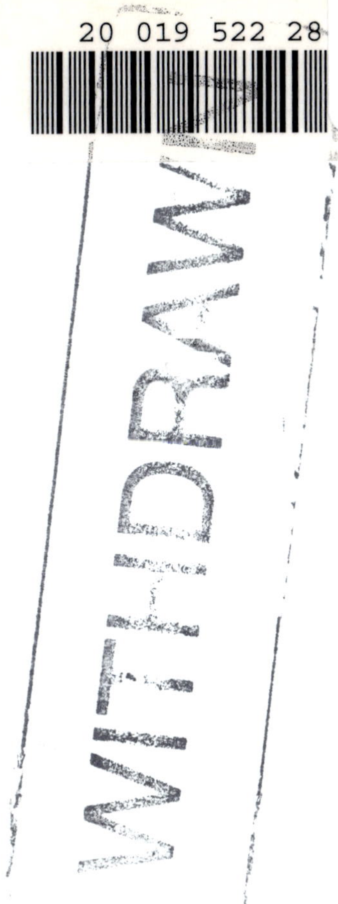
WITHDRAWN

Thin Films

Author and Subject Cumulative Index
Including Tables of Contents, Vols. 1–24

VOLUME 25

Serial Editors

Honorary Editor

Thin Films

Author and Subject Cumulative Index
Including Tables of Contents, Vols. 1–24

VOLUME 25

Preface by

Stephen M. Rossnagel
IBM Watson Research Center
Yorktown Heights, New York

Indexes compiled by

Janet Perlman
Southwest Indexing
Phoenix, Arizona

ACADEMIC PRESS
San Diego London Boston New York Sydney Tokyo Toronto

This book is printed on acid-free paper. ∞

Academic Press
525 B Street, Suite 1900, San Diego, CA 92101-4495, USA
1300 Boylston Street, Chestnut Hill, MA 02167, USA
http://www.apnet.com

Academic Press Limited
24–28 Oval Road, London NW1 7DX, UK
http://www.hbuk.co.uk/ap/

International Standard Serial Number 1079-4050
International Standard Book Number 0-12-533025-1

Printed in the United States of America
98 99 00 01 02 MV 9 8 7 6 5 4 3 2 1

Contents

Preface vii
Author Index. 1
Subject Index 289
Appendix: Tables of Contents, Vols. 1–24. 459

Preface

"Today, thin films of metals, semiconductors, and dielectrics have become of increasing importance for fundamental studies in many fields of physics, electronics, and chemistry, and are employed in numerous practical applications. The tremendous progress in the field of thin film physics has been stimulated and widened to the largest degree by the recent development of the most efficient high and ultra-high vacuum systems . . ."

With those words, the founding series editor, Georg Haas, initiated the Physics of Thin Films series over thirty-five years ago. It is interesting and indeed quite surprising that the original introduction written so long ago in the infancy of both the age of space exploration as well as the so-called information revolution is still quite accurate. It is rare to find a highly technical field that survives in an identifiable form for one decade, let alone almost four.

It is also quite interesting to look at the list of editors and contributors over the years. The list reads like a Who's Who of thin film science and technology. Many names will be familiar, particularly to the more senior readers, and the younger readers should realize that these authors constitute the founders of this field.

The original concept of the Thin Film series was to provide a forum for publishing state-of-the-art review articles on the most relevant and important aspects of thin film science and technology. At the time there were few journals devoted to these areas. The American Vacuum Society was still in its infancy and was not specifically focused on thin film areas. Other physics journals published occasional thin film articles, but they were more science oriented, less practical, and often lost within the large numbers of papers in other fields.

The Thin Films series was originally intended to be an annual publication, but as will be evident in the listings below, this aggressive schedule could not be met over the years. The series currently stands at Volume 24, published in early 1998, and hopefully over the next several years the production rate will return to roughly an annual schedule. Essentially all of the articles published are invited, review articles, and while many authors will readily agree to contribute such an article, they tend to take much longer to write than originally anticipated.

The scope of the Thin Films series has varied over the years, perhaps mirroring the trends of the times and the interests of the editors. The earliest volumes had a distinct emphasis on both evaporative deposition as well as optical characterization and applications. More recently, the scope of the series has moved to cover the areas of semiconductor-based film technology along with the separate area of organic thin films. This index, which includes a complete compilation of the technical indices of all twenty-four volumes, is intended to be a resource for finding information across this broad spread of years and topics. While many if not all of the early volumes are out of print, many of the larger technical libraries in universities as well as such labs as IBM Research, Bell Labs, and Xerox PARC generally have complete or near-complete sets.

The original and founding editor for the Physics of Thin Films series was Georg Haas, of the Night Vision Lab, U.S. Army Electronics Command in Fort Belvior, Virginia. In early volumes he was joined by Rudolf Thun of the IBM Research Lab in Yorktown Heights, New York. In the early 1970s Maurice Francombe of Westinghouse Electric Corporation Research Labs in Pittsburgh began contributing to the series, first as a guest editor with R. Hoffman of Case Western, and then as a series editor with Haas, joined shortly thereafter by Hoffman. With the retirement of Haas in the early 1980s, John Vossen of RCA-Princeton joined as a series editor. This continued until the mid-1990s when John unfortunately passed away. At about that time, Abraham Ulman of Polytechnic University, Brooklyn, New York, joined as series editor for Organic Films, and Steve Rossnagel of IBM Research in Yorktown Heights became series editor, concentrating primarily on non-organic films (with the exception of semiconductor applications of organic films). Maurice Francombe became editor emeritus and still plays a vital role in the development of volumes for the series.

Overview of Volumes 1–24

VOLUME 1, 1963 (G. HAAS, EDITOR)

The introductory volume contains a fascinating article by H. Caswell on the vacuum and deposition technology of the time. Although most of the pumps (such as the ''Evapor-Ion pump'') and fixtures (such as gold o-rings) will be completely unfamiliar to many current readers, this overview touches on many topics and problems still of importance today such as outgassing and vapor pressure, gauging, and mass spectroscopy. Also included in this volume are extensive discussions of optical film measurements and some very basic thin film science. In Thun's chapter on structure, the basic nucleation, growth, and epitaxial issues are discussed that should still be a basic part of the training for

anyone in thin film areas. Also included are chapters on low temperature and magnetic films which are basic, interesting starting points for fields that have changed significantly over the years. In general, Volume 1 would function well as an introductory textbook, and I imagine it was used as one over the years.

VOLUME 2, 1964 (G. HAAS AND R. THUN, EDITORS)

This volume contains three chapters that deal with film fundamentals and four chapters that look at coating applications. The first chapter is a fundamental study of structural disorder by C. Neugebauer of the GE Research labs. A second chapter describes the interaction of electron beams with thin film material, which is key to understanding TEM results. The third basic paper looks at optical constants of thin films. The applied chapters cover electronic components (thin film transistors and components and circuits) as well as a detailed study of antireflection and light-absorbing coatings. These latter two chapters are in many ways still quite relevant to current thin film technology. The chapters on electrical components are perhaps less so, although they show the early directions of the field that have evolved extensively over the past thirty-four years.

VOLUME 3, 1966 (G. HAAS AND R. THUN, EDITORS)

This volume branches out from the concentration on evaporative deposition of Volumes 1 and 2. The first chapter (K. Behrndt) describes thickness and rate monitoring devices and techniques needed for accurately reproducing film properties. A second chapter by L. Maissel describes, for the first time in the series, the field of sputtering as it existed in the 1960s, based on dc bias and early rf-diode systems. Following this is the first series article on ''gas phase deposition,'' which is a field now known as chemical vapor deposition (CVD). The chapter, by L. Gregor, focuses on the deposition of dielectrics, which is something that is currently done worldwide for semiconductor manufacturing. Later chapters include the topics of II-VI detector compounds, an interesting paper by R. Hoffman on mechanical properties, and a short paper on lead-salt detectors, the precursor for many of today's II-VI detectors.

VOLUME 4, 1967 (G. HAAS AND R. THUN, EDITORS)

This volume starts with a discussion of precision measurements in thin film optics, followed by a group of articles on growth, nucleation, and structure. The thin film optics paper, which predates the application of lasers, nevertheless describes the basic physics of such areas as ellipsometry, interference, absorption,

and reflection. The nucleation and growth chapter is a basic, fundamental study that is still quite accurate. Next follows an interesting paper on evaporated single crystal films, a field we now know as MBE. A paper by Lawless describes electrodeposition, which might be considered a somewhat ''old'' field in thin films. However, at the time of this writing, the major semiconductor manufacturers of the world are rapidly moving toward the electrodeposition of Cu for semiconductor wiring. In addition, the volume contains an interesting early paper by Crowell and Sze on hot electron transport and electron tunneling in materials, another field that has broadly evolved into today's semiconductor manufacturing realm.

VOLUME 5, 1969 (G. HAAS AND R. THUN, EDITORS)

This volume contains articles on interference photocathodes, multilayer interference filters, and the deposition of primarily optical films from organic solutions. The volume also contains an article by M. Francombe and J. Johnson on the current developments in semiconductor films, followed by an article on CVD technology.

VOLUME 6, 1971 (M. FRANCOMBE AND R. HOFFMAN, GUEST EDITORS)

The first article on anodic oxidation describes the basics of this technology, followed by an article by D. Larson on size-dependent electrical conduction that perhaps anticipates the current trends toward near-quantum size technologies. This is followed by a mostly theoretical chapter on optical properties of metal films followed by chapters on interactions in multilevel films and diffusion. The multilayer chapter (A. Yelon) describes some novel effects in exchange coupling, which have become the basis of current magnetoresistive thin-film magnetic read sensors.

VOLUME 7, 1973 (G. HAAS, M. FRANCOMBE, AND R. HOFFMAN, EDITORS)

The first chapter, by D. Dove, examines the status of research with amorphous films, and is followed by a chapter by W. Hunter on metal film optical filters for the extreme ultraviolet. Next is a chapter on liquid phase epitaxy, focused primarily on semiconductor and magnetic oxide materials. The fourth chapter is a seminal paper on electromigration effects in thin films by F. d'Heurle and R. Rosenberg. The final article covers the topic of built-up molecular films, which was a small area at the time but has become widespread in the past ten years.

Volume 8, 1975 (G. Haas, M. Francombe, and R. Hoffman, Editors)

The first two chapters in this volume continue the earlier series trend with studies of optical films, the first chapter on antireflection coatings and the second on inhomogeneous optical films. The third article (Z. Meiksin) describes electrical transport in discontinuous and cermet films, followed by a chapter on electrical conduction in nonmetallic amorphous films. The final chapter describes novel techniques for fabrication of unusual topographical shapes, primarily in Si. The latter has become more important recently with the widespread development of Micro-Electrical-Mechanical Systems, or MEMS, and the chapter provides an interesting early look at the field.

Volume 9, 1977 (G. Haas, M. Francombe and R. Hoffman, Editors)

This volume deals primarily with optical effects and characterization of films. The first article, by J. Vossen, reviews the field of transparent conducting films. This is followed by a paper on metal-dielectric interference filters describing some unusual and interesting effects. A third chapter is a detailed look at surface plasma oscillations and applications by H. Raether. The final paper, by P. Chaudhari, describes what was a very interesting and exciting field of magnetic bubble memory films. Although this field initially was quite promising, today it remains primarily an interesting effect.

Volume 10, 1978 (G. Haas and M. Francombe, Editors)

This volume focuses primarily on optical films and application. The first chapter describes a topic of great interest at the time of photothermal solar energy conversion. The second chapter describes the use of films for ultraviolet astronomy as well as thermal coatings on satellites. Later chapters describe scattering effects, integrated optics and optical communications, and the final chapter describes the use of films for correction of optical elements.

Volume 11, 1980 (G. Haas and M. Francombe, Editors)

The first article by W. Hunter and G. Haas describes the development of reflectance coatings for diffraction gratings in the extreme ultraviolet. This is followed by a wide-ranging article by C. Wood on MBE of III-V compounds. The third chapter by H. Holloway describes developments in IV-VI compound films, used as infrared detectors, and follows earlier work in Volume 3 on lead-salt

detectors. This is followed by an article by A. Jonscher on theoretical work relating to dielectric properties.

VOLUME 12, 1982 (G. HAAS, M. FRANCOMBE AND J. VOSSEN, EDITORS)

This volume also focuses primarily on optical topics. The first article reviews recent progress in the development of metal coatings and protective layers for front-surface mirrors. The second chapter describes the field on photoemissive materials and the physics of photoemission. The third article by K. Chopra et al. explores the field of spray pyrolysis as a deposition technique for low cost, organic films. The final chapter gives an updated look at the field of plasma-enhanced chemical vapor deposition, used widely now for semiconductor manufacturing.

VOLUME 13, 1987 (M. FRANCOMBE AND J. VOSSEN, EDITORS)

This volume explores some of the less conventional film deposition and etching techniques that were beginning to emerge in the mid-1980s. The first chapter, by T. Takagi, is on ionized cluster beam deposition, which showed early promise and unique capabilities. The next chapter by R. Bunshah and C. Deshpandy is on a variation of evaporation known as activated reactive evaporation, followed by a chapter by U. Gibson on ion beam deposition, and a paper by Ashby on laser-induced etching. The last chapter, by Woodall, Braslau, and Freeouf, describes the broad area of contacts to GaAs devices and the future directions of the field.

VOLUME 14, 1989 (M. FRANCOMBE AND J. VOSSEN, EDITORS)

This volume follows Volume 13 in describing contemporary film deposition technology. The first chapter, by W. Westwood, covers the area of reactive sputtering, a technique that has become widely used. This is followed by an article on plasma oxidation and another deposition article on a technique known as arc-based plasma deposition (P. Johnson). The last article, by J. Vossen, describes the general topic of surface preparation by means of glow discharge exposure.

VOLUME 15, 1991 (M. FRANCOMBE AND J. VOSSEN, EDITORS)

This volume moves toward novel solid state uses of thin films with a long article on microwave magnetostatic wave phenomena in epitaxial magnetic oxide films (J. Adams et al.), which is followed by another article in the general area of

magnetic alloy films by B. Krusor and G. Connell. The second article focuses on magneto-optic materials, such as Gd-FeCo and Tb-FeCo for recording applications. A third article explores quantum-well semiconductor devices (D. Coon and K. Bandara), and the fourth article describes work on Fourier transform infrared spectroscopy.

VOLUME 16, 1992 (M. FRANCOMBE AND J. VOSSEN, EDITORS)

This book describes films for emerging applications and covers the recent work in high temperature superconducting films (N. Dhere) and thin film permanent magnet films (F. Cadieu). A third chapter by K. Reddy describes work on lateral diffusion and electromigration in films, a topic that was first covered in Volume 7 and will be revisited as the topic for a complete, upcoming volume. The fourth article describes the failure and cracking of films that are deposited on highly deforming or elongating surfaces, such as might be present in the deposition of films onto thin polymer films that are deformed or rolled.

VOLUME 17, 1993 (M. FRANCOMBE AND J. VOSSEN, EDITORS)

This volume focuses primarily on mechanical and dielectric properties of films. It begins with an article by S. Barnett on the mechanical properties of thin film superlattice films and is followed by a paper by J. Musil, J. Vyskocil, and S. Kadlec on hard coatings prepared by sputtering and arc-based deposition. Dielectric properties are explored in an article by Krishnaswamy et al. on piezoelectric films, and this is followed by an interesting, broad article by Francombe on ferroelectric thin films for integrated electronics. The final chapter by C. Granqvist explores electrochromic tungsten-oxide-based films that have quite unique optical properties.

VOLUME 18, 1994 (M. FRANCOMBE AND J. VOSSEN, EDITORS)

This volume describes plasma technologies for thin film deposition and modification. The first chapter by M. Lieberman and R. Gottscho describes high-density plasma technology, used primarily for etching applications. This is followed by a paper by O. Popov on electron-cyclotron-resonance plasmas (ECR), which are also used mostly for etching. A third chapter by S. Rohde is on a variant of magnetron sputter deposition technology known as unbalanced magnetrons in which large ion fluxes can be drawn to the sample during deposition. The final chapter by C. Steinbruchel describes the phenomena and problems of particle formation in plasmas.

Volume 19, 1994
(K. Vedam, Guest Editor; M. Francombe and J. Vossen, Series Editors)

This volume reports on the optical characterization of real surfaces and films. This series of papers describes a variety of experimental techniques for optical characterization, with the emphasis on real or experimental results. The papers cover a technique known as ''reflectance anisotropy'' (B. Drevillon and V. Yakovlev), real-time spectroscopic ellipsometry (chapters by R. Collins et al. and by H. Nguyen et al.) and then two characterization studies using ellipsometry. The first is on inhomogeneous transparent films (P. Chindaudom and K. Vedam) and the second on ferroelectric films by S. Trolier-McKinstry et al.

Volume 20, 1995 (A. Ulman, Editor)

This volume is the first to focus exclusively on organic films. Starting with this volume, a significant component of the volumes in the series will concentrate in this field under the direction of Ulman. Non-organic film topics will be covered by Rossnagel along with the support of Emeritus Editor M. Francombe. This volume covers a wide range of topics on organic films for the late 1990s: seventeen chapters in all, and is the starting point for future volumes.

Volume 21, 1995 (M. Francombe and J. Vossen, Editors)

This volume covers the topic of homojunction and quantum-well infrared detectors. The first chapter by A. Perera discusses the area of heavily doped silicon homojunctions as well as Ge and GaAs based junctions. This is followed by a chapter on SiGe/Si quantum wells by R. Karunasiri et al. and a chapter on quantum-well infrared photodetectors based on III-V compounds. There are two additional chapters on quantum-well studies, one by K. Choi on 3-terminal multi-quantum-well infrared photodetectors and one focused on solar cell applications.

Volume 22, 1996 (S. Rossnagel, Editor)

This volume focuses on the computer-based modeling of thin film deposition, and in particular, its application to the somewhat challenging topographies present in current-day semiconductor wafers. The papers range from Monte Carlo-based models to almost completely analytic, although the vast majority of the cases examined are in two dimensions rather than three.

VOLUME 23, 1998 (M. FRANCOMBE AND J. VOSSEN (DECEASED), EDITORS)

This volume covers topics similar to Volume 21 in addition to other, more electronic applications. The first chapter by D. Greve describes work with Si-Ge epitaxial growth, from experimental work to high performance circuit applications. The second and third articles examine various detector films, primarily in the infrared. The fourth chapter describes the development and optimization of quantum-well infrared photodetectors, and this is followed by an article on semiconductor infrared sources, also based on quantum-well structures.

VOLUME 24, 1998 (A. ULMAN, EDITOR)

Self-assembled monolayers (SAMs) of thiolates on gold are probably the most studied group of SAMs. Vol. 24 provides the reader with up-to-date information on this important subject, from the preparation of thiols and their SAMs (Paul Laibinis, MIT), to their utilization in microcontact printing, a technology that aims at replacing photolithography (George Whitesides, Harvard). The chapters, written by the most recognized experts in their respective fields, while offering a broad overview, give in-depth critical analysis of every topic. Vol. 24 should be on the desk of every expert and novice who are interested in this exponentially growing research area.

VOLUME 26, 1998 (R. POWELL AND S. ROSSNAGEL, EDITORS)

This volume differs from earlier volumes in that it is an authored, single-topic book based on sputter deposition for semiconductor applications, and covers conventional sputter deposition, followed by a number of variations that have been developed for the unique topographies of semiconductor features. The variations include collimated sputtering, high temperature or reflow sputter deposition, and also ionized PVD, or the condensation of thin films from metal plasmas fed by sputtering.

Upcoming Volumes

At the time of this writing, at least four new additions to the Thin Film series are in progress. These volumes cover topics such as ultra-thin oxides for semiconductor device applications, a more complete look at the deposition of films from ions, a wide-ranging study of electromigration phenomena and results, and also a look at the low-k dielectric materials that are needed for high speed semiconductor applications.

In the next several years, the Thin Film series will tend to focus on primarily semiconductor applications along with organic films. The early focus of the series on optical films and characterization has built a tremendous base of understanding in this area. However, current industrial trends are such that much of the intriguing new topics in thin film science and technology will be driven by the relentless miniaturization and wide-ranging applications of semiconductor-based technologies.

Organic thin films are very important in many applications, from anticorrosion coatings to the design of surfaces, and to biosensors. They are prepared by a variety of techniques, such as spin-coating, high vacuum evaporation, self-assembly and Langmuir-Blodgett. Recently, organic thin films have been prepared on nanoparticles, opening new opportunities for the design of new materials with novel properties.

As part of the Thin Films series, we plan to address important issues concerning organic thin films. The recent volume on self-assembled monolayers of thiols on gold brings the most up-to-date information that allows both the novice and the expert to read about this exciting research area. For the future we are planning volumes on nanoparticles, other self-assembly developments, and applications.

We encourage our readers to suggest topics for future volumes. Please send suggestions via e-mail to:

Abraham Ulman: aulman@duke.poly.edu

Stephen Rossnagel
February 1998
Yorktown Heights, NY

Author Index

Numbers in parentheses (except Vols. 16 & 17) are reference numbers; the complete citations can be found at the end of each chapter.

A

Aarnink, A. M., **16:**111, 117, 142
Abarenkov, I., **12:**105(178), 162
Abbott, N. A., **24:**229(35)
Abbott, N. L., **24:**21(104), 26(104), 27(104), 28(104), 44(59), 45(59), 229(43)
Abbott, P. R., **21:**106(45), 107(45), 108(45), 112
Abbott, R. R., **21:**103(39), 107(39), 112, 118(14), 121(21), 123(21), 128(14), 133(14, 44), 137(14), 141(14), 149(14), 150(14), 160(14), 172(14), 193(14), 194(21), 195(21), 209(14), 221(136), 232–233, 237, 244(21), 249(25), 261(21), 284(25), 309
Abe, A., **14:**52(169), 77
Abe, H., **16:**86–87, 138; **19:**35
Abe, T., **13:**64(28), 65, 68, 88(28), 91(28), 105; **17:**111, 142
Abeledo, C., **2:**31(55), 61
Abeles, B., **10:**60(136, 139), 69; **12:**72(73), 79(73), 160; **14:**82(1), 116(1), 125
Abelès, F., **1:**70(7), 87(7), 88, 91, 120, 143, 185; **2:**210, 214, 228, 237, 238; **4:**77(192, 193, 194), 78(194), 95; **5:**53(7), 57, 59, 85, 114(18), 121(18), 141; **6:**100, 145, 183, 188, 203; **9:**13, 66, 75, 76(12), 78(12), 143, 240(232a), 261; **19:**50
Abeles, J. H., **23:**290(110)
Abelson, J., **14:**99(41), 110(41), 126
Abelson, J. R., **19:**52, 103
Abernathy, D., **24:**3(17), 4(17), 184(43)
Abkovitz, M., **11:**256(60, 62, 64, 69), 257(60, 62, 64, 69), 260(69), 315, 316
Aboaf, J. A., **5:**303, 321; **9:**31(114), 35(114), 67
Abraham, D., **3:**180(37), 209
Abraham, F. F., **13:**8(18), 57
Abraham, K. M., **13:**97(95), 106
Abraham, W. G., **12:**143(313), 166
Abrahams, M. S., **7:**126(16), 128, 129, 130, 131(20), 132(21), 133(28), 136(15, 28), 137(34, 35), 138, 139(35), 147(34), 157(34, 73), 193(163), 230(35), 244(269), 247, 248, 249, 252, 255
Abrahams, S. C., **1:**205(78), 230; **9:**264, 267, 276, 287, 294; **17:**229, 230, 279, 295
Abrain, I., **23:**255(78)
Abramson, M. B., **7:**330, 367(61), 394

Abruña, H. D., **24:**21(101), 23(101), 24(101), 27(101), 82(16), 86(16)
Abstreiter, G., **11:**45(57), 97; **21:**78(4), 111, 290(43), 310; **23:**254(75)
Abt, N., **17:**253, 255, 297
Acero, A., **24:**184(37)
Acher, O., **19:**2, 4, 7–9, 29–31, 38, 39, 41
Ackermann, K. P., **17:**111, 142
Ackermann, P. W., **5:**138, 141; **9:**33(118), 67
Ackley, D. E., **10:**41(80), 67
Acloque, P., **10:**237, 258
Acosta-Ortiz, S. E., **19:**9
Acrivos, A., **24:**72(114)
Adachi, H., **14:**20(78), 75; **16:**2, 17–19, 30, 33–34, 36, 124, 126, 129–130; **17:**228, 232, 248, 249, 250, 251, 264, 296, 297
Adachi, M., **17:**163, 221, 228, 232, 243, 296, 297
Adachi, R., **17:**87, 140
Adachi, S., **19:**97
Adalsteinsson, D., **22:**270(131), 276
Adam, A., **24:**153(14)
Adam, H., **1:**37(100), 66
Adam, J. D., **15:**14, 18, 20, 22, 25, 40, 43–44, 47–48, 48, 54, 59, 62, 64, 68, 72–76, 78; **16:**107, 141; **17:**146, 148, 186, 206, 219, 221, 222
Adam, N. K., **7:**312, 314, 315, 320, 362(3), 392
Adamo, W., **17:**172, 173, 222
Adams, A., **5:**273, 317
Adams, A. C., **12:**263, 264(98), 292; **13:**99(105), 107; **15:**293, 299–300, 311; **22:**61(62), 78, 204(63), 273
Adams, C. T., **7:**68(59), 113
Adams, J. B., **22:**127(38), 172, 254(98), 261(98), 262(98), 274
Adams, P. M., **21:**93(21), 94(21), 111
Adams, S. E., **17:**241, 287, 297
Adams, T. E., **14:**82(4), 103(4), 125
Adamsky, R. F., **5:**197(172), 234
Adamson, A. W., **7:**312(10), 392; **24:**220(81)
Adamson, Peggy L., **4:**139, 187
Adar, F., **13:**180(105), 195
Adar, R., **21:**40(86), 74
Adda, Y., **7:**258, 260(2), 305; **16:**234, 266
Addiss, R. R., **5:**164(84), 172(84), 184, 231
Addiss, R. R., Jr., **1:**35, 66; **3:**169, 174, 181, 208
Adelman, S. A., **22:**136(60), 173
Adkins, L. R., **15:**18, 37, 39, 64, 72, 75, 137–139
Adler, D., **7:**34(141), 40
Adler, S. L., **4:**329, 369
Adriaens, M. R., **10:**84, 164
Adrian, F. J., **16:**59, 79, 135, 137
Aegerter, M. A., **17:**361, 370
Aers, G. C., **15:**250, 263; **21:**170(73, 75), 178(73), 234
Affinito, J., **14:**11(41), 44–45(157), 53(41), 54(41), 74, 77; **17:**111, 116, 142
Affronte, M., **16:**27, 128
Agar, A. W., **1:**198, 229; **2:**83, 94, 96(36), 97, 98, 109(82), 130(82), 132(150), 141, 142, 145; **3:**34, 52(166), 57
Agarwal, M. S., **14:**141(39), 142–143(39), 196
Agarwal, P., **13:**94, 106
Agarwal, V., **13:**79(78), 99(78), 106
Agarwal, V. K., **7:**364(165), 373(188, 189, 190, 191, 192), 374(188,

189), 376(189), 377(190), 378(191, 192), 396, 397
Aggarwal, P. S., **3:**174, 208
Aggarwala, B. N., **7:**276(72), 289(96), 308, 309; **16:**240, 267
Agius, B., **17:**315, 316, 329, 330, 365
Agnello, P. D., **23:**24(97), 49–50(178–180)
Agnihotry, S. A., **17:**304, 312, 364, 365
Agostinelli, E., **16:**59, 79, 135, 137
Agostinelli, J. A., **16:**97–98, 140
Agranovich, V. M., **19:**10
Agrawal, A., **17:**312, 365
Agrawal, D. C., **16:**331, 340
Agrawalla, B. S., **13:**190(160), 191(160), 197
Agreev, A. N., **15:**140
Aguilar, M., **24:**163(24)
Agule, B. J., **5:**283, 319
Agusta, B., **4:**320(166), 324
Aharoni, A., **6:**232(117a, 117b), 235(117), 262, 297, 299
Aharoni, S. M., **9:**8(38), 65
Aheam, J. S., **21:**241(11), 282(11), 290(49), 309–310
Ahilea, E., **9:**5(19), 65
Ahn, B. H., **7:**154(66), 219(205), 249, 253
Ahn, K. Y., **6:**209(18), 210, 226, 291(215), 295, 300; **9:**266, 295
Ahonen, R. G., **13:**135(104), 141(104), 150
Ahrenkiei, R. K., **16:**97, 99, 140
Aichison, R. E., **3:**179(32), 209
Aikawa, Y., **17:**318, 366
Aiki, K., **10:**228(7, 8), 251(46), 252(47), 258, 259
Ainger, F. W., **17:**237, 238, 296
Ainslie, N., **7:**268, 269(61), 287(94), 290, 294(99), 308, 309, 310
Ainslie, N. G., **16:**250, 268
Airey, R. W., **12:**147(318, 319), 166
Airy, G. B., **1:**71, 121; **9:**16, 66
Aisenberg, S., **12:**246(31, 32), 247, 253(32), 290; **13:**111, 112(30, 32), 115(32), 124(73), 142, 147, 148, 149, 150
Aita, C. R., **13:**68, 105; **14:**11(42), 16(60), 31(119), 66(218), 74, 76, 78; **17:**153, 220
Aitchison, R. E., **2:**389(114), 422; **9:**9, 31(43), 33(43), 35(43), 36(43); **12:**234
Aitkozhin, S. A., **5:**166(116), 177, 232
Aizaki, N., **23:**55(9)
Aizenberg, J., **24:**240(70)
Ajakawom, A., **24:**33(134)
Ajioka, T., **15:**295, 311
Akamine, S., **24:**84(21)
Akari, S. D., **24:**95(73)
Akasaka, H., **15:**204, 206, 217
Akasaki, I., **5:**219(201), 234, 282, 318; **7:**163(88, 90), 164(90), 236(249c), 249, 254
Akasaki, M., **5:**282, 318
Akashi, K., **16:**51, 133
Akbar, S. A., **16:**99, 140
Akhiezer, A. I., **15:**26, 137; **18:**147
Akhtar, M., **17:**359, 361, 369
Akiyama, T., **24:**33(134)
Akiyama, Y., **22:**188(40), 272
Akovali, G., **12:**248(40), 251(40), 290
Akram, H., **17:**313, 324, 329, 365, 366
Akselrad, A., **9:**288, 294
Aksenov, I. I., **14:**163(113–115), 170(123, 124), 176(140, 141), 177(145, 147, 148), 185(113–115), 197, 198
Al-Jamaily, G. A., **13:**121(64, 65), 123(65), 130(92), 131, 132(65, 102), 134(65, 92), 139(119),

Al-Jamaily, G. A., (*continued*) 140(119), 142(119), 145(64, 65), 148, 149, 150
Aladjem, A., **6:**4(15), 74
Alarco, J. A., **16:**67, 136
Alavi, K., **15:**229, 233–235, 262; **21:**241(16), 309; **23:**221(9), 224(9)
Albella, J.-M., **14:**112(98), 128
Alben, R., **15:**185, 216
Albers, M. W., **24:**249(96)
Albert, K., **24:**151(5), 170(40), 173(40, 42)
Albert, L., **4:**199(55), 200(55), 201(55), 202(55, 78), 203(55), 204(55), 206(55), 233(55), 252
Albin, D., **16:**97, 99, 140
Albiston, S. D., **15:**161–162, 171, 185–186, 215–216
Albon, N., **5:**261(58), 315
Albrecht, M. G., **11:**198, 203
Albrecht, T. R., **24:**84(21)
Albright, H. T., **10:**129(64), 130(64), 147(64), 151(64), 165
Albritton, J., **21:**117(9), 128(9), 232
Alcock, C. B., **4:**283(108b), 322
Alder, B. J., **24:**207(21)
Alderson, R. H., **2:**387(105), 388(105), 389(105), 422
Aldiner, G. K., **13:**117(41), 148
Aleck, B. J., **16:**317, 336, 339
Aleksandrov, B. N., **6:**92, 118, 119, 120, 121, 122, 123, 129, 131(168), 135(168, 180), 145, 147, 148
Aleksandrov, V. I., **19:**239–241
Aleksanyan, I. T., **3:**255, 257, 258(96), 260, 263, 264(96), 265, 268, 273
Alemany, L. B., **24:**153(12)
Aleskoskii, V. B., **3:**176(25), 209
Alessandrini, E. I., **1:**279(3), 331; **6:**337(44), 347
Alexander, A. E., **7:**312(13), 350, 354, 355, 358(126), 393, 395
Alexander, E. G., **5:**265, 269, 316
Alexander, F. B., **9:**50(228), 70; **12:**263(98), 264(98), 292; **22:**121(20), 172
Alexander, J., **24:**98(94)
Alexander, J. H., **6:**37(133), 77; **12:**261(92), 262(92), 271(158), 292, 293
Alexander, L., **7:**11(24), 37
Alexander, L. E., **1:**205(89), 230; **4:**218(97), 222(97), 253
Alexander, R. W., **9:**237(223), 261
Alexander, W. B., **7:**263(27), 306
Alexopoulos, K., **9:**219(176), 259
Alfano, R. R., **4:**83(211), 96; **21:**330(25), 333(25), 367
Alferov, Zh. I., **7:**155(68), 186(146, 148, 149), 187(148), 217(200), 234(248), 246(276, 279), 247(281, 282), 249, 251, 253, 254, 255, 256; **13:**162(44), 164(44), 193
Alfieri, I., **11:**113(51), 200
Alger, T. D., **24:**153(12)
Ali, T., **21:**334(29), 341(29), 342(29), 355(29), 367
Alibert, C., **9:**22(93), 67; **23:**221(10), 225(10)
Alien, J. A., **3:**219, 271
Alien, J. E., Jr., **18:**16
Alien, S. D., **17:**121, 143
Alien, T. H., **15:**290, 311
Aliev, F. I., **7:**36(160), 41
Alikhodzhaeva, M. A., **10:**64(145), 69
Alivisatos, A. P., **24:**36(147), 222(105)
Allam, D. S., **3:**133(16), 162
Allara, D., **15:**303, 311

Allara, D. L., **12:**277(212), 279(212), 280(212), 295; **14:**103(49), 126; **19:**258, 282, 288, 290; **23:**48(161); **24:**3(10, 11, 12, 15, 18), 4(14, 18), 5(12), 10(35, 40), 14(35), 17(10, 40), 18(11, 35), 32(15), 33(137), 44(1, 2, 4, 8, 14, 15, 37), 45(15), 46(14), 55(14), 56(14), 66(4, 8), 70(14, 15), 82(7, 19), 86(19), 87
Allegretti, J. E., **5:**256(45), 266, 315, 316
Allen, D. E., **15:**54–55, 138
Allen, D. H., **16:**302, 339
Allen, E. A., **9:**5(22), 65
Allen, F. G., **1:**8, 36, 64, 66; **9:**47(223), 70; **12:**259(81), 292
Allen, G. A., **7:**242(264), 255; **12:**124(253), 164
Allen, H. A., **7:**225(224), 253
Allen, J. A., **1:**279(2), 331; **2:**48(80, 81), 49(81), 51, 62
Allen, J. T., **19:**32
Allen, J. W., **7:**210(191), 252; **15:**176, 191, 216–217
Allen, M. P., **22:**127(32), 128(32), 134(51), 135(58), 142(51), 172–173; **24:**207(19), 208(19)
Allen, R., **15:**154, 168, 176, 186–187, 214–216
Allen, R. P., **16:**193, 230
Allen, S. D., **13:**190(160), 191(160), 197
Allen, S. J., **15:**229–231, 259, 262; **21:**98(34), 100(34), 112, 184(81), 235
Allen, S. J., Jr., **15:**259, 264; **23:**238(43, 44), 239(43), 243(43, 44), 244(43, 44), 251(44)
Allen, T. H., **13:**111, 114, 122, 130, 131(18, 98), 132, 133, 134, 135, 149
Allenson, M. B., **6:**247(153), 298
Allerman, A. A., **23:**278(101)
Allgaier, R. S., **3:**277, 278, 300; **11:**114(60), 120, 121(72), 201
Allinson, D. L., **4:**149(34), 156(34), 171(34, 122), 172, 173, 175(34), 188, 190
Allinson, H., **24:**21(91)
Allis, W. P., **18:**137
Allison, D. P., **24:**98(92)
Allison, J. F., **5:**146(14), 229
Allison, R., **4:**68(163), 94
Allpress, J. G., **4:**152(71, 72), 161, 180(71, 72), 189
Almasi, G. S., **6:**209(18), 210, 226, 295; **9:**266, 295
Almassy, R. J., **11:**56(93), 57(93), 98
Almeleh, N., **7:**187, 209, 220, 222(150), 251
Almén, O., **3:**70, 73(34), 76(34), 125
Almer, F. H. R., **1:**51(126), 67
Alperin, H. A., **9:**273, 274, 296; **15:**147, 214
Alpert, D., **1:**14(27), 26, 28, 33, 35, 42, 64, 65, 66; **3:**4, 51(26), 53
Als-Nielsen, J., **24:**116(36), 117(36), 180(10, 11), 181(12, 15), 185(10), 186(10, 44), 192(12), 199(12, 50)
Alsford, R. W., **9:**5(17), 65
Alstrup, O., **5:**267, 316
Alt, C. E., **3:**30(149), 38(149), 51(149), 52(183), 57, 58
Alt, L., **2:**370(53), 421
Alt, L. L., **3:**138(34), 163; **4:**263(56), 264, 319(56), 321; **5:**287(206); **12:**259, 292
Alt-Anthony, C., **3:**30(150), 31(150), 40, 51(150), 57
Alterovitz, S., **13:**79(80), 106

Alvarez, M., **24:**222(104)
Alvarez, M. M., **24:**151(10), 154(10), 156(10), 163(10)
Alves, C. A., **24:**44(20, 31, 34), 83(20), 88(46), 90(20), 91(46), 112(5), 128(60)
Alves, E., **15:**249, 263
Alwitt, R. S., **6:**68, 79
Aly, S. H., **16:**168, 177, 194–201, 213, 230
Amador, J., **16:**2, 124
Amal, S., **17:**361, 370
Amano, J., **13:**112(29), 115, 145(33), 148
Amano, O., **11:**254, 255
Amano, S., **11:**36(13), 96
Amans, R. L., **9:**59(266), 71
Amberseley, M. D., **17:**153, 220
Ameen, M. S., **16:**14, 17, 126; **17:**229, 234, 235, 263, 296
Amelinckx, S., **2:**20, 60; **6:**12, 75
Amelincky, S., **4:**144(26), 155(26), 162(94), 182(26), 184(26), 187, 189, 195(21), 199(69), 225(129), 251, 252, 254
Amelio, G. F., **23:**109(61)
Amemiya, H., **18:**124
Ament, W., **15:**32, 135, 137
Amer, N. M., **16:**110, 142; **24:**94(66)
Ames, I., **1:**20(48), 65; **3:**139(39), 163; **7:**265, 274(53), 278(73), 283; **16:**250, 268
Amick, J. A., **5:**247(29), 260(54), 261(54), 274(120), 279, 282, 315, 317, 319; **7:**127(18), 247
Amiguet, J., **13:**78(63), 79(63), 106
Amini, A., **22:**132(47), 172
Amm, D. T., **17:**241, 290, 297
Amsel, G., **6:**6, 25, 27, 74, 76; **12:**292; **14:**109(61), 126
Amundsen, T., **6:**82(24, 28), 144
Amzallag, E., **21:**21(43), 72
An, H., **17:**22, 74
An, I., **19:**53–58, 64, 70, 83, 84, 129, 148, 152, 154, 157, 159, 160, 163, 165–167, 171, 175, 181, 182, 269, 280
Anastasio, T. A., **3:**34, 52(165), 57
Andant, A., **3:**13, 51(58), 52(58), 54
Andelin, J. P., Jr., **9:**136(50), 144
Anderegg, J. W., **24:**44(34)
Anderegg, M., **9:**193(93), 257
Anders, F. J., **1:**226(175), 232; **3:**245(73), 272
Andersen, H. C., **5:**247(20), 314; **22:**134(49, 51), 152(51), 172
Andersen, H. H., **22:**110(23), 115
Andersen, P. H., **16:**32, 129
Anderson, A. C., **15:**60, 139
Anderson, A. H., **1:**319(75), 333; **13:**187(126), 196
Anderson, D. A., **12:**255(58), 256(58), 272(164), 273, 291, 293, 296; **19:**72
Anderson, D. B., **10:**227(1), 228, 258
Anderson, D. E., **3:**88(68), 122(68), 126; **4:**353, 371
Anderson, D. P., **17:**3, 49, 61, 62, 73
Anderson, E. H., **13:**163(48), 193
Anderson, G. B., **16:**78, 137
Anderson, G. S., **2:**367(23), 420; **3:**66(24), 102, 125, 127; **4:**269(80); **13:**110(10), 111(11), 147; **14:**2(1), 22(1), 73, 203(8), 217(8), 237
Anderson, H. M., **18:**300–301
Anderson, J., **12:**115(214), 163
Anderson, J. C., **11:**189(152), 203; **13:**96, 106
Anderson, J. S., **6:**334(37), 346
Anderson, L. P., **12:**248(43, 44, 45), 251(43, 44, 45), 253(45), 291

Anderson, M. R., **24:**44(47)
Anderson, O., **17:**332, 333, 366
Anderson, R. J., **21:**117(8), 128(8), 221(8), 223(8), 232, 251(26), 309
Anderson, R. L., **5:**272, 275, 317, 318
Anderson, T. G., **13:**190(155), 191(155), 197
Anderson, W. T., **13:**212(39), 224
Anderson, W. W., **14:**36–37(139), 76; **21:**15(39), 72
Andersson, A. M., **17:**358, 361, 369, 370
Andersson, J. Y., **21:**117(11), 128(32), 216(119, 121), 232–233, 236, 272(35); **23:**213(20, 21)
Andersson, T., **16:**92, 139
Andersson, T. G., **21:**207(106, 107), 235
Ando, E., **17:**352, 368
Ando, H., **12:**186(93), 197(93), 200(93), 234, 264(107), 292
Ando, K., **13:**172(77, 78), 173(77, 78), 177(77, 78), 190(77, 78), 191(77); **14:**91(23), 125; **15:**103, 113, 140
Ando, T., **15:**222, 261; **21:**97(31), 98(31), 99(31), 100(31), 112, 122(27), 233, 303(52), 310
Andoh, Y., **17:**83, 128, 129, 139
Andosca, R. G., **18:**218
Andra, W., **1:**293(40), 332; **6:**263, 268, 282, 288(170, 198), 299
Andrade, E. N. Da C., **4:**120, 135, 152(68), 189
Andraden, J. D., **24:**206(2)
Andre, E., **7:**236(249f), 254
Andreev, A. A., **14:**177(143, 144), 178(150), 198
Andreev, V. M., **7:**186(146), 217(200), 246(276, 279), 247(281), 251, 253, 256
Andres, R. J., **5:**273, 317
Andrew, E. R., **6:**103, 112, 124(96), 129(96), 135(96), 145
Andrew, J. E., **13:**154(16, 20), 172(16), 181(16, 20), 182(16), 183(16, 20), 184(20), 192
Andrew, S. R., **21:**128(34), 233
Andrews, A. M., **11:**157(129), 171, 200, 202, 203
Andrews, E. R., **1:**248(43), 251(43), 274
Andrews, J. M., **14:**220(72), 238
Andrews, P. V., **6:**114(145, 146), 146
Andrews, R. A., **10:**237(25), 258
Andrievski, A. I., **7:**27(113), 29(123), 30, 34, 35(126), 40
Androshuk, A., **5:**298, 321
Andrushko, A. F., **3:**172(10), 174(10), 208; **5:**164(90), 231
Anemogiannis, K., **17:**217, 224
Anfinogenov, V. B., **15:**64, 80–81, 139
Ang, C. Y., **5:**270(102), 317
Angel, D. W., **1:**149, 173, 180(68), 186; **4:**6(16), 50(101), 91, 93; **7:**45(14), 47(14, 20), 48(14, 27), 49, 54, 55, 56, 58, 61(20), 63(27), 66(20), 70(14), 74, 75, 76, 81, 90(20), 111, 112; **10:**77(15), 164
Angilello, J., **16:**5, 124
Angilello, L. J., **14:**65(214), 78
Angst, X., **24:**169(38)
Angus, J. C., **12:**246(27, 28, 29), 290; **19:**61, 105
Anlage, S. M., **16:**12, 68, 101, 125, 136
Annoh, K., **12:**246(22), 254(22), 290
Anselmetti, D., **24:**127(51)
Ansermet, J.-Ph., **24:**154(18)
Anstead, R. J., **10:**120(52), 165
Antal, J. J., **1:**204(75), 230; **3:**32(178), 38, 40(212), 52(178), 58
Antcliffe, G. A., **11:**121(73), 201

Antell, G. R., **5:**276, 279, 318
Anthony, T. C., **15:**171, 215
Antippa, A. F., **19:**281
Antognazza, L., **16:**19, 28, 126, 128
Antreasyan, A., **23:**55(8)
Anttila, A., **13:**142(127), 150
Antypas, G. A., **7:**170, 171(110), 181(132), 189, 190(152), 191, 192(155), 196, 197(170), 198(170, 173), 243(267), 244(270), 250, 251, 252, 255; **9:**46(206), 70; **12:**114(191, 192, 194), 123(251), 124(252), 125(256), 126(263, 264), 128(256, 264), 132, 162, 165
Anzai, H., **15:**300, 311
Anzlowar, M., **16:**53, 74, 134, 137
Aoki, K., **24:**12(52), 15(52)
Aoki, M., **13:**99(100), 107
Aoki, S., **16:**88, 138
Aoki, T., **19:**97
Aoki, Y., **15:**210–211, 213, 217
Aoyagi, E., **16:**88, 139
Aoyama, T., **23:**47(158)
Apfel, J., **10:**168(8), 225
Apfel, J. H., **5:**85, 86; **9:**105(20), 119(29), 132(39), 143(54), 144
Apfelstedt, I., **16:**9, 125
Apker, L., **12:**60(34), 101(34), 108(34), 159, 162
Apling, A., **3:**169(6), 174, 183(15), 208
Appert, J. R., **7:**244(269), 255
Appleyard, E. T. S., **1:**248(42), 251(42), 274; **2:**22(22), 23, 60; **6:**104, 146
Apsley, N., **15:**234–235, 262; **21:**117(12, 13), 152(13), 153(13), 232
Arabski, J., **16:**30, 32, 129–130
Arai, F., **14:**103(47), 126
Arai, K., **23:**144(35)
Arai, T., **9:**34(157, 161), 35(157), 68
Arakawa, E. T., **2:**124(132c, e), 144; **9:**176(40, 41, 42, 43, 44, 45), 178(41), 189(79), 220(180), 221(183), 225(191), 226(191), 227(205), 228(205), 229(205), 237(222), 243(240), 256, 257, 260, 261
Arakawa, H., **12:**292
Araki, Y., **15:**148, 214
Aranovich, J., **12:**176(54), 177(54), 178(54), 179(54), 233
Aratani, K., **15:**171, 198, 216
Arau, B. A., **21:**362(44), 368
Araujo, C. A., **17:**259, 298
Araujo, G. L., **21:**313(6), 319(11), 320(11), 321(13), 341(6), 342(6), 343(40), 344(6), 351(13, 40), 355(6), 357(13), 359(13, 42), 360(6, 40), 361(6, 40), 362(40), 366, 368
Archard, J. F., **2:**199, 237
Archer, J. L., **5:**313(295, 297–299), 314(298), 322; **9:**282, 287, 296; **15:**8, 136
Archer, R. J., **1:**162, 186; **2:**223, 238; **4:**79(200, 201), 84(216), 95, 261; **23:**91(22), 101(32)
Archibald, P. C., **6:**215(43), 296
Arendt, F., **5:**118(21), 141
Arendt, M. F., **24:**12(52)
Arendt, P. N., **16:**37, 48, 130, 132
Argall, F., **11:**226(21), 314
Argyle, B. E., **9:**275, 288, 290, 292, 294, 295
Argyrakis, P., **13:**153(5), 182(5), 183(5), 192
Argyres, P. N., **1:**330(111), 334
Arias, J. M., **21:**40(84, 88), 74
Arifova, M. U., **6:**137(213), 139(213),

140(226), 141(226), 142(213), 148
Arikado, T., **13:**176(91), 177(91), 180(104), 185(91, 104), 187(91, 104), 195
Aristotle, **24:**180(1)
Arkhipenko, V. A., **18:**137
Armand, M., **17:**359, 369
Arminen, E., **17:**134, 144
Armington, A. F., **5:**161, 230
Armstrong, J., **7:**225(226), 253
Armstrong, S. R., **19:**2, 10, 15
Armstrong, W. J., **4:**262(47), 321; **5:**255(41), 315
Arnaki, R., **7:**315(22), 393
Arnason, S., **16:**42, 131
Arnaud D'Avitoya, F., **12:**116(223), 118(227), 163
Arndt, D. P., **19:**192, 239, 241
Arnell, R. D., **18:**273, 275–276, 278, 279
Arni, R., **12:**272(170, 171), 274(170, 171), 294
Arnold, H., **12:**270(154), 293
Arnold, J. C., **22:**270(129), 275
Arnon, O., **12:**7, 36(63), 50, 51
Arnone, C., **13:**171(74), 186(74), 190(74), 194
Arnquist, W. N., **3:**276, 289(12), 299
Arntz, F. O., **17:**306, 345, 347, 364, 367
Aronovich-Magran, J., **12:**235
Arons, R. R., **16:**24–25, 127
Aronson, A., **22:**52(79), 79, 254(96), 274
Aronson, B. S., **5:**310, 322
Arrowsmith, H. W., **16:**193, 230
Arthur, J. R., **11:**36(1), 42(1, 38), 44(1, 38), 56(1), 57(1), 58(1), 61, 63(111), 65(1), 66(1, 38), 67(1), 73(1), 79(1), 86, 89(1), 96, 97, 99
Arvin, M. J., **9:**28(106), 67
Arwin, H., **19:**94, 158, 169, 182
Asada, K., **14:**103(47), 126
Asahara, Y., **15:**15, 18, 137
Asahi, H., **11:**67(133b), 80(177), 100, 101
Asai, H., **21:**121(20), 123(20), 233
Asai, K., **21:**31(62), 32(62), 73
Asakawa, K., **13:**185(139), 188(139), 196
Asano, H., **14:**240
Asano, K., **6:**36(116), 38(116), 76
Asano, M., **15:**172–173, 216
Asano, T., **16:**2, 123
Asao, A., **12:**71, 75, 76, 78, 79, 159, 160
Asaoka, T., **16:**263, 269
Asch, A. E., **11:**158(131), 164, 165, 191, 192, 193, 194, 195, 196, 197, 203
Ascolese, M., **16:**32, 129
Aselage, T. L., **16:**4, 81, 124
Asg, Y. W., **16:**254, 268
Ashbum, J. R., **16:**2, 123
Ashby, C. I. H., **13:**176(88, 89, 90), 177(90), 178(88, 89), 179, 186(88, 89, 90), 189(88, 89, 90), 195
Ashby, M. F., **16:**288, 338
Ashcroft, N. W., **19:**129, 131, 134, 135, 155
Ashford, A., **7:**174(122), 250
Ashley, E. J., **2:**285(33), 304, 315(15, 16), 334(16), 336(16), 359; **4:**3(9), 11(9, 35), 26(30), 39(74), 46(82), 90, 91, 92; **6:**132(197), 148; **10:**130(69), 132(69), 165, 207(40), 226; **12:**8(33), 28(55), 36(55), 37(67), 50, 51
Ashley, J. C., **2:**124(132f, g), 144; **4:**329, 365(21), 369; **9:**167(27),

Ashley, J. C., (*continued*)
174(34), 179(54), 256, 257
Ashley, K. L., **7:**142(48), 158(78a), 248, 249
Ashok, S., **12:**235; **14:**232(118), 234(146–148), 234–235(147), 235(148), 239, 240
Ashrit, P. V., **17:**355, 356, 368
Ashworth, F., **2:**387(105), 388(105), 389(105), 422
Ashworth, J. L., **3:**137, 162, 172(15); **4:**261(33), 263(33), 314, 320; **5:**288, 320
Askew, R., **16:**123, 143
Aslam, M., **12:**147(318), 166
Asmar, N. G., **16:**4, 124
Asmussen, J., **14:**98(38), 118(38), 126; **18:**124, 154
Asom, M. T., **21:**117(7), 128(7, 30), 129(30), 130(30), 131(30), 132(30), 135(30), 136(30), 137(30), 138(30), 139(30), 140(30), 141(30), 142(30), 143(30), 144(30), 145(30), 146(30), 147(30), 148(30), 149(30), 150(30), 160(30), 170(76, 77), 171(76), 172(76), 173(76), 174(76), 175(76, 77), 176(76), 177(77), 178(77), 179(77), 180(77), 221(135, 137, 139), 222(135, 137, 139), 232–234, 236–237
Asomoza, R., **22:**7(13), 77
Aspnes, D. E., **14:**82(4), 103(4), 125; **16:**13, 125; **19:**2, 4, 6, 9–19, 23, 26, 27, 32–37, 41–43, 50–53, 59, 94, 129, 139, 148; **21:**327(21), 367
Asthana, A., **16:**99, 140
Astle, M. J., **24:**182(24)
Astoin, N., **7:**45(9), 111
Astridge, R. A., **12:**135(291), 165
Atailyan, Y. J., **15:**88, 140
Atalla, M. M., **2:**150, 190, 370(44), 420; **4:**300(120, 121), 301(117), 323, 326, 341(50), 342(50), 345(50), 347(50), 360, 369, 370
Atkins, P. W., **24:**46(92), 57(92), 58(92)
Atkinson, A., **14:**97(37), 115(37), 116(37), 126
Atkinson, J. T. N., **6:**71(209), 79
Atkinson, R. H., **4:**245, 255
Atre, S. V., **24:**33(137), 44(37), 218(77)
Attardo, M. J., **7:**265, 273, 274(72), 276(51, 72), 284, 286(51), 307, 308; **16:**240, 267
Atwood, J. L., **16:**100, 140
Atzei, A., **10:**163, 166
Aubert, A., **14:**14(49), 74; **17:**111, 142
Aubrey, J. E., **6:**116, 117, 132(164), 147
Auciello, O., **13:**116(38), 126(38), 148; **14:**229(93–95), 239; **16:**14, 17, 57, 60, 126, 134; **17:**229, 234, 235, 263, 296
Auerbach, D. I., **24:**240(74)
Augustine, F., **12:**181(69, 70), 190(69, 70), 234
Augustyn, W. H., Jr., **10:**293(8), 308
Auld, B. A., **15:**105, 140
Aurentz, D., **24:**172(42), 173(42)
Ausserr, D., **24:**169(35)
Austerman, S. B., **9:**276, 294; **13:**48(58), 58
Austin, I. G., **11:**290(120), 317
Austin, W. E., **7:**45(10, 13), 103(13), 107(84), 111, 114
Autranen, R., **3:**179(29), 209
Auvert, G., **13:**178(101), 195

Auwarter, M., **3:**11, 51(50), 54; **13:**60, 61, 62(10), 104
Averbach, B. L., **4:**221, 230, 253; **7:**27(110), 39
Avery, D. G., **2:**201, 237
Aveston, J., **16:**287, 316, 319, 334, 338
Avouris, P., **24:**99(101), 106(101)
Awender, H., **3:**24(100), 51(100), 55
Axe, J. D., **24:**44(19)
Axelrod, N., **1:**131(12), 184
Axelrod, N. N., **6:**54(148), 69, 70, 71(148), 77; **9:**209(132), 258
Aylesworth, K. D., **16:**80, 137–138, 212, 216–218, 230–231
Aylett, M. R., **13:**186(150), 189(150), 197
Ayling, S. G., **2:**369(37), 420
Aylsworth, J. W., **5:**240(9), 314
Ayoub, G. T., **19:**291
Ayres, C. F., **17:**130, 143
Ayusawa, M., **17:**325, 366
Azbel, M. Ya., **6:**82(15, 33), 91, 97, 127, 143, 144, 145, 190, 203
Aziz, R. A., **1:**244(31), 274
Aziza, A., **21:**21(43), 72
Azoulay, J., **16:**53, 134
Azouz, I. B., **24:**169(35, 36), 228(20)
Azuma, H., **14:**232(130), 239
Azzam, R. M. A., **19:**2, 5, 10, 42, 43, 50, 51, 55, 148, 150, 158, 169, 196, 203, 280, 281, 288, 305

B

Baar, D. J., **17:**228, 247, 297
Baars, J., **11:**111(39), 200
Baba, A., **17:**234, 270, 296
Baba, M., **17:**345, 367
Baba, N., **17:**318, 366
Baba, S., **11:**68, 69, 100; **16:**262, 269, 288, 338
Babaev, V. O., **13:**18(24), 20(24), 57
Babalola, I. A., **12:**114(207), 163
Babenko, Yu. E., **12:**178(64), 234
Babiskin, J., **6:**82(22), 144
Bacal, M., **14:**86(13), 113(67), 125, 127, 212(29), 237
Bach, B. W., **11:**3(1), 4(1), 5(1), 6(1), 8(1), 14(1), 19(1), 33
Bach, H., **5:**101(10a), 105(16), 119(22), 141; **14:**231(109), 239
Bachmann, K. J., **10:**252(68), 259; **17:**238, 239, 297
Bachmann, K. M., **11:**47(60), 66(60), 67(60), 97
Bachmann, L., **1:**189(11), 228; **3:**217, 271; **4:**120, 136, 136(51)
Bachmann, P. K., **17:**110, 111, 119, 141
Bachner, F. J., **9:**116(26), 144; **10:**52(111), 53(111), 54(111), 68
Bachrach, R. Z., **7:**196(167), 252
Back, R., **24:**161(21), 169(21)
Bacon, D. D., **15:**147, 150, 152, 161–162, 171, 214–215; **16:**28–29, 36–37, 129–130; **17:**118, 142; **19:**129, 139, 155
Bacon, R., **3:**246, 272
Baczewski, A., **5:**282, 319
Bädeker, K., **9:**1, 64
Bader, G., **17:**355, 356, 368
Bader, H. P., **14:**226(89), 239
Bader, M., **3:**74(52), 75(52), 126
Badertscher, G., **13:**213(49), 224
Badia, A., **24:**151(8), 154(8), 155(8b, 8c, 19), 156(8, 19), 157(8b, 8c), 158(8b), 159(8b, 8c), 161(8b, 21), 169(21), 222(103)
Badilenko, G. F., **17:**64, 65, 77
Badoz, P. A., **23:**161(60–62)
Badzian, A. R., **17:**110, 111, 119, 141; **19:**105

Badzian, T., **17:**110, 111, 119, 141; **19:**105
Baer, E., **4:**121(54), 136
Baer, Y., **9:**217(166), 259; **15:**196, 217
Baerg, W., **16:**252, 268
Baeri, P., **13:**166(60), 194
Baert, K., **23:**2(4)
Baez, A. V., **4:**50(117), 93
Bagchi, A., **9:**203(121), 258
Bagchi, S. M., **2:**110, 143
Bagg, J., **7:**330, 367(61), 394
Bagghi, S. N., **7:**2, 3(9), 37
Bagley, B. G., **16:**18, 42, 45, 126, 131; **19:**103
Baglin, J. E. E., **13:**124(74), 126(74), 149
Bahl, I., **17:**149, 219
Bahl, S. K., **7:**34(137), 40
Bahr, G. F., **1:**191(20), 228; **2:**97, 142
Bai, G. R., **16:**91, 139
Bailey, A. I., **4:**3(5), 90; **7:**324(39), 336, 337, 389, 389(104), 393
Bailey, D., **17:**171, 221
Bailey, D. S., **17:**174, 178, 222
Bailey, J. E., **16:**287, 338
Bailey, P., **13:**158(34), 169(34), 193
Baillargeon, **23:**272(96)
Bailyn, M., **6:**97, 145
Bain, C. D., **24:**2(1), 3(11, 13), 10(37), 12(13, 37), 13(13), 14(13, 37), 15(13, 37, 64), 16(13, 37, 64, 70), 17(13, 72, 74), 18(11, 13), 20(13), 21(92), 30(13, 92), 32(37, 70), 44(12, 13, 14, 48, 57), 46(12), 55(12, 14), 56(12, 14), 65(12), 70(14, 105), 74(121), 75(121, 122), 82(18), 86(18), 89(51, 55), 92(55), 112(3), 150(3), 151(8), 154(8), 156(8), 206(3), 212(57, 58, 64), 218(3, 70, 71), 220(57), 228(4, 17)
Bain, J. A., **17:**58, 76
Baines, M. J., **6:**97, 98, 145
Baird, C. S., **3:**113, 127
Baird, D. C., **1:**244(31), 274
Baird, J. D., **1:**194, 229
Baird, M. E., **6:**57(163), 77; **11:**211(7), 314
Baird, S. S., **4:**267(74), 322
Baise, A. I., **13:**153(7), 154(7), 172(7), 173(7), 182(7), 183(7), 192
Baitinger, W. E., **14:**230(104), 239
Bajars, G., **17:**318, 319, 320, 354, 366
Bajpai, S. N., **15:**34, 40, 64, 72, 81, 137–139
Bak, P., **21:**52(101), 74
Bakai, A. S., **18:**147
Baker, A. G., **5:**312, 322
Baker, F. A., **1:**28(64), 65
Baker, J. E., **21:**287(39), 310; **23:**48(162)
Baker, J. M., **14:**212(27), 220(74), 237, 238
Baker, M., **12:**282(233), 283(233), 295; **14:**96(36), 114(36), 122(36), 126
Baker, R. F., **2:**125, 144
Baker, S. P., **17:**49, 63, 77
Baker, W. E., **5:**275, 317
Baklanov, M. R., **13:**185(136), 188(136), 196
Balachander, N., **24:**11(42)
Balachandran, U., **16:**32, 129
Balagopal, R., **15:**308, 312
Balanski, M., **11:**132(94), 133(94), 142(94), 202
Balde, J. W., **2:**369(34), 420
Baldo, J., **12:**266(137), 267(137), 268(137), 293
Baldrey, J. A., **5:**269, 316
Baldwin, F. L., **2:**331(45), 332(45), 360

Baldwin, K W., **16:**53, 74, 134, 137
Balerna, A., **17:**310, 365
Bales, J. W., **23:**243(63)
Balestrino, G., **16:**99, 140
Baliga, B. J., **14:**234(146–148), 234–235(147), 235(148), 240
Baliga, S., **16:**28, 128
Balk, P., **3:**135(25), 136(25), 162; **4:**260(19), 301(19, 128), 320, 323
Balkanski, M., **21:**21(43), 72
Ball, D. J., **14:**27(105), 75
Ball, G. J., **12:**30, 36, 51
Ball, L. T., **14:**21(85), 75
Ball, R. C., **22:**5(7), 77
Ballantine, D. S., **17:**210, 212, 223
Ballard, K. H., **2:**392(122), 423
Ballard, S. S., **4:**75(188), 95
Ballato, A., **17:**188, 204, 206, 222, 223
Ballentine, P. H., **16:**109, 142
Ballman, A. A., **10:**256(59), 259; **13:**164(56, 57), 165(56, 57), 166(56), 167(57), 176(56, 57), 186(56, 57), 189(56, 57), 194
Balluffi, R. W., **7:**282(77), 286(93), 309; **16:**236–237, 267
Balon, J. R., **11:**146(116), 149(116), 151(116), 202
Balooch, M., **16:**60, 135; **24:**96(75), 97(75), 98(75)
Baltensperger, W., **6:**247, 298
Baltz, A., **4:**159(88), 171(121), 186, 189, 190; **6:**113(136), 146
Balz, W. E., **16:**41, 131
Balzarotti, A., **16:**13, 125
Bamberg, E., **24:**21(91, 102), 26(102), 85(24), 88(45), 89(50), 90(24, 45), 99(24), 212(65)
Bamer, J. B., **16:**63, 106, 112–114, 135, 141–142
Ban, I., **23:**48(170)
Ban, V. S., **16:**81, 138
Banavar, J. R., **21:**38(74, 75), 73
Banda, Y., **5:**313(304), 322
Bandara, K. M. S. V., **15:**222, 229, 233, 238–239, 241–243, 249–251, 253–254, 257–258, 261, 263–264; **21:**4(12), 8(27), 72, 113–237, 119(15), 121(26), 167(72), 168(72), 169(72), 170(76, 77, 78), 171(76), 172(76), 173(76), 174(76), 175(76, 77), 176(76), 177(77), 178(77), 179(77), 180(77, 78), 181(78), 189(82)
Bando, Y., **16:**25, 51, 73, 127, 132; **17:**229, 236, 287, 290, 296
Bandy, S. G., **11:**100
Banerjea, A., **17:**56, 76
Banerjee, A., **12:**169, 170(27), 172, 173(27), 174(27), 177(54a), 178(27), 179(27, 65), 180(65), 183(25, 27, 28, 73), 185(30, 54a, 73), 186(30, 54a, 73), 187(54a, 73), 189(27, 28, 29, 30), 190(27, 30), 192(30), 193(27), 194(27), 195(24), 196(24, 27), 197(27, 73), 198(30, 73), 199(29, 30, 73), 225(28), 228(30), 233, 234
Banerjee, B. C., **1:**226, 232; **4:**212, 253
Banerjee, I., **17:**24, 45, 48, 49, 51, 52, 56, 57, 74
Banfield, J. R., **14:**223(76), 238
Bange, K., **17:**310, 332, 333, 348, 356, 357, 365–367, 369
Banks, C., **15:**47, 55, 138–139
Banks, E., **17:**306, 364
Banning, M., **1:**135, 147, 149(24), 155(24), 185; **3:**9(36), 51(36), 54
Banning, S. A., **16:**84, 90, 138
Bansal, N. P., **16:**98, 140
Banse, B. A., **19:**35
Bao, Z. L., **16:**41, 131
Bar, G., **24:**97(78), 98(78, 84)

Baraff, G. A., **4:**332, 370; **6:**140(228), 149
Barakat, N., **4:**86(227), 96
Baral, D., **17:**43, 48, 49, 50, 51, 62, 75, 77
Baranov, A. K., **6:**225(86), 297
Baratoff, A., **24:**127(51), 222(102)
Barbee, T., **17:**153, 220
Barbee, T. W., Jr., **17:**8, 12, 66, 74, 77
Barber, D. J., **22:**113(28), 115
Barber, J. A., **13:**8(18), 57
Barbero, C., **17:**326, 366
Barbonneau, F., **17:**317, 365
Barbour, J. C., **18:**209
Barboux, P., **16:**13, 56, 63, 125, 134
Bard, A., **24:**27(125), 44(25, 50), 68(50), 82(16), 86(16), 230(44)
Bard, A. J., **17:**323, 366
Bardas, D., **12:**143(313), 166
Bardasis, A., **6:**298
Bardeen, J., **1:**255(57), 256(57), 258(57), 274; **4:**339, 370; **11:**132, 202; **13:**201, 202(6), 223
Bardell, P., **4:**320(166), 324
Bardell, P. H., Jr., **4:**370
Bardini, N. G., **6:**2, 73
Bardos, L., **12:**276, 294; **18:**216
Barean, J. P., **24:**206(7)
Barefield, E. K., **16:**92, 139
Bareiss, M., **3:**63, 125
Barenblatt, G. I., **16:**279, 337
Barger, W. R., **17:**210, 212, 223
Barker, A., **14:**99(41), 110(41), 126
Barker, A. S., Jr., **9:**154(12), 236(218), 238(226, 227), 256, 261; **10:**32(59), 66; **11:**84(189, 193), 101, 102
Barker, D. L., **13:**157(31), 193
Barker, R. A., **18:**70
Barlett, J. T., **4:**152(70), 189
Barlow, K.-J., **14:**92(91), 127
Barna, G. G., **17:**354, 368
Barnard, B. R., **16:**31–32, 129
Barnard, G. P., **1:**40, 66
Barner, B. J., **24:**66(104)
Barnes, B. T., **10:**36(64), 67
Barnes, C. R., **3:**137, 162; **5:**288, 320
Barnes, J., **21:**313(5), 334(5), 336(5), 348(38), 355(38), 356(38), 366–367
Barnes, J. M., **21:**332(28), 334(29), 338(28), 341(29), 342(28), 355(28, 29), 367
Barnes, P. A., **13:**218(61), 224
Barnes, S. C., **4:**199(58, 60), 200, 204(58, 60), 206, 209, 210, 211, 228, 233(58, 84), 252, 253, 254
Barnett, D. M., **17:**43, 47, 75
Barnett, S. A., **13:**109(5), 116(5), 147; **14:**61(198), 65(209), 68(209), 78; **17:**4, 5, 6, 7, 10, 13, 18, 19, 23, 24, 25, 27, 28, 32, 34, 47, 49, 53, 63, 64, 65, 71, 73–77, 131, 143
Barnham, K., **21:**325(18), 351(18), 353(18), 367
Barnham, K. W. J., **21:**312(3, 4), 313(5), 322(3), 323(4), 324(4, 15), 325(15), 329(15), 330(27), 331(27), 332(27), 333(3, 15), 334(5, 15, 29), 336(5, 15), 341(29), 342(29), 348(38), 353(27), 355(27, 29, 38), 356(38), 357(3), 361(43), 366–368
Barnoski, M. K., **10:**255(55), 259
Barns, R. L., **5:**156(149), 185, 233; **9:**264, 267, 276, 287, 294
Barrat, A., **24:**169(37)
Barrell, H., **4:**85(221), 96
Barrer, R. M., **6:**304(6), 343(61), 346, 347
Barrera, J., **11:**53(88), 98
Barrera, R. G., **19:**11

Barrett, C. S., **4:**142(20), 187, 227, 253; **7:**33(138), 40
Barrett, D. L., **5:**284(194), 319
Barrett, J. H., **17:**22, 74
Barrett, M. A., **4:**84(217), 96
Barron, A. C., **13:**121(64), 130(92, 94), 131(92), 133(94), 134(92, 94), 135(94), 145(64), 148, 149
Barrow, A. C., **13:**132(102), 150
Barry, C., **16:**80, 137–138
Barsony, I., **11:**244(51), 245, 315; **15:**300, 311
Bart, W. L., **1:**207(98), 227(98), 230; **3:**253(88), 255(88), 257(88), 273; **16:**332, 339
Barteau, M. A., **24:**240(75)
Bartelink, D. J., **12:**60(32), 75(32), 159
Bartell, L. S., **7:**15(35), 38
Bartels, G., **9:**276, 284, 297; **15:**14, 137
Barth, J., **19:**280
Barth, N., **2:**44(73), 62
Barthell, B. L., **17:**4, 64, 73
Bartholomew, C. S., **16:**57, 60, 134
Bartholomew, R. F., **9:**33(120), 68
Bartlett, R. S., **6:**104, 146
Bartoe, J.-D. F., **10:**91(27a), 164
Bartoli, F. J., **23:**279(36, 102)
Barton, E., **7:**11(22), 37
Barton, J. L., **4:**200(76), 252
Barton, R., **16:**42, 131
Barton, R. W., **16:**2, 69, 123, 136; **17:**209, 223, 272, 290, 293, 298
Barton, S. W., **24:**184(35, 41), 188(41), 192(35)
Bartram, M. E., **22:**209(67), 273
Barua, A. K., **15:**281, 310
Bar'yakhtar, V. G., **15:**26, 137
Baselgia, L., **15:**95, 140
Baseman, R. J., **16:**57, 61–62, 71–72, 134–136
Bashara, N. M., **3:**132, 162; **4:**39(70), 92; **19:**2, 5, 42, 43, 50, 51, 55, 148, 150, 158, 169, 195, 203, 280, 281, 288, 291
Basharov, R., **14:**196
Baskes, M. I., **22:**127(36), 172
Basol, B., **13:**79(81), 95(90), 106
Bass, J., **6:**97, 103(82), 113, 119, 124(82), 127, 129, 130, 131(195), 135(82), 145, 148
Bass, J. C., **7:**240(254b), 255
Bass, S. J., **7:**223(218, 219), 253; **12:**130(275), 165
Bassani, F., **12:**57(20), 58(20), 68(20), 159
Basseches, H., **2:**397(155, 156), 424; **3:**108(106), 115, 127, 128; **14:**2(13), 73
Bassett, D. W., **11:**111(40), 119(40), 200
Bassett, G. A., **1:**195(35), 196(36, 42), 218, 229, 230, 280(5), 330, 331, 334; **2:**13, 14(10), 18, 19, 22(6a), 60, 132(153), 134, 145; **3:**32(160), 57, 212(4, 7), 213(4), 218, 257(106), 270, 273; **4:**118, 119(42, 43), 120, 128(69), 133(43), 135, 136, 144(28), 150(4, 38, 42, 61), 152(42), 155, 156(38), 160, 165, 166, 169, 175, 187, 188, 189, 226(133), 254; **9:**4(9, 10), 65
Bastard, G., **15:**222, 261; **21:**328(22, 23), 367; **23:**243(66)
Basu, P. K., **15:**222, 261
Batchelder, D. N., **24:**11(46), 21(46), 33(46), 34(46), 228(31)
Batdorf, R. L., **4:**332(29), 370
Bate, R. T., **6:**92, 121, 123, 136, 145; **11:**169, 203
Bateman, J. B., **2:**216(19b), 238;

Bateman, J. B., (*continued*) **3:**18(83), 51(83), 55; **7:**333(75), 336(99), 342, 354, 358, 367(113), 394, 395
Bates, B., **9:**132(40), 144; **10:**117, 165
Bates, C. W., **7:**34(142), 40
Bates, C. W., Jr., **12:**75(87), 160
Bates, R. I., **18:**275–276
Batey, J., **15:**227, 244, 261
Bath, H. H. A., **3:**19(117), 24(117), 26(117), 51(117), 56
Batke, E., **21:**96(27), 111, 155(64), 234
Batlogg, B., **16:**3, 124
Batra, I. P., **19:**253
Batty, W., **23:**228(30)
Baucom, K. C., **23:**231(16)
Baud, C., **12:**121(236), 164
Baudry, P., **17:**361, 370
Bauer, E., **1:**196, 229; **4:**223, 253; **5:**193, 233; **9:**215(153), 217(153), 259; **16:**7, 125
Bauer, G., **2:**240, 303
Bauer, G. H., **12:**289(269), 296
Bauer, H. J., **2:**30(40), 61; **14:**219(56), 238
Bauer, J., **12:**270(155), 293
Bauer, V. G., **13:**110(7), 147
Bäuerle, D., **16:**79, 137
Baughman, R. H., **16:**99, 140
Baughman, R. J., **16:**54, 134
Baumann, R. C., **17:**228, 260, 298
Baumeister, P., **1:**70(22), 120; **4:**50(102), 93; **5:**60(13), 69, 70, 71, 72, 86, 134(37), 141; **9:**74(9, 10), 79(10), 82(10), 86(10), 102(10), 123(9), 124(9), 127(9), 141(53), 142(53), 143(53), 144; **12:**7, 36(63), 50, 51; **19:**237, 238–240
Baumeister, P. W., **3:**2(20), 53; **10:**168(5), 175(27), 187, 204(35), 225, 226
Baur, G., **17:**350, 367
Bauser, E., **11:**45(57), 97
Baxter, R. D., **7:**170(101), 250; **14:**231(108), 239
Bay, H. L., **22:**110(23), 115
Bayard, R. T., **1:**26, 28, 65; **3:**4, 51(26), 53
Bayer, E., **24:**151(5)
Bayer, G., **5:**106(17), 141
Baylander, E. G., **12:**259(82), 292
Bayley, A. R., **14:**226(83), 238
Bayly, A. R., **13:**126(79), 149
Bayne, M. A., **14:**31(124), 32(124), 76
Bazin, C., **2:**228, 238; **6:**183, 184, 188, 203
Bazyura, R. I., **6:**339(56), 347
Beach, D. B., **23:**30(115), 32(115)
Beadle, C. W., **2:**179(61), 192
Beaglehole, D., **6:**199(63), 204; **9:**226(193), 260; **19:**280
Beagley, B., **24:**135(62)
Beale, H., **13:**99(110), 107
Beale, P. D., **17:**241, 259, 287, 297, 298
Bealty, D. C., **16:**250, 268
Beams, J. W., **1:**207(96, 97, 98, 99), 227, 230; **2:**53, 54, 56, 58, 62; **3:**253; **16:**332, 339; **17:**43, 75
Bean, C. P., **1:**315, 324(92), 333, 334; **2:**31(51, 52), 61; **6:**15, 75, 206(3, 5); **16:**11
Bean, J. C., **11:**36(29), 97; **14:**232(115), 239; **15:**222, 261; **17:**26, 75; **21:**79(6, 7), 111; **23:**2(2), 3(13), 55(8)
Bean, K. E., **5:**296, 320
Beard, D. W., **7:**390(209), 397
Beasley, J. K., **10:**59(129), 68
Beasley, M. R., **16:**2, 12, 42, 50, 54,

68, 101, 123, 125, 131, 134, 136; **17:**2, 73
Beattie, J. R., **2:**339, 360; **4:**84(212), 96; **10:**132(70), 143, 165
Beauchamp, K. M., **16:**51, 132
Beauchamp, W. T., **10:**22(35), 56(35), 57(35), 66
Beaumont, D., **2:**315(9), 316(9), 359; **4:**60(140), 94; **12:**6(26), 50
Bebb, H. B., **7:**157(70), 249
Bebker, W., **16:**51, 133
Bechmann, R., **3:**24(122, 125), 51(122, 125), 56
Beck, A. H., **12:**88, 89(128), 99(128), 161
Beck, N., **3:**168, 208
Beck, T. L., **24:**33(133), 75(123), 211(52)
Beck, W. A., **11:**155(127), 198, 199, 202, 203; **21:**117(9), 128(9), 137(49), 180(49), 191(87), 212(49), 221(87), 232, 234, 235, 257(32), 309
Becke, H., **3:**140(50), 163
Becker, A., **2:**68(11), 141
Becker, E. J., **1:**27(58), 65
Becker, F. S., **15:**299, 301–302, 311
Becker, G. E., **14:**232(115), 239
Becker, J. A., **1:**27, 65, 205(83), 230
Becker, J. J., **1:**316(70), 333; **2:**31(53), 61; **15:**185, 216
Becker, P. C., **21:**193(91), 235
Becker, R., **1:**283(21), 332; **4:**98(11), 99(11), 134, 193, 251
Becker, S., **16:**79, 137
Becker, W. M., **5:**274, 317
Beckerman, M., **2:**379(84), 390(120), 391(120), 392(84, 120), 422, 423; **3:**4(27), 5(27), 51(27), 53
Beckey, H. D., **1:**205, 230
Beckman, W. A., **10:**11(16), 14(16), 23(16), 65
Beckmann, P., **10:**168(13), 173(13), 176, 225
Becquerel, A. E., **3:**275, 299
Bedeaux, D., **19:**128, 129, 139, 140, 173
Bednorz, J. G., **16:**2, 123
Bedo, D. E., **7:**46(18), 48, 70(26), 112
Beebe Jr., T. P., **24:**36(145)
Beeck, O., **2:**48(78), 50, 51, 52, 62; **3:**219, 271
Beeghly, H. F., **3:**39(193), 52(193), 58
Beeman, D., **22:**172, 312(45)
Beer, A. C., **5:**222, 235; **7:**170(101), 250
Beery, J. G., **16:**42, 47, 131
Beesley, J. P., **6:**54(153), 77
Behm, R. J., **23:**32(119)
Behrens, G., **3:**32, 39(158), 51(158), 57
Behringer, A. J., **3:**176(23), 209
Behringer, R. E., **1:**315(66), 333; **13:**161(39), 182(39), 183(39), 193
Behrisch, R., **3:**62, 128; **14:**217(45, 46), 229(46), 230(46), 237, 238
Behrish, R., **17:**83, 84, 140
Behrndt, K., **7:**18(53), 38
Behrndt, K. H., **1:**32(77), 58, 59, 60, 65, 67, 316(68), 333; **3:**4(27), 5(27), 9(40), 10(40), 13(60), 24(104, 106, 107, 112, 113), 25(113), 26(107, 113), 27(131, 136), 28(136), 29(136), 31(106, 151), 32(40, 106, 154), 34(154, 167), 35, 40(40, 106), 42(1); **4:**11, 21, 23(36), 91; **5:**197(172), 234; **12:**12, 50
Behrndt, M. E., **1:**252(49), 254(49), 274

Beischer, D. E., **7:**330(63), 334(63, 93, 94), 348, 374(194), 394
Beischer, D. V., **5:**306, 321
Beister, G., **14:**36(138), 48(162), 55(162), 76, 77; **17:**111, 116, 142
Beitat, U., **13:**117(40), 148
Bek-Bulatov, I. Kh., **14:**145(65), 196
Bekebrede, W. R., **9:**270, 296
Belan, R. M., **16:**16, 25, 126
Belder, G. F., **24:**95(73)
Belford, G. G., **4:**339, 370
Belitsky, V. F., **16:**237, 267
Belk, J. A., **3:**9(37), 10, 36(169), 49(37), 51(37), 52(169), 54, 57
Belk, R., **4:**26(62), 92
Bell, A. T., **12:**238(6), 248(40), 251(40), 290; **13:**67(39), 105; **14:**84(6), 85(6), 125
Bell, C. M., **24:**12(52), 15(52)
Bell, D. G., **3:**277, 300
Bell, H. B., **4:**265(60), 314(60), 319(60), 321
Bell, J. A., **17:**46, 49, 51, 53, 76
Bell, J. M., **17:**318, 365
Bell, R. J., **9:**237(223), 261
Bell, R. L., **7:**243(267), 244(270), 255; **12:**113(187), 114(191, 194), 123(251), 125, 126(263, 264), 128(264), 132(258), 135(297), 137, 138, 162, 163, 164, 165
Bell, R. O., **5:**154(42), 230
Bellamy, W. C., **11:**42(47), 89(232), 97, 103
Bellan, P. M., **15:**45, 138
Bellemans, A., **22:**134(51), 142(51), 172; **24:**209(35)
Bellevance, D. W., **11:**132(90), 201
Belohoubek, E., **16:**106, 141
Belous, V. A., **14:**170(123, 124), 198
Belser, M., **16:**123, 143
Belser, R. B., **2:**368(31, 32), 393(32), 420; **3:**24(108), 51(108), 55, 96, 108, 109, 126; **6:**101(90), 145, 311, 346
Belt, R. F., **15:**113, 141
Beltram, F., **21:**133(44), 233, 249(25), 284(25), 309
Belvinger, A. J., **5:**283, 319
Belyaeva, N. N., **12:**204(97), 217(97), 234
Belyakov, L. V., **13:**162(43), 163(47, 49), 164(43), 165(43), 183(49), 186(47), 189(47), 193
Ben Daniel, D. J., **18:**133
Ben-Dov, I., **24:**19(83), 20(83)
Benacka, S., **16:**52, 133
Benbenek, J. E., **1:**37(98), 66
Bendre, S. T., **16:**70, 136
Beneking, H., **7:**153(62), 248
Benettar, J. J., **24:**169(35)
Benferhat, R., **19:**8
Beni, G., **17:**304, 364
Beniere, F., **16:**242–244, 253–254, 265, 267
Benjamin, P., **1:**207(101), 209, 230, 231; **2:**408(192), 424; **3:**96, 126; **6:**312, 346
Bennet, A. J., **9:**159(15), 256
Bennett, B. L., **16:**37, 130
Bennett, H. E., **2:**285, 304, 315, 334(16), 336, 359; **4:**4(10, 13), 9(26), 10(32), 11(35), 12(10, 32), 13, 16(32), 17, 18(40), 19(10), 38(69), 39(69, 74), 41(69), 46(10, 82), 48(88), 49(88), 50(13, 26, 98), 58(127), 60(37, 127), 62(98), 64(326.27), 76, 132, 148, 152(2), 177, 202; **7:**54(37), 112; **9:**21, 67, 227(201), 260; **10:**35, 37(74), 67, 130(69), 132(69), 165, 169(14), 207(40), 225, 226; **12:**6
Bennett, H. S., **15:**191–192, 217

Bennett, J., **24:**229(42)
Bennett, J. A., **1:**279(1), 331; **3:**19, 51(90), 55; **13:**61(12), 104
Bennett, J. M., **2:**235(40), 238, 285, 304, 315(15), 359; **4:**2(3), 3(3), 6(15), 9(26), 10(32), 12(32), 16(32), 25(60, 61), 26(15, 30), 28(15), 34(15, 60, 61), 36(15), 39(74), 46(82), 50(26), 55(123), 64(32), 85(15), 86(15, 60, 61), 88(15, 60, 61); **6:**27, 76, 132(197), 148, 152(2), 188, 202; **7:**54(36, 37), 66(51), 112, 113; **9:**21, 67; **10:**41(78), 67, 97(32a), 164, 207(40), 226; **12:**6, 28(55), 36(64), 37(67), 38(64), 50, 51; **13:**121(63), 148; **19:**199, 305
Bennett, L. H., **16:**59, 79, 135, 137
Bennett, P., **6:**106(123), 107, 131(123), 146
Bennett, R. J., **5:**152(34), 186(34), 199(34), 230, 269(100), 272(100), 317
Bennett, R. S., **14:**220(73), 238
Bennett, V., **13:**126(78), 149
Bennett, W. R., **17:**49, 51, 53, 76
Bennewitz, C. D., **14:**27(109), 75
Bensaoula, A., **21:**325(17), 330(17), 332(17), 338(17), 350(17), 362(17), 367
Benson, D. K., **17:**315, 316, 365
Bensoussan, M., **13:**180(103), 195
Bentini, G. G., **14:**122(86), 127
Benz, M. E., **24:**71(111)
Benzig, W. C., **12:**266(137), 267(137), 268(137), 293
Berberich, P., **16:**52, 133
Berbezier, I., **23:**24(96), 30(96)
Berckhemer, H., **2:**44(76), 62
Berdyshev, A. A., **6:**247(150, 151), 298
Berenbaum, L., **7:**264, 265, 266(46), 267(71), 270(66, 67), 274, 275(71), 283(66), 284(66, 71), 290(97), 293(46, 71), 299, 301(71), 304(97), 307, 308, 310
Berendsen, H. J. C., **22:**132(48), 134(54), 136(54), 172–173; **24:**220(85)
Berera, G., **17:**357, 358, 369
Berezin, A. B., **16:**52, 133
Berg, O. E., **7:**102(81), 114
Berg, R. S., **10:**58(128), 68
Berg, S., **12:**248(43, 44, 45), 251(43, 44, 45), 253(45), 291; **14:**14(50, 51), 74; **17:**111, 112, 116, 135, 142, 144
Berge, B., **24:**181(12), 192(12), 199(12)
Berger, H., **2:**175, 191; **3:**171(16, 87), 172, 176(16), 208, 210; **5:**218, 234
Berger, H. H., **11:**88(222), 102; **13:**213(50), 214(50), 224
Berger, V., **21:**136(46), 233
Berggren, K. K., **24:**230(44)
Bergh, A. A., **7:**158(77), 205(182), 249, 252
Berghezan, A., **1:**223, 232
Berghouse, B., **12:**286, 295
Berglund, C. N., **12:**81, 82, 160; **21:**21(48), 72
Bergman, J. G., **17:**279, 299
Bergmann, E., **17:**83, 95, 128, 130, 139, 141
Bergmann, F., **3:**11(54), 51(54), 54
Bergmark, T., **7:**45, 48(16), 111
Bergstresser, T. K., **7:**173(118), 250
Bergstrom, D. B., **22:**240(87), 274
Berkenblit, M., **5:**274(128), 275(136), 317, 318
Berkey, W. E., **14:**150(81, 82), 197

Berkeyheiser, J. E., **7:**169(96), 250; **13:**51(60), 52(60), 53(60), 58
Berkley, D. D., **16:**51, 133
Berkovits, V. L., **19:**2
Berkowitz, A. E., **6:**214(38), 215, 223, 296
Berkowitz, H. L., **15:**222, 226, 229, 261
Berkowitz, J., **17:**307, 364
Berkowitz, M. L., **24:**220(87)
Berkowski, M., **16:**5, 124
Berkstresser, G. W., **16:**101, 141
Berkstresser, R., **5:**257(47), 315
Berlincourt, D. A., **13:**48(58), 58
Berman, L. E., **24:**184(36, 43)
Bernal, J. D., **2:**110, 143
Bernard, W. J., **6:**6(27), 7(27), 26, 33(27), 74
Bernasek, S., **24:**112(5), 117(15), 128(15)
Berndt, W., **2:**23(24), 46, 60, 62
Berne, B. J., **24:**207(23)
Bernert, B., **9:**188, 257
Bernett, M. K., **7:**329(49), 393
Bernieri, E., **17:**310, 365
Berning, J. A., **1:**117(49), 121, 171(67), 186; **2:**246, 304
Berning, P. H., **1:**70(20), 92(40), 104(28), 117(49), 120, 121, 148, 154(53), 171(67), 175(53), 178(53), 182(53), 185; **2:**242(13), 246, 265(25), 273, 276, 304, 338(58), 360; **3:**10(49), 51(49), 54, 124, 128; **4:**39(72), 40(72), 92; **5:**53(9), 85; **9:**16(75), 66, 74, 75, 76(11), 78(11), 79(8), 82, 91(8), 95, 116, 143; **10:**54(115), 68, 77(20), 118, 164; **12:**10(37), 48(37), 50
Bernland, L. G., **19:**168
Bernstein, A. I., **19:**195
Bernstein, J. L., **17:**229, 230, 279, 295
Bernstein, S., **7:**354, 355(136, 139, 142), 356(136, 139), 358(136), 395
Berréhar, J., **16:**292, 334, 340
Berreman, D. W., **19:**282, 288
Berrera, G., **17:**306, 345, 347, 364
Berry, A. D., **16:**85, 138
Berry, B. E., **5:**158, 230
Berry, B. S., **17:**49, 76; **22:**120(17), 121(17), 172
Berry, J., **11:**78(166), 83(166), 101
Berry, R. W., **2:**371(63), 379(85), 393(85, 125, 142), 397(63), 398(85), 405, 406(125), 407(125), 409(85), 421, 422, 423; **3:**108(104), 109, 127, 133(18), 162; **4:**258, 282, 320; **6:**4(17), 5(17), 7(17), 74; **14:**2(7, 12), 29(12), 30(116), 73, 76
Bersey, S., **16:**112–114, 142
Berson, B., **7:**239, 240(254a), 255
Bertaut, F., **4:**222, 224(111), 253
Bertelsen, B. I., **2:**405(183), 424
Berti, M., **14:**122(86), 127
Bertilsson, L., **24:**19(79), 20(79)
Bertness, K. A., **21:**322(14), 337(14), 367
Bertozzi, C. R., **24:**23(112), 27(112), 30(112), 44(11), 45(11), 219(80)
Berz, M., **21:**121(18), 123(18), 232
Besoke, M. K., **24:**88(45), 90(45), 212(65)
Besselov, V. N., **19:**2
Besser, P. J., **9:**276, 282, 287, 294, 296; **15:**8, 136
Besson, J., **13:**76, 79(61), 105
Besson, J. M., **11:**132(94), 133(94), 142(94), 202
Bessonova, E. S., **12:**176(52), 233

Bessonowa, E. S., **5:**94(6), 121(6), 130(6), 141
Best, G., **2:**328(34), 359
Best, P. E., **2:**100(61, 62), 142
Betarbet, S., **21:**62(116), 63(116), 74
Betarbet, S. R., **21:**39(82), 65(126), 66(126), 67(126), 68(126), 69(126), 70(126), 73, 75
Beterov, I. M., **13:**185(136, 137), 188(136, 137), 196
Bethe, H., **4:**351, 357(76), 371
Bethe, H. A., **1:**198(47), 229; **2:**68, 118, 124(143), 141, 143, 144
Bethea, C. G., **15:**222, 226, 229–230, 233–234, 236–239, 261–262; **21:**103(39), 106(45), 107(39, 45), 108(45), 112, 116(5), 117(6, 7, 10), 118(14), 121(21), 123(5, 21), 124(5), 125(5), 128(7, 10, 14, 38), 129(5, 39, 40), 133(14, 44), 136(5, 39, 40), 137(14, 47), 141(14, 50)
Bethell, D., **24:**154(15, 16)
Bethge, H., **1:**194, 229; **4:**150(46), 152(46), 188
Bett, R. F., **17:**153, 154, 220
Betts, F., **7:**32, 33(134, 135), 34(142), 40
Betts, J. F., **7:**333(76), 394
Betz, D. F., **17:**313, 365
Betz, G., **6:**11, 13(43), 15, 74; **14:**9(31), 16(31), 74
Betz, J., **16:**81, 138
Bevacqua, S. R., **5:**277(156), 318
Bevans, J. T., **2:**330, 360; **4:**60(153), 94
Bewilogua, K., **12:**246(35), 248(35), 251(62, 63), 256, 257(65), 290, 291; **13:**99(108), 107, 111(20), 143(20), 147
Beyeler, M. U., **17:**310, 314, 315, 365
Beyer, W. H., **24:**182(24)
Beyler, C. A., **19:**12, 32, 35
Beynon, J., **15:**289, 311
Bezak, V., **6:**137, 139, 148
Bezuidenhout, D. F., **19:**237
Bhadra, R., **16:**52, 133; **17:**44, 48, 49, 50, 51, 58, 59, 76
Bhalla, A. S., **17:**241, 297
Bhan, M. K., **15:**286, 310
Bhandage, G. T., **16:**98, 140
Bharagava, R. W., **7:**162(84, 86), 211(84), 249
Bhasin, K. B., **16:**5, 43, 50, 107, 124, 131–132, 137, 141
Bhat, B. M., **12:**71(64), 72(64), 85(101), 160
Bhat, R., **19:**2, 16, 17, 27, 32–37; **21:**327(21), 367
Bhatia, T. B., **12:**148(320), 166
Bhatt, H. J., **7:**296, 302(109), 310
Bhattacharjee, T., **15:**33, 137
Bhattacharya, P. K., **21:**207(108), 235
Bhattacharya, R. N., **16:**97, 99, 140
Bhattacharya, S., **19:**161
Bhattachayra, P., **15:**222, 260
Bhide, G. K., **12:**85(101), 148(320), 160, 166
Bhola, S. R., **5:**270(108), 271, 317
Bhushan, M., **14:**55(172), 77
Biard, J. R., **7:**158(78a), 214(196), 249, 253
Biberian, J. P., **13:**185(131), 187(131), 196
Biberman, L. M., **2:**94, 141
Bibik, V. F., **12:**71(60), 74(60), 78(60), 80(60), 81(60), 84(60), 119(60), 159
Bicelli, L. P., **4:**199(65, 66, 67, 68), 200(65, 66, 67, 68), 204, 207, 209(65, 67, 68), 210, 211(68), 228, 236(66), 252

Bickley, W. P., **2:**369(41), 420; **3:**110, 127
Bickmann, A. F., **24:**19(81), 20(81, 89)
Bicknell, R. W., **4:**150(59), 188; **5:**152(34), 186(34), 187(153), 199(34), 230, 233, 269(100), 270(104, 105), 272(100), 317; **7:**28(120), 40
Biebel, E. M., **17:**217, 224
Biebl, H., **21:**121(17), 123(17), 152(17), 232
Biebuyck, H. A., **24:**2(2), 10(39), 14(39, 60), 15(60, 64), 16(64), 17(73), 18(88), 20(88), 21(39, 92, 93), 23(39), 30(92, 93), 44(48, 58), 89(49), 98(83), 106(111), 119(42), 121(42), 150(3), 212(58, 63, 64), 219(63), 228(25), 229(35, 38), 230(47), 235(58), 236(60), 237(62), 238(62, 66), 245(84), 246(84), 248(89, 90)
Biefield, R. M., **13:**176(90), 177(90), 186(90), 189(90), 195; **15:**222, 260; **21:**364(47), 368; **23:**231(16), 278(101)
Bieganski, P., **19:**180
Biegelsen, D. K., **15:**278–279, 310; **23:**49(30)
Bienenstock, A., **7:**32, 33(134, 135), 34(142), 40
Bierig, R., **1:**43(120), 66
Biersack, J. P., **14:**6(22), 73; **22:**32(43), 78, 164(85), 173
Bigan, E., **19:**8
Bigelow, W. C., **24:**2(6), 3(6, 9), 17(6, 9), 181(16)
Bikel, W., **7:**67(55), 113
Bikerman, J. J., **7:**327(44), 343(44, 115), 393, 395
Bilaniuk, N., **15:**115, 117, 141
Bilberger, M., **22:**54(58), 78
Bilby, B. A., **3:**268, 273
Bilger, G., **12:**289(269), 296
Billat, R., **17:**310, 355, 365, 368
Billgren, P. E., **17:**136, 144
Billings, W. C., **17:**184, 222
Billington, C., **3:**17, 18, 51(76, 77), 55
Billow, H., **1:**202(70), 229
Bills, D. G., **1:**6, 29(11), 36, 46, 66
Bilous, O., **4:**320(167), 324
Bimberg, D., **19:**40, 41
Bindels, J. F. M., **13:**187(125), 196
Binder, K., **19:**253
Binh, L. N., **13:**135(105), 136(105), 139(105), 140(105), 141(105), 150
Binias, G., **9:**178(51), 257
Binnig, G., **24:**82(1, 2, 3), 84(22)
Binsma, J. J. M., **23:**226(11), 227(11), 228(11), 229(11), 232(11)
Biondi, M. A., **1:**14, 15, 64, 255, 265(60), 275; **4:**60(139), 94
Bionta, R. M., **22:**154(75), 155(75), 173
Biragnet, F., **6:**230(109), 233, 234, 235, 243(140), 244(140), 245(140), 250(109, 118), 277(140), 297, 298
Birch, J., **17:**15, 16, 17, 24, 31, 74
Birchenall, C. E., **4:**123(59), 127(59), 136; **6:**345, 347
Bird, R. B., **13:**8(17), 57
Bird, V. M., **9:**227(197), 260
Birecki, H., **15:**171, 215
Birkhoff, R. D., **2:**67(7), 70(13a), 124(132b, c, e), 128(7), 140, 144; **9:**176(40, 43, 44, 45), 256, 257
Birks, L. S., **1:**205(92), 216(128), 230; **3:**39, 52(189), 58
Birli, R., **24:**21(98), 22(98)
Birnbaum, A., **7:**350, 395

Birnbaum, M., **13:**185(138), 188(138), 196
Birtcher, R. C., **6:**124(190), 127(190), 148
Birtwistle, J. K., **3:**19(91), 51(91), 55
Biryukov, S. V., **19:**252
Bis, R. F., **11:**109, 111, 116(66), 200, 201
Bisanti, P. B., **17:**49, 51, 52, 76
Bischow, K. H. J., **15:**167
Bishop, A. R., **24:**150(2)
Bishop, C. A., **14:**44(155), 77
Bishop, F. W., **3:**19(89), 51(89), 55
Bishop, S. G., **11:**50(73), 98
Bisset, D. C., **7:**354, 355(143), 395
Biste, L., **12:**270(154, 155), 293
Biswas, D., **15:**222, 260
Biswas, S., **13:**96(92), 106
Bitner, L. R., **14:**43(152), 77
Bito, Y., **17:**361, 362, 370
Bittman, C. A., **7:**255, 257
Biunno, N., **16:**57, 60, 134
Bjonholm, T., **24:**98(95)
Bjorck, E., **3:**112, 127; **4:**267(73), 322
Bjork, R. H., **9:**244, 261
Bjorkholm, J. E., **13:**153(14), 182(14), 183(14), 192
Blachman, A. G., **14:**66(215), 78; **22:**119(10), 121(10), 171
Black, A., **24:**240(70)
Black, G. M., **11:**111(41), 119(71), 132(41), 165(71), 166(71), 186(41), 187, 200, 201
Black, J. R., **7:**265(52, 54), 279(75), 282(52), 284, 286(54), 307, 308; **16:**240, 267; **19:**149
Black, P. W., **10:**39(75), 40(75), 41(75), 43(75), 67
Blackburn, H., **1:**63, 67; **2:**365(9), 403(180), 420, 424; **3:**6, 51(32), 52(32), 53, 215(15), 220(30, 31), 225, 229(30), 234, 235(30, 31), 236(30), 237, 238(15, 30), 239, 243, 270, 271, 272
Blackburn, L. D., **7:**18(59), 38
Blacknall, D. M., **7:**142, 147(47), 248
Blackwell, H. E., **4:**68(165), 94
Blackwell, J., **21:**117(7), 128(7), 221(139), 222(139), 232, 237; **24:**163(26)
Blades, J. D., **1:**55, 67
Blair, I. C., **12:**272(177), 294
Blair, J. C., **7:**265(50), 269, 270(62, 63), 286(50), 307, 308; **16:**240, 254, 267–268
Blakely, J. M., **3:**216, 255, 256, 257, 258(17), 261(17), 262, 263, 264(17), 268; **16:**236–237, 267; **22:**72(65), 78
Blakemore, J. S., **3:**278
Blakeslee, A. E., **5:**261(57), 315; **17:**12, 29, 74, 75; **21:**341(34), 364(46), 367–368
Blamire, M. G., **17:**22, 74
Blanc, J., **7:**128, 129, 130, 131(20), 133(28), 247
Blanc, P., **24:**163(26)
Blanchard, G. J., **24:**21(96), 23(96), 44(27), 45(27, 86, 87), 46(27, 86, 87, 93), 53(27), 56(27), 57(27), 60(27), 61(27), 64(27), 70(27), 71(108, 109, 111), 93(60), 94(60)
Blanco, J. R., **19:**139, 150
Blanco, M., **24:**228(10)
Bland, R. D., **13:**111(14), 112(14), 145(14), 147; **14:**135(13), 136(13), 195, 219(58), 220(58), 238; **22:**160(76), 173
Blandenet, G., **9:**39, 69
Blank, J. M., **5:**269, 317
Blank, S. L., **9:**271, 276, 283, 284, 288,

Blank, S. L., (*continued*) 292, 294, 295, 296; **10:**256(61), 259; **15:**9–11, 13, 136–137
Blank-Lapierre, A., **2:**23(28, 31), 60
Blankenship, B. A., **16:**8, 125
Blasenbrey, S., **11:**289(113), 317
Blatt, F. J., **6:**82(21), 96, 120, 129(172, 195), 130, 131(195), 135(172), 144; **11:**132(92), 202
Blatt, J. M., **6:**137(221), 148
Blattner, R. J., **19:**149
Blayo, N., **19:**45
Blazey, R., **10:**169(23), 226
Bleakney, W., **1:**41, 66; **14:**34(132), 76
Blech, I. A., **7:**265(47, 48), 307; **16:**240, 247–248, 250, 253–254, 267–268
Bleil, G. E., **3:**179(30), 209
Blenkinsop, S. E., **16:**52, 133
Blevin, W. R., **4:**60(137, 149), 94
Blickensderfer, R., **10:**43(89), 67
Bloch, A. N., **9:**240(233), 261
Bloch, F., **1:**235, 273, 323, 333
Bloch, J. M., **24:**184(35), 192(35)
Blocher, J. M., **9:**31(111), 67; **12:**271(157), 293
Blocher, J. M., Jr., **5:**238(1), 240(14), 261, 289(224), 314, 316–318, 320
Block, J. B., **17:**171, 221
Blodgett, K. B., **2:**240, 303; **5:**87, 140; **7:**312(5, 6), 325(5, 6), 328(6), 335(5, 6), 336(97), 337(5), 338, 342(6, 114), 343(5, 6), 348(5, 6), 350(5, 6), 351(5, 97), 356(5, 6), 360(5), 387(204), 388(6), 392, 395, 397; **24:**180(6)
Bloem, J., **3:**134(21), 137(21), 162; **4:**263(53), 264, 319(53), 321; **5:**288, 320
Bloembergen, N., **15:**92, 140
Blois, M. S., **1:**317, 333
Blok, J., **3:**285, 296(69), 300, 301
Blom, G. M., **7:**169(100), 250
Blom, H.-O., **14:**14(50, 51), 74, 233(134), 240; **17:**111, 112, 116, 142
Blonder, G. E., **13:**144(134), 150
Blonder, R., **24:**19(81), 20(81, 89)
Bloom, J. H., **23:**87(6)
Bloomberg, D. S., **15:**144, 177, 182–183, 190, 192, 196, 200, 207–210, 212, 214, 216–217
Bloomer, I., **19:**61
Bloomer, R. N., **1:**16, 19(42), 64
Bloss, W. L., **21:**98(35), 112, 128(33), 162(69), 233–234
Blossey, D. F., **11:**256(62), 257(62), 315
Bloxsom, J., **2:**406(187), 424
Blum, F. D., **24:**151(6)
Blum, S. E., **7:**163(88a), 220(210, 211), 221(212), 225(230), 249, 253; **16:**13, 125
Blum, W., **4:**192, 227(137), 251, 254
Blumberg, R. H., **1:**239(7), 242(25), 244(7), 252(49), 254(49), 258(71), 273, 274, 275; **3:**227, 229(50), 242, 272
Bluzer, N., **21:**219(126), 236
Board, J., **22:**134(51), 142(51), 172
Board, K., **11:**76(156a), 88(156a), 100; **13:**219(64), 224
Boardman, A. D., **9:**159(17), 256
Boatgo, G., **24:**114(31), 115(31)
Boatner, L. A., **16:**43, 48, 131; **24:**98(92)
Bobb, L. C., **2:**381(89), 422; **3:**110(114), 127; **5:**154(41), 155(41), 156(41), 191, 192, 204, 228(41), 230, 233; **6:**109, 110, 131(127, 128, 129), 132(127, 128, 129), 146; **7:**133(26), 247

Bobbin, S. E., **17:**49, 53, 76
Bobeck, A. H., **9:**264, 265, 267, 271, 275, 276, 287, 294, 296
Boccara, A. C., **16:**13, 125
Bockemuehl, R. R., **2:**150, 190
Bockenhoff, E., **21:**136(46), 161(67), 233–234
Bockris, J. O'M., **4:**191, 192, 195(2), 196, 197(24, 25, 40), 198, 199(36, 39, 64), 200(36, 39, 76), 201(36), 202, 203(2), 204(36, 39, 64), 206(36, 39, 64), 209(36, 39, 64), 210(64), 211(36, 64), 251, 252
Bode, D., **11:**106(4), 108(4), 199
Bode, D. E., **3:**286, 288(48), 291(53), 293, 296, 298(48), 300; **12:**169, 201, 230(33), 233
Bode, H., **1:**143, 185
Boden, E. G., **22:**227(80), 274
Bodesheim, J., **9:**254(251), 261
Boeckner, C., **10:**143, 166; **12:**29(59), 51
Boeffel, C., **24:**153(14)
Boehm, H.-P., **5:**118(21), 141
Boelens, S., **14:**170(122), 198
Boerio, A. H., **12:**157(335), 166
Boerio, F. J., **24:**33(133), 211(52)
Boers, A. L., **14:**208(20), 237
Boersch, H., **1:**189(9), 228; **6:**198(60), 203; **7:**16(42), 38; **9:**174, 219(177), 220(179), 256, 259
Boeters, K. E., **5:**164(94), 231
Boettcher, A., **1:**199(61, 62), 211(120), 214(123), 221, 224(120, 151, 169), 229, 231, 232; **6:**311, 346
Boettner, E. A., **4:**47(86), 92
Bogardus, E. H., **14:**219(56), 238
Bogenschütz, A. F., **3:**11, 51(54), 54
Bogoliubov, N. N., **1:**255(58), 274
Bogorad, A., **16:**50, 132
Bohac, P., **17:**263, 298
Bohandy, J., **16:**59, 79, 135, 137
Bohm, D., **1:**126(4), 184; **2:**99, 103(56), 142; **19:**298
Böhm, F., **17:**356, 357, 369
Böhm, G., **21:**290(43), 310
Böhm, H. J., **24:**210(41)
Bohmman, D. Y., **16:**76, 137
Bohn, H. G., **16:**237, 267
Bohnke, C., **17:**359, 361, 370
Bohnke, O., **17:**320, 359, 361, 366, 370
Bohr, N., **2:**67, 140
Bohr, T., **21:**52(101), 74
Bohrer, J. J., **2:**386(102), 389(102), 422
Boicy, J. H., **9:**32(116), 33(116), 67
Boiko, B. T., **3:**216(18), 256(104), 257(104), 258(104), 264(104), 271, 273; **6:**110, 111, 132(130), 146; **7:**22, 23(75, 76), 39
Boileau, A. R., **12:**91(140), 161
Bois, D., **13:**178(101), 195
Bois, P., **21:**161(67), 170(74), 175(74), 234
Bokshtein, B. S., **16:**234, 266
Bokshtein, S. Z., **16:**234, 266
Bolanovski, B., **14:**196
Bolduc, P. E., **3:**213(10a), 270
Boles, J., **9:**133(45), 143(45), 144
Boley, M. S., **16:**4, 124
Bolger, D. E., **5:**158, 230; **7:**236(249e), 254
Bolland, T. K., **9:**194(97), 258
Bolle, H. J., **4:**86(231), 96
Bollmann, W., **1:**227(184), 232; **2:**14(12), 60
Bolnedi, S., **22:**266(115), 270(125), 275–276
Boloshin, G. A., **10:**36(62), 67
Boltzmann, L., **6:**303, 346

Bommarito, G. M., **24:**181(12), 192(12), 199(12)
Bonacci, E. L., **10:**265(5), 308
Bondarev, V. V., **16:**71, 79, 136
Bondi, A., **2:**8, 60
Bongers, P. F., **15:**9, 136
Bongianni, W. L., **15:**8, 36, 72, 74, 136–137
Bongors, P. F., **9:**276, 296
Bonner, T., **24:**94(71), 95(71), 222(102)
Bonner, W. A., **9:**275, 276, 294, 297
Bonnet, M., **9:**51(252), 52, 62(252), 71
Bonnet, W. A., **15:**66, 139
Bontempo, A., **24:**208(28, 29)
Bontemps, N., **16:**13, 125
Boog, H. C., **3:**296(69), 301
Booker, G. R., **3:**218(22), 271; **4:**150(60), 169(108, 110, 111), 172, 186, 188, 190; **5:**163(72), 168, 170, 188(157), 196(72, 169), 197, 198, 199(169), 205(72, 178), 231, 233, 234, 249(34), 315
Boone, T., **16:**2, 124
Booth, J. G., **15:**110–111, 140–141
Booth, R. C., **15:**113, 141
Bootsma, G. A., **19:**280, 281
Booty, M. J., **6:**188, 203
Bor, J., **2:**199, 237
Borburgh, J. C. M., **9:**244(243), 261
Bordes, N., **16:**50, 132
Bordewijk, P., **11:**288, 317
Bordure, G., **9:**22(93), 67
Bordwell, F. G., **24:**9(30)
Borel, J.-P., **19:**168
Boreman, G. D., **15:**238, 262
Boresh, J. P., **16:**51, 132
Borghs, G., **21:**287(40), 310
Borgogno, J. P., **19:**192, 239, 240, 280
Borie, B., **4:**40, 92, 144(27), 187, 230, 254
Borisova, I. I., **12:**176(55), 233
Borisova, T. I., **6:**64(187), 78
Borkan, H., **2:**150(22, 24, 25), 153(24), 154(24), 156(39), 157(40), 159(39), 163(39), 168(22, 39), 173(40), 179(61), 183(24), 189(67), 191, 192; **5:**219(200), 234
Bormann, R., **16:**19–20, 126
Born, M., **1:**72(32), 121; **2:**105(72), 142; **5:**53(8), 57, 62, 85; **6:**34(97); **10:**171(25), 226; **19:**281, 287, 288
Bornstein, **19:**237
Borodin, V. P., **13:**63(27), 105
Borrelli, N. F., **9:**22(97, 98, 99), 23(97, 98, 99), 26, 67
Borukhov, M. Yu., **14:**145(65), 196
Borzyak, P. G., **12:**71(60), 74, 78, 80, 81, 84, 94, 95, 101(172), 103, 119(60), 159, 161, 162
Bos, J. G., **9:**33(134), 68
Bos, L. W., **19:**131
Bos, P. B., **10:**7(13), 65
Bosch, B. G., **23:**55(185)
Bosch, W., **14:**179(155), 186(155), 198
Bose, A., **22:**243(92), 274
Bose, J. C., **3:**275, 299
Bose, T. K., **11:**211(8), 232(8), 314
Bösenberg, J., **9:**186(67), 187(67), 190(67), 197(104, 105), 236(220, 221), 237(221), 240(221), 241(220, 221), 257, 258, 261
Bosi, S., **16:**26, 127
Bosio, L., **24:**184(42), 188(42)
Bosnell, J. R., **9:**56(261), 71
Bostner, L. A., **16:**10, 48, 125
Bosvieux, C., **7:**259, 260(7), 306; **16:**254, 269
Boswell, F. W., **17:**60, 77
Boswell, F. W. C., **2:**42(69), 61

Boswell, R. W., **18:**40
Bothe, W., **2:**80, 141
Bott, S. G., **16:**100, 140
Böttcher, C. J. F., **11:**288, 317
Bottka, N., **4:**58(128), 94
Botvinkin, O. K., **12:**176(55), 233
Bouanani, A. M., **16:**98, 140
Bouchard, M. C., **6:**185, 203
Bouchoule, A., **18:**297–299
Boudart, M., **5:**274(124, 125), 317
Bougnot, J., **12:**169(19), 232
Bouhet, A. A., **7:**323(36), 393
Bouillon, F., **4:**150(55), 188
Bouilov, L. L., **19:**104
Bounden, J. E., **6:**51(145a), 77
Bourdieu, L., **24:**169(37), 181(13)
Bouree, J. E., **19:**38
Bourgeois, P., **5:**166(117), 177, 232
Bourgoin, J. C., **14:**231(112), 239
Bourgon, M., **7:**31(130), 40
Bourne, H. C., Jr., **15:**154, 215
Bourne, L, C., **16:**106, 117, 121, 141–143
Bourneix, J., **7:**174(121), 250
Bousquet, P., **1:**70(13), 120; **4:**85(226), 96; **5:**101(11), 141; **6:**201, 202(78), 204
Bouteville, A., **14:**7(24), 8(24), 74; **17:**111, 142
Bouvier, J., **16:**13, 125
Bouwman, W. G., **24:**181(12, 15), 192(12), 199(12)
Bovard, B., **13:**124(71), 127(71), 149
Bovey, E., **12:**71(65), 160
Bowden, E. F., **24:**18(77), 19(77), 45(80)
Bowden, F. P., **7:**289(207), 390(208), 397
Bower, H. K., **15:**292, 311
Bower, J. E., **16:**60, 135
Bower, K., **12:**14, 50
Bowerman, E. R., **2:**393(134), 423; **3:**109, 127
Bowlden, H. J., **2:**202, 237
Bowling, R. A., **15:**300–301, 311
Bowman, D., **5:**281, 318
Bowman, R. C., Jr., **21:**93(21), 94(21), 111
Boxman, R. L., **14:**175(134), 176(139), 177(149), 198; **17:**93, 140
Boyce, J. B., **16:**69, 78, 136–137
Boycott, B. B., **21:**66(124), 75
Boyd, D. R., **9:**51(248, 249), 71
Boyd, E. L., **1:**286(27), 287(27), 289(27), 332; **3:**214(12), 270
Boyd, E. P., **16:**83–84, 88, 94, 138, 140; **17:**238, 297
Boyd, G. D., **17:**354, 368
Boyd, I.-W., **14:**122(85), 127
Boyer, L. L., **17:**119, 142, 234, 251, 252, 296, 297
Boyer, R. Q., **3:**28(145), 51(145), 57
Boyle, W. S., **14:**139(32), 195
Boynton, R. J., **14:**38(144), 64(205), 76, 78
Bozada, C. A., **15:**244, 263
Bozdemir, S., **11:**293(124), 302(139), 317
Bozler, C. O., **10:**250(42), 259; **13:**157(31), 159(31), 183(31), 191(31), 193
Bozorth, R. M., **1:**292(39), 332; **6:**211(28), 295, 297(190), 299; **15:**180, 261
Bozovic, I., **16:**2, 53–54, 123, 134
Brackley, G., **3:**169(9), 208
Bradford, A., **11:**111(47), 200
Bradford, A. P., **1:**223(162), 232; **2:**319, 346(22), 348(22), 359; **3:**17(75), 55; **4:**39(71), 92, 282(100, 103), 322; **7:**55(37a),

Bradford, A. P., (*continued*) 112; **10:**129(61, 62, 63, 65), 131, 134(65), 135(62, 65), 136(62, 65), 137(65), 138(65), 139(65), 151(65), 153(61), 156(84), 165, 166; **12:**2(5, 7), 3(14), 10, 12(39), 29(57, 58), 49, 50, 51
Bradley, A., **3:**156(81), 157, 164; **5:**245, 314
Bradley, D. C., **5:**304, 321
Bradley, D. E., **1:**188(4), 228
Bradley, D. J., **9:**132(40), 144; **10:**117, 165
Bradley, R. C., **3:**65, 70, 125
Bradley, R. R., **5:**148, 196(169), 199(169), 229, 233, 271, 317
Bradshaw, A. M., **9:**217(167, 168), 259
Bradsky, M. H., **22:**123(25), 172
Bradwell, A., **6:**72(215), 79
Brady, G. W., **7:**35(154), 41
Brady, M. J., **17:**153, 154, 220
Brag, A. R., **6:**13(56), 75
Braganza, C., **12:**288(257), 295
Bragg, W. L., **2:**110, 143; **4:**140(19), 187
Braggins, T. T., **16:**22, 127
Braginski, A. I., **16:**8, 11–12, 18, 52, 87, 101, 108, 121, 125–126, 133, 138, 141–143
Brailsford, A. D., **11:**184(147), 203
Brain, R., **22:**264(109), 275
Brambring, J., **6:**197(56), 203; **9:**186(65, 66), 195(98), 257, 258
Bran, M., **7:**34(140), 40
Brand, W., **16:**97–98, 140
Brandenberg, W. M., **4:**60(138, 148), 94
Brander, R. W., **5:**284, 319
Brandes, H., **4:**198, 252
Brandes, R. G., **1:**27(58), 65, 205(83), 230
Brandle, C. D., **9:**281, 294; **16:**101, 107, 141
Brändli, G., **6:**86, 89, 100, 144
Brandli, G., **6:**190, 203
Brandmayr, R. J., **3:**139(42), 163
Brandon, D. G., **6:**4(15), 74
Brandow, S. L., **24:**98(91)
Brandt, G., **17:**350, 367
Brandt, G. E., **11:**10(4), 33
Brandt, J. A., **2:**330, 360
Brannon, J. H., **13:**153(7), 154(7), 172(7), 173, 182(7), 183(7), 190(158, 161), 191(158, 161), 192, 197
Brannon, P. J., **13:**190(163), 191(163), 197
Bransky, I., **6:**223(76a), 296
Bransky, J., **6:**223, 296
Braren, B., **13:**153(10), 170(71), 171(71), 172, 182(10, 71), 183(10, 71), 192, 194
Braski, D. W., **7:**64(46), 65, 66, 69, 72, 74(50), 113
Braslau, A., **24:**181(13)
Braslau, N., **13:**209(27, 28), 210(32), 211, 212, 213, 215, 216, 217, 218, 223, 224
Bratt, P. R., **21:**4(8), 41(8), 71
Brattain, W. H., **2:**157(41), 191, 284(42), 304
Bratton, R. J., **9:**28(105a), 67
Brauer, G., **12:**88(132), 161
Brauer, P., **2:**240(9a), 303
Brauer, W., **2:**125(141), 144
Braun, B., **21:**312(4), 323(4), 324(4, 15), 325(15), 329(15), 333(15), 334(15), 336(15), 366–367
Braundmeier, A. J., **19:**149
Braundmeier, A. J., Jr., **9:**176(41, 42), 178(41), 221(183), 243(240), 254(253), 256, 260, 261

Braunsfurth, G., **3:**19(86), 51(86), 55
Braunstein, G. H., **16:**97, 140
Bravman, J. C., **16:**53–54, 106, 134, 141; **17:**43, 47, 75
Bray, A. R., **4:**300(120), 323
Bray, J. W., **16:**101, 141
Breazeale, J. B., **1:**207(98), 227(98), 230; **3:**253(88), 255(88), 257(88); **16:**332, 339
Brebee, G., **7:**261, 262(12), 306
Brebrick, R. F., **3:**277, 300; **11:**108(22, 23), 200
Breidenbach, R., **17:**110, 141
Breitling, G., **2:**44(76), 62; **7:**20(63, 64), 21(69), 24(63, 86), 27(105), 38, 39
Breitling, H. M., **7:**270(65), 308; **16:**246, 255, 258, 267, 269
Bremer, J. W., **1:**253(51, 52), 263(77), 264(51, 52), 274, 275
Bren, V. G., **14:**177(146), 177(147), 198
Brendel, C., **12:**269(153), 293
Brendel, R., **19:**149
Brennan, K. F., **15:**222, 226, 261
Brennan, S., **17:**58, 76
Brenneman, A. E., **1:**251(48), 274; **4:**302(139), 306(139), 309(139), 323
Brennemann, A. E., **3:**146(66), 148(66), 164
Brenner, A., **3:**222, 271; **4:**192, 251
Brenner, D. W., **24:**221(100, 101)
Brenner, S. S., **2:**58(92, 93), 62; **4:**118(40), 135
Bresse, J. F., **12:**176(50), 179(50), 180(50), 183(50), 186(50), 187(50), 190(50), 197(50), 200(50), 233
Brett, M. J., **14:**44–45(157), 45(158), 77; **22:**1, 3(63), 7(48), 9(19), 10(19), 11(20, 21), 12(20), 15(23), 17(25), 18(23, 25, 26), 20(19), 22(31), 23(31, 32, 33), 24(20, 33), 25(33, 34), 26(33), 28(33), 30(37, 38, 48), 32(46, 48), 34(26), 35(26), 37(46), 38(46), 39(49), 40(48, 49), 41(49), 42(48), 44(49), 45(25, 26, 48, 49, 51, 52), 46(25, 53), 47(25), 49(53), 50(19, 56), 52(56), 53(56), 54(56, 58), 55(59), 56(56), 59(48), 60(59, 60), 61(61, 63), 62(59), 64(59), 66(61), 67(63, 64), 68(61), 70(63), 72(63), 74(64), 75(64, 66), 76(64, 66–74), 77–79, 123(27), 172, 227(81, 82), 232(85), 274
Breuch, R. A., **10:**140(75), 166
Breuer, H. D., **19:**142
Breuer, K., **12:**251(62), 256(62), 291
Brewer, J. A., **16:**16, 25, 126
Brewer, J. W., **17:**153, 172, 220
Brewer, P., **13:**169(68), 177(96, 97), 185(96, 97, 134, 141), 186(68), 187(134), 188(96, 97, 141, 142), 189(68), 194, 195, 196, 197
Brewer, P. D., **23:**234(37)
Brierley, C. J., **17:**237, 238, 296
Briggs, G. S., **12:**157(336), 166
Briggs, H. B., **2:**284(42), 304
Bright, T. B., **23:**48(161); **24:**10(40), 17(40), 44(15), 45(15), 70(15), 82(19), 86(19), 112(4), 128(58)
Bril, A., **3:**201(76), 210
Brillouin, L., **2:**65(3), 120(3), 140
Brillson, L. J., **13:**202, 223; **14:**234(145), 240
Brinckman, F. E., **12:**269(146), 293
Bringans, R. D., **23:**49(30)
Brink, J., **13:**185(128), 187(128), 196
Brinker, C. J., **17:**239, 297

Brinkerhoff, J. M., **3:**108(168), 128
Brinkman, J. A., **12:**246(25), 290
Brinlee, W. R., **15:**83, 140
Brinsmaid, S., **13:**60, 62(9), 104
Briones, F., **19:**10, 20–22
Brishka, A. A., **17:**355, 356, 368
Bristow, J. R., **1:**248(42), 251(42), 274; **2:**22(22), 23, 60
Brito Cruz, C. H., **21:**193(91), 235
Broachard, J., **1:**127, 184
Broadbent, K. D., **1:**321(85), 322(85), 333; **6:**224(82), 291(82), 297
Broadbent, R. H., **6:**54(152), 77
Broadhurst, M., **11:**271(93), 316
Broadley, J. S., **5:**252(36), 315
Broberg, K. B., **16:**279, 337
Brochier, A., **9:**282, 294
Brock, A. J., **6:**61, 78
Brock, D., **13:**79(81), 106
Brock, J. D., **24:**169(39)
Brocklesby, W. S., **16:**110, 142
Brockway, L. O., **7:**15(35), 38
Brodbeck, D., **24:**98(85)
Brodie, D. E., **16:**53, 133
Brodie, I., **14:**208(17), 237
Brodsky, M. B., **17:**44, 48, 49, 50, 51, 52, 76
Brodsky, M. H., **7:**26(98), 39; **10:**41(79, 81), 67; **14:**59(189), 77; **15:**277, 279, 310; **22:**5(8), 77
Brodsky, S., **22:**119(12), 161(12), 167(12), 171
Brody, T. P., **5:**216(192), 228(192), 234
Broers, A. N., **6:**103(102a), 145
Brogan, W. T., **5:**197(172), 234
Broida, H. P., **9:**198(111), 258
Bröker, F. J., **6:**116(158), 147
Brokman, A., **12:**287, 295
Bronco, C. J., **4:**50(119), 93
Bronnes, R. L., **7:**162(86), 249
Bronson, R. P., **7:**333(82), 334(82, 91), 394
Brooke, L., **16:**253, 268
Brooks, H., **4:**139, 187, 328, 369
Brooks, R. E., **15:**55, 138
Brooks, T. A., **15:**294, 311
Broom, R. T., **1:**272(98), 275
Broquet, P., **6:**106(124), 131(124), 146
Brorsson, G., **16:**67, 136
Brosl, N., **14:**177(149), 198
Broszeit, E., **17:**130, 143
Brotzen, F. R., **17:**43, 47, 49, 62, 63, 75, 77
Broudy, R. M., **6:**13(54, 55), 75
Brown, A. E., **3:**139(42), 163
Brown, A. S., **7:**126, 228, 229(16), 247
Brown, B. W., **17:**306, 364
Brown, C. A., **24:**112(21,22)
Brown, D., **16:**37–38, 130
Brown, D. M., **4:**260(21), 301(21), 320
Brown, D. R., **16:**37, 130
Brown, E. R., **15:**244–246, 262–263; **23:**243(63)
Brown, F., **6:**24(84, 86), 75
Brown, G. A., **4:**305(145), 309(145), 323
Brown, G. R., **24:**151(8), 154(8), 156(8)
Brown, H. R., **16:**292, 334, 340
Brown, I. G., **17:**94, 141
Brown, J. D., **14:**27(109), 75
Brown, K. W., **15:**245–246, 263
Brown, L. C., **6:**315(29), 326(33), 328(34), 339(53), 340(57, 58), 346, 347
Brown, M. A. C. S., **7:**170(102), 250
Brown, M. D., **12:**265(113), 292
Brown, P. J., **3:**39(198), 52(198), 58
Brown, R., **2:**371(61), 372(61), 373(61), 421; **3:**33(163), 52(163),

57; **13:**125(77), 149; **16:**106, 141; **24:**9(34)
Brown, R. R., **10:**162, 166
Brown, S. C., **12:**238(3), 289; **18:**122–145
Brown, W. J., **4:**60(149), 94
Brown, W. R., **6:**195, 203; **9:**176, 256
Brownell, R. B., **3:**5(29), 51(29), 53
Brownson, J., **5:**154(40), 230
Brubaker, W. M., **1:**19(45), 65
Bruce, C. F., **4:**85(219), 96
Bruce, G., **3:**70, 73(34), 76(34), 125
Bruce, R., **14:**51(167), 77
Bruchez Jr., M. P., **24:**36(147)
Bruchhaus, R., **17:**291, 299
Brück, L., **1:**216, 231
Bruck, L., **4:**159(85), 189
Bruckner-Lea, C., **24:**183(32, 33), 184(33), 192(32)
Brueck, S. R. J., **16:**75, 137
Brueckner, G. E., **10:**91(27a), 164
Bruening, M., **24:**33(135)
Brug, J., **15:**171, 215
Bruggeman, D. A. G., **19:**41, 43, 59, 60, 72, 82, 97, 99, 107, 110, 139, 155–157, 206, 210
Brum, J. A., **15:**222, 261; **21:**328(23), 367
Brun, G., **16:**98, 140
Brundle, C. A., **14:**230(107), 239
Brundle, L. K., **15:**35, 137
Brunner, O., **16:**19, 28, 126, 128
Bruns, R., **9:**153(9), 242(9), 255
Brust, D., **12:**57, 58, 68(20), 159
Brust, M., **24:**154(15, 16)
Bruun, H., **7:**323(35), 393
Bruyére, J. C., **3:**24(109), 51(109), 55; **6:**206(9), 210, 211, 212(24, 25), 225(24), 226(93), 241, 242, 244(24, 25), 249(157), 250(24), 253, 254, 255(24, 25), 256, 257, 258, 259, 260(24, 157, 160), 269(171), 271, 277, 278(189), 291, 292(9, 22), 293(9), 295, 297, 298, 299
Bruynseraede, Y., **16:**52, 133
Bryant, A. M., **7:**36(164), 41
Bryce, P., **13:**112(29), 148
Brycki, B., **16:**106, 141
Brymner, R., **1:**33(82), 65
Brynestad, J., **16:**3, 124
Bryskin, V. V., **17:**350, 367
Brzoska, J. B., **24:**169(35)
Buachidze, Z. E., **13:**186(154), 190(154), 197
Bubar, S. F., **6:**4, 74
Bubay, H., **17:**154, 155, 182, 222
Bube, R. H., **2:**172(50), 176(57), 177(50), 178(50), 191, 192; **3:**188(48), 197(48), 209; **5:**223(211), 225(211), 235; **12:**169(12, 13, 14, 15, 16), 176(54), 177(54), 178(54), 179(54), 183(16, 18), 189(14), 190(14), 193(14), 194(13, 14), 196(12, 14), 232, 233, 235; **17:**153, 154, 221
Bubenzer, A., **15:**307, 312
Bublick, A. I., **7:**18(52), 38
Bubulac, L. O., **12:**137(309), 165; **21:**40(88), 74
Bucek, H., **12:**286, 295
Bucelot, T. J., **21:**84(17), 111
Buch, F., **12:**169(15, 17), 232
Buchal, Ch., **16:**5, 28, 121, 124, 128, 143
Buchan, N. I., **19:**35
Buchanan, D. A., **18:**124, 165
Buchanan, D. N. E., **17:**318, 366
Buchanan, M., **14:**48(163), 77; **15:**250, 263; **17:**153, 220; **21:**5(7), 32(7), 33(7), 41(7), 71, 81(12), 101(37),

Buchanan, M., (*continued*) 111, 112, 128(35), 170(73, 75), 178(73), 179(79), 209(35), 233, 234, 235, 287(38), 310
Bucher, H., **7:**385(201), 397
Bucher, J. P., **17:**111, 142
Bucher, J.-P., **24:**44(42), 86(33)
Bucher, K., **21:**312(1), 313(1), 321(1), 322(1), 344(1), 366
Buchholz, S., **24:**163(28), 240(73)
Buchner, S. P., **11:**153(126), 155, 199, 202, 203
Buchta, R., **14:**233(134), 240
Buchwald, C. E., **7:**368(175, 176), 396
Buck, M., **24:**45(77), 65(77), 92(57, 58)
Buck, T. M., **1:**8(16), 64; **4:**7(18), 91, 309, 323; **9:**47(223), 70; **12:**259(81), 292
Buck, D. A., **3:**146(60), 163
Buckel, W., **1:**202(65, 70), 211(121), 222, 229, 230, 231, 243, 244(26); **19:**168
Buckle, H., **6:**329, 346
Buckley, W. D., **22:**118(3), 171
Buckman, A. B., **19:**50
Budai, J. D., **16:**10, 43, 48–49, 70, 125, 131–132, 136
Budd, K. D., **17:**240, 297
Budden, K. G., **18:**36
Buddhabadana, S., **11:**279(102), 280, 316
Buder, R., **6:**273(179, 180), 274, 275(179), 299
Budewski, E., **4:**202(79), 203(79), 252
Budhani, R. C., **13:**96(93), 106, 143(129), 150; **16:**51, 133
Budiansky, B., **16:**288, 338
Budnick, J. I., **1:**1(1), 63, 241(11), 273
Budnikov, V. N., **18:**137
Budo, Y., **1:**61(149), 62(149), 63(149), 67; **2:**403(176, 177, 178), 404(177), 424; **3:**140(45), 148(45), 150, 163, 164, 221(36), 229(65), 236(61), 237(36), 238(36), 242(65), 246(65), 250(36), 251(36), 271, 272; **4:**282(106), 322
Bueche, F., **9:**8(39), 65
Buerger, M. J., **2:**110, 143
Buess-Herrman, C., **24:**183(29)
Buff, F. P., **24:**184(40), 190(40)
Buffat, P., **19:**168
Bugenis, C., **3:**38(181, 182, 184), 44, 52(181, 182, 184), 58
Buhaenko, D. S., **19:**32
Buhay, H., **17:**229, 236, 237, 270, 271, 272, 275, 276, 277, 278, 280, 281, 282, 283, 284, 285, 286, 290, 292, 294, 296, 298–300
Buhl, L. L., **10:**228(9), 254(9), 258
Buhl, R., **1:**191(28), 229; **14:**135(14), 195
Buhrman, R. A., **16:**51, 61–62, 66, 70, 107, 133, 135–136, 141; **19:**113
Buiocchi, C., **21:**221(137), 222(137), 237
Buiocchi, C. J., **7:**126(15), 128, 129, 130, 131(20), 136(15), 137(34, 35), 138, 139(35), 147, 157(34), 230(35), 247, 248
Bukhshtab, E. I., **6:**118(167), 141, 147, 149
Bulat, E. S., **18:**218
Bulgakov, Y. V., **3:**70, 125
Bullard, R. L., **2:**379(84), 392(84), 422
Bullington, J., **17:**271, 298
Bullock, D. C., **16:**34–35, 130
Bullot, J., **19:**61, 85, 87
Bulman, J. B., **16:**119, 121, 143
Bulow, H., **7:**18, 21(51), 38

Bumble, B., **16:**21, 116–117, 122, 127, 142
Bunding Lee, K. A., **24:**27(123), 29(123)
Bundy, F. P., **7:**23(82), 39
Bunkin, F. V., **13:**169(70), 185(70), 187(70), 194
Bunner, J., **12:**270(156), 293
Bunshah, R. F., **13:**60, 62(1), 63(27), 64(2, 29), 65, 66, 67, 68, 75(1, 2, 30, 58, 59), 76, 78, 79, 80(1, 50, 59), 83(30, 60, 70, 85), 84(29, 72, 85), 87, 89(76), 90, 91(83), 92, 93, 94(70, 89), 95, 96(93), 97(93), 98(97); **14:**132(9, 10), 134(9), 195; **17:**9, 60, 64, 65, 77, 81, 87, 91, 92, 100(111), 101(111), 102(64, 114, 115, 116), 103, 104, 105, 106, 107, 139, 140
Burattini, E., **17:**310, 365
Buratto, W. S., **24:**12(55), 14(55)
Burbank, R. D., **4:**171, 190
Burch, J. F., **16:**119, 121, 143
Burdiyan, I. I., **7:**177(118d), 250
Burge, D. K., **4:**38(69), 39(69), 41(69), 68(160), 81(69), 84(69), 92; **12:**36, 38(64), 51
Burge, R. E., **2:**65(2), 82, 85(2), 86, 87(38), 88(38), 89, 91, 92, 93, 97, 140, 141, 142
Burger, R. M., **5:**190(162), 233
Burger, W. R., **14:**232(132), 239
Burggraf, C., **7:**16(44), 38
Burgin, T. P., **24:**33(137), 44(37)
Buritz, R. S., **1:**42, 66
Burke, E. R., **9:**265, 296
Burke, M. G., **17:**237, 275, 276, 277, 280, 296
Bürker, U., **9:**179(52), 220, 257, 260
Burkett, R. H. W., **2:**376(74), 389(115), 421, 422
Burkhard, W. J., **3:**146(64), 147, 163
Burkhardt, E. G., **16:**107, 141
Burkhardt, P. J., **3:**132(14), 141(52), 162, 163; **4:**260, 301(19), 302(137), 305(137), 306(137), 320, 323
Burkland, W. A., **17:**201, 223
Burlamacchi, P., **10:**238(30), 258
Burmeister, R. A., **5:**282, 318
Burmester, A., **6:**120(172), 129(172), 130(172), 135(172), 147
Burnham, R. D., **7:**234(247, 248), 254
Burns, F., **13:**153(7), 154(7), 172(7), 173(7), 182(7), 183(7), 192
Burns, J., **1:**14(28), 64; **4:**68(163), 94; **6:**224, 291(85), 297
Burns, J. R., **5:**146(14), 229
Burns, J. W., **7:**169(99), 250
Burns, S. G., **17:**160, 198, 201, 221, 223
Burns, W. K., **10:**254(52), 259
Burr, A. A., **6:**310(16), 346
Burrell, G. J., **11:**142(113), 143(113), 202
Burrell, M. C., **16:**41, 131
Burridge, J. C., **2:**334(52), 360
Burrows, P., **24:**181(20), 182(20)
Burrus, C. A., **7:**222(216), 253; **11:**78, 101
Burstein, E., **3:**277(26, 27), 278(27), 300; **9:**154(11), 240(232), 256, 261
Burstein, G. T., **14:**230(106), 239
Burt, M. G., **12:**126(265, 266), 128(265, 266), 164
Burton, J., **5:**2(9), 45
Burton, J. A., **12:**101(171), 102(171), 137(304), 162, 165
Burton, J. J., **16:**234, 266
Burton, R. L., **16:**57, 60, 134–135
Burton, S., **15:**291, 297–299, 311

Burton, W. K., **4:**192, 193, 194, 195, 251
Burwell, D. A., **24:**164(31), 168(31)
Busch, G., **6:**247(152), 298; **9:**217(166), 259
Buschert, R. C., **7:**27(109), 39
Buschor, F. W., **17:**111, 142
Buschow, H. H. J., **15:**168, 215
Buschow, K. H. J., **15:**167–168, 215; **16:**213, 230
Bush, H., **17:**172, 173, 222
Busham, R. D., **10:**251(45), 259
Bushik, A. I., **14:**147(74), 196
Buss, R. J., **22:**186(32), 209(32), 272
Butcher, D. N., **12:**277(199), 294
Butcher, P. N., **11:**290(117), 317
Butera, R. A., **24:**3(19)
Butler, C. P., **2:**329, 330(38), 359
Butler, D., **17:**241, 287, 297
Butler, G., **7:**60, 65(41), 112
Butler, J. F., **3:**277(31, 32), 300; **11:**158, 203
Butler, J. K., **7:**234(242), 254
Butler, N. S., **17:**86, 140
Butorina, L. N., **5:**163(70), 170, 231
Butt, H.-J., **24:**21(91, 102), 26(102), 85(24), 88(45), 89(50), 90(24, 45), 94(67), 95(67), 98(97), 99(24), 106(97), 212(65)
Butter, D. C., **22:**119(13), 120(13), 161(13), 171
Buttner, F. H., **4:**139(13), 187
Button, C., **21:**312(4), 323(4), 324(4), 330(27), 331(27), 332(27), 348(38), 353(27), 355(27, 38), 356(38), 366–367
Buttry, D. A., **24:**27(124), 28(124), 29(124), 30(124)
Buwalda, E. P., **2:**312(5), 358
Bybocas, J., **16:**12–13, 101, 103, 105, 125
Byer, N. E., **11:**153(125, 126), 155(127), 198(157), 199(127, 157), 202, 203
Bykov, J. V., **13:**18(24), 20(24), 57
Bylander, E. G., **5:**265, 266, 316; **9:**47(224), 70; **11:**108, 200
Byram, E. T., **10:**78(22), 164
Byungsung, O., **15:**222–224, 229, 233, 237–239, 241, 243, 250, 261–262; **21:**8(27), 39(82), 72, 73, 121(26), 185(82), 189(82), 206(26), 233, 235

C

Caballero, D. L., **1:**70(3), 71(3), 86(3), 120
Cabanski, W. A., **23:**101(36)
Cabelli, S. A., **21:**221(134), 236
Cabn, J. W., **2:**2, 59
Cabrera, N., **2:**58(91), 62; **4:**138(6), 146, 147, 157, 158, 160(29), 161; **14:**114(69), 127
Cachet, H., **11:**253, 255
Cadieu, F. J., **16:**149, 153, 159–161, 164, 168–169, 177, 186, 189–216, 219–220, 225–227, 229–231
Cadoff, I. B., **5:**199, 234, 270(106), 306(272), 317, 321
Cadoret, R., **19:**52
Cafarella, J. H., **15:**100, 140
Caffrey, M., **24:**119(40)
Caffrey, R. E., **5:**288(216), 320
Cafrrey, R. E., **4:**261(32), 263(32), 264, 320
Cagle, E. H., **7:**56(39), 112
Cahay, M., **15:**226, 250, 261
Cahn, J. B., **17:**10, 74
Cahn, J. W., **4:**108, 135; **17:**35, 40, 68, 75, 77
Cai, S.-M., **24:**27(128), 31(130)

Cairns, R. B., **7:**47(21), 112
Calawa, A. R., **3:**277(31, 32), 300; **11:**36(18, 23), 49(67), 57, 58, 65(95), 67(67), 81(185a), 96, 97, 98, 99, 101, 128(78, 80), 155(78), 201
Calawa, D. R., **23:**237(33)
Calbick, C. J., **2:**72(20), 73(24), 96(49), 109(24, 81), 130(24), 132(154), 141, 142, 145, 371(62), 373(62), 394(147), 395(147), 397(152), 421, 423, 424
Calderon, L., **21:**240(4), 246(4), 248(4), 252(4), 278(4), 308
Calderon, M. A. G., **18:**151
Caldwell, K., **24:**183(32, 33), 184(33), 192(32)
Caldwell, W. B., **24:**112(24), 129(24)
Cale, T. S., **22:**4(3), 77, 175, 177(1, 2, 4–6), 178(8–12), 180(17), 181(21, 22), 183(28, 29), 185(28, 29), 188(8–10, 35, 36), 193(8–10, 43–46, 48, 49, 51), 194(52, 53), 196(35), 201(55), 202(8, 17, 43, 59–61), 205(43), 211(60, 68, 69), 212(68, 69), 215(75, 76), 216(60, 61, 69, 77), 220(60, 69), 226(48, 49, 51, 68), 227(52, 53, 83, 84), 228(48, 52), 229(52), 232(28, 29, 53, 75, 83, 84), 233(53), 235(83, 84), 238(53), 241(88), 245(60, 61, 68,77), 253(55, 97), 254(97), 262(97), 265(111), 266(112, 113, 115), 269(113, 118–120), 270(121–127, 130), 271–276
Calhoun, B. A., **9:**275, 294, 295
Calistri-Yeh, M., **24:**169(39), 170(39)
Callaghan, M. J., **1:**20(49), 65
Callaway, J., **13:**49(59), 58
Callen, H., **9:**288, 294
Callender, R. E., **11:**119(67), 131, 168, 201, 203
Callott, T. A., **9:**217(162), 259
Calvert, J. M., **24:**228(27)
Calvert, L. D., **6:**12, 75
Camack, W. G., **2:**312(4), 358
Cambel, A. B., **9:**46(204), 70
Cambey, L. A., **1:**43(115), 66
Cameron, J. F., **3:**39(206), 52(206), 58
Camillone III, N., **24:**4(17, 69, 72, 73), 44(17, 46), 82(12), 83(12), 86(12, 29), 93(12), 94(61), 112(7, 8, 9, 10, 11, 12, 13, 15), 117(7, 11, 13, 15), 119(9, 11, 13), 121(12), 122(13), 123(13), 125(9), 128(15), 142(11), 143(11), 144(11), 206(14), 212(14), 216(68)
Camley, R. E., **15:**37, 138
Camlibel, I., **17:**229, 260, 279, 280, 281, 282, 298, 299
Cammarata, R. C., **17:**49, 53, 54, 56, 58, 62, 76, 77
Campagna, M., **17:**318, 366
Campbell, D. J., **24:**112(24), 129(24)
Campbell, D. R., **7:**186(145), 251, 263(31), 307; **16:**236, 241, 266
Campbell, D. S., **1:**63, 67; **2:**365(9), 369(41), 403(180), 420, 424; **3:**6, 51(32), 52(32), 53, 110, 114, 127, 215, 220(30, 31), 225, 229(30), 234, 235(30, 31), 236(30), 237, 238(15, 30), 239, 243, 270, 271, 272; **5:**161, 196, 230, 238(2), 314; **6:**94, 145; **9:**7(34), 8, 31, 65, 67; **13:**125, 149; **14:**27(108), 32(108), 75; **17:**111, 141
Campbell, I. E., **3:**133(20), 162
Campbell, I. H., **13:**180(105), 195
Campbell, J. J., **6:**12, 75
Campbell, L. J., **16:**73, 137

Campbell, N. R., **12:**54(3), 71, 158
Campbell, R. B., **5:**157, 230, 284, 319
Campbell, S. D., **14:**71(232), 79
Campidelli, Y., **23:**161(61, 62)
Campisano, S. U., **13:**166(60), 194
Canfield, L. R., **1:**134, 135, 149(25, 64), 153(25), 154(25), 155(25), 157(25), 158(25), 159(62), 161(62), 162(62), 163(62), 165(62), 167(64), 170(25, 62, 64), 182(23), 185, 186; **2:**315, 359; **4:**41(78), 60(135), 68(78), 92, 94; **9:**227(195), 260; **10:**76(11), 77(16), 92(29, 30), 94(30), 95(29), 96(29, 30), 164; **12:**5, 10(38), 50
Canillas, A., **19:**45, 280
Cannady, S. S., **10:**157(86, 87), 162(91), 166
Cannon, R. M., **16:**288, 338
Cantagrel, M., **14:**226(84), 238; **22:**113(29), 115
Cantella, M. J., **23:**101(33)
Cao, G., **24:**163(25), 164(25)
Cao, G. H., **24:**182(23), 199(23)
Cao, H. C., **16:**291, 340
Cao, H. L., **16:**36, 130
Cao, L., **16:**41, 131
Capasso, F., **21:**184(80), 194(80), 235; **23:**239(46), 240(54), 243(65), 262(65, 80, 83, 84, 85, 87), 263(65), 265(85), 266(46), 268(46), 269(92), 271(54), 272(54, 83, 96), 273(86), 274(87), 279(83), 291(80), 298(80)
Capella, L., **5:**165(109), 174(109), 232
Capelli, M. A., **22:**182(25), 188(25), 193(25), 272
Capio, C. D., **12:**263(98), 264(98), 292; **13:**99(105), 107; **22:**204(63), 273
Caplan, P. J., **14:**232(118), 239
Capone, B. R., **23:**87(6)
Caprile, C., **16:**252, 268
Capriulo, A. J., **5:**298(250), 321
Caracciolo, R., **16:**91, 139
Carchano, H., **11:**278(97), 316
Carcia, P. F., **14:**42(150), 77
Card, F. E., **3:**169, 208
Card, H. C., **15:**293, 311
Cardona, A. H., **16:**117, 121, 142–143
Cardona, M., **3:**277(25), 300; **15:**277, 279, 286, 310; **17:**45, 76; **19:**10, 73, 280
Cardwell, D. A., **16:**107, 123, 141, 143
Careem, M. A., **11:**256(68), 257(68), 263(83), 265, 266, 267, 275(96), 316
Carey, G. P., **4:**361(80), 362(80), 364(80), 367(80), 371
Carey, R. I., **24:**10(39), 14(39, 58), 21(39), 23(39)
Cargill, G. S., **7:**13(29), 30(127), 38, 40; **9:**277, 294; **15:**156, 159–160, 185, 215–216
Carl, D. A., **18:**124, 125, 159, 184, 185–186, 217
Carlile, R. N., **18:**308, 312
Carlson, D. E., **9:**12(61), 33(60, 151, 152), 60(273), 66, 68, 71; **12:**238(2), 289
Carlson, H. G., **4:**305(145), 309, 323
Carlson, O. W., **7:**264(43), 307
Carlsson, J.-O., **16:**92, 139
Carlston, C. E., **3:**77(61), 79, 126
Carmichael, H., **3:**28(146), 51(146), 57
Carnahan, B., **17:**66, 77
Caron, L. G., **17:**279, 299
Carpenter, F. E., **7:**67(57), 113
Carpenter, J. A., **16:**237, 267
Carpenter, M. K., **17:**304, 318, 361, 362, 364, 366, 370

Carr, P. H., **15:**63, 66, 139; **17:**187, 222
Carr, P. W., **24:**163(26)
Carr, W. N., **7:**214(196), 220(207), 253
Carrejo, J. P., **24:**98(94)
Carroll, J. A., **5:**184, 233
Carroll, R. W., **24:**249(96)
Carroll, W. N., **2:**147(2), 178(2), 190, 365(3), 366(3), 376(77), 419, 421
Carruthers, G. R., **10:**75(8, 9), 164
Carruthers, J. R., **10:**255, 259
Carson, R. A., **11:**241(50), 244(50), 315
Carter, C. B., **16:**61–62, 66, 70, 135–136
Carter, D. L., **11:**169(142), 203
Carter, F., **15:**157, 215
Carter, G., **1:**27(57), 29(57), 65; **6:**102(94), 145; **12:**238(9), 290; **14:**217(39), 237; **22:**113(26, 27, 30), 115
Carter, R. L., **15:**51–52, 64, 67, 75, 77, 83, 88, 138–140
Carter, W. B., **16:**92, 139
Cartwright, C. H., **2:**240, 303
Caruso, A. J., **11:**24(18), 25(18), 27(18), 28(18), 29(18), 34
Carve, G. E., **12:**176(56), 233
Carver, T. E., **24:**84(21)
Casais, M. T., **16:**2, 124
Casan, C. J., **7:**170, 171(110), 250
Cascales, C., **16:**2, 124
Cascarano, F. M., **22:**186(33), 209(33), 272
Case, T. W., **3:**275, 299
Case, W. E., **13:**121(58), 148
Casey, H., **2:**392(123), 423
Casey, H. C., **11:**75(148), 78(148), 79(148), 100
Casey, H. C., Jr., **7:**149, 154(59, 67), 155(67a), 162, 164(82), 166(92a), 186(142), 225(228), 248, 249, 251, 253; **10:**228(6), 251, 258; **11:**51(80, 81), 54(80), 66(12 1), 79(169, 171), 81(184), 89, 98, 99, 101, 103
Casey, J. J., **5:**307, 322
Casey, M., **2:**379(82), 393(144), 397(82, 144), 411(144), 422, 423
Cashman, R. J., **3:**276, 299; **12:**168(5), 232
Casimir, H. B. G., **1:**247(39), 249(39)
Caskey, G., **17:**313, 365
Cassie, A. B. D., **24:**15(68), 16(68), 17(68)
Castagna, T. J., **18:**105, 124
Castaing, R., **2:**125, 144
Castellano, R. N., **14:**24(93), 75
Castera, J. P., **15:**84, 140
Castner, D. G., **24:**15(65), 34(139, 140), 35(139, 140), 130(61)
Castro, A., **16:**2, 124
Castrucci, P. P., **4:**284, 320(166), 321, 324
Caswell, H. L., **1:**22, 23, 25(53), 34, 37(103), 43(116), 48, 52(128), 53, 54, 55(53, 128, 130), 56, 60(116), 61, 62(149), 63(149), 65, 66, 67, 241(14), 273; **2:**178(60), 192, 365(5), 403(175, 176, 177), 404(177), 419, 424; **3:**140(45), 148(45), 150, 163, 164, 213, 220(33), 221(36), 229(65), 235(33), 236(61), 237(36), 238(33, 36), 242, 246, 250(36), 251(36), 270, 271, 272; **4:**282(106), 322; **14:**38(143), 76
Catalan, A. B., **17:**229, 300
Catalan, L. A., **2:**258, 304
Catalano, A., **15:**286–287, 310
Catchpole, C. E., **12:**147(318), 166
Cathcrart, J. V., **4:**144(27), 187
Catherine, Y., **12:**272(163, 165, 166),

Catherine, Y., (*continued*)
273, 289(266, 268), 293, 296; **15:**303, 306, 312
Catlett, D. C., **17:**4, 64, 66, 73
Catlett, D. S., **17:**64, 65, 77
Catlin, A., **3:**253(90), 255(90), 257, 258(90), 259, 260, 263, 264(90), 265(90), 273; **4:**171(123), 181, 190; **5:**163(66, 67), 168(66, 67), 188, 199(66, 67), 231
Caudle, G. F., **1:**227(185), 232
Cauge, T. P., **5:**288(217), 320; **12:**266(127), 293
Caugherty, P. D., **3:**39(196), 52(196), 58
Caughman, J. B. O., II, **18:**182
Caugley, D. M., **21:**103(38), 107(38), 112
Cava, R. J., **16:**3, 124
Cave, E. F., **5:**260, 267, 273, 316
Cavese, J. M., **19:**72
Cavin, O. B., **16:**3, 124
Cawthorne, C., **9:**167(24), 256
Cawthron, E. R., **14:**208(18), 237
Cech, R. E., **1:**210(119), 231
Celis, J. P., **17:**110, 141
Celis, R., **2:**23(26), 60
Celler, G. K., **22:**160(78), 173
Celli, V., **9:**245(245), 261
Cercignani, C., **22:**189(41), 194(41), 272
Cerdeira, F., **21:**79(7), 111
Ceren, E., **12:**39(68), 51
Cerenkov, P. A., **2:**123, 144
Cerny, R., **17:**83, 89, 90, 119, 121, 123, 126, 127, 128, 131, 139, 140, 143
Cerullo, M., **23:**18(67)
Cervinka, L., **7:**36(162), 41
Chaara, M. B., **22:**181(22), 193(43), 202(43, 59), 205(43), 270(126), 272–273, 276
Chabal, Y. J., **23:**47(159)
Chabot, R., **12:**29, 246(32), 247, 253(32); **13:**111, 112(30, 32), 115(32), 142(16), 147, 148
Chadi, D. J., **12:**115(215), 163; **19:**21; **23:**16(61)
Chadwick, J. E., **24:**44(41)
Chaffin, R. J., **21:**364(47), 368
Chahine, R., **11:**211(8), 232(8), 314
Chailapakul, O., **24:**44(54, 67, 68), 46(67, 68), 86(36)
Chakoumakos, B. C., **24:**98(92)
Chakrabarty, J., **16:**302, 311, 339
Chakravarti, B. N., **5:**304(264), 321
Chakraverty, B. K., **4:**111, 112(31), 120, 127(31), 135, 152, 189
Chaldyshev, V. A., **12:**105(176, 177), 106(176), 162
Challa, A., **22:**177(1–3), 271
Chamberlain, M. B., **16:**250, 268
Chamberlain, R. R., **3:**197(62), 209
Chamberlain, R. V., **24:**44(53)
Chamberlin, R. R., **12:**168, 169(2, 3, 4, 10, 11), 176(4), 183(4), 190(4), 194(11), 232
Chambers, A., **17:**27, 31, 75
Chambers, R. G., **6:**82(14), 86, 113, 133(132), 135(132), 143, 146, 190, 203
Chan, K. C., **24:**33(138), 228(32)
Chan, S-W., **16:**13, 42, 48, 63, 71, 125, 131–132, 135–136
Chan, W. K., **17:**271, 298
Chan, W. S., **11:**119(68), 201
Chance, R. R., **24:**73(118, 119)
Chand, N., **15:**230, 233–234, 262; **21:**137(48), 139(48), 154(48), 158(65), 160(48), 179(48), 188(84), 189(84), 190(84), 234–235, 244(20), 309

Chandra, A., **11:**51(74), 77(163), 98, 101
Chandra, S., **2:**181(62), 192; **17:**304, 312, 364, 365
Chandrasekhar, S., **1:**205(77), 230
Chandross, E. A., **10:**242(33), 245(35, 36), 258, 259; **24:**206(2)
Chang, C., **15:**294–295, 311; **22:**209(65), 273
Chang, C. A., **11:**67(131), 68, 70(137), 71, 73(137, 145), 74, 79(137), 99; **13:**212(46), 217(46), 220(70), 224, 225; **16:**28, 41–42, 51, 128, 131–132
Chang, C. C., **12:**277(209, 210, 213), 279, 280(212), 282(210), 294, 295; **13:**99(104), 107; **14:**103(46, 49), 126, 232(66), 238; **15:**303, 311; **16:**56, 60, 63, 134–135; **23:**48(161)
Chang, C. Y., **13:**209(29), 210(29), 215(29), 223; **21:**24(54), 73
Chang, C.-P., **22:**211(72), 274
Chang, F., **21:**241(9, 12, 13), 256(9), 269(9), 300(12), 308, 309
Chang, G. K., **3:**154, 164
Chang, H., **6:**224, 249(84, 158), 257, 259, 260(159, 163, 164), 263(164), 264, 265(162, 163), 267, 268, 270, 271, 272, 281(172), 282, 283, 288(208), 289(208), 290(208), 291(85, 208c, 210, 211, 213, 214), 292, 297; **9:**298, 299
Chang, H. C., **5:**284(192), 319
Chang, I. C., **15:**55, 138–139; **17:**182, 222
Chang, I. F., **17:**304, 352, 364, 367
Chang, J., **7:**24(93), 34(139), 39, 40
Chang, J.-J., **16:**106, 141
Chang, K. W., **15:**37, 64, 72–73, 86–87, 101, 138–140
Chang, L. L., **7:**155(67a, 67b), 166(67b), 249; **11:**36, 41, 42(40), 44(40, 54), 47(3), 58(3), 66, 67, 68, 70(137), 71(137), 73(130, 137, 145), 74(137, 145), 79(3, 137), 84, 85(130, 197, 198), 86(204, 205), 87(214), 89(3), 96, 97, 99, 100; **15:**245, 263; **17:**2, 8, 10, 11, 12, 15, 35, 37, 39, 42, 73–75; **21:**81(11), 111
Chang, M. S., **10:**238(30), 258
Chang, N. S., **15:**37, 137
Chang, P., **14:**39(145), 48(145), 49(145), 67(145), 71(233), 76, 79, 226(97), 238
Chang, P. C., **23:**279(22, 102)
Chang, R. K., **13:**205(18), 223
Chang, R. P. H., **12:**276(207, 215), 277(206, 208, 209, 210, 213), 279; **14:**82(1, 4), 87(15), 92(27), 100(43), 103(4, 46, 49), 113(66), 116(1), 125, 126, 127; **15:**303, 311; **16:**43–44, 46–47, 131; **17:**7, 12, 73
Chang, S., **17:**210, 223
Chang, S. J., **21:**93(21), 94(21), 111
Chang, S.-C., **24:**23(115), 32(115), 33(115), 44(44)
Chang, T. H. P., **6:**103(102b), 145
Chang, W., **21:**121(25), 233, 240(4), 241(8, 10, 18), 245(18), 246(4), 248(4), 252(4), 256(8, 10, 27), 259(10), 260(8), 265(27), 278(4), 308, 309
Chang, W. H., **21:**211(114), 236, 241(9, 11, 13), 256(9, 29), 267(29), 269(9), 282(11, 36), 290(49), 308, 309, 310
Chang, W. S. C., **13:**127(87), 149

Chang, Y. C., **19:**2, 9, 11, 16–18, 19, 23, 32; **21:**96(28), 112, 155(63), 234
Chang, Y.-K., **22:**227(84), 232(84), 235(84), 274
Chanin, G., **1:**241(9), 273
Channin, D. J., **10:**257(66, 67), 259; **17:**153, 154, 220
Chantre, A., **13:**178(101), 195
Chao, I., **24:**23(115), 32(115), 33(115), 44(44)
Chao, M. S., **7:**330(65), 394
Chao, S. S., **15:**287–288, 310–311
Chapin, J., **17:**116, 142
Chapin, P. C., **15:**244–245, 263
Chaplart, J., **11:**42(43), 97; **13:**219(65), 224
Chapman, B., **13:**72, 105; **14:**2(5), 73; **17:**81, 92, 139
Chapman, C. M., **2:**370(47), 421
Chapman, D. W., **17:**238, 253, 297
Chapman, G. H., **13:**187(126), 196
Chapman, R., **17:**347, 357, 358, 367, 369
Chapman, R. A., **11:**132(89, 90), 146(117), 148(117), 149(117), 201, 202
Chapman, R. L., **11:**81(185a), 101
Chaporon, A., **11:**254, 255
Chappell, T. I., **13:**176(87), 185(87), 188(87), 189(87), 195
Char, K., **16:**2, 121, 123, 143
Charap, S. H., **6:**222, 296
Charig, J. M., **4:**150(59), 188; **5:**187, 233, 266, 267, 270(104, 105), 316, 317
Charles, R. G., **5:**302(259), 321; **16:**87, 138
Charles, R. J., **11:**271(92), 316
Charlton, J. J., **3:**139(42), 163
Charmet, J. C., **19:**280
Charsan, S. S., **3:**105(96), 127
Charschan, S. S., **2:**369(33, 34), 420
Chartier, J. L., **14:**124(89), 127
Charych, D. H., **24:**5(21), 106(107), 221(93, 99), 228(8)
Chase, E. W., **16:**56, 60, 63, 71, 134–136
Chase, W. R., **9:**45(199), 46(199), 69; **12:**259(83), 292
Chasmer, R. P., **4:**50(111, 112, 113, 114, 115), 93
Chatenay, D., **24:**169(37), 181(13)
Chau, K., **15:**55, 139
Chau, Y. C., **12:**196(94), 197(94), 234
Chaudhari, P., **9:**271, 273, 275, 278, 280, 284, 285, 290, 291, 292, 294, 295; **14:**59(189), 77; **15:**147, 214; **16:**11, 41–42, 71, 118–122, 125, 131, 136, 143; **22:**5(8), 77, 123(25, 26), 172
Chaudhari, S. M., **16:**70, 136
Chaudhry, S., **16:**116, 142
Chaudhuri, S., **21:**38(77), 39(83), 73
Chaudhury, M. K., **24:**233(53)
Chauhan, S. P., **12:**246(29), 290
Chauley, J. D., **7:**330(60), 394
Chaurasia, H. K., **9:**43(195), 44, 69
Chavira, E., **16:**4, 124
Chayahara, A., **15:**293, 311
Chazalviel, J.-N., **17:**318, 366
Chckowski, D. H., **12:**146(317), 147(317), 166
Chebotaev, V. P., **13:**185(136, 137), 188(136, 137), 196
Cheesman, D. F., **7:**316(25), 393
Chelikowsky, J. R., **12:**115(216), 163; **13:**202(9), 223; **19:**102
Chemseddine, A., **17:**317, 365
Chen, A., **24:**151(9), 154(9), 156(9)
Chen, C. H., **12:**246(24), 290; **16:**36–37, 55–56, 130, 134

Chen, C. J., **17:**228, 229, 260, 287, 298, 299; **24:**237(63)
Chen, C. K., **23:**146(43), 165(43)
Chen, C. L., **1:**7(14), 64; **14:**162(112), 163(112), 197; **17:**93, 140
Chen, C. S., **24:**248(91), 249(91, 95)
Chen, C. Y., **13:**96(93), 97(93), 106; **16:**63, 135
Chen, C.-H., **24:**151(9), 154(9), 156(9)
Chen, D., **6:**272, 299; **13:**31, 58
Chen, D. R., **11:**76, 100
Chen, D. X., **16:**12, 125
Chen, F. F., **17:**107, 141; **18:**40, 46, 50, 80, 150
Chen, H., **24:**228(10)
Chen, J., **19:**269; **21:**117(8), 128(8), 221(8), 223(8), 232
Chen, J. B., **16:**91, 139
Chen, J. M., **11:**153(125, 126), 202
Chen, K., **16:**149, 177, 186, 189, 192, 203–205, 214–216, 220–225, 229–231; **24:**112(24), 129(24)
Chen, K. C., **22:**54(58), 78
Chen, K. Y., **12:**253(67), 291
Chen, L., **17:**318, 319, 366
Chen, M., **21:**161(68), 234
Chen, Q., **19:**12, 32, 35
Chen, S. H., **24:**46(90), 57(90)
Chen, T. C., **9:**265, 295
Chen, W. P., **9:**240(232), 261
Chen, W.-X., **14:**232(123), 239
Chen, Y. J., **9:**240(232), 261
Chen, Y. K., **23:**228(31), 229(31)
Chen, Y. L., **16:**96, 140
Chen, Y. Y., **17:**197, 198, 222
Chen, Y.-L., **17:**229, 300
Chen, Z. G., **21:**207(106), 235
Chen, Z. J., **16:**53–54, 134
Chency, G. T., **5:**303, 321
Chenery, D. H., **15:**276, 310
Cheney, M. E., **15:**244, 263
Cheng, A., **24:**119(40)
Cheng, C. C., **23:**27(100)
Cheng, J., **13:**160, 170(38), 186(38), 193; **24:**44(40)
Cheng, J.-Z., **24:**27(128), 31(130)
Cheng, K. H., **17:**320, 366
Cheng, L.-J., **21:**88(18), 111
Cheng, L.-Y., **22:**30(36), 77, 235(86), 274
Cheng, S. C., **13:**5(9), 21(9), 56, 112(31), 116(35), 148
Cherki, C., **6:**6(29), 27(93), 54(158), 55, 56, 57(158, 164, 165), 58, 63, 74, 76, 77, 78
Chern, C. H., **21:**81(14), 84(15, 16), 86(14), 87(14), 100(36), 111, 112
Chern, C. S., **16:**91, 139
Chernayev, V. N., **9:**39(169), 40(169), 69; **12:**177(61), 234
Cherreton, M., **16:**254, 268
Cheseldine, D. M., **6:**26(90), 76
Chesley, F. G., **3:**276(11), 299
Chester, L. A., **12:**277(203), 278, 294
Cheung, D. T., **21:**117(8), 128(8), 221(8, 134), 223(8), 232, 236
Cheung, N. H., **16:**65, 136
Cheung, T. D., **16:**164, 168, 177, 190–191, 193–202, 206–211, 213–214, 230
Chevalier, G., **6:**269(171), 299
Chevallier, J., **7:**196(168), 252
Chew, N. G., **16:**52, 133
Chewnpu, J., **15:**33, 137
Chezel, P., **6:**244, 298
Chhabra, D., **7:**268, 269(61), 308
Chi, C.-C., **16:**17, 41, 126, 131
Chi, G. C., **13:**176(84), 185(84), 186(84), 189(84), 195
Chi, K. C., **13:**79(80, 82), 95(82), 106
Chi, L. F., **24:**98(96)

Chi, M. C., **15:**185, 216
Chi-yi-Kao, **16:**240, 251, 267–268
Chia, P. S., **11:**146(116), 149(116), 151(116), 202
Chialvo, A., **22:**128(41, 42), 172
Chiang, C., **18:**222
Chiang, C. K., **16:**59, 135
Chiang, S., **24:**82(5), 86(27), 98(81)
Chiang, Yu. N., **6:**124(191), 148
Chiarotti, G., **12:**56(17), 158
Chiba, K., **9:**47(222a), 70
Chicotka, R. J., **7:**171(111), 177(179), 193(157), 196(169), 203(180), 204(179), 250, 251, 252
Chidambaram, K., **12:**176(53), 233
Chidsey, C. E. D., **24:**6(24), 10(40), 12(47, 50), 14(55, 57, 59), 15(50, 57), 17(40), 20(59), 23(112), 27(112, 122), 29(122), 30(112), 44(10, 11, 15, 17, 18), 45(11, 15, 17, 18, 69, 71, 72, 73), 70(15), 82(11, 12, 16, 17, 19), 83(11, 12), 86(11, 12, 16, 17, 19, 25, 29), 89(47), 112(4, 6, 7, 8, 9, 12, 20), 117(7, 20), 119(9, 20), 121(12), 125(6, 9, 49), 127(6), 128(20, 58), 129(20), 199(49), 206(14), 212(14), 217(68), 218(78), 219(80)
Chien, W. Y., **11:**81(184), 101
Chikawa, J., **12:**88(131), 96(131), 161
Chikazumi, S., **1:**281(13, 14), 293, 331; **6:**221(66), 227(96), 296, 297; **12:**94(148), 161
Childs, C. B., **4:**50(104), 93
Childs, P. E., **17:**354, 368
Chin, D., **16:**28, 128
Chin, F. K., **19:**148
Chin, J., **12:**253(67), 291
Chin, Y. H., **24:**154(18)
Chindaudom, P., **19:**199, 206, 213, 254, 263
Chinn, S. R., **11:**36(18), 96
Chinnok, E., **10:**168(3), 169(3), 225
Chiou, C., **1:**241(10), 273, 274
Chirigos, J. N., **4:**113(32), 120, 135
Chisholm, M. F., **16:**17, 21, 23, 43, 48, 126–127, 131
Chistyakov, N. S., **6:**238(127), 298
Chiu, L. C., **15:**229, 261; **21:**116(2), 155(60), 232, 234
Chiu, R. C., **16:**97, 140
Chiu, T., **21:**287(41), 310
Chlipala, J. D., **13:**187(125), 196
Chludzinski, J. W., **23:**237(33)
Cho, A. Y., **10:**250, 259; **11:**36, 42, 44(1), 50, 51(75, 80, 81), 53(89), 54(80, 89), 55(89), 56(1), 57(1, 75), 58(99, 102), 59(75, 102, 104), 61(1), 65(1), 66, 67, 73(1), 75(46, 148), 76, 77(157, 159, 161), 78(148, 168, 172, 173), 79, 80; **13:**218(59, 61), 221(59), 224; **15:**229–230, 261–262; **21:**116(2), 155(60), 184(80), 194(80), 196(94, 95), 207(105), 232, 234, 235; **23:**221(9), 224(9), 239(46), 240(54), 243(65), 262(65, 80, 83, 84, 85, 87), 263(65), 265(85), 266(46); **24:**114(29)
Cho, H. C., **22:**186(33), 209(33), 272
Choa, F. S., **23:**228(31), 229(31)
Choe, J.-W., **21:**4(12), 8(25, 26, 27), 9(25), 10(25), 11(25), 17(25), 28(26), 29(26), 31(26, 61), 36(26), 37(25), 72, 73, 185(82, 83), 186(83), 189(82), 206(102), 235; **23:**239(45), 243(45), 247(45), 251(45), 261(81), 291(81, 111)
Choi, C.-H., **17:**34, 75

Choi, H. K., **23:**229(12), 230(13, 14), 230–231(34), 232(17, 18, 35), 237(33), 278(100)
Choi, K. K., **15:**222, 226, 229–230, 234, 236–237, 239, 260–262; **21:**116(5), 117(10), 120(16), 121(16, 25), 122(16), 123(5), 124(5), 125(5), 126(29), 128(10, 38), 129(5, 39, 40), 136(5, 39, 40), 137(47), 141(50), 146(5, 39, 40), 149(5, 39), 209(10, 40), 211(114), 215(10), 232; **23:**190(12)
Choi, S.-K., **24:**247(88)
Chollet, L., **17:**44, 75
Chopra, K. L., **2:**381(89), 422; **3:**110, 127; **4:**166, 189; **6:**63, 78, 82, 101(18), 106(122a), 107, 109, 110, 124(189), 127(189), 131(122a, 127, 128, 129), 132(127, 128, 129), 143, 146, 148; **7:**1(4), 34(137), 37, 40, 372(183), 396; **9:**4(12, 13), 5(15, 16), 6(12, 23), 42(181), 65, 69, 277, 295; **12:**71(67), 160, 169, 172, 176(26, 49, 51, 53), 177(54a), 179(65), 180(65), 183(25, 28, 73), 185(30, 49, 54a, 73), 186(26, 30, 49, 54a, 73), 187(54a, 73, 87, 88), 189(28, 29, 30, 88, 89), 190(30, 87, 88), 191(88), 192(30), 194, 195(24), 196(24), 197(26, 73), 198(30, 49, 73), 199(29, 30, 49, 73), 200(30, 54a, 73), 203(42), 204(43), 205(42, 43), 207(42), 208(42), 210(42), 211(42), 214(42), 215(42), 217(40, 43), 217(104), 218(99, 111), 219(106, 111, 112), 220(112), 223(106, 113), 224(120), 225(28), 226(122, 124, 125), 227(124), 228(30), 229(106, 111), 230(99, 106), 233, 234, 235; **13:**18(25), 19(25), 57, 79(78), 99, 100(111), 101(111), 106, 107, 117(39), 119(48), 127, 131, 143(129), 145(96), 148, 149, 150; **16:**4, 124, 236, 267
Chorey, C. M., **16:**76, 137
Chorin, A. J., **22:**98(18), 115
Chorney, P., **18:**65
Chou, C. H., **22:**217(79), 274
Chou, N. J., **9:**31(114), 35(114), 67
Chou, P., **16:**91, 139
Chou, S. F., **15:**287, 310
Chow, D. H., **23:**234(20, 37), 235(20), 281(108)
Chow, E. T-K., **4:**22, 23, 91
Chow, J. T., **24:**98(84)
Chow, S. L., **9:**6(24), 65
Chow, T. P., **14:**234(146–148), 234–235(147), 235(148), 240
Chow, T. S., **16:**282–284, 317, 338
Chowdari, B. V. R., **17:**359, 369
Choyke, W. J., **11:**10, 33; **12:**76, 160
Chretien, M. L., **16:**99, 140
Chrisey, D. B., **16:**65, 78, 136–137
Christen, D. K., **16:**10, 43, 48–49, 70, 125, 131–132, 136
Christensen, H., **5:**266(79), 273, 316, 317
Christensen, R. L., **1:**20(48), 65
Christiansen, J., **16:**51, 133
Christie, E. A., **10:**43(85), 44(85), 45(85), 58(85), 67
Christman, S. B., **12:**121(238), 164
Christou, A., **13:**210(35), 212(39), 224
Christov, S. G., **6:**22, 75
Christy, R. W., **2:**370, 421; **3:**146(61, 63), 160, 163; **5:**243(16), 311, 314, 322; **9:**236(219), 261
Chromik, S., **16:**52, 133

Chrzan, D., **17:**15, 74
Chu, C. S., **16:**239, 267
Chu, C. W., **16:**2, 123
Chu, J. C., **12:**137(309), 165
Chu, S. N. G., **23:**243(65), 262(65, 80, 85, 87), 263(65), 265(85), 272(96), 273(86), 274(87), 291(80), 298(80)
Chu, T. K., **18:**151
Chu, T. L., **4:**316, 324; **5:**154(45), 157, 160, 161, 199(174), 202(174), 230; **12:**265(112), 292; **17:**153, 154, 220
Chu, W. K., **17:**22, 74
Chu, W. W. L., **3:**40(212), 58
Chu, X., **18:**283
Chuang, R. J., **24:**72(116)
Chuang, T. J., **13:**153, 171(75, 76), 181(108), 182(108), 183(108), 184(75, 76), 185(75, 76, 129), 187(75, 76, 129, 132), 190(158), 191(76, 158), 192, 194, 195, 196, 197; **14:**230(107), 235(150), 239, 240
Chubachi, N., **14:**20(81), 27(81), 75; **17:**153, 219
Chubb, T. A., **10:**78(22), 164
Chudnovskii, F. A., **17:**328, 344, 366
Chun, C. S., **17:**24, 46, 49, 51, 53, 56, 58, 75
Chun, S. K., **21:**80(10), 97(10), 111
Ch'un-ha, L., **7:**31(129), 40
Chung, C., **24:**3(16), 4(16), 32(16), 44(28), 53(28), 70(28)
Chung, D. W., **16:**21, 127
Chung, T. C., **24:**181(13)
Chupka, W. A., **17:**307, 364
Churchill, D., **7:**331, 333(72), 394
Chwalek, J. M., **16:**107, 141
Chye, P. W., **12:**74(85), 78, 114(206, 207), 115(212, 217), 116(212, 222), 117(212), 160, 163
Chynoweth, A. G., **12:**60(29, 31), 159
Chyung, L. J., **17:**58, 76
Ciccotti, G., **22:**123(24), 172
Ciddor, P. E., **4:**85(219), 96
Cikmach, P. D., **17:**310, 337, 338, 349, 350, 365, 367
Cillessen, J. F. M., **17:**209, 223, 292, 299
Cima, M. J., **16:**97, 100, 140
Cinader, G., **19:**41
Cirkler, W., **6:**104, 105, 146
Claassen, W. A. P., **15:**297, 311
Clabes, J. G., **14:**232(129), 239
Claeson, T., **16:**67, 136
Claesson, Y., **17:**131, 143
Clampitt, R., **13:**3(8), 56
Clans, H., **16:**4, 124
Clark, A. F., **16:**11–12, 125
Clark, C. L., **3:**39, 52(188), 58
Clark, G. J., **13:**135(108), 136(108), 137(108), 150
Clark, G. L., **7:**354(137, 138), 355(146), 356(137), 395, 396
Clark, G. W., **1:**316(68), 333
Clark, J. P., **16:**226, 231
Clark, M. R., **24:**151(9), 154(9), 156(9)
Clark, R. A., **24:**45(80)
Clark, R. J., **10:**252(68), 259
Clarke, J., **16:**57, 75–77, 114, 117, 134, 137, 142; **24:**106(106)
Clarke, J. E., **11:**171(144), 203
Clarke, J. H. R., **22:**134(51), 142(51), 172
Clarke, K. D., **19:**237
Clarke, P. J., **17:**87, 140
Clarke, R. C., **17:**173, 222
Class, W., **17:**111, 142
Clausen, E. M., Jr., **19:**241

Clausen, O. W., **4:**60(138), 94
Clauss, J., **24:**153(14)
Clawson, A. R., **5:**184, 185, 233; **7:**201(175d), 252; **11:**57(97), 77(164), 99, 101
Clegg, B., **11:**59(96), 76(96), 99
Clegg, P. L., **2:**199(5), 237; **3:**27(135), 28, 51(135), 56
Clegg, W. W., **6:**237(124b), 298
Clem, P. G., **24:**243(81)
Clemens, B. M., **17:**24, 46, 49, 51, 53, 56, 58, 75, 76
Clemens, H. J., **12:**114(210), 163
Clerc, G., **6:**211(23, 24, 25), 212(24, 25), 225(24), 230(109), 233(109), 234(109, 118), 235(109), 242(23, 24, 25), 244(24, 25), 249(157), 250(24, 109, 118), 254(24, 25), 255(24, 25), 256(24), 257(24), 258(24), 259(24), 260(24, 157), 271(24, 157), 295, 297, 298
Clernets, A. N., **17:**153, 219
Cleveland, C., **24:**151(10), 154(10), 156(10), 163(10), 222(104)
Climent, A., **14:**90(90), 105(55), 106(55), 112(97–99), 124(90), 126, 127, 128
Clough, R. B., **5:**282, 319; **7:**170(105), 193(162), 250, 252
Clougherty, E. V., **7:**18(58), 38
Clouston, J. G., **5:**312, 322
Clow, H., **6:**238, 239(129), 241(128), 298
Clunie, J. S., **7:**355(147), 396
Cobine, J. D., **3:**80(63), 88(63), 126; **6:**37, 77; **14:**139(30), 156(89), 175(131), 195, 197, 198; **17:**95, 141
Coblentz, W. W., **3:**275, 299; **4:**60(145), 94
Coburn, J. W., **14:**16(61, 62), 17(61), 28(112), 74, 76, 208(21), 209(21), 234(136, 138, 139, 141), 235(150), 237, 240
Cochran, J. F., **6:**123, 129(182, 183, 184), 131(182, 183, 184), 147
Cochrane, W., **4:**223, 232, 234(149), 235(149), 236(149), 253, 254
Cockcroft, J. D., **4:**122, 123(58a), 127(58a), 136
Cocks, F. H., **12:**258(73, 74), 291
Codling, K., **7:**48, 63(27), 112
Cody, G. D., **10:**60(136), 69; **12:**71, 79(72), 160; **19:**73, 75
Coehlo, R., **6:**54(158), 55(158), 56, 57(158, 164), 58(158, 164), 63(158), 77, 78
Coeure, P., **9:**282, 294
Coey, J. M. D., **15:**144, 214; **16:**215, 231
Coffin, L. F., **5:**263, 316
Coffman, B., **2:**373(64), 421
Cogan, S. F., **17:**347, 353, 359, 360, 367–369
Coghill, H. D., **3:**195(57), 209; **11:**111, 200
Cohen, C., **14:**122(86), 127
Cohen, D., **16:**59, 65, 68, 135
Cohen, J., **23:**91(22), 101(32)
Cohen, J. B., **4:**222(112), 228(141), 253, 254; **17:**29, 47, 75
Cohen, J. M., **22:**254(98), 261(98), 262(98), 274
Cohen, L., **7:**240(256), 255
Cohen, L. K., **24:**208(27)
Cohen, M., **4:**223(116), 253; **7:**18(59), 38; **16:**50, 132
Cohen, M. H., **6:**140(227), 148; **7:**34(141), 40; **10:**40(77), 67; **19:**139
Cohen, M. L., **12:**114(203), 115(216), 163; **13:**202(9), 223; **19:**102

Cohen, M. S., **1:**285(24), 286(28, 29, 30), 287, 288, 289(33), 297(46), 301(54), 332; **3:**214(11, 13), 247(11), 250, 255(87), 270, 273; **6:**228(106), 297
Cohen, R. W., **10:**60(136), 69; **12:**72(73), 79(73), 160
Cohen-Boulakia, J., **24:**33(135)
Cokell, C. M., **17:**209, 223
Colas, E., **15:**259, 264; **19:**2, 27, 32–34, 36, 37; **21:**241(15), 309; **23:**238(43, 44), 239(43), 243(43, 44), 244(43, 44), 251(44), 292(114)
Colbert, W. H., **9:**43(189, 192), 69
Colbow, K., **17:**315, 318, 328, 347, 365, 366
Colby, J. W., **14:**220(72), 238
Colclough, M. S., **16:**117, 121, 142–143
Coldren, L. A., **15:**66, 139
Cole, R. C., **21:**162(69), 234
Cole, R. H., **11:**222(17), 227, 228, 293(127, 130), 314, 317
Cole, B. E., **13:**135, 141(104), 150
Cole, K. S., **11:**222(17), 227, 314
Coleman, H. S., **6:**310(16), 314, 327, 346
Coleman, J. J., **11:**66(122), 99; **12:**277(210, 211), 279(210, 211), 282(210), 295
Coleman, M. V., **7:**24(92), 35(152), 36(159), 39, 41
Coleman, P. D., **15:**244–245, 263
Coleman, R. V., **6:**82, 103(101), 112(30), 144, 145
Colen, M., **13:**79(70), 83(70), 92, 93, 94(70), 106
Collard, D. M., **24:**19(86), 20(86), 27(86), 30(86), 44(52)
Colligan, J. S., **12:**238(9), 290; **14:**217(39), 237
Collignon, J. S., **17:**82, 139
Colligon, J. S., **22:**113(26, 27), 115
Collins, D. A., **11:**77(164), 101; **13:**205(20), 223
Collins, D. M., **11:**36, 63, 64, 67(110), 96, 99, 100
Collins, F. C., **4:**260(17), 320
Collins, G. J., **15:**295–296, 311; **17:**153, 220
Collins, J. H., **12:**269(149), 293; **15:**8, 47–48, 48, 53, 57, 82, 101, 136, 138–140
Collins, L. E., **1:**218(136), 231
Collins, R. E., **4:**353, 371
Collins, R. H., **1:**10, 11, 12, 64
Collins, R. J., **21:**94(24), 111
Collins, R. W., **19:**50–52, 54–56, 72, 94, 103, 104, 110–115, 129, 135, 148, 152, 154, 157, 159, 160, 163, 165–167, 171, 175, 176, 181, 182, 196, 203, 205, 206, 280, 310
Collinson, M., **24:**18(77), 19(77)
Collman, J., **16:**84, 90, 138
Colmenares, C. A., **14:**202(2), 236
Colmet, R., **17:**109, 141
Colombani, A., **2:**30(41), 61; **6:**118(166), 147
Colombini, E., **14:**63(204), 64(204), 66(204), 78; **17:**153, 220
Colson, D., **24:**181(13)
Colson, J. P., **4:**79(202), 84(202), 95
Coltrin, M. E., **22:**186(31, 32), 209(32), 272
Comas, J., **11:**66(120), 99
Combemale, Y., **10:**228(4), 253(4), 254(4), 258
Combet, H. A., **6:**141, 149
Comer, J. J., **5:**297, 321
Comfort, J. H., **14:**232(131), 239

Comizzoli, R. B., **12:**258(77), 291
Compton, D. J. M., **5:**275, 317
Compton, K. T., **14:**150(80), 196
Compton, V. B., **6:**211(28), 295
Comsa, G., **24:**117(38)
Condas, G. A., **1:**54, 67; **12:**85(102), 160
Condon, C. R., **17:**116, 142
Cong, Y., **19:**106, 108, 116
Conger, R. L., **1:**317, 333; **6:**224(78), 291(78), 297
Conjeand, P., **4:**150(43), 152(43), 188
Conklin, J. B., **3:**277(24), 300
Conklin, R. E., **2:**390(119), 423
Connell, E. M., **5:**211(220), 235
Connell, G. A. N., **15:**144–145, 150, 154, 159, 168, 176–177, 182–183, 186–187, 190; **16:**69, 78, 136–137; **17:**209, 223, 272, 290, 293, 298
Connell, R. A., **1:**245(33), 274; **3:**156, 157(83), 164
Connelly, W. G., **23:**124(16)
Connolly, J., **21:**324(15), 325(15, 18), 329(15), 333(15), 334(15), 336(15), 337(32), 340(32), 348(38), 351(18), 353(18), 355(38), 356(38), 367
Connolly, J. C., **23:**230(15), 231(15)
Connors, M. K., **23:**278(100)
Conrad, E. E., **4:**266(68), 267(68), 270(68), 271(95, 96), 310(68), 322; **12:**259(84), 292
Conrady, A. E., **4:**55, 93
Conragen, J., **5:**226(216), 235
Conroy, J., **17:**154, 155, 182, 222; **24:**183(32, 33), 184(33), 192(32)
Consoli, T., **18:**122, 159
Constant, E., **11:**254, 255
Constant, G., **13:**99(98), 107; **17:**121, 143
Conty, E., **15:**55, 138
Conway, B. E., **4:**196, 197(24, 25, 41), 251, 252
Conway, E. D., **22:**85(6), 114
Conwell, E. M., **4:**328(12), 331(27), 332, 369, 370; **9:**240(234), 261
Coogan, C. K., **2:**284(41), 304
Cook, C. L., **5:**312, 322
Cook, E. L., **6:**3, 74
Cook, H. C., **4:**212, 253
Cook, H. D., **7:**334(83), 394; **9:**56(260), 57, 58, 62(260), 71
Cook, H. E., **6:**335(41), 347; **17:**24, 30, 35, 36, 37, 38, 74, 75
Cook, J. M., **18:**60
Cook, M., **11:**293(125), 317
Cook, M. J., **24:**21(99), 24(99), 25(99)
Cook, P. W., **13:**183(118), 184(118), 196
Cook, R. F., **16:**5, 124
Cook, R. L., **5:**121, 141
Cook, W. R., Jr., **19:**250
Cooke, D. W., **16:**37, 48, 130, 132
Cooke, M. J., **22:**30(35), 77
Cooley, K. W., **12:**287(254), 288(254), 295
Coon, D. D., **15:**222, 226–227, 229–234, 238–239, 241–243, 245–246, 248–254, 256–259, 260, 264; **21:**4(10, 18), 5(20), 6(18, 20), 7(18), 8(27), 9(18), 22(18, 20), 27(20), 28(18), 29(20, 58), 31(61), 32(18, 64), 35(68), 38(74, 75, 77, 78, 81), 39(78, 82, 83), 40(78), 41(58, 68, 89, 90, 94), 42(94), 45(20), 49(68, 78, 89, 97), 50(98), 51(94), 52(94), 53(90), 55(90), 56(90), 59(94, 107, 108), 60(109), 61(95, 107, 108, 114); **23:**144(36), 239(45), 243(45), 247(45), 251(45)

Coon, P. A., **23:**35n(138), 41(145)
Coons, S. B., **16:**119, 121, 143
Cooper, C. B., **3:**70, 125
Cooper, E. I., **16:**5, 124
Cooper, G., **2:**418(196), 425
Cooper, H. W., **4:**265(62), 319(62), 321
Cooper, J. N., **1:**264(79, 80), 275
Cooper, L. N., **1:**255(57), 256(57), 258(57), 272(102), 274, 275
Cooper, R., **6:**68(204), 73(218), 78, 79
Copel, M., **23:**43(74)
Copeland, J. A., **7:**240(255), 255
Coppens, R., **6:**127(192), 148
Coppola, A. J., **3:**137(32b), 163; **4:**315(159), 324
Corbett, J. M., **17:**60, 77
Corbett, J. W., **14:**231(112), 239
Corbitt, T. S., **24:**99(103), 100(103), 106(103)
Corboy, J. F., **5:**276(146), 278, 280, 318
Cördoba-Torresi, S. I., **17:**326, 327, 366
Coren, R. L., **3:**120(148), 128; **6:**277, 299
Corenzwit, E., **6:**209(19), 295
Corgan, J. M., **7:**97(72), 113
Corkhill, J., **14:**32(125), 76
Corkish, R., **21:**325(16), 351(16), 357(16, 41), 359(16), 367, 368
Corl, E. A., **4:**274, 322
Corliss, L. M., **1:**205(80), 230
Corn, R. M., **24:**18(75), 19(75), 46(83, 84), 66(104)
Cornely, R., **9:**6(25), 43(25), 65
Corning, J. J., **4:**258(10), 319(10), 320
Cornog, D. G., **14:**36–37(139), 76
Coronell, D. G., **22:**178(13), 182(26), 271–272
Corrigan, D. A., **17:**304, 364
Corrsin, L., **3:**176(23), 209
Corruccini, R. J., **2:**329(36), 359
Corsi, C., **11:**113, 200
Cortes, R., **24:**184(42), 188(42)
Corts, R. W., **16:**328, 339
Cossar, B. C., **24:**8(28), 9(28)
Cosslett, V. E., **2:**73, 125(138), 141, 144; **4:**224(123), 253
Costable, J., **14:**57(177), 77
Costanzo, C., **15:**238, 262
Costard, E., **21:**136(46), 161(67), 233, 234
Costich, V. R., **9:**98(14), 110(24), 111(24), 123(14), 124(14), 135(14), 136(46), 143, 144
Cotterell, B., **16:**279, 337
Cotterell, D. L., **14:**208(18), 237
Cottey, A. A., **6:**86, 144
Cotti, P., **6:**82, 86, 89, 100, 106(85), 113, 115, 124(85), 135(85, 140), 136, 144, 145, 146, 190, 203
Cottington, R. L., **7:**331(67), 394
Cotton, P., **1:**101, 121
Cotton, R. J., **17:**334, 367
Cottrell, A. H., **4:**147(30), 187; **7:**116, 119(4), 247
Coudere, P., **15:**306, 312
Cougulin, J. P., **12:**73(80), 160
Coulter, J. K., **10:**76(12), 128(12), 164, 262(3a), 308; **12:**8(35), 15(44), 40(44), 50
Courths, R., **17:**332, 366
Courtney, W. J., **10:**133, 166
Courtney-Pratt, J. S., **4:**3(5), 90; **7:**336(102, 103, 104), 337(102, 103, 104), 389(104), 395; **12:**148(321), 166
Courvoisier, J. C., **5:**163(68), 168, 231
Coutts, M., **7:**158(79), 249
Coutts, M. D., **10:**60(136), 69; **12:**72(73), 79(73), 160

Coutts, T. J., **14:**59(184), 77
Covington, B. C., **21:**121(19), 123(19), 232
Covington, D. W., **11:**56(91, 92, 93), 57(93), 77(158, 160), 88, 98, 100, 102
Covington, E. J., **7:**342, 354, 358(113), 395
Covy, J., **13:**103(119), 107
Cowache, P., **12:**226(113), 235
Cowan, J. J., **9:**225(191), 226(191), 260
Cowden, W. G., **22:**215(76), 274
Cowher, M. E., **9:**282, 295
Cowley, J. M., **1:**194, 196(40), 198, 229; **2:**118(106, 107, 110), 143; **7:**15(36), 38, 334(89), 394
Cox, G. A., **5:**160(51), 230
Cox, H. L., **16:**319, 339
Cox, J. N., **15:**302, 311
Cox, J. T., **1:**148, 149(49), 169(49), 170(49), 185; **2:**262, 263(21, 22, 23), 265(24b), 284(22, 36), 286(21), 289(20), 290(24b), 294(22), 295(21, 22), 296(21), 302(23), 304, 334(51), 346(51), 360; **3:**2, 53; **4:**9(24), 91, 282(101), 285(101), 322; **5:**123(28), 141; **9:**19(78), 66; **10:**77(17, 18, 19), 78(19), 79(19), 80(19), 83(18), 84(18), 97(32), 98(32), 99(32), 100(33), 101(33), 102(33), 103(33), 104(33), 106(33), 107(34)
Cox, P. A., **15:**196, 217
Cox, R. E. L., **9:**45, 69
Cox, R. H., **13:**210(33), 213, 224
Cox, W. P., **4:**263(55), 264(55), 319(55), 321
Cozine, H. T., **3:**291(55), 300
Cracknell, A., **3:**266, 273
Craford, J., **7:**209(185a), 252
Craford, M. G., **7:**169(95), 238, 250, 254
Craievich, A., **24:**182(22)
Craig, A. G., Jr., **10:**43(90), 44(90), 67
Craig, D. R., **1:**197(46), 229
Craig, G., **16:**84, 90, 138
Craig, R., **3:**28(145), 51(145), 57
Craigen, D., **17:**315, 365
Craighead, H. G., **13:**153(14), 161(39), 182(14, 39), 183(14, 39), 192; **19:**129; **24:**229(40, 41)
Cram, L. S., **9:**220(180), 260
Crandall, M. G., **22:**85(7–9), 114
Crandall, R. S., **17:**304, 318, 319, 320, 350, 352, 364, 366, 367
Crane, G. R., **17:**279, 299
Crangle, J., **6:**211(29), 295
Crank, J., **6:**304(7), 346; **16:**236, 266
Crawford, J. L., **4:**185(144), 186, 190; **6:**337(43), 347
Crawley, R. L., **7:**186(144a), 251; **11:**109(29), 200
Crawshaw, D. D., **12:**142, 165
Creager, S. E., **24:**23(111), 27(111), 44(51)
Creasey, C. J., **6:**117, 147
Cree, F., **5:**259(48), 315
Creedy, F., **14:**150(79), 196
Creighton, J. R., **19:**35
Cremer, E., **2:**346(72, 74), 360, 402(170), 424; **10:**129(59), 135(59), 165
Cremers, C., **16:**52, 133
Creuzburg, M., **9:**210, 212, 214(146), 218(174), 259
Crittenden, E. C., Jr., **1:**58, 67, 226(175, 179, 180), 232, 264(79, 80), 275, 282, 324(97), 325(97), 326, 331, 334; **2:**30(37), 61; **3:**19, 51(88), 55, 220(29, 63), 229(63),

Crittenden, E. C., Jr., (*continued*) 234(63), 240(63), 241(63), 243(63), 245(73), 271, 272; **6:**116(152), 147; **22:**120(15), 172
Croce, P., **4:**23(57), 24, 92; **6:**188(30, 31), 203; **10:**169(18), 225
Crocker, A., **1:**36(93), 66
Croft, M., **16:**60, 135
Croll, I. M., **6:**214(35), 295
Crombeen, J. E., **12:**122(247), 164
Cromer, D. T., **9:**12(56), 66
Cronemeyer, D. C., **5:**275, 317; **9:**275, 280, 281, 287, 288, 295, 296
Crook, A. W., **1:**70(5), 71(5), 120; **3:**27(135), 28, 51(135), 56
Crooks, R. M., **24:**12(50), 15(50), 18(50, 76), 19(50, 76), 33(138), 44(54, 55, 56, 63, 64, 65, 66, 67, 68), 46(63, 64, 65, 66, 67, 68), 70(55, 63), 86(36), 92(56), 94(64, 65), 99(102, 103), 100(103), 106(102, 103), 161(23), 162(23), 221(88), 228(32)
Croset, M., **12:**277(197), 280(197), 292, 294; **14:**100(42–44), 117(76), 120(76), 121(44), 126, 127
Cross, C., **14:**232(114), 239
Cross, K. B., **5:**197, 198, 234; **13:**61(12), 104
Cross, L. E., **17:**241, 297; **19:**253
Cross, P. C., **15:**266, 309; **19:**299
Cross, R. W., **15:**111, 141
Crossley, I., **7:**120(6), 247
Croteau, A., **19:**250
Crouch, P. E., **22:**269(118–120), 275
Croutch, M. L., **5:**265(69), 316
Crowe, J. W., **1:**271(97), 275
Crowe, K. R., **12:**70(55), 159
Crowell, C. R., **4:**330, 331(24), 332, 333(25), 334(36), 335(36), 337(25), 338, 339, 340, 341, 342(50), 343, 345, 346, 347(50), 349, 354(51), 356(75), 357, 358(75), 359, 360, 361(80), 362(79, 80), 363, 364(80); **12:**84, 160; **23:**88(16)367, 369, 370, 371
Crowell, J., **9:**198, 219(175), 229(209), 258, 259, 260
Crowell, J. E., **22:**186(33), 209(33), 272
Crowley, J. M., **17:**29, 47, 75
Crowther, T. S., **1:**302(51), 319(75), 332, 333
Cruchon-Dupeyrat, S., **24:**94(63)
Cruzan, P. D., **14:**226(81), 238; **22:**119(6), 171
Csihas, L., **7:**68(59), 113
Cuage, T. P., **5:**298(247), 321
Cuccia, L. A., **24:**151(8), 154(8), 155(8b, 8c), 156(8), 157(8b, 8c), 158(8b), 159(8b, 8c), 161(8b), 222(103)
Cuchiaro, J. D., **17:**259, 298
Cuff, K. F., **3:**277(29), 300; **11:**142(110), 202
Cui, G. J., **16:**121, 143
Cui, Y. D., **16:**41, 131
Cukauskas, E. J., **16:**85, 138
Cullen, G. W., **5:**152(36), 153(36), 230, 272(113), 306(269), 317, 321
Cullen, J. D., **17:**318, 365
Cullis, A. G., 13.3(8), 13.56, 13.158(34), 13.169(34), 13.193, 23.24(89), 23.30(114), 23.37n(114), **23:**38(114), 39(114)
Cullity, B. D., **4:**218(98), 253; **16:**157, 229; **17:**17, 74
Culp, J., **15:**70, 100, 139
Culp, J. W., **2:**393(134), 423; **3:**109(110), 127

Culver, H. E., **2:**393(133), 423
Cummings, K. D., **13:**176(84), 185(84), 186(84), 189(84), 195
Cummins, D. O., **5:**160(51), 230
Cummins, S. E., **17:**229, 266, 267, 268, 269, 298
Cunneen, T. J., **24:**9(32)
Cunningham, B. B., **3:**27(129), 51(129), 56
Cunningham, J. E., **15:**249–250, 263; **21:**193(91), 235, 287(41), 310; **23:**290(109)
Cunningham, R., **14:**179(153), 180(153), 186(153), 187(153), 192(153), 198
Cunningham, R. D., **9:**22(90), 23(90), 27, 67
Cuomo, J. J., **9:**51(242), 54(242), 70, 271, 273, 274, 277, 278, 280, 282, 285, 294, 295, 296; **10:**58(126), 68; **12:**282(237), 283(237), 295; **13:**74(56), 105, 109(4), 147; **14:**8(28), 9(33), 18(72), 23(89), 25(98), 26(98, 101), 61(197), 65(98, 214), 67(98), 68(98), 74, 75, 78, 203(11), 209(11), 212(11), 217(11), 228(91), 229(96), 237, 239; **15:**147, 154, 164, 214–215, 277, 279, 310; **16:**14, 15, 17, 23, 53, 78, 125–126, 133, 137; **17:**82, 87, 139, 140, 153, 158, 220, 234, 236, 296; **22:**30(39), 32(39), 39(47), 45(47), 78, 161(79), 173
Cuppens, R., **17:**228, 240, 256, 257, 298
Curcio, J. A., **7:**67(57), 113
Curzon, A. E., **3:**212(5), 270
Cusano, D. A., **3:**194(53, 54), 197(61), 201, 205, 209, 210; **5:**282(178, 179), 319; **7:**209(187), 252
Cushman, D., **17:**201, 223
Cuthbert, J. D., **10:**236(16), 258
Cuthbert, J. O., **7:**162, 211(85), 249
Cutiss, C. F., **13:**8(17), 57
Cutler, P. H., **4:**339, 370
Cutshaw, C., **16:**123, 143
Czaja, W., **5:**188, 233; **7:**162(83), 249
Czanderna, A. W., **3:**27(142), 28(142), 51(142), 56
Czekala, L., **16:**123, 143
Czerny, M., **2:**321, 359
Czorny, B. R., **5:**260, 267, 273, 316

D

Da Silva, E. M., **3:**31(153), 51(153), 57, 146(65), 147, 156(82), 163, 164
Daalder, J. E., **14:**139(29), 140(29), 141(38), 145(68), 149(77), 150(77), 151(38), 155(84), 157(93, 94), 158(84, 93), 159(93), 160(106), 162(94), 163(94), 195, 196, 197; **17:**93, 140
Daban-Haurou, J. L., **14:**96(35), 126
Dabizha, E. V., **17:**60, 77
Dacey, G. C., **2:**150, 190
Dachraoui, M., **12:**226(113), 235
Dafesh, P. A., **21:**162(69), 234
Dahlgren, D. A., **24:**228(30)
Dahlgren, S. D., **14:**223(75), 238; **16:**193, 230
Dahmen, K-H., **16:**87, 138
Dai, B. T., **13:**190(160), 191(160), 197
Dai, U., **16:**50, 132
Daillant, J., **24:**181(13)
Dakss, M. L., **10:**235(11), 236(17), 258
Dalal, V., **12:**137(307), 165
Dalal, V. L., **7:**170(104), 250; **23:**91(24)
Dalby, J. L., **19:**6

Dale, E. B., **5:**164(96), 172(96), 231
Dale, J. R., **13:**210(30), 223
Dallesasse, J. M., **21:**287(39), 310
Dalton, J. V., **4:**309, 315, 323, 324; **5:**296, 300(242), 320; **14:**226(86), 238
Dalven, R., **11:**106(8), 199
Dalvie, M., **22:**83(1), 84(1–3), 85(1), 92(1), 93(1), 97(1), 98(2), 101(2, 3), 102(2), 105(2), 114, 188(39), 270(129), 272, 275
Daly, K., **16:**106, 141
Daly, K, P., **16:**119, 121, 143
Dam, B., **16:**47, 50, 132
Damany, M. H., **1:**148(51), 149(51), 159(51), 160(51), 185
D'Amico, C., **3:**97(82), 126
Damjanovic, A., **4:**191, 192, 195(2), 196(2, 34, 35, 36, 37, 38, 39), 198, 199(36, 38, 39, 64), 200(36, 38, 39), 201, 202, 203(2), 204, 206, 207(38), 209(36, 38, 39, 64), 210, 211, 233(38), 251, 252
Damon, R. W., **15:**26, 35, 92, 137, 140
Danbert, B. F., **7:**318(28, 29), 393
Dandl, R. A., **18:**122, 124, 151, 152, 155
Daneu, V., **13:**162(46), 163(46), 186(46, 145, 153), 189(46, 145), 190(46, 145, 153), 193, 197
Daniel, M. F., **17:**329, 366
Daniel, M. R., **15:**18, 37–38, 40, 47–48, 54, 59, 62, 72–76, 78, 81, 132, 137–139; **17:**146, 148, 186, 219
Daniels, F., **10:**43(87), 44(87), 67
Daniels, J., **9:**202(116), 205(129), 206, 214(147), 258, 259
Daniels, R. D., **1:**226(176), 232, 282(18), 331; **3:**220(63), 229(63); **22:**120(15), 172
Danielson, G. C., **6:**104, 105, 131(110), 146
Danilina, T. l., **14:**43(152), 77
Dankov, P. D., **12:**73, 160
Danroc, J., **14:**14(49), 74; **17:**111, 142
Danso, K. A., **16:**240, 267
D'Antonio, C., **3:**254, 255(95), 256, 257, 258(95), 263, 264(95), 265(95), 267, 269, 273
D'Antonio, P., **15:**156–157, 215
Dao, G., **16:**252, 268
Dao, L. H., **17:**359, 369
Dapkus, P. D., **16:**81, 138; **19:**2, 12, 24, 25, 28, 32, 35
Darack, S., **14:**103(46), 126; **15:**303, 311
Darken, C. S., **7:**181(133), 251
Darken, L. S., **6:**303(3), 345, 346, 347
Darmois, G., **2:**23(30), 60
Darnell, P. S., **2:**384(94), 422
Das, D., **16:**169, 193, 202–203, 207, 230
Das, D. N., **3:**174(21), 208
Das, S. B., **6:**128(193), 148
Das, S. K., **12:**253(67), 291; **16:**28–29, 129
Das, S. R., **12:**169(25, 28), 172(25), 183(25, 28), 189(28), 225(28), 233
Das Gupta, D. K., **11:**230(29), 315
Das Gupta, S., **11:**254, 255
D'Asaro, L. A., **7:**234(244), 254
Dasgupta, S., **24:**211(49)
Dash, W. C., **4:**169(107), 190; **5:**187(154), 233
DaSilva, E. M., **1:**30(70), 65
Datlov, J., **18:**143
Daude, A., **9:**178(60), 227(204), 257, 260
Daugherty, M. J., **21:**128(33), 233
Daughton, J. M., **6:**249(158), 298

Daunt, J., **1:**265(83), 275
Dautremont-Smith, W. C., **17:**304, 364
Dautreppe, D., **1:**281(12), 331; **6:**280(192), 299
Davazoglou, J. A., **17:**314, 315, 365
Davem, W., **5:**287(205), 291(205), 319
Davern, W., **3:**138(35), 163; **12:**260(88), 261, 292
Davey, J. E., **1:**215, 231; **3:**241, 272; **5:**163(64, 65), 165(105, 106, 111); **12:**76, 160; **13:**212(39), 224
Davey, W. B., **6:**310(16), 314(16), 346
David, E., **6:**201, 204; **10:**60(135), 69
David, J., **2:**23(28), 60
David, J. P., **5:**165(109), 174, 232
Davidenkov, N. N., **3:**223, 271
Davidse, P. D., **3:**76, 102, 113, 127, 129; **4:**261(37), 269, 271(37), 272(37); **13:**111(12), 147
Davidson, B. A., **16:**17, 36–37, 126, 130
Davidson, D. W., **11:**228, 236(37), 249(53), 251(53), 253, 255, 314, 315
Davidson, E. B., **14:**238
Davies, D. D., **6:**211(28), 295
Davies, D. K., **14:**162(112), 163(112), 197; **17:**93, 140
Davies, G., **11:**49(67), 67(67), 98
Davies, H., **10:**168(9), 225
Davies, J. A., **6:**24, 75
Davies, J. E., **15:**13, 25, 137
Davies, J. T., **7:**312(11), 316(25), 393
Davies, M., **11:**254, 255
Davies, P. B., **24:**17(74)
Davis, B. M., **17:**7, 12, 34, 43, 44, 49, 51, 73, 75
Davis, E. A., **11:**226(20), 256(20), 258, 290(20), 314
Davis, E. M., **2:**413(195), 425; **4:**258(10), 319(10), 320
Davis, F., **24:**23(107)
Davis, G. D., **14:**51(164), 77
Davis, G. M., **13:**153(5), 182(5), 183(5), 192; **17:**287, 299
Davis, G. P., **13:**181(107), 185(107), 188(107), 195
Davis, J., **21:**5(1), 71
Davis, J. C., **4:**339, 370
Davis, J. K., **2:**389(113), 422
Davis, L. W., **4:**353, 371
Davis, M., **5:**272(114), 274, 317
Davis, M. F., **16:**54, 134
Davis, N. O., **2:**124(132c), 144; **9:**176(43), 256
Davis, P., **16:**252, 268
Davis, R. F., **17:**238, 239, 298
Davis, R. H., **1:**16(34, 35), 64
Davis, R. W., **24:**228(28)
Davis, W. A., **15:**88, 140
Davis, W. D., **1:**40, 66; **14:**34(136), 76, 151(62), 158(62), 161(62), 196
Davison, W. H. V., **3:**39(201), 52(201), 58
Davisson, C. J., **1:**203, 230
Daw, M. S., **22:**127(36), 172
Dawson, B., **2:**110(100), 143
Dawson, D., **17:**197, 202, 223
Dawson, L. R., **15:**222, 260; **21:**364(47), 368; **23:**231(16)
Dawson, P. M., **6:**38(136), 51(13 6), 77
Dawson, R. M. A., **19:**83
Dawson-Elli, D. F., **17:**87, 140
Day, D., **23:**130(23)
Day, G. F., **7:**140, 141, 142(36), 248
Day, P., **16:**2, 124
Day, W. B., **11:**77(160), 100
Dayananda, M. A., **7:**261(15), 306
Dayem, A. H., **16:**42, 131
De, M. L., **2:**94, 142
de Boeij, **19:**2

de Croes, M., **3:**39(202), 40(202), 52(202), 58
de Frésart, E., **23:**60(202)
De Gasperis, P., **15:**81, 140
de Gennes, P. G., **19:**280; **24:**228(13)
de Graaf, A. M., **6:**247, 298
de Jong, A. N., **3:**18, 51(82), 55
de Jong, J., **21:**132(43), 133(43), 137(, 43), 145(52), 147(52), 149(43, 52), 150(43, 52), 213(52), 219(43), 233
de Jonge, F. A., **9:**266, 295
de Klerk, J., **5:**165(110), 176, 232
de Kluizenaar, E. E., **19:**281
de Leeuw, D. M., **17:**209, 223, 292, 299
De Levie, R., **24:**183(28)
de Lodyguine, A., **5:**240(10), 314
de Lozanne, A. L., **16:**17, 52, 74, 126, 133
De Luccia, F. J., **21:**128(33), 233
de Man, H. J., **21:**10(31), 32(31), 72
de Nijs, J. M. M., **19:**193, 203
de Pablo, J. J., **24:**208(34)
De, S. C., **15:**281, 310
De Smet, D. J., **19:**281, 282
de Vries, A. E., **14:**15(54), 74
De Vries, J. W. C., **16:**97–98, 140
de Waard, P. J., **16:**237, 267
de Wael, J., **7:**329(48), 334(48, 87), 359(87), 393, 394
De Wames, R. E., **15:**32, 135–136, 137
de Winter, J. C., **14:**113(68), 127
de With, G., **17:**110, 141
Deaderick, J., **2:**397(151), 408(151), 409(151), 423, 424
Deal, B. E., **4:**260, 261, 295(113), 301(25, 113, 124), 302(140), 304(140); **14:**114(70), 127; **17:**274, 298
Dean, P. J., **7:**162, 171(111), 205(182), 210(189), 211(85, 189), 249, 250, 252
Dearborn, E. F., **9:**275, 296
Deardorff, D. K., **10:**43(89), 67
Deb, S. K., **15:**284–285, 310; **17:**303, 307, 336, 344, 355, 364, 368
Debbar, N., **15:**222, 260
Debe, M. K., **15:**266, 309; **19:**281, 293
DeBell, G., **10:**168(6), 225
Debenedetti, P. G., **22:**128(41, 42), 172
DeBoer, J. H., **2:**23(23), 60; **5:**247, 315
Debye, P., **7:**2(6), 37; **11:**223(19), 314; **24:**72(112)
Decius, J. C., **15:**266, 309; **19:**299
Deck, R. J., **11:**114(53), 201
Decker, A., **5:**9(16), 45
Deckert, C. A., **13:**181(113), 195
Decroux, M., **16:**19, 126
DeFay, R., **7:**180(131), 251
defontaine, D., **17:**10, 35, 40, 74, 75
deForest, A. V., **16:**272, 276, 285, 337
Defour, M., **19:**2, 9, 29–31, 38, 41
Degenhart, H. J., **2:**371(57), 373(65), 376(65), 387(57), 388(57), 389(57), 393(134), 404(181), 405(181), 421, 423, 424; **3:**109(110), 127
DeGraff, P. D., **13:**182(116), 183(116), 196
DeGrath, P. D., **14:**232(119), 239
DeGreve, L., **2:**23(26), 60
Deguchi, T., **15:**172, 216
Dehoff, R. T., **16:**245, 267
DeHoog, F. J., **3:**296(69), 301
Deitch, R. H., **7:**238(252), 254; **9:**54, 71
Deiter, R. H., **3:**241, 272
Dejewski, S. M., **23:**161(63)
Dekeyser, W., **4:**195(21), 251
Dekker, A. J., **2:**125(141), 144; **12:**64, 101(170), 159, 162

DeKlerk, J., **13:**61, 104; **17:**153, 219
del Alamo, J. A., **21:**32(66), 73
del Castillo, H. M., **23:**146(46)
Del Sole, R., **19:**2
Delacourt, D., **21:**96(29), 112
Delahaye-Vidal, A., **17:**304, 364
deLaitre, S., **21:**161(67), 170(74), 175(74), 234
Delamarche, E., **24:**2(2), 18(88), 20(88), 44(43), 89(49), 106(111), 127(51), 218(75), 230(47), 236(60), 237(62), 238(62, 66)
DeLano, R. B., Jr., **1:**251(48), 274
Delavaux, J. M., **13:**127(87), 149
Delavignette, P., **4:**162, 189; **6:**12, 75
Deleuil, R., **4:**85(226), 96
Delichere, P., **17:**314, 315, 316, 329, 330, 348, 365, 367
Deline, V., **16:**18, 21, 126–127
Deline, V. R., **13:**217(55), 224
Dell'Oca, C. J., **6:**12(52a), 13(52a), 17, 18, 75; **12:**275(186), 294; **14:**120(78), 127
Delord, J. F., **4:**260(18), 301(18), 320
Demchishin, A. V., **13:**62(25), 79(25), 81, 82, 83, 92, 93(25), 104; **14:**63(200), 78, 136(16), 195; **17:**64, 65, 77, 119, 143, 156, 220; **19:**194, 216; **22:**2(1), 77, 122(23), 172
Demeny, M., **7:**315(22), 330(60), 393
Demers, L., **24:**222(103)
Demers, L. M., **24:**151(8), 154(8), 155(8b, 8c, 19), 156(8, 19), 157(8b, 8c), 158(8b), 159(8b, 8c), 161(8b)
Demikhovskii, V. Ya., **6:**137(216, 223), 148
Demiryont, H., **13:**118(46), 131(46, 100, 101), 132, 134(100, 101), 139(100, 101, 117), 140(100, 101, 117), 148, 149, 150
Demny, J., **2:**15, 60
Demota, K., **24:**86(35), 237(65), 238(65)
Demoz, A., **24:**183(30), 192(30)
Dempster, A. J., **1:**39, 66
Demske, D. L., **11:**119(71), 165(71), 166(71), 201
Den Baars, S. P., **19:**12, 32, 35
den Engelsen, D., **19:**281, 293
den Hartog, E. A., **18:**79
Denard, L., **7:**327(45), 346, 393, 395
DeNatale, J. F., **16:**28–29, 129
Denberg, D. L., **17:**153, 219
Denbigh, P. N., **7:**10(20), 13(28), 16(39), 23(78), 37, 38
Denda, M., **23:**105(55)
Denes, L. J., **17:**154, 155, 182, 222
Deneuville, A., **17:**310, 332, 344, 355, 365–368
Deng, L., **24:**14(61), 25(61), 106(112), 249(13)
Denhoff, M. W., **16:**54, 134
Denicolo, I., **24:**182(22)
Denison, D. R., **13:**190(155), 191(155), 197
Denisov, F. P., **3:**32(179), 38(179), 52(179), 58
Denmark, H. S., **3:**276(11), 299
Denney, D. J., **11:**229(25), 315
Dennison, B. J., **9:**12(58), 33(58, 130), 66, 68
Dennison, D., **9:**74, 132(12), 143
Dennison, D. M., **2:**208(11), 237; **5:**21, 45
Dent Glasser, L. S., **9:**11(52), 66
Deori, K. L., **11:**261(79, 81), 263(79, 8 1), 270(79), 272, 273(79), 316
Deppe, D. G., **21:**287(39), 310
Deppert, K., **19:**2, 23, 26–29

Derbali, Y., **24:**163(24)
Dereniak, E. L., **21:**25(56), 73
Derevyanchenko, A. S., **7:**22, 23(75), 39
Derick, L., **4:**300(118), 323; **7:**158(75), 249
Derjaguin, B. V., **19:**104
Derkits, G. E., **21:**38(74, 75, 77), 39(83), 73
Dermit, G., **1:**218(135), 231
Dermody, D. L., **24:**18(76), 19(76), 44(65, 66), 161(23), 162(23)
Dernier, P. D., **11:**86(215), 102
Deroo, D., **17:**359, 361, 369, 370
deRoode, W. H., **15:**13, 137
DeRosa, F., **15:**259, 264; **21:**241(15), 309
DeRosa, R., **23:**238(43, 44), 239(43), 243(43, 44), 244(43, 44), 251(44)
DeRose, J. A., **24:**237(64)
Derrien, J., **12:**116(223), 118(227), 163
Desai, A., **15:**101, 140
DeSandre, L., **13:**124(71), 127(71), 149
DeSantolo, A. M., **16:**2, 60, 124, 135
Desbat, B., **17:**329, 366
DeSchepper, L., **17:**110, 141
Deshotels, W. J., **12:**181, 190(69, 70), 234
Deshpandey, C., **13:**67, 79(78, 79, 83), 87(86, 87), 91(83), 92(83), 96(93), 97(93), 98(97), 99(78), 100(111), 101(111), 105, 106, 107
Deshpandey, C. V., **17:**64, 65, 77, 81, 139
Desmaison, J., **17:**109, 141
Desorbo, W., **1:**241(12), 273
Despic, A. R., **4:**196(33), 198(33), 251
Desu, S. B., **22:**186(34), 203(34), 206(34), 210(34)
Deutch, J., **24:**218(74)
Deutsch, M., **24:**3(17), 4(17), 180(10, 11), 181(12), 182(23), 184(34, 36, 39, 40, 43), 185(10), 186(10), 187(34), 188(34), 190(34, 40), 192(12), 199(12, 23)
Deutsch, T. F., **11:**198, 203; **13:**164(53), 167(62), 169(22), 182(114, 116, 117), 183(114, 116, 117, 119), 184(119), 185(53, 62, 143), 186(53, 143), 187(62), 188(53, 143), 189(53), 193, 194, 195, 196, 197
Deutscher, G., **16:**50, 53, 132, 134; **19:**161
Deutscher, K., **5:**1(6), 2(11), 11(11), 12(11), 14(19), 40(19), 45; **12:**86(114), 161
Dev, S. B., **11:**256(57), 257(57), 315
Devanathan, M. A. V., **4:**196(40), 197, 252
Devant, G., **6:**188(31), 193(47), 203; **10:**169(18), 225
Devaty, R. P., **21:**4(18), 5(2), 6(18), 7(2, 18), 9(18), 17(2), 22(18), 27(2, 57), 28(2, 18, 26, 57), 29(26), 31(26), 32(2, 18, 57, 64), 35(2), 36(2, 26), 71, 72, 73
Devaty, R. R., **23:**144(36, 37)
DeVaux, L. H., **11:**146, 149(116), 151, 202; **21:**41(92), 74
Deveaud, B., **11:**49(71), 98
Devenyi, J., **6:**230(109), 233(109), 234(109, 118), 235(109), 250(109, 118), 277, 279, 297, 299
Devine, R. A. B., **13:**190(156), 191(156), 197
Devine, R. L. S., **15:**250, 263
Devlen, W. J., **13:**218(58), 224

Devlin, J., **11:**76(156), 88(156, 225), 100, 102
Devoldre, P., **11:**42(43), 47(65), 48(65, 66), 86(217, 218), 97, 98, 102
Dew, R. J., **2:**379(79), 418(79), 421
Dew, S., **22:**23(33), 24(33), 25(33), 26(33), 28(33), 46(53), 49(53), 77–78
Dew, S. K., **16:**28–30, 33, 129–130; **22:**1, 3(63), 7(48), 9(19), 10(19), 15(22, 23), 17(25), 18(23, 25, 26), 20(19, 29), 25(34), 30(37, 38, 48), 32(46, 48), 34(26), 35(26), 37(46), 38(46), 39(49), 40(48, 49), 41(49, 50), 42(48), 44(49, 50), 45(25)
Dew-Hughes, D., **16:**90, 139
DeWald, J. F., **2:**158, 191; **6:**9, 15, 16(38), 22(42), 74, 75
DeWames, R. E., **21:**40(84, 88), 74, 117(8), 128(8), 221(8, 134), 223(8), 232, 236, 251(26), 309
DeWit, J. H. W., **9:**28, 67
DeWitt, L., **24:**71(111)
DeWolf, T. N., **16:**272, 285, 337
Dexter, D. L., **4:**339, 370; **5:**211(198), 218, 234
Dexter, R. N., **4:**340(48), 358(48), 370
Dey, S. K., **17:**228, 240, 242, 297, 300; **19:**250
Dhaka, V. A., **4:**320(168), 324
Dhanabalan, M., **16:**244, 257–258, 265, 267, 269
Dheer, P. N., **6:**115(150), 121(150), 133(150), 135(150), 147
Dhere, A. G., **16:**3, 124
Dhere, N. G., **16:**6, 16, 28–32, 38, 125–126, 129
Dhere, R. G., **16:**28–32, 129
d'Heurle, F. M., **4:**302(139), 306(139), 309(139), 323; **7:**265(53, 55), 274(53), 278(73), 284(53), 287(73, 94), 288(73, 95), 289(95, 99), 290, 291(95), 292(100, 102), 293(95, 104, 105), 294(95, 99, 102, 105), 295(105), 296(55), 297(102, 105), 298(114), 307; **14:**86(14), 98(14), 105(14), 125; **16:**234, 239–240, 248, 250, 254, 266–268; **17:**108, 126, 141
Dhez, P., **17:**2, 8, 12, 73, 74
Dhirani, A.-A., **24:**33(136), 128(57)
Dhote, A. M., **16:**70, 136
Di Domenico, Jr., M., **7:**225(227), 253
Diaye, A. N., **12:**169(16), 183(16), 232
Diaz, J., **12:**277(197), 280(197), 294
Dichi, J. R., **17:**307, 364
Dickens, P. G., **17:**305, 320, 364, 366
Dickey, J., **12:**60(34), 101(34), 108(34), 159
Dickey, J. E., **16:**258, 269
Dickinson, L., **24:**155(19), 156(19), 164(30), 165(30), 167(30), 168(33), 172(43), 173(43)
Dickinson, P. H., **16:**84, 90, 138
Dickson, E. W., **4:**150(40, 63), 156(40), 173, 188, 223(121), 235(155), 243, 244, 253, 254
Dictrich, D., **12:**256(61), 257(65), 291
Dicumegard, D., **12:**277(197), 280(197), 294
DiDomenico, M., **17:**229, 279, 280, 281, 282, 299
Diefendorf, R. J., **5:**262, 263, 316; **14:**31(120), 76
Diehl, D. A., **12:**263(99), 264(99), 292
Diehl, T., **16:**297–298, 339
Dieleman, J., **14:**232(126), 239
Dieter, G. E., **16:**295, 339
Dietrich, D., **13:**99(108), 107, 111(20), 143(20), 147

Dietrich, M. R., **16:**8, 125
Dietrich, O., **16:**285, 338
Dietrich, W., **1:**310(61), 311, 333
Dietsche, W., **16:**52, 133
Dietz, R. W., **4:**2(3), 3(3), 90; **7:**54(36), 112; **10:**97(32a), 164
Dietze, H. D., **1:**312(63), 315(63), 333
Dietze, H.-J., **16:**79, 137
Dietzsch, W., **3:**39(203, 207), 52(203, 207), 58
Digacomo, G., **16:**261, 269
Diggle, J. W., **6:**2, 74
Dignam, M. J., **6:**3, 11, 13, 16, 17(66), 21, 22, 27(72), 31, 34, 74, 75; **19:**281
Dijkkamp, D., **16:**60, 62–63, 135
Dike, L. E., **24:**248(91), 249(91, 94)
Dillinger, J. F., **1:**292(39), 332
Dillon, J. F., **15:**185–186, 216
Dillon, J. R., Jr., **17:**17, 74
Dillon, Jr., J. F., **15:**161–162, 171, 215
Dillon, R. O., **13:**79(80), 106
DiLorenzo, J. V., **11:**42(46), 50(46), 75(46), 76(46, 155), 88, 91(233), 97; **13:**218(59), 221(59), 224
Dilorio, M. S., **16:**41, 111, 115, 117, 122, 131, 142
Dima, I., **5:**164(90), 231
DiMaria, D. J., **14:**92(28, 29), 93(30), 125, 126
DiMasi, E., **24:**184(39)
Dimigen, H., **12:**253(55), 254(55), 291; **14:**67(225), 78, 219(59), 238
DiMilla, P., **24:**248(89)
Dimitriou, P., **14:**94(33), 100(42), 103(33), 106(56), 118(33), 124(56), 126
Dimitrova, P. P., **12:**204(98), 234
Dimmey, L. J., **12:**258(73, 74), 291
Dimmler, K., **17:**241, 287, 297
Dimmock, J. O., **11:**106(5), 132, 143, 199, 202; **21:**37(73), 73
Dimos, D., **16:**11, 71, 119, 125, 136
DiNardo, R. P., **9:**179(55), 257
Dineen, J. J., **2:**369(34), 420
Dinger, T. R., **16:**21, 110, 127, 142
Dingle, R., **11:**57(94), 79(176), 84, 85, 86(207), 99, 100, 101, 102; **15:**221, 260
Dingle, R. B., **3:**19, 51(86), 55; **4:**328, 369; **6:**87, 94, 101(88), 118, 144, 145, 190, 203
DiNola, A., **22:**134(54), 136(54), 173
Dion, D. T., **17:**228, 239, 240, 241, 297
DiPaola, A., **17:**317, 365
DiQuarto, F., **17:**317, 365
Dirks, A. G., **14:**60(190), 77; **15:**161, 163, 167–168, 215; **22:**123(28, 29), 172, 262(108), 275
Dischler, B., **15:**307, 312; **21:**121(17), 123(17), 152(17), 232
DiSimone, D., **11:**51(79), 54(79), 57(98), 60(98), 62(98), 76(154), 98, 99, 100
Dissado, L. A., **11:**305, 306(143), 317
Ditchbum, R. W., **1:**72(31), 121, 131(13), 149(13), 185
Ditchburn, R. W., **2:**222, 238; **3:**98(84, 85), 126
Dite, A. F., **16:**71, 79, 136
Ditlefsen, E., **6:**87, 144
Divatia, A. S., **1:**16(34, 35), 64
DiVita, S., **3:**139(42), 163
Dixit, G. A., **22:**119(13), 120(13), 161(13), 171
Dixit, K. R., **1:**216(130), 219, 231
Dixon, E. O., **1:**118(50), 121
Dixon, J. R., **11:**111(36), 200
Dixon, R. W., **11:**79(169), 101
Djakov, B. E., **14:**144(42), 145(63, 67), 146(67), 196

Djoglev, D. H., **12:**204(98), 234
Dobberstein, P., **9:**174(32), 220(179, 181), 221(185, 187), 227(200), 229, 256, 259, 260
Dobi sgov , J., **17:**126, 143
Dobierzewska-Mozrzymas, E., **19:**180
Dobrott, R. D., **7:**36(164), 41
Dobrowolski, J. A., **5:**60(14), 86
Dobson, C. D., **7:**236(250), 254
Dobson, P. J., **19:**168, 173, 179
Dobson, P. S., **7:**286(89), 309
Dockerty, R. C., **9:**22(86), 23(86), 24, 25, 67
Dodson, B. W., **17:**26, 28, 56, 58, 75, 76
Doerbeck, F. H., **7:**241(258a), 255
Doerbeck, G. G., **2:**396(149), 423
Doerner, M. F., **17:**46, 47, 49, 62, 63, 76, 77
Doerr, H. J., **13:**79(79, 83), 91(83), 92(83), 94(89), 103(120), 106, 107; **17:**60, 64, 65, 77
Doetsch, H., **15:**95, 115, 117, 140–141
Doherty, J. T., **1:**285(23), 328(104), 332, 334; **6:**273(178), 299
Dohler, G. H., **11:**59, 67(105), 82, 83(187), 99, 101
Doi, K., **9:**47(218), 48(218), 70
Doi, T., **16:**28–29, 129
Doi, Y., **3:**235(57), 236, 272; **13:**60(6), 78, 104; **14:**132(12), 195
Dolby, R. M., **2:**125, 144
Dolginov, L. M., **7:**223(217), 253
Dolique, G., **23:**255(78)
Doll, J. D., **22:**136(60), 173
Dollinger, R., **14:**140(35), 195
Dolmer, P., **24:**184(35), 192(35)
Domashevskaya, E. P., **17:**344, 367
Domeji, B., **6:**24, 75
Dona dalle Rose, L. F., **14:**122(86), 127
Donaldson, E. E., **9:**46(206), 70
Donaldson, W. R., **16:**109, 142
Donavan, T. M., **15:**306, 312
Dong, D. W., **14:**93(30), 126
Dongen, T. V., **23:**226(11), 227(11), 228(11), 229(11), 232(11)
Donnadieu, A., **6:**201, 202, 204; **17:**304, 314, 315, 364, 365
Donnelly, J. P., **10:**250(42), 253(69), 259; **11:**106(7), 115(7), 130(85, 86), 131(7, 85), 171(143); **13:**182(114), 183(114), 195
Donnelly, V. M., **13:**68(46), 105; **14:**234(142), 240
Donohue, P. C., **10:**31(53), 66
Donohue, T., **13:**184(124), 196
Donovan, R. P., **4:**301(123), 323; **5:**255(40), 315
Donovan, T. M., **4:**3(8, 9), 7(8), 8, 11(9, 35), 90, 91; **10:**41(78), 67; **15:**278–279, 310
Dontje, J. H., **14:**180(159), 199
Doo, V. Y., **3:**137(31), 162; **4:**270(87), 314(87), 322; **5:**287(209), 288; **12:**265(111), 292
Doodie, A. F., **2:**119(107), 143
Doormann, V., **15:**112, 141
Doremus, R. H., **6:**201(74), 204; **9:**15(66, 67), 66, 198(109), 258; **10:**60(133), 69; **12:**72(74), 79(74), 160; **19:**139
Dorier, J. L., **18:**299
Döring, W., **1:**283(21), 323, 325, 326, 332, 334; **4:**98(11), 99(11), 134, 193, 251
Dorn, F. W., **12:**87, 161
Dorn, R., **9:**217(160), 259
Dorodnov, A. M., **13:**79, 106; **14:**138(23), 138(27), 195; **17:**92, 93, 94, 95, 140

Doron, B., **10:**16(28), 25(28), 28(28), 43(28), 47(28), 66
Dosanjh, S., **21:**348(38), 355(38), 356(38), 367
Dosanjh, S. S., **21:**334(29), 341(29), 342(29), 355(29), 367
Dota, K., **10:**236(14), 258
Dou, S. X., **16:**4, 124
Doucet, J., **24:**182(22)
Dougal, S. M., **24:**135(63)
Doughty, D. W., **3:**13(60), 47(60), 50(60), 51(60), 54
Douglas, L., **6:**208(15), 295
Douglas, R. W., **4:**304, 305, 323
Douglass, D. H., Jr., **1:**271(95, 96), 275
Doumani, T. F., **24:**8(26)
Doupont, E. B., **21:**96(29), 112
Dousmanis, G. C., **7:**230(234), 254
Dove, D. B., **6:**225(90), 297; **7:**10(20), 13(28), 16(39), 18(56), 22(77); **11:**42, 44, 97
Dovek, M. M., **24:**86(26)
Dovzhenko, A. V., **15:**115, 118, 141
Dowling, J. E., **21:**66(124), 75
Dowman, A. A., **12:**88, 89, 99, 161
Downie, T. C., **6:**2(7), 74
Downing, R. G., **15:**302, 311
Downs, M. J., **19:**199, 238, 254
Doyle, B. L., **13:**132(102), 150
Doyle, N. J., **17:**206, 222, 229, 271, 272, 276, 278, 280, 281, 282, 283, 284, 285, 286, 290, 292, 298, 299
Doyle, W. T., **9:**198, 258
Dozier, C. M., **14:**239
Dozier, W. D., **16:**119, 121, 143
Drabeck, L., **16:**100, 106, 141
Dragieva, I. D., **12:**204(98), 234
Drago, F., **7:**240(256), 255
Drake, B., **24:**84(23), 94(67), 95(67)
Drake, R., **16:**110, 142
Drake, R. M., **2:**344(67), 360
Draper, P. H., **6:**6(28), 12, 74, 75
Drapkin, I., **14:**71(231), 79
Drawhorn, R. A., **24:**44(59), 45(59)
Drechsler, M., **1:**209, 231
Dreeben, A., **7:**171(113), 250
Dreicer, H., **18:**151
Dreiner, R., **6:**11(44), 12, 21, 33, 58, 75
Dreith, F., **2:**312(6), 359
Drescher, K., **22:**205(64), 273
Dresner, J., **2:**150(32), 170, 181(31, 64), 191, 192; **3:**171(22, 40), 175, 176(22), 181(40), 208, 209; **5:**164(85), 172(85), 209(85), 218(85), 219(85), 231
Dresselhaus, G., **4:**340(48), 358(48), 370
Drévillon, B., **19:**2, 4, 6, 8, 9, 29–31, 38, 41, 45, 52, 280, 287
Drexhage, K. H., **7:**372(182), 386(202), 390(182), 391(212), 396, 397
Dreyer, Karl, **1:**36(94), 66
Dreyfus, B., **6:**246, 298
Driehuys, B., **16:**32, 129
Drier, M., **24:**106(114)
Drigo, A., **2:**30(36), 61
Drigo, A. V., **14:**122(86), 127
Driscoll, M., **16:**107, 141
Driscoll, M. M., **17:**153, 154, 156, 181, 182, 196, 197, 200, 201, 220–223
Driscoll, T. J., **14:**17(66), 74
Drobeck, J., **5:**288(216), 296, 300(242), 320
Drory, M. D., **16:**288–290, 338–339
Drouet, M. G., **14:**141(36), 195
Drowley, C. I., **22:**215(76), 274
Drucker, J., **22:**177(1, 2), 271
Drude, P., **1:**72(29), 78(35), 121, 142, 159(61), 185, 234, 273; **2:**223,

238; **3:**16, 51(68), 54; **6:**177, 202; **7:**44(5), 111, 363(161), 396
Druilhe, R., **19:**38
Drum, C. M., **4:**169(115), 190; **5:**294, 320
Drumheller, C. E., **2:**401(166), 424
Drummeter, L. F., Jr., **2:**307(1), 312(1), 358; **3:**10(48), 51(48), 54; **4:**282, 322; **7:**93, 98(68), 113; **10:**22, 35(33), 54(33), 66, 72, 129(2), 131, 146(78), 153(2), 164, 166; **12:**42(71), 44(71), 51
Drummeter, L. R., **12:**29(56), 51
Drummond, A. J., **10:**5(6), 65
Druon, C., **11:**236(40), 244, 315
Drutnev, V. A., **3:**172(10), 174(10), 208
Druyvesteyn, W. F., **6:**114, 146; **9:**266, 295
Druzhinina, N. N., **14:**155(87), 197
Dryden, J. S., **11:**253, 254, 255
Du, Q., **24:**221(99)
du Chatenier, F. J., **16:**237, 267
Du Mond, J., **6:**310, 335, 346
du Prè, F. K., **6:**62, 78
Dube, D. C., **11:**261(76), 262, 316
Dubey, R. C., **16:**98, 140
DuBois, A., **9:**56(264), 58(264), 62(264), 71
Dubois, J. C., **10:**230(10), 238(10), 242(10), 258
Dubois, L. H., **24:**2(3), 10(35), 14(35), 18(35), 44(3, 4, 5, 6, 9), 45(6), 47(94), 55(94), 64(94), 66(4, 6, 9), 70(5, 6), 82(10), 83(10), 86(10, 40), 87(41), 88(41), 89(41), 90(41), 91(41), 93(10), 112(1), 114(26), 121(44), 125(50), 206(4), 212(55, 62), 213(62), 217(62), 228(16), 237(16)
Dubouloz, P., **7:**331(69), 394
Ducharme, D., **19:**281
Ducommun, J. P., **14:**226(84), 238; **22:**113(29), 115
Dudas, J., **14:**66(222), 70(222), 78
Dudding, R. W., **3:**20(96), 21(96), 51(96), 55
Dudkevich, V. P., **19:**251, 253
Dudley, S. C., **15:**244, 263
Dudonis, J., **13:**111(21), 134, 140(21), 141(21), 144, 147, 150
Duevel, R. V., **24:**18(75), 19(75), 46(83)
Duffie, J. A., **10:**11(16), 14(16), 23(16), 61(141), 65, 69
Duffield, T., **15:**229–231, 259, 262; **21:**184(81), 235
Duffy, C. M., **4:**261(41), 262(47), 321
Duffy, D. C., **24:**17(74)
Duffy, M. C., **5:**255(41), 315
Duffy, M. T., **5:**304, 321; **10:**257(66, 67), 259; **17:**153, 154, 220
Dufour, C., **3:**5, 10(41), 17, 49(228), 51(30, 41, 42, 78), 53, 54, 55, 59; **5:**75, 78(24), 86; **9:**74, 132(5), 143
Dufour, G., **14:**98(94, 96), 101(94), 113(94), 114(96), 127, 128
Dugdale, J. S., **6:**104(116), 146
Duggal, V. P., **6:**110, 112, 132(131), 141, 142, 146, 149
Duggan, G., **21:**312(3), 322(3), 329(24), 333(3), 341(24, 35), 357(3), 366, 367
Dugger, D., **21:**207(109), 235
Duh, K., **11:**111(38), 114(38), 200
Duiker, H. M., **17:**241, 259, 287, 297, 298
Duke, C. B., **9:**203(121, 123, 123a), 258; **12:**115(218), 163
Düker, H., **1:**191(26), 228

Dumesnil, M. E., **2:**392(122), 423; **4:**292, 311, 322
Dumin, D. J., **3:**224, 271; **5:**146(15), 152, 153(39), 186(35), 207, 209(35, 39), 214(15, 35), 215, 227(35), 229, 230, 234, 275, 317
Dumke, W. P., **7:**193(157), 246(278), 251, 255
Dummer, G. W. A., **2:**411(193), 425
DuMond, J., **17:**2, 73
Dumont, B., **17:**304, 364
Duncan, W. M., **19:**45
Dungan, D. F., **16:**96, 140
Dunkle, R. V., **2:**315(7), 317, 326(32), 344, 359, 360; **4:**60(151, 153), 94; **10:**2, 14(2), 28(2), 65, 132, 165
Dunkleman, L., **1:**131(15, 16), 185
Dunlap, C. J., **24:**163(26)
Dunlap, H. L., **14:**231(110), 239
Dunlap, H. R., **23:**234(20), 235(20)
Dunlap, W. C., **5:**145(2), 229, 275, 317
Dunlop, G. L., **21:**207(107), 235
Dunn, B., **13:**98(97), 106
Dunn, C. G., **6:**4, 74
Dunn, C. N., **11:**77(161), 100
Dunn, S. T., **4:**60(150), 94
Dunnaway, D. J., **24:**44(26), 45(26), 46(26, 65, 66), 60(26)
Dunse, J. U., **7:**146, 219(53), 248
Dünstadter, H., **3:**116, 128
Dunstädter, H., **9:**50(232), 70
Dupouy, G., **2:**72, 141
Duprat, M., **24:**163(24)
Dupuie, J. L., **17:**153, 154, 221
Durasova, Y. A., **3:**102(92), 127
Durbin, M. K., **24:**112(24), 129(24)
Durelli, A. J., **16:**272, 281, 285–287, 332, 334, 337
Durham, J. A., **18:**301, 310
Durig, J. R., **13:**48, 58
Dürig, U., **24:**221(89), 222(89)
Dürrschnabel, W., **2:**393(140), 423
Dus, R., **24:**53(102)
Dusek, V., **18:**125
Dushman, S., **1:**9, 51(127), 64, 67; **2:**329(35), 359, 387(111), 422; **5:**162, 230
Dutartre, D., **23:**24(96), 30(96)
Dutault, F., **12:**174(46, 47), 175(46), 180(46), 231(46), 233
Dutchak, Ya. V., **7:**29(124), 30, 34, 35(126), 40
Dutcher, J. R., **17:**49, 51, 76
Duthler, C. J., **9:**198(111), 258
Dutta, B., **16:**55, 58, 63, 111, 134–135, 142
Dutta, M., **21:**240(4), 241(7), 246(4), 248(4), 252(4), 256(28), 267(28), 277(28), 278(4), 308, 309
Dutta, P., **17:**15, 74; **24:**112(24), 129(24), 180(11)
Dutta, S., **15:**226, 250, 261
Dutta, V., **12:**226(124, 125), 227(124), 235
Dutton, R. W., **22:**270(132), 276
Duval, M., **12:**248(39), 251(39), 290
Duwez, P., **2:**42, 61, 62; **7:**28(115), 33(136), 40
Dvoinin, V. I., **12:**204(97), 217(97), 234
Dwight, H. B., **3:**44(221), 59
Dwir, B., **16:**26–28, 127–128
Dyatlovitskaya, B., **5:**2(10), 45
Dyatlovitskaya, B. I., **12:**101(172), 162
Dybwad, G. L., **17:**153, 219
Dye, R., **16:**37, 130
Dye, R. C., **16:**58, 61, 75, 135, 137
Dyer, P. E., **13:**153(15), 154(15, 16, 17, 20), 172(16), 181(16, 20), 182(15, 16, 17), 183(15, 16, 17, 20), 184(20), 192

Dykaar, D. R., **16:**107, 141
Dyment, J. C., **7:**234(243, 244), 254
Dynes, R. C., **16:**74, 137
Dyson, F., **2:**234, 238
Dyson, F. J., **1:**324, 334
Dyson, J., **3:**2, 53; **4:**22(46), 91
Dzurko, K. M., **19:**12, 32, 35

E

Eaglesham, D. J., **23:**18(67)
Earl, H., **10:**168(3), 169(3), 225
Early, B. F., **7:**64(46), 113
Earnshaw, J. C., **24:**182(23), 199(23)
Eastman, D. E., **9:**217(169), 259; **12:**114(196), 163
Eastman, J. M., **10:**175(27), 181, 186(31), 187, 188, 196, 203, 204(35), 211, 226
Eastman, L. F., **11:**10, 51(74, 79), 54(79, 90), 56(90), 57(98), 60(98), 62(98), 66(128), 70(140a), 76(154, 156, 156a), 77(162, 163), 78(166), 83(188a), 85(162), 88(156, 156a, 225), 98, 99, 100, 101; **13:**12, 164(55), 185(55), 189(55)
Easton, B. C., **12:**81(185), 82(185), 101; **21:**211(114), 236, 241(11), 256(30), 276(30), 282(11, 36), 309
Easton, E. C., **14:**150(79), 196
Eastwood, H. K., **14:**27(107), 75
Eastwood, J. W., **22:**132(44), 172
Eaton, S., **17:**241, 287, 297
Eaves, E., **15:**249, 263
Eaves, L., **15:**249–250, 263
Ebara, Y., **24:**10(38), 23(38), 24(38)
Ebata, Y., **17:**197, 201, 222, 223
Eberhardt, A., **24:**44(45), 45(74), 82(14), 83(14), 86(14, 39), 112(17, 18), 117(17, 18), 119(17, 18, 39), 127(17, 18), 132(39), 135(17), 136(17), 137(17), 139(66), 150(3), 206(16), 212(16), 213(16), 214(16)
Ebersbach, U., **12:**291
Eberstein, A., **4:**2(1), 90
Ebert, J., **13:**61, 62, 79(24a), 101, 104, 111, 112, 130, 131(17), 133, 135, 139, 140(17), 144, 147, 149
Ebert, W., **16:**60, 135
Ebihara, B. T., **16:**5, 76, 124, 137
Ebina, A., **12:**103, 104, 105, 162
Eccleston, W., **14:**91(25), 92(91), 103(25), 116(25), 117(25), 118(25), 125, 127
Eckart, F., **12:**82, 101(97), 102, 160
Ecker, G., **3:**96, 126; **14:**138(18), 195, 196, 219(63), 238
Eckert, E. R. G., **2:**330(41), 344, 360; **10:**14(22), 57(124), 65, 68
Eckert, H. U., **18:**52
Eckhardt, G., **13:**213, 224
Eckhardt, H., **16:**99, 140
Eckstein, J. N., **16:**53–54, 134
Eckstein, W., **22:**32(43), 78
Economou, N. A., **4:**199(56, 61), 200(56, 61), 204(56, 61), 206, 209, 242(61), 252
Economu, E. N., **9:**146(4a), 255
Eddols, D. V., **11:**166(120), 202
Eddy, M. M., **16:**81, 100, 102–105, 106, 117, 121, 138, 141–142
Edeleanu, C., **16:**281–282, 337
Edelman, F. L., **13:**99(112), 100(112), 107
Edelman, V. S., **6:**133(198), 148
Edelstein, A. S., **17:**49, 62, 77
Eden, R. C., **12:**59(22), 137(309), 159, 165
Edenhofer, B., **12:**286, 295

Edenholm, B., **3:**36(171), 37, 52(171), 57

Edgecumbe, J., **3:**88(68), 122, 126; **7:**243(267), 244(270), 255; **12:**114(191, 194), 125(256), 128(256), 132(281), 162, 163, 164, 165

Edinger, K., **24:**86(35), 237(65), 238(65)

Edlinger, J., **19:**195

Edwards, A. G., **1:**43, 66

Edwards, D., **12:**121(244), 164

Edwards, D. F., **12:**253(66), 291

Edwards, D. K., **2:**312(4), 315, 358, 359; **4:**60(144, 152), 94; **10:**14(26, 27), 43, 44(26), 45(26), 66

Edwards, H. H., **1:**263(77), 275

Edwards, J. A., **16:**52, 133

Eesley, G. L., **17:**24, 46, 49, 51, 52, 53, 56, 75, 76

Efeoglu, I., **18:**278, 279

Effer, D., **5:**150(31), 230, 276, 279, 318

Efferson, K. R., **6:**82(26), 144; **16:**92, 139

Efimov, V. M., **13:**99(112), 100(112), 107

Efremow, N. N., **13:**182(114), 183(114), 195

Efthimiopoulos, T., **13:**153(5), 182(5), 183(5), 192

Egami, T., **15:**186, 216

Egan, J. J., **12:**286, 295

Egan, W. G., **1:**330(112), 334; **19:**262

Egerton, R. F., **11:**114(54), 153(123), 201, 202

Eggarter, E., **18:**16

Eglash, S. J., **15:**230, 239, 262; **21:**116(4), 117(4), 232, 240(3), 308; **23:**229(12), 230(13, 14), 230–231(34)

Egner, B., **16:**74, 137

Egorov, V. V., **6:**9(41), 74

Eguchi, K., **16:**51, 133

Eguchi, N., **17:**135, 144

Eguiluz, A., **9:**159(16), 256

Ehlbeek, H. W., **2:**396(149), 423

Ehler, K. W., **12:**269(151), 293

Ehlers, K. W., **6:**43(141), 77

Ehlert, L. B., **13:**135(104), 141(104), 150

Ehlert, R. C., **7:**354, 355(144), 358(150), 396

Ehrenberg, W., **3:**201(73), 210; **5:**221(204), 234

Ehrenreich, H., **4:**370; **6:**153, 202; **10:**30, 32(52), 66; **12:**79, 80, 160

Ehrhardt, C., **3:**31(152), 32(152), 51(152), 57

Ehrlich, D. J., **13:**155(30), 157(31), 159(31), 161(42), 163(50, 51), 164(53), 165(30), 167(62), 168(42), 169(30, 62, 69), 170, 171(74), 181(30), 182(50, 51), 183(31, 50, 119), 184(119, 120, 121), 185(53, 62, 69, 143), 186(53, 74, 143, 145), 187(62, 69), 188(53, 143), 189(53, 145), 190(74, 145), 191(31), 193, 194, 197

Ehrlich, G., **1:**209, 223, 230; **2:**12, 13, 14(6), 60; **4:**107(24), 135

Ehrlich, S., **24:**180(11)

Ehrlick, D. J., **11:**198(159), 203

Eibl, O., **17:**291, 299

Eibschutz, M., **17:**229, 279, 280, 281, 282, 299

Eicher, S., **14:**51(167), 77

Eichhorn, G., **12:**270(154), 293

Eichkorn, G., **4:**219(107), 228(107), 231, 253
Eichner, L., **13:**153(14), 182(14), 183(14), 192
Eikum, A., **7:**286(90, 92), 309
Eilers, J. E., **24:**32(131), 44(49), 64(49), 88(44), 114(27), 128(59), 210(46)
Einstein, A., **16:**237, 267; **21:**5(22), 72
Eisenbeiss, A., **15:**227, 244, 261
Eisenberg, A., **24:**94(72), 95(72)
Eisenberger, P., **24:**44(45, 46), 45(72, 74, 76), 82(13, 14), 86(13, 14, 39), 89(47), 93(13, 14), 94(61), 112(7, 10, 11, 15, 16, 17, 18, 19, 20), 114(29), 115(29), 116(29), 117(7, 11, 15, 16, 17, 18, 20), 119(11, 16, 17, 18,19, 20), 125(16), 127(16, 17, 18), 128(15, 20), 129(20), 130(16), 131(16), 133(16, 18), 135(17), 136(17), 137(17), 138(16), 139(66), 142(11), 143(11), 144(11), 150(3), 181(20), 182(20), 199(47), 206(15, 16), 211(15), 212(15, 16), 213(16), 214(16), 215(15), 216(15), 218(78)
Eisenbud, L., **15:**228, 261
Eisenhutt, O., **6:**310(12), 346
Eisenstein, A., **3:**39, 52(192), 58
Eisenthal, K. B., **24:**72(116)
Eisert, F., **24:**92(58)
Eisner, E., **4:**86(228), 96
Ejiri, A., **9:**186(68), 187(70), 257
Ekenburg, U. E., **23:**228(30)
Ekholm, R., **7:**315(21), 393
Ekin, J. W., **16:**8–9, 11, 125
Ekvall, R. A., **2:**393(139), 423
Ekwall, P., **7:**315(21), 393
El-Assal, K., **14:**57–58(180), 77
Elabd, H., **23:**101(33)
Elam, W. T., **16:**53, 134
Elbert, R., **24:**89(53), 92(53)
Elbott, R. J., **11:**223, 314
Eldert, C., **5:**70, 86
Eldridge, H. B., **2:**124(132a), 144
Eldrige, H. B., **9:**174(33), 256
Elgan, D. L., **5:**287(210), 320
Elich, J. J. P., **17:**352, 368
Elings, V. B., **24:**94(67), 95(67)
Elinson, M. I., **6:**137(213), 139(213), 140(225, 226), 141(225, 226), 142(213), 148
Eliseev, P. G., **7:**216(197a), 253
Elkins, P. E., **9:**276, 294
Elleman, A. J., **11:**107(9), 200
Ellett, M. R., **3:**277(29), 300
Ellingson, R. J., **21:**365(48), 368
Elliott, A., **3:**276, 289(13), 299
Elliott, C. T., **6:**73(218), 79
Elliott, D., **9:**35(165), 36(165), 37(165), 69; **12:**190(91), 234
Elliott, L. T., **7:**142(46), 248
Elliott, M. T., **15:**9, 14–16, 136–137
Elliott, N., **1:**205(80), 230; **16:**37, 130
Elliott, R. P., **6:**337(45), 347
Elliott, S. R., **11:**258(71), 316
Elliott, T. C., **23:**130(23)
Ellis, B., **11:**142(113), 143(113), 202
Ellis, G., **16:**272, 276, 285, 337–338
Ellis, J. W., **7:**363(155), 396
Ellis, P. D., **24:**154(18)
Ellis, S. G., **5:**166(118), 177, 232
Ellis, T., **4:**218, 219, 232(103), 253
Ellis, W. C., **3:**243(66), 272; **4:**174, 190; **5:**156(149), 185, 233, 268, 316
Ellis, W. P., **2:**214, 238
Ellison, A. H., **7:**331, 359, 363(71), 394
Ellsworth, D. L., **14:**234(144), 240
Elman, B., **21:**207(109), 235

Elman, J. F., **24:**212(60)
Elpat'evskaya, O. D., **5:**209(185), 234
Elsharkawi, A. R., **19:**281, 293, 294
Elshazly-Zaghloul, M., **19:**281, 305
Elsner, O. V., **7:**385(201), 397
Elson, J. M., **9:**154(11), 245(245), 256, 261
Elsworth, L., **1:**35(86), 66
Eltoukhy, A. H., **14:**231(111), 239; **17:**8, 11, 34, 74
Ely, B. W., **3:**169(6), 197(68), 208, 210
Ely, K., **17:**149, 219
Emberson, R. M., **4:**83(207), 95
Emde, F., **3:**44(221), 59
Emeleus, K. G., **3:**96, 126; **14:**219(63), 238
Emeny, M. T., **15:**234–235, 262; **21:**117(12, 13), 128(36), 152(13), 153(13), 232, 233
Emeric, A., **6:**201, 204; **9:**190(85), 257
Emeric, N., **6:**201, 204; **9:**190(85), 257
Emerling, M. R., **21:**60(112), 65(112), 74
Emerson, A. B., **24:**14(59), 20(59)
Emerson, L. C., **2:**124(132b, c, f), 144; **9:**167(27), 176(44), 179(54), 256, 257
Emery, K., **21:**312(1), 313(1), 321(1), 322(1), 344(1), 366
Emiliani, G., **19:**192
Emmen, J. H. P. M., **16:**47, 50, 132
Emtage, P. R., **3:**132(7), 146(7), 162; **11:**142, 143(109), 146, 147, 148, 202; **15:**37–38, 40–41, 43–44, 50, 75–76, 81, 95–100, 107–109, 111, 132, 138; **17:**146, 148, 186, 219
Enard, J.-P., **14:**86(14), 90(90), 95(34), 98(14), 105(14), 112(99), 119(64), 122(34), 124(90), 125, 126, 127, 128
Enck, R. S., **12:**132(281), 143(313), 165, 166
Endo, N., **17:**279, 299
Endo, T., **17:**356, 369
Endo, U., **16:**4, 33, 124
Endoh, Y., **17:**22, 74
Endres, G., **16:**55, 58, 61, 64, 72–74, 135, 137
Endriz, J. G., **9:**227(203), 228(203), 260
Engelhart, G., **24:**150(4), 165(4)
Engelman, J. H., **3:**39(200), 42, 52(200), 58
Engelsen, D., **7:**333, 363, 364, 365(73), 394
Engelsrath, A., **4:**73(172), 95
Engemann, D., **12:**274(172), 275(172), 294
Engemann, J., **14:**23(92), 75
England, P., **15:**259, 264; **16:**61, 63, 135; **23:**238(43, 44), 239(43), 243(43, 44), 244(43, 44), 251(44)
Engle, G. M., **12:**242, 290
Engler, D. M., **24:**98(98, 99), 106(98, 99)
English, A. T., **16:**250, 252, 268
English, J., **3:**11(56), 12(56), 13(56), 19(117), 24(117), 26(117), 49(56), 51(56, 117), 54, 56
English, J. H., **14:**162(111), 197; **17:**93, 94, 140
Englman, R., **6:**91, 144
Engquist, I., **24:**22(106)
Engström, A., **2:**125(138), 144
Engstrom, P. W., **12:**150(322), 166
Enjouji, K., **14:**55(174), 77
Enke, K., **12:**253, 254, 291
Ennen, H., **21:**121(17), 123(17), 152(17), 232
Ennos, A. E., **3:**146, 163, 236, 272;

4:4(4), 90; **6:**106, 131(120), 146; **10:**273, 308; **19:**264
Enokihara, A., **16:**111, 142
Enomoto, Y., **16:**25, 73, 127
Enstrom, R. E., **7:**193(163), 244(269), 246(277), 252, 255; **12:**118(229), 120(229), 126(229), 127(229), 163
Entenberg, A., **22:**119(11), 154(11), 156(11), 171
Entine, G., **12:**176(48a), 233
Enyo, M., **4:**196(29, 30), 198(29), 251; **17:**315, 365
Enz, W., **9:**264, 295
Eom, C. B., **16:**12, 22, 68, 78, 101, 125, 127, 136–137
Epanechnikov, V. A., **15:**37–38, 138
Ephrath, L. M., **14:**220(73), 234(140, 143), 238, 240
Epstein, A. L., **4:**265(60, 61), 314(60, 61), 319(60, 61), 321
Epstein, A. S., **7:**163, 164(89), 249
Epstein, H. T., **7:**333(79), 394
Epstein, L. I., **1:**70(17), 88, 92, 120, 171(66), 186; **2:**275(28), 276, 304; **5:**60(10), 66, 67, 80(26), 85, 86
Epworth, R. W., **13:**161(39), 182(39), 183(39), 193
Era, M., **16:**111, 116, 142
Erb, R. A., **4:**120, 136
Erbil, A., **16:**82–84, 88, 94, 138, 140; **17:**238, 297
Erdelen, C., **24:**15(65), 34(141), 35(141)
Erdelyi, A., **11:**214(10), 314
Erdey-Gruz, T., **4:**199(49), 204(49), 252
Erdman, W. C., **5:**298, 321
Eremenko, V. V., **6:**124(191), 148
Ergun, S., **7:**11(24), 37
Erhardt, H. G., **23:**101(33)
Eriott, N. E., **16:**37, 48, 130, 132
Erlandsson, R., **24:**98(81)
Erlbach, E., **1:**248(45), 251(45), 274
Erler, H., **12:**246(35), 248(35), 290
Erler, H. J., **12:**257(65), 291; **13:**99(108), 107, 111(20), 143(20), 147; **14:**25(99), 26(99), 75
Ermanis, F., **7:**142(45), 162, 164(82), 248, 249
Errafi, N., **24:**184(42), 188(42)
Errede, L. A., **3:**157(85), 164
Erturk, E., **14:**174(125), 198
Erukhimov, M. Sh., **6:**137(215), 148
Esaki, L., **11:**36(3), 41(3, 34), 44(53), 45(53, 61), 47(3), 47(53, 61), 58(3), 67(130, 131), 68(131, 137), 70(137), 71(137), 73(130, 137, 145), 74(137, 145), 79(3, 137), 81(185a), 85(130, 185b, 197, 198), 86(202, 203, 205); **12:**115(213), 163; **15:**87(214), 89(3), 96, 97, 99, 100, 101, 102, 245, 252, 263; **21:**81(11), 111, 116(1), 232
Escadier, P., **6:**273(179, 181), 274(179, 181), 275(179), 299
Esch, R. P., **4:**261(26), 271(26), 288, 320
Escher, J., **12:**122(249), 123(249, 251), 126(249), 129(249), 131(249), 139, 164, 165
Escher, J. S., **12:**54(9), 118(229), 120(229), 126(229), 127(229), 130; **21:**14(38), 72
Escoffery, C. A., **3:**172, 174(12), 208
Escudero, R., **16:**4, 124
Eshbach, J. R., **15:**26, 35, 137
Espinosa, G. P., **9:**276, 284, 297
Esposito, B. J., **23:**124(15)

Este, G., **13:**140(120), 150; **14:**8(26), 36(26), 39(147), 41(149), 42(149), 46(159), 47–48(147), 67(227, 228), 68(228), 69(228), 70(228), 74, 76, 77, 78
Estermann, I., **4:**121, 136
Estier, R. C., **16:**58, 61, 75, 135
Etienne, R., **11:**36(9), 41(35), 42(43), 45(55), 47(65), 48(65), 86(216, 218), 96, 97, 98, 102
Etler, P. J., **5:**152(34), 153(38), 186(34), 199(34), 209(38), 214(38), 230, 269(100), 272(100), 317
Etsel, T. H., 14.91(24);, **14:**116(24), 117(24), 119(24), 121(24), 125
Ettenberg, M., **7:**134(29a), 171(113, 114), 186(143), 232, 233(239); **12:**113(190), 129(272), 162, 165
Etzel, K., **5:**247(26), 315
Etzold, K., **6:**314(26), 346
Etzold, K. F., **17:**234, 236, 296
Evall, J., **24:**3(13), 12(13), 13(13), 14(13), 15(13), 16(13), 18(13), 20(13), 30(13), 44(12), 46(12), 55(12), 56(12), 65(12), 82(18), 89(51), 112(3), 212(57), 218(70), 220(57), 228(17, 18), 237(18)
Evans, A. G., **16:**288–291, 333–334, 338–339
Evans, C. A., Jr., **13:**217(55), 224
Evans, C. C., **1:**279(2), 331; **2:**48(81), 49(81), 51(81), 62; **3:**219(25), 271
Evans, D. F., **24:**221(94)
Evans, D. M., **3:**218, 271
Evans, G. B., **7:**242(261), 255; **12:**112(184), 118(184), 162
Evans, J., **17:**271, 298
Evans, J. A., **7:**174(118a), 177(118c), 250
Evans, J. H., **9:**167(25), 256
Evans, K., **23:**60(202)
Evans, K. R., **15:**244, 263; **21:**121(22, 23), 123(22, 23), 194(22), 195(22), 233, 303(53), 310
Evans, L. C., **22:**85(8), 114
Evans, M. V., **10:**43(88), 44(88), 67
Evans, P. E., **12:**261, 292
Evans, P. R., **5:**150(31), 230
Evans, R. J., **4:**260(24), 301(24), 320
Evans, S. D., **24:**11(46), 12(53), 15(53), 21(46, 91), 23(114), 31(114), 33(46), 34(46), 44(22, 23, 24), 70(22), 220(84), 228(31)
Evans, T., **5:**167(134), 203(134), 232
Evans, U. R., **7:**60(40), 112
Evers, W., **16:**24–25, 127
Eversole, W. G., **12:**246(26), 290
Eversteijn, F. C., **4:**265(64), 266, 314, 321
Evett, A. A., **1:**6, 29(11), 64
Evetts, J. E., **17:**22, 74
Evitts, H. C., **4:**265(61, 63), 314(61), 319(61, 63), 321
Ewald, A. W., **2:**201(8), 237
Ewald, P. P., **2:**117, 143
Ewing, W. S., **21:**219(128), 236; **23:**106(57)
Excher, J. S., **12:**165
Exharos, G. J., **13:**123(68), 149
Eyring, E. M., **4:**107(23), 135
Eyring, H., **4:**107(23), 135

F

Faber, T. E., **1:**254(53, 54), 274
Fabian, W., **13:**218(62), 221(62), 224
Fabré, D., **1:**145(40), 149(40), 185, 186
Face, D. W., **16:**28, 33, 116, 128, 142
Factor, B. J., **24:**181(13)

Fagan, E. A., **12:**272(160), 298
Fagen, E. A., **7:**34(141), 40
Faget, J., **2:**110, 143
Fagot, M., **2:**110(92), 143
Fahlman, A., **7:**45, 48(16), 111
Fahrenbruch, A. L., **12:**169(15, 17, 18), 183(18), 232
Fahrenfort, J., **4:**75(176), 95
Fahrinre, T. O., **11:**115, 201
Faininger, K. H., **5:**272(113), 317
Fairbank, R. W., **24:**228(7)
Faist, J., **23:**239(46), 240(54), 243(65), 262(65, 80, 83, 84, 85, 87), 263(65), 265(85), 266(46), 268(46), 269(92), 270, 271(54), 272(54, 83, 96), 273(86), 274(87), 279(83), 291(80), 298(80)
Faith, T. J., **14:**235(152), 240
Faith, W. N., **9:**203(122), 215(152), 258, 259
Falaras, P., **17:**315, 316, 329, 330, 348, 365, 367
Falco, C. M., **16:**43, 47, 131; **17:**24, 45, 46, 48, 49, 51, 52, 53, 56, 57, 58, 74, 76
Falconer, I. S., **14:**21(85), 75; **22:**146(69), 173
Falk, D. S., **6:**298
Fal'ko, I. I., **7:**264(45), 307
Falkovskii, L. A., **6:**140(229), 149
Falter, L. M., **16:**99, 140
Falvo, M., **24:**151(9), 154(9), 156(9)
Fan, H. Y., **21:**21(45), 72, 94(24), 111
Fan, J. C., **11:**81(185a), 101
Fan, J. C. C., **9:**116(26), 144; **10:**52(105, 111), 53(111), 54(1 1 1), 68
Fan, L. C., **17:**210, 223
Fang, C. J., **15:**286, 310
Fang, C.-C., **22:**117, 129(43), 141(63), 142(43), 145(43), 152(74), 160(78), 161(81, 82), 165(82, 86), 172–173, 194(54), 273
Fang, D. Y., **14:**148(75, 76), 149(75), 196
Fang, F., **5:**216(196), 234, 277, 318
Fang, Y. K., **13:**209(29), 210(29), 215(29), 223
Fankuchen, I., **4:**23(56), 92; **7:**354, 355(140), 395
Fano, U., **9:**227, 260; **12:**134(288), 165
Fantini, F., **16:**252, 268
Farabaugh, E. N., **11:**116(66), 201
Faraday, M., **10:**37(66), 67; **19:**127
Faris, S. M., **16:**110, 142
Farley, C. W., **21:**117(8), 128(8), 221(8), 223(8), 232, 241(9, 12, 13), 251(26), 256(9), 269(9), 300(12), 308, 309
Farmer, J. C., **19:**52
Farnsworth, H. E., **4:**143, 187; **5:**190, 233
Farnsworth, P. T., **12:**152(325), 166
Farrall, G. A., **14:**175(131), 198
Farrell, C. E., **16:**28, 42, 51, 128, 131–132
Farrell, D. E., **16:**98, 140
Farrell, H. H., **16:**236–237, 266; **19:**2, 9, 11, 16–19, 21, 23
Farrow, L. A., **16:**63, 112, 135–136
Farrow, R. C., **11:**36, 42(6), 48(32), 49(68), 66(125, 126), 67(125), 79(6); **16:**3, 36–37, 51, 124, 130, 132
Farrow, R. L., **13:**205(18), 223
Fartash, A., **17:**43, 49, 53, 75, 76
Faruoki, R. T., **22:**83(1), 84(1), 85(1), 92(1), 93(1), 97(1), 114, 188(39), 272
Faska, T. S., **21:**117(9), 128(9), 191(87), 221(87), 232, 235, 257(32), 309

Fastie, W. G., **10:**74(4), 164
Fathauer, R. W., **23:**49(37), 109(59), 161(63, 64)
Fathipour, M., **14:**234(144), 240
Fathy, A., **16:**106, 141
Fattow, R. F. C., **13:**3(8), 56
Fatunmbi, H. O., **24:**228(7)
Faucher, J. A., **7:**364(163), 396
Fauchet, P. M., **13:**180(105), 195
Faughnan, B. W., **17:**304, 318, 319, 320, 350, 352, 364, 366, 367
Faulkner, K. R., **12:**135(291), 165
Faust, J. W., **4:**153, 156(75), 189
Faust, J. W., Jr., **5:**186(150), 187, 208(150), 233
Faust, R. C., **4:**86(232), 87(232), 96; **12:**71(69), 160
Favenec, P. N., **11:**79(71), 98
Fawcett, E., **6:**91(63, 64), 113, 122(63), 135(63, 207), 136(207), 144, 148
Fawcett, R. W., **7:**30(127), 40
Fayers, F. J., **3:**246(79), 272
Fechtig, H., **7:**102(80), 113
Feder, J., **4:**98(12), 99(12), 134
Feder, R., **6:**115(143), 132(143), 146; **7:**133(27a), 247; **13:**124(74), 126(74), 149
Federighi, T., **7:**297(111), 310
Fedorov, G. V., **3:**11(53), 51(53), 54, 261(112), 265(112), 267(112), 273
Feenstra, R., **16:**10, 43, 48–49, 125, 131–132
Feenstra, R. M., **23:**281(107)
Fehling, H., **14:**196
Fehlner, F. P., **14:**82(5), 88(5, 18), 116(5), 117(5), 125
Fehrenbacher, M., **23:**32(119)
Fei, X., **16:**81, 138
Feibelman, P. J., **9:**159(17, 18), 203(121), 256, 258; **22:**254(104, 105), 275
Feidenhans'l, R., **24:**114(30), 115(30), 116(30)
Feigelson, R. S., **12:**169(12, 16), 182(72), 183, 193(72), 196(12), 232, 234
Feillade, G., **6:**6(29), 74
Feinberg, I., **4:**320(167), 324
Feinleib, J., **10:**33(60), 34(60), 66
Feinstein, L. G., **14:**24(93), 75
Feist, W., **4:**349, 354, 367, 370; **5:**300(256), 321
Feist, W. M., **9:**31, 36(113), 37(113), 39(113), 67; **12:**258(76), 259(76), 265(76), 271, 272(76), 291
Felcher, G. P., **17:**24, 46, 49, 51, 52, 53, 56, 58, 75, 76
Felder, R. J., **16:**21, 28–29, 127, 129
Feldkeller, E., **3:**1(2), 53
Feldman, C., **2:**367(16), 384(91), 393(133), 420, 422, 423; **3:**168, 171(1, 75), 180, 200(1), 201(75), 202(75), 208, 210
Feldman, J. L., **15:**157, 215
Feldman, L. C., **12:**277(208), 279(208), 294; **23:**2(2)
Feldmann, W. L., **16:**42, 45, 131
Feldmesser, K., **1:**30(71), 65
Feldtkeller, E., **1:**198, 202(71), 229, 230, 302(48), 306(57), 332, 333; **2:**42, 61; **6:**206(7), 226, 228(107), 229, 230(92), 231, 234, 235, 237, 240(136), 249, 250(92), 260(165), 292(156), 295, 297, 298, 299
Felix, C. L., **23:**279(22, 102)
Fell, J., **15:**276, 310
Fellenzer, H., **9:**202(141), 259
Feller, F. R. G., **14:**31(117), 76
Felt, R. M., **9:**32(116), 33(116), 67

Felton, L. E., **16:**251–252, 268
Femandez, M., **16:**123, 143
Fencht, D. L., **5:**276(140), 318
Fendler, R., **14:**55(173), 77
Feng, D., **17:**24, 49, 51, 53, 56, 75
Feng, G., **19:**52, 103
Feng, G. F., **19:**102, 135
Feng, I., **17:**171, 221
Feng, S., **23:**241(57), 291(57)
Feng, T., **12:**226(117), 235
Feng, Y., **24:**228(9)
Fenner, D. B., **16:**69, 136; **17:**209, 223, 272, 290, 293, 298; **23:**49(30)
Fensham, P. J., **5:**274(125), 317
Fenske, G. R., **17:**110, 141
Fenter, P., **24:**44(45), 45(72, 74), 82(13, 14), 83(13, 14), 86(13, 14, 39), 89(47), 112(7, 15, 16, 17, 18, 19, 20), 117(7, 15, 16, 17, 18, 20), 119(16, 17, 18, 19, 20), 125(16), 127(16, 17, 18), 128(15, 20), 129(20), 130(16), 131(16), 133(16, 18), 135(17), 136(17), 137(17), 138(16), 139(66), 150(3), 181(20), 182(20), 199(47), 206(15, 16), 211(15), 212(15, 16), 213(16), 214(16), 215(15), 216(15), 218(78)
Ferguson, E. T., **1:**281(11), 282, 292(11), 331
Ferguson, W. F. C., **2:**284(40), 304
Fermi, E., **4:**344(56), 370; **5:**9(14), 45
Fernandez, V. M., **24:**18(78), 19(78)
Ferrand, B., **9:**282, 294
Ferrari, L. A., **18:**137, 151
Ferrari, M. J., **16:**75–77, 117, 137, 142
Ferrario, M., **22:**134(51), 142(51), 172
Ferraris, J. P., **24:**98(84)
Ferré, J., **2:**110(92), 143
Ferrell, R. A., **2:**100, 142; **6:**193, 195, 197(48), 203, 298; **9:**146(2), 150(2), 173, 210(133), 255, 256, 258
Ferretti, A., **10:**33(60), 34(60), 66
Ferrier, F., **2:**72, 141
Ferrieu, E., **12:**263(104), 264(104), 292; **13:**61, 104
Ferris, N., **24:**23(114), 31(114), 44(23)
Ferry, D. K., **15:**222, 249, 254, 261, 263
Fert, C., **2:**110(92), 143
Feshbach, H., **9:**84(16), 143
Fetisow, E. P., **9:**192(87), 257
Fetterman, H. R., **16:**121, 143
Feucht, D. L., **5:**155(46), 227(46), 230, 276(141), 279(160), 318
Feuerbacher, B., **9:**178(80), 186(69), 187(69), 188(77), 189(78), 190(80); **12:**8(34), 50
Feuerbacher, B. F., **10:**76, 77(10), 84, 164
Feuersanger, A. E., **3:**139(41), 163; **5:**305, 321; **17:**235, 296
Feuerstein, M., **7:**102(80), 113
Feurer, R., **13:**99(98), 107
Fichman, M., **7:**330, 367(61), 394
Fickenscher, P., **16:**51, 133
Ficker, J., **4:**238(160), 254
Field, D. R., **19:**281, 293
Field, R. J., **17:**153, 154, 220
Fielding, W., **5:**283, 284, 319
Fields, D. L., **24:**8(28), 9(28)
Figlarz, F., **17:**304, 364
Fijimura, S., **16:**149, 229
Filby, J. D., **5:**147, 152, 183(20, 141), 185(20, 141), 199, 229, 233, 261, 268, 315, 316
Filion, A., **17:**279, 299
Fillard, J. P., **13:**121(60), 148
Finch, G. I., **1:**218, 231; **4:**234(152), 235(153), 254

Finch, R. H., **4:**169(106), 190; **5:**249(33), 260(53), 315
Fincham, D., **22:**132(47), 172
Findikoglu, A., **16:**106, 141
Fine, J., **9:**46(208), 70
Finegan, J. D., **1:**280(6), 283, 289(6, 20), 331; **3:**220(34, 41), 221, 222(38); **22:**120(16), 172
Fink, J., **24:**154(16)
Finkelnburg, W., **14:**196
Finkenrath, H., **3:**124, 128
Finklea, H. O., **24:**21(95), 27(126), 28(126), 29(126)
Finlayson, K. M., **14:**219(60), 220(60), 221(60), 222(60), 223(60), 238
Finn, D., **11:**66(122), 99
Finnegan, J. J., **2:**387(112), 422
Finstad, T. G., **17:**22, 74
Finzi, L. A., **6:**236, 237, 259(162), 264(162), 265(162), 288(208), 289(208), 290(208), 298, 299, 300
Fiori, C., **13:**190(156), 191(156), 197
Fiory, A. T., **16:**17, 126; **21:**79(7), 111; **23:**2(2)
Firment, L. E., **24:**144(68)
Fisanick, G. J., **16:**96, 140
Fischer, **16:**19, 28, 126, 128
Fischer, A., **9:**33(131), 35(131), 36(131), 37, 68; **12:**187(75), 234, 294
Fischer, A. C., **11:**42(45), 44(49), 45(45, 49, 56, 59), 59(49), 83(188), 89(230, 231), 97, 101, 103
Fischer, B., **9:**238(228, 229), 261
Fischer, D. W., **21:**121(22), 123(22), 194(22), 195(22), 233, 303(53), 310
Fischer, F. H., **13:**187(125), 196
Fischer, H., **4:**199(52, 55, 56), 200(52, 55, 56), 201(55), 202(55, 78), 203(55), 204(52, 55, 56), 206, 211, 219(107), 227, 228(107), 231, 233(55), 251, 252, 253
Fischer, J., **24:**92(62)
Fischer, K., **16:**70, 136
Fischer, R., **12:**274(172), 275(172), 294, 296; **15:**244–245, 263
Fischer, R. F., **9:**275, 294
Fischer, R. J., **3:**139(42), 163
Fischer-Colbrie, A., **19:**40, 41
Fischetti, M., **14:**94(31), 126
Fisher, A. D., **15:**111, 113–116, 118, 122, 141
Fisher, A. W., **12:**292
Fisher, B., **16:**59, 65, 68, 135
Fisher, D. G., **7:**244(269), 255; **12:**100(166), 113, 118(229), 120(229), 125, 126, 127, 128(185), 129(185, 270, 271), 131, 133, 134(185), 162
Fisher, G. B., **12:**62, 159
Fisher, J. C., **4:**192, 203(81), 252; **6:**15, 75
Fisher, M. E., **24:**184(40), 190(40)
Fisher, R. A., **3:**43, 49(218), 59
Fisher, R. M., **16:**288, 338
Fisher, T. F., **3:**99, 126
Fishman, C., **12:**226(117), 235
Fiske, M. D., **1:**241(15), 273
Fistul, V. I., **9:**51(243, 244), 70, 71
Fitch, R. K., **14:**23(91), 75
Fitton, B., **9:**178(80), 190(80), 193(93), 257
Fitts, R. W., **12:**150(322), 166
Fitzgerald, D. J., **4:**301(134), 323
Fiyimori, S., **12:**246(19), 290
Fizeau, H., **3:**2, 53
Flam, R. P., **10:**237(26), 258
Flament, C., **24:**181(15)
Flamm, D. L., **13:**68(46), 105;

14:234(142), 240; **15:**294–295, 311; **18:**60; **22:**61(62), 78, 211(70, 71), 212(73), 273–274
Flamme, C., **16:**58, 61, 75, 135
Flanagan, T. P., **3:**19, 51(90), 55
Flanders, J. M., **5:**153(38), 209(38), 214(38), 230
Flannery, R. E., **11:**168(141), 203
Flaschen, S. S., **4:**261(41, 42), 310(155), 321, 324
Flatley, D., **5:**290, 320
Flecken, F., **1:**10, 43(118), 64, 66
Fleet, S. G., **3:**214, 215, 216, 270
Fleig, G. J., **10:**124(53a), 165
Fleisch, H., **24:**163(24)
Fleischer, H. J., **5:**164(94), 231
Fleischmann, M., **4:**191, 233(3), 235(3), 251
Fleming, R. M., **16:**2–3, 51, 53, 124, 132, 134; **17:**14, 36, 38, 74, 75
Flengas, S. N., **14:**91(24), 116(24), 117(24), 119(24), 121(24), 125
Flesner, L. D., **21:**128(37), 233
Fletch, K., **22:**119(11), 154(11), 156(11), 171
Fletcher, N. H., **4:**139, 187
Flevaris, N. K., **17:**77
Flicker, H., **2:**134, 145
Flint, B., **10:**115, 165
Flitsch, R., **23:**47(160)
Flom, E. B., **24:**184(37, 41), 188(41)
Flonio, G. C., **17:**153, 154, 220
Flood, J. J., **3:**116(133, 170), 128, 129; **5:**167(132, 133), 193(133), 194(133), 203(132, 133), 232
Florescu, N. A., **3:**97(80), 126
Florez, L. T., **19:**2, 6, 9, 11–19, 23, 26, 27, 32–37; **21:**241(15), 309
Florinskaya, V. A., **4:**271(92), 322
Flörke, O. W., **5:**106(17), 141
Flory, F., **19:**195
Floyd, R., **15:**64, 67, 139
Fluit, J. M., **3:**73(57), 126
Flükiger, R., **16:**9, 125
Flur, B. L., **3:**107, 118, 119, 127, 128; **6:**237(124a), 272(174), 298, 299
Fochs, P. D., **3:**169, 208
Focht, M. W., **21:**221(135, 137), 222(135, 137), 236, 237
Foex, M., **12:**168, 232
Fofanov, G. M., **12:**169(37), 201(37), 233
Fogarassy, E., **14:**98(94–96), 101(94, 95), 113(94), 114(96), 127, 128; **16:**76, 78, 137
Fogdall, L. B., **10:**157(86, 87), 162(91), 166
Foglietti, V., **16:**117, 122, 142
Fokkink, L. G. J., **24:**21(105), 26(105), 27(105), 28(105), 86(37, 38)
Folch, A., **24:**98(88), 221(90), 222(90)
Folda, T., **24:**89(53), 92(53)
Foley, G., **17:**347, 357, 358, 367, 369
Foley, G. H., **9:**116(26), 144
Folkers, J. P., **24:**5(22), 14(58), 17(22), 112(23), 121(23), 163(28), 218(72, 73, 74), 230(43), 240(73), 248(89)
Folkes, P. A., **15:**222, 260; **21:**289(42), 310
Folks, J. R., **12:**126(261), 128(261), 132(261), 164
Foltyn, S., **16:**58, 61, 75, 135
Fonash, S. J., **12:**235; **14:**232(118, 122), 232(120), 232(127), 239
Fonstad, C. G., **9:**27, 67
Fontenille, J., **12:**272(168), 273(168), 294
Forbes, D. H., **5:**204(176), 234, 270(106, 107), 272, 306(272), 317, 321
Ford, C. W., **21:**362(44), 368

Ford, J. F., **24:**44(61)
Ford, N. C., **1:**316, 333
Ford, W. K., **3:**267(114), 273
Foreman, W. T., **4:**50(96), 68(96), 93
Forester, D., **15:**157, 215
Forgensen, F., **16:**149, 229
Fork, D. K., **16:**68–69, 78, 136–137; **17:**209, 223, 271, 272, 290, 293, 298
Fork, R. L., **21:**193(91), 235
Forlani, F., **6:**66, 78; **7:**373(185, 193), 378(185), 397
Forman, P. F., **10:**265(5), 308
Formigoni, N. P., **17:**232, 233, 234, 237, 265, 271, 296
Forouhi, A. R., **19:**61
Forrest, S. R., **23:**221(9), 224(9); **24:**181(20), 182(20)
Forrester, A. T., **1:**255(60), 265(60, 88), 275
Forrester, M. G., **16:**8, 14, 18, 22, 52, 101, 108, 125–127, 133, 141–142
Forsbergh, P. W., Jr., **19:**253
Forse, R. J., **16:**81, 102–105, 121, 138
Forster, D., **13:**154(16), 172(16), 181(16), 182(16), 183(16), 192
Forster, J., **18:**90, 169, 187
Försterling, K., **6:**186, 203
Forstmann, F., **9:**192(89), 257
Forsvoll, K., **6:**82(16, 25), 113, 115, 122, 128, 135(138, 149, 179), 143, 144, 146
Forsyth, D. L., **7:**56(39), 112
Forsythe, W. E., **2:**320(25), 359
Fort, T., Jr., **7:**330(58), 394
Forte, A. R., **17:**315, 365
Forty, A. J., **2:**132(153), 134(153), 145; **4:**152(69), 189, 195, 251
Foster, L. M., **7:**163(88a), 177(179), 182(137), 203(178a), 204(179); **11:**73(146, 146a), 100; **13:**204(17), 206(17), 223
Foster, N. F., **5:**164(83), 172(83), 231; **17:**153, 159, 219, 220
Foster, T. J., **15:**249–250, 263
Foster, W. J., **24:**181(12), 192(12), 199(12)
Fotakis, C., **13:**153(5), 182(5), 183(5), 192
Foti, G., **13:**166(60), 194
Fotiadis, L., **21:**241(8, 10), 246(24), 256(8, 10, 27), 259(10), 260(8), 265(27), 308, 309
Fotland, R. A., **3:**146(64), 147, 163
Fouad, M., **6:**116(155), 147
Fourcade, R., **17:**314, 365
Fournier, J. O., **24:**8(28), 9(28)
Foutell, A., **17:**134, 144
Fowler, A. B., **2:**176, 192; **21:**122(27), 233, 303(52), 310
Fowler, A. O., **2:**385(100), 422
Fowler, G. O., **14:**240
Fowler, G. W., **12:**129(271), 164
Fowler, P., **2:**315, 336, 359; **12:**6, 50
Fowler, R. H., **4:**342(53), 351, 370, 371; **12:**54(10), 158; **21:**22(53), 73; **23:**88(15)
Fowler, W. B., **1:**131(16), 185
Fox, G. R., **19:**251, 265, 267, 272
Fox, M. A., **24:**20(90), 44(36, 52), 45(36), 218(72)
Foxon, C. T., **11:**36, 64(115), 66(124), 70, 89(4), 96, 97, 99; **19:**15; **21:**30(60), 73, 312(4), 313(5), 323(4), 324(4, 15), 325(15), 329(15), 333(15), 334(5, 15), 336(5, 15), 366, 367; **23:**254(76)
Foy, D. W., **5:**255(41), 315
Foy, P. W., **7:**234(245), 254
Foy, R W., **11:**75(148), 78(148), 79(148), 86(210), 100, 102

Foyt, A. G., **11:**106(7), 115(7), 130(85), 131(7, 85), 171(143), 199, 201, 203
Fragnito, H. L., **21:**193(91), 235
Fragstein, C. B., **6:**155, 202
Fragstein, C. v., **19:**134
Frait, Z., **1:**288(32), 332
Francis, A. B., **1:**19(46), 65
Francis, E. W. R., **1:**19(42), 64
Francis, G., **3:**80(62), 90(62), 126; **6:**42, 77
Francis, M. D., **24:**163(24)
Francis, S. A., **7:**331, 359, 363(71), 394
Francis, S. M., **19:**32
François, M., **16:**2, 124
Francombe, M. H., **2:**381(89), 422; **3:**62, 100(145), 103(146), 110(114), 115, 116(170), 118(145, 146), 120, 125, 127, 128, 129; **5:**163(75), 166, 167(132, 133, 135), 170, 176(209), 179(121, 122), 180(121, 122), 193(133), 194(133), 195, 196(75), 197, 199, 200, 203(132, 133, 135), 205, 206, 207, 208(121, 135), 222(209), 227; **6:**102, 109(127), 131(127), 132(127), 145, 146, 317(30), 325(32), 346; **11:**171(145), 183(145), 203; **15:**222, 229, 233–234, 238–239, 241, 243, 250, 261–262; **17:**92, 111, 140, 146, 148, 153, 156, 157, 162, 163, 187, 219–221, 226, 228, 229, 230, 232, 233, 234, 235, 236, 237, 248, 252, 265, 266, 267, 268, 269, 270, 271, 272, 273, 274, 276, 278, 280, 281, 282, 283, 284, 285, 286, 287, 290, 292, 294, 295–300; **21:**4(12), 5(2), 7(2), 8(25, 26), 9(25), 10(25), 11(25), 14(37), 17(2, 25), 20(37), 27(2, 57), 28(2, 26, 57), 29(26), 31(26, 61), 32(2, 57), 35(2), 36(2, 26), 37(25), 62(116), 63(116), 71, 72, 73, 74, 121(26), 185(82, 83), 186(83), 189(82), 206(26), 233, 235; **23:**112(37), 239(45), 243(45), 247(45), 251(45), 261(81), 291(81, 111)
Francon, M., **2:**233(35), 238; **3:**18, 51(81), 55
Frank, C. W., **24:**46(90), 57(90)
Frank, F. C., **1:**218, 231; **2:**132(150), 145; **4:**139, 142(22), 152(69), 157(15), 162(15), 166, 175(15), 184(22, 134), 187, 189, 190, 193, 194(16), 195, 203, 207, 208(86), 251, 253; **7:**125, 133(12), 247; **22:**113(28), 115
Frank, G., **7:**244(268), 255; **12:**124(254), 164
Frank, I., **2:**123, 144; **9:**174, 256
Frank, K., **16:**51, 133
Frank, L. A., **10:**159(89), 166
Frankenthal, R. P., **15:**150, 169–170, 214
Frankl, D. R., **4:**295(111), 296(111), 323; **6:**141, 142, 149
Franklin, B., **24:**180(4)
Franklin, T. D., **23:**124(16)
Franks, A., **2:**125(134), 144; **11:**21(14), 33
Franks, J., **3:**201(73), 210; **7:**236(249a, 250), 254
Franson, K. D., **1:**207, 230
Frantsevich, I. W., **7:**264(42), 307
Frantz, P., **24:**221(99)
Frantz, V. L., **23:**101(33)
Fraser, D., **16:**252, 268
Fraser, D. B., **9:**56(260), 57, 58,

Fraser, D. B., (*continued*) 62(260), 71; **14:**22(86, 87), 33(86, 87, 128), 66(86, 87), 75, 76; **16:**23, 127; **18:**22
Fraser, J. W., **12:**257(68), 291
Fraser, K. A., **24:**188(45), 195(45)
Frawley, J. J., **3:**179(29), 209
Fray, S. J., **4:**60(136), 94
Frechette, V. D., **7:**3(16), 37
Frederking, T., **6:**123, 147
Freedman, A., **24:**14(63), 15(63), 16(63)
Freedman, J. F., **1:**279(3), 286(27), 287(27), 289(27), 331, 332; **3:**214(12), 226, 228, 243(48), 270, 271; **6:**291(215), 300
Freedman, N. J., **15:**35, 137
Freeman, C. F., **12:**145(316), 152(316), 153, 166
Freeman, E. C., **15:**277, 279, 310
Freeman, G. H. C., **4:**67(158), 94
Freeman, J. H., **12:**248, 271, 290
Freeman, R. H., **6:**129(195), 131(195), 148
Freeman, T. L., **24:**11(46), 21(46), 33(46), 34(46), 44(24), 228(31)
Freeman, W. G., **16:**31–32, 129
Freeouf, J. L., **13:**203(13), 204(17), 206, 208(26), 219(72), 220(72), 221(72), 223, 225
Freese, R. P., **15:**168, 215
Frei, E. H., **6:**262(168a), 299
Freiburg, C. H., **16:**24–25, 127
Freik, D. M., **6:**116(161), 147
Freiser, M. J., **15:**192, 217
Freller, H., **11:**69(139), 100; **14:**180(157), 198
Freltoft, T., **16:**66–67, 136
Fremont, C., **3:**10(43, 44), 51(43), 54
Frenkel, A., **16:**110, 142
Frenkel, D., **24:**207(20), 208(20, 30, 33)
Frenkel, J., **1:**209, 231; **2:**2, 23(25), 59, 60; **4:**110, 135, 192, 251
Frensley, W. R., **15:**252–253, 264
Frerichs, R., **2:**367(27, 28), 420; **3:**96, 121, 126, 128, 169, 208; **5:**180, 232
Fresia, J. F., **6:**291(208b), 300
Freundlich, A., **21:**325(17), 330(17), 332(17), 338(17), 345(36), 346(36), 350(17), 362(17), 367
Frey, B. L., **24:**46(84)
Freyer, J., **13:**210(37), 212(37), 224
Friaut, R. J., **22:**134(55), 173
Fricke, R., **2:**13, 60
Fridmann, S., **15:**291, 311
Fridrich, E. G., **9:**59(267), 71
Friedel, J., **7:**123(10), 247, 259, 260(7), 306; **15:**178, 216; **16:**254, 269
Friedel, P., **14:**82(3), 90(3), 94(33), 103(3, 33), 118(33), 125, 126, 212(30), 237
Friedertshauser, P. E., **6:**116(160), 147
Friedlaender, F. J., **6:**230(110), 231(110, 114, 115), 232, 235(114, 115), 237, 297
Friedman, A. N., **6:**116, 117, 132(162), 147
Friedman, D. J., **21:**322(14), 337(14), 367
Friedman, H., **3:**39, 52(189), 58; **10:**78(22), 164
Friedman, H. L., **4:**302(139), 306(139), 309(139), 323
Friedman, L., **24:**11(41)
Friedman, N., **6:**277, 278, 299
Friedmann, G., **4:**87(239), 96
Friedrich, L., **22:**67(64), 74(64), 75(64), 76(64), 78

Friedrich, L. J., **22:**74(66), 76(66), 78
Frieser, R., **6:**37(133a), 77
Frieser, R. G., **4:**267(75), 322; **5:**266, 268, 316; **12:**292
Frisbie, C. D., **24:**19(84, 85), 20(84, 85), 28(127), 97(79), 98(79, 80), 228(29)
Frischat, G. H., **5:**140(44), 141
Frish, M., **14:**8(28), 74
Fritz, I. J., **15:**222, 260
Fritz, M. C., **24:**21(98), 22(98), 94(70), 95(70), 96(76), 98(76)
Fritze, H., **3:**122(156), 128
Fritzsche, C. R., **6:**36, 38, 76
Fritzsche, D., **9:**174(32), 219(177), 220(179), 256, 259
Fritzsche, H., **12:**296
Frohlich, H., **2:**400(160), 424; **6:**34(99), 51(99), 56(99), 76; **11:**288(109), 317; **12:**54(14), 158
Froissart, C., **6:**185, 203
Frolich, H., **4:**338(37), 370
Frolov, V. D., **7:**216(197a), 253
Froment, M., **14:**98(94, 96), 101(94), 113(94), 114(96), 127, 128; **17:**315, 316, 329, 330, 365
Fromherz, T., **21:**241(16), 309
Fromhold, A. J., Jr., **12:**282(233), 283(233), 295
Fromhold, A. T., **6:**3, 74; **14:**115(72, 73), 116(72, 73), 118(72), 127, 212(27), 212(28), 237
Fromhold, A. T., Jr., **14:**96(36), 114(36), 120(83), 121(83), 122(36), 126, 127
Fromm, E., **17:**135, 144
Frommer, J., **24:**82(6), 89(54), 92(54), 94(69, 71), 95(69, 71), 218(79)
Froome, K. D., **14:**196
Frosch, C. J., **4:**300(118), 323; **5:**150, 230, 277, 279, 318
Frossati, G., **11:**249(55), 252, 315
Frost, M. S., **11:**222, 256(67), 257(67), 270, 271, 314, 316
Frova, A., **6:**22, 75
Fry, H., **16:**332, 339
Frye, R. C., **16:**107, 141
Fryer, E. M., **6:**113(140), 135(140), 136(140), 146
Fryer, R. E., **4:**53(122), 54, 93
Fu, C. L., **24:**122(47)
Fu, C. Y., **14:**99(41), 110(41), 126
Fu, Y., **23:**146(45)
Fuchs, B. A., **19:**210
Fuchs, C., **16:**76, 78, 137
Fuchs, E., **6:**226(92), 230(92), 234(92), 250(92), 297
Fuchs, H., **24:**98(96)
Fuchs, K., **2:**381(87), 422; **3:**19, 51(85), 55; **6:**84, 85, 104(42), 144, 190, 203
Fuchs, R., **6:**195(51), 203; **9:**159, 162(19), 163(19), 167(26), 192(91), 198, 255, 256, 257, 261
Fueki, K., **14:**20(80), 75; **16:**3, 8, 96, 124, 140
Fuerbacher, B. P., **10:**84, 164
Fuhs, W., **12:**296
Fujicka, J., **17:**209, 223
Fujihara, M., **24:**94(69, 71), 95(69, 71), 218(79)
Fujii, K., **13:**20(34), 21(34), 30(34), 57; **17:**83, 139
Fujii, T., **15:**147, 154, 163–164, 171, 185–186, 214, 216–217
Fujii, Y., **17:**9, 10, 15, 74, 153, 154, 220
Fujiki, Y., **3:**39(187), 52(187), 58
Fujime, S., **7:**18(54, 55), 21(70, 72), 28(110, 119), 29(54, 55, 72), 38, 39, 40
Fujimori, S., **13:**142(124), 150

Fujimoto, F., **1:**198(51), 229; **2:**119(109), 143; **9:**166(21), 218(21, 173), 256, 259
Fujimoto, K., **17:**287, 299
Fujimoto, M., **3:**277(80), 300
Fujinaga, M., **22:**181(18), 271
Fujino, K., **17:**361, 362, 370
Fujioka, J., **17:**290, 299
Fujishima, A., **24:**27(128), 231(51), 233(51)
Fujishima, S., **17:**169, 221
Fujita, F. E., **4:**238(161), 239(161), 254
Fujita, I., **14:**66(220), 78
Fujita, J., **16:**30, 32, 129
Fujita, M., **17:**361, 370
Fujita, N., **17:**290, 299
Fujitsuka, M., **13:**71(52), 105
Fujiwara, H., **6:**227, 297
Fujiwara, K., **2:**119(108), 134, 143, 145; **9:**60(270), 71
Fujiwara, R., **17:**355, 369
Fujü, T., **11:**50(72), 70(72), 98
Fujü, Y., **11:**67(130a), 99
Fujuki, T., **11:**70(141), 71(141), 72(141), 100
Fukai, F., **11:**92(238), 103
Fukamichi, K., **15:**168, 177, 215–216
Fukano, Y., **1:**221, 231
Fukaro, Y., **7:**11, 18(26), 38
Fukatsu, S., **23:**10–11(38)
Fuks, M. Ya., **3:**216(18), 256(104), 257(104), 258(104), 264(104), 271, 273
Fukto, M., **24:**181(14)
Fukuda, Y., **1:**220(146), 231; **14:**7(23), 69(23), 73
Fukuhara, S., **12:**92(141), 161
Fukukawa, Y., **12:**292
Fukumoto, H., **14:**105(54), 126; **16:**27, 127
Fukumoto, Y., **13:**47(55), 58
Fukunishi, S., **14:**31(121), 76; **17:**261, 298
Fukushima, K., **13:**20(33), 21(33), 28(33), 57
Fukuta, M., **12:**278(206), 294
Fukutomi, M., **13:**71, 105; **16:**2, 123
Fukuyoshi, K., **17:**354, 368
Fulcomer, E., **6:**222, 259, 264(162), 265(162), 296, 299
Fulk, M. M., **2:**330(40), 360
Fullenbaum, M. S., **6:**298
Fuller, C. R., **3:**113, 127; **4:**267(74), 305(145), 309(145), 322; **7:**269; **16:**254, 268
Fuller, H. W., **1:**295(43), 296(44), 322(86), 332, 333; **6:**228(100, 101, 104), 230(104), 238(100, 101, 104), 294(100), 297
Fuller, W. D., **2:**393(127), 406(127), 423
Fullerton, E., **17:**49, 51, 53, 76
Fullman, R. L., **4:**203(81), 252
Fullowan, T. R., **21:**221(135), 222(135), 236
Fulton, E. J., **9:**167(24), 256
Fulton, R. L., **11:**293(131), 317
Fulton, T. A., **16:**13, 125
Funabashi, M., **14:**20(80), 75
Funatogawa, . Z., **1:**324(96), 326(96), 334
Fung, A., **10:**168(12), 169(17), 225
Funk, E. R., **4:**139(12a), 187
Funkenbusch, E. F., **24:**163(26)
Funkenbusch, P. D., **14:**66(218), 78; **17:**153, 220
Fuoss, R. M., **11:**229(24), 315
Furman, S. A., **9:**136(49), 144
Furnas, Jr., T. C., **7:**390(209), 397
Furois, P. C., **3:**92(181), 110(181), 129
Furrey, J. T., **15:**149, 154, 165, 168, 171, 214

Furst, O., **7:**24(85), 39
Furtado, C. S., **13:**96(92), 106
Furuichi, J., **11:**253, 255
Furukawa, K., **7:**3(13), 23(80), 37, 39
Furukawa, Y., **9:**35(163a), 68
Furusaki, T., **17:**318, 365
Furuuchi, D., **9:**47(221), 48(221), 62(221), 70
Furuuchi, S., **5:**125(29), 141
Furuya, K., **10:**236(19), 258
Fuschillo, N., **3:**217(19a), 244(19a), 271; **9:**6(25), 45(25), 65
Fusco, F. A., **24:**3(12), 5(12)
Fussell, W. B., **10:**132, 166

G

Gaber, B. P., **24:**98(91)
Gabla, L., **3:**11(52), 51(52), 54
Gabor, D., **2:**101, 110(98), 142, 143
Gabor, J., **7:**334(85), 394
Gabor, T., **5:**261, 316
Gabriel, C. T., **18:**98–99
Gabriele, P. C., **22:**154(75), 155(75), 173
Gabrusenoks, J. V., **17:**337, 349, 350, 367
Gaddipati, A. R., **16:**9, 125
Gaerttner, M. R., **13:**112(25), 120(25), 147
Gafni, H., **6:**66(198), 78
Gagne, S., **21:**62(117), 74
Gaidai, Y. A., **15:**115, 118, 141
Gaide, A., **6:**115, 124(148), 135(148), 147
Gaidukov, Yu. P., **6:**103(100), 122, 131(100), 132(100), 145
Gaines, G. L., **7:**313, 316, 319, 320, 321, 322, 323(14), 325, 326(43), 327(43, 47), 330(43, 47, 56, 57, 62), 334(47, 62), 335(14, 47), 336, 337(105), 349(47), 359(47), 366(14), 389(43), 391(14), 393, 394, 395; **24:**180(7)
Gaiser, R. A., **9:**33(122), 68
Gal, M., **15:**283, 310
Gale, B., **11:**21(14), 33
Gale, R. J., **17:**43, 47, 75
Galen, J. J., **3:**169, 208
Gallagher, A., **19:**72, 76, 80
Gallagher, C. J., **14:**139(30), 195
Gallagher, D. M., **7:**368(175, 176), 396
Gallagher, P. C. J., **4:**167(104a), 190
Gallagher, P. K., **15:**147, 214; **16:**3, 124
Gallagher, W. J., **16:**5, 21, 110, 116–118, 122, 124, 127, 142–143
Gallaway, W. S., **9:**227(197), 260
Galle, P., **17:**153, 184, 219, 222
Gallet, F., **24:**181(15)
Galli, G., **5:**283, 319
Galliorenzi, T. G., **10:**237, 258
Gallois, B., **16:**91, 139
Gallone, L. M., **5:**166(114), 176(114), 203(114), 232
Galloway, M. D., **16:**43, 48, 131–132
Gallus, G., **9:**20(80a), 67
Galyamov, B. Sh., **17:**318, 366
Gambino, R. J., **9:**51(242), 54(242), 70, 271, 273, 274, 277, 278, 280, 282, 285; **13:**109(4), 147; **14:**18(70, 72), 20(70), 61(197), 75, 78; **15:**147–149, 154, 164, 190, 197, 214–215, 217; **16:**21, 41–42, 110, 127, 131, 142
Gambke, T., **17:**304, 356, 357, 364, 369
Gamble, F., **16:**106, 141
Gammel, P. L., **16:**13, 125
Gamo, K., **13:**166(61), 167(64, 66), 168(61), 169(65), 180(64), 185(61, 64, 65, 66, 140),

Gamo, K., (*continued*) 186(149), 188(61, 64, 65, 66, 140), 189(149), 194, 197
Gandais, M., **4:**23(57), 24(57), 92; **6:**188(30, 31), 203
Gandhi, S. K., **17:**153, 154, 220
Gandy, T. H., **22:**178(9), 188(9), 193(9), 211(68), 212(68), 215(75), 226(68), 232(75), 245(68), 253(97), 254(97), 262(97), 270(121), 271, 273–275
Gang, O., **24:**182(23), 199(23)
Gangulee, A., **7:**290(99), 292(102), 293(104, 105), 294(99, 102, 105); **15:**174, 180, 216; **16:**234, 250, 254, 266, 268
Ganguly, A. K., **15:**37, 40–42, 47, 50, 137–138
Gans, F., **3:**188, 198, 209
Gans, R., **6:**201(65), 204
Ganter, C., **24:**8(29)
Gantmakher, V. F., **6:**82(35, 38), 144
Gao, J., **16:**22, 111, 117, 127, 142
Gao, L., **16:**2, 123
Gao, W., **24:**151(8), 154(8), 155(8b), 156(8b), 158(8b), 159(8b), 161(8b), 164(30), 165(30), 167(30), 168(33), 170(41), 171(41), 172(43), 173(43), 222(103)
Gao, Z., **24:**228(9)
Garbauskas, M. F., **16:**48, 52, 132–133
Garbe, S., **1:**19(41, 47), 27(47), 28(41), 64, 65; **7:**242(262), 244(268); **12:**114(193), 121(193), 124(254), 126(260), 162, 164
Garber, E. M., **17:**201, 223
Garber, J. A., **11:**230(32), 315
Garber, M., **11:**132(91), 156(91), 202
Garbuny, M., **12:**91(139), 161
Garbuzov, D. Z., **7:**155(68), 186, 187(148), 249, 251; **23:**230(15), 231(15)
Garcia, A. R., **16:**58, 61, 75, 135
Garcia, J. L., **24:**18(78), 19(78)
Garcia, N., **6:**117, 118, 141, 142, 147, 149
Garcia-V zquez, V., **16:**43, 47, 131
Gard, G. A., **12:**248(36), 271(36), 290
Gardner, I. C., **1:**146, 185
Gardner, N. C., **12:**246(28, 29), 290
Gardner, R. N., **15:**168, 215
Gardner, T. J., **24:**240(74)
Gardner, W. J., **13:**48(57), 58
Garfield, B. R. C., **12:**86(116), 87, 89, 97(116), 98, 99, 161
Garfinkel, H. M., **9:**33(120), 68
Garfinkel, M., **1:**241(18), 255(60), 265(60), 273, 275
Garg, S. K., **24:**48(98)
Garie, R., **17:**329, 366
Garmire, E., **10:**247(37), 248, 259; **13:**127(88), 149
Garmon, L. B., **4:**150(58), 153, 188, 212, 213, 233(95), 234(156), 238(156, 162, 163), 238(162, 163), 240(162), 242(95), 243(156, 163), 245(156, 163), 253, 254
Garnache, R. R., **5:**307(277), 322
Garnaes, J., **24:**98(95)
Garner, C. M., **12:**115(217), 116(222), 163
Garner, P. W., **12:**114(206), 163
Garoff, S., **24:**206(2)
Garone, P. M., **23:**41(113), 55(187, 188)
Garrell, R. L., **24:**44(41)
Garrett, C. G. B., **2:**157(41), 191
Garrett, J. P., **7:**143, 145, 146(49), 237, 238(49), 248
Garrett, J. R., **11:**65(119), 99

Garrett, K. W., **16:**287, 338
Garriott, O. K., **7:**56(39), 112
Garrison, B. J., **13:**153(2, 3), 154(3), 182(2, 3), 183(2), 192
Garrison, S. M., **16:**121, 143
Garrod, R. I., **1:**191(23, 24), 228
Garton, C. G., **11:**288(108), 317
Garton, G., **6:**62, 78
Garver, M. M., **22:**243(92), 274
Garverick, L. M., **14:**232(131), 239
Garvin, H. L., **10:**248(40), 259
Garwin, R. L., **1:**248(45), 251(45), 274
Garzon, F. H., **16:**42, 47, 131
Gasiot, J., **13:**121(60), 148
Gaskill, D. K., **16:**85, 138
Gasparov, V. A., **16:**71, 79, 136
Gast, P. R., **10:**5(9), 6(9), 65
Gastaud, A., **4:**85(226), 96
Gates, D. M., **2:**315, 316(9), 359; **4:**60(140), 94; **12:**6, 50
Gates, J. L., **23:**124(16)
Gates, S. M., **23:**30(115), 32(115)
Gatin, M., **24:**44(47)
Gatos, H. C., **3:**179(28), 209; **5:**285(201), 319
Gaub, H. E., **24:**94(68, 70), 95(68, 70), 96(76), 98(76)
Gault, W. A., **21:**221(134, 135, 137), 222(135, 137), 236, 237
Gautherin, G., **13:**109(3), 111(3), 142(3), 147
Gautier, H., **17:**172, 221
Gauzzi, A., **16:**26–28, 127–128
Gavaler, J. R., **16:**8, 18, 22, 52, 125–127, 133
Gaviola, E., **3:**46(225), 59; **10:**262, 308
Gavrilovskaia, E. M., **14:**196
Gavrilyuk, A. I., **17:**328, 344, 366
Gawehn, H., **2:**367(22), 420; **3:**90, 126
Gay, J. C., **9:**282, 294
Gay, J. G., **12:**284(241), 285(241), 295
Gayer, B., **12:**269(146), 293
Gayler, M. L. V., **6:**313, 346
Gaylord, T. K., **19:**282, 288
Gaynor, E. S., **15:**111, 113–116, 118, 141
Gazard, M., **10:**230(10), 238(10), 242(10), 258
Gazi, S., **16:**52, 133
Gear, C. W., **22:**132(46), 172
Geballe, T. H., **11:**256(58), 257(58), 315; **16:**2, 12, 22, 42, 50, 54, 63, 68–69, 78, 84, 90, 101, 123, 125; **17:**209, 223, 271, 272, 290, 293, 298
Gebhard, E., **2:**393(140), 423
Gebhardt, J. J., **5:**259(48), 315
Gee, J. M., **13:**190(157), 191(157), 197
Gee, M. L., **24:**15(69), 17(69)
Geerk, J., **16:**23–24, 74, 127, 137
Geesner, C. R., **3:**137, 162; **5:**288, 320
Geffcken, W., **9:**74, 132(1), 141, 143, 144
Geffcker, W., **5:**94(7), 123, 137(39), 141
Geib, K., **13:**139(117), 140(117), 150
Geier, H. J., **16:**245, 254, 259, 261–263, 267, 269
Geiger, J., **9:**204(127), 210, 212(134), 258
Geiling, S., **7:**19(61), 38
Geis, M. W., **13:**182(114, 116, 117), 183(114, 116, 117), 195, 196
Geisler, M., **18:**155
Geissinger, E. L., **3:**106, 127
Geittner, P., **12:**262(95), 292
Gelber, R. M., **9:**116(27), 132(39), 144
Gelfand, M. P., **24:**184(40), 190(40)
Geller, M., **6:**72(214), 79
Geller, S., **9:**275, 276, 295
Gennser, U., **21:**84(17), 111

Genovese, E. R., **6:**272(174), 291(210), 292(210), 299
Genzel, L., **4:**50(103), 93
George, I. A., **14:**196
George, S. M., 23.35n(138), **23:**41(145)
George, T., **23:**14(50), 109(59), 161(63)
George, T. D., **7:**36(164), 41
George, T. H., **5:**190(162), 233
Georgescu, V., **9:**33(156), 34(156), 68
Georgiadis, R., **24:**92(59), 93(59)
Georgopoulos, P., **24:**180(11)
Geppert, D. V., **2:**148(5), 190; **4:**326, 360, 369
Gerand, B., **17:**304, 332, 364
Gérard, P., **17:**310, 344, 355, 365–368
Gerasimova, L., **9:**51(247), 71
Gerber, C., **24:**18(88), 20(88), 44(43), 82(1, 2), 89(49), 106(111), 127(51), 218(75), 238(66)
Gerber, E. A., **3:**24(124), 51(124), 56; **17:**188, 222
Gerber, J., **1:**55, 67
Gerdy, J. J., **24:**127(53), 206(18), 211(18), 213(18), 214(18)
Gerenser, L. J., **24:**12(53), 15(53)
Gereth, R. A., **12:**266, 267, 293
Gergis, I. S., **16:**28–29, 111, 116, 129, 142
Gerhard-Multhaupt, R., **17:**229, 263, 264, 298
Gerhardt, R., **16:**45, 132
Gering, J. M., **15:**245, 263
Gerl, M., **7:**261(10), 306
Gerlach, E., **17:**350, 367
Gerlach-Meyer, U., **14:**234(139), 240
Gerloff, U., **7:**102(81), 114
Germer, L. H., **1:**203, 230; **2:**110, 143; **7:**1(1), 11(21), 37, 334, 359(86), 394
Germer, L. M., **14:**139(32), 195
Gerritsen, A. N., **6:**128(193), 148
Gerritsma, G. J., **16:**111, 117, 142
Gershenzon, M., **7:**210, 211(189), 252
Gershoni, D., **21:**197(98), 235
Gerstenberg, D., **2:**393(128, 130, 138), 394(146, 147), 395(147), 397(146), 398(146), 423; **3:**122, 128
Gertner, E. R., **11:**107(18), 171(18, 144), 200, 203; **21:**40(84), 74
Gesell, T. F., **9:**189(79), 227(205), 228(205), 229, 257, 260
Gesi, K., **11:**255
Gesser, H., **4:**211(90), 253
Gethin, D. T., **17:**132, 143
Getty, W., **5:**300(256), 321
Gettys, W. B., **6:**103(99), 122(99), 132(99), 145
Geusic, J. E., **9:**276, 294
Gevers, M., **2:**400(165), 424; **6:**62, 78
Ghandi, S. K., **14:**59(182), 77
Ghannam, M., **23:**2(4)
Ghate, P. B., **7:**265(50), 269, 270(63), 286(50), 307, 308; **16:**240, 250–251, 254, 267–268
Ghazali, A., **21:**13(35), 72
Ghez, R., **9:**283, 284, 295; **14:**115(74), 127
Ghezzo, M., **15:**299–300, 311
Ghidini, G., **23:**45–46(156)
Ghiti, A., **23:**228(30)
Ghosh, C., **12:**59(25), 60(39), 64(47), 68(52), 70(54), 74(82), 85(104), 86(52, 112, 115), 89(52), 90(52), 91, 96, 97, 98, 99(52), 100, 101, 103(39, 115), 105, 107, 108(39), 109(104, 175, 180), 110(180), 111, 112, 145(315), 148(320), 158, 159, 160, 161, 162, 166, 226(117), 235

Ghosh, P. K., **17:**241, 297
Ghosh, S. K., **2:**178(59), 192
Ghoshtagore, R. N., **11:**171(145), 183(145), 203
Gia-Russo, D. D., **10:**237(22), 258
Giacomo, P., **1:**127(7), 184; **3:**2(20), 13(63), 14, 15, 46(227), 49(63); **10:**169, 226
Giaever, I., **1:**267(90, 91), 268(94), 269, 275
Giallorenzi, T. G., **9:**54(256), 71
Giannelis, P., **19:**263
Gianola, U. F., **9:**275, 295
Gibbings, C. J., **23:**13(47)
Gibbons, D. F., **3:**243(66), 272
Gibbons, E. F., **18:**143
Gibbons, J. F., **13:**218(63), 224
Gibbs, D., **24:**127(56)
Gibbs, G. B., **6:**307(9), 346
Gibbs, J. W., **4:**100(17), 113(17), 119(17), 134, 138(5), 150(5), 187, 192, 251
Gibbs, P., **11:**255
Gibson, A. F., **3:**293, 300; **11:**223, 314
Gibson, D. W., **2:**331, 360
Gibson, J. M., **21:**79(7), 111
Gibson, U., **13:**137, 138(109, 110, 111, 116), 139(44, 109, 110, 111, 116), 148, 150
Gibson, W. M., **10:**238(27), 258
Giedd, G. R., **1:**31(74), 32(74), 65, 252(45, 50), 254(49), 274; **2:**365(7), 419; **3:**3, 5(23), 51(23), 53
Gier, J. T., **2:**315(10), 326(32), 359; **4:**60(144, 153), 94; **10:**2, 14(2, 26), 28(2), 43(26), 44(26), 45(26), 65, 66
Gierlowski, P., **16:**5, 97, 124, 140
Giess, E., **16:**17, 23, 126
Giess, E. A., **9:**264, 275, 276, 281, 283, 284, 287, 288, 294, 295, 296; **15:**13, 25, 137; **16:**5, 71, 124, 136
Giessen, B. C., **17:**2, 8, 10, 11, 12, 15, 35, 37, 39, 42, 73–75
Giessen, W., **7:**20(606), 38
Gifford, F. E., **12:**287(254), 288(254), 295
Giglia, R. D., **17:**352, 358, 368, 369
Gijsbers, J. R., **15:**168, 215
Gilbert, B. L., **17:**351, 352, 367
Gilbert, G. B., Jr., **5:**182, 233
Gilbert, S. L., **7:**134(29a), 171(113), 247, 250
Gilbert, T. L., **1:**323, 334
Gilchrist, J. le G., **11:**241(47, 48), 243, 249(55), 252(55), 253, 254, 255, 315
Gilder, H. M., **7:**262(20), 306; **16:**254, 269
Giles, G. E., **13:**166(59), 194
Gilgen, H. H., **13:**155(27), 162(45, 46), 163(45, 46), 164(27, 45), 165(58), 169(68), 171(72, 73), 174(58, 72), 178(72), 181(27), 186(46, 58, 68, 72), 189(45, 46, 58, 68, 72), 190(46), 192, 193, 194
Gilham, E. J., **3:**109, 127; **9:**9, 43(188), 62(40, 41), 65, 69
Gilinskii, I. A., **15:**40, 44, 138
Gill, W. E., **3:**91, 129
Gillam, E., **3:**99, 126
Gillaspy, J. D., **24:**230(44)
Gillen, G., **24:**229(42)
Gillery, F. H., **9:**50(229), 70
Gillery, P., **14:**196
Gilles, J., **3:**180, 181(38, 39), 209
Gilles, J. C., **14:**107(59), 126
Gilles, J. F., **9:**42(179), 69
Gilles, J. M., **2:**183(66), 192; **5:**184, 233

Gillet, R., **14:**14(49), 74; **17:**111, 142
Gillett, C. M., **7:**174(118a), 177(118c), 250
Gillette, R. B., **2:**353(78), 361; **7:**90(64), 113; **10:**14(21), 43(21), 65, 125(54), 165
Gillham, E. J., **6:**106, 110(119), 146
Gillhausen, H. F. V., **7:**374(194), 397
Gilmartin, H. R., **23:**124(15)
Gilmer, G. H., **16:**236–237, 266; **22:**123(29), 172
Gilmour, A. S., Jr., **14:**140(35), 195
Gilstrap, P., **13:**131(99), 133(99), 134(99), 135(99), 139(99), 140(99), 149
Gimpl, M. L., **3:**217, 244, 271
Gingrich, N. S., **7:**3(12), 37
Ginley, D. S., **16:**54, 123, 134, 143
Ginsberg, D. M., **1:**265(89), 275
Ginsburg, V. L., **1:**256, 258(63), 259(63), 275
Ginther, R., **10:**237(25), 258
Gintherodt, H. J., **24:**94(69, 71), 95(69, 71), 106(114), 127(51), 218(79), 222(102)
Ginzberg, V. L., **16:**22, 127
Ginzburg, V. L., **18:**127, 137; **19:**10
Ginzburg, W., **9:**174, 256
Girard, R., **6:**244(141), 298
Giri, A. P., **17:**120, 121, 143, 314, 365; **19:**194, 195, 230
Giron, I., **4:**199(62), 204(62), 227, 252
Girouard, F. E., **17:**355, 356, 368
Giroud, M., **16:**42, 45, 131
Gittleman, J. I., **10:**60(139), 69; **17:**351, 367
Giusto, M. B., **5:**271(110), 317
Givens, M. P., **1:**149, 186
Gizeli, E., **17:**212, 223
Gizeman, O. L. J., **19:**280, 281
Gjonnes, J., **2:**114(112), 119, 143; **7:**14(31), 38
Gjostein, N. A., **16:**236, 267; **22:**254(102, 103), 256(103), 275
Gladel, G., **17:**109, 141
Glang, R., **1:**216(132), 231; **3:**93(181), 110, 129, 171(24), 176(24), 209, 224, 271; **4:**258, 320; **5:**146(11), 149, 164(89), 172, 229, 231, 266, 267, 316; **7:**1(5), 37; **9:**42, 69, 133, 144, 277, 296; **14:**51(166), 77; **16:**4, 40, 124, 131; **17:**86, 91, 126, 140, 232, 295
Glarum, S. H., **11:**293(123, 126), 317; **16:**3, 124
Glaser, W., **2:**122(124), 144
Glasper, J. L., 23.24(89), 23.30(114), 23.37n(114), **23:**38(114), 39(114)
Glass, A. M., **13:**164(56, 57), 165(56, 57), 166(56), 167(57), 176(56, 57); **15:**66, 139; **17:**227, 230, 295
Glass, E., **24:**3(9), 17(9)
Glass, H. L., **9:**276, 294; **15:**9, 14–18, 37, 39, 64, 75, 136–139
Glass, S. J., **1:**58, 67, 323, 334
Glatz, A. C., **7:**390(211), 397
Glauber, R. J., **11:**302, 317
Glauche, E., **5:**313(303), 322
Glazer, A. A., **6:**216, 220(53), 269, 276, 277, 296, 299
Gleason, K. K., **19:**93
Gleason, P. R., **1:**146, 185; **10:**74(6), 91(6), 164
Gleim, P. S., **5:**296(40), 320
Gleiter, H., **17:**65, 77
Glenn, R. W., **3:**105(96), 127
Glimm, J., **22:**98(17), 115
Glocker, R., **2:**123, 144; **4:**254, 327
Gloge, D., **10:**168, 169, 225
Glogovsky, K. G., **21:**117(7), 128(7), 221(131, 134, 135, 137, 139),

222(131, 135, 137, 139), 232, 236, 237
Glossop, A. B., **4:**243, 255
Glover, R. E., III, **1:**257(68), 264(78), 265(85, 87), 275
Glowienski, L. D., **7:**274, 283(70), 308
Gmitter, T., **23:**48(161)
Gmitter, T. J., **19:**2
Gnaedinger, R. J., Jr., **4:**261(41, 42), 321
Gnall, R. P., **4:**262(45), 270(45), 272(45), 273(45), 288, 321
Goad, D., **6:**21, 27(72), 75
Gobbert, M. K., **22:**270(122–124), 275
Gobeli, G. W., **4:**79(201), 95; **6:**177, 202
Gobrecht, H., **5:**164(94), 231
Gobrecht, R., **12:**96(160), 97(160), 162
Goddard III, W. A., **24:**127(53), 206(18), 211(18, 47, 49), 213(18), 214(18)
Goddard, N. J., **17:**212, 223
Goddard, W. A., **24:**150(3)
Godwin, R. P., **9:**188(77), 257
Godyak, V. A., **18:**18, 22, 80, 147, 148
Goebloed, J. J., **11:**135(98), 202
Goeckner, M. J., **22:**7(14), 77
Goedeke, S., **15:**245, 263
Goedicke, K., **14:**34(133), 76, 204(13, 15, 16), 233(13), 237
Goeke, H., **16:**168, 192, 230
Goell, J. E., **10:**236, 258
Goemans, A. H., **15:**297, 311
Goerbitz, C. H., **24:**211(48)
Goercke, P. J., **3:**188(46), 198, 209, 210
Goering, H. L., **5:**154(44), 158(44), 159(44), 230
Goerlich, P., **5:**1(1), 44
Goetting, L. B., **24:**240(72), 241(72)
Goetz, G. W., **12:**157(335), 166
Goetzberger, A., **4:**301(125), 323
Gogadze, G. A., **6:**137(219), 148
Gogol, C. A., **17:**116, 142
Goh, M. C., **24:**98(93), 100(93)
Göhre, H., **3:**27(141), 28(141), 29(141), 30(141), 51(141), 56
Goisa, S. N., **12:**121(243), 164
Gokieli, B., **12:**265(109), 292
Goland, A. N., **1:**204(75), 230
Goland, A. W., **9:**179(55), 257
Golant, V. E., **18:**149
Golay, M. J. E., **9:**129(35), 144
Gold, R. B., **13:**218(63), 224
Goldak, J., **7:**33(138), 40
Goldberg, A. C., **21:**191(87), 221(87), 235, 257(32), 309
Goldberg, A. J., **3:**277, 300
Goldberg, H. S., **21:**124(28), 233
Goldberg, L., **23:**279(102)
Goldberg, P., **3:**203(82, 83), 205, 210; **5:**164(86), 231
Golden, W. G., **15:**273, 310
Goldfarb, R. B., **15:**154, 215; **16:**11–12, 125
Goldie, H., **15:**68, 99, 139
Goldman, A. M., **16:**16, 51, 126, 132–133
Goldman, D. T., **3:**69, 125
Goldman, V. J., **15:**249–250, 263; **23:**290(109, 110)
Goldner, E. L., **17:**347, 367
Goldner, R. B., **10:**52(112), 53(112), 68; **17:**306, 345, 347, 357, 358, 364, 367, 369
Goldsmith, A., **4:**371
Goldsmith, N., **4:**261(34), 263(34), 274, 320; **5:**277, 279, 280, 290(231), 318, 320
Goldsmith, S., **14:**175(134), 177(149), 198

Goldstein, B., **12:**114(201), 118(226), 119, 121(201, 240), 163, 164
Goldstein, M., **11:**254, 255
Goldstein, Y., **2:**157(43), 191
Goldsztaub, S., **7:**16(44), 38
Golovashkin, A. I., **6:**154(7), 202; **11:**137(101), 202
Golovcenco, I., **9:**33(137), 68; **12:**187(77), 234
Golub, J. E., **23:**292(114)
Golzhauser, A., **24:**86(35), 237(65), 238(65)
Gomer, R., **1:**21, 65; **9:**34(158), 68
Gomez, M., **15:**284–285, 310
Gomi, Y., **17:**162, 221
Gommel, G., **7:**20(62), 38
Gomzales, A., **7:**290(98), 310
Gonda, S., **11:**42(44), 45(58), 67(130a, 133), 73(123, 144), 97, 99, 100; **18:**91–92, 96
Gondo, Y., **1:**324(96), 326, 334; **2:**30(46), 61
Gonser, B. W., **3:**133(20), 162; **5:**306, 321
Gonsov, L., **6:**36(104), 76
Gonzalez, D. L., **21:**52(102), 74
Gonzalez, O. D., **7:**264(41), 307
Good, A., **4:**200(72), 252
Good, E., **16:**42, 131
Good, R. H., **1:**205(82), 230
Goodenough, J. B., **1:**312(64), 327, 333, 334; **9:**275, 295; **10:**52(105), 68; **17:**333, 366
Goodhue, W. D., **15:**244–246, 248, 251, 262–263; **23:**243(63)
Goodier, J. N., **16:**280, 337
Goodman, A., **2:**170(48), 191
Goodman, A. M., **5:**221(206), 234; **6:**65, 73, 78; **23:**88(17)
Goodman, B. B., **1:**255(59), 265(59), 274
Goodman, F. O., **14:**208(19), 237
Goodman, J., **3:**156(80), 164; **9:**60(272), 71
Goodman, J. W., **10:**172, 226; **15:**207, 217
Goodman, L. A., **9:**3(4), 64
Goodnick, S. M., **14:**234(144), 240
Goodwin, A. R., **4:**60(136), 94; **7:**236(250), 254
Goodwin, C. A., **11:**36(26), 97
Goodwin, H. A., **24:**24(116)
Goodwin, M. W., **15:**245, 263
Goodyear, S. W., **16:**52, 133
Gooedicke, K., **14:**35(137), 38–39(137), 76
Goorsky, M. S., **23:**24(97)
Goos, F., **3:**122, 128
Goosen, K. W., **21:**215(117), 236
Goossen, K. W., **15:**229, 233–235, 262; **16:**101, 141
Goppert-Berarducci, K. E., **24:**12(53), 15(53)
Goral, J. P., **16:**6, 28–29, 125, 129
Goranchev, B. G., **14:**10(37), 51(165), 66(224), 74, 77, 78
Goranson, R. W., **7:**367(173), 396
Gorbacher, A. A., **1:**329(109), 334
Gorbatkin, S. M., **18:**124, 125, 159, 184, 194, 203–205, 208
Gorbunova, K., **4:**199(51), 204(51), 252
Gorden, R. L., **10:**207(39), 226
Gordon, G. D., **2:**332, 860
Gordon II, J. G., **24:**112(21, 22)
Gordon, J., **7:**236(249e), 254
Gordon, J. E., **2:**53, 62; **3:**256(108), 273
Gordon, P., **7:**294(107), 310
Gordon, R. G., **22:**178(11), 271
Gordon, W. L., **16:**5, 76, 107, 124, 137, 141

Gore, G., **4:**191(1), 251
Goree, J. A., **22:**7(14), 77
Gorenstein, A., **17:**326, 327, 361, 366, 370
Gorkov, L. P., **1:**256(64, 65), 275
Gorkun, Yu. I., **6:**91, 145
Gorlich, P., **12:**54(4), 85(4), 92(4), 158
Gorman, C. B., **24:**177(28), 240(73), 245(84), 246(84)
Gorman, G., **16:**21, 127
Gornik, E., **21:**78(4), 111, 290(43), 310
Gorochov, O., **16:**28–29, 98, 129, 140
Gorodetsky, G., **13:**154(18), 182(18), 183(18), 192
Gorshunov, B. P., **16:**5, 124
Gorski, D. A., **11:**130, 158(131), 164(137), 192, 196(81), 201, 203
Gorsuch, P. D., **2:**59, 62
Gorter, C. J., **1:**247(39), 249(39), 274; **2:**23(34), 25, 60
Goruk, W. S., **6:**2(4), 22(4), 63, 64(4), 73
Goryachev, D. N., **13:**162(43, 44), 163(47, 49), 164(43, 44), 165(43), 183(49), 186(47), 189(47), 193
Goshen, R., **12:**282(219), 295; **14:**103(50), 124(50), 126
Goss, C. A., **24:**5(21)
Gossard, A. C., **11:**53(89), 54(89), 55(89), 78(165), 79(174, 176), 81(178, 179), 84(189, 192, 193), 85(165, 195, 196, 199, 200, 201), 86(179, 207), 98, 101, 102; **15:**221, 244, 269–261, 277; **17:**14, 36, 38, 74, 75; **21:**155(61), 234
Gossenberger, H., **5:**247(29), 274(120), 315, 317
Gossett, C. R., **16:**65, 136
Gossman, H.-J., **23:**18(67)
Goswami, A., **3:**174, 208
Goto, E., **1:**321, 333; **6:**208(16), 209, 262, 263, 264, 265, 266, 272, 284, 286, 295, 299, 300
Goto, N., **17:**315, 316, 317, 329, 331, 365
Goto, T., **16:**30, 129; **17:**234, 270, 296
Gotoh, S., **16:**14, 125
Gotoh, Y., **17:**354, 368; **24:**94(69), 95(69), 218(79)
Gottesfeld, S., **19:**50
Gottlieb, E. G., **5:**152(36), 153(36), 230, 272(113), 276(146), 278, 280, 317, 318
Gottlieb, M., **15:**115, 118, 141; **17:**182, 222
Gottlieb, W., **17:**154, 155, 182, 222
Gottsche, H., **4:**156(79), 159(79), 175(79), 189
Gottscho, R. A., **13:**181(107), 185(107), 188(107), 195; **18:**124, 159, 165, 194
Goudswaard, B., **6:**54(154), 77
Gough, S. R., **11:**236(37), 249(53), 251(53), 253, 255, 315
Gould, R. W., **18:**65
Gould, S. A. C., **24:**94(67), 95(67), 98(89)
Goulding, C. W., **6:**2(7), 74
Goulding, P. A., **19:**32
Gound, P. R., **2:**387(112), 422
Goureaux, G., **2:**30(41), 61
Gourley, P. L., **15:**222, 260
Gourrier, S., **14:**82(3), 86(13), 90(3), 94(33), 100(42), 103(3, 33), 113(67), 118(33), 125, 126, 127, 212(29, 30), 237
Goutzoulis, A., **17:**182, 222
Gover, A., **10:**250(44), 259
Govil, M., **22:**215(75), 232(75), 274

Govindarajan, N., **12:**71(64), 72(64), 75(89), 160
Gower, M. C., **13:**153(5), 182(5), 183(5), 192; **17:**287, 299
Grabe, B., **16:**250, 268
Grabowski, K. S., **16:**65, 78, 136–137; **17:**209, 223
Grace, J. M., **16:**32, 129
Graczyk, J. F., **7:**12, 13(27), 16(46), 24, 25(27), 38
Grader, G. S., **16:**36–37, 130
Gradmann, U., **4:**143, 187
Gradmann, V., **4:**249(171), 255
Graettinger, T. M., **17:**229, 234, 263, 296
Graf, K., **24:**98(96)
Graham, C. D., **1:**284, 332
Graham, C. T., **2:**384(95), 422
Graham, J. J., **24:**98(94)
Graham, R. L., **6:**24(84), 75; **24:**14(63), 15(63, 64), 16(63, 64), 119(42), 121(42)
Grainger, D. W., **24:**34(139, 140), 35(139, 140), 130(61), 221(94)
Grajewski, V., **17:**135, 144
Gran, B. E., **1:**294(41), 332
Graner, F., **24:**181(15)
Grange, J. D., **11:**68, 100
Granier, J., **13:**76, 79(61), 105
Granqvist, C. G., **17:**303, 304, 350, 351, 355, 356, 358, 361, 364, 367, 369; **19:**60, 113, 129, 139, 180
Grant, A. J., **13:**3(8), 56
Grant, D. M., **24:**153(12)
Grant, J., **4:**50(105), 93
Grant, P., **16:**54, 134
Grant, P. M., **6:**188, 203; **15:**53, 57, 138–139
Grant, R. J., **11:**256(59), 257(59), 315
Grant, R. W., **11:**157(129), 202
Grant, T. C., **16:**240, 251, 267–268
Gras-Marti, A., **14:**21(84), 75
Gravé, I., **21:**220(130), 236, 290(44), 310
Graves, B. B., **1:**37(99), 66
Graves, D. B., **18:**306, 308
Graves, R. H. W., **19:**281
Gray, E., **16:**37, 130
Gray, E. R., **16:**37, 48, 130, 132
Gray, K. E., **16:**17, 28, 32, 36, 126, 128–130
Gray, P. V., **4:**260(21), 301(21), 320; **12:**265(113), 292
Gray, R. B., **3:**39, 52(191), 58
Gray, S., **2:**170(49), 179(61), 181(49), 191, 192
Graybeal, J., **16:**116, 142
Graybeal, J. M., **16:**28, 33, 128
Grcar, J. F., **22:**180(16), 271
Greaves, J. C., **22:**213(74), 215(74), 221(74), 222(74), 274
Grebennik, I. B., **6:**339(56), 347
Grechanuk, N. I., **17:**60, 77
Green, A. J., **16:**92, 139
Green, A. K., **15:**306, 312
Green, D., **17:**339, 367
Green, D. C., **17:**318, 365
Green, J. B., **15:**60, 139; **17:**172, 173, 222
Green, J. J., **15:**92, 140
Green, J.-B. D., **24:**12(48), 46(85), 86(32), 98(82), 101(82), 105(82), 221(92)
Green, M., **11:**198, 203; **17:**320, 354, 358, 366, 368, 369; **21:**357(41), 368
Green, M. A., **21:**312(1), 313(1), 321(1), 322(1), 344(1), 366
Green, M. J., **24:**66(104)
Green, R. W., **16:**52, 133
Green, S. J., **24:**151(9), 154(9), 156(9)
Green, W. J., **16:**59, 79, 135, 137

Greenaway, D. L., **3:**277(25), 300
Greenberg, C. B., **17:**359, 361, 369
Greenberg, K. E., **13:**155(29), 183(29), 184(29), 193
Greenburg, S. A., **10:**140(75), 166
Greene, J. E., **10:**41(82), 67; **12:**275(181), 294; **13:**109(5), 116(5), 147; **14:**59(186), 60(193), 61(198), 65(209), 68(209), 77, 78, 219(55), 231(111), 238, 239; **17:**7, 8, 11, 34, 63, 65, 73, 74, 119, 122, 130, 131, 143; **18:**245; **22:**121(22), 172, 240(87), 274
Greene, L. C., **3:**179(26), 209
Greene, L H., **16:**18, 42, 45, 126, 131
Greene, P. D., **7:**153(63, 64), 154(64), 156(63, 64, 67c), 180, 181, 189; **13:**186(147), 189(147), 197
Greene, P. E., **7:**128, 131(19), 198, 199, 201(175), 247, 252
Greene, R. F., **5:**219, 234; **6:**89, 144
Greene, R. L., **16:**14, 125
Greenhouse, H. M., **7:**50(32), 112
Greenland, K. M., **3:**2(9), 17(9), 18, 51(9, 76, 77), 53, 55; **4:**21, 91
Greenlief, C. M., **23:**29(106), 30(115), 32(115)
Greenough, R. E., **13:**154(20), 181(20), 183(20), 184(20), 192
Greer, A. L., **17:**35, 37, 39, 42, 75, 77
Greer, J. A., **16:**74, 137
Greer, L., **17:**66, 77
Gregg, S. J., **7:**349, 395
Greggi, J., **16:**8, 18, 125–126
Gregor, H. P., **7:**330, 367(61), 390(210, 211), 394, 397
Gregor, L. V., **2:**365(4), 370(4), 380(4), 389(4), 390(4), 419; **3:**132(14), 139(39), 146(56, 66), 148(66), 149(74), 151(73, 74, 75), 152(56, 73, 74), 153(56, 74, 76), 154(77), 155(75), 156, 157(83), 158(87), 159(74), 160(90), 161(90), 162, 163; **4:**258, 260, 301(19), 316(162), 320, 324; **5:**238(5), 245, 289(5), 293(235), 314, 320; **12:**265(124), 293; **14:**221(64), 222(64), 238
Gregory, P. E., **12:**54(9), 74(85), 78, 114(206), 115(217), 117(221), 118, 120(235), 121(230), 130(276), 136(9), 138, 139, 158, 163, 164, 165
Gregory, S., **19:**51
Greiner, J. H., **6:**214(35, 36, 38), 215, 223, 295, 296, 297; **12:**282(230, 231), 283, 295; **14:**10(35), 74, 214(33), 214(34), 215(34), 237
Grenier, C., **1:**242(23), 274
Grenier, C. G., **6:**82(26), 144
Grenning, D. A., **7:**127, 131(17), 247
Gresh, M., **2:**397(151), 406(190), 407(190), 408(151), 409(151), 423, 424; **6:**71(208), 79
Gress, R. W., **9:**31(115), 67
Gretz, R. D., **4:**100(18), 106(18), 110(29), 111(29), 115, 117, 118(18, 37), 122(58), 124(58), 125(58), 127(18, 37), 128, 134, 135, 136; **5:**260, 315
Greve, D. W., **16:**26, 127; **23:**163(47, 48)
Gridley, J. C., **21:**219(127), 236
Grieco, M. J., **5:**257, 269, 273, 297, 306(274), 315, 316, 320, 322
Griesbaum, K., **24:**9(33)
Grieser, J. L., **16:**57, 60, 134
Griffin, A., **9:**159(17), 256
Griffin, R., **10:**154(80), 166
Griffith, A. A., **16:**276, 331, 337
Griffith, R. W., **12:**289(264), 296
Griffiths, C. H., **5:**163(77), 171, 231

Griffiths, J. E., **13:**205(19), 223
Griffiths, P. R., **15:**271, 310
Grigorovici, R., **7:**3, 23(14), 24(87), 36(163), 37, 39, 41; **19:**73
Grigson, C. W. B., **7:**11(22), 16(37, 38, 40), 17(40), 18(56), 37, 38
Grill, A., **12:**272(170, 171), 274(170, 171), 294
Grill, C., **12:**169(21), 232
Grim, P. C. M., **24:**95(73)
Grimmeiss, H. G., **7:**171(112), 210(188), 250, 252
Grimminger, E. H., **7:**27(108), 39
Grimsditch, M., **17:**24, 34, 43, 44, 45, 46, 48, 49, 50, 51, 52, 53, 56, 57, 58, 74–76
Grindle, A. W., **3:**179(31), 183(31), 209
Grinevich, G. P., **7:**264(36, 37, 40), 307
Gritsenko, V. A., **13:**99(112), 100(112), 107
Grobe, E., **7:**143(51), 248
Grodkiewicz, W. H., **9:**264, 267, 275, 276, 287, 288, 294, 295, 297; **10:**256(62), 259
Groff, R. P., **6:**103(102), 145
Grolleau, B., **12:**289(266), 296
Gronau, M., **16:**168, 192
Grone, A. R., **7:**258, 262(6), 263(23), 306; **16:**238, 267
Gronet, D. M., **23:**24(91)
Groppe, J. V., **23:**101(33), 124(14)
Gross, E. P., **6:**72(214), 79
Gross, P., **5:**246(18), 247(18), 314
Gross, R., **16:**118–122
Grossklaus, W., **13:**76(60), 79(60, 73), 83(60), 95(73), 105, 106
Grosskreutz, J. C., **16:**279, 292, 337
Grossman, J. J., **5:**273, 317
Groth, R., **9:**33(141), 35, 37, 62(141), 68; **12:**187(82), 234
Grove, A. S., **4:**260, 261, 295(113), 301(25, 113, 124, 134), 302(140); **14:**114(70), 127; **17:**274, 298
Grove, W. R., **3:**61(1), 124; **14:**2(11), 73
Grovenor, C. R. M., **17:**121, 143
Grover, N. B., **2:**157(43), 191
Groves, S. H., **11:**36(18, 23), 96, 97, 107(19), 200
Groves, W. O., **7:**169(95), 209(185a), 238, 250, 252; **11:**66(122), 99
Grozinger, C., **24:**164(30), 165(30), 167(30), 168(33)
Grube, W. I., **12:**284(241), 285(241), 295
Grübel, G., **24:**3(17), 181(15), 184(43)
Gruber, G. A., **5:**199(174), 202(174), 234, 287, 296(241), 297(241), 320; **12:**265(112), 292
Grudkowski, T. W., **16:**56–57, 134
Gruen, D. M., **14:**6(21), 73
Grünbaum, E., **1:**219(143), 231; **4:**131, 136, 138, 140(9), 159, 160(9), 161(92), 162, 176(88a), 178, 179, 187, 189, 190; **6:**102(93b), 145
Grunberg, G., **6:**244(141), 291(209), 298, 300
Grunberg, P., **15:**37–38, 138
Grundmann, M., **19:**40, 41
Grundy, P. J., **15:**171, 216
Gruner, G., **16:**100, 106, 141
Grunes, R. L., **3:**269, 273
Gruninger, H., **7:**27(108), 39
Grunthaner, F. J., **23:**49(37)
Grunthaner, P. J., **23:**49(37)
Gruntz, K. J., **15:**286, 310
Grunwell, J. R., **24:**25(120)
Grunze, M., **24:**45(77), 65(77), 86(35),

92(57, 58), 127(55), 237(65), 238(65)
Grushkevitz, Y., **5:**221(203), 234
Gruss, L. L., **3:**194(52), 209
Grützmacher, D. A., **23:**50(180)
Gu, H., **16:**32, 130
Gu, P., **22:**163(83), 173
Gu, Y., **24:**3(19)
Gualtieri, G. J., **13:**205(19), 223
Guan, W. Y., **16:**23, 41, 127, 131
Guarnieri, C. R., **14:**65(214), 78; **16:**78, 137
Gubanov, A. I., **6:**64(183), 78
Gudat, W., **10:**116(45), 117(45), 165
Gudkow, L. V., **10:**64(145), 69
Guellil, A. M., **22:**127(38), 172
Guentert, O. J., **5:**262, 300(256a), 316, 321
Guenther, K. H., **13:**121, 124(70), 148; **19:**194, 195, 230, 243, 251
Guerci, C. F., **9:**276, 295
Guerri, S., **14:**112(99), 128
Guestblom, B., **11:**211(9), 314
Gugan, D., **6:**104(116), 146
Guggenheim, E. A., **7:**181(135), 251
Guggenheim, H. J., **5:**311(284), 322; **17:**229, 279, 280, 281, 282, 299
Guggi, D., **16:**23, 127
Guha, S., **19:**61
Guibault, G. G., **24:**48(96)
Guichar, G. M., **12:**114(211), 163
Guidi, R. L., **11:**132(89, 90), 201
Guien, Z., **16:**4, 124
Guild, J., **2:**330, 860
Guile, A. E., **14:**138(20), 157(98), 195, 197
Guillard, M., **10:**117, 165
Guillemet, C., **10:**237, 258
Guinier, A., **2:**125, 144; **7:**2, 3(10), 37; **17:**12, 15, 74
Guivarc'h, A., **12:**272(168), 273(168), 294
Gulari, E., **17:**153, 154, 221; **21:**207(108), 235
Gulbransen, E. A., **3:**28, 51(147, 148), 57
Guldner, W. G., **2:**397(154), 424; **3:**86, 126
Gullberg, J. E., **3:**28(145), 51(145), 57
Gummel, H. K., **4:**327(8), 369
Gumnick, J. L., **12:**70(55), 159
Gunapala, S., **21:**137(48), 139(48), 154(48), 160(48), 179(48), 234
Gunapala, S. D., **21:**14, 38(78, 81), 39(78), 40(78), 49(78), 73, 94(25), 111, 113–237, 128(30), 129(30), 130(30, 42), 131(30), 132(30), 133(42), 135(30), 136(30), 137(30), 138(30), 139(30), 140(30), 141(30), 142(30), 143(30); **23:**2, 102(43), 145(30), 146(30), 146(46), 146(46)4(30), 147(30), 148(30), 149(30), 150(30, 54, 55), 151(42), 153(42), 154(42), 155(59), 158(59), 159(59), 160(30), 163(70), 164(70), 165(70), 166(70), 181(55), 188(84), 189(84), 190(84), 192(89), 193(89), 194(89), 195(89), 197(97, 98), 199(97), 200(97), 201(97), 202(99, 100), 203(99), 204(99, 101), 205(101), 207(97, 99, 105), 208(112), 209(112), 210(112), 211(42), 211(112)
Gundlach, K. H., **6:**73(219), 219; **9:**218(171), 259
Gundry, P. M., **12:**114(202), 121(241, 242), 163, 164
Gunn, J. B., **7:**239(254), 255; **13:**210(31, 32), 212(32), 223, 224

Gunshor, R. L., **17:**172, 221
Günther, K. G., **5:**165(102, 103), 173, 174, 175(103), 209(182), 214, 232, 234
Günther, K.-G., **1:**44, 67
Guntherodt, G., **17:**45, 76
Güntherschulze, A., **6:**11, 13(43), 15, 74
Guntherschulze, A., **14:**196, 219(62), 238
Guntzmann, G., **12:**87, 88, 161
Gupta, A., **16:**5, 57, 61–62, 64, 71–72, 78, 110, 118–120, 122, 124, 134–137, 142–143, 292, 340
Gupta, A. K., **16:**116, 118, 142–143
Gupta, D., **7:**285, 295, 296, 297, 300, 301, 302, 303, 304(108), 310; **16:**236, 241, 254–255, 266, 268–269
Gupta, P., 23.35n(138)
Gupta, R. P., **13:**210(37), 212(37), 224
Guptill, M. T., **21:**36(70), 73
Gurev, H. S., **10:**37(73), 45(94), 46(73), 67
Gurevich, L. E., **6:**117(165b), 147
Gurevich, S. A., **13:**162(44), 164(44), 193
Gurevich, V. L., **6:**82(23), 144
Gurley, J. A., **24:**94(67), 95(67)
Gurney, R. W., **9:**9, 65
Gurp van, G. J., **7:**296(110), 310
Gururaja, R., **19:**272
Gurvitch, M., **16:**17, 126
Gurzhi, R. N., **6:**97, 99, 127, 145
Guseva, M. B., **13:**18(24), 20(24), 57
Gust, D., **24:**33(132)
Gustafson, K., **12:**282(238), 295
Gutan, V. B., **5:**221(204), 234
Guth, G., **21:**221(135), 222(135), 236
Guthric, A., **3:**107(100), 127
Gutierrez, D. A., **21:**162(69), 234
Gutierrez, W. A., **2:**151(35), 176(35), 191; **5:**216(195), 234
Gutmann, R., **17:**229, 263, 298
Gutsche, E., **3:**171(16), 172(16), 176(16), 208
Guy, A. G., **7:**264(44), 307
Guy, D. R. P., **21:**128(36), 233
Guyot-Sionnest, P., **24:**33(136), 128(57)
Guzman, A. M., **17:**334, 367
Gwathmey, A. T., **4:**227, 254
Gwilliams, G. F., **12:**143(313), 166
Gyorgy, E. M., **9:**288, 295; **15:**147, 150, 152, 161–162, 169–171, 185–186, 214–216; **17:**17, 74
Gyurk, W., **2:**375(73), 421
Gzowski, O., **11:**256(65), 257(65), 316

H

Haacke, G., **9:**58, 59(265a), 62(265a), 71; **12:**169(7), 186(93), 197(93), 200(93), 232, 234; **17:**358, 369
Haaijman, P. W., **9:**9(44), 65
Haak, J. R., **22:**134(54), 136(54), 173
Haaland, C. M., **1:**36(95), 66
Haanstra, H. B., **22:**21(30), 77
Haarpaintner, G. H., **21:**313(5), 325(18), 334(5), 336(5), 348(38), 351(18), 353(18), 355(38), 356(38), 366, 367
Haas, G., **17:**126, 143
Haas, G. A., **1:**28(63), 65
Haas, L. D., **11:**119(70), 130(87), 131(70), 201
Haas, O., **17:**326, 366
Haas, T. E., **17:**306, 345, 347, 357, 358, 364, 367, 369
Haas, W., **2:**406(186, 187), 424

Haase, G., **1:**211(120), 214(123), 224(120), 231
Haase, O., **3:**3, 51(22), 53
Haayman, P. W., **9:**33(138), 35, 68
Haaymaü, P. W., **12:**187(78), 234
Haber, M. D., **7:**330, 367(61), 394
Haberecht, R. R., **3:**139(43), 163; **5:**299(253, 254), 321
Haberey, F., **11:**294(133), 317
Habermeier, H.-U., **16:**53, 60, 69, 134–136
Habib, M. A., **17:**318, 361, 362, 366, 370
Habibi, M., **17:**312, 365
Hablanian, M. H., **1:**13(23), 64
Habraken, F. H. P. M., **19:**280, 281
Hachenberg, O., **2:**125(141), 144
Hack, R. A. A., **22:**262(108), 275
Hackerman, N., **4:**210(89), 253
Hackett, Jr., W. H., **7:**166, 167(93), 225(226), 249, 253
Hackforth, H. L., **2:**326(33), 359
Hacman, D., **6:**116(159), 147; **10:**82(23), 164; **12:**48(77), 51
Hacskaylo, M., **4:**78, 95; **13:**121(61), 148
Hada, T., **17:**153, 220
Haddenhorst, H G., **1:**70(12), 120
Hadjipanayis, G. C., **16:**212, 216–218, 230–231
Hadley, C. P., **3:**20(95), 21(95), 51(95), 55
Hadley, L., **3:**9(38), 54; **9:**13, 66, 74, 100, 102(18), 113, 115, 132(2), 143
Hadley, L. N., **2:**208(11), 217, 237, 238, 336, 360; **5:**21, 45
Hadziioannou, G., **24:**95(73)
Haefer, R., **3:**11(50), 51(50), 54
Haegel, N. M., **21:**5(19), 72
Haelbich, H.-J., **10:**116, 117, 165
Haensel, R., **7:**46(17), 48(25), 112
Haering, R. R., **2:**151(37, 37a, 37b), 176(37), 191
Haertling, G. H., **17:**249, 297; **19:**259
Haessler, H., **14:**180(157), 198
Hafl, E. L., **16:**43–48, 131–132
Hagar, C. K., **2:**418(197), 425
Hagberg, D. S., **17:**241, 297
Hagberg, J., **16:**100, 140
Hage-Ali, M., **14:**232(118), 239
Hagedom, F. B., **2:**379(78), 421
Hagedorn, F. B., **6:**213, 214(39), 262(168), 263, 265, 267, 269(39), 292(169), 294, 295, 296, 299; **9:**270, 287, 295, 296
Hägele, G. Z., **24:**164(31), 168(31)
Hagemann, H.-J., **10:**116(45), 117(45), 165
Hagen, S. J., **16:**14, 125
Hagena, O. F., **13:**8(14), 54(63), 57, 58
Hagenlocher, A., **5:**276, 318
Hagenlocher, A. K., **2:**367(18), 420; **17:**235, 296
Hagenlocker, A. K., **3:**139(41), 163
Haggmark, L. G., **14:**6(22), 73; **22:**164(85), 173
Haghiri-Gosnet, A. M., **17:**121, 143
Hagino, M., **12:**87, 95, 135(295), 161, 162, 165
Hagström, S. B. M., **9:**188(75), 257
Hagstrum, H. D., **3:**97(82), 126
Hahn, M. R., **16:**42, 50, 131
Hahn, R., **9:**141(52, 53), 142(53), 143(52, 53), 144
Hahn, R. B., **12:**202(95), 234
Hahn, R. E., **10:**37(71), 45(93), 67; **12:**42(72), 45(72), 51
Hahner, G., **24:**45(77), 65(77), 92(57)
Haidinger, W., **5:**163(68), 168(68), 231
Haigh, J., **13:**186(150), 189(50), 197
Haile, J. M., **22:**134(51), 142(51), 172

Haine, M. E., **1:**19(42), 64, 188(1), 228; **2:**83, 94, 96(36), 97, 98, 141
Hairu, T., **12:**296
Haisler, W. E., **16:**302, 339
Haisty, R. W., **5:**279(165), 318; **7:**185(140), 251
Haitjema, H., **17:**352, 368
Hajicek, D. J., **14:**228(90), 239
Hakansson, G., **14:**60(193), 78
Hkansson, G., **17:**89, 99, 119, 123, 124, 129, 130, 140, 143
Hakim, M., **11:**238(42), 315
Hakki, B. W., **5:**226, 235; **7:**174(123), 193, 194, 195(160), 196(160, 167), 208(186), 251, 252
Halaby, S., **3:**133(19), 162
Halaby, S. A., **2:**365(4), 370(4), 380(4), 389(4), 390(4), 419
Halbrecht, R. R., **2:**370(55), 421
Halbritter, J., **16:**8, 125
Halder, N. C., **7:**28(117), 40
Hale, A. P., **3:**113(128), 127
Hale, G. M., **12:**46(75), 51
Hale, J., **21:**68(127), 75
Hale, M. E., **1:**295, 296(44), 332
Hales, M. C., **7:**170, 191(106), 250
Halford, J. H., **19:**148
Halherbe, A., **9:**132(42), 144
Hall, A. C., **19:**50
Hall, C. E., **1:**188(3), 228; **2:**93, 110(42), 112(42), 141; **4:**253
Hall, D. B., **10:**256(63), 257(63), 259
Hall, D. G., **9:**254(253), 261
Hall, E., **7:**290(98), 310
Hall, E. O., **17:**69, 77
Hall, J. A., **12:**151(323), 152, 154(327), 166
Hall, J. F., Jr., **2:**284(40), 304
Hall, L. A., **7:**51(33), 112
Hall, L. D., **1:**18, 64
Hall, L. H., **11:**132(92), 202
Hall, M. J., **4:**190
Hall, P. M., **2:**379(78), 393(138), 421, 423; **3:**122, 128; **6:**4(17), 5(17), 7(17), 74; **14:**30(116), 76
Hall, R., **3:**140(49, 50), 163
Hall, R. N., **4:**330, 369, 370; **7:**120(4a), 247; **21:**348(39), 368
Hallais, J., **19:**31, 56
Halle, S., **13:**177(97), 185(97), 188(97, 142), 195, 197
Haller, B., **3:**112(178), 129
Haller, E. E., **21:**5(3), 18(3), 36(69, 72), 37(69), 71, 73
Haller, I., **2:**132, 145; **3:**24(118), 27(118, 127), 51(118, 127), 56, 146, 147; **12:**289(267), 296; **13:**124(74), 126(74), 149
Halliday, J. S., **2:**97, 142; **3:**226, 229(47), 234, 235(47), 238, 244, 246, 271
Hallmark, V. M., **24:**15(65)
Hallock, R. B., **16:**100, 140
Halperin, A., **13:**99(106), 107
Halpern, V., **11:**290(119), 317
Halteman, E. K., **3:**246, 272
Ham, F. S., **6:**92, 145
Hamaguchi, S., **22:**81, 83(1), 84(1–5), 85(1), 92(1), 93(1), 97(1), 98(2, 4), 101(2–4), 102(2), 105(2), 111(4), 112(4), 113(4), 114, 114(4, 5), 188(39), 242(91), 270(129), 272, 274–275
Hamakawa, Y., **9:**35(163b), 68; **14:**31(122), 76; **17:**290, 299; **19:**250
Hamaker, H. C., **21:**327(20), 367
Hamann, D. R., **16:**4, 124
Hamasaki, T., **15:**293, 311
Hamberg, I., **17:**350, 367
Hamburger, L., **2:**37, 61; **3:**19(86), 51(86), 55

Hamdi, A. H., **16:**95–96, 140
Hamill, D. W., **5:**285(200), 319
Hamilton, D. R., **5:**156(47), 230
Hamilton, M. C., **10:**236(18), 258
Hamilton, T. M., **5:**313(298), 314(298), 322; **15:**8, 136
Hamilton, T. N., **9:**276, 294
Hamisch, H., **1:**189(9), 228
Hamm, R., **21:**197(97, 98), 199(97), 200(97), 201(97), 207(97), 213(116), 214(116), 216(116), 235, 236
Hamm, R. A., **21:**150(54), 202(99, 100), 203(99), 204(99), 207(99), 234
Hamm, R. N., **9:**225(191), 226(191), 227(205), 228(205), 229(205), 237(222), 260, 261
Hammad, F. H., **4:**239(166), 255
Hammer, D. C., **2:**124(132b), 144; **9:**176(44), 256
Hammer, J. M., **10:**253, 254, 257(66, 67), 259; **17:**153, 154, 220
Hammer, L. H., **14:**232(124), 239
Hammer, W. N., **13:**124(74), 126(74), 149
Hammes, J. P., **3:**156(81), 157, 164; **5:**245, 314
Hammond, B., **16:**101, 103, 105–107, 141
Hammond, R., **16:**106, 141
Hammond, R. B., **16:**12–13, 81, 100–106, 121, 125, 138, 141
Hammond, R. H., **14:**65(214), 78; **16:**2, 19–20, 42, 50, 123, 126, 131
Hamon, B. V., **6:**57(162), 77
Hampden-Smith, M. J., **22:**182(27), 272; **24:**99(103), 100(103), 106(103)
Hampe, A., **9:**227(200), 260
Hampe, W., **6:**201(71), 204
Hampshire, D. P., **16:**45, 48, 132
Hampy, R. E., **12:**269(144), 293
Hamrin, K., **7:**45, 48(16), 111
Han, J.-U., **16:**250, 268
Han, P. D., **16:**99, 140
Han, W., **24:**33(132)
Hanak, J. J., **5:**306, 313(302), 321, 322; **14:**18(71), 75
Hanakawa, Y., **12:**275(174), 294
Hanau, R., **3:**96, 99, 126
Hanawa, T., **7:**22(74), 39
Hanbrucken, M., **17:**30, 75
Hanbucken, M., **19:**70
Handelman, E. T., **5:**163(76), 166, 168, 196(76), 231
Handy, R. M., **4:**350(66), 352(66), 353, 370; **7:**335, 369, 370, 375, 379, 381(96), 394
Haneji, N., **14:**103(47), 126
Haneman, D., **12:**122(246), 164
Hang, K. W., **9:**12(60), 33(60, 152), 66, 68
Hanigofsky, J. A., **16:**92, 139
Hanken, D. G., **24:**46(84)
Hänlein, W., **1:**44(124), 67
Hannay, N. B., **9:**30(110), 67
Hansen, E. E., **11:**108(24), 112(24), 200
Hansen, J. R., **12:**91(139), 161
Hansen, P., **15:**112, 137, 141, 180, 182, 197, 206, 216–217
Hansen, R. C., **16:**101, 141
Hansen, W. N., **4:**75(177, 178, 179, 180, 181), 95
Hanshew, D. D., **24:**27(126), 28(126), 29(126)
Hansma, P. K., **24:**84(23), 94(67), 95(67), 106(113)
Hanson, M., **7:**364(162), 396
Hanson, W. F., **9:**176(40), 256
Hansson, G. V., **23:**55(10)

Hantsche, H., **17:**332, 333, 366
Hantzehe, E., **14:**138(22), 195
Hantzpergue, J. J., **14:**7(24), 8(24), 74; **17:**111, 142
Haq, K. E., **3:**114(174), 129; **5:**167(126), 170, 180, 196(126), 232; **13:**61, 104
Haque, C. A., **4:**143, 187
Hara, H., **21:**21(42), 72
Hara, K., **3:**15(67), 51(67), 54
Hara, M., **24:**150(3)
Hara, T., **5:**219(201), 234, 282(171, 172), 318; **7:**163, 164(90); **18:**96–97
Harada, K., **16:**37–38, 130
Harada, M., **15:**154–155, 163, 214
Harada, R. T., **3:**281, 300
Haraki, H., **3:**225, 233, 241, 271
Harame, D. L., **23:**55(11)
Haranou, T., **17:**309, 365
Harbeke, G., **21:**121(18), 123(18), 232
Harber, D., **1:**189(8), 228
Harbison, J., **15:**229–231, 259, 262; **21:**184(81), 235
Harbison, J. P., **19:**2, 6, 9, 11–19, 21, 23, 26, 27, 32–37; **21:**241(15), 309
Harding, G. L., **10:**57(120), 68; **17:**313, 365; **18:**271, 272
Harding, W. E., **2:**413(195), 425; **4:**258(10), 319(10), 320
Hardwick, D. A., **17:**42, 43, 47, 75
Hardy, A. C., **4:**50(89), 75(174), 92, 95
Hardy, H. K., **7:**297(112), 310
Hardy, S. C., **4:**117(38), 118(38), 135
Harford, M. Z., **16:**53, 134
Hargis, P. J., Jr., **13:**190(157), 191(157), 197
Haring, H. E., **2:**393(135), 405(135), 423
Haring, R. A., **14:**15(56), 74
Hariu, T., **16:**251, 268
Hariz, A., **19:**12, 32, 35
Harker, A. B., **16:**28–29, 111, 116, 129, 142
Harkins, W. D., **7:**312, 315(8), 392
Harkness, A. C., **6:**4(16), 74
Harlow, F. E., **7:**90(65), 113
Harlow, J. D., **17:**313, 365
Harman, T. C., **3:**277(31, 32), 300; **11:**36(18), 96, 106(6, 7), 115(7), 128(78, 80), 130(85), 131(7, 85), 142, 146, 155(78), 166(6), 199, 201, 202
Harmer, A. L., **12:**70(56), 121(245), 131(245), 159, 164
Harmon, L. D., **21:**52(104), 60(110, 111), 74
Harms, H., **6:**249(156), 292(156), 298
Harper, J. M. E., **12:**282(237), 283(237), 295; **13:**109(4, 6), 147; **14:**9(33), 18(70, 72), 20(70), 23(89), 25(98), 26(98, 101), 61(197), 65(98–214), 67(98), 68(98), 74, 75, 78; **15:**148–149, 164, 214; **17:**153, 158, 220, 234, 296
Harper, M. E., **18:**245
Harper, W. J., **12:**76, 160
Harra, D. J., **14:**10–11(38), 74
Harrah, L. A., **13:**153(9, 11), 182(9, 11), 183(9, 11), 192
Harrick, N. H., **15:**273–275, 310
Harrick, N. J., **4:**75(182, 183, 184, 185), 76(189), 95
Harrington, D. B., **1:**44, 66
Harrington, D. J., **2:**367(29), 420
Harriott, L. R., **13:**176(84), 185(84), 186(84), 189(84), 195
Harris, E. P., **19:**264
Harris, G., **22:**30(35), 77; **23:**48(170)
Harris, J., **9:**159(17), 256; **10:**16(28),

25(28), 28(28), 43(28), 47(28), 66; **24:**184(35), 192(35)
Harris, J. E., **1:**5, 64; **7:**286(87, 88, 91), 309
Harris, J. H., **10:**238(28), 258
Harris, J. J., **11:**51(77), 57(77), 58(77), 59(77), 60(77), 65(77), 77(77), 88(77), 98
Harris, J. M., **5:**285, 319; **24:**74(120)
Harris, J. S., **7:**140, 141, 142(36), 248; **12:**137(309), 165
Harris, J. S., Jr., **21:**160(66), 234
Harris, Jr., J. S., **15:**229, 231, 239, 261–262; **16:**53–54, 134
Harris, L., **2:**208, 237, 315, 336, 359; **10:**59(129, 130), 68; **12:**6, 31, 50, 51
Harris, L. A., **1:**14, 64; **3:**132(6), 162
Harris, L. P., **14:**161(110), 197
Harris, M. T., **6:**4(17), 5(17), 7(17), 74; **14:**30(116), 76
Harris, M. W., **7:**333(75), 394
Harris, N. N., **5:**121, 141
Harris, R., **15:**185–186, 216–217
Harris, R. K., **24:**164(31), 168(31)
Harris, S. E., **17:**182, 222
Harrison, A. K., **24:**72(115)
Harrison, D. E., **3:**67, 73(46), 125; **14:**15(53), 74, 218(47), 238; **22:**119(14), 120(14), 127(14), 172
Harrison, D. H., **9:**132(41), 141(53), 142(53), 143(41, 53), 144; **10:**117, 165
Harrison, D. J., **24:**183(30), 192(30)
Harrison, G. R., **4:**50(94), 93
Harrison, L. G., **16:**237, 267
Harrison, M. J., **9:**192(90), 193, 257
Harrison, M. R., **16:**31–32, 129
Harrop, P. J., **6:**54(160), 63, 77
Harshavardhan, K. S., **13:**127(86), 149
Harsta, A., **16:**92, 139
Hart, C. W., **3:**198, 210
Hart, D., **16:**63, 112, 136, 142
Hart, D. L., **16:**71, 136; **17:**236, 237, 270, 271, 296
Hart, D. M., **3:**41(214), 58
Hart, E. W., **16:**237, 267
Hart, Jr., H. R., **16:**9, 125
Hart, P. B., **5:**153(38), 166(114), 176(114), 203(114), 209(38), 214, 230, 232
Hart, R. R., **14:**231(110), 239
Harte, K. J., **6:**235(119), 298
Hartel, E., **6:**36(115), 76
Hartemann, P., **15:**79–80, 84, 139–140
Härter, R., **24:**248(89, 90)
Hartl, M., **6:**13(53), 75
Hartley, N. E. W., **14:**217(42), 237
Hartline, H. K., **21:**62(115), 63(115), 74
Hartman, C. S., **17:**167, 221
Hartman, J. A., **6:**236, 237, 298
Hartman, R. E., **7:**395
Hartman, R. L., **11:**79(169), 101
Hartman, T. E., **4:**23(53), 91, 351, 371; **6:**64(191), 78; **7:**269, 270(62); **12:**272(177), 294
Hartmann, M., **15:**171, 197, 217
Hartmann, R. E., **2:**215, 216(19b), 238
Hartmann, R. S., **2:**216(19b), 238
Hartmann, W., **16:**51, 133
Hartnagel, H. L., **12:**277(193), 294
Hartnett, J. P., **10:**14(22), 65
Hartree, O. P., **1:**194(33), 229
Hartsough, L. D., **13:**190(155), 191(155), 197
Hartstein, A., **9:**240(232), 261
Harvey, G. A., **9:**14, 66
Harvey, J., **6:**12, 75; **14:**32(125), 76
Harvey, J. A., **11:**66(124), 99

Harwit, A., **15:**229, 231, 239, 261–262; **21:**160(66), 234
Harzallah, B., **24:**184(42), 188(42)
Hasegawa, F., **23:**166(90)
Hasegawa, H., **4:**340(47), 370; **12:**277(193), 294
Hasegawa, R., **6:**238(129), 239(129), 298
Hasegawa, T., **14:**20(80), 75; **16:**3, 8, 96, 124, 140
Hasenberg, T. C., **23:**234(20), 235(20)
Hash, M., **16:**4, 124
Hashiba, M., **12:**277(217), 295
Hashimoto, H., **1:**196(41), 229; **2:**132(152), 135, 137, 145; **4:**155, 189, 212, 253
Hashimoto, K., **17:**173, 222
Hashimoto, M., **5:**282(172), 318; **7:**163(88), 249
Hashimoto, N., **5:**307, 322
Hashimoto, S., **17:**334, 367
Hashimoto, Shin, **23:**161(64)
Hasimoto, K., **24:**231(51), 233(51)
Haskal, H. M., **10:**52(112), 53(112), 68
Haskins, C. P., **7:**368(175, 176), 396
Hasko, S. E., **7:**158(77), 249
Hasnain, G., **15:**229–230, 233–234, 238, 262; **21:**106(45), 107(45), 108(45), 112, 117(6), 118(14), 121(21), 123(21), 128(14), 133(14), 137(14, 48), 139(48), 141(14), 149(14), 150(14), 154(48), 158(65), 160(14, 48), 172(14), 179(48), 193(14), 194(21), 195(21)
Hasnian, G., **15:**51–52, 138
Hasper, A., **22:**215(75), 232(75), 274
Hass, G., **1:**54, 60, 61, 62(146), 67, 125(2), 135(25), 147, 148, 149(22, 25, 48, 50, 56), 150(22), 152(22), 153(25, 48, 50), 154(25, 53), 155(25), 157(25), 158(25), 172, 173(56), 175(25, 50, 53); **2:**26(50), 160, 161, 162), 177(50), 178(53), 180(70), 182(50, 53), 184, 184(50), 185, 186, 188(5, 6), 197(46), 199, 214, 220, 223(155, 156, 157, 158, 159), 224(56), 226(30), 228, 229, 231, 232, 238, 262(20), 263(21, 22), 265(24b), 284(22, 35, 36, 37, 39), 285(32, 34), 286(21), 289(20), 290(24b, 35), 294(22), 295(21, 22, 43), 296(21), 297(39), 303, 304, 307(1), 312(1), 324, 334(51, 52, 54), 336, 338(57, 58), 339, 340, 346(22, 51, 69, 70), 348(22), 353(66), 358, 359, 360, 401(167, 168), 402(167), 424; **3:**2, 9(38, 39), 10(47, 48, 49), 17(75), 51(47, 48, 49), 53, 54, 55; **4:**9(24), 10(33, 34), 39, 40, 50(106, 107), 60(34), 84(215), 91, 92, 93, 96, 282, 285(101), 286(99), 322; **5:**123(28), 141; **6:**158, 188, 202, 203; **7:**45(7), 47(24), 55(37a), 57, 58(24), 72, 73(63), 93(68), 94(69), 98(68), 111, 112, 113; **9:**13, 19(78), 66, 74, 100, 102(18), 113, 115, 129(6), 130, 131, 143, 144, 227(195), 260; **10:**22, 35(33), 54(33), 57(122), 66, 68, 72, 74(5), 76(11, 12), 77(14, 15, 16, 17, 18, 19, 20), 78(19), 79(19), 80(14, 19), 82(24), 83(18), 84(18), 86(27), 87(27), 88(27), 89(27), 92(30), 94(30), 95(31), 96(30), 97(31, 32), 98(32), 99(32), 100(33), 101(33), 102(33), 103(33), 104(33), 106(33), 107(34), 108(34), 109(34), 110(36),

111(36), 112(36), 113(36), 118(20), 119(50), 120(52a), 121(52a), 123, 124(5), 128(12, 24, 56, 58), 129(2, 58, 61, 63, 64, 65, 66), 130(56, 64), 131(66), 134(65), 135(24, 62, 65), 136(62, 65), 137(65), 138(65), 139(65), 141(66), 142(66), 143(66), 145(66), 146(66), 147(57, 64), 149(79), 150(79), 151(64, 65, 79), 152(56, 79), 153(2, 56, 61), 154(79), 156(84), 164, 165, 166, 262(3a, b), 308; **12:**2(1, 3, 4, 5, 7, 8, 9, 10, 11, 12), 3(11, 12, 14, 15), 4(16), 8(16, 31, 32), 10(37, 38, 39), 12(39), 15(44, 45), 16, 18(3, 47), 22(48), 23(48), 24(49), 25(50), 26(51, 52), 28(54), 29(56, 57, 58), 31(8, 10, 48,60), 33(11), 35(12), 37(48), 38(48), 40(44, 69), 41(47), 42, 44(47, 71), 48(37, 49), 49, 50, 51
Hasted, J. B., **11:**253, 255, 270(87), 316; **17:**92, 140
Hastings, J. M., **1:**205(80), 230
Hasunuma, H., **3:**229(54, 55), 230(54, 55), 234(55), 272
Hasuo, H., **12:**282(234), 295
Haszko, S. E., **12:**266(138), 268(138), 293
Hata, T., **17:**153, 220; **18:**249
Hatakeyama, I., **15:**171, 198, 216
Hatamian, M., **16:**107, 141
Hatcher, A. P., **7:**148, 149(57), 248
Hathcock, K. W., **24:**24(118), 26(118)
Hathorn, F. G. M., **14:**118(77), 127
Hatta, S., **16:**30, 129
Hatta, S.-I., **16:**33, 36, 130
Hattendorff, H. D., **9:**176(46), 177(46), 178(46), 179(46), 257
Hattori, M., **9:**182(63), 186(63), 257
Hattori, S., **14:**29(113), 76
Hattori, T., **16:**88, 138
Hatwar, T. K., **15:**171–172, 216
Hatzakis, M., **9:**264, 295
Hauchecome, G., **16:**76, 137
Hauert, R., **17:**229, 263, 298
Hauge, P. S., **19:**264
Haupfear, E. A., **22:**209(66), 273
Hauschild, H., **2:**223, 238
Hauschildt, D., **12:**296
Hauser, B., **16:**22, 111, 127
Hauser, J. J., **2:**367(30), 396(30), 420; **3:**87, 121(153, 154), 126, 128; **12:**246(21), 290
Hauser, V. E., **4:**316(163), 324; **5:**288(214), 320
Häussling, L., **24:**11(46), 21(46, 100), 25(121), 33(46), 34(46, 141), 35(141), 86(34), 221(89), 222(89), 228(31)
Hauth, W. F., Jr., **2:**386(102), 389(102), 422
Hautman, J., **21:**98(33), 112; **24:**75(124), 88(42, 43), 90(42, 43), 98(43), 139(65), 206(5, 6, 7, 8a, 8b), 209(5, 6, 7, 8), 210(5, 6, 7, 8), 212(5, 6, 7, 8), 213(5), 214(5), 215(5, 6), 216(5, 6), 217(5), 220(8)
Havinga, E., **7:**329(48), 334(48, 87), 359(87), 363(156), 393, 394, 396
Havriliak, S., **11:**229(23), 315
Hawkes, P. L., **7:**375(196), 397
Hawkins, R. E., **11:**249(53), 251(53), 315
Hawrylo, F. Z., **7:**137(34), 142(41), 143(50), 146(53), 157(34), 186(147), 187(150), 188(151), 209(150), 215(246), 218(202, 203, 246), 219(53), 220(150),

Hawrylo, F. Z., (*continued*) 221(213), 222(150), 230(234), 234(241, 242), 248, 251, 253, 254
Hawryluk, A. M., **13:**163(50, 51), 182(50, 51), 183(50), 193
Haya, H., **15:**281, 310
Hayakawa, H., **10:**256(57), 259; **16:**92, 139; **17:**153, 154, 161, 162, 219, 220, 221, 261, 263, 298
Hayakawa, S., **12:**274(184), 294; **14:**63(199), 78; **17:**153, 218, 219
Hayasaka, N., **13:**185(135), 187(135), 196
Hayashi, I., **7:**142(45), 214(198), 217(198, 200a), 234(240, 245), 248; **11:**58(99), 59(104), 99; **14:**65(208), 78
Hayashi, K., **15:**185, 216
Hayashi, M., **18:**16
Hayashi, N., **6:**208(16), 209(16, 17), 262(16), 263(16), 264(16), 265(16), 266(16), 272(173), 282(195), 283(195), 284, 286(16), 295, 300
Hayashi, Y., **17:**309, 365; **19:**116
Hayden, J. D., **2:**389(117), 390(117), 422
Hayes, J. R., **15:**222, 260
Hayes, R. E., **3:**6, 51(33), 54; **9:**5(17), 65
Hayes, T. R., **21:**240(6), 308
Hayman, C. C., **19:**61, 105
Haynes, D., **3:**19(117), 24(117), 26(117), 51(117), 56
Haynes, J. T., **17:**180, 181, 190, 222
Haynes, R. W., **13:**164(55), 185(55), 189(55), 194
Haywood, C. T., **7:**265, 269, 270(63), 286(50), 307, 308; **16:**240, 254, 267–268
Haywood, W. H., **14:**10–11(38), 74
Hazelwood, F. J., **12:**275(182, 183), 294
Heaney, J. B., **10:**57(122), 68, 125, 126(54b), 128(56), 129(65, 66), 130(56), 131(66), 134(65), 135(65), 136(65), 137(65), 138(65), 139(65), 141(66), 142(66, 77), 143(66), 145(66), 146(66), 149(79), 150(79), 151(65, 79); **12:**28(54), 29(57, 58), 31(60), 152(56, 79), 153(56), 154(79, 81, 82), 157(81, 82), 162(94), 165, 166
Heany, J. B., **7:**95, 98(70), 113
Heathcote, V. A., **1:**33(81), 65
Heaton, L., **1:**204(74), 230; **7:**27(109), 39
Heavens, O. S., **1:**71, 120, 121, 129(9), 184, 218(136), 231, 330(110), 334; **2:**236(42), 238, 242(11, 11a), 304; **3:**2(10, 11, 19), 17(10), 41(214), 51(10), 53, 59, 235(60), 236, 246, 272; **4:**21, 22(48), 25, 26(48), 77(195), 78(196, 198), 86(238), 91, 95, 96; **5:**49(2), 85, 163(77), 171, 231; **6:**27, 76, 152(3), 202, 312, 346; **7:**365(106), 396; **9:**16(74, 76)
Hebard, A. F., **13:**144(134), 150
Hebb, H. M., **12:**64, 159
Hebb, M., **5:**9(15), 45
Hecht, H. G., **10:**210, 226
Hecht, R. J., **2:**397(153), 406(153), 424; **3:**107(102), 127
Hecq, A., **14:**16(63), 74
Hecq, M., **9:**50, 56(264), 58, 62(264), 70, 71; **14:**16(63), 74
Hedgecock, N. E., **9:**6(24), 65
Hedin, L., **21:**98(32), 112
Hedman, J., **7:**45, 48(16), 111
Hefft, K., **3:**11(51), 49(51), 51(51), 54

Hegde, H., **16:**149, 159–161, 164, 168, 177, 180, 186, 189, 192–193, 203–205, 214–216, 220–225, 229–231
Hegde, M. S., **16:**63, 106, 111, 135, 141–142
Hegner, M., **24:**6(23), 106(109, 114)
Hegyl, I. J., **3:**196(59), 209; **5:**283, 319; **7:**232(236a), 254
Hehenkamp, Th., **7:**262(19), 306
Heiblum, M., **12:**282(237), 283(237), 295; **13:**212(46), 217(46), 220(70), 224, 225; **15:**222, 260; **21:**240(5), 308
Heide, H. G., **2:**130(145, 146), 144
Heidemann, W. A., **17:**253, 289, 297
Heiden, C., **16:**23, 121, 127, 143
Heidenreich, R. D., **1:**1(2), 63, 191, 228; **2:**100, 107(77), 108(66), 109(81), 112, 118, 119(113), 122(66), 142, 143; **4:**171, 190, 225(128), 254; **7:**23(81), 39
Heidrich, P. F., **10:**235(11), 236(17), 258
Heil, O., **2:**149(14), 190
Heiland, W., **13:**117(40), 148; **14:**218(48), 238
Heilman, R. K., **24:**181(14)
Heilner, E. J., **9:**288, 295
Heim, R. C., **15:**287, 310
Heim, U., **7:**170(108, 109), 250
Heiman, D., **21:**155(61), 234
Heiman, F. P., **2:**140(10, 12), 162, 165(12), 190; **4:**300(122), 323; **5:**146(14), 229
Heiman, N., **15:**147, 180, 182, 214
Heiman, W., **12:**119(234), 164
Heime, K., **13:**212(40), 224
Heine, V., **6:**82(34), 144; **12:**105(178), 162
Heinen, V. O., **16:**5, 76, 107, 124, 137, 141
Heinrich, B., **17:**328, 366
Heinrich, H., **11:**130(87), 145, 201, 202; **15:**227, 244, 261
Heinrich, J. P., **16:**193, 230
Heinz, D. M., **9:**276, 294
Heisenberg, W., **15:**125, 141
Heisig, U., **13:**67(32), 105; **14:**11(39), 34(133), 35(137), 36(138), 38–39(137), 44(39), 48(162), 51(138), 55(163, 173), 74, 76, 77, 204(13–16), 233(13), 237
Heising, R. A., **3:**21(98), 51(98), 55
Heising, U., **17:**111, 116, 142
Heisinger, L., **1:**224(165), 232
Heitman, W., **14:**26(104), 75
Heitmann, D., **9:**179(56), 180(58), 221(189), 223(189a), 257, 260; **21:**96(27), 111, 155(64), 234
Heitmann, W., **12:**28, 35, 36(62), 48(78), 50, 51, 73, 74, 76(81), 77; **13:**61, 104, 111, 112, 130, 133, 147
Helbig, W., **24:**251(97)
Heleg-Shabtai, V., **24:**19(81), 20(81, 89)
Helg, A. G., **24:**230(44)
Hellenthal, W., **2:**30(42, 45), 36, 61
Heller, A., **24:**19(81), 20(81, 89)
Heller, J., **14:**10(36), 74; **17:**111, 142
Hellman, E. S., **16:**54, 134
Hellmer, R. P., **16:**54, 134
Helm, C. A., **24:**180(11)
Helm, M., **15:**259, 264; **21:**241(15, 16), 309; **23:**238(43, 44), 239(43), 243(43, 44), 244(43, 44), 251(44), 292(114)
Helmersson, U., **16:**100, 140; **17:**7, 63, 65, 73, 122, 136, 143, 144
Helms, C. R., **12:**74(84), 75(84), 81(84), 160

Helwig, G., **1:**224(166), 232; **3:**156, 164; **9:**50(231), 70
Hembach, R. J., **10:**22(34), 66
Hemingway, D. J., **10:**169, 226
Hemmerdinger, L. H., **10:**22(34), 66
Hemminger, J. C., **24:**228(30)
Hempstead, C. F., **16:**50, 132
Hems, J., **22:**119(13), 120(13), 161(13), 171
Hench, L. L., **7:**24, 25(94), 39
Henck, S. A., **19:**45
Henderson, D., **7:**24, 25(96), 39; **14:**59(189), 77; **22:**5(8), 77, 123(25, 26), 172
Henderson, G., **4:**85(225), 86(225), 96
Henderson, G. A., **7:**225(224), 253
Henderson, T., **23:**255(77)
Hendriks-Van Itterbeek, W., **6:**127(192), 148
Hendus, H., **7:**27(103), 39
Henein, G. H., **17:**43, 49, 54, 75, 76
Heneman, D., **15:**283, 310
Henini, M., **15:**249–250, 263
Henisch, H. K., **3:**197(64, 65), 210; **5:**145, 219(3), 229
Henke, B. L., **7:**342, 395
Henke, S., **16:**72–73, 137
Henneberger, J., **14:**35(137), 38–39(137), 51(138), 76
Hennes, J., **1:**131(16), 185
Hennessy, P., **15:**54–55, 138
Henning, C. A. O., **17:**60, 77
Henninger, E. H., **7:**27(109), 39
Henninger, J. H., **10:**132(72), 166
Henon, S., **19:**281
Henrich, V. E., **2:**150(25), 157(40), 173(40), 191; **5:**219(200), 234
Henry, C. H., **7:**162, 211(85), 249; **11:**84(190), 101; **21:**312(2), 315(2), 319(2), 321(2), 322(2), 351(2), 366
Henry, D., **14:**98(40), 126; **18:**40
Henry, J. E., **21:**193(91), 235
Henry, R. A., **17:**318, 366
Henry, R. L., **16:**85, 138
Henry, W. M., **12:**157(336), 166
Henschke, E. B., **3:**65, 66, 125
Hensel, B., **16:**72–73, 137
Hensel, J. C., **4:**340(47), 370
Hensler, D. H., **10:**236(16), 258
Hentzell, H. T. G., **14:**25(98), 26(98), 65(98), 67(98), 68(98), 75; **17:**110, 121, 135, 141, 143, 144, 153, 158, 220
Heraga, K., **16:**149, 229
Herak, T. V., **18:**218
Herb, R. G., **1:**16, 18(38, 39), 64
Herd, S. R., **9:**4(14), 5(14), 65, 275, 285, 295; **15:**161, 167, 215
Herickhoff, R. J., **9:**176(40, 45), 256, 257
Herickoff, R. J., **2:**124(132e), 144
Hering, R., **10:**169(15), 225
Heritage, M. B., **7:**11, 18, 19(25), 23(78), 34(137), 37, 39, 40
Herkert, R., **11:**112(46), 200
Herman, E, S., **9:**6, 7(30), 65
Herman, F., **7:**24, 25(96), 39
Hermann, A. M., **16:**2, 9, 13, 81, 123, 125, 138
Hermann, W. C., **13:**112(23), 121(65), 123(65), 126, 127(80), 130(94), 131(65), 132(65), 133(94), 134(65, 94), 135(94), 137(23), 138(23), 139(23), 142(23), 144(23, 82), 145(65), 147, 148, 149
Hermans, J., **24:**220(85)
Hermansen, A., **3:**16(69, 73), 17(73), 49(73), 51(69, 73), 54; **9:**74, 143
Hernandez, J. P., **12:**59(24), 109(179), 111(179), 112(179), 159, 162

Herndon, T. O., **1:**319(75), 333
Heroux, L., **1:**130(11a), 184
Herpin, A., **1:**87(37), 91, 121; **5:**75, 78(24), 86
Herr, B. R., **24:**23(110), 27(110), 112(24), 129(24)
Herre, F., **7:**24(89), 27(106), 39
Herrick, C. S., **5:**247(22, 23), 315; **13:**60, 104
Herring, C., **4:**331(26), 341, 370
Herriott, D. R., **3:**36, 57; **12:**7, 50
Herrmann, K., **16:**121, 143
Herrmann, K. H., **11:**147, 202
Herrmann, R. I., **19:**195
Herron, N., **17:**337, 367
Herschitz, R., **16:**50, 132
Hershinger, L. W., **5:**164(87), 231
Hershkowitz, N., **18:**80
Hertle, H., **21:**78(4), 111
Hertz, H., **12:**54, 158
Herzig, Ch., **16:**254, 269
Herzig, H., **10:**57(122), 68, 125(54b), 126(54b), 142(77), 165, 166; **12:**14(41), 50
Herzog, A. H., **7:**127, 131(17), 169(95), 209(185a, 238), 247, 250, 252, 254
Herzog, H.-J., **23:**2(1)
Hess, D. W., **13:**68(44), 105; **15:**294, 311
Hess, H. F., **16:**52, 133
Hess, K., **15:**222, 261
Hess, N., **16:**50, 132
Hess, R. B., **4:**58(128), 94
Hesseborn, H., **14:**220(71), 238
Hester, W. A., **17:**153, 154, 156, 197, 220, 222
Hetherington, G., **4:**266(67), 321
Heumann, F. K., **12:**265(113), 292
Heuvel, H. J., **14:**174(125), 198
Hewett, W. A., **24:**9(30)
Hewitt, B. S., **9:**276, 283, 284, 288, 292, 294, 295; **11:**42(46), 50(46), 75(46), 76(46), 97; **15:**13, 137
Hewitt, L. R., **14:**219(57), 238
Hewitt, S. J., **3:**246(79), 272
Hexter, R. M., **15:**266, 309
Heyman, P. M., **17:**350, 352, 367
Hezel, R., **19:**149
Hhner, G., **24:**21(98), 22(98)
Hibbard, R. R., **10:**14(20, 25), 65, 66
Hibbs, A. R., **2:**312(5), 358
Hibi, T., **2:**109(84), 143
Hibiya, T., **15:**11, 18, 113, 137, 141
Hichwa, B. P., **17:**313, 365
Hicinbothem, W., **7:**170(104), 250
Hickam, W. M., **1:**34(84), 66; **5:**302(259), 321
Hickernell, F. S., **14:**32(127), 76; **17:**150, 153, 154, 172, 173, 220–222
Hickey, J. R., **10:**5(5), 65
Hicklin, W. H., **2:**368(31, 32), 393(32), 420; **3:**96, 108, 109, 126; **11:**77(158), 100
Hickman, J., **17:**315, 365
Hickman, J. J., **24:**21(103), 23(103, 113), 26(103), 27(103, 113), 30(113), 240(74)
Hickmott, T. W., **1:**25(59), 49, 65; **3:**132(5), 162; **6:**64(180), 66, 78
Hicks, H. G. B., **7:**153, 154, 236(64, 249d), 249, 254
Hicks, R. F., **15:**308, 312
Hidber, H. R., **24:**98(85), 251(97)
Hiehaus, W. C., **11:**88(223), 102
Hiesinger, L., **3:**110, 124, 127; **14:**175(126), 198
Hietala, V. M., **16:**123, 143
Hietel, B., **6:**202, 204
Higaki, K., **16:**32–33, 129
Higashi, N., **24:**34(142), 35(142)

Higashino, H., **16:**111, 142
Higgins, J. A., **11:**107(18), 171(18), 200
Higgins, T., **15:**54, 138
Higuma, Y., **14:**31(122), 76
Hijiya, S., **12:**292, 293
Hikichi, K., **11:**253, 255
Hikita, M., **17:**217, 224
Hilber, W., **21:**241(16), 309
Hildebrand, P. B., **10:**207(39), 226
Hildebrandt, A. F., **10:**25(42), 66
Hildenbrand, D., **16:**84, 90, 138
Hill, C., **13:**158(33), 169(33), 193; **15:**249, 263
Hill, C. G., **22:**181(23), 272
Hill, D. E., **7:**142(43), 169(95), 248, 250
Hill, D. N., **16:**92, 139
Hill, F. A., **15:**154, 214
Hill, G., **15:**249–250, 263; **21:**348(38), 355(38), 356(38), 367
Hill, G. N., **11:**77(160), 100
Hill, J. E., **12:**168(3), 169(3), 232
Hill, P. S., **16:**302, 339
Hill, R. M., **4:**86(235), 96; **6:**64(178), 78, 310(10), 311(10), 313(22), 346; **9:**6(27, 28), 65; **11:**206(1), 207(1), 209(5), 251, 253, 256, 258, 259, 278(5), 290(72), 291, 305, 306(143), 314, 315, 316, 317
Hille, P. F., **6:**92(71), 121(71), 123(71), 136(71), 145
Hilliard, J. E., **6:**335(41), 336(42), 347; **17:**2, 10, 24, 30, 35, 36, 37, 38, 40, 41, 43, 47, 48, 49, 50, 51, 54, 62, 73–75
Hillier, J. H., **2:**110(87), 125, 143, 144
Hilsch, R., **1:**198, 202(54), 227(54), 229, 243, 244(26), 274; **2:**38, 40, 43, 44, 45, 61; **6:**113(134), 146; **7:**21(50), 38
Hilsum, C., **7:**174(122), 193(161), 240(254c), 250, 252, 255
Hilton, A. R., **7:**34(140), 36(164), 40
Hilton, G. C., **16:**99, 140
Himes, R. C., **5:**261(58), 315
Himes, W., **2:**366(19), 411(19), 420
Himpsel, F. J., **12:**114(196), 163
Hindley, N. K., **15:**191, 217
Hinds, K., **24:**130(61)
Hines, R. L., **3:**76, 102, 126
Hink, W., **3:**39(186), 52(186), 58
Hinneberg, H. J., **12:**257(65), 291; **13:**99(108), 107, 111(20), 143(20), 147
Hinnen, C., **24:**45(82)
Hino, I., **13:**203(16), 207(16), 208(16), 223
Hinson, D. C., **4:**67, 94
Hintenberger, H., **3:**281, 300; **11:**106(1), 199
Hinteregger, H. E., **7:**51(33), 112; **10:**24(39), 66
Hinterlegger, H. E., **1:**130(11a), 137, 184, 185
Hintermann, H. E., **4:**238, 255; **17:**109, 130, 141, 143
Hinton, R. W., **4:**228(141), 254
Hipple, J. A., **1:**41, 42, 43(113), 66
Hiraga, K., **16:**88, 139
Hirai, T., **16:**30, 84, 86, 88–92, 111, 116, 129, 138–139, 142; **17:**234, 270, 296
Hiraki, A., **13:**205(22), 223; **15:**281, 310
Hirano, S., **17:**241, 297
Hirao, T., **16:**17, 19, 126; **18:**209, 213, 215
Hirata, K., **17:**229, 236, 287, 290, 296
Hiratsuka, H., **12:**246(22), 248(40), 251(40), 254(22), 290
Hirayama, H., **23:**55(9)

Hirayama, M., **13:**99(103), 107
Hirita, K., **16:**25, 51, 73, 127, 132
Hirobe, K., **14:**232(130), 239
Hirochi, K., **14:**20(78), 75; **16:**2, 17–19, 30, 33–34, 36, 124, 126, 129–130
Hirofuji, Y., **11:**67(133), 100
Hirohashi, R., **17:**361, 362, 370
Hirohava, Y., **12:**277(217), 295
Hirose, F., **23:**30(116), 32(116)
Hirose, M., **11:**89(231), 103; **12:**294; **13:**164(54), 188(54), 190(54, 159), 191(54, 159), 193, 197
Hirsch, A. A., **6:**207(12, 13, 14), 223(76a), 273(12), 277, 278, 295, 296, 299; **9:**5(19), 65
Hirsch, D., **16:**79, 137
Hirsch, E. H., **13:**3(5), 56, 117(42), 118(42), 119, 144(42, 135), 148, 150
Hirsch, J., **21:**54(106), 59(106), 74
Hirsch, P. B., **1:**227, 232; **2:**14, 60, 111(101), 113(101), 114(101), 118, 121(101), 143; **4:**142(21), 182(21), 187, 225, 226(130), 254; **6:**228(105), 297
Hirschberg, K., **5:**2(11), 11(11), 12(11), 45
Hirschelder, J. D., **13:**8(17), 57
Hirschhorn, J., **3:**254(95), 255(95), 256, 257(95), 258(95), 263(95), 264(95), 265(95), 273
Hirschhorn, J. S., **6:**116(153), 147
Hirschlaff, E., **1:**265(82), 275
Hirschwald, W., **7:**263(28), 306; **22:**20(27), 77
Hirshorn, H., **4:**371
Hirth, J. P., **4:**97(3, 4, 5, 6, 8), 99(13), 100(4, 5, 6, 8), 102(3), 103, 104(20), 106(3, 6, 20), 107(3), 108(3, 27), 109(3, 4, 6), 110, 113, 114(3 3), 117(5), 119(3, 48), 120, 121, 122(58), 123(3, 27), 124(5); **16:**4, 124; **17:**6678), 125(27, 58), 127(27), 131(3, 6), 133(4), 134, 135, 136, 137(3), 150, 151, 152, 154(63a), 155(3, 63a), 159, 160, 164, 184, 187, 188, 189, 190, 193, 251; **22:**254(102), 275
Hiruta, Y., **17:**339, 344, 367
Hirvonen, J. P., **13:**142(127), 150
Hisano, K., **15:**277, 310
Hisatsugu, T., **11:**70(141), 71(141), 72(141), 100
Hitchon, W. N. G., **22:**178(14), 271
Hittorf, W., **18:**52
Hiwatari, Y., **22:**145(68), 173
Hiyakawa, H., **11:**67(130a), 99
Hiyaniizu, S., **11:**50(72), 70, 71(141), 72, 98, 100
Hiziya, K., **2:**132, 145
Hjortsberg, A., **13:**121(56), 148
Hlady, V., **24:**183(33), 184(33)
Hmiel, A. F., **14:**55(171), 77; **17:**116, 142
Ho, C., **17:**326, 366
Ho, F., **13:**155(28), 183(28), 184(28), 192
Ho, J. C., **16:**79, 137
Ho, K. M., **24:**122(47)
Ho, K.-C., **17:**359, 361, 369
Ho, M., **24:**24(117), 26(117), 27(117)
Ho, P., **21:**190(86), 191(88), 235, 290(46), 310; **22:**186(32), 209(32), 272
Ho, P. S., **7:**262, 263(22, 30), 274(70), 282(22), 283(70), 306, 307, 308; **16:**234, 236, 238, 241, 248, 250, 253–254, 266–268
Ho, V. A., **14:**107(57), 126
Ho, V. Q., **12:**282(245), 295; **14:**109(62), 126

Hoai, T. X., **11:**147, 202
Hoang, H. H., **16:**241, 253, 267
Hoar, T. P., **14:**230(106), 239
Hoatson, G. L., **24:**153(13)
Hobel, M., **16:**23, 127
Hobson, J. P., **1:**21, 30(66), 65
Hobson, W. S., **21:**128(30), 129(30), 130(30), 131(30), 132(30), 135(30), 136(30), 137(30), 138(30), 139(30), 140(30), 141(30), 142(30), 143(30), 144(30), 145(30), 146(30), 147(30), 148(30), 149(30), 150(30), 160(30), 233
Hockett, L. A., **24:**44(51)
Hockings, E. F., **7:**174(118b), 250
Hockney, R. W., **22:**132(44), 172
Hodge, I. M., **11:**256(59), 257(59), 315
Hodgkinson, I., **19:**281
Hodgson, R. T., **13:**185(127), 187(127), 190(127), 191(127), 196
Hodul, D., **22:**45(52), 46(53), 49(53), 76(72), 78–79
Hoekstra, B., **15:**9, 136
Hoekstra, J. P., **1:**37(102), 38, 66; **3:**153(76), 164
Hoene, E. L., **12:**72, 73(75, 81), 74(81), 76(81), 77(81), 83(81), 84(81), 89, 99, 119(234), 160, 161, 164
Hoene, J. V., **5:**302(259), 321
Hoener, C. F., **22:**227(80), 274
Hoenig, R. E., **11:**61(108), 99
Hoenig, S. A., **9:**45(203), 69
Hoerni, J. A., **4:**300(119), 323
Hoesler, W., **17:**291, 299
Hofer, E., **4:**219(105, 106), 228(105, 106), 230, 253
Hofer, R., **24:**98(85)
Hoffman, C. A., **23:**279(22, 36, 102)
Hoffman, D., **13:**63(26), 104, 112(25), 120, 147
Hoffman, D. G., **4:**260(18), 301(18), 320
Hoffman, D. M., **2:**384(96, 97), 385(97), 386(97), 422; **9:**43(193), 44, 45(201), 46(201), 47(197), 48(197), 49(197), 69; **14:**217(36), 237
Hoffman, D. W., **14:**65(211, 212), 65(213), 68(211), 78; **22:**46(54), 78, 119(8, 9), 120(18, 19), 121(18, 19), 122(18), 151(18), 154(8, 19), 171
Hoffman, F., **6:**274(183a), 299
Hoffman, H., **1:**325(99), 334; **2:**30(48), 61
Hoffman, H. J., **13:**176(85, 87), 185(85, 87), 187(85), 188(87), 189(87), 195
Hoffman, R. A., **6:**141, 142, 149
Hoffman, R. G., **3:**19(91), 51(91), 55
Hoffman, R. W., **1:**58, 59, 67, 226(175, 176, 177, 178, 179, 180), 232, 280(6), 282(17, 18), 283, 289(6, 20), 324(97), 325(97), 326, 331, 332, 334; **2:**30(37), 61; **3:**19, 51(88), 55, 213(10), 214(10), 217, 220(29, 32, 34, 41, 63), 221, 223, 225(44), 228(41), 229(32, 41, 53, 63), 230, 234(32), 240, 241, 243, 244(67), 245(32, 73, 74), 247, 250(41, 67); **4:**230(142), 254; **6:**116(152), 147; **16:**317, 339; **17:**126
Hoffmann, C. L., **24:**15(65)
Hoffmann, I., **9:**190(81), 195(101), 197(101), 257, 258
Hoffmann, K. W., **9:**218(172), 259
Hoffmann, R., **12:**48(76), 51
Hoffmann, W., **5:**106(17), 141

Hofman, H. H., **12:**86(114), 161
Hofmann, H., **5:**12(18), 45
Hofmann, S., **17:**108, 135, 141, 144
Hofstein, S. R., **2:**149(10, 12), 161, 165(12), 190; **4:**300(122), 303, 305, 323
Hogaboom, G. B., **4:**192, 251
Hogarth, C. A., **15:**288, 290, 311
Hoh, J. H., **24:**106(173)
Hohkawa, K., **16:**41, 131
Höhler, A., **16:**23, 127
Holmke, D. K., **11:**108(25), 109(25, 29), 115(25, 65), 119(25), 120(25); **14:**12(45), 74; **17:**111, 112, 142
Holber, W., **13:**159(35), 174(35, 79, 81), 175, 185(35, 79, 81, 133, 134, 141); **16:**71–72, 136
Holber, W. M., **18:**90, 124, 165, 169, 182, 187, 194
Holden, B. J., **6:**22, 75
Holden, J., **1:**35(85), 66; **4:**87(241), 88, 96
Holden, J. T., **3:**19(117), 24(117), 26(117), 51(117), 56
Holden, W. S., **11:**78(168), 101
Holder, G. N., **16:**28, 128
Holder, S. L., **16:**28, 128
Holiahan, J. R., **13:**190(155), 191(155), 197
Holister, G. S., **16:**319, 339
Hollahan, J. R., **12:**263, 264, 292; **14:**84(6), 85(6), 125
Holland, L., **1:**16, 18(37), 31(72), 35, 64, 65, 66, 127, 184, 207(100), 224, 230, 232; **2:**285(31), 304, 307(2), 333(2), 358, 365(1), 366(1), 367(1), 368(1), 384(1), 387(1), 390(1), 419; **3:**2(12), 11(55), 12(55), 13(55), 19(117), 24(117), 26(117), 32(12), 43(12), 44(12, 219), 49(55), 51(55, 117), 53, 54, 56, 59, 62, 76, 97(81), 100, 101, 103, 107, 110, 124; **4:**2, 9(23), 22(49), 91, 258
Holland, O. W., **16:**57, 60, 134
Hollands, E., **14:**27(108), 32(108), 75; **17:**111, 141
Hollands, K. G. T., **10:**57(123), 68
Holleck, H., **17:**101, 102, 103, 104, 105, 106, 109, 110, 141
Holley, C., **7:**351, 354(133, 139), 355(133, 139), 356(139), 358(149), 395, 396
Holliday, J. E., **2:**69, 71, 141
Hollins, R. A., **17:**318, 366
Hollis, M., **13:**203(15), 223
Hollohan, J. R., **12:**238(6), 239(12), 241(12), 290
Holloway, H., **5:**150, 154(41), 155, 156(41), 190, 191, 192, 204, 228(41), 230, 233; **7:**133(26), 247; **11:**107(12, 13, 14), 108(17), 109, 114(58), 115, 119(64), 120(31, 64), 121, 128(13), 130(14, 82, 86), 131(13, 82), 135(13, 99), 142, 151(122), 155(13), 158(131), 163(136), 164(137), 166(63, 139, 140), 173, 178
Holloway, P. H., **16:**23, 126
Holm, R., **4:**351, 371; **14:**230(105), 239
Holm, R. T., **16:**85, 138
Holman, R., **17:**260, 263, 298
Holmen, J. O., **1:**294(41), 332
Holmes, R., **14:**141(39), 142–143(39), 144(42), 145(63, 67), 146(63), 147(71), 148(71), 196
Holmes, R. E., **5:**313(299), 322
Holmquist, W. R., **1:**285(23),

Holmquist, W. R., (*continued*) 328(104), 332, 334; **6:**273(178), 299
Holmwood, R. A., **3:**92(181), 110(181), 129, 224(42a), 271
Holnyak, N., **11:**66(122), 99
Holonyak, N., **21:**287(39), 310
Holonyak, N., Jr., **5:**277, 318; **7:**234(238, 247, 248), 254
Holst, K., **9:**239(231), 241(231), 242(231), 261
Holst, P., **9:**276, 284, 297; **15:**14, 137
Holstein, T., **6:**154, 202
Holt, A. S., **7:**331(68), 394
Holt, D. B., **5:**164(91), 231; **7:**123, 124(11), 247
Holt, L., **7:**117, 368, 370, 373, 374(117), 395
Holtom, G. R., **24:**71(110)
Holtom, R., **12:**121(242, 245), 126(261), 128(261), 132(261), 135(245, 292), 164, 165
Holton, R., **12:**70(56), 114(202), 121(241), 159, 163, 164
Holverson, P. J., **22:**119(13), 120(13), 161(13), 171
Holwech, I., **6:**82(16, 19, 25), 113, 115, 124(139), 127, 128, 135(138, 139, 149), 143, 144, 146, 147
Holzl, J., **9:**218(172), 259
Holzmann, R. T., **12:**257(68, 69), 291
Hom-ma, Y., **14:**226(85), 238
Homak, L. A., **16:**107, 141
Homma, H., **16:**52, 133
Honcia, G., **9:**119(30), 144
Honda, K., **17:**361, 370
Honda, S., **15:**150, 154–155, 161, 162, 163, 195, 214–215
Hong, C. S., **15:**283, 310
Hong, H.-G., **24:**12(51, 52), 15(51, 52), 27(51), 29(51), 163(25), 164(25)
Hong, J. D., **13:**217(55), 224
Hong, J. P., **16:**121, 143
Hong, M., **15:**147, 150, 152, 154, 161–162, 169–171, 185–186, 214–216; **16:**2, 13, 17, 21, 28–29, 36–37, 51, 74, 110, 124–127, 129–130, 132; **21:**163(70), 164(70), 165(70), 166(70), 234; **23:**60(202)
Hong, S., **17:**49, 63, 77
Hong, S. C., **21:**207(108), 235
Hongbao, L., **16:**4, 124
Honick, K. R., **2:**389(116), 422
Honig, J. M., **3:**27(142), 28(142), 51(142), 56
Honig, R. E., **1:**14(31), 21, 35, 37(98), 64, 66; **2:**365(13), 380(13), 405(13), 420; **3:**69(31), 70, 73(31), 75(31), 76(31), 77(31), 78(31), 125; **9:**47(222), 70
Honjo, G., **1:**202, 229
Honke, D. K., **11:**36(21), 41(21), 96
Honma, N., **6:**272(173), 284(173), 299
Hood, D., **11:**81(185), 82(185), 101
Hoogendoorn, A., **3:**73(44), 75(44), 76(44), 77(44), 125
Hoogendoorn, C. J., **17:**352, 368
Hook, H. O., **1:**14(31), 21, 64
Hooper, W. W., **4:**298(115), 323; **7:**255 (257)
Hoover, W. G., **22:**123(24), 139(61), 172–173
Hoper, J. H., **6:**240(137), 249(137), 298
Hopf, E., **22:**85(6), 114
Hopfield, J. J., **11:**295(135), 317
Hopkins, B. E., **4:**37(66), 92
Hopkins, D. C., **1:**241(22), 274
Hopkins, G. P., **12:**133(283), 165
Hopkinson, H. R., **10:**115, 165

Hopwood, J., **18:**125, 194; **22:**32(40, 41), 78, 101(19, 20), 111(19, 20), 114(19, 20), 115, 242(89, 90), 248(89, 90), 249(89, 90), 274
Hor, P. H., **16:**2, 123
Horak, J., **7:**174(121), 250
Horan, D. M., **7:**101(74), 113
Horenstein, M. N., **15:**180, 186–188, 216
Hori, Y., **9:**47(218), 48(218), 70
Horiai, N., **15:**171, 216
Horibe, T., **13:**20(44), 24(44), 51(44), 57
Horie, M., **13:**20(39), 22(39), 40(39), 57, 143(131), 150
Horigome, S., **15:**285–286, 310
Horiike, Y., **13:**176(91), 177(91), 180, 185(91, 104, 135), 187(91, 104, 135), 195, 196
Horikoshi, H., **3:**224(43), 229(55), 230, 232, 233, 234, 271, 272
Horikoshi, Y., **19:**10, 12, 15, 19–22
Horioka, K., **13:**180(104), 185(104), 187(104), 195
Horita, H., **11:**68(138), 69(138), 100
Horiuchi, S., **7:**375, 379, 383(195), 397
Horl, E. M., **1:**202, 230
Horn, K., **9:**217(160), 259
Hornauer, D., **9:**241(236), 242, 243(239), 244, 254(252), 261
Hornbogen, E., **17:**65, 77
Horne, D. E., **9:**278, 297; **14:**7–8(25), 69(25), 74
Horne, J. C., **24:**71(108, 109)
Horne, R. W., **1:**227(183, 184), 232; **2:**14(11), 60
Hornreich, R. M., **6:**225, 291(88), 297
Hornung, J., **11:**111(39, 40), 119(40), 200
Horose, H., **13:**96(91), 106
Horsley, B., **9:**33(132), 35, 37, 68
Horstmann, C., **9:**252(247a), 261
Horstmann, R., **7:**278, 288(73), 308; **16:**250, 268
Horton, J. A., **4:**75(178), 95
Horton, W. J., **17:**212, 224
Horvath, E., **12:**177(63), 234
Horwitz, J. S., **16:**65, 78, 136–137
Horwitz, S. S., **17:**199, 200, 202, 203, 204, 205, 206, 207, 223, 224
Hosaka, T., **16:**333, 339
Hosemann, R., **2:**110, 143; **7:**2, 3(9), 36(162), 37, 41
Hosford, C. D., **17:**64, 77
Hoshi, Y., **14:**42–43(151), 77; **17:**88, 91, 140
Hoshino, S., **17:**197, 222
Hosokawa, N., **4:**218(103), 219(103), 232(103), 253; **12:**246(22), 254(22), 290; **14:**34(135), 76
Hosono, K., **13:**20(43), 23(43), 49(43), 57
Hoss, P. A., **5:**282, 318
Hostetler, M. J., **24:**151(7, 9), 154(7, 9), 156(9)
Hothersall, A. W., **4:**227(136), 254
Hottel, H. C., **10:**23(37), 43(86), 44(86), 66, 67
Hottier, F., **19:**31, 52, 56, 59, 196, 203, 280
Hou, M., **22:**32(42), 78
Houchens, A., **10:**169(15), 225
Houde, A. L., **3:**27(143), 51(143), 56
Houghton, D. C., **14:**8(26), 36(26), 74; **21:**81(12), 111; **23:**13(47)
Houle, F. A., **13:**171(76), 176(92, 93, 94, 95), 181(112), 184(76), 185(76, 92, 93, 94, 95), 187(76, 92, 93, 94, 95), 190(93), 191(76, 93), 194, 195
Houlton, R. J., **16:**37, 130

Housekeeper, W. G., **1:**37(97), 66
Houska, C. R., **16:**237, 267
Houston, B. B., **11:**114(60), 201
Houston, J. E., **9:**22(87), 23(87), 67; **14:**219(58), 220(58), 238; **24:**44(56), 94(64, 65), 221(88)
Houston, W. V., **1:**235, 273
Houtz, C. C., **2:**406(185), 424
Hovel, H. J., **5:**282, 319; **7:**246(280), 256; **12:**176(57), 234; **21:**318(9), 366
Hovis, W. A., Jr., **5:**85(31), 86
Howard, A. J., **23:**231(16)
Howard, J. K., **7:**273(69), 292, 302(101), 308, 310; **16:**238, 248, 250, 253–254, 267–268
Howard, L., **24:**94(71), 95(71)
Howard, P. J., **16:**31–32, 129
Howard, R., **16:**107, 141
Howard, R. E., **13:**153(14), 161(39), 182(14, 39), 183(14, 39), 192, 193; **16:**42, 74, 101, 110, 131, 137, 141–142
Howard, R. W., **5:**154(53), 230
Howard, W. E., **11:**36(3), 41(3), 47(3), 58(3), 79(3), 85(198), 89(3), 96, 102
Howarth, L. E., **4:**341(51), 343(51), 345(51), 349(51), 354(51); **12:**84(99), 160; **23:**146(16)
Howe, A. T., **17:**354, 368
Howe, R. T., **17:**210, 217, 223; **24:**237(61)
Howell, G. M., **12:**282(223), 295
Howell, J. R., **10:**11(17), 13(17), 36(17), 65
Howells, G. M., **6:**45(145), 48(145), 77
Hower, P. L., **7:**255
7.(257)
Howie, A., **2:**111(101), 113(101), 114(101), 119(116), 121(101), 136, 143, 145; **4:**142(21), 169(110), 182(21), 187, 225(130), 226(130), 254; **6:**228(105), 297
Howitt, D. O., **16:**28–29, 129
Howorth, J. R., **12:**70(56), 121(245), 126(261), 128(261), 132(261, 282), 133(283), 135(291, 292), 159, 164, 165
Howson, R. P., **5:**165(107), 174, 232; **14:**44(154, 155), 56(175), 62(175), 77; **17:**111, 116, 142; **18:**275
Hrbek, **14:**9(32), 74
Hren, J. J., **19:**241
Hruska, S. J., **4:**97(6), 100(6), 106(6), 108(26), 109(6, 28), 123, 124(28, 63a), 131(6), 134, 135, 136, 151(67a), 189
Hseih, S. J., **21:**103(39), 107(39), 112, 221(136), 237
Hshieh, J. J., **22:**161(81), 173
Hsia, S. T., **22:**54(58), 78
Hsiang, T. Y., **16:**107, 141
Hsiau, Z., **22:**270(132), 276
Hsieh, J. J., **10:**253(69), 259; **22:**117, 161(80), 173, 178(11), 188(38), 193(38), 211(70–72), 271–274
Hsieh, K. C., **21:**287(39), 310
Hsieh, K. H., **13:**203(15), 223
Hsieh, K. Y., **17:**238, 239, 297
Hsieh, S. J., **21:**118(14), 121(21), 123(21), 128(14), 133(14, 44), 137(14), 141(14), 149(14), 150(14), 160(14), 172(14), 193(14), 194(21), 195(21), 209(14), 232, 233, 244(21), 249(25), 261(21), 284(25), 309
Hsieh, T., **16:**74, 137
Hsu, J. W. P., **16:**42, 50, 131

Hsu, L. S., **13:**123(68, 69), 131, 132(69), 149
Hsu, N. J., **16:**91, 139
Hsu, R., **15:**302, 311
Hsueh, F. L., **23:**124(15)
Hsung, R. P., **24:**33(136), 128(57)
Hu, A., **17:**24, 49, 51, 53, 56, 75
Hu, B. H., **21:**121(19), 123(19), 232
Hu, C., **10:**238(30), 258
Hu, C. M., **24:**72(113)
Hu, H., **3:**217, 271; **19:**265
Hu, H. L., **9:**264, 276, 295
Hu, H.-K., **14:**7(23), 69(23), 73
Hu, J., **24:**221(93), 228(8)
Hu, M., **21:**128(37), 233
Hu, M. S., **16:**288–291, 333–334, 338–339
Hu, Q., **23:**241(57), 291(57)
Hu, R., **16:**28, 119, 121, 128, 143
Hu, S. M., **3:**137, 163; **4:**270(88), 297(88), 314, 316, 322, 324; **12:**265(124), 292, 293
Huang, C. I., **15:**244, 263
Huang, F. H., **9:**50(225), 70
Huang, J., **16:**91, 99, 139–140; **17:**45, 76
Huang, J. Y., **24:**228(30)
Huang, K., **2:**105(72), 142; **6:**34(97), 76
Huang, K. G., **24:**112(24), 127(56), 129(24)
Huang, K. H., **17:**22, 74
Huang, S., **24:**249(95)
Huang, T., **16:**21, 127
Huang, T. C., **16:**34–35, 130
Huang, Z. J., **16:**2, 123
Huang, Z. Q., **16:**55–56, 64, 134
Hubbard, C. R., **15:**156–157, 215
Hubbard, K., **16:**37–38, 130; **17:**47, 49, 53, 63, 76, 77
Hubbard, P., **21:**5(1), 71
Huber, A. M., **11:**49, 50(70), 98
Huber, D. L., **9:**275, 295
Huber, E. E., **1:**285(24), 286(28, 29), 304(54), 312, 332, 333
Huber, E. E., Jr., **3:**214(13), 250(87), 255(87), 270, 273
Huber, F., **2:**393(129), 406(186, 188), 423; **5:**287(208), 319; **6:**64(185), 78
Huber, W., **11:**142(111), 144, 145(115), 202
Huber, W. K., **1:**42(109), 66
Huberman, B., **21:**54(106), 59(106), 74
Huberman, M. C., **23:**14(50)
Huberman, M. L., **17:**57, 76
Hubner, H. J., **23:**254(75)
Hubsch, H., **12:**253(55), 254(55), 291
Huchital, D. A., **12:**118(225), 131(225), 163
Hudis, M., **12:**286, 287, 295
Hudock, B. F., **3:**109(110), 127
Hudock, P., **11:**111, 200
Hudock, P. F., **2:**393(134), 423
Hudson, J. B., **4:**123, 124(61), 127(60, 61), 136
Hudson, M. D., **17:**237, 238, 296
Huebener, R. P., **6:**82(10), 116(156), 143, 147
Huebner, D., **5:**164(96), 172(96), 231
Hues, S. M., **17:**49, 51, 76
Huet, P., **2:**30(41), 61; **6:**118(166), 147
Huff, E. A., **16:**28, 128
Huffman, J., **21:**5(6), 71
Huffman, J. E., **21:**36(70, 71), 73
Hug, H. J., **24:**98(85)
Huggins, R. A., **17:**318, 326, 359, 366, 369
Hughes, C. J., **24:**182(23), 199(23)
Hughes, D., **11:**256(63), 257(63), 315
Hughes, G., **23:**103(48)

Hughes, H. L., **14:**231(108), 239
Hughes, O. H., **15:**249–250, 263
Hugo, G., **21:**207(108), 235
Hugot-LeGoff, A., **17:**314, 315, 316, 329, 330, 348, 365, 367
Hui, C. Y., **24:**233(53)
Hui, S., **21:**94(25), 111, 150(55), 155(59), 158(59), 159(59), 181(55), 234
Huijer, P., **2:**387(107), 422
Huijser, A., **12:**114(205, 208), 163; **13:**201(2), 223
Huisman, B.-H., **24:**23(108)
Hulburt, E. O., **4:**60(132), 94
Hull, D., **16:**273, 337
Hull, G. W., **16:**18, 126
Hull, R., **21:**79(7), 111
Hull, R. W., **6:**36(101), 38(101), 45(101), 48, 49, 76
Hulliger, J., **17:**229, 263, 298
Hultgren, C. T., **16:**4, 124
Hultman, L., **17:**4, 5, 6, 10, 13, 15, 16, 17, 18, 19, 23, 24, 25, 27, 28, 31, 32, 34, 63, 64; **18:**269, 284
Hultquist, A. E., **12:**257, 291
Hum, D. M., **3:**277(23), 300
Hummel, R. E., **7:**270(65), 308; **16:**240–241, 245–246, 250–251, 253–255, 258–259, 261–263, 267–269
Humphrey, D. A., **21:**204(101), 205(101), 235
Humphrey, F. B., **6:**216(60), 238, 239, 240(133, 134), 268, 296, 298
Humphrey, J. N., **3:**289, 300
Humphreys, R. G., **16:**52, 133
Humphreys-Owen, S. P. F., **2:**199, 237; **4:**69, 94
Humphries, R. D., **2:**370(55), 421; **3:**139(43), 163; **5:**299(253, 254), 321
Humphris, R. R., **5:**163(66, 67), 168(66, 67), 188, 199(66, 67), 231
Hunderi, O., **9:**190(86), 257; **19:**60, 129, 139, 168, 180
Hung, L. S., **16:**97–98, 140; **23:**48(162)
Hung, S. H., **15:**185, 216
Hunn, J. D., **16:**28, 128
Hunsperger, R. G., **10:**248(40), 259
Hunt, B. D., **16:**41, 131
Hunt, G. H., **5:**85(32), 86
Hunt, J. D., **17:**197, 198, 222
Hunter, W. R., **1:**130, 134, 147(22, 48), 148(49), 149(22, 48, 49), 150(22), 152(22), 153(48), 169(47), 170(49), 173(56), 180(70), 184, 185, 186; **2:**227, 238, 285(34), 304, 324(29), 334(53), 359, 360; **3:**10(47), 51(47), 54; **4:**10(33, 34), 50(101), 60(34), 68(164), 69(167), 70, 72, 73, 74, 75, 91, 93, 94, 95; **7:**44(6), 45(7, 14), 47(14, 23, 24), 48(14, 27), 57, 58(24), 63(27), 70(14), 71(6), 72, 73(6, 63), 90(65), 103(6), 111, 112, 113
Huntington, H. B., **7:**258(4, 6), 261(4), 262(6, 9, 22), 263(22, 29, 31); **9:**50(225), 70; **16:**234, 238–239, 254, 265, 266–267, 269
Hurada, S., **14:**226(85), 238
Hurle, D. T. J., **7:**142(37), 248
Hurley, K. H., **15:**303, 311
Hurley, M. D., **11:**36(21, 22), 41(21), 96, 97, 115(65), 121(76); **14:**12(45), 74; **17:**111, 112, 142
Husinsky, W., **14:**9(31), 16(31), 74
Hussein, M. A., **18:**194
Husstad, B., **6:**97, 98, 128, 145
Hutcheson, E. T., **7:**64(45a), 112;

10:76(12), 77(18, 19), 78(19), 79(19), 80(19), 83(18); **12:**8, 50
Hutchinson, A. L., **23:**239(46), 240(54), 243(65), 262(65, 80, 83, 84, 85, 87), 263(65), 265(85), 266(46), 268(46), 269(92), 271(54), 272(54, 83, 96), 273(86), 274(87), 279(83), 291(80)
Hutchinson, J. W., **16:**288, 338
Hutchinson, T. E., **4:**156(82), 189
Hutchison, J. E., **24:**24(118, 119), 26(118, 119), 151(9), 154(9), 156(9)
Huth, B. G., **15:**187, 310
Hutley, M. C., **9:**227(197), 260; **10:**58(127), 68; **11:**9(3), 10(3), 19(3), 21, 23(17), 24(17), 25(17), 26(17), 27(17), 33, 34
Hwang, C. J., **7:**158(76, 77), 249
Hwang, D. M., **16:**45, 63, 132, 135
Hwang, H. J., **23:**261(81), 291(81)
Hwang, H. L., **15:**283, 310
Hwang, T., **15:**295–296, 311
Hyder, S. B., **12:**54(9), 130(276, 280), 136(9), 138(9), 139(9), 158, 165, 271, 293

I

Iafrate, G. J., **15:**222, 260–261; **21:**121(25), 233, 240(4), 241(7, 8, 10, 18), 245(18), 246(4, 24), 248(4), 252(4), 256(8, 10, 27, 28), 259(10), 260(8), 265(27), 267(28), 277(28), 278(4), 308, 309
Iampas, F. J., **12:**289(264), 296
Iams, H. A., **12:**154(329, 330), 166
Ianelli, J. M., **23:**161(64)
Ianno, N. J., **16:**80–81, 137–138
Ibach, H., **9:**215, 216, 217(156, 157, 160), 259
Ibaev, Yu. K., **6:**225(86), 297
Iball, J., **7:**354, 355(143), 395
Ibara, Y., **16:**98, 140
Ibbotson, D. E., **13:**68(46), 105; **15:**294–295, 311; **22:**211(70, 71), 212(73), 273–274
Ibers, J. A., **2:**83, 141
Ibl, N., **6:**23, 75
Ibrahim, M. M., **19:**280
Ibukuro, S., **10:**236(19), 258
Ichibara, M., **14:**105(54), 126
Ichihara, M., **15:**281, 310
Ichikawa, F., **15:**295, 311
Ichikawa, S., **17:**354, 368
Ichikawa, Y., **16:**2, 17, 19, 30, 33–34, 36, 124, 126, 129–130
Ichinokawa, T., **3:**39(211), 40(211), 52(211), 58
Igarashi, H., **16:**30, 32, 53, 129, 134
Igarashi, T., **15:**221, 260
Igarishi, Y., **14:**175(135), 198
Iglesias, I., **5:**298, 321
Ignacio, J., **22:**5(5), 77
Ignat'ev, A. A., **15:**9, 136
Ihantola, H. K. J., **2:**159, 161, 191
Ihashi, T., **15:**210–211, 213, 217
Ihmori, T., **3:**27(128), 51(128), 56
Iida, S., **9:**287, 295
Iijima, K., **16:**51, 132; **17:**228, 229, 232, 233, 236, 237, 242, 243, 245, 246, 247, 283, 287, 290; **19:**250
Iijima, T., **15:**171, 198, 216
Iizuka, S., **18:**91–92, 96
Ijiro, K., **24:**36(144)
Ikeda, M., **11:**67(133b), 100
Ikeda, T., **17:**345, 367
Ikeda, Y., **17:**354, 368
Ikegami, Y., **23:**228(32)
Ikonopisov, S., **6:**22, 75

Ikuzo, I., **9:**47(218), 48(218), 70
Ilegems, H., **7:**180(126, 129), 181(134), 183(126, 129), 184, 185(126), 201, 203(134), 204, 205(126), 251
Ilegems, M., **11:**53(89), 54(89), 55(89), 57(94), 65, 66(117, 120, 121), 79(171), 86(208, 209, 210, 210a, 211, 212), 91, 98, 99, 101, 102, 103
Iley, R., **5:**283, 319
Il'inskii, A. I., **3:**256(101), 257, 261(112), 264(101), 265(101, 112), 266, 267(112), 273
Ilinskii, A. I., **17:**3, 60, 73
Illegems, M., **10:**228(6), 251(6), 258
Imahori, H., **24:**33(134)
Imai, I., **9:**34(157), 35(157, 164), 36(164), 37(164), 68, 69
Imaishi, N., **22:**188(40), 272
Imamoto, H., **15:**244, 263
Imamura, N., **15:**171–172, 198–199, 207, 216
Imamura, S., **12:**95(154, 155), 96, 162
Imamura, T., **12:**282(234), 295
Imbrie, P., **16:**302, 339
Imura, T., **4:**218, 253; **14:**105(54), 126; **15:**281, 310; **16:**98, 140
Inaba, H., **17:**353, 368
Inagaki, M., **7:**23(79), 39
Inagawa, K., **13:**60(5), 64(28), 65(28), 68(28), 78(5), 79(5), 88(28); **17:**110, 141
Inam, A., **16:**55–56, 58, 60–61, 63, 106, 110–114, 134–136, 141–142; **17:**236, 237, 270, 271, 296, 298
Inamura, N., **15:**154, 174–175, 180–181, 214
Inamura, S., **14:**66(220), 78
Inata, T., **15:**222, 261
Ing, D. W., **5:**154(45), 160(45), 161(45), 230, 303(263), 321; **12:**269(150), 293; **17:**153, 154, 220
Ing, S. W., Jr., **2:**370(53), 421; **3:**138(34, 35), 141(53), 142(53), 144(53), 163; **4:**263(56), 264(56), 319(56), 321; **5:**279, 287, 291, 318, 319; **12:**259(87), 260, 261(88), 292
Inga, J. B., **15:**113, 141
Ingber, D. E., **24:**228(6), 229(6), 248(91), 249(6, 91, 94, 95)
Ingelstam, E., **3:**36(171, 173), 37, 52(171), 57
Ingersoll, K. A., **13:**126, 149; **14:**226(86), 238
Inghram, M. G., **17:**307, 364
Ingraham, A. P., **7:**276(72), 308; **16:**240, 267
Ingram, M. D., **11:**256(59), 257(59), 315
Ingrey, S. J., **14:**24(94), 64(205), 75, 78
Inkson, J. C., **12:**126(265, 266), 128(265, 266), 164
Inman, M. C., **4:**160(91a), 189
Inn, E. C., **2:**329, 330(38), 359
Inn, E. C. Y., **4:**68(161), 94
Innes, D., **7:**33(138), 40
Ino, S., **4:**150(49), 156(49, 81), 159(81), 160(81), 164(49), 165, 172, 175(49), 178(49), 188
Ino, T., **7:**22(74), 39
Inokana, H., **13:**112(31), 148
Inokawa, H., **13:**5(9), 20(33), 21(9, 33), 28(33), 56, 57, 116(35), 148
Inokuchi, H., **24:**27(128)
Inokuti, M., **19:**131, 133, 153, 155
Inoue, E., **17:**355, 356, 368, 369
Inoue, J., **7:**198(174), 252

Inoue, K., **13:**20(42), 23(42), 57; **15:**150, 195, 214
Inoue, M., **16:**92, 139, 251, 268
Inouye, T., **2:**93, 141
Ioffe, A. F., **6:**64(184), 78
Iogansen, L. V., **2:**195, 237; **6:**137(218), 148
Iordanis, P. C., **12:**275(182), 294
Iqbal, Z., **12:**238(2), 270(156), 289, 293; **16:**99, 140
Ireland, C. L. M., **15:**115, 118, 141
Irie, A., **16:**111, 116, 142
Irland, M. J., **3:**13, 51(57), 54
Irvan, R. S., **14:**235(152), 240
Irvine, T. F., Jr., **2:**330(41), 860; **10:**14(22), 65
Isaac, E. D., **6:**286, 287(206), 300
Isaac, M., **21:**162(69), 234
Isaacson, G. C., **14:**23(90), 75
Isaeva, R. V., **6:**103(97), 119, 132(97), 145
Isaki, L., **9:**217(158), 259
Isard, J. O., **14:**120(79), 127
Isebeck, K., **4:**150(52), 188
Ishak, W. S., **15:**86–87, 140
Ishibashi, T., **13:**44(53), 58
Ishida, A., **12:**282(235, 236), 295
Ishida, H., **15:**279, 310; **17:**361, 370
Ishida, K., **9:**218(173), 259; **14:**19(76), 75
Ishida, T., **13:**20(32), 21(32), 57
Ishida, Y., **11:**238(43), 239, 240, 247, 255, 315
Ishiguro, K., **9:**34(157), 35(157), 68
Ishii, M., **24:**99(104), 100(104), 101(104)
Ishikawa, H., **12:**293; **13:**212(47), 224; **14:**103(48), 126
Ishikawa, Y., **1:**205(81), 230
Ishitani, A., **14:**65(208), 78
Ishiyama, S., **13:**38(51), 58
Ishkin, A. I., **3:**27(116), 51(116), 56
Isin, A., **6:**103(101), 145
IslamRaja, M. M., **22:**182(25), 188(25), 193(25), 209(65), 272–273
Ismeurt, M., **13:**185(131), 187(131), 196
Ison, H. C. K., **7:**60, 65(41), 112
Israelachvili, J. N., **24:**2(5), 15(69), 17(69), 89(52), 92(52), 169(37), 206(2), 221(95), 222(95), 240(76)
Ito, A., **24:**119(41), 121(41)
Ito, H., **9:**49(224a), 70; **15:**152–153, 214
Ito, K., **2:**110(89), 143; **17:**345, 367
Ito, R., **23:**10–11(38)
Ito, T., **12:**292, 293; **14:**103(48), 126; **23:**47(158)
Itoh, A., **11:**67(130a), 99; **14:**11(43), 74; **15:**152, 214; **17:**110, 111, 141, 142, 153, 154, 219, 220
Itoh, H., **13:**99(101), 107
Itoh, S., **13:**20(40), 23(40), 45(40), 57
Itoi, Y., **16:**97, 140
Itozaki, H., **16:**32–33, 37–38, 129–130; **17:**43, 49, 51, 53, 75
Ittner, W. B., III, **1:**57(136), 67, 245(32), 257(69), 258(69), 262(76), 272(99), 274, 275; **3:**120, 128, 131(3), 162
Itzhak, D., **12:**272(170, 171), 274(170, 171), 294
Ivanov, R. D., **3:**117(141), 118, 128
Ivanov, Z. G., **16:**67, 136
Ivanova, M. K., **1:**148, 149(52), 185
Ivey, H. F., **3:**197(66), 203, 210
Iwade, S., **23:**105(55)
Iwai, N., **23:**228(32)
Iwaki, H., **17:**83, 128, 129, 139
Iwaki, M., **17:**82, 139
Iwakura, M., **24:**36(143)

Iwami, M., **13:**205(22), 223
Iwamura, S., **14:**18(74), 19(74, 82), 20(82), 30(74), 75
Iwanaga, M., **2:**135(158), 137(158), 145
Iwasa, K., **17:**352, 368
Iwasaki, H., **4:**260(23), 301(23), 320; **16:**86, 88–90, 92, 138–139
Iwata, M., **3:**15(67), 51(67), 54
Iwauchi, S., **12:**264(105), 292
Iyer, S. S., **16:**250, 268; **21:**84(17), 111
Izawa, A., **17:**355, 356, 368
Izotova, T. P., **6:**238(126, 127), 298

J

Jaccodine, R. J., **4:**169(115), 190
Jack, K. H., **4:**266(67), 321; **12:**87, 88, 161
Jack, M. A., **15:**57, 139
Jackel, L. D., **13:**161(39), 182(39), 183(39), 193
Jackman, R., **24:**229(37), 234(37), 235(37)
Jackman, R. J., **24:**235(55), 242(55)
Jackman, T. E., **23:**13(47)
Jackson, C. S., **4:**122, 124(58), 125(58), 136
Jackson, D. A., **12:**135(299), 165
Jackson, D. C., **17:**27, 31, 75
Jackson, D. M., Jr., **5:**154(43), 230
Jackson, N. F., **6:**36(120), 76
Jackson, P. A., **3:**207(85), 210
Jackson, T., **13:**208(26), 219(72), 220(69, 72), 221(72, 73), 223, 224, 225
Jackson, W. B., **15:**191, 217
Jackson, W. H., **2:**371(63), 393(143), 397(63), 421, 423
Jacob, H., **5:**247(26), 315
Jacobi, K., **12:**115(219), 122(248), 163, 164; **14:**98(39), 126
Jacobi, R., **5:**131(33), 141
Jacobs, E. E., **7:**331(68), 394
Jacobs, I. S., **2:**31(51), 61; **6:**206(5), 215(5, 47), 220, 222(47), 295, 296
Jacobs, J. E., **3:**198, 210
Jacobs, J. T., **6:**124(190), 127, 148; **19:**253
Jacobs, M. H., **3:**32(161), 57, 212(6), 213(6), 216(6), 218(6), 225(6), 243(6), 270, 271; **4:**118(45, 46), 119(45), 120(45), 132(75), 133, 135, 150(41, 62, 63), 155(41), 156(41), 165, 167, 168, 170, 171(102), 174, 175, 188, 189, 190, 235(155), 243(1), 243(155, 168)
Jacobs, P. M. W., **6:**13(56), 75
Jacobs, R. M., **5:**303(262), 321
Jacobs, S. D., **13:**138(116), 139(116), 150
Jacobson, A. J., **17:**320, 366
Jacobson, B., **13:**79(72, 76, 83), 83, 84, 89, 91, 92(83), 106
Jacobson, D. C., **16:**74, 137
Jacobsson, R., **3:**24(120), 26(120), 51(120), 56
Jacobus, G., **1:**149(64), 167(64), 170(64), 186
Jacobus, G. F., **2:**263(22), 284(22), 294(22), 295(22), 304; **10:**92, 94(31), 95(29, 31), 96(29, 31), 97(31), 164
Jacquemain, D., **24:**181(12), 192(12), 199(12, 50)
Jacquemin, J. L., **9:**22(93), 67
Jacques, A., **10:**242(32), 258
Jacquinot, P., **3:**13, 51(62), 54
Jadus, D. K., **5:**279, 318

Jaeger, F., **18:**33, 147
Jaffe, B., **19:**250
Jaffe, H., **19:**250
Jaffe, J. H., **4:**86(237), 96
Jaffe, M. S., **9:**59(267), 60(268), 71
Jagannath, C., **21:**207(109), 235
Jaggi, R., **1:**248(46), 251(46), 274, 329(108), 334
Jahnke, E., **3:**44(221), 59
Jahoda, F. C., **1:**143, 185
Jain, A. L., **16:**28, 128
Jain, J. D., **16:**98, 140
Jain, K., **13:**159(36), 160(36), 181(109), 182(109, 115), 183(109, 115), 193, 195
Jain, M. K., **22:**183(28, 29), 185(28, 29), 215(75), 232(28, 29, 75), 265(110, 111), 266(112, 113), 269(113), 270(121), 272, 274–275
Jain, S. C., **21:**10(30), 12(30), 15(30), 72; **23:**3(14), 5(14), 145–146(42), 146(45)
Jain, V., **24:**96(77), 98(77), 106(108)
Jaininger, J., **6:**47(143), 77
Jake, R. A., **16:**92, 139
Jaklevic, R. C., **6:**36(107), 76
Jakob, M., **2:**332(49), 360; **10:**155(83), 156(83), 166
Jalfe, B., **17:**243
Jalochowski, M., **17:**15, 74
Jamal, T., **13:**102, 107
James, B. D., **3:**113(128), 127
James, B. W., **22:**146(69), 173
James, C., **6:**116(164), 117(164), 132(164), 147
James, E. A., **14:**223(78), 238
James, G. H., **16:**302, 339
James, H. M., **3:**294(61), 300
James, J. H., **16:**26–28, 127–128
James, L. W., **7:**196, 197, 198(170), 243(265, 267), 244(270), 252, 255; **12:**76(92), 77, 81(92), 114(191, 192, 194), 118(92, 231), 123(251), 124(252), 125(257), 126(263, 264), 128, 130(277), 137(308), 138(308), 160, 162, 163, 164, 165
James, R. B., **21:**96(28), 112, 155(63), 234
James, T. W., **16:**100, 106, 141
Jan, Q., **17:**22, 74
Jan, W., **21:**287(41), 310
Janacek, T., **22:**18(26), 34(26), 35(26), 45(26, 51), 76(76), 77–79
Janak, J. F., **6:**114(144), 132(144), 146, 247(154), 298
Janata, J., **24:**183(32, 33), 184(33), 192(32)
Janda, K. C., **23:**29(102, 103), 38(102, 103)
Jandus, J., **9:**47(216), 70
Janeff, W., **3:**110(116), 127
Jang, S.-M., **23:**35(134)
Janisch, F. R., **6:**293(226), 300
Jankowski, A. F., **17:**49, 51, 54, 55, 56, 61, 63, 76, 77; **22:**154(75), 155(75), 173
Jannitti, E., **14:**122(86), 127
Janocko, M. A., **16:**8, 52, 125, 133
Janousek, B. K., **21:**128(33), 233
Jansen, M., **15:**9, 136
Jansen, P., **21:**287(40), 310
Janssen, A. P., **14:**229(97), 239
Janssen, G. C. A. M., **22:**267(117), 275
Janssen, G. H., **9:**33(138), 35(138), 68; **12:**187(78), 234
Janssen, J. E., **2:**353(79), 357, 361; **10:**55(118), 68; **12:**45, 51
Jansson, U., **16:**92, 139
Jantsch, W., **11:**119(70), 130(87), 131(70), 201
Janus, A. R., **4:**270(89), 322

Jaquinot, P., **1:**127(7), 184
Jardinier-Offergeld, M., **4:**150(55), 188
Jares, A. L., Jr., **2:**320(26), 359
Jarezebski, Z. M., **10:**52(106), 68
Jarosik, N., **21:**126(29), 233
Jarrett, A. H., **3:**10(46), 51(46), 54
Jarrett, D. N., **19:**141
Jarrold, M. F., **16:**60, 135
Jarzebski, Z. M., **9:**30, 67
Jaschke, M., **24:**21(91, 102), 26(102), 88(45), 89(50), 90(45), 98(97), 106(97), 212(65)
Jasper, M., **11:**132(91), 156(91), 157(128), 202
Jasperson, S. N., **9:**227(199), 228(199), 260; **10:**61(142), 69; **19:**6
Jaswal, S. S., **15:**185, 216
Javet, P., **4:**219(105, 106), 228(105, 106), 230, 253
Jaworski, B. M., **2:**94(45), 141
Jayaraman, A., **7:**174(123), 251
Jayson, J. S., **7:**225(228), 253
Jedamzik, D., **16:**31–32, 129
Jedlicka, E., **9:**47(216), 70
Jehn, H. A., **17:**108, 135, 141, 144
Jelatis, J. G., **6:**72(214), 79
Jelen, R. A., **17:**196, 222
Jelinski, L. W., **24:**153(13)
Jelley, J., **2:**123(131), 144
Jellinek, H. H. G., **13:**154(19), 183(19), 192
Jellison, Jr. G. E., **19:**199, 201, 212
Jellyman, P. E., **4:**271(93), 322
Jen, S., **15:**60, 139
Jenkin, II, K. R., **16:**13, 42, 45, 125, 131
Jenkins, F. A., **3:**2(20), 53
Jenkins, G. I., **7:**351, 395
Jenkins, G. M., **12:**246(23), 254(23), 290
Jenkins, J. E., **14:**147(71), 148(71), 196
Jenness, J. R., Jr., **10:**14(23), 65
Jennings, L. D., **1:**241(17), 273
Jennings, T. A., **6:**36(118), 40, 43, 76, 77
Jennings, V. J., **5:**284(192), 318
Jenny, F. F., **2:**376(77), 421
Jensen, H. H., **1:**324(98), 326, 334
Jensen, J. D., **11:**111(41, 42), 113(52), 119(42, 71), 132(41), 135(52); **12:**218(110), 219(110), 235
Jensen, J. T., Jr., **1:**28(63), 65
Jensen, K. F., **22:**178(13), 182(26), 271–272
Jensen, M. H., **21:**52(101), 74
Jentzsch, J., **3:**11(57), 51(54), 54
Jeon, N., **24:**14(60), 15(60)
Jeon, N. L., **24:**240(72), 241(72), 243(80, 81), 245(80)
Jeon, Y., **16:**60, 63, 135
Jeong, W. G., **19:**12, 32, 35
Jeppesen, J., **6:**113, 124(139), 127, 128, 135(139), 146
Jeppesen, M. A., **19:**257
Jeppesen, R. H., **3:**213, 270
Jeppesen, S., **19:**2, 26, 29
Jepsen, D., **7:**268, 269(61), 308
Jepsen, D. W., **3:**172(14), 208
Jepsen, R. L., **1:**19(43, 44, 46), 20(49), 65; **14:**208(17), 237
Jeric, S., **12:**73(81), 74(81), 76(81), 77(81), 83(81), 84(81), 160
Jervis, B. W., **5:**153(38), 209(38), 214(38), 230
Jervis, T., **17:**47, 49, 53, 63, 76, 77
Jesion, G., **11:**189(149), 191(149), 203
Jesser, W. A., **4:**139, 145, 146,

150(12), 158, 162, 187, 235(154), 243(154), 245, 254; **7:**18(57), 38
Jezequel, G., **9:**178(60), 257
Jia, C. L., **16:**28, 121, 128, 143
Jiang, C., **16:**28, 128
Jiang, Q. D., **16:**41, 131
Jiang, S.-S., **17:**22, 74
Jijima, K., **16:**25, 73, 127
Jillson, D. C., **5:**277(156), 318
Jimenez, J. R., **23:**166(91)
Jin, B. Y., **17:**2, 73
Jindal, P. C., **12:**284(240), 295
Joannopoulos, J. D., **12:**114(203), 163
Jochems, P. J. W., **5:**163(68), 168(68), 231
Joffrin, J., **11:**312(144), 317
Jogai, B., **15:**245–246, 263
Johansen, O. A., **14:**180(159), 199
Johansson, B. O., **14:**65(209), 68(209), 78; **17:**131, 136, 143, 144
Johansson, G., **7:**45, 48(16), 111
Johansson, M., **16:**100, 140
Johari, G. P., **11:**240(46), 246, 253, 254, 255, 315
John, F., **22:**87(11), 115
John, H. F., **5:**186(150), 208(150), 233
John, S., **17:**337, 367
Johnson, A. W., **13:**153(9, 11), 155(29), 174(80), 178(98), 182(9, 11), 183(9, 11, 29), 184(29), 186(80, 98, 144), 189(80, 98, 144), 190(162), 191(162), 192, 193, 194, 195, 197
Johnson, B. R., **16:**51, 132–133
Johnson, C. S., Jr., **24:**151(9), 154(9), 156(9)
Johnson, C. Y., **7:**102(79), 113
Johnson, D., **5:**313(302), 322
Johnson, D. B., **6:**339(53), 340(57), 347
Johnson, E. W., **6:**133(199), 148
Johnson, F., **5:**271(110), 317
Johnson, F. A., **4:**60(136), 94
Johnson, F. S., **1:**131(14), 136, 185; **2:**308(3), 358; **7:**51, 94(34), 112; **11:**33, 34
Johnson, G. R., **4:**199(70), 200(70), 206(70), 252
Johnson, H., **5:**216(194), 228(194), 234
Johnson, H. H., **6:**133(199), 148
Johnson, J. E., **2:**151(36), 191; **5:**165(108), 174, 176(209), 222(209), 232, 235
Johnson, J. L., **16:**96, 140
Johnson, J. R., **2:**370(45), 406(45), 420
Johnson, L. A., **17:**260, 263, 298
Johnson, L. E., **3:**277, 300
Johnson, L. F., **13:**126, 149; **14:**226(86), 238
Johnson, L. H., **15:**168, 215
Johnson, M. C., **6:**36(117), 76
Johnson, M. R., **11:**146, 148, 149(117), 202
Johnson, N., **14:**99(41), 110(41), 126
Johnson, P. B., **9:**236(219), 261
Johnson, P. C., **14:**175(133), 179(153), 180(153, 158), 181(158), 182(158), 183(133), 185(133), 186(153, 158), 192(153, 168), 193(168), 197(153), 198, 199; **17:**102, 103, 141
Johnson, R. I., **16:**78, 137
Johnson, R. L., **19:**280
Johnson, S. E., **9:**198(111), 258
Johnson, T. H., **3:**291(54), 300; **12:**169, 201, 233
Johnson, W. J., **7:**186(144a), 251; **11:**130(83), 131(83), 190(153), 201, 203
Johnsson, B., **24:**19(80), 20(80)
Johnsson, K. P., **24:**36(147)

Johnston, A. R., **4:**82(205), 95
Johnston, L. H., **13:**48(57), 58
Johnston, R. G., **4:**16(39), 65(39), 91
Johs, B., **16:**80–81, 137–138
Jona, F., **5:**166, 230, 247(25), 315
Jonath, A. D., **14:**36–37(139), 76
Jones, B. P., **22:**194(52), 227(52), 228(52), 229(52), 273
Jones, C. E., **7:**34(140), 36(164), 40, 41
Jones, C. K., **12:**286(246), 295
Jones, D. I., **12:**296
Jones, E. C., **16:**70, 136
Jones, E. W., **23:**14(50), 102(43), 109(59), 146(46), 161(63)
Jones, F., **14:**57(177), 77; **22:**117, 129(43), 141(63), 142(43), 145(43), 152(74), 160(78), 161(81), 172–173, 194(54), 273
Jones, F. E., **4:**50(111, 112, 113, 114, 115), 93
Jones, G. A., **16:**28–29, 129
Jones, G. R., **13:**3(8), 56
Jones, H., **4:**338(40), 365(81), 366(81), 370, 371; **7:**260(8), 306; **9:**7(36), 65
Jones, J. P., **14:**229(97), 239
Jones, L., II, **24:**33(137), 44(37)
Jones, M. L., **19:**169, 180, 281, 291
Jones, P. L., **12:**258(73, 74), 291
Jones, R. A., **3:**9(40), 10(40), 32(40), 40(40), 44(40), 51(40), 54
Jones, R. C., **3:**278, 300
Jones, R. E., **3:**121(149), 128, 150, 154(77), 164; **12:**259, 263(85), 264(85), 292
Jones, R. L., **15:**244, 263
Jones, S. J., **11:**253, 255
Jones, T. H., **12:**88(128), 89(128), 99(128), 161
Jones, T. T., **3:**151(69), 164
Jones, W. E., **24:**9(34)
Jones, W. K., **16:**4, 124
Jonker, H. D., **15:**11, 137
Jonscher, A. K., **6:**63, 64, 78; **11:**206(1), 207(1, 2), 208(2), 217(13, 14), 221(16), 222, 226(21), 230(26, 27), 232(33, 34, 35), 234, 236, 238, 239, 240, 241, 244, 245, 247, 248, 249, 250, 252, 256(2, 67, 68), 257, 258, 259, 261(76, 77, 78, 79, 80, 81, 82)
Jönsson, J., **19:**2, 23, 26–29
Joo, S.-K., **17:**318, 359, 366, 369
Joosten, J., **11:**135(98), 202
Jordan, A. S., **7:**166(92), 249
Jordan, E. L., **2:**370(52), 421; **3:**134(24), 135(24), 136(24), 141, 143(24), 162; **4:**263, 321; **5:**289(222), 320
Jordan, L. K., **9:**215(150), 259
Jordan, N. A., **5:**225(214, 215), 235
Jordan, R. C., **10:**7(12), 65
Jorgensen, G. V., **7:**49, 50, 54, 55, 56(30), 112
Jorgensen, P. S., **6:**2(1), 73
Jorgensen, W. L., **24:**209(36, 37, 38), 210(36, 37, 38), 211(36), 220(36)
Jorgenson, G. V., **2:**400(159), 424; **4:**269(81), 322
Jorke, H., **23:**14(51)
Jorritsma, J., **24:**86(37)
Joseph, A. S., **5:**150(28), 230
Joseph, D. M., **15:**308, 312
Joseph, R. I., **15:**6, 8, 136
Josephs, R. M., **6:**114, 123, 146
Joshi, M., **7:**35(155), 41
Joshi, R. V., **22:**117, 119(12), 161(12, 81), 165(86), 167(12), 171, 173, 188(38), 193(38), 272
Joshi, V., **22:**161(80), 173
Joshnson, R. A., **22:**127(37), 172

Jossem, E. L., **1:**11(21), 64
Josserand, Y., **19:**8
Jost, R., **9:**37(165a), 69
Jost, W., **6:**304(4), 307, 308(4), 310(14, 15), 346; **16:**234, 236, 266
Jotatis, A., **13:**144(132), 150
Joubert, J. C., **9:**282, 294
Jovanovic, S., **3:**253(91), 255(91), 257, 258(91), 259, 273
Joy, D. C., **12:**246(33), 248(33), 252(33), 290; **13:**142(123), 150
Joyce, B. A., **5:**147, 148, 149(18), 152, 186(34), 187(153), 196, 199, 229, 230, 233, 238(3), 266, 267, 269–272, 314, 316, 317; **11:**36, 41(36), 42(36, 39, 42), 45, 48(36), 59(36), 60(36), 61, 64(115), 66(124), 70, 89(4), 96, 97, 99; **19:**10, 15
Joyce, J. B., **17:**209, 223, 272, 290, 293, 298
Joyce, R. J., **6:**37(133), 77
Joyce, S. A., **24:**94(64), 221(88)
Joyee, B. A., **4:**150(59), 188
Joyee, R. J., **12:**261, 262, 271(158), 292, 293
Joyner, K., **11:**230(29), 315
Jozwikowska, A., **21:**40(87), 74
Jozwikowski, K., **21:**40(87), 74
Juang, C., **23:**291(82)
Judaprawira, S., **11:**57(98), 60(98), 62(98), 76(154), 83(188a), 99, 100, 101
Judeinstein, P., **17:**361, 370
Juenker, D. W., **4:**74(173), 95; **10:**91, 103(28), 106, 164
Juhasz, C., **11:**153(123), 202
Jull, D. W., **2:**101, 142
Jung, G., **16:**97, 140
Jung, T. A., **24:**98(85)
Jungblut, F., **17:**100, 110, 141
Junge, A. E., **9:**33(145, 147, 150), 68
Jungling, K., **13:**121(64), 130(92), 131(92), 134(92), 145(64), 148, 149, 155(29), 183(29), 184(29), 193
Junno, B., **19:**23, 27, 28
Jupnik, H., **2:**272, 304
Jura, G., **2:**384(92), 422
Juranek, H. J., **9:**245(245), 261
Juretschke, H. J., **1:**330(112), 334; **4:**339, 370; **6:**86, 144, 243(140), 244(140, 142), 245(140), 247(155), 277, 298, 299
Jurgensen, C. W., **18:**71; **22:**113(31), 115
Jursa, A. S., **1:**18, 130(11), 184
Jushida, K., **19:**253
Justi, E., **6:**82(6), 143
Juttner, B., **14:**139(33), 147(74), 157(98), 195, 196, 197
Jutzi, W., **6:**293(224, 225), 300

K

Kabanov, V. A., **3:**158(88), 164
Kabayashi, A., **3:**277(28), 300
Kabius, B., **16:**28, 72–73, 128, 137
Kadano, K., **17:**83, 139
Kadereit, H. G., **6:**107, 108, 109, 110(126), 116(126), 131(126), 146
Kadin, A. M., **16:**109, 142
Kadlec, S., **14:**14(48), 74; **17:**83, 88, 89, 90, 91, 97, 98, 99, 111, 112, 114, 115, 116, 117, 118, 121; **18:**263–264, 265, 266, 267, 269, 271, 278
Kadlecova, I., **6:**103(100), 122, 131(100), 132(100), 145
Kadota, T., **12:**275(175), 294
Kaga, E., **17:**355, 369

Kagan, M. B., **7:**246(279), 256
Kaganov, M. I., **6:**91, 92, 123, 144, 145, 190, 203; **15:**26, 137
Kagdis, W., **5:**271, 317
Kageyama, Y., **16:**21, 127
Kagiwada, R. S., **17:**153, 154, 220
Kagonovich, E. B., **12:**223(114), 229(114), 230(114), 235
Kahan, G. J., **1:**251(48), 274
Kahle, W., **3:**171(16), 172(16), 176(16), 208
Kahng, D., **4:**300(121), 323, 326, 334(36), 335(36), 360, 369, 370; **5:**266, 316; **7:**149(60), 248
Kaintha, R. C., **12:**169(41, 42), 179(65), 180(65), 203, 205(42), 207(42), 208(42), 210(42), 211(42, 99), 212(99, 102, 105, 106), 213(105, 106), 214, 215(42), 216(102), 217(102), 218(99, 102), 219(102, 106, 112), 220(102, 112), 221(102), 222(102), 223(102, 106, 113), 226(121a), 229(41, 102, 106), 230(99, 102, 106), 233, 234, 235
Kainuma, Y., **2:**106, 142
Kaischew, R., **1:**210(117), 231; **4:**202, 203, 252
Kaise, M., **16:**28, 33, 128
Kaiser, S. W., **11:**32(22), 97, 108(25), 109(25), 115(25), 119(25), 120(25), 200
Kaiser, W., **5:**294(237), 320; **12:**292; **21:**94(24), 111
Kaito, C., **17:**310, 365
Kajiyama, K., **9:**35(163a), 68; **13:**220(71), 221, 225
Kakagi, T., **17:**153, 154, 160, 220
Kakinoki, J., **7:**22(74), 39
Kalil, F., **2:**70(13a), 141
Kalinkin, I. P., **3:**176(25), 209
Kalinovich, D. F., **7:**264(39), 307
Kalisher, M. H., **17:**162, 221, 261, 298
Kallander, D. A., **4:**261(41), 321
Kalnins, I. L., **4:**310(155), 324
Kalokitis, D., **16:**106, 141
Kalt, S., **16:**60, 135
Kalter, H., **14:**234(137), 240
Kamada, T., **16:**17, 19, 30, 126, 129
Kamagaki, K., **16:**51, 132
Kamangar, F. A., **21:**62(118), 74
Kamasz, S., **22:**76(68), 78
Kamat, S. V., **17:**66, 77
Kamath, G. S., **5:**281, 318
Kambersky, V., **1:**288, 289(34), 332
Kamehara, N., **16:**34, 130
Kamei, K., **11:**77(162), 85(162), 101
Kamimori, T., **17:**308, 309, 318, 352, 353, 364, 366, 368
Kamimura, K., **11:**76(149), 100
Kaminar, N. R., **21:**362(44), 368
Kaminow, I. P., **10:**253(48), 254, 255, 259
Kaminsky, G., **7:**210, 211(189), 252
Kaminsky, M., **3:**62, 129; **12:**253(67), 291; **14:**217(37), 237; **22:**160(77), 173
Kamiya, I., **19:**2, 15–17, 27, 32–36
Kamiya, Y., **4:**171, 190
Kamiyama, M., **13:**19(27), 57
Kammerdiner, L., **17:**256, 257, 298
Kammerer, O. F., **3:**39, 52(190), 58
Kammura, W., **13:**205(22), 223
Kamo, M., **19:**104
Kampas, F. J., **13:**68(40), 72(40), 105
Kamprath, N., **16:**164, 168, 193, 202, 208–211, 214, 230–231
Kampschoer, G. L. M., **17:**228, 240, 256, 257, 297, 298
Kampwirth, R. T., **14:**38(140), 76; **16:**17, 28, 32, 36, 126, 128–130
Kan, E. C., **22:**270(132), 276

Kan, H., **12:**135(295), 165
Kanai, M., **16:**79, 137
Kanamaya, M., **14:**175(135), 198
Kanamori, J., **6:**215, 296
Kanamori, S., **14:**66(223), 78
Kanamori, T., **16:**86, 138
Kanata, T., **17:**161, 222, 228, 229, 262, 298
Kanatzidis, M. G., **24:**71(111)
Kanaya, H., **23:**166(90)
Kanaya, K., **1:**191(25), 228; **2:**110(89), 143; **17:**84, 85, 140
Kanayama, S., **16:**263, 269
Kanazama, K., **12:**277(218), 282(218), 295
Kanazawa, H., **9:**158(14a), 256
Kanazawa, K., **14:**104(51), 126
Kandel, E. R., **21:**61(113), 62(113), 65(113), 74
Kane, E. O., **12:**68(53), 159; **21:**94(22), 111; **23:**226(25, 26)
Kane, J., **9:**39(171, 172), 40, 41(170, 171, 172, 174), 62(171, 172), 69; **12:**177(59, 62), 234
Kane, M. J., **15:**234–235, 262; **21:**117(12, 13), 128(36), 152(13), 153(13), 232, 233
Kane, W. M., **2:**148(4), 190; **4:**325(2), 351(2), 369; **5:**164(87), 231
Kanedo, M., **11:**253, 255
Kanekama, N., **13:**119(49), 126(49), 148
Kaneko, H., **17:**313, 314, 336, 345, 346, 365, 367
Kaneko, K., **16:**96, 140
Kaneko, M., **15:**112–113, 115–116, 141
Kaner, E. A., **6:**82(33), 144
Kanetkar, S. M., **16:**70, 136
Kanev, V., **12:**87(118, 119), 161
Kang, C., **7:**120, 121(5), 247
Kang, C. S., **7:**153, 156, 235(63), 249
Kang, D., **24:**28(127)
Kang, H., **24:**44(43), 127(51)
Kang, J. H., **16:**14, 17, 28, 36, 125–126, 128–130
Kang, K., **17:**354, 368
Kangro, A., **12:**96(159), 162
Kanicki, J., **19:**61
Kannewurf, C. R., **16:**57, 60, 87, 93–94, 134–135, 138–140
Kano, G., **13:**176(86), 188(86), 189(86), 195
Kano, T., **5:**277(152), 318
Kansky, E., **12:**73(81), 74(81), 77(81), 83(81), 84(81), 99(164), 119(234), 160, 162, 164
Kantelhardt, D., **3:**39(209), 52(209), 58
Kanter, H., **2:**126(142), 127, 128(142), 129(142), 144; **4:**349, 352(61), 353; **21:**128(33), 233
Kantsel, V., **14:**156(90), 197
Kao, K. C., **15:**293, 311; **19:**281, 293, 294
Kao, Y. H., **6:**82(17), 117, 118, 143, 147
Kapin, A. T., **14:**151(83), 152–153(83), 154(83), 156(83), 158(83), 197; **17:**93, 94, 140
Kapitulnik, A., **16:**42, 50, 131
Kapitza, H., **9:**237(225), 238(225), 243(225, 239), 244(239), 261
Kaplan, D. R., **11:**168(141), 203
Kaplan, L., **3:**140(46), 163
Kaplan, M. L., **12:**246(24), 290
Kaplan, R., **16:**85, 138
Kaplow, R., **7:**27(110), 36(156), 39, 41
Kapoor, V. J., **16:**36, 130
Kapp, D. S., **7:**354, 355(145), 396
Kaputnik, A., **16:**2, 123
Kar, S., **19:**161

Karaborni, S., **24:**208(29), 210(43, 44)
Karakashian, A. S., **9:**244(242), 261
Karakostas, Th., **17:**77
Karasawa, N., **24:**211(49)
Karavaev, G. F., **12:**105(176), 106(176), 162
Kard, P. G., **1:**101, 107, 121; **5:**60(17), 86
Kargin, V. A., **3:**158(88), 164
Karim, A. A., **13:**79(83), 91(83), 92(83), 106
Kariya, T., **12:**176(48), 224(48), 233; **14:**31(122), 76
Karkut, M. G., **16:**19, 28, 126, 128
Karlik, S., **2:**406(185), 424
Karlsson, S. E., **7:**45, 48(16), 111
Karmarkar, M. M., **16:**96, 140
Karpenko, A. N., **15:**115, 118, 141
Karpenko, B. V., **6:**247, 298
Karpovich, D. S., **24:**21(96), 23(96), 44(27), 45(27, 86, 87), 46(27, 86, 87), 53(27), 56(27), 57(27), 60(27), 61(27), 64(27), 70(27, 105), 74(121), 75(121), 93(60), 94(60)
Karunasiri, R. P. G., **15:**229, 232–234, 239, 259, 261; **21:**4(17), 38(78, 81), 39(78), 40(78), 48(96), 49(78), 72, 73, 74, 77–112, 78(1, 2, 3), 81(13, 14), 84(15), 86(14), 87(14), 88(18), 91(2, 19, 20), 95(20, 26), 97(26), 100(36), 101(3), 103(41), 104(43); **23:**240(51)
Kasai, I., **11:**111(39, 40), 119(40), 200
Kasai, T., **14:**240
Kasami, A., **7:**158, 225(78), 249
Kase, T., **17:**353, 368
Kaseman, P. W., **12:**157(336), 166
Kashiwagi, K., **17:**153, 219
Kashyap, S. C., **15:**286, 310
Kasner, W. H., **17:**229, 236, 237, 270, 271, 272, 276, 278, 296, 298
Kaspaul, A. F., **2:**412(194), 425
Kaspaul, E. E., **2:**412(194), 425
Kasper, E., **11:**36(30), 97; **23:**2(1), 3(15), 55(185)
Kasper, J. S., **7:**23(82), 39
Kasprzak, L., **9:**4(14), 5(14), 65
Kastalsky, A., **15:**229–231, 245, 259, 262–263; **21:**184(81), 235; **23:**290(110)
Kasukawa, A., **23:**228(32)
Katada, K., **7:**22(73, 74), 39
Katama, Y., **11:**36(28), 97
Kataoka, S., **17:**345, 367
Katardjiev, I. V., **22:**113(32, 33), 115
Katayam, S., **12:**176(48), 224(48), 233
Katayama, H., **15:**168, 172, 215–216
Katayama, T., **15:**154, 156, 158, 164–165, 183, 185, 215, 285–286, 310
Katelaar, J., **7:**36(161), 41
Katellus, H. P., **14:**66(217), 78
Katiyar, R. S., **15:**284–285, 310
Katnani, A. D., **14:**234(145), 240
Kato, H., **3:**229(54), 230, 272
Kato, I., **14:**103(48), 126
Kato, K., **17:**241, 297
Kato, N., **2:**120, 143
Kato, R., **1:**148(54, 55), 169, 186
Kato, T., **16:**28–29, 98, 129, 140
Kato, Y., **14:**236(155), 240
Katoh, Y., **14:**236(154), 240
Katrich, G. A., **4:**349, 354(60), 367, 370
Katsube, S., **3:**15(67), 51(67), 54
Katsumo, H., **12:**135(295), 165
Katsura, T., **24:**36(143)
Katter, M., **16:**214, 230
Katto, K., **12:**264, 292
Katz, E., **24:**19(81), 20(81, 89)

Katz, H. E., **24:**12(55), 14(55, 59), 20(59), 163(27)
Katz, H. W., **16:**162, 230
Katz, L., **2:**69(13), 70(13), 71, 141
Katz, L. E., **12:**282(221), 295
Katz, L. F., **14:**220(72), 238
Katz, M., **12:**272(170, 171), 274, 294
Katz, W., **14:**234–235(147), 234(147, 148), 235(148), 240; **23:**91(20), 102(42)
Katzir, A., **10:**228(7), 258; **13:**99(106), 107
Kauer, E., **10:**30(51), 66
Kauffman, J.-M., **24:**21(106)
Kauffman, W. M., **6:**48(144), 77
Kaufherr, N., **17:**110, 141
Kaufman, F., **12:**238(8), 290; **14:**85(9), 125
Kaufman, H. R., **13:**109(4), 112, 114, 147; **14:**18(70), 20(70), 23(88, 89), 26(101, 102), 33(130); **17:**84, 140, 234, 296
Kaufman, J., **13:**153(7), 154(7), 172(7), 173(7), 182(7), 183(7), 192
Kaufman, L., **7:**18(58, 59), 38
Kaufman, R. L., **12:**277(208), 279(208), 294
Kaufmann, H., **17:**263, 298
Kaupp, E., **4:**227, 254; **6:**310(12), 346
Kauppila, J. E., **2:**150, 190
Kaur, I., **12:**169(43), 204(43), 205, 208, 212(103), 214(103, 103a), 217(43, 103), 218(103), 222(103, 103a), 229(103), 230(103), 233, 235
Kavanagh, K. L., **13:**220(69), 224
Kavassalis, C. A., **13:**161(41), 162, 176(41), 186(41), 190(41), 193
Kaw, P., **9:**15(69), 66
Kawabata, A., **14:**65(210), 78; **17:**153, 154, 219, 220, 228, 232, 243, 296, 297
Kawabe, H., **16:**88, 90, 139
Kawabi, K., **5:**160(51), 230
Kawaguchi, T., **17:**249, 297
Kawahara, H., **17:**361, 362, 370
Kawai, N. J., **11:**86(205), 102
Kawai, S., **16:**79, 137; **17:**209, 223, 290, 299
Kawai, T., **16:**79, 137; **17:**209, 223, 290, 299
Kawakami, K., **17:**352, 368
Kawakami, T., **7:**186, 187(144), 251
Kawakatsu, H., **1:**191(25), 228; **2:**100(89), 143
Kawamoto, A., **15:**170, 174, 210, 215, 217
Kawamoto, E. H., **24:**181(13), 184(36)
Kawamura, K., **12:**246(23), 254(23), 290
Kawamura, N., **4:**260(23), 301(23), 320
Kawamura, T., **18:**146
Kawamura, Y., **11:**67(133b), 100; **13:**153(8), 154, 182(8), 183(8), 192; **21:**121(20), 123(20), 233
Kawamure, N., **13:**217(57), 224
Kawana, A., **14:**31(121), 76
Kawasaki, M., **14:**20(80), 75; **16:**118–120, 122, 143
Kawashima, M., **19:**12, 15
Kawata, Y., **17:**288, 299
Kawaziri, K., **17:**355, 356, 368
Kawi, T., **16:**4, 33, 124
Kay, D. H., **4:**253
Kay, E., **2:**367(21), 368(21), 420; **3:**62, 91, 92, 97, 100(173), 101(5), 118, 120, 125, 126, 128, 129; **9:**282, 296; **14:**16(61), 17(61), 28(112), 74, 76, 208(21), 209(21),

Kay, E., (*continued*)
234(136, 139), 237, 240; **17:**86, 140, 153, 219
Kaye, Wilbur., **4:**46(85), 92
Kaygorodov, V. N., **7:**292(103), 310
Kayser, W., **6:**221(71), 240(71, 137), 249(137), 296, 298
Kazakov, G. T., **15:**64, 80–81, 139
Kazakov, V. A., **14:**175(136), 198
Kazan, B., **17:**304, 364
Kazarinov, R. F., **23:**238(42), 240(42, 53)
Kazmazovskaya, T. S., **7:**34(145), 40
Kazyaka, T. G., **13:**154(18), 182(18), 183(18), 192
Keahl, G. T., **5:**85(29, 30), 86
Kear, B. H., **16:**91, 139
Keating, D. T., **3:**39, 52(190), 58; **7:**9(19), 37
Keating, P. N., **5:**225(212), 235
Kechouane, M., **13:**178(101), 195
Keck, P. H., **5:**171(82), 231
Keddie, J. L., **19:**263
Kee, R. J., **22:**180(16), 186(30, 31), 271–272
Keefe, G. E., **9:**266, 271, 295, 296
Keefer, K. D., **16:**4, 81, 124
Keeley, T. C., **1:**265(83), 275
Keem, J. E., **15:**288, 311
Keenan, W. J., **13:**60(9), 62(9), 104
Kehler, H., **1:**188(5), 228
Kehoe, R. B., **4:**156(80), 159(80), 175(80), 189
Kein, W., **12:**177(59), 234
Keister, F. Z., **3:**20(94), 51(94), 55
Keitel, G., **7:**46(17), 112
Keith, H. D., **3:**246, 272
Keith, V., **12:**277(192), 294; **14:**86(11), 106(11), 125
Kelbanikov, N. S., **12:**71(63), 160
Kel'bert, S. L., **14:**145(65), 196
Keler, G., **16:**70, 136
Kelkar, G. N., **12:**148(320), 166
Keller, A., **2:**132(150, 151), 145
Keller, C., **16:**9, 125
Keller, H., **7:**27(102), 39
Keller, M., **1:**191(27), 228
Keller, P., **17:**288, 299
Kellett, B. J., **16:**26–28, 127–128
Kelley, E. F., **13:**61, 104
Kellog, G. J., **24:**181(13)
Kelly, A., **16:**287, 316, 319, 334, 338–339
Kelly, E. F., **5:**165(110), 176, 232
Kelly, J. C., **2:**236(41), 238; **3:**41(214), 59; **4:**26(63), 78(196), 92, 95; **6:**27, 76; **14:**217(42), 237
Kelly, M., **14:**234(145), 240
Kelly, M. J., **12:**71(59), 73, 159
Kelly, M. K., **16:**13, 125; **19:**2, 11, 15, 16, 26
Kelly, P. M., **4:**225, 254
Kelly, R., **13:**116(38), 117(43), 118, 119, 126(38), 127(43, 51, 84), 135(43); **14:**8(30), 8–9(30), 9(34), 17(30, 65), 74, 229(93), 230(102, 103), 239
Kelly, W. A., **5:**85(31), 86
Kelly, W. C., **5:**271(110), 317
Kelso, S. M., **15:**191, 217; **21:**327(21), 367
Kember, D., **15:**276, 310
Kendall, G., **16:**43, 132
Kendall, J. T., **5:**284, 319
Kenigsberg, N. L., **17:**153, 219
KenKight, C. E., **3:**76(55), 126
Kennedy, D. I., **9:**5(17), 65; **16:**258, 269
Kennedy, D. L., **7:**268(58), 308
Kennedy, D. R., **2:**369(36), 420; **3:**112, 127; **4:**267(72), 322; **6:**67(199), 78

Kennedy, T. N., **18:**221
Kennemore, C., III, **13:**137, 138(109, 110, 111, 114, 116), 139(109, 110, 111, 116), 150
Kenney, B. T., **3:**137(32b), 163; **4:**315(159), 324
Kenny, M. J., **17:**318, 365
Kent, A. D., **16:**28, 128
Kent, C. V., **2:**199, 237
Kent, M. S., **24:**181(13)
Kenty, J. L., **4:**126(65), 136
Kenyon, B. A., **10:**125(54), 165
Kenyon, B. H., **7:**90(64), 113
Keodam, M., **3:**88(167), 128
Kepley, J., **24:**44(63, 64), 46(63, 64), 70(63)
Keramidas, V. G., **19:**2, 36
Kerchner, H. R., **16:**10, 43, 48–49, 125, 132
Kerestes, J. A., **15:**20, 62, 137
Kerherve, F., **16:**76, 137
Kern, E. L., **5:**285(200), 319
Kern, K., **24:**44(42), 86(33)
Kern, R., **3:**11(51), 49(51), 51(51), 54
Kern, W., **9:**39(170, 171, 172), 40(170, 171, 172), 41(170, 171, 172), 62(170, 171, 172), 69; **12:**177(62), 234, 258(77), 268, 291, 292, 293; **13:**181(113), 195; **15:**287, 310; **16:**4, 81, 124, 138, 153, 229; **17:**81, 83, 84, 139, 232, 296
Kerner, P., **24:**106(109)
Kerns, R. C., **15:**284–285, 310
Kerr, D. R., **3:**141(52), 163; **4:**298(114), 299, 302, 303, 304, 305, 306, 307, 308, 310, 311, 313, 315, 316(162), 323, 324; **5:**287(209), 288, 293(209), 295, 296, 297, 299(209), 319; **12:**258(75), 291; **17:**273, 274, 298
Kerr, J. T., **1:**9(17), 64
Kersch, A., **22:**178(15), 271
Kerwin, D. B., **13:**105, 131(101), 134(101), 139(101), 140(101), 150
Kesaiev, I. G., **17:**92, 140
Kesan, V. P., **21:**84(17), 111, 193(93), 235
Kessler, V. G., **17:**318, 366
Kester, D. J., **14:**18(73), 19(73), 75
Kestigian, M., **9:**270, 296
Ketcham, R. S., **17:**160, 198, 201, 206, 221, 223
Ketcham, T. D., **16:**70, 136
Ketchen, M. B., **16:**118–120, 143
Ketterson, J. B., **16:**43–44, 46–47, 131; **17:**2, 7, 12, 34, 43, 44, 45, 48, 49, 50, 51, 52, 53, 62, 73, 75, 76; **24:**180(11)
Keuch, T. F., **19:**35
Keune, D. L., **11:**66(122), 99
Keve, E. T., **17:**229, 230, 279, 295
Key, P. H., **13:**154(16, 20), 172(16), 181(16, 20), 182(16), 183(16, 20), 184(20), 192
Keyes, R. J., **21:**25(55), 73
Keyes, R. W., **21:**65(122), 75
Keywell, F., **3:**67, 75(54), 125, 126
Khadzhai, G. Ya., **7:**264(45), 307
Khaikin, M. S., **6:**82(37), 133(198), 134(200), 144, 148
Khan, I. H., **5:**164(87), 231
Khan, J. M., **3:**32(157), 51(156, 157), 57
Khan, M. R., **17:**24, 46, 49, 51, 53, 56, 58, 75
Khang, D., **13:**61, 104
Khanna, A. P. S., **16:**102–105, 141

Khanna, U., **7:**362(152), 371(178, 179, 180), 372(178, 180, 184), 396
Khare, M., **16:**16, 38, 126
Khare, N., **16:**116, 142
Khasanov, S. S., **16:**71, 79, 136
Khasieva, R. V., **19:**2
Khokle, W. S., **16:**98, 140
Khoroshikh, V. M., **14:**158(99), 161(109), 163(115), 176(140), 177(147), 185(115), 197, 198
Khotkevich, V. I., **7:**264(45), 307
Khoury, J. T., **24:**151(10), 154(10), 156(10), 163(10), 222(104)
Khuri-Yakub, B. T., **17:**153, 172, 219, 220, 222
Kibbel, H., **11:**36(30), 97; **23:**14(51), 55(185)
Kibbler, A. E., **21:**322(14), 337(14), 367
Kichuchi, Y., **14:**20(81), 27(81), 75
Kicinski, F., **11:**106(3), 108(3), 199
Kidder, M., **12:**135(293), 165
Kidou, H., **17:**318, 365
Kidson, G. V., **6:**305(8), 346
Kiehl, W., **16:**81, 138
Kienel, G., **6:**339(54), 347; **9:**20(80a), 67
Kiermasz, A., **14:**91(25), 92(91), 103(3–25), 116(25), 117(25), 118(25), 125, 127
Kierstead, F. H., **4:**60(143), 94
Kies, F. K., **3:**269(121), 273
Kiessig, H., **3:**38, 40(185), 52(185), 58; **6:**187, 203
Kietzman, B. E., **1:**19(46), 65
Kiin, H. S., **16:**65, 136
Kikuchi, K., **21:**33(67), 34(67), 73
Kikuchi, M., **10:**237(21), 258; **14:**183(162), 199; **19:**93
Kikuchi, S., **7:**236(249b), 254
Kikuchi, Y., **17:**153, 219
Kikuta, K., **17:**241, 297
Kiledjian, M. S., **21:**129(41), 233
Kilipatrick, J., **10:**168(2), 225
Kim, B. F., **16:**59, 79, 135, 137
Kim, C., **17:**49, 62, 77
Kim, C. K., **7:**174(123), 251
Kim, D. H., **16:**16, 126
Kim, D. M., **13:**155(25), 166(25), 192
Kim, E., **24:**44(60, 62), 228(25), 229(33, 34, 35), 230(33, 48), 235(33), 240(48, 69, 71), 245(83), 246(34, 83, 85, 86), 251(97)
Kim, H., **9:**31(114a), 67
Kim, H. G., **19:**250
Kim, J., **17:**49, 51, 76
Kim, K. S., **14:**230(101, 104), 239
Kim, M. W., **24:**181(13), 182(23), 199(23)
Kim, S., **13:**205(22), 223; **22:**123(26), 172
Kim, S. W., **17:**135, 144
Kim, S. Y., **19:**206, 216, 280
Kim, T., **24:**12(50), 15(50), 18(50, 76), 19(50, 76), 33(138), 44(66), 46(66), 228(32)
Kim, T. W., **16:**121, 143
Kim, Y., **24:**36(148), 98(90)
Kim, Y. B., **16:**50, 132
Kim, Y. H., **16:**16, 126
Kim, Y. S., **4:**274(97), 322
Kim, Y.-T., **19:**52, 54, 55, 129, 148, 280; **24:**44(25, 50), 68(50)
Kimata, M., **11:**67(133), 68(134), 100; **23:**88(14), 103(47), 105(55), 106, 124(17)
Kimball, W. A., **7:**318, 329(26), 333(80), 334(92), 393, 394
Kimbara, A., **9:**190(84), 257
Kimblin, C. W., **14:**139(34), 156(92, 101), 160(101), 161(101), 195, 197; **17:**93, 95, 141

Kimel, G., **16:**59, 65, 68, 135
Kimerling, L. C., **12:**282(221), 295
Kimmel, R. J., **24:**183(32), 192(32)
Kimmerling, L. C., **11:**52(83), 98
Kimura, H., **11:**146(116), 149(116), 151(116), 202
Kimura, S., **14:**88(17), 89(19), 109(19), 117(19), 125
Kimura, T., **10:**253(49), 254(49), 259
Kinbara, A., **3:**220(35), 225, 229(45), 232, 233, 241, 271; **11:**68(138), 69(138), 100; **16:**262, 269, 288, 338; **19:**128, 139, 140
Kinch, M. A., **21:**103(40), 112, 133(45), 233
Kinder, H., **16:**52, 133
King, C. N., **11:**268(85), 269, 316
King, D. E. N., **3:**201(74), 210
King, D. L., **21:**312(1), 313(1), 321(1), 322(1), 344(1), 366
King, E. L., **12:**259(83), 292
King, J. H., **10:**159(88), 166
King, Jr., H. E., **24:**182(22)
King, L. L. H., **17:**238, 239, 297
King, P., **2:**265(24a), 289(34a), 304
King, R. J., **19:**199, 238, 254
King, R. M., **11:**66(127), 68(135, 136), 99, 100
King, W., **10:**43(90), 44(90), 67
Kingdon, K. H., **3:**62, 125; **12:**71(57), 114(197), 159, 163
Kingery, W. D., **12:**294; **15:**292, 311
Kingon, A. I., **16:**14, 17, 126; **17:**226, 228, 229, 232, 234, 235, 236, 238, 239, 241, 248, 250, 251, 253, 255, 258, 260, 263, 264, 291, 295–299
Kingsbury, N. G., **17:**162, 221
Kingston, J. J., **16:**57, 75–77, 117, 134, 137, 142
Kingston, R. H., **4:**289(109), 322; **21:**219(124), 236
Kinn, T. P., **1:**127(5), 184
Kinney, J. H., **24:**96(75), 97(75), 98(75)
Kino, G. S., **17:**86, 140, 153, 167, 172, 173, 184, 219, 221, 222
Kinoshita, H., **22:**39(47), 45(47), 78, 161(79), 173
Kinosita, K., **1:**283(19), 331; **3:**229(56), 230, 232, 236(56), 272
Kinsbron, E., **16:**247, 250, 252, 267–268; **19:**129, 139, 155
Kip, A. F., **4:**340(48), 358(48), 370; **6:**82(36), 144
Kippenham, B. W., **5:**149(24), 229, 246(19), 255(43), 273(19), 314, 315
Kirby, P. A., **7:**218(203a), 253
Kircher, C. J., **2:**367(28), 420; **3:**121, 128; **19:**264
Kirchgessner, J., **16:**106, 141
Kirchner, F., **4:**171(119), 190
Kirchner, Fritz, **1:**62, 67
Kirchner, P. D., **11:**54(90), 56(90), 98; **13:**219(72), 220(72), 221(72, 73), 225
Kirenskii, L. V., **6:**238(126, 127), 298
Kiridena, W., **24:**96(77), 98(77)
Kirk, E. C. G., **17:**22, 74
Kirk, P. L., **3:**28(145), 51(145), 57
Kirk, R., **10:**237(25), 258
Kirkaldy, J. S., **16:**234, 266
Kirkpatrick, S., **9:**273, 274, 296
Kirkup, L., **17:**318, 365
Kirkwood, J. G., **11:**229(24), 315
Kirtley, J. R., **14:**93(30), 126
Kiseler, V. P., **12:**105(177), 162
Kishiga, F., **12:**277(198), 294
Kishimoto, A., **17:**318, 366
Kishio, K., **14:**20(80), 75; **16:**3, 8, 96, 124, 140
Kisilev, V. A., **19:**2

Kislova, G. K., **3:**118(142), 128
Kiss, J. G., **13:**169(70), 185(70), 187(70), 194
Kistemaker, J., **3:**73(57), 126
Kita, H., **4:**196(30, 32), 198(32), 251
Kitabatake, M., **14:**20(78), 26(100), 65(100), 75; **16:**2, 17, 19, 33–34, 124, 126
Kitaev, G. A., **12:**169, 201, 203, 204, 205(38), 210, 217(97), 233, 234
Kitaigorodskii, A. F., **24:**114(25), 123(25), 125(25), 133(25), 134(25)
Kitamura, N., **1:**202(67), 229; **15:**152–153, 214
Kitao, M., **17:**313, 324, 329, 339, 344, 365–367
Kitatani, T., **21:**330(26), 331(26), 340(26), 343(26), 367
Kitayama, T., **17:**279, 299
Kitazawa, K., **14:**20(80), 75; **16:**3, 8, 96, 124, 140
Kittel, C., **1:**316(69), 333; **6:**245(143), 298; **15:**135, 141; **22:**123(31), 135(57), 172–173
Kiu, J. K., **17:**153, 154, 220
Kiuchi, M., **12:**274(185), 275(175), 294; **17:**83, 139; **18:**209
Kiyono, H., **15:**281, 310
Kiyota, K., **17:**355, 356, 357, 368, 369
Kizel, V. A., **19:**199
Kizilyalli, H. M., **11:**260(75), 261, 316
Kjr, K., **24:**180(10, 11), 181(12, 14, 15), 185(10), 186(10), 192(12), 199(12, 50)
Klaassen, F. M., **3:**285, 296, 300, 301
Klabunde, C. E., **16:**10, 43, 48–49, 125, 132
Klages, C.-P., **15:**112, 141
Klapetzky, A., **16:**37–38, 130
Klapetzky, A. J., **16:**37, 130
Klatte, S. J., **24:**75(123)
Klauder, J. R., **21:**12(34), 13(34), 72
Klaumunzer, S., **16:**72–73, 137
Klazes, R. H., **19:**73
Klee, M., **16:**97–98, 140
Kleeman, W., **19:**259
Kleimack, J., **5:**266(79), 316
Kleimack, J. J., **4:**332(29), 370
Klein, B. J., **16:**254, 269
Klein, B. M., **17:**119, 142
Klein, D. J., **5:**204(176), 234
Klein, D. L., **24:**106(106)
Klein, G. P., **6:**26, 76
Klein, H. M., **7:**36(164), 41
Klein, J., **23:**101(33)
Klein, M. J., **1:**58, 67, 323, 326, 333, 334; **2:**30(50), 61
Klein, M. L., **24:**75(124), 88(42, 43), 90(42, 43), 99(43), 127(54), 139(65), 206(5, 6, 7, 8a, 8b), 207(24), 209(5, 6, 7, 8, 39, 40), 210(5, 6, 7, 8, 40, 42), 211(51), 212(5, 6, 7, 8), 213(5, 40), 214(5, 40), 215(5, 6, 40), 216(5, 6, 40), 217(5), 220(8)
Klein, M. V., **23:**255(77)
Klein, N., **6:**66, 67, 69(196), 70, 72(196), 73, 78; **7:**373(186, 187), 397; **16:**107, 141
Klein, U., **16:**107, 141
Klein, Z., **12:**39(68), 51
Kleinman, D. A., **15:**222, 227, 229–230, 234, 239, 244, 261–262; **21:**117(10), 128(10), 196(94), 209(10), 215(10), 232, 235
Kleinsasser, A. W., **16:**21, 127
Klement, W. Jr., **2:**42(72, 72), 61, 62
Klemm, W., **12:**87, 161
Klemperer, O., **2:**68(12), 71, 92(12), 141; **9:**190(82), 257

Kleperis, J. J., **17:**337, 338, 349, 350, 355, 356, 367, 368
Klerer, J., **2:**370(51), 421; **3:**135(26), 162; **4:**261(31), 263, 319(52), 320, 321; **6:**4, 7(20), 74
Kliewer, K. L., **6:**195(51), 203; **9:**159, 162(19), 163(19), 192(91), 255, 256, 257, 261
Klimenko, G. K., **7:**264(36), 307
Kliment, V., **16:**52, 133
Klimin, A. I., **12:**105(176, 177), 106(176), 162
Kline, G. R., **17:**197, 198, 199, 201, 204, 206, 222, 223
Klokholm, E., **3:**146(65), 147, 163, 220, 223(32a), 228, 229(52), 230, 245; **16:**288, 309, 338; **22:**120(17), 121(17), 172
Kloos, T., **9:**150(7), 201(115), 202(120), 203(120), 206(7), 207(120), 208, 209(7), 255, 258
Klopfer, A., **1:**19(41, 47), 27(47), 28, 43(1 14), 46, 47, 51(114), 64, 65, 66
Klopman, B. B. G., **16:**111, 142
Klose, S., **12:**257(65), 291; **13:**99(108), 107, 111(20), 143(20), 147
Klotsman, S. M., **7:**267, 270(56, 57), 292(103), 308, 310
Klug, D. A., **19:**10
Klug, H. P., **1:**205(89), 230; **4:**218(97), 222(97), 253
Kluksdahl, N., **15:**249, 263
Kluksdahl, N. C., **15:**222, 254, 261
Klyarfel'd, B. N., **14:**155(87), 197
Klyavin, Ya. K., **17:**355, 356, 368
Kmetz, A. R., **17:**304, 364
Knab, O. D., **7:**216(197a), 253
Knapp, J. A., **12:**114(196), 163
Knecht, J., **12:**274(172), 275(172), 294
Kneer, G., **6:**210, 295
Kneller, E., **1:**330, 334
Kneller, E., **6:**231(114), 235(114), 237(114), 297
Knewstubb, P. E., **6:**38, 51(136), 77
Knight, J. C., **12:**296; **15:**297, 311
Knight, J. R., **5:**150, 230; **7:**170, 191(106), 250
Knight, S., **9:**283, 292, 295
Knights, J. C., **12:**264(103), 292; **14:**99(41), 110(41), 126; **15:**277–279, 310
Knobler, C. M., **24:**181(12), 192(12), 199(12)
Knoch, L., **23:**60(202)
Knoedler, C. M., **15:**222, 260; **21:**240(5), 308
Knoll, W., **24:**25(121), 150(3)
Knolle, W. R., **15:**272, 292, 311
Knoop, A. N., **12:**292
Knorr, K., **12:**277(192), 294; **14:**86(11), 106(11), 125
Knorr, T. G., **1:**286(31), 332
Knotek, O., **14:**179(155, 156), 186(155), 198; **17:**110, 141
Knott, G., **7:**350, 354, 358(131), 395
Knowles, J. E., **1:**106, 121
Knudsen, M., **4:**121(55), 136
Ko, W. H., **4:**305(146), 309(146), 323
Koba, R., **15:**288–289, 311
Kobayashi, A., **12:**94(148), 161
Kobayashi, K., **16:**96, 140; **17:**228, 229, 262, 298, 299
Kobayashi, K. L. I., **11:**36(28), 97
Kobayashi, K. W., **17:**201, 223
Kobayashi, M., **13:**60(6), 78, 104; **14:**132(12), 195; **15:**172–173, 216
Kobayashi, N., **16:**86, 88–92, 138–139; **17:**361, 362; **19:**12, 19
Kobayashi, S., **15:**172, 216

Kobayashi, T., **9:**271, 296; **15:**154, 171, 174–176, 179–181, 195–199, 204, 214, 216–217
Kobayashi, Y., **17:**161, 221, 287
Kobel, E., **14:**196
Kobiyama, S., **13:**1(1), 56
Kobliska, R. J., **15:**174, 180, 216
Kobos, P., **15:**111, 141
Kobrin, P. H., **16:**28–29, 111, 116, 129, 142
Koçak, H., **21:**68(127), 75
Koch, D. J., **22:**178(14), 271
Koch, F. B., **14:**232(113), 239
Koch, G. E., **13:**60(9), 62(9), 104
Koch, H., **9:**34(159), 35(159, 163), 68
Koch, J. F., **6:**82(36), 133, 134(201, 202, 203), 144, 148
Koch, R. H., **16:**21, 41–42, 116–118, 122, 127, 131, 142–143, 252, 268
Koch, S. M., **19:**2, 9, 29–31, 38, 41
Köck, A., **21:**290(43), 310
Kock, H. G., **6:**116(158), 147
Kodaira, K., **17:**318, 370
Kodama, M., **13:**25(47), 27(47), 58, 102(116), 103, 107
Kodami, K., **19:**35
Kodas, T. T., **22:**182(27), 272
Koedam, M., **3:**73(44), 75(44), 76(44), 77(44), 125; **9:**5(21), 65
Koeh, H., **12:**49(80), 51
Koehler, J. S., **17:**66, 67, 68, 77
Koehler, W. E., **2:**315, 359; **12:**6, 50
Koehler, W. F., **3:**36, 57; **4:**2(1), 4(11, 12), 25(60), 34(60, 64), 58(127); **10:**204(37), 226
Koenig, H., **3:**110, 124(161), 127, 128, 156, 164
Koenig, S. H., **6:**116, 117, 132(162), 147
Koening, J., **12:**181(70), 190(70), 234
Koenings, J., **12:**262(94), 292
Koester, C. J., **4:**85(218), 96
Koestner, R. J., **15:**245, 263
Koffyberg, F. P., **9:**22(94), 30, 67
Kofink, W., **1:**70(10), 120
Kofstad, P., **9:**9, 10(45), 66; **14:**116(75), 127
Koga, Y., **5:**302(257), 303(260), 307, 321, 322; **12:**264, 292
Kogan, V. G., **16:**73, 137
Kogelnik, H., **10:**228(5), 238(29), 251, 258
Kogut, A. N., **7:**31, 34(131), 40
Koh, P. K., **3:**39(196), 52(196), 58
Kohane, T., **15:**8, 136
Kohayashi, K., **2:**135(158), 137(158), 145
Kohiki, S., **16:**30, 33–34, 129–130
Kohl, P. A., **13:**160, 164(56, 57), 165(56, 57), 166(56), 167(57), 170(38), 176(56, 57), 178(100), 186(38, 56, 100), 188(100), 189(56), 193, 194, 195
Kohler, H., **16:**5, 124
Kohler, M., **6:**82(6), 143
Kohlschütter, V., **4:**199(48, 50), 200(72), 204(48, 50), 252
Kohn, E., **13:**212(40), 224
Kohn, E. S., **7:**244(272), 246(275), 255; **12:**135, 165; **23:**97(10)
Kohn, G., **6:**293(225), 300
Kohn, W., **2:**400(161), 424; **7:**372(181), 396; **9:**146(6), 255; **15:**221, 260; **21:**97(30), 112
Kohnke, E. E., **9:**22(87), 23(87), 67
Kohno, O., **16:**88, 138
Koidi, P., **21:**121(17), 123(17), 152(17), 232
Koidl, P., **15:**307, 312
Koike, Y., **17:**355, 356, 357, 368, 369
Koikeda, T., **6:**227(96), 297
Koinkar, V. N., **16:**70, 136

Koinuma, H., **14:**20(80), 75
Kojima, T., **15:**244, 263
Kokada, H., **17:**355, 369
Kokai, H., **17:**153, 219
Koki, A., **17:**201, 223
Kokkinakis, T., **9:**219(176), 259
Kokorish, N. P., **5:**265(70), 273(70), 316
Kokoropoulos, P., **2:**353(77), 361; **10:**43(87, 88), 44(87, 88), 67
Kola, R. R., **22:**160(78), 173
Kolb, G. A., **7:**135(30), 248
Kolb, G. H., **5:**247, 315
Koleshko, V. M., **16:**237, 267
Kolesova, V. A., **4:**271(94), 322
Kolk, A., **6:**208(15), 295
Koller, L. R., **3:**195, 209; **4:**47(87), 50, 92; **11:**111, 200; **12:**54(2), 71, 158
Kollmorgen, F., **2:**240, 303
Kolomiets, S. I., **7:**177(118d), 250
Kolomoets, N. V., **13:**32(49), 58
Koltun, M. M., **10:**57(119), 68
Koma, A., **11:**44(54), 48(64), 84(194), 97, 98, 102
Komaki, K., **9:**166(21), 218(21, 173), 256, 259
Komar, A. P., **1:**205(86), 230
Komarek, E. E., **6:**36(111), 76
Komatsubara, K. F., **11:**36(28), 97
Komem, Y., **16:**255, 269
Kometani, T. Y., **16:**2, 124
Kominiak, G. J., **10:**28(49), 58(128), 61(140), 66, 68, 69; **12:**265(115), 292; **13:**111(13, 14), 112(14), 145(14), 147; **14:**135(13), 136(13), 195, 210(24, 25), 211(24), 212(24), 213(24), 220(68), 221(24, 25), 223(77), 233(135), 237, 238, 240; **22:**160(76), 173
Komiya, S., **13:**60(5), 78, 79(5, 62), 104, 106
Komnik, Y. F., **9:**2, 64
Komnik, Yu. F., **1:**209, 230; **6:**118(167), 141, 147, 149
Komolova, L. F., **17:**327, 335, 366
Komori, A., **18:**45, 51
Komuma, M., **12:**287(252, 253), 295
Konagai, X., **24:**99(104), 100(104), 101(104)
Kondakov, S. F., **16:**71, 79, 136
Kondo, H., **1:**283(19), 331; **3:**229(56), 230(56), 236(56), 272
Kondo, K., **15:**221, 260
Kondo, M., **13:**119(49), 126(49), 148; **14:**67(226), 78
Kondo, N., **13:**20(34, 35), 21(34), 22(35), 30(34), 31(35), 57
Kondo, R., **9:**60(271), 71
Kondo, S., **17:**309, 324, 325, 327, 328, 329, 365, 366
Kondratiev, V. N., **12:**238(10), 290
König, H., **1:**59(143), 67, 224, 225, 232
Konig, H., **7:**24(84), 39
Konig, U., **11:**36(30), 97
König, U., **13:**212(40), 224
König, W., **1:**142(27), 185
Konishi, A., **12:**246(22), 254(22), 290
Konkle, K. H., **13:**187(126), 196
Konnert, J. H., **15:**156–157, 215
Konnerth, K., **7:**142, 218, 219(38), 248
Konno, H., **1:**324(96), 326(96), 334; **2:**30(46), 61
Kono, T., **1:**282(16), 331
Konopka, A., **16:**97, 107, 140–141
Konopka, J., **16:**97, 107, 140–141
Konstantinova, F., **19:**281
Konuma, K., **21:**31(62), 32(62), 73; **23:**144(35)
Koo, J. Y., **17:**19, 20, 21, 32, 74

Koo, M., **24:**172(42), 173(42)
Kooi, C. F., **1:**285, 328(104), 332, 334; **6:**247(155), 273, 274(179, 180, 181), 275(179), 298, 299
Kooi, E., **4:**260(22), 262(46), 301(22, 129), 305(22), 320, 321, 323
Koomen, M. J., **7:**104(82), 114
Koonce, S. E., **3:**112(123), 127, 139(44), 163; **4:**267(77), 322; **5:**310(282), 322; **9:**51(239), 70
Koop, S. A., **4:**123(61), 124(61), 127(61), 136
Kooy, C., **9:**264, 295
Kopf, R. F., **21:**192(89), 193(89), 194(89), 195(89), 235
Kopp, W., **15:**244–245, 263
Koppe, H., **2:**77, 141
Koppelmann, G., **2:**214, 238; **3:**32(177), 36(176), 37, 52(177), 57; **12:**48(78), 51
Koppelmann, Von G., **10:**168(4), 169, 225, 226
Korb, H. W., **5:**303(262), 321; **7:**234(248), 254
Koren, G., **13:**153(6), 155(28), 167(63, 67), 181(6), 182(6), 183(6, 28); **16:**57, 59, 61–62, 64–65, 68, 71–72, 78, 110, 134–137, 142
Korenblit, I. Ya., **6:**117(165b), 147
Korenic, E. M., **24:**44(6), 45(6), 66(6), 70(6), 86(40), 125(50), 212(62), 213(62), 217(62)
Korhonen, A. S., **17:**119, 143
Koriufskii, A. D., **12:**126(262), 164
Korndorfer, C., **14:**36(138), 48(162), 55(162), 76, 77
Korndörfer, C., **17:**111, 116, 142
Korneeva, T. I., **6:**183, 203
Kornfeld, A., **21:**40(86), 74
Kornmann, M., **13:**78(63), 79(63), 106
Korol'kov, V. I., **7:**217(200), 246(276), 253, 255
Korotkikh, V. L., **12:**126(262), 164
Kortan, A. R., **16:**36–37, 51, 130, 132
Korzo, V. F., **9:**39, 40, 69; **12:**177(60, 61), 234
Koschel, W. H., **11:**50(73), 98
Kosevich, A. M., **6:**137(208), 148
Kosevich, V. M., **7:**21(66), 38
Koshida, N., **17:**345, 367
Koshiga, F., **12:**277(205), 294
Koshizuka, N., **13:**172(78), 173(78), 177(78), 190(78), 191(78), 194; **15:**103, 113, 140; **16:**14, 125
Kosicki, B. B., **13:**61, 104
Koskinen, J., **13:**142(127), 150
Köslin, H., **10:**163, 166
Koslover, R., 18.78n
Kosonocky, W. F., **21:**4(15), 19(15), 72; **23:**101(33), 124(14)
Kosowsky, S. D., **24:**112(23), 121(23)
Kossel, D., **5:**1(5), 2(12), 31(22), 40(24), 45
Kossel, W., **1:**199(58), 210, 229, 231
Kost, A. R., **23:**234(20), 235(20)
Köster, W., **2:**387(109), 422
Köstlin, H., **9:**37, 69
Kostopoulos, D., **16:**242–245, 253–254, 257–260, 265, 267
Kotani, N., **22:**181(18), 271
Koteles, E. S., **21:**207(109), 235
Kotrubenko, B. P., **7:**170(103), 250
Kotthaus, J. P., **21:**96(27), 111, 155(64), 234
Kötz, R., **17:**326, 366
Kotzebue, K. L., **15:**98, 140
Koury, C. J., **13:**94(89), 106
Koutsky, J. A., **4:**121, 136
Koutsos, V., **24:**95(77)
Kouvel, J. S., **6:**215(44, 45, 47, 50), 217, 218(44, 50), 220, 222, 296

Kouyumdjiev, C. N., **16:**317, 336, 339
Koval, I. F., **12:**121(243), 164
Koval'chuk, V. S., **7:**262(18), 264(33), 306, 307
Kovalev, V. I., **6:**137(213), 139(213), 142(213), 148
Kovener, G. S., **9:**237(223), 261
Kovenskii, I. I., **7:**264(39, 42), 307
Kowalchik, M., **7:**148, 164, 165, 166(56), 224(222), 248, 253
Kowalczyk, S. P., **9:**217(165, 169), 259; **15:**177, 216
Kowne, I. A., **2:**94(45), 141
Koyama, S., **16:**4, 33, 124
Koyanagi, K., **13:**20(37), 22(37), 33(37), 57
Koyanagi, T., **13:**20(38), 22(38), 34(38), 57
Koza, M. A., **19:**2, 36
Kozawa, S., **4:**83, 92
Kozlov, G. V., **16:**5, 124
Kozlova, E. K., **13:**169(70), 185(70), 187(70), 194
Kozlowski, L. J., **21:**117(8), 128(8), 221(8, 134), 223(8), 232, 236, 257(31), 258(31), 309
Kraak, H. H., **2:**23(23), 60
Krabach, T. N., **21:**36(70), 73; **23:**161(63)
Kraitchman, J., **5:**289, 320; **6:**37(126), 42, 43, 49, 76
Krajewski, J. J., **16:**3, 124
Krall, H. R., **12:**143(314), 144, 166
Kramer, B. M., **17:**103, 104, 141
Kramer, C., **21:**322(14), 337(14), 367
Kramer, D. A., **9:**47(222), 70; **11:**61(108), 99
Kramer, E. J., **24:**169(39), 170(39), 233(53)
Kramer, G., **11:**189(150), 203
Kramerinko, G. S., **12:**71(60), 74(60), 78(60), 80(60), 81(60), 84(60), 119(60), 159
Krane, K. J., **9:**203(120a), 258
Kraselov, A. V., **5:**265(70), 273(70), 316
Krasilov, Y. I., **19:**199
Krassner, L., **7:**246(277), 255
Kraus, C. J., **2:**147(2), 178(2), 190, 365(30), 366(3), 419
Kraus, H., **5:**167(136), 182, 233
Kraus, T., **2:**346(72), 360, 402(170), 424; **10:**129(59), 135(59), 165
Krausbauer, L., **14:**226(79), 238
Krause, G. O., **7:**133(24, 27), 247
Krauss, A. R., **14:**6(21), 73; **16:**14, 17, 126; **17:**234, 235, 296
Krebs, H., **7:**20(65), 27(101), 38, 39
Krebs, K., **2:**214, 238; **4:**78(197), 95; **9:**119(30), 144
Kreibig, U., **9:**190(83), 198(110), 218(110), 257, 258; **19:**102, 128, 129, 134, 142, 178
Kreines, N. M., **15:**95, 140
Kreismanis, V. G., **13:**164(55), 185(55), 189(55), 194
Kreitchman, J., **12:**276, 294
Kremer, F. J. B., **24:**21(102), 26(102), 89(50)
Kren, J. G., **3:**171(24), 176(24), 209; **5:**164(89), 172(89), 231
Kreplin, R. W., **10:**78(22), 164
Kress, K. A., **10:**108, 165
Kressel, H., **7:**137(34, 35), 138, 139(35), 142(41, 44), 143(50), 146(52, 53), 147(34, 52), 153(65), 157(34), 166(91), 171(114), 177(177), 186(141, 147), 187(150), 188(151), 192, 193(156), 201, 202(177), 205, 207(183); **12:**125(294), 213(193), 214(195, 197), 215(246),

Kressel, H., (*continued*)
216(193, 197), 217(197), 218(201, 202, 203, 246), 219(53), 220(150), 221(213), 222(150), 223, 224(91), 226(232, 233), 227(232, 233), 230(35), 231, 232(236), 234(241, 242), 243(266), 244(272), 245(273, 274), 246(277), 248, 249, 250, 251, 252, 253, 254, 255, 1659(150, 183)
Kretschmann, E., **9:**153(10), 155(13), 177(47), 181, 187(71, 73), 196, 197(103), 223(189d), 229(210), 231(47, 215), 233, 235(10), 242(102), 245(103, 245), 246(212, 246, 247), 247(103, 249), 250(103), 251, 252, 253, 254, 256, 257, 258, 260, 261
Kretschmar, G. G., **5:**165(104), 174, 232
Krieble, J. G., **5:**247(23), 315
Krikorian, E., **3:**103(175), 115, 129; **5:**167(127, 128), 181, 193, 196(127, 128), 197, 198, 199, 232; **11:**113, 200
Kriman, A. M., **15:**222, 249, 254, 261, 263
Krimmel, E. F., **13:**185(128), 187(128), 196
Krinchik, G. S., **1:**329(109), 334
Krischer, C., **17:**171, 221
Krishnamurthy, B. S., **9:**15(68), 66
Krishnan, K., **15:**303, 311
Krishnan, S., **19:**168
Krishnaswamy, J., **16:**65, 136
Krishnaswamy, S. V., **17:**153, 154, 155, 156, 157, 159, 160, 189, 190, 191, 192, 193, 197, 199, 200, 201, 202, 203, 204, 205, 206, 207, 215, 220–224, 226, 248, 294, 295, 297, 299
Krivoglaz, M. A., **7:**262(21), 306
Kroeger, D. M., **16:**3, 124
Kroemer, H., **7:**135(32), 248; **11:**81(184), 101
Kröger, E., **9:**174(35), 176(35), 179, 197(103), 201, 210, 223(189d), 229(210), 245(103), 247(103), 250(103), 256, 258, 259, 260
Kröger, F. A., **3:**172, 208; **7:**142(39), 248
Kroger, H., **7:**47(22), 112
Krohn, K. E., **13:**182(116), 183(116), 196
Krokhina, A. I., **3:**102(92), 127
Krolikowski, W. F., **4:**329, 369; **12:**60(43), 159
Krongelb, S., **4:**261(38), 263(38), 264(38), 321; **5:**303, 321
Kronig, R., **7:**44(4), 111
Kronmüller, H., **6:**222(75), 296
Kruangam, D., **19:**61
Krueger, H. A., **13:**48(58), 58
Kruer, M. R., **21:**219(127), 236
Krug, J., **15:**64, 139
Krug, W., **3:**36(172), 57
Krüger, G., **6:**66, 78
Kruger, H., **4:**271(91), 322
Kruger, J., **4:**37, 38, 92
Kruglikov, E. P., **14:**196
Kruglova, A. V., **9:**33(142, 143), 68
Krumme, J.-P., **15:**112, 141
Krupanidhi, S. B., **14:**57–58(180), 77; **17:**153, 220, 228, 242, 288, 297; **19:**251, 265, 272
Kruppa, E. F., **12:**286(243), 295
Kruse, F. H., **7:**35(150), 41
Kruse, P. L., **10:**11(19), 65
Kruse, P. W., **11:**123(77), 181(77), 201
Krutter, H., **7:**1(3), 37

Kryder, M. H., **6:**216(60), 268, 296; **9:**266, 295; **15:**154, 197, 204–205, 215, 217
Krysinski, P., **24:**44(53)
Kryszewski, M., **15:**291, 311
Kryzhanovskii, B. P., **9:**22(82), 27(82), 33(142, 143, 144), 47(222b), 67, 68, 70
Kryzhanoyskii, B. P., **12:**187(83), 234
Ksendzov, A., **23:**14(50), 161(63)
Ku, H. Y., **2:**400(163), 424
Ku, S. M., **5:**161, 230
Kuan, C. H., **21:**241(9, 13), 256(9), 269(9), 308, 309
Kuan, R., **21:**256(28), 267(28), 277(28), 309
Kuan, T. S., **13:**217, 224; **15:**222, 261; **23:**24(97)
Kuang, A. X., **19:**250
Kubaschewski, O., **4:**37(66), 92
Kubo, H., **19:**51
Kubo, T., **17:**345, 367
Kubo, Y., **16:**53, 134, 262, 269
Kubono, T., **14:**157(95–97), 197
Kubota, J., **17:**228, 229, 262, 298, 299
Kubota, K., **17:**161, 221, 287, 299
Kucera, J. T., **16:**28, 33, 116, 128, 142
Kuchibhatla, D., **13:**127(83), 142(83), 149
Kuchieko, S. I., **17:**318, 366
Kudman, I., **7:**147(55), 173(119), 174(119), 179(119), 248, 250
Kudo, T., **17:**318, 365, 366
Kudryashova, M. D., **10:**32(55), 66
Kuech, T. F., **15:**222, 261
Kueny, A., **17:**24, 45, 48, 49, 51, 52, 56, 57, 74
Kuglin, C. D., **3:**277(29), 300
Kühl, C., **17:**109, 110, 141
Kuhlman, G., **13:**94(89), 106
Kuhlmann-Wilsdorf, D., **3:**255, 256, 273; **4:**139, 158(86), 162(86), 187
Kuhn, H., **2:**334(52), 360; **4:**60(133), 94; **7:**372(182), 379(199, 200), 383, 384(199), 385(200), 386(199, 200, 202, 203), 390(182), 391(213), 396, 397
Kuhn, L., **10:**228(7), 236(17), 258
Kuhn, M., **12:**282(224), 295
Kuhn-Kuhnenfeld, F., **13:**178(99), 186(99), 189(99), 195
Kuhnert, K., **9:**178(51), 257
Kuindersma, P. I., **23:**226(11), 227(11), 228(11), 229(11), 232(11)
Kuiper, A. E. T., **15:**297, 311
Kukimoto, H., **12:**294; **16:**93, 139
Kukuk, H. S., **2:**365(11), 420; **3:**24(110), 51(110), 55
Kukuyev, V. I., **17:**327, 335, 344, 366, 367
Kula, W., **16:**97, 99, 140
Kulak, A. I., **17:**317, 365
Kulakovskii, V. D., **21:**207(106), 235
Kulcke, W., **7:**27(104), 39
Kulik, I. O., **6:**137(210, 214, 219), 148
Kulik, L. A., **6:**142, 149
Kulkarni, A. K., **17:**241, 287, 297
Kuma, S., **24:**89(50)
Kumagai, H. I., **14:**28(111), 76
Kumagai, S., **6:**27, 76
Kumagai, T., **16:**28–29, 129
Kumakura, H., **16:**9, 125
Kumar, A., **16:**91, 139; **24:**17(73), 21(102), 26(102), 228(6, 25, 26), 229(6, 35, 38), 230(26), 231(26), 234(26), 240(26), 247(87), 248(90), 249(6)
Kumar, C. S., **10:**237(22), 258
Kumar, K., **16:**169, 193, 202–203, 207, 230

Kumar, S., **16:**101, 141; **17:**49, 53, 76
Kumar, S. N., **12:**226(122), 235
Kumar, V., **19:**265
Kume, M., **14:**18(75), 20(75), 75
Kümmel, R., **6:**137(220), 148
Kun, Z., **17:**182, 222
Kunchenko, V. V., **14:**177(146), 198
Kung, A., **17:**24, 46, 49, 51, 53, 56, 58, 75
Kunig, H. E., **5:**214, 216(192), 217, 228(192), 234
Kunikiyo, T., **22:**181(18), 271
Kunimatsu, K., **17:**315, 316, 329, 331, 365
Kunin, L. L., **4:**139(14), 156(14), 187
Kunioka, A., **9:**50(236a), 70
Kunori, M., **13:**1(1), 56
Kuntz, U. E., **3:**137(32a), 163
Kunz, C., **7:**46(17), 48(25), 112; **9:**202(119), 204(125, 126), 205(126), 209(126), 211(138, 139), 258, 259; **10:**115, 116(44, 45), 117(45), 165
Kunz, W. E., **15:**86–87, 140
Kunze, C., **12:**85(107), 102, 161
Kunzel, H., **11:**83(188), 101
Kunzler, J. E., **6:**122, 147
Kuo, C. C., **6:**134(201), 148
Kuo, C. Y., **9:**60(269), 71
Kuo, J. M., **21:**94(25), 111, 128(30), 129(30), 130(30), 131(30), 132(30, 43), 133(43), 135(30), 136(30), 137(30, 43), 138(30), 139(30), 140(30), 141(30), 142(30), 143(30), 144(30), 145(30, 52), 146(30), 147(30, 52), 148(30), 149(30, 43, 52), 150(30, 43, 52, 55), 155(59), 158(59), 159(59), 160(30), 170(78), 180(78), 181(55, 78), 182(78), 183(78), 184(78), 206(104), 213(52), 219(43), 233, 234, 235
Kuo, P. K., **24:**96(77), 98(77)
Kuo, Y. S., **17:**91, 92, 140
Kuper, A. B., **4:**305(146), 309(146), 323
Küpfer, H., **16:**9, 125
Kupperian, J. E., **10:**78(22), 164
Kuppermann, A., **4:**339(45), 370
Kuppers, D., **12:**262(94, 95, 96), 292
Küppers, J., **9:**215, 259
Kuptsis, J. D., **6:**337(44), 347; **9:**276, 283, 284, 295; **14:**18(72), 75; **15:**13, 25, 137
Kurabayashi, T., **19:**35
Kurger, F., **22:**118(1), 171
Kurita, S., **19:**61, 105
Kuriyama, Y., **13:**20(36), 22(36), 32(36), 57
Kurnick, S. W., **2:**297(45), 304(45), 420
Kuroda, K., **12:**282(235, 236), 295; **14:**19(77), 66(219), 75, 78; **16:**41, 52, 131, 133
Kuroda, R., **6:**272(173), 284(173), 299
Kurogi, Y., **17:**279, 299
Kurosawa, H., **16:**84, 86, 88–90, 92, 111, 116, 138–139, 142
Kurov, G. A., **1:**218(134), 231
Kurtin, S., **13:**202, 223
Kurtz, R. L., **16:**28, 128
Kurtz, S. K., **7:**162(86), 249
Kurtz, S. R., **21:**322(14), 337(14), 367; **23:**231(16), 278(16)
Kurtzig, A. J., **9:**287, 296
Kurumizawa, T., **24:**169(37)
Kuryla, M. S., **21:**362(44), 368
Kus, P., **16:**23, 127
Kuse, N., **21:**290(44), 310
Kushida, K., **17:**162, 221; **19:**260
Kushihashi, A., **9:**60(270), 71

Kushman, D., **17:**199, 222
Kushner, M. J., **13:**68, 105; **17:**87, 140; **18:**124, 194
Kushner, R. A., **14:**220(67), 231(67), 238
Küster, H., **13:**61, 101(24), 104, 130, 149
Kusuda, T., **15:**150, 154–155, 161, 162, 163, 195, 214–215
Kutzner, J., **14:**141(36 37), 156(91), 195, 197
Kuvas, R. L., **11:**77(161), 100
Kuwabara, K., **17:**354, 368
Kuwahara, K., **2:**30(47), 61; **14:**67(226), 78
Kuwahara, M., **17:**288, 299
Kuwano, H., **14:**240
Kuwano, Y., **12:**267, 293; **19:**61
Kuzel, R., Jr., **17:**83, 89, 90, 119, 121, 123, 126, 127, 128, 131, 139, 140, 143
Kuz'menko, P. P., **7:**262(18), 264(33, 36, 37, 38, 40), 306, 307
Kuzmin, A., **17:**310, 365
Kuznetsov, A. Y., **9:**33(142, 143, 153), 68
Kuznetsova, E. G., **6:**64(187), 78
Kuznetzoff, P., **4:**301(133), 323
Kuznetzov, B. A., **24:**18(78), 19(78)
Kuzuo, R., **15:**170, 174, 215
Kwak, B. S., **16:**83–84, 88, 94, 138, 140; **17:**238, 297
Kwak, J. F., **16:**54, 134
Kwo, J., **16:**2, 13, 17, 28–29, 36–37, 51, 110, 124–126, 129–130, 132, 142
Kwo, J. R., **16:**74, 137
Kwok, H. S., **16:**55–56, 64–65, 134, 136
Kwok, N. L., **12:**186(94), 197(94), 234
Kwok, T., **16:**234, 250, 266, 268
Kwong, D. L., **13:**155(25), 166(25), 192
Kwong, Y. H. K. D. L., **15:**295, 299, 311
Kyle, N., **5:**273, 317
Kyono, C. S., **21:**193(93), 235

L

L dakss, M., **10:**235(11), 258
L. M. Simard, Inc., **15:**152–153, 214
Labate, E. E., **4:**341(51), 343(51), 345(51), 349(51), 354(51), 361(80); **12:**84(99), 160; **23:**88(16)
Labdi, S., **16:**13, 32, 125, 130
LaBrie, J. J., **4:**265(62), 319(62), 321
Labrunie, G., **10:**257(65), 259
Labun, A. H., **22:**76(68), 78, 243(93), 245(93), 274
Labunov, V., **14:**94(32), 103(32), 106(32), 118(32), 126
Lacey, E. T. M., **15:**171, 216
LaChapelle, T. J., **5:**238(4), 261, 270, 314
Lacharme, J. P., **14:**120(79), 127
Lackey, W. J., **16:**92, 139
Lacklison, D. E., **15:**113, 141
Lacoe, R. C., **21:**162(69), 234
Lad, R. J., **14:**66(218), 78; **17:**153, 220
Ladany, I., **7:**133(29), 166(91), 177(177), 201, 202(177), 205, 207, 209(183), 219(204), 223(91, 219), 224(91), 225(225), 226(232, 233), 227(232, 233), 247, 249, 252, 253, 254
Ladd, G. O., **5:**279(160), 318
Laderman, S. S., **16:**2, 22, 123, 127
Ladwig, H., **9:**33(136), 68; **12:**187(76), 234

Laegreid, N., **3:**73(42), 74(51), 77(42), 85(42), 125, 126
Laendle, K. W., **2:**370(53), 421; **3:**138(34), 163; **4:**263(56), 264(56); **12:**259(87), 292
Lafferty, J. M., **1:**27(61, 62), 30(62, 68), 65; **14:**138(19), 195
Lagally, M. G., **24:**99(100), 106(100)
Lagarde, Y., **9:**39(172a), 69
Lagier, G., **13:**121(59), 148
Lagzdons, J. L., **17:**354, 368
Lahav, M., **24:**180(11), 181(12, 15), 192(12), 199(12)
Lahaye, J., **12:**174(47), 233
Lai, Y. P., **9:**297
Laibinis, P. E., **19:**282, 288, 310; **24:**3(15, 20), 4(15), 5(22), 9, 10(37), 11(20), 12(20, 37), 13(20), 14(20, 37), 15(20, 37, 64), 16(20, 37, 64), 17(22), 21(103), 23(103, 113), 26(103), 27(103, 113), 30(113), 32(15, 37, 71), 44(8, 16), 66(8)
Laibowitz, R., **9:**4(14), 5(14), 65
Laibowitz, R. B., **16:**21, 41–42, 57, 71, 117–118, 127, 131, 134, 136, 143
Laidler, K. J., **24:**46(91), 57(91), 62(91)
Laikin, M., **1:**144(39), 169(39), 185
Lairson, B. M., **16:**106, 141
Laitinen, H. A., **7:**330(65), 394; **9:**31(114a), 35(165), 36(165), 37(165), 67, 69
Laitiner, H. A., **12:**190(91), 234
Lakatos, A. I., **11:**256(62, 64, 69), 257(62, 64, 69), 315, 316
Lakhani, A. A., **15:**245, 263
Lakin, K. M., **17:**153, 154, 197, 198, 204, 206, 220–223
Lakin, K. W., **17:**198, 222
Lakin, L. R., **6:**228(104), 230(104), 238(104), 297
Lakshmanan, T. K., **2:**369(39), 387(106), 388(106), 393(39), 420, 422; **3:**116, 124(165), 128; **5:**167(137), 233; **9:**50(235), 70
Lalak, J., **17:**227, 295
Lally, V., **2:**371(60), 421
Lamb, D. R., **6:**63, 78
Lamb, J. L., **14:**68(229), 70(229), 78
Lamba, J. J. S., **24:**33(137), 44(37)
Lambe, J., **6:**36(107), 76
Lambeir, R., **2:**23(26), 60
Lambert, V. L., **11:**121(75), 201
Lammers, U. H., **15:**55, 138
Lamont, L. T., **14:**71(232), 79, 208(17), 237
Lamp, B. D., **24:**44(33), 53(33)
Lampe, D. R., **17:**229, 237, 271, 272, 275, 276, 277, 278, 280, 281, 282, 283, 285, 286, 290, 294, 296, 298–300
Lampert, C. M., **15:**304–305, 312; **17:**304, 364
Lampert, M., **2:**150(28), 191
Lampert, M. A., **5:**145, 229
Lampkin, C. M., **12:**172, 233
Lamport, C. B., **5:**285, 319
Land, C. E., **17:**249, 297; **19:**259
Landau, L., **9:**270, 296
Landau, L. D., **11:**215(11), 314; **15:**103, 113, 128, 130, 140–141; **16:**22, 127; **22:**147(72), 173
Landau, L. V., **1:**256, 258(63), 259(63), 275
Lander, J. J., **4:**40(77), 92; **5:**190, 233
Landermann, J., **9:**282, 295
Landford, W. A., **14:**103(49), 126
Landfors, A. A., **1:**13(23), 64
Landgren, G., **21:**117(11), 232

Landheer, D., **15:**250, 263; **21:**81(12), 111, 170(73), 178(73), 234
Landin, A. R., **17:**153, 197, 198, 201, 220, 222, 223
Landman, U., **24:**151(10), 154(17), 156(10, 17), 163(10), 211(53), 222(53, 104)
Landmann, U., **9:**203(123), 258
Landorf, R. W., **9:**283, 296; **15:**9–10, 136
Landsberg, P. T., **7:**207(184), 252
Lane, E., **15:**303, 311
Lanford, W. A., **12:**269(139), 277(212), 279(212), 280(212), 293, 295; **13:**135(108), 136(108), 137(108), 150; **15:**291, 311
Lang, A. R., **5:**191(164), 233
Lang, C. A., **24:**86(26)
Lang, D. V., **11:**51(81), 52(82, 83), 53(89), 54, 55, 89(227), 98, 103
Lang, G. A., **5:**247(30), 315
Lang, J. K., **15:**196, 217
Lang, N. D., **9:**146(6), 255
Lang, W., **2:**101, 142
Langberg, E., **3:**66, 125
Langdon, J. L., **4:**257(2), 320, 320(167), 324
Lange, B., **3:**275, 299
Lange, V. N., **7:**170(103), 250
Lange, W. J., **1:**33(83), 36(92), 42(108), 66
Langell, M. A., **14:**17(67), 74
Langenburg, D. N., **3:**277(26, 27), 278(27), 300
Langendam, W. T., **2:**387(107), 422
Langer, C., **5:**240(12), 314
Langford, R. C., **2:**379(80), 421
Langier, J., **1:**281(12), 331
Langkowski, J., **9:**203(120b), 258
Langley, B. W., **16:**12, 101, 125
Langley, R. C., **10:**59(131), 60(131), 68
Langlois, J.-M., **17:**279, 299
Langmuir, D. B., **2:**72(21), 125(21), 141
Langmuir, I., **1:**209, 230; **3:**62, 125; **7:**312(4, 6), 315, 316, 317(19), 319(30), 320(19), 323(4), 324(38), 325(6, 42), 328(6), 330(42), 335(6), 337(6, 19, 108), 338(19, 109), 341(19), 342(6, 19), 343(6, 19), 347(118), 348(6), 350(6, 19, 128, 129, 130); **12:**114, 163; **13:**60, 104; **24:**180(5)
Langmuir, J., **5:**87, 140
Lanir, M., **11:**160, 203
Lankard, J. R., **13:**153(7), 154(7), 172(7), 173(7), 182(7), 183(7), 192
Lanson, J. D., **17:**153, 219
Lanza, C., **13:**203(13), 220(69), 223, 224; **21:**193(92), 235
Lapersonne-Meyer, C., **16:**292, 334, 340
Lapeyre, G. J., **10:**108, 165; **12:**115(214), 163
Lapin, A., **10:**169(16), 225
Laponsky, A. B., **12:**133(287), 165
Lappe, F., **3:**116, 117, 128; **9:**50(233), 70
Lappin, T. A., **10:**61(141), 69
Larabee, G. B., **15:**300–310, 311
Larbalestier, D. C., **16:**45, 48, 132
Larden, M., **14:**135(14), 195
Lardon, M. A., **14:**226(89), 239
Lark-Horovitz, K., **3:**281(43), 300
Lark-Horowitz, K., **5:**274, 317
Larnois, H., **17:**121, 143
LaRoy, B., **6:**120(172), 129(172), 130(172), 135(172), 147
Larsen, I. D., **15:**55, 138

Larsen, N. B., **24:**230(47), 237(62), 238(62)
Larsen, P. K., **17:**228, 240, 256, 257, 297, 298
Larson, D. C., **6:**82, 110, 111, 112(30), 124, 125, 126, 127, 132(130), 144, 146; **9:**7(32), 8(32), 65
Larson, K., **2:**216(19b), 238
Larson, T. M., **16:**9, 125
Larsson, T., **14:**14(50, 51), 74; **17:**111, 112, 116, 142
Lasaponara, L., **13:**131(97), 140(97), 141(97), 149
Lashkari, D. A., **24:**228(28)
Lasko, W. R., **2:**373(66), 420
Laso, M., **24:**208(34)
Lassegues, J. C., **17:**329, 366
Lasser, M. E., **4:**370
Lastras-Martinez, A., **19:**9
Lathlaen, R., **12:**263(99), 264(99), 292
Lathrop, D. K., **16:**51, 70, 107, 133, 136, 141
Latil, J. P., **2:**331(44), 360
Lau, E., **3:**36(172), 57
Lau, K. F., **17:**153, 154, 201, 220, 223
Lau, S. S., **14:**104(53), 126, 232(123), 239; **23:**48(162)
Laudan, H., **16:**5, 124
Laude, L., **5:**165(109), 174(109), 232
Laue, E. G., **10:**6(10), 7(10), 65
Laugal, R. C. O., **16:**96, 140
Laughlin, D. E., **15:**154, 215; **16:**26, 127
Laugier, A., **7:**196(168), 252
Laugier, J., **6:**280(192), 299
Laurence, G., **19:**56
Laurenson, L., **1:**18(37), 35(85, 86), 64
Laurent, A., **14:**86(14), 95(34), 97(16), 98(14, 16), 103(16), 105(14, 16), 109(16), 112(99), 113(16), 119(64), 120(16), 122(34), 125, 126, 127, 128
Lauritzen, J. I., **11:**289(114), 317
Lautenschlager, P., **19:**99, 100, 107, 109
Lautz, G., **6:**82(6), 143
Laux, S. E., **13:**208(26), 223
Lave, E. G., **2:**331(45), 332(45), 360
Lavernhe, B., **14:**90(90), 124(90), 127
Laville-Saint-Martin, B., **3:**112(178, 179), 129
Lavine, J. M., **4:**327, 351(9), 369
Lavine, M. C., **3:**179(28), 209
Laviron, M., **13:**219(65), 224
Lavrich, D., **24:**139(66)
Lavrinets, D. A., **6:**36(119), 76
Law, B., **19:**280
Law, D., **12:**277(195), 294
Law, H. B., **12:**154(331), 166
Law, J. T., **1:**8(16), 64; **9:**47(223), 70; **12:**259(81), 292
Law, T. J., **3:**32(161), 57, 212(6), 213(6), 216(6), 218(6), 225(6), 243(6), 270; **4:**118(45, 46), 119(45), 120(45), 133(45), 135, 150(41), 155(41), 156(41), 165(41), 175, 190, 243(168), 255; **16:**281–282, 337
Lawless, K. R., **4:**150(54, 58), 153, 180(130), 181(131), 188, 190, 199(63), 204(63), 212, 223(115), 233, 234(63, 156), 238, 239, 240(162), 243, 245, 249, 252, 253, 254, 255
Lawley, A., **3:**255, 256, 257, 258, 259(109), 264(109), 273
Lawrence, G., **13:**11(20), 57
Lawson, J., **2:**199, 237
Lawson, R. P. W., **13:**112(29), 148
Lawson, W. D., **3:**276, 300

Lax, B., **4:**340(48), 358(48), 370; **18:**122
Lax, M., **11:**290(116), 317; **13:**155(21, 22), 166(23), 192; **21:**38(76), 73
Lax, P. D., **22:**85(10), 87(10), 89(10), 91(10), 92(10), 93(10), 114
Laxhuber, L. A., **24:**180(11)
Lay, K. W., **16:**70, 136
Layer, E. H., **1:**53(129), 67, 226(180), 232; **2:**370(46, 47, 48), 420, 421
Layton, C. K., **5:**197, 198, 234
Lazarev, B. G., **1:**244(27, 28), 274
Lazarides, B., **19:**192, 239, 240, 280
Lazarus, D., **7:**262(20), 306; **16:**254, 269
Laznovsky, W., **5:**287(208), 319
Le Bas, J., **9:**5(18), 65
Le Bihan, R., **14:**124(89), 127
Le Blanc, F. J., **1:**130(11), 184
Le Claire, A. D., **6:**345, 347
Le Fur, P., **7:**137, 138, 139(35), 142(41), 230(35), 248
Le, H. Q., **15:**244, 263; **23:**232(17)
Le, P., **17:**174, 194, 222
Le Traon, J. Y., **6:**141, 149; **16:**242–244, 253–254, 265, 267
Leadbeater, M. L., **15:**249–250, 263
Leamy, H. J., **14:**60(190), 77; **15:**161, 163, 167, 215; **22:**123(28, 29), 172
Learn, A. E., **1:**264(79), 275
Learn, A. J., **6:**113(135), 146; **7:**284(85), 309; **13:**61, 104
Leary, P. A., **14:**8(28), 74
Leavitt, A. J., **24:**36(145)
Leavitt, J. A., **13:**122(66), 148; **16:**43, 47, 131
Leavitt, R. P., **21:**256(30), 276(30), 309
Lebedeva, M. D., **5:**164(95), 231
Leben, L., **7:**289(207), 397
LeBlanc, L. J., **10:**91(28), 103(28), 106(28), 164
LeBlanc, R., **21:**191(87), 221(87), 235
Leblanc, R. M., **19:**281
Lebrun, A., **11:**254, 255
Lech, J., **1:**43(119), 66
Lechter, W. L., **16:**53, 134
Leckenly, R. E., **13:**8(13), 57
LeClaire, A. D., **7:**258(3), 305
LeComber, P. G., **11:**256(60), 257(60), 315; **12:**296
LeContellec, M., **12:**272(168), 273, 294
Lecoy, G., **6:**36(104), 76
LeCraw, R. C., **9:**288, 296
Leder, F., **2:**393(134), 423; **3:**109(110), 127
Leder, L., **1:**126(3), 184
Leder, L. B., **2:**99(54), 100(54), 101(54, 70), 102(54), 103(54, 70), 104(54, 70), 105(70), 106(70), 108(54), 123, 142, 144; **4:**370
Lederey, W. C., **14:**17(66), 74
LeDue, J. M., **7:**236(249f), 254
Lee, A., **10:**255(55), 259
Lee, A. B., **10:**254(52), 259
Lee, A. E., **16:**119, 121, 143
Lee, B. W., **12:**115(218), 163
Lee, C., **21:**78(5), 111
Lee, C. A., **4:**332(29), 370; **12:**277(195), 294; **23:**48(170)
Lee, C. C., **21:**121(19), 123(19), 232
Lee, C. H., **4:**316(161), 324; **5:**287(213), 297, 298(241), 320; **12:**265(112), 292
Lee, C. P., **10:**250, 259
Lee, C. Y., **21:**211(114), 236, 241(11), 256(30), 276(30), 282(11, 36), 290(49), 309, 310
Lee, D., **15:**234–235, 262; **21:**117(12), 128(36), 232, 233

Lee, E. L., **3:**213(10a), 270
Leé, F., **3:**11(52), 51(52), 54
Lee, H., **23:**230(15), 231(15)
Lee, H. Q., **15:**244–246, 262
Lee, H. Y., **19:**250
Lee, J., **15:**252, 264; **21:**207(109), 235
Lee, J. K., **13:**8(18), 57
Lee, J. N., **15:**111, 113–116, 141
Lee, J. W., **15:**154, 215
Lee, J.-W., **16:**26, 127
Lee, K., **15:**147, 180, 182, 214; **17:**318, 365
Lee, L. K., **24:**44(62), 245(83), 246(83)
Lee, L. T., **24:**181(13)
Lee, M., **16:**2, 123
Lee, R., **16:**52, 133
Lee, R. E., **22:**110(24), 115, 244(94), 274
Lee, S., **15:**295, 299, 311; **17:**49, 51, 76; **24:**184(36)
Lee, S. J., **16:**43–44, 46–47, 131
Lee, S. K., **15:**295, 299, 311
Lee, S. S., **17:**228, 242, 300
Lee, S. T., **16:**97, 140
Lee, S. Y., **16:**28, 128
Lee, T., **24:**106(106)
Lee, T. F., **21:**72
Lee, T. G., **21:**9(28)
Lee, T. H., **14:**196
Lee, T. P., **11:**78(168), 79(175), 80, 101
Lee, T. R., **24:**10(39), 14(39), 21(39), 23(39)
Lee, V., **14:**19(79), 20(79), 42(79), 75; **16:**18, 21, 126–127
Lee, V. Y., **16:**34–35, 130
Lee, W. L., **6:**36(122), 37(128, 142b), 38(122), 45(122, 128), 46(122, 128), 47, 48, 49(122, 128), 50, 76, 77
Lee, W. W., **22:**121(21), 154(21), 172
Lee, W. Y., **14:**19(79), 20(79), 42(79), 75; **16:**18, 21, 34–35, 116–117, 122, 126–127, 130, 142
Lee, W.-T. P., **6:**212, 260, 261, 293(31), 295
Lee, Y. H., **14:**232(125), 239; **18:**203, 204
Leek, J. H., **1:**27(57), 29(57), 65
Lefkow, A. R., **14:**178(151), 183(151), 198; **17:**87, 140
Legat, W. H., **4:**263(55), 264(55), 319(55), 321
Leger, P., **3:**49(229), 59
LeGoff, E., **24:**71(111)
LeGoves, F. K., **16:**71, 119, 136
Legrand, J.-F., **24:**3(17), 181(12, 15), 184(43), 192(12), 199(12)
Legros, R., **13:**186(151), 190(151), 197
LeGuillerm, J., **6:**244(141), 298
Leheny, R. F., **7:**254; **14:**113(68), 127
7.(237)
Lehman, H. S., **3:**135, 137, 162; **4:**257(1), 261(27, 29, 37), 263(29, 54), 264, 269(37), 271(27, 37), 272, 273(27, 29), 274, 275(29, 37), 277(29, 37), 280(37), 282(29), 283, 284(29), 285(37), 286, 287, 288, 301(132), 319(54); **5:**290, 320; **12:**291
Lehman, H. W., **17:**153, 159, 219
Lehmann, H. W., **9:**51, 52, 70; **14:**226(79), 238
Lehmpfuhl, G., **2:**108(79), 109, 142
Lehoczky, S. L., **16:**250, 268; **17:**47, 60, 61, 66, 68, 69, 76
Lehovec, K., **6:**61, 78
Lehr, E., **16:**285, 338
Lehrer, W., **7:**255
7.(257)

Leibenguth, R., **23:**55(8)
Leibenguth, R. E., **21:**117(7), 128(7), 167(72), 168(72), 169(72), 170(76), 171(76), 172(76), 173(76), 174(76), 175(76), 176(76), 216(122, 123), 221(131, 135, 137, 139), 222(131, 135, 137, 139), 232, 234, 236, 237
Leibowitz, D., **9:**45(201), 46(201), 69
Leiby, C. C., Jr., **1:**7(14), 64
Leidheiser, H., Jr., **4:**227, 238(162), 239(162), 240(162), 254
Leigh, W. J., **13:**153(12), 182(12), 183(12), 192
Leighton, P. A., **5:**312, 322
Leisegang, S., **1:**202(68), 229; **2:**75, 122(125, 126), 123(126), 130(144), 141, 144
Leiser, C. F., **2:**374(70), 421
Leiserowitz, L., **24:**180(11), 181(12, 15), 192(12), 199(12, 50)
Leite, R. C., **10:**247(38), 259
Lejus, A. M., **14:**107(59), 126
Lekner, J., **19:**286, 305
LeLievre, R. A., **13:**161(41), 162(41), 176(41), 186(41), 189(146), 190(41), 193, 197
Lely, J. A., **2:**387(107), 422; **5:**283, 319; **9:**33(134), 68
Lemberg, H. L., **9:**240(233), 261
Lemons, R. A., **15:**66, 139; **16:**42, 47, 131
Lemperiere, G., **13:**74, 105; **14:**12(44, 46), 74
Lempérière, G., **17:**111, 112, 128, 142, 143
Lems, W., **9:**37(165a), 69
Lénard, **2:**68(10), 141
Lenard, P., **21:**5(21), 72
Lengfellner, H., **16:**81, 138
Lenham, A. P., **4:**84(214), 96
Lenk, T. J., **24:**15(65)
Lennox, R. B., **24:**151(8), 154(8), 155(8c, 19), 156(8, 19), 157(8c), 159(8c), 161(21), 169(21)
Lenz, F., **1:**189(14, 15), 191, 228; **2:**71, 75(26, 27), 80(26), 81(26), 82(26), 83, 91, 93, 94, 141
Leon, L. M., **16:**4, 124
Leonard, R. A., **4:**263(54), 264(54), 319(54), 321
Leonberger, F. J., **10:**250, 259
Leong, W. Y., 23.30(114), 23.37n(114), **23:**38(114), 39(114)
Leonhardt, H., **5:**281, 318
Leonhardt, R., **7:**21, 24(67), 38
Leont'ev, N. I., **1:**44, 66
Lepage, J., **16:**292, 340
Lepestkin, A. N., **15:**9, 136
LePhat Vinh, A., **10:**25(41), 66
Lepie, M. P., **5:**298(250), 305, 321
Lepoole, J. B., **2:**125, 144
Leppavuori, S., **16:**100, 140
Leppla, P. W., **7:**354(137, 138), 356(137, 138), 395
Lercel, M. J., **24:**229(40, 41)
Lerch, P., **16:**57, 76–77, 117, 134
LeRiche, R., **1:**26(54), 65
Leskela, M. A., **16:**17, 126
Leslie, J. D., **12:**277, 294; **14:**86(11), 106(11), 125
Lesnikovskaya, V. E., **13:**99(112), 100(112), 107
Lesovoy, M. V., **17:**327, 335, 366
Lessor, A. E., **2:**379(83), 392(83), 422
Lestrade, J. C., **11:**253, 255
LeSueur, E. J., **11:**274(95), 316
Leta, D. P., **24:**94(72), 95(72)
Letsinger, R. L., **24:**36(146)
Lettington, A. H., **4:**84(211a), 96; **12:**30, 36, 51
Leung, O. M., **24:**98(93), 100(93)

Leung, T. Y. B., **24:**44(46), 94(61), 112(10, 11, 13), 117(11, 13), 119(11, 13), 122(13), 123(13), 139(66), 142(11), 143(11), 144(11)
Leupold, H. A., **16:**226, 231
Leurgans, P., **2:**256(17a), 289(17a), 304
Leurgans, P. J., **1:**105, 121
Leusink, G. J., **22:**266(115), 267(117), 275
Levanoni, M., **17:**153, 154, 220
Levarsky, J., **16:**52, 133
Levedakis, D. A., **22:**202(61), 216(61, 77), 245(61, 77), 273–274
Leveiller, F., **24:**181(12), 192(12), 199(12, 50)
Levelut, A., **11:**312(144), 317
Levenson, J. A., **23:**255(78)
Levenson, L. L., **1:**13, 64; **22:**262(107), 275
Levenstein, H., **1:**209(107), 230
Leveque, G., **17:**315, 365
Lever, R. F., **5:**247(25), 274, 315, 317
Leverenz, H. W., **3:**193(51), 209
Leverett, V., **12:**121(241), 164
Levi, A. F., **16:**53, 134
Levi, A. F. J., **15:**222, 260; **16:**74, 110, 137, 142; **21:**240(6), 308
Levi, D. L., **5:**246(18), 247(18), 314
Levin, E. M., **9:**11(49, 50, 51), 20(51), 66
Levin, R. M., **15:**300, 311
Levine, B. F., **15:**222, 226, 229–230, 233–234, 236–239, 261–262; **23:**211(18)
Levine, E. R., **7:**158(79), 249
Levine, J. D., **9:**46, 70; **12:**121, 164
Levine, M. A., **11:**189(150), 203
Levine, O., **7:**331(66), 394
Levine, P. A., **23:**124(15)
Levine, S. N., **3:**131(1), 162
Levinson, M., **16:**28–29, 129
Levinstein, H., **3:**276(17), 277, 286, 288(48), 293, 298(48), 300
Levinstein, H. J., **9:**271, 283, 294, 296; **12:**266(138), 268(138), 293; **15:**9–10, 136; **22:**121(20), 172
Levoska, J., **16:**100, 140
Levy, D., **14:**90(90), 124(90), 127
Levy, J. L., **3:**296, 301
Levy, M., **17:**171, 221
Lew, A. Y., **23:**281(108)
Lewandowski, S. J., **16:**5, 97, 107, 124, 140–141
Lewicki, G., **6:**37(132), 77; **12:**269(152), 293
Lewin, R. W., **17:**111, 116, 142
Lewis, B., **4:**263(49), 285, 321; **5:**161, 196, 230
Lewis, C. H., **5:**271, 317
Lewis, C. R., **21:**362(44), 368
Lewis, C. W., **2:**30(35a), 60, 384(93), 387(104), 391(121), 422, 423; **6:**116(154), 147
Lewis, D. W., **10:**162(93), 166
Lewis, H. W., **1:**260(75), 275
Lewis, L. B., **23:**29(102, 103), 38(102, 103)
Lewis, N., **16:**48, 132; **24:**44(38), 45(38), 70(38)
Lewis, N. S., **24:**228(14)
Lewis, T. J., **13:**216(53), 224
Ley, J. M., **15:**115, 118, 141
Ley, L., **9:**217(165, 169), 259; **15:**177, 216, 286, 310
Ley, R. P., **9:**28(104), 29, 30, 67
Leyendecker, T., **14:**179(155, 156), 186(155), 198
Lezec, H. J., **13:**163(48), 193
Lhermitte-Sibire, I., **17:**109, 141

Li, C. H., **5:**147, 148, 149(17), 196(17), 229, 252(35), 315
Li, C. Y., **16:**41, 131
Li, D. Q., **16:**41, 131
Li, D. X., **16:**43–44, 46–47, 131
Li, H. C., **16:**23, 127
Li, J., **24:**45(72, 75), 112(7, 13, 15, 19), 117(7, 13, 15), 119(13), 122(13), 123(13), 128(15)
Li, L., **21:**5(7), 32(7), 33(7), 41(7), 71
Li, M., **24:**46(89), 60(89), 150(3), 184(37)
Li, P., **22:**3(63), 61(63), 67(63), 70(63), 72(63), 78
Li, P. C., **5:**298(250), 321
Li, Q., **16:**23–24, 127
Li, S., **12:**296; **24:**33(132)
Li, S. S., **21:**103(42), 112, 190(85, 86), 191(88), 235, 290(46), 298(51), 310
Li, Y. M., **19:**54, 66, 68, 69, 75, 78–81
Li, Y. Q., **16:**91, 139
Liang, K. C., **5:**284(199), 319
Liang, K. S., **16:**282, 338; **24:**45(72, 75, 76), 82(13), 86(13, 39), 112(7, 14, 18, 19), 117(7, 18), 119(16, 18, 19), 121(14), 122(13), 123(13), 125(16), 127(16, 18), 128(15), 130(16), 131(16), 133(16, 18), 138(16), 181(20), 182(20), 199(4)
Liao, H., **22:**177(4, 5–7), 178(7, 11, 12), 181(7, 21, 22), 271–272
Liao, J.-H., **24:**71(111)
Liao, K., **23:**35(134)
Liarokapis, E., **13:**155(23), 166(23), 192
Liau, Z. L., **23:**232(35)
Liaw, J. H. W., **15:**14–15, 137
Libov, L. D., **7:**223(219), 253
Licata, T. J., **16:**42, 131
Licht, S., **15:**9–10, 136
Licht, S. J., **9:**283, 288, 296
Lichtenstein, T., **13:**125(76), 149
Lichtman, D., **1:**50(125), 67; **9:**46(207), 70
Liddel, H. M., **13:**121(55), 148
Liddell, H. M., **4:**78(198), 95
Lide, D. R., **24:**54(103), 64(103)
Lieber, C. M., **24:**36(148), 97(79), 98(79, 80, 90)
Lieberman, M. A., **18:**62, 124, 147, 159
Lieberman, M. L., **9:**50, 51(237), 70
Lieberman, S., **17:**153, 220
Liebert, C. H., **10:**14(20), 65
Liebert, L., **17:**288, 299
Liebhafsky, H. A., **3:**39(194, 195), 52(195), 58
Liebhofsky, H. A., **1:**205(91), 206(91), 230
Liebowitz, D., **13:**63(26), 104
Liebson, M., **1:**37(101), 66
Liedberg, B., **24:**19(79), 20(79), 21(106), 218(76, 77)
Liehr, M., **23:**29(106)
Lienert, **19:**40, 41
Liesk, W., **6:**226(92), 230(92), 234(92), 235(120), 250(92), 297, 298
Lievscu, G., **21:**221(135), 222(135), 236
Liezhao, C., **16:**4, 124
Lifshitz, E., **9:**270, 296
Lifshitz, E. M., **11:**215(11), 314; **15:**103, 113, 128, 130, 140–141; **22:**147(72), 173
Lifshitz, I. M., **6:**137(208), 148
Ligensjo, J., **3:**24(105), 51(105), 55
Ligenza, J. R., **4:**261, 321; **6:**37(125), 41(125), 42, 43, 49, 76, 77;

Ligenza, J. R., (*continued*) **12:**282(222, 224), 295; **14:**90(22), 125
Ligeon, E., **12:**272(168), 273(168), 294
Light, T. B., **5:**274, 317; **7:**28(117, 118), 40, 133(27a), 247
Lighthill, M. J., **4:**207, 253
Lighty, P. E., **4:**199(71), 200(71), 209(71), 218, 220, 231, 252, 253
Likharev, L K., **16:**101, 141
Lile, D. L., **11:**77(164), 101; **13:**205(20), 223
Lilienfeld, J., **2:**149(13), 190
Lilienfeld, J. E., **9:**174(32), 256
Lilienthal, H., **9:**273, 274, 296
Lilienthal, H. R., **16:**51, 132
Lim, G., **14:**19(79), 20(79), 42(79), 75; **16:**18, 21, 126
Lim, T. C., **17:**162, 221, 261, 298
Limansky, I., **12:**151(324), 166
Limoge, Y., **16:**253–254, 268
Limpaecher, R., **14:**86(12), 125
Lin, B., **24:**180(11)
Lin, B. J., **13:**159(36), 160(36), 182(115), 183(115), 193, 195
Lin, C., **16:**110, 142; **19:**281
Lin, C. H., **6:**291(208b), 300
Lin, C.-H., **23:**279(22, 102)
Lin, H., **17:**228, 260, 292, 299
Lin, J. H., **10:**64(146), 69
Lin, P., **13:**79(79), 106
Lin, R. J., **13:**87(86, 87), 106; **16:**25, 36, 127, 130
Lin, S. C. H., **7:**35(155), 41
Lin, S. Y., **15:**288, 311
Lin, T. L., **21:**4(16), 72, 106(44), 112, 208(112), 209(112), 210(112), 211(112), 212(112), 213(112), 236; **23:**14(50), 49(37), 95, 102(43), 109(59), 144(41), 146(46)
Lin, T. X., **16:**32, 130
Lin, Y. F., **15:**222, 229, 234, 238–239, 241, 243, 250, 261–262; **21:**4(12), 72, 121(26), 185(82, 83), 186(83), 189(82), 206(26), 233, 235
Lin, Y. J., **15:**233, 238, 262
Lin, Y. S., **6:**260(163, 164), 263(164), 264(163, 164), 265(163), 267(163, 164), 268, 272, 292(221), 299, 300; **9:**271, 296
Lin, Z., **22:**193(48–51), 194(50), 226(48–51), 228(48), 273; **24:**3(19)
Lin-Chung, P. J., **19:**282, 288
Linares, R. C., **5:**313(293, 294), 322; **9:**281, 296
Lincke, R., **4:**50(97), 93
Lincoln, R. L., **10:**43(89), 67
Lindau, I., **9:**188(75), 193(92), 194, 257, 258; **12:**114(206), 115(212, 217, 220), 116, 117(212), 128(268), 163; **13:**201(4), 202(4), 203(16), 207(4, 16), 208(4, 16), 223
Lindberg, B., **7:**45, 48(16), 111
Lindberg, V., **22:**119(11), 154(11), 156(11), 171
Lindberg, V. L., **3:**13, 51(57), 54
Lindemer, T. B., **16:**43, 132
Linden, K. J., **7:**220(209), 253
Lindenfeld, P., **6:**82(11), 143
Linder, B., **10:**154(80), 166
Lindfors, P. A., **14:**175(132), 183(132), 184(132), 198
Lindgren, I., **7:**45, 48(16), 111
Lindgren, R. A., **6:**246, 298
Lindhard, J., **17:**350, 367
Lindle, J. R., **23:**279(102)

Lindley, W. T., **11:**106(7), 115(7), 131(7), 171(143), 199, 203
Lindmayer, J., **6:**58, 60, 78
Lindner, R., **4:**295(112), 300(120), 323
Lindquist, R. E., **2:**201(8), 237
Lindsay, R. B., **4:**23(55), 92
Lindsay, S. M., **24:**33(132), 237(64)
Lines, M. E., **17:**227, 230, 295
Lineweaver, J. L., **1:**9(17), 64
Ling Yang, **2:**2(2), 59
Ling, Z., **16:**4, 124
Lingensjo, J., **1:**32(76), 65; **2:**365(12), 420
Linger, K. R., **12:**272(169), 274, 294
Linh, N. T., **13:**219(65), 224
Linholm, L. W., **13:**214(51), 224
Link, N. T., **11:**36(9), 38(31), 41(35), 42(43), 45(55), 47(65), 48(65, 66), 49(71), 86(216, 217, 218), 96, 97, 98, 102
Linke, R. A., **11:**77(159), 100
Linker, G., **16:**23–24, 127
Linker, H. C., **16:**74, 137
Linnett, J. W., **22:**213(74), 215(74), 221(74), 222(74), 274
Lins, S. J., **2:**365(11), 420; **3:**24(110), 51(110), 55
Lions, P. L., **22:**85(7–9), 114
Liou, S. H., **16:**17, 80, 110, 126, 137–138, 142
Lipari, G., **24:**21(95), 72(117)
Lippel, P. H., **24:**86(27)
Lippert, W., **1:**191(19), 228; **2:**94, 130(147), 131, 141, 145
Lis, S., **12:**176(48a), 233
Lischka, K., **11:**119(70), 130(87), 131(70), 142(111), 144, 145(115), 201, 202
Lisina, G. A., **12:**92(144), 161
Lisle, D., **11:**189(152), 203
Lissberger, P. H., **3:**15, 49(64), 51(64), 54
Little, J. W., **21:**117(9), 128(9), 191(87), 221(87), 232, 235, 257(32), 309
Litton, C. W., **11:**56(93), 57(93), 58(101), 98, 99
Litton, F. B., **5:**247(20), 314
Liu, B. Y. H., **10:**7(12), 65
Liu, C., **22:**254(98), 261(98), 262(98), 274
Liu, C. A., **16:**282, 284, 338
Liu, D., **22:**17(25), 18(25, 26), 34(26), 35(26), 39(49), 40(49), 41(49), 44(49), 45(25, 26, 49, 51), 46(25), 47(25), 77–78, 227(81), 232(85), 274
Liu, G., **24:**199(49)
Liu, G.-Y., **24:**18(16, 18, 69, 71, 72, 73), 44(17, 18, 21), 45(17, 18, 69, 71, 72, 73), 82(11, 12), 83(11, 12), 86(11, 12, 28, 29), 89(47), 90(28), 94(63), 96(77), 98(77, 86, 87), 106(107, 108), 112(6, 7, 8, 9, 12), 117(7), 119(9), 121(12), 125(6, 9), 127(6), 128(20), 129(20), 206(14), 212(14), 217(68), 218(78), 221(91), 222(91)
Liu, H. B., **16:**79, 137
Liu, H. C., **15:**222, 227, 229, 234, 239, 245–246, 248, 250–254, 256, 260–263; **21:**5(7), 32(7), 33(7), 41(7), 49(97), 50(98), 71, 74, 81(12), 101(37), 111, 112, 128(35), 143(51), 170(73, 75), 178(73), 179(79), 209(35), 233, 234, 235, 287(38), 310; **23:**241(49), 243(49), 251(49, 71)
Liu, H. K., **16:**4, 124
Liu, J., **18:**76

Liu, J. K., **21:**208(112), 209(112), 210(112), 211(112), 212(112), 213(112), 236
Liu, J. X., **16:**51, 132
Liu, L. Z., **15:**229, 232–233, 259, 261; **21:**103(41), 112
Liu, N. C., **16:**164, 168, 193, 230
Liu, X., **22:**163(83), 173
Liu, Y., **24:**94(72), 95(72), 221(94)
Liu, Y. H., **15:**95, 140; **16:**81, 138
Liu, Y. Z., **12:**130(279), 165
Liu, Z.-F., **24:**27(128), 31(130)
Livage, J., **17:**318, 361, 365, 370
Livesay, R. G., **9:**33(117), 67
Liveseu, G., **21:**221(137), 222(137), 237
Livingston, J. D., **1:**324(92), 334; **2:**31(52), 61; **16:**168, 230
Lizet, J., **10:**257(65), 259
Lloyd, J. C., **1:**316(68), 333
Lloyd, J. M., **11:**162, 203
Lloyd, J. R., **16:**240, 251–252, 255, 267–269
Lloyd, O., **14:**158(100), 159(100), 160(100), 197
Lloyd, P., **2:**369(38), 420; **3:**111, 127
Lo, D. S., **6:**227(99), 233(99), 260(99), 297
Loane, C. M., **7:**334(83), 394
Loar, H., **5:**266(79), 316
Lobb, C. J., **16:**14, 125
Lobotka, P., **16:**249, 268
Loch, L. D., **5:**310(281), 322; **9:**22(84), 27(84), 28, 67
Lock, C., **1:**131(15), 185
Lock, J. M., **1:**241, 248(19), 251(19), 274
Locker, L. D., **6:**36(112), 40, 45(142), 46, 76, 77; **7:**102(76), 113
Lockhart, L. B., **5:**88, 140
Lockhart, L. B., Jr., **2:**265(24, 24a), 289(34a), 304
Lockwood, A. H., **11:**146(116), 149(116), 151(116), 202
Lockwood, H. F., **7:**186(141, 147), 192, 193(156), 214(195), 215(246); **15:**252, 264
Locquet, J. P., **16:**52, 133
Lodan, F. R., **17:**121, 143
Lodder, A., **16:**239, 267
Lodding, A., **7:**263, 302(26), 306
Lodge, J. I., **4:**268, 322
Loeb, A. L., **2:**208, 237
Loehr, J. P., **21:**121(24), 123(24), 233
Loewen, E. G., **11:**14(8), 33
Loewenstein, E. V., **4:**73(172), 95
Löfs, S., **24:**19(80), 20(80)
Loferski, J. J., **2:**134, 145; **23:**146(45)
Löffler, K. H., **1:**189(9), 228
Logan, J. S., **3:**141(52), 163; **4:**302, 305(137), 306(137), 323; **14:**57(177), 77
Logan, M. A., **22:**186(33), 209(33), 272
Logan, R. A., **4:**332(29, 30), 370; **11:**52(82, 83), 79(174, 176), 81(178, 182), 86(207), 98, 101, 102; **12:**277(194), 294; **13:**164(56), 165(56), 166(56), 176(56), 186(56), 189(56), 194; **17:**14, 36, 38, 75; **21:**204(101), 205(101), 215(118), 235, 236, 272(34), 309, 327(21), 367
Logan, R. M., **7:**142(37), 148, 164, 165, 166(56), 169(98), 210, 211(190), 223(221), 225, 226, 227, 228(231), 237, 248, 250, 252, 253, 254
Logothetis, E. M., **11:**107(12, 13, 14), 115(12, 61), 119(61), 121, 128(13), 130(14, 83), 131(13),

135(13), 155(13), 166(63, 139), 200, 201, 203
Loh, E., **13:**46(54), 58
Loh, K. W., **17:**173, 222
Lohmann, A., **4:**86(230), 96
Lohner, T., **19:**233
Loiacono, D. N., **24:**6(24), 12(47), 14(57), 15(57), 44(10), 86(25), 125(49)
Lokoshko, V. V., **14:**177(146), 198
Lommel, J. M., **1:**284, 332
Lomonosa, T. N., **1:**148(52), 149(52), 185
Loncar, G., **12:**276(189), 294
London, A., **17:**172, 173, 222
London, F., **1:**208(105, 106), 230, 247, 248, 250, 255(40), 259(40), 274
London, H., **1:**247, 248(42), 251(42), 265, 274, 275
Londono, J. D., **24:**151(9), 154(9), 156(9)
Longden, A. C., **6:**103, 145
Longendorfer, D. H., **13:**161(41), 162(41), 176(41), 186(41), 190(41), 193
Longeway, P. A., **13:**68(43), 105
Longini, R. L., **5:**155(46), 219, 227(46), 230, 234
Longo, J. M., **9:**275, 295
Longo, J. T., **11:**107(18), 157(129), 171(18, 144), 200, 202, 203
Longo, T. A., **5:**279, 299(161), 318
Longshore, R., **11:**157(128), 202
Lonsdale, K., **2:**106(71, 72), 142
Loofbourow, J. R., **4:**50(94, 109), 93
Loper, G. L., **13:**159(37), 161(40), 164(40), 184(122, 123), 185(122, 123, 124a, 130), 187(122, 123, 130), 190(123, 130), 191(130), 193, 196
Lopez, G., **24:**248(89)
Lopez, G. P., **24:**228(6), 229(6), 248(90), 249(6, 96)
Lopez-Ottero, A., **11:**111, 112, 114(37, 59), 119, 121, 130(87), 131(70), 145(115), 200, 201, 202
Lopez-Rios, T., **9:**240(232a), 261
Loposzko, M., **19:**250
Lord, R. C., **4:**50(94), 93; **13:**48(57), 58
Lord Rayleigh, **6:**221, 296
Loree, T. R., **13:**157(31), 193
Lorentz, H. A., **1:**234, 273
Lorentz, R. D., **16:**28–29, 129
Lorenz, J., **22:**181(20), 271
Lorenz, M. R., **7:**163(88b), 171(111), 174(120), 176(124), 177(179), 178(118c), 193(157), 204(192), 209(120), 210(192), 222(124), 223(192), 225(223), 249, 250, 251, 252, 253
Lorenz, W., **4:**198, 252
Lorkovic, I. M., **24:**28(127)
Lormand, G., **16:**254, 268
Losche, M., **24:**181(14)
Lossy, R., **14:**23(92), 75
Lostis, P., **3:**24, 51(102), 55
Lothe, J., **4:**97(9), 98(12), 99(12, 14), 103, 109(9, 14), 134; **6:**87, 97, 98, 128, 144, 145
Lotspeich, J. F., **10:**255(55), 259
Loucks, T., **4:**339, 370
Louie, S. G., **13:**202(9), 223
Lounsbury, J. B., **14:**219(50), 238
Love, J., **5:**1(8), 45
Love, R. B., **12:**14, 50
Love, R. W., **1:**32(77), 65; **3:**24(107, 113), 25(113), 26(107, 113), 44(113), 51(107, 113), 55, 56
LoVecchio, P., **11:**132(91), 156(91), 157(128), 202
Lovell, A. C. B., **6:**84, 104, 144, 146

Lovell, R., **11:**215(12), 314
Lovett, R. A., **24:**184(40), 190(40)
Low, F. J., **21:**5(1, 5), 71
Lowe, C. R., **17:**212, 223
Lowengrub, M., **16:**279, 337
Lowenhardt, P. K., **18:**51
Lowenheim, F. A., **4:**192, 251
Loweth, C. J., **24:**36(147)
Lowndes, D. H., **13:**181(106), 195; **16:**10, 48, 70, 125, 136
Lowney, J. R., **11:**111(36), 200; **21:**13(36), 72
Lozowski, E. P., **22:**76(75), 79
Lu, Y., **19:**93
Lu, Z.-N., **17:**24, 49, 51, 53, 56, 75
Lubecka, M., **6:**285, 286(205), 300
Lubezky, I., **12:**39, 51
Lubig, A., **16:**5, 124
Lubinsky, A. R., **12:**115(218), 163
Lubitz, P., **15:**112, 141; **16:**53, 134
Luborsky, F., **2:**31(54), 61
Luborsky, F. E., **6:**215(48), 296; **15:**149, 154, 165, 168, 171, 214–215; **17:**121, 143
Luby, S., **16:**249, 268
Luc, F., **21:**161(67), 170(74), 175(74), 234
Lucas, A. A., **9:**167(23), 213(142, 143), 256, 259
Lucas, F. B., **14:**150(79), 196
Lucas, M. S. P., **2:**365(6), 419; **3:**110, 127; **6:**86, 106, 107, 131(121, 122), 144, 146; **9:**42(180, 182), 43(182), 69; **14:**175(128, 129), 198
Lucero, A. B., **10:**159(88), 166
Lucovsky, G., **7:**27(111, 112), 39, 40; **15:**159, 215, 277, 288, 290, 297, 310–311; **19:**77
Lucy, J. E., **14:**57(177), 77
Ludeke, R., **5:**166(119), 178, 232; **9:**217(158), 259; **11:**36(3, 15), 41(3), 42, 43, 44(40, 53), 45(41, 53, 61), 47(3), 48(63, 64), 58(3), 67(41, 131), 68, 70(137), 71(137), 73(137, 145), 74, 79(3, 137), 85(197, 198), 87, 89(3), 96, 97, 98, 99; **12:**115(213), 163; **13:**220(70), 225
Ludington, C. E., **23:**87(6, 8), 101(38), 102(40)
Ludlow, J. H., **17:**245, 297
Ludmirsky, D. B., **16:**5, 124
Ludorf, W., **16:**79, 137
Lueas, R. C., **4:**346, 347(59), 349, 354(59), 367, 370
Luedtke, W. D., **24:**151(10), 154(10, 17), 156(10, 17), 163(10), 211(53), 222(53, 104)
Lugijjo, E., **12:**292, 293
Luine, J. A., **16:**119, 121, 143
Luitjens, S. B., **16:**149, 229
Lujan, R. A., **15:**278–279, 310
Lukaszek, T., **17:**188, 204, 206, 222, 223
Luke, C. L., **7:**148, 164, 165, 166(56), 248
Lukens, W. E., **7:**28(117), 40
Lukes, F., **10:**169(19), 226
Lukovtsev, P. D., **6:**6(33), 9, 74
Luk'yanchuk, B. S., **13:**169(70), 185(70), 187(70), 194
Lum, R. M., **13:**164(56, 57), 165(56, 57), 166, 167, 176(56, 57), 186(56, 57), 189(56, 57), 194
Lum, W. Y., **11:**50(73), 57(97), 98, 99
Lummis, F. L., **2:**175(55), 192; **3:**285, 300
Lunden, A. B., **12:**169(35), 201(35), 203(35), 204(35), 233
Lundqvist, B. I., **21:**98(32), 112
Lundqvist, L., **21:**128(32), 216(119,

121), 233, 236, 272(35), 309; **23:**213(20, 21)
Lundstrom, M. S., **15:**226, 250, 261; **21:**337(30), 367
Lundy, R., **1:**21(52), 65
Luneberg, R. K., **1:**70(2), 120
Lunev, V. M., **14:**158(99), 161(109), 197
Lunt, S. R., **24:**228(14)
Luo, H. L., **16:**28, 128
Luo, W. M., **16:**91, 139
Lupatkin, W. L., **7:**162(86), 249
Luryi, S., **15:**228, 245, 261, 263
Luscher, P. E., **11:**36, 96
Lusis, A., **17:**310, 318, 319, 320, 337, 338, 349, 350, 354, 355, 356, 365–368
Lutes, O. S., **1:**257(66), 275
Lüth, H., **9:**217(160), 259
Luth, W., **13:**213(49), 224
Luther, K., **17:**236, 237, 270, 271, 296
Luther, K. E., **16:**22, 127
Luther, L. C., **5:**281, 318; **21:**221(135, 137), 222(135, 137), 236, 237
Lüthi, B., **6:**96, 145
Luthi, R., **24:**94(71), 95(71)
Luthje Philips, H., **14:**219(59), 238
Lutsch, A. G. K., **13:**185(128), 187(128), 196
Lutskii, V. N., **6:**137(213), 139(213), 140(225, 226), 141(225, 226), 142, 148, 149
Lutsko, J. F., **17:**59, 76, 77
Luttinger, J. M., **15:**221, 260
Lutwyche, M. I., **16:**57, 134
Lutz, C. P., **24:**98(98, 99), 106(98, 99)
Lux, R. A., **15:**222, 226, 229, 261
Luzet, D., **24:**181(13)
L'vov, V. S., **15:**95, 140
L'vova, T. N., **19:**2
Ly, H., **24:**46(89), 60(89), 150(3)
Lyashenko, S. P., **9:**34(160), 35(160), 36(160), 68
Lydtin, H., **12:**262(94, 95, 96), 292
Lye, R., **5:**9(16), 45
Lye, R. G., **7:**269(64), 308; **12:**64, 101(170), 159, 162
Lyford, E., **9:**33(117), 67
Lynch, D. W., **19:**131
Lynch, R. T., **7:**201(176, 178), 202(176), 203(178), 221(212), 252, 253
Lynds, L., **16:**56–57, 134
Lynton, E. A., **1:**241(8, 9), 273
Lyo, I., **24:**99(101), 106(101)
Lyon, S. A., **15:**229, 233–235, 262; **21:**215(117), 236
Lyons, F., **24:**24(116)
Lyons, V. J., **5:**276, 279, 318
Lytle, F. W., **7:**27(100), 31(133), 32(100), 39, 40
Lytle, W. O., **9:**12(59), 33(59, 145, 147, 150), 66, 68; **12:**187(84), 234
Lyubimov, G. A., **14:**138(21), 144(21), 160(102–105), 161(102–104), 195, 197

M

M, X., **16:**4, 124
Ma, Q. Y., **16:**42, 131
Ma, S. N., **21:**41(94), 42(94), 51(94), 52(94), 59(94), 74
Ma, Y. Y., **12:**169(14, 15), 189(14), 190(14), 193(14), 194(14), 196(14), 232
Maa, B. Y., **19:**2, 12, 24, 25, 28, 32, 35
Maa, J.-S., **14:**240
Mabbitt, A. W., **7:**193(164), 252
Maboudian, R., **24:**237(61)
McAfee, K. B., **12:**60(27, 28, 29), 159

McAleer, W. J., **5:**267(85), 316
McAlister, A. J., **9:**182(62), 186(62), 257
McAlister, E. D., **4:**60(155), 94
McAloon, K. T., **24:**135(62)
McAuley, J. W., **12:**187(80), 234
McAvley, J. W., **9:**33(140), 68
McAvoy, B. R., **15:**67, 139; **16:**107, 141; **17:**153, 155, 156, 157, 159, 160, 171, 180, 181, 187, 190, 191, 192, 193, 195, 220–222, 248, 294, 299
McBride, W. R., **4:**46(81), 50(98), 62(98), 92, 93
McCaldin, J. O., **12:**117(224), 119(224), 163; **13:**202, 206(8), 223
McCammon, R. D., **11:**253, 255
McCamont, J. W., **6:**116(160), 147
McCamy, J. W., **16:**70, 136
McCann, M. P., **16:**55–56, 134
McCann, W. H., **6:**338(46), 347
McCarley, R. L., **24:**19(87), 20(87), 27(87), 28(87), 30(87, 129), 33(129), 34(129), 44(25, 26, 39, 50), 45(26), 46(26), 60(26), 68(50), 70(39)
McCarroll, B., **4:**107(24), 135
McCarroll, W. H., **12:**85(109), 86, 87, 88, 89, 93(111), 94, 96, 99, 103, 161, 162
McCarron, K. T., **17:**206, 223
McCarthy, D. W., **4:**50, 92
McCarthy, G. T., **7:**34(143), 40
McCarthy, J. P., **5:**210, 234
McCarthy, K. A., **4:**75(188), 95
McCarthy, S. L., **9:**236(219a), 261
McCarthy, T. J., **24:**206(2)
McCaughan, D. V., **14:**220(67), 231(67), 238
MacChesney, J. B., **5:**311, 322
McClanahan, E. D., **14:**31(124), 32(124), 76, 223(75), 238
McClelland, G. M., **24:**98(81)
McClelland, J. J., **24:**230(44)
McCloud, H. A., **9:**20(80), 66
McClure, D., **13:**169(68), 177(96), 185(96), 186(68), 188(96), 189(68), 195
McClure, G. W., **14:**155(86), 197
MacColl, L. A., **4:**338, 370
McCombe, B. D., **11:**50(73), 98; **21:**287(37), 310
McConnell, M. D., **16:**168, 230
McCormick, L. D., **14:**17(66), 74
McCoy, G. L., **11:**56(93), 57(93), 98
McCracken, G. M., **14:**217(41), 237
McCrackin, F. L., **3:**16(72), 51(72), 54; **4:**79(202), 83, 84(202), 84(206), 95
McCrum, N. G., **11:**235(36), 238(36), 315
McCulloch, D. J., **12:**288(255), 295
McCullough, J. D., **7:**35(150), 41
McCulluch, D., **12:**251(52), 252(52), 254(52), 255(52, 57), 269(52), 291
McCumber, D. E., **16:**112, 119, 142
McCune, E., **15:**55, 138
McDermott, C. A., **24:**86(32), 221(92)
McDermott, M. T., **24:**12(48), 46(85), 86(32), 98(82), 101(82), 105(82), 221(92)
McDonald, D. B., **16:**32, 129
MacDonald, D. K. C., **1:**239(6), 273; **2:**40(67), 61; **6:**82(12, 13), 87, 119, 143
McDonald, G. E., **10:**49(101), 68
MacDonald, H. E., **2:**176(57), 192
MacDonald, I. A., **6:**71(209), 79
McDonald, I. R., **24:**206(9, 10, 11, 12), 208(26), 209(9, 10, 11, 12),

210(9, 10, 41, 42), 212(9, 10, 11, 12), 213(10), 214(10), 216(67), 217(12), 218(12), 219(9, 10), 221(11), 222(11)
Macdonald, J., **9:**116, 144
MacDonald, J. C., **24:**44(58), 98(83)
McDonald, J. E., **3:**97, 126
MacDonald, J. R., **3:**220(49), 226, 229(49), 272; **11:**288(110, 111, 112), 292(121, 122), 317
MacDonald, K. A., **17:**318, 365
McDonald, N. A., **24:**44(41)
McDonie, A. F., **12:**100(166), 162
MacDougall, J. E., **24:**98(89)
McElfresh, M. W., **16:**110, 142
McEuen, P. L., **24:**106(106)
McFadden, R. S., **17:**359, 360, 369
McFarland, D. G., **1:**14(29), 64
McFarland, M., **4:**282(103), 322; **10:**129(62, 63), 135(62), 136(62), 165; **12:**2(7), 3(14), 49, 50
McFarland, R. H., **1:**14(29), 64
McFarlane, J. C., **6:**54, 77
McFarlane, S. H., III, **5:**272(113), 317; **7:**134(29a), 137, 138, 139(35), 223(219), 230(35), 247, 248, 253
McFarlane, S. J., **5:**152(36), 153(36), 230
McFee, J. H., **11:**45(48), 47, 58(103), 66(60), 67, 80(103), 97, 99; **17:**279, 299
McFeely, F. R., **9:**217(165, 169), 259; **15:**177, 216
McGarr, D. M., **14:**39(145), 48(145), 49(145), 67(145), 76
McGaughan, D. V., **14:**232(113), 239
McGee, H. L., **3:**146(56), 151(73, 75), 152(56, 73), 153(56), 155(75), 163, 164
McGee, J. D., **12:**147, 166
McGill, T. C., **12:**117(224), 119(224), 163; **13:**202(7, 8, 12), 206(8), 223; **21:**9(28), 72
McGinn, J. T., **14:**232(124), 239
McGinnies, R. T., **10:**59(130), 68
McGlauchlin, L. D., **10:**11(19), 65; **11:**123(77), 181(77), 201
McGraw, R. B., **5:**313(293, 294), 322
McGreer, K. A., **16:**51, 132
McGregor, P. J., **15:**302, 311
McGuire, G. E., **13:**123, 149
McGuire, I. R., **15:**190, 197, 217
McGuire, T. R., **9:**275, 288, 295; **16:**11, 41, 51, 78, 125, 131–132, 137
Machajdik, D., **16:**52, 133
Machi, T., **16:**34, 130
Machida, K., **15:**15, 18, 137; **17:**315, 365; **18:**221
Machiko, G. J., **16:**87, 138
Mchl, W., **4:**196(26, 27), 198(26), 251
Machlin, E. S., **3:**267, 273
McIlraith, A. H., **14:**23(91), 75
McIntosh, A. B., **5:**252(36), 315
McIntosh, K. A., **23:**243(63)
McIntyre, B. J., **24:**106(105)
McIntyre, D., **14:**60(193), 78; **17:**119, 143
McIntyre, L., **16:**43, 47, 131
McIntyre, P. C., **16:**97, 140
McKay, K. G., **12:**60(27, 28), 159
MacKenna, E. L., **4:**263(55), 264, 319(55), 321
McKenney, D. B., **10:**22(35), 56(35), 57(35), 66
Mackenzie, D. R., **13:**127(91), 143(91), 144(133), 149, 150
McKenzie, D. R., **14:**21(85), 75; **22:**146(69), 173
MacKenzie, J. D., **3:**138, 163; **4:**263(57, 58), 264, 319(57, 58),

MacKenzie, J. D., (*continued*) 321; **12:**260, 261, 292; **17:**228, 229, 260, 287, 298, 299
MacKenzie, J. K., **4:**164, 189, 265(63), 319(63), 321
Mackenzie, J. O., **5:**292, 320
McKenzie, K. R., **14:**86(12), 125
McKeown, D., **3:**24(111), 27(111), 51(111), 56, 65, 125; **4:**60(138), 94
McKeown, P., **24:**34(140), 35(140)
Mckernan, S., **16:**66, 136
McKetta, J. J., Jr., **9:**12(55), 66
Mackey, H. J., **6:**82(27), 144
McKinley, A., **12:**114(209), 163
McKinstry, H. A., **5:**119(23), 141
Mackintosh, A., **17:**315, 365
Mackintosh, W. D., **22:**118(3), 171
Mackor, A., **16:**82–84, 138
Macksey, H. M., **7:**234(248), 254; **13:**212(42), 224
MacLachlan, D. F. A., **2:**389(116), 422
McLauchlan, T. A., **4:**22(50), 91; **7:**352(135), 395
McLaurin, R. C., **2:**223, 238
McLean, B. N., **3:**291(54, 55), 300
Maclean, B. N., **12:**169, 201, 233
McLean, D., **7:**293, 293(106), 310
McLean, D. A., **2:**374(68), 375(68), 376(68), 379(88), 393(124, 126, 136), 397(68), 400(164), 405(136), 406(126), 407(68), 408(68), 409(68), 411(124), 421, 423, 424; **3:**108(105), 109(105), 127, 138(37), 163; **5:**311, 322
McLean, W., **14:**202(2), 236
McLennan, M., **15:**226, 250, 261
McLennan, W. D., **3:**5(29), 51(29), 53
Macleod, H. A., **9:**136, 144; **10:**54(116), 68; **13:**120(54), 122(66), 124(54, 71), 127(71, 89), 128; **19:**251, 281
Macleod, J. E. S., **4:**86(235), 96
McLevige, W. V., **11:**66(120), 99
McLintock, I. S., **12:**246(20), 290
McMahon, H. O., **7:**368, 375, 379(171), 396
McMahon, R., **17:**111, 116, 142
McMahon, R. W., **14:**44–45(157), 77
McMahon, T. J., **10:**61(142, 143), 69; **11:**119(69), 201
McMarr, P. J., **19:**196, 205, 232, 280
McMaster, A. D., **3:**217(19a), 244(19a), 271
McMaster, H. A., **9:**33(121), 68
MacMillan, H. F., **21:**362(44), 368
McMillan, L. D., **17:**241, 259, 287, 297, 298
McMurdie, H. F., **9:**11(49, 50, 51), 20(51), 66
McNally, J. J., **13:**121(65), 123(65), 131(65), 132(65), 134(65), 139, 140, 142(119), 145(65), 148, 150
McNeil, J. R., **13:**112(23), 121, 123(65), 126, 127, 130, 131(65, 92), 132(65, 102), 133, 134(65, 92, 94), 135, 137(23), 138(23), 139(23, 118, 119), 140(119), 142(23, 119), 144(23, 82), 145(64, 65), 147, 148; **17:**234, 251, 252, 296, 297
McNeil, M. B., **16:**279, 292, 337
McNeill, W., **3:**194(52), 209; **6:**36(118), 40, 43(118), 76, 77
MacNeille, S. M., **1:**118(50), 121
McNichol, J. J., **6:**294, 300
McPhedran, R. C., **9:**227(198), 260
McQuistan, R. B., **10:**11(19), 65
McQuistan, R. D., **11:**123(77), 181(77), 201

MacRae, A. U., **2:**65, 140; **4:**150(57), 185
McTaggart, F. K., **12:**73(79), 160, 238(4), 289
McVittie, J. P., **18:**98–99; **22:**5(5), 30(36), 77, 182(25), 188(25, 37), 193(25, 37), 209(65), 235(86), 270(132), 272–274, 276
McWhan, D. B., **17:**8, 10, 11, 12, 14, 15, 17, 36, 38, 74, 75
McWhirter, J. T., **16:**62, 135
MacWilliains, K. P., **23:**72(234)
McWilliams, R., **18:**78
Maczek, C., **9:**241(235), 261
Madakson, P., **16:**17, 126
Madakson, P. B., **16:**17, 23, 126
Madar, R., **13:**11(20), 57
Madarsz, F. L., **21:**18(41), 72
Madden, K., **7:**185(140), 251
Madden, P. A., **24:**210(41)
Madden, R. P., **1:**133(19), 134, 135, 148(53), 149(23, 25, 64), 153(25), 154(25, 53), 155(25), 157(25), 158(25), 159(62), 161(62), 162(62), 163(62), 165(62), 167(64), 170(25, 62, 64), 175(53), 178(53), 182; **2:**185, 186, 315, 338(58), 359, 360(23, 53); **3:**10(49), 51(49), 54; **4:**16(39), 37, 39(72), 40(72), 41(78), 50(108, 116), 60(135), 65(39, 67, 156, 157), 68, 91, 92, 93, 94; **7:**48, 63(27), 112; **10:**72
Maddocks, F. S., **1:**316(68), 333; **2:**367(14), 373(14), 404(14), 405(14), 420
Madelung, O., **7:**149, 204(61), 248
Mader, S., **4:**138(7), 187; **5:**178, 232; **6:**101(89), 145; **7:**28(114), 40; **9:**7(35), 65
Maderic, B. P., **17:**254, 255, 256, 298
Madey, T. E., **16:**28, 128
Mador, I. L., **5:**312, 322
Madoz, J., **24:**18(78), 19(78)
Madura, J. D., **24:**209(36), 210(36), 211(36), 220(36)
Maecker, H., **14:**196
Maeda, A., **16:**54, 134
Maeda, H., **16:**2, 123
Maeda, K., **7:**158, 225(78), 249
Maekawa, S., **11:**36(13, 14), 50(72), 70(72, 141), 71(141), 72(141), 96, 98, 100
Maeno, Y., **15:**172–173, 216
Maesler, M. A., **12:**289(265), 296
Maffei, N., **14:**57–58(180), 77
Magee, C., **16:**60, 135
Magee, T. J., **13:**217, 224
Magnuson, G. D., **3:**79, 80(61), 126
Magnussen, O. M., **24:**3(17), 4(17), 182(23), 184(34, 36, 39, 43), 187(34), 188(34), 190(34), 199(23)
Mahadev, V., **22:**175, 241(88), 270(130), 274, 276
Mahan, A. H., **19:**93
Mahan, A. I., **1:**72(33), 121
Mahan, G. D., **9:**217(169), 259; **11:**295(134), 317; **21:**11(32), 72
Mahan, J. E., **12:**296
Maher, D. M., **23:**48(170)
Maher, G. H., **14:**31(120), 76
Maheswari, S. P., **17:**318, 361, 362, 366, 370
Mahlman, G. W., **3:**281, 294, 300
Mahoney, G. E., **11:**91(233), 103
Mai, K. A., **17:**201, 223
Mai, Y.-W., **16:**279, 337
Maier, R. L., **9:**127(31, 32), 143(31, 32), 144
Mailloux, R. J., **15:**62, 139

Maissel, L., **5:**179(123), 180(123), 232; **17:**86, 91, 126, 140, 232, 295
Maissel, L. I., **2:**367(24, 25), 393(24, 25, 144), 397(144, 153, 157, 158), 406(153), 408(191), 411(144), 420, 423, 424; **3:**76, 92(73), 96(73), 102, 103(95), 104(95), 105(95), 107, 108(107), 109(107, 109), 113, 126, 127, 129, 132(13), 162; **4:**9(21), 91, 258, 261(37), 267, 269, 271(37), 272(37), 275(37), 277(37), 280(37), 285, 288(37), 314, 320, 321, 322; **6:**54, 69(151), 70, 77, 82, 143; **7:**1; **14:**2(2, 6), 26(2), 27(2), 73, 203(10), 217(10, 38), 221(64), 222(64), 237, 238
Maivald, P., **24:**94(67), 95(67)
Maiz, J. A., **16:**251–252, 268
Majda, M., **24:**5(21)
Majda, M. M., **24:**44(53)
Majinmdar, A., **24:**98(94)
Majkrzak, C. F., **17:**22, 74
Major, S., **12:**177(54a), 185(54a), 186(54a), 187(54a), 200(54a), 233
Majumdar, A. J., **5:**119(23), 141
Majumdar, D., **15:**171, 216
Majumder, K. S., **19:**149
Makhov, A. F., **2:**71, 141
Maki, K., **3:**232(60b), 236(60b), 272
Makida, S., **16:**98, 140
Makimoto, T., **19:**12, 19
Makino, Y., **3:**299, 301
Makishima, S., **9:**22(85), 23(85), 67
Makita, V., **11:**73(123), 99
Makita, Y., **11:**36(13, 14), 96
Makovsky, L. E., **17:**307, 364
Makowieki, D. M., **14:**202(2), 236
Makowski, M. P., **12:**181(70), 190(70), 234
Maksimov, M. Z., **1:**44(121), 66
Maksymowicz, A., **6:**285(205), 286(205), 300
Maksymowicz, L., **6:**285, 286(205), 300
Malarkey, E. C., **15:**115, 122, 141
Malé, D., **1:**101, 121; **6:**183, 203
Malek, Z., **1:**288(32), 293, 332
Maley, N., **19:**52, 103
Malherbe, A., **10:**115, 117, 165
Malhotra, L. K., **12:**169(23), 176(53), 194(23), 226(122), 233, 235; **13:**127(86), 149; **15:**286, 310
Malik, A., **24:**112(24), 129(24)
Malik, R. J., **15:**222, 226, 229–230, 234, 236–239, 261–262; **21:**116(5), 117(6, 10), 123(5), 124(5), 125(5), 126(29), 128(10, 38), 129(5, 39, 40), 136(5, 39, 40), 137(47), 141(50), 146(5, 39, 40), 149(5, 39, 53), 161(53), 162(53), 196(94), 209(10, 40), 215(10, 118), 232
Malin, J. I., **23:**279(22, 102)
Malin, M., **19:**288
Malinowski, S., **4:**202(79), 203(79), 252
Malitson, I. H., **11:**135(96), 202; **12:**51; **19:**59, 148, 206, 207–209, 212, 214, 217, 236, 238, 241, 257
Malkin, O. A., **14:**196
Malliaris, A., **9:**8(37), 65
Mallone, D. P., **14:**140(35), 195
Mallouk, T. E., **24:**12(51, 52), 15(51, 52), 27(51), 29(51), 163(25), 164(25)
Malloy, F., **3:**139(42), 163
Maloff, I., **2:**371(60), 421
Maloney, T. J., **12:**165
Malozemoff, A. P., **9:**271, 296; **15:**177, 216

Mameno, K., **13:**20(45), 24(45), 53(45), 58
Man, A., **5:**240(8), 314
Man, L. I., **5:**164(98), 172(98), 193(98), 232
Man, P., **21:**94(23), 111
Manaila, R., **7:**24(87), 36(163), 39, 41
Manako, T., **16:**30, 129
Manasevit, H. M., **5:**147, 151, 152, 153(32), 154, 186(33), 199, 209(33), 229; **11:**108(20), 121(20), 200; **12:**129(273), 165
Manasreh, M. O., **21:**121(22, 23, 24), 123(22, 23, 24), 194(22), 195(22), 233, 303(53), 310
Mandeville, P., **15:**250, 263
Mandich, M. L., **16:**60, 135
Mandler, D., **24:**183(31), 192(31)
Mandlich, M. L., **16:**2, 124
Manfra, M. J., **23:**278(100)
Manghi, F., **19:**2
Mani, T., **17:**358, 369
Manifacier, J. C., **12:**176(50), 179(50), 180(50), 183(50); **13:**121(60), 148
Manini, P., **16:**17, 126
Maniv, S., **13:**67(33, 34), 105; **14:**3(15), 5(16), 8(15), 9(16), 33(129), 35(16), 36(16), 38(141), 39(129, 141), 41(15), 44(153, 156), 45(153–156), 46(156), 63(204), 64(204), 66(204), 68(129), 73, 76, 77, 78; **17:**111, 142, 153, 219, 220
Mankiewich, P. M., **16:**42, 74, 101, 107, 110, 131, 137, 141–142
Manley, D. F., **7:**236(249d), 254
Mann, B., **7:**379(199, 200), 383, 384(199), 385(200), 386(199, 200), 387(199), 391(200), 397
Mann, C. K., **24:**44(61)
Mann, H. T., **3:**146(62), 160, 163
Mann, J. B., **1:**33(80), 65
Mannhart, J., **16:**11, 71, 119, 125, 136
Mannhart, J. D., **16:**17, 126
Manning, J. R., **7:**261(14), 306; **16:**234, 266
Manohara, H., **23:**24(92)
Mansingh, A., **17:**241, 242, 288, 297
Mansingh, M., **11:**256(66), 257(66), 316
Mansuripur, M., **15:**180, 186–188, 203, 207, 216–217
Mantei, T. D., **18:**163
Mantese, J. V., **16:**95–96, 140; **17:**229, 300
Mantz, T. F., **13:**187(125), 196
Manus, G. M., **7:**364(163), 396
Many, A., **2:**157(43), 191; **5:**221(203), 234
Manz, R. C., **5:**266(81), 316
Mao, G., **24:**34(140), 35(140)
Mao, X. L., **16:**79, 137
Maoz, R., **24:**181(17)
Mapother, D. E., **1:**241(18, 22), 273, 274
Mar, H. Y. B., **10:**48(98), 64(146), 68, 69
Mar, R. W., **9:**22(95), 27, 67
Mar, W., **24:**127(54), 206(7, 8b), 209(8, 40), 210(8, 40), 212(8), 213(40), 214(40), 215(40), 216(40), 220(8)
Marayama, A., **17:**308, 309, 364
Marbot, R., **19:**6
Marchal, M., **9:**51(252), 52, 62(252), 71; **22:**113(29), 115
Marchand, J. F., **1:**55, 67, 241(13), 245(32), 273, 274; **2:**393(137), 423
Marcotte, V. C., **9:**31(114), 35(114), 67

Marcu, V., **13:**186(152), 190(152), 197
Marcus, P. M., **1:**242(24), 249(47), 274
Marcus, R. B., **6:**116(157), 147
Marcus, S. M., **10:**31(53), 66
Marcuse, D., **9:**254, 261
Mareschal, J., **9:**282, 294
Marey, H. O., **16:**57, 60, 87, 93–94, 134–135, 138–140
Marfaing, Y., **13:**186(151), 190(151), 197
Margalit, S., **15:**229, 261; **21:**116(2), 155(60), 232, 234
Margaritondo, G., **12:**121(238), 164
Marhic, M. E., **14:**11(42), 74
Mariani, J. L., **6:**54(158), 55(158), 57(158, 165), 58(158, 165), 63(158), 77, 78
Mariano, A. N., **3:**179(28), 209
Mar'in, A. V., **14:**43(152), 77
Marinace, J. C., **5:**275, 276, 317, 318
Marinero, E. E., **15:**170, 215
Marinov, M., **13:**119(50), 127(50), 145, 148
Maris, O., **12:**169(21), 232
Marjin, M., **12:**169(21), 232
Mark, A., **5:**265, 316
Mark, G., **6:**36(103b), 76
Mark, H. F., **9:**12(55), 66
Mark, J. T., **1:**36(94), 66
Mark, P., **6:**64(191), 78; **12:**115(218), 163, 194(92), 195(92), 199(92), 234
Markelz, A. M., **16:**4, 124
Markert, L. C., **17:**7, 63, 65, 73
Marks, T. J., **16:**87, 93–94, 138–140; **17:**287, 299
Marley, J. A., **9:**22(86, 99), 23(86, 99), 24, 25, 67
Marra, W. C., **24:**114(29), 115(29), 116(29)
Marraud, A., **4:**23(57), 24(57), 92; **6:**188(30, 31), 203
Marriott, J. G., **9:**32(116), 33(116), 67
Marruchi, J., **1:**218(137), 231
Marsh, D. M., **1:**226, 232; **2:**54, 62; **3:**229(93), 254, 255(93), 256, 257(107), 268(119), 273
Marsh, J. H., **16:**3, 124
Marsh, O. J., **14:**231(110), 239
Marsh, P., **16:**2, 53, 124, 134
Marsh, S., **16:**28, 128
Marshall, A. F., **16:**2, 22, 42, 50, 53–54, 63, 123, 127, 131, 134–135
Marshall, G. W., **24:**96(75), 97(75), 98(75)
Marshall, N., **9:**238(228, 229), 261
Marshall, R. V., **2:**189(68), 192
Marshall, S. J., **24:**96(75), 97(75), 98(75)
Martens, G., **16:**53, 134
Martens, J. S., **16:**123, 143
Martens, U., **17:**310, 365
Mrtensson, J., **24:**3(18), 4(18)
Marth, P. T., **1:**27(60), 65
Marti, A., **21:**313(6), 319(11), 320(11), 321(13), 341(6), 342(6), 343(40), 344(6), 351(13, 40), 355(6), 357(13), 359(13, 42), 360(6, 40), 361(6, 40), 362(40), 366, 368
Marti, O., **24:**84(23)
Martienssen, W., **1:**198, 202(54), 227(54), 229
Martin, B., **6:**92(71), 121(71), 123(71), 136(71), 145
Martin, B. A., **17:**213, 224
Martin, C. R., **10:**91(28), 103(28), 106(28), 164
Martin, G. M., **11:**52(86), 56, 98
Martin, J., **16:**37–38, 58, 61, 75, 130, 135

Martin, P. J., **13:**109(1), 112(24), 120, 124(54, 72), 127(72, 91), 128, 131, 132(54), 134(54), 135(54, 72, 105, 106, 107, 108), 136, 137, 138, 139(105), 140(105), 141(72, 105, 122), 143, 144, 145(138), 147, 148; **14:**26(103), 61(103, 194, 196), 75, 78, 147(72), 167(72), 176(72), 182(72), 196; **19:**195; **22:**118(2), 150, 171
Martin, R., **21:**79(8), 111
Martin, R. J., **9:**169(28), 256; **10:**227(3), 230(3), 234(3), 235(12), 236(3, 16), 241(31); **11:**58(103), 80(103), 99
Martin, S. J., **17:**172, 221
Martin, S. W., **12:**286(246), 295
Martin, W. E., **10:**256(63), 257(63), 259
Martin-Binachon, J. C., **9:**282, 294
Martindale, J. G., **4:**120, 135, 152(68), 189
Martinelli, R. U., **12:**54(8), 113, 114(8, 195), 124(185), 125, 128(8, 185); **23:**230(15), 231(15)
Martinet, E., **21:**161(67), 234
Martinez-Duart, J.-M., **14:**105(55), 112(97–99), 126, 128
Martins, R., **12:**296
Martinuzzi, S., **5:**165(109), 174(109), 232
Marton, C., **2:**101(70), 103(70), 104(70), 105(70), 106(70), 142
Marton, J. P., **9:**22(90), 23(90), 27(90), 67; **10:**52(106), 60(137), 68, 69; **16:**250, 268
Marton, L., **1:**126(3), 184, 202, 230; **2:**99, 100, 101(54), 102(54), 103(54), 104(57, 70), 105(70), 106(70), 108(54), 123, 142, 144; **7:**16(43), 38
Martyna, G. J., **24:**207(23, 24)
Marucchi, J., **12:**169(21), 232
Marusa, H. P., **13:**51(60), 52(60), 53(60), 58
Maruska, H. P., **7:**169(96, 97), 170(105), 250
Maruyama, M., **7:**236(249b), 254
Maruyama, S., **11:**44(51), 74, 97, 100
Marvin, A., **9:**245(245), 261
Marwuis, F., **15:**206, 217
Marx-Tibbon, S., **24:**19(83), 20(83)
Maselko, J., **21:**50(99), 52(103), 74
Maserjian, J., **5:**183, 233; **21:**4(16), 72, 106(44), 112; **23:**109(59), 144(41)
Masetti, E., **19:**238, 240
Mashtakova, G. S., **14:**177(146), 198
Masing, L., **2:**405(184), 424; **6:**7(35), 23, 27(83), 54, 66(35), 74, 75
Maskalick, N. J., **2:**30(35a), 60
Maskil, N., **24:**184(36, 37)
Maskus, M., **24:**21(101), 23(101), 24(101), 27(101)
Mason, A., **16:**97, 99, 140
Mason, A. R., **16:**6, 28–30, 125, 129
Mason, P. R., **11:**253, 255, 260(75), 261, 316
Mason, R. C., **14:**150(81), 197
Mason, W. P., **3:**21(97), 51(97), 55; **17:**195, 222
Massenet, O., **6:**206(8, 9), 210, 211(22, 23, 24, 25, 26, 27), 212, 214(40, 41), 215(40, 41, 42), 216(41), 217(42), 219(42), 220(42), 225(24), 226, 230(109), 233(109), 234(109, 118), 235(109), 241(22, 93, 139), 242(22, 23,24, 25), 244(24, 25, 139), 245, 247(155), 250(24, 40, 41, 109, 118), 253(22, 93), 254(24, 25, 40, 41), 255(24, 25),

Massenet, O., (*continued*) 256(24), 257(24), 258(24), 259(24), 260(24), 271(24), 273(179, 181, 183), 274, 275(179), 277, 279, 292(9), 293(9), 295, 296, 297, 298, 299, 300, 2243
Masserjian, J., **6:**63, 78
Massey, H. S. W., **2:**64, 67, 68, 74(1), 99(55), 100, 140, 142
Massies, J., **11:**36, 41(35), 42(43), 45(55), 47, 48, 86, 96, 97, 98, 102; **13:**219(65), 224
Masson, D., **15:**279, 281–283, 310
Masters, B. C., **7:**286(87, 88, 91), 309
Masterson, K. D., **10:**19(32), 45(95), 51(103), 66, 67, 68
Masuda, Y., **17:**234, 270, 296, 308, 312, 353, 364, 368
Masui, A., **17:**309, 365
Masui, S., **15:**176, 179, 195–199, 216
Masumoto, H., **16:**30, 86, 129, 138; **17:**234, 270, 296
Mataga, P. A., **16:**291, 340
Matakura, Y., **3:**137(28), 141(28), 162; **5:**289(221), 320
Matano, C., **6:**303, 346; **16:**236, 266
Mate, C. M., **24:**98(81)
Mateika, D., **9:**276, 284, 297; **16:**5, 124
Matheson, L. A., **1:**191, 228
Mathew, R., **14:**71(231), 79
Mathews, A., **17:**80, 139
Mathews, H., **17:**167, 221
Mathews, J. W., **9:**290, 296
Mathewson, A. G., **19:**153, 155
Mathieu, H. J., **24:**18(88), 20(88), 106(111)
Mathieu, X., **10:**228(4), 253(4), 254(4), 258
Mathur, B., **13:**187(126), 196
Mathur, D. R., **11:**131(88), 168, 201
Matijasevic, V., **16:**19–20, 126; **17:**2, 73
Matisoo, J., **1:**286(27), 287(27), 289(27), 332
Matossi, F., **4:**271(90, 91), 322
Matovich, E., **5:**273, 317
Matsik, S., **21:**7(23), 41(23, 93), 42(23), 45(23), 46(23), 47(93), 57(93), 58(93), 59(93), 72, 74
Matsik, S. G., **21:**65(126), 66(126), 67(126), 68(126), 69(126), 70(126), 75; **23:**261(81), 291(81)
Matson, R. J., **16:**28–29, 129
Matsubara, K., **13:**2(2), 20(28, 29, 34, 35, 36, 37, 38, 43, 44, 45), 21(34), 22(35, 36, 37, 38), 23(28, 29, 43), 24(28, 44, 45), 25(47), 27(47), 30(34), 31(35), 32(36), 33(37), 34(38), 38(51), 45(29), 47(55), 48(56), 49(43); **17:**51(44), 53(45), 56, 57, 58, 153
Matsubara, S., **16:**27, 127; **17:**247, 290, 293, 299
Matsubosa, T., **14:**65(208), 78
Matsuda, A., **10:**237, 258; **19:**76, 80
Matsuda, H., **13:**143(130), 150; **22:**145(68), 173
Matsuda, R., **15:**172, 216
Matsuda, S.-P., **16:**28–29, 129
Matsudaira, N., **17:**354, 368
Matsuhiro, K., **17:**308, 309, 312, 353, 364, 368
Matsui, A., **1:**148(54), 169(54, 55), 186
Matsui, T., **5:**306(273), 321; **10:**237(20), 258; **16:**111, 142
Matsui, Y., **17:**290, 299; **22:**182(24), 272
Matsumara, K., **17:**153, 154, 160, 220
Matsumaru, H., **10:**236(14), 258

Matsumoto, H., **15:**204, 206, 217
Matsumoto, K., **24:**99(104), 100(104), 101(104)
Matsumoto, O., **12:**287(252, 253), 295
Matsumoto, S., **19:**104
Matsumoto, T., **13:**119(49), 126(49), 148
Matsumoto, Y., **15:**244, 263
Matsumura, K., **12:**264(107), 292; **14:**91(23), 125
Matsumura, S., **22:**188(40), 272
Matsunaga, N., **11:**64(114), 99
Matsunami, H., **9:**49(224a), 70; **12:**277(218), 282(218), 295; **14:**104(51), 126
Matsuno, Y., **12:**296
Matsunobu, Y., **24:**36(144)
Matsuo, M., **11:**239(44), 247(44), 254, 255, 315
Matsuo, S., **18:**122, 124, 125, 152, 159, 167, 168, 178, 209, 217, 218
Matsuo, Y., **15:**37, 137
Matsuoka, M., **14:**42–43(151), 77; **17:**88, 91, 140, 334, 367; **18:**81, 82, 84, 124, 165, 194
Matsuoka, S., **11:**238(43), 254, 255, 315
Matsushima, Y., **11:**42(44), 45(58), 67, 68(134), 73, 97, 99, 100
Matsushita, K., **12:**296
Matsushita, M., **5:**302(257), 303, 321
Matsushita, T., **16:**88, 139; **17:**318, 365
Matsuura, K., **24:**10(38), 23(38), 24(38)
Matsuura, M., **16:**149, 229
Matsuzaki, T., **17:**228, 232, 243, 296, 297
Mattheiss, L. F., **16:**4, 124
Matthews, A., **13:**78(65), 106; **14:**132(8), 134(8), 195; **17:**100, 132, 141, 143
Matthews, H., **17:**167, 220
Matthews, H. I., **4:**200(75), 252
Matthews, J. W., **2:**14, 16, 17, 20, 60; **3:**244, 272; **4:**128, 131, 133, 136, 139(11, 12), 143, 144(11), 145, 146, 147(11), 148, 149(31, 34), 150(12, 35, 36, 47, 50, 52), 152(47, 50), 153, 155, 156(34, 35, 36, 50), 157, 158, 159, 160(47, 50, 91); **6:**102(92), 145, 337(43), 347; **7:**18(57), 38; **17:**26, 27, 28, 29, 31, 75; **21:**341(34), 367; **23:**225(23)
Matthews, M. D., **13:**79(71), 106
Matthews, N., **17:**196, 222
Matthiessen, A., **4:**327, 369
Mattis, D. C., **6:**92, 145
Mattox, D. M., **3:**97, 126; **9:**53(253b), 71; **10:**22, 23(36), 24(36), 28(49), 44(92), 45(92), 49(36, 100), 50(36, 100), 57(36), 61(140), 63(92, 144), 64(92), 66, 67, 68, 69; **13:**111(13, 14), 112(14), 145(14), 147; **14:**132(5, 6), 135(13), 136(13), 195, 201(1), 202(6, 7), 233(6, 135), 236, 237, 240; **17:**82, 87, 121, 139, 140; **22:**160(76), 173
Mattson, D. R., **15:**271, 310
Mattson, E., **4:**196(28), 251
Mattson, J., **17:**34, 43, 44, 49, 51, 53, 75, 76
Mattuck, R. D., **3:**18, 51(83), 55; **7:**336(98, 99), 351(98, 99), 395
Matuda, N., **16:**288, 338
Matukura, Y., **4:**265(65), 314(65), 321
Matyash, A. A., **12:**126(262), 164
Mau, H. J., **14:**196
Maung, M., **16:**111, 115, 117, 122, 142

Maurain, C., **6:**241, 298
Maurel, P., **15:**125, 141
Maurer, R. D., **4:**46(84), 92
Maverick, E. F., **5:**265(69), 316
Maxfield, M., **16:**99, 140
Maximov, V. N., **18:**18
Maxwell Garnett, J. C., **12:**71, 79(72), 160
Maxwell, K. H., **5:**154(41), 155(41), 156(41), 228(41), 230
Maxwell, L. R., **7:**1(2), 37
Maxwell-Garnett, J. C., **10:**60, 68; **19:**43, 127, 139–141, 160
Maxwell-Garnett, J. G., **6:**201, 204
May, R. M., **21:**53(105), 74
Mayadas, A. F., **6:**102(95), 114(143, 144), 132(143, 144), 145, 146; **7:**285; **19:**135, 164
Mayer, A., **5:**270(108), 271, 317
Mayer, E. H., **2:**394(146), 397(146), 398(146), 423
Mayer, H., **1:**71, 120, 208, 220, 222, 227(104), 230; **2:**30(49), 61, 242(10), 303, 381(90), 422; **3:**2(6), 17(6), 27(136, 139, 141), 28(136, 139, 141), 29(136, 139, 141), 30(139, 141), 51(136, 139, 141), 53, 56; **5:**1(3), 45, 49(3), 85, 101(12), 141; **6:**82, 104, 119, 129(169), 130, 131(111, 112, 113, 114), 143, 146, 147, 151, 202, 204; **9:**218(172), 259
Mayer, J. W., **13:**220(69), 224; **14:**230(100), 239; **16:**234, 266; **23:**48(162)
Mayer, S. E., **5:**271, 317
Mayer, T. M., **14:**9(33), 74; **18:**70
Mayer, W., **3:**72(40), 125
Mayer, W. N., **2:**367(23), 420; **3:**102(93), 127; **4:**269(80), 322; **13:**111(11), 147
Mayeux, C., **17:**121, 143
Mayfield, J. R., **1:**304(53), 332
Maynard, R., **6:**246, 298
Mayne-Banton, V., **13:**153(13), 181(13), 182(13), 183(13), 192
Maynes-Banton, V., **16:**55, 134
Mayo, S. L., **24:**211(47)
Maystre, D., **9:**227(198), 260; **11:**21(15), 29(19), 30(19), 34
Mazaki, H., **16:**25, 51, 73, 127, 132
Mazer, J. A., **13:**214(51), 224
Maziar, C. M., **21:**193(93), 235, 337(30), 367
Mazor, A., **19:**76, 82
Mazur, J. H., **23:**49(37)
Mazzeo, N. J., **6:**291(208c), 300
Meaburn, J., **9:**136(48), 144
Mead, C. A., **2:**148(3), 190, 400(162), 424; **3:**131(4), 162; **4:**325, 349, 350, 354, 367, 369, 370; **5:**222(208), 234, 286(203), 319; **6:**36(106), 37(132), 65, 76, 77, 78; **7:**379(198), 397; **12:**117(224), 119(224), 163, 269(152), 272(177), 293, 294; **13:**201(5), 202, 203(5), 206(8), 219(66), 223, 224; **21:**60(112), 65(112), 74
Meade, R. H., **5:**313(298), 314(298), 322
Meakin, P., **16:**288, 338; **22:**5(7), 77
Meakins, R. J., **11:**236(38), 253, 254, 255, 315
Mealmaker, W. E., **12:**186(93), 197(93), 200(93), 234
Mears, R., **3:**179(31), 183(31), 209
Meca, F., **11:**249(54), 252, 275(54), 277, 281(54, 103), 282(103), 283(103), 315, 316
Mecartney, M. L., **16:**51, 132
Meckel, B., **3:**74(51), 126

Medelci, N., **21:**325(17), 330(17), 332(17), 338(17), 350(17), 362(17), 367
Medernach, J. W., **13:**155(29), 183(29), 184(29), 193
Medicus, G., **3:**65(17), 125
Medrano, F. J., **24:**18(78), 19(78)
Medrud, R. C., **9:**50, 51(237), 70
Mee, C. H. B., **12:**81, 100, 105, 109(105), 110(183), 160, 162
Mee, J. E., **5:**260(55), 313(201, 295, 297, 298), 314(298, 301), 315, 322; **15:**8, 136
Meek, P. R., **17:**161, 221
Meek, R. L., **14:**232(113), 239
Meeks, E., **22:**180(16), 271
Meeks, E. L., **11:**56(91, 92), 77, 88, 98, 100, 102
Mees, B. M., **5:**287(211), 320
Meessen, A., **19:**128, 139
Megerle, K., **1:**268, 269, 275
Meggett, B. T., **11:**66(127), 68, 99, 100
Megtert, S., **16:**30, 32, 129–130
Meguellati, F., **16:**98, 140
Mehal, E. W., **5:**146(13), 154(13), 229, 279, 318
Mehl, R. F., **6:**314(27), 345, 346, 347
Mehrotra, A. K., **16:**98, 140
Mehrotra, B. N., **15:**303, 311
Mehta, B. R., **12:**176(49, 51), 185(49), 186(49), 187(87), 190(87), 198(49), 199(49), 233, 234
Mehta, R. R., **5:**298(247), 321; **9:**50(236), 51(236), 52, 70; **12:**266(127), 293
Mei, L., **10:**41(82), 67
Meier, W., **6:**201(69), 204
Meier-Hirmer, R., **16:**9, 125
Meieran, E. S., **7:**133(25), 247, 265(48, 52), 282(52), 307; **16:**240, 250, 267–268
Meiklejohn, W. H., **1:**285(25), 294(25), 332; **6:**206(4), 215, 216, 217, 218; **15:**144, 214
Meiksin, Z. H., **10:**60(134), 69
Meilus, O., **13:**144(132), 150
Meinel, A. B., **10:**2(3), 7(3), 11(3), 14(3), 19, 21(3), 22(3, 35), 28(3, 48), 35(3), 37(48), 44(3), 56, 57(35), 58(3), 62(48), 65, 66
Meinel, M. P., **10:**2(3), 7(3), 11(3), 14(3), 19, 21(3), 22(3), 28(3), 35(3), 44(3), 58(3), 65
Meinema, H. A., **16:**82–83, 138
Meiners, L. G., **13:**205(20), 223
Meissner, H., **1:**212, 275; **6:**121, 122, 147
Meissner, K. W., **3:**281(43), 300; **4:**85(220), 96
Meissner, W., **1:**247(38), 274
Melcher, R. L., **13:**154(18), 182(18), 183(18), 192
MeLennan, W. D., **5:**210, 212, 213, 234
Melius, C. F., **22:**186(32), 209(32), 272
Mell, H., **19:**75
Melliar-Smith, C. M., **14:**217(44), 220(69), 226(69, 82), 234(82), 237, 238
Mellor, J. W., **5:**240(7), 314
Melman, P., **21:**207(109), 235
Melngailes, I., **11:**106(5, 6), 142, 146, 166(6), 199, 202
Melnick, B. M., **17:**259, 298
Melnik, P. V., **12:**121(243), 164
Melnyk, A. R., **9:**192(90), 193, 257
Melville, H. W., **3:**146(55), 151(55, 69), 163, 164
Memaninsch, R. J., **15:**297, 311
Memmert, U., **23:**32(119)
Memzes, C. A., **12:**235
Mena, A. A., **5:**283, 319

Menda, M., **17:**361, 370
Mendel, E., **9:**281, 290, 296
Mendelson, A., **16:**290, 302, 311, 339
Mendelson, S., **4:**169(112, 113), 173, 186, 190; **5:**187, 205(152), 233
Mendelssahn, K., **1:**265(83), 275
Mendenhall, H. E., **9:**35(162), 68
Mendlowitz, H., **1:**126(3), 145(41), 184, 185; **2:**99(54), 100(54), 101(54, 70), 102(54), 103(54, 70), 104(54, 70), 105(70), 106(70), 108(54), 142
Mendolia, M. S., **16:**287, 291, 334, 336, 338
Menezes, S., **17:**3, 49, 61, 62, 73
Meng, C. C., **15:**245–246, 263
Meng, R. L., **16:**2, 123
Meng, W. J., **17:**49, 52, 76
Menke, H., **7:**2(8), 37
Menna, A. A., **5:**154(42), 230
Menna, R. J., **23:**230(15), 231(15)
Menter, J. W., **1:**195(34, 35), 196(36, 42), 226, 227, 229, 230, 280(5), 330(113), 331, 334; **2:**13(6a), 14, 18(6a), 22(6a), 53, 54, 56, 58(83), 59(83), 60, 62; **3:**212(4), 213(4), 256(100), 257(100), 258(100), 261, 263(100), 264(100), 266, 267(100), 268, 270, 273; **4:**128(69), 136, 144(25, 21), 155(25, 28, 78), 160(28), 166(4), 182(25), 184(25), 187, 189, 254; **7:**334, 389(90), 394; **16:**332
Menzel, D., **9:**217(167), 259
Menzer, E., **1:**70(10), 120
Menzer, G., **4:**171, 190
Meray, G. M., **23:**101(33), 124(15)
Meray-Horvath, L., **2:**150(25), 189(67), 191, 192
Mercandall, L. M., **12:**277(197), 280(197), 294
Mercandalli, L. M., **14:**102(45), 126
Mercer, S. L., **1:**19(43), 20(49), 65
Mercy, J.-M., **21:**287(37), 310
Mergerian, D., **15:**115, 122, 141
Merisov, B. A., **7:**264(45), 307
Merker, D. R., **7:**318(28, 29), 393
Merten, U., **7:**148, 149(57), 248
Mertens, F. P., **2:**216, 238; **4:**83(210), 96
Mertens, G., **16:**60, 69, 135–136
Mertens, R., **23:**2(4)
Mertens, R. P., **21:**10(30, 31), 12(30), 15(30), 32(31, 65), 72, 73
Mertz, J. L., **11:**79(172, 174), 81(172, 178, 179, 182), 84(189, 193), 86(179), 101, 102
Merwin, L. H., **24:**164(31), 168(31)
Merz, J. L., **7:**201, 202(176), 252
Merz, K. M., **5:**284, 319
Meservey, R., **19:**165
Meskauskas, A., **13:**144(132), 150
Mesker, O. R., **14:**223(78), 238, 240
Messerly, M. J., **13:**122(66), 124(71), 127(71), 148, 149
Messiaen, A. M., **18:**146
Messick, L., **11:**77(164), 101
Messier, R. A., **14:**18(73), 19(73), 64(202, 203), 75, 78; **17:**110, 111, 119, 120, 121, 141, 143, 310, 314, 365; **18:**246, 248; **19:**61, 105, 194, 195, 230
Methfessel, S., **1:**292(38), 294, 304(55), 306(55), 309, 312, 313, 329(108); **16:**168, 192, 230
Metropolis, N., **24:**208(25)
Metselaar, R., **17:**110, 141
Metz, B., **17:**304, 364
Metze, G. M., **11:**78(166), 83(166), 99, 101; **13:**164(55), 185(55), 189(55), 194
Metzger, G. W., **17:**251, 252, 297

Meunier, J., **19:**281
Meunier, P., **6:**292(217), 300
Meunier, P.-L., **15:**125, 141
Meunier, R., **14:**95(34), 96(35), 122(34), 126
Meuris, M., **21:**287(40), 310
Meyer, D., **16:**80, 137–138
Meyer, D. T., **9:**42, 43(183), 69
Meyer, E., **24:**89(54), 92(54), 94(69, 71), 95(69, 71), 218(79)
Meyer, F., **19:**281
Meyer, G., **2:**226, 238; **24:**94(66)
Meyer, J. R., **23:**279(22, 36, 102, 103), 280(103)
Meyer, N. I., **12:**60(32), 75(32), 159
Meyer, O., **16:**23–24, 74, 127, 137
Meyer, R. B., **17:**288, 299
Meyers, A. M., **18:**271, 284
Meyers, J. H., **11:**114(53), 201
Meyerson, B., **12:**256(59, 60), 291
Meyerson, B. S., **23:**24(84), 34(125), 48(84, 85)
Meystre, P., **15:**206, 217
Mezei, F., **17:**2, 73
Mezey, E. G., **5:**238(6), 314
Mezin, A., **16:**292, 340
Micalski, A., **12:**265(109), 292
Miccoli, G., **15:**81, 86–87, 140
Miceli, P. F., **16:**60, 62–63, 135
Michael, J. R., **16:**70, 136
Michalske, T. A., **24:**44(56), 94(64, 65), 221(88)
Michalski, A., **12:**242(17), 243(17), 265(125), 290, 293
Michel, B., **24:**2(2), 18(88), 20(88), 44(43), 86(34), 89(49), 106(112), 127(51), 218(75), 221(89), 222(89), 230(47), 236(60), 237(62), 238(62, 66)
Michel, C., **7:**162(86), 249
Michel, D., **14:**126
Michel, E., **4:**50(105), 93
Michel, P., **6:**310(11), 346
Michel, S., **16:**27, 128
Micheletti, B., **12:**264(110), 292
Micheletti, F. B., **12:**194(92), 195(92), 199(92), 234
Micheli, A. L., **16:**95–96, 140; **17:**229, 300
Michels, D. J., **7:**90, 101(66), 113; **11:**17(11), 18, 19(11, 13), 32(11), 33
Michikami, O., **14:**232(117), 236(154, 155), 239, 240
Middelhock, A., **6:**4(24), 5(24), 74
Middelhoek, S., **1:**296, 302(49, 52), 303(49, 52), 304(55, 56), 306(55), 308(56, 59), 309(55), 312(65), 313(65), 314(49), 332, 333; **6:**206(6), 221(69, 70), 228, 230, 231, 232(111), 235, 236(6, 111, 122), 238, 240, 295, 296, 297, 298
Mie, G., **9:**198(106), 258
Miedler, L. J., **4:**47(86), 92
Mielenz, K. D., **1:**91, 121; **3:**2(21), 53
Mierop, H. P., **7:**218(201), 253
Miescher, E., **3:**5, 51(31), 53
Mieville, L., **16:**28, 128
Migliorato, P., **6:**22, 75
Migliuolo, M., **16:**16, 25–26, 126–127
Mihama, K., **1:**202(67), 229; **2:**132(152), 145
Mii, Y. J., **21:**91(19), 111, 155(62), 161(68), 234; **23:**240(51)
Mika, K., **15:**37, 138
Mikalsen, D., **17:**87, 140; **22:**39(47), 45(47), 78, 161(79), 173
Mikalsen, D. J., **16:**17, 23, 53, 126, 133
Mikes, T. L., **7:**72, 73(63), 113; **10:**120(52, 52a), 121(52a), 165; **11:**14(10), 15(10), 16(10), 17(11), 18(11), 19(11), 32(11), 33

Mikhailov, G. K., **14:**160(104), 161(104), 197
Mikhailov, Y. G., **1:**244(29), 274
Mikkelsen, J. C., Jr., **14:**99(41), 110(41), 126
Mikkola, D. E., **17:**287, 299
Miklich, A. H., **16:**57, 76–77, 117, 134, 142
Mikolaichuk, A. G., **5:**164(101), 232; **7:**29(124), 31, 34(131), 40
Mikolajczak, P., **17:**15, 74
Mikoshiba, M., **17:**161, 221, 261, 263, 298
Mikoshiba, N., **15:**37, 137
Miksic, M. G., **1:**245(37), 274
Miksie, M. C., **2:**151(37, 37a), 176(37), 191
Mikulyak, R. M., **11:**81(183), 86(210, 210a), 101, 102
Milano, U., **15:**92, 140
Miles, D. C., **24:**44(41)
Miles, J., **1:**272(100), 275
Miles, J. L., **2:**370(43), 420; **6:**2(2), 36, 37(108), 38, 39(108), 45, 73, 76; **7:**368, 375, 379(177), 396
Miles, M. H., **17:**318, 366
Miles, R. H., **23:**234(20, 37), 235(20), 281(108)
Milgram, A., **1:**149, 186
Millany, M. H., **11:**281(103), 282(103), 283(103), 316
Millard, J. P., **10:**162(92), 166
Millard, M., **12:**249(37), 290
Miller, A., **5:**147, 152(33), 186(33), 199(33), 209(33), 229, 230, 238(4); **16:**28–29, 129, 302, 339
Miller, B. A., **21:**128(34), 233
Miller, B. I., **7:**222(216), 234(243), 253; **10:**252(68), 259
Miller, B I., **11:**45(48), 47, 58(103), 66(60), 67, 78(167), 80(103), 97, 99, 101
Miller, C. J., **24:**44(40)
Miller, D. B., **18:**122, 143, 152, 159, 169
Miller, D. C., **9:**281, 294; **13:**210(34), 214(34), 216(34), 217, 224; **15:**170, 215
Miller, D. J., **12:**122(246), 164; **16:**32, 129
Miller, D. P., **4:**169(109), 190
Miller, F. D., **12:**91(140), 161
Miller, H. C., **14:**151(62), 158(62), 161(62, 108), 196, 197
Miller, J. A., **22:**186(30), 272
Miller, J. C., **19:**167, 168
Miller, J. N., **19:**40, 41
Miller, Jr., J. H., **16:**28, 128
Miller, K. J., **5:**257, 269, 273, 306, 315, 316, 322
Miller, N. D. J., **15:**36, 137
Miller, P. B., **1:**258, 275
Miller, R., **23:**101(33)
Miller, R. C., **11:**79(176), 86(207), 101, 102; **15:**222, 227, 244, 261
Miller, R. E., **3:**156(82), 164
Miller, R. F., **1:**330(110), 334
Miller, R. J., **16:**254, 268
Miller, R. L., **17:**174, 178, 222
Miller, S. E., **5:**154(44), 158(44), 159(44), 230; **10:**227(2), 258
Miller, S. L., **4:**332(28), 370
Miller, W. A., **22:**261(106), 275
Miller, W. E., **14:**232(114), 239
Miller, W. H., **21:**62(115), 63(115), 74
Miller, W. R., Jr., **9:**209(132), 258
Miller, W. T., **24:**106(106)
Milleron, N., **1:**13(24), 14(30), 16, 64; **3:**41(214), 58
Millidge, S., **21:**128(36), 233
Mills, D., **6:**4, 74

Mills, D. L., **6:**298; **15:**37, 138
Mills, R. E., **23:**124(16)
Milner, C. J., **1:**43(115), 66
Milner, T., **21:**5(1), 71
Milnes, A. G., **5:**276(140), 282, 318, 319
Miloslavskii, V. K., **3:**124, 128; **9:**34(160), 35(160), 36(160), 68
Milton, A. F., **10:**254(52), 259
Mimura, T., **12:**278(206), 294
Mimura, Y., **15:**154, 174–175, 180–181, 214
Mina, R. T., **6:**82(37), 144
Minahan, J. A., **4:**323
Minakata, M., **14:**20(81), 27(81), 75; **17:**153, 219
Minakata, S., **17:**209, 223, 290, 299
Minamisono, T., **13:**167(64), 180(64), 185(64), 188(64), 194
Minamiya, A., **5:**276(143), 318
Minden, H. T., **3:**281, 294(63), 300, 301; **5:**279, 318; **7:**121(8), 247
Minemura, J., **21:**330(26), 331(26), 340(26), 343(26), 367
Miner, C., **13:**67(33, 34), 105; **14:**44(153), 45(153), 77
Miner, C. J., **14:**44(156), 45(156), 46(156), 77
Ming, Y., **15:**95, 140
Mingazin, T. A., **12:**88(130), 161
Minn, S., **14:**175(127), 198
Minn, S. S., **2:**23(32), 24, 60; **9:**6(26), 43(184, 187, 191), 65, 69
Minnaja, N., **6:**66, 78; **7:**373(185, 193), 378(185, 193), 397
Minomo, S., **18:**165, 184, 185
Minowa, Y., **13:**134(103), 150
Mir, J. M., **16:**98, 140
Miralles, N., **24:**163(24)
Miranda, F. A., **16:**5, 76, 107, 124, 137, 141
Miranda, R., **14:**213(31), 237
Mircea, A., **11:**52(84, 86), 56(86), 98; **14:**86(13), 113(67), 125, 127
Mircea-Roussel, A., **13:**11(20), 57
Mirkarimi, P. B., **17:**4, 5, 6, 13, 18, 19, 24, 25, 47, 49, 53, 63, 71, 73, 76, 77
Mirkin, C. A., **24:**23(110), 27(110), 36(146), 112(24), 129(24)
Misawa, S., **11:**67(130a), 99; **17:**153, 154, 219, 220
Mischel, P., **7:**220(208), 253
Misener, A. D., **1:**248(42), 251(42), 274
Misiano, C., **13:**131(97), 140(97), 141(97), 149; **14:**220(70), 238
Missiroli, G., **3:**39(205), 52(205), 58
Missler, J. M., **21:**62(118), 74
Misumi, T., **14:**34(135), 76
Mita, Y., **13:**181(110), 185(139), 188(139), 195, 196
Mitch, A., **12:**135(293), 165
Mitchell, D., **16:**54, 134
Mitchell, D. F., **4:**150(54), 188, 223(115), 253
Mitchell, D. L., **3:**299, 301
Mitchell, E. N., **1:**290, 292, 319, 332, 333; **6:**114, 115, 124(141), 129, 132(141), 135(141), 146, 224(80), 292(80), 297
Mitchell, G. R., **3:**277, 300
Mitchell, I. V., **22:**118(3), 171
Mitchell, J., **5:**287(208), 319
Mitchell, J. M., **3:**124(165), 128; **5:**167(137), 233
Mitchell, J. P., **4:**301, 323
Mitchell, J. W., **1:**279(2), 331; **2:**48(80, 81), 49(81), 51(81), 62; **3:**219(25), 271; **4:**152(70), 162, 189
Mitchell, K., **12:**54(11, 12, 13), 158

Mitchell, K. W., **12:**169(15), 232; **21:**364(46), 368
Mitchell, M. A., **16:**54, 134
Mitoff, S. R., **11:**271(92), 316
Mitra, N., **16:**13, 125
Mitra, S. S., **3:**39(210), 52(210), 58
Mitsuhada, T., **7:**122(9), 247
Mitsui, S., **11:**51(76, 78), 52(78), 60(76), 76(78, 152), 98, 100
Mitsui, Y., **11:**76(152), 100
Mitsuishi, A., **4:**50(95), 93
Mitsuyu, T., **16:**17, 19, 33–34, 111, 126, 130, 142; **17:**163, 221, 228, 248, 297
Mittall, K. L., **16:**274, 282, 337–338
Mitterauer, J., **14:**141(40), 196
Mittmann, M., **24:**228(28)
Mittonneau, A., **11:**52(84, 86), 56(86), 98
Mitzumo, J., **4:**238(161), 239(161), 254
Miura, N., **24:**99(104), 100(104), 101(104)
Miura, S., **16:**27–30, 57, 127, 129, 134; **17:**247, 290, 293, 299
Miura, Y., **3:**137(28), 141(28), 162; **4:**260(20), 265(65), 301(20), 314, 320, 321; **5:**289(221), 320
Miyake, K., **12:**94(147), 95(147), 161; **14:**88(17), 89(19), 109(19), 117(19), 125; **17:**313, 314, 336, 345, 346, 365, 367; **18:**216
Miyake, S., **2:**119, 143; **14:**240
Miyake, T., **15:**168, 215
Miyamoto, N., **12:**277(200), 294; **23:**30(116), 32(116)
Miyamoto, T., **17:**353, 368
Miyamura, H., **17:**83, 139
Miyanchi, T., **7:**171(116), 250
Miyano, K., **17:**24, 45, 48, 49, 51, 52, 56, 57, 74
Miyano, T., **17:**310, 365
Miyaoka, S., **15:**210–211, 213, 217
Miyasaka, Y., **16:**27–30, 57, 127, 129, 134; **17:**197, 223, 247, 290, 293, 299
Miyashita, K., **17:**288, 299
Miyashita, T., **6:**208(16), 209(16), 262(16), 263(16), 264(16), 265(16), 266(16), 272(173), 284(173), 286(16), 295, 299
Miyata, N., **6:**215(46), 216, 296; **23:**47(158)
Miyatake, H., **15:**147, 214
Miyauchi, K., **17:**357, 369
Miyazaki, M., **15:**154, 156, 158, 185, 215–216
Miyazawa, H., **12:**92(141), 94(148), 161
Miyazawa, S., **10:**256(58), 259; **16:**52, 133; **17:**261, 298
Miyoshi, T., **17:**352, 368
Mizerov, M. N., **13:**162(44), 163(47), 164(44), 186(47), 189(47), 193
Mizoguchi, T., **15:**185, 216
Mizugaki, S., **13:**20(42), 23(42), 57
Mizuhashi, M., **17:**318, 352, 353, 366, 368
Mizuno, H., **10:**237(21), 258
Mizuno, K., **16:**111, 142
Mizuno, M., **17:**356, 369
Mizuno, O., **7:**236(249b), 254
Mizuno, S., **15:**279, 281, 310
Mizuno, Y., **17:**324, 366
Mizushima, Y., **13:**220(71), 221(71), 225
Mizutani, F., **24:**36(143), 89(48)
Mizutani, S., **15:**172, 216
Moali, J., **11:**261(79), 263(79), 270(79), 273(79), 316
Moazed, K. L., **4:**97(8), 103(20), 104(20), 106(20), 108(27), 115,

118(35, 39, 41), 123(27), 125(27), 127(27, 35), 134, 135, 150, 151(63a), 152, 154(63a), 155(63a), 159, 160, 188

Mobis, A. J., **7:**355(147), 396

Mobius, D., **7:**385(201), 386(203), 391(213), 397

Moch, P., **5:**166(117), 177, 232

Mochan, W. L., **19:**11

Mochel, J. M., **9:**33(133), 68

Mochizuiki, S., **15:**290, 311

Mochizuki, K., **19:**35

Mochrie, S. G. J., **24:**128(56)

Modine, F. A., **19:**212

Moeckly, B. H., **16:**61–62, 66, 135–136

Moerel, M., **7:**36(161), 41

Moerkirk, R. P., **21:**256(28), 267(28), 277(28), 309

Moest, R. R., **5:**150, 155(27), 230, 277, 318

Moffat, D., **16:**106, 141

Moffat, H. K., **22:**180(16), 186(32), 209(32, 67), 271–273

Mogab, C. J., **12:**265(114), 292, 294; **14:**226(82), 234(82), 238; **22:**61(62), 78

Mogami, T., **14:**226(88), 238

Mogro-Campero, A., **16:**41, 43–48, 50, 52, 101, 131–133, 141

Mogyorosi, P., **13:**169(70), 185(70), 187(70), 194

Mohaidat, J. M., **21:**330(25), 333(25), 367

Mohammed, K., **21:**184(80), 194(80), 235

Mohapatra, S. K., **17:**321, 354, 366, 368

Mohnke, V. A., **14:**66(221), 78

Mohr, T. O., **6:**216(58), 283(58), 293(225), 296, 300, 338(48), 339(49, 50, 51), 347

Mohwald, H., **24:**116(36), 117(36), 180(11)

Moisan, M., **15:**283, 310; **18:**65

Moison, J. M., **13:**180(103), 195

Moizhes, B. Ya., **14:**161(107), 197

Mokhtar, S., **4:**86(227), 96

Mokrushin, S. G., **12:**169(34, 36), 201(34), 233

Molarius, J. M., **17:**119, 143

Molenbroek, A. M., **19:**2

Molenstedt, G., **7:**16(41), 38

Moliere, G., **1:**191, 228

Moliére, G., **22:**127(34), 172

Molière, K., **2:**108(79), 109, 142

Molinary, E., **19:**2

Moll, A., **2:**367(15), 420

Moll, E., **14:**135(14), 195; **17:**95, 141

Moll, J. L., **4:**326(7), 343(54), 344(54), 346(54), 347(54), 348(54), 349(54), 354(54), 367(54), 369, 370; **5:**221(205), 234; **7:**142(40), 243(265), 248; **12:**60(32), 75(32), 84(98), 128, 130(279), 159, 160, 164, 165

Mollenkopf, H. C., **6:**82(27), 144

Möllenstedt, G., **1:**189(10, 13), 191, 199(58, 59, 60), 228, 229; **2:**122, 144

Moller, W. M., **6:**273(178), 299

Molnar, B., **3:**116, 129; **5:**167(133), 193, 194, 203(133), 232

Molnar, B. J., **7:**30, 31(132), 34(139), 36(157), 40, 41

Molozemoff, A. P., **16:**90, 139

Molzen, W. W., **9:**51(253), 52, 62(253), 71; **10:**52(108), 68

Monaco, S. F., **1:**70(15), 120

Monaghan, D., **18:**273, 276, 280

Mönch, G. C., **3:**8(35), 9(35), 51(35), 54

Monch, G. C., **6:**338(47), 347

Mönch, W., **2:**38(63), 39(63), 40, 61
Monch, W., **12:**114(210), 163
Monchamp, R. R., **5:**267, 316
Mond, L., **5:**240(11, 12), 314
Monemar, B., **7:**171(112), 250
Monin, J., **12:**60(42), 63, 159
Monkowsk, J. R., **15:**288–289, 311
Monroe, D., **16:**110, 142
Montariol, F., **6:**119, 122, 135(170, 178), 147
Montero, I., **14:**112(98, 99), 128
Montgomery, H. C., **7:**161(81), 249
Montgomery, M. D., **5:**163(65), 168(65), 209(65), 231
Montmory, R., **6:**206(2), 209(2), 210(22), 211(22, 23, 24, 25), 212(24, 25), 214(40, 41), 215(40, 41, 42), 216(41), 217(42), 219(42), 220(42), 224, 225(24), 226, 230(109), 233(109), 234(109), 235(109), 241(22, 93, 139), 242(22, 23, 24, 25), 243(140), 244(24, 25, 140), 245(140), 250(24, 40, 41), 253(22, 93), 254(24, 25, 40, 41), 255(24, 25), 256(24), 257(24), 258(24), 259(24), 260(24), 271(24), 277(140), 285(203), 286(203), 295, 296, 297, 298, 300
Moodie, A. F., **1:**196(40), 198, 229
Mooij, J. E., **17:**77
Mooij, J. H., **15:**190, 217
Moojui, G. C. A. M., **24:**208(33)
Moon, P., **10:**5(7), 6, 7(7), 21, 65
Moon, R. L., **7:**243(267), 255; **12:**114(191, 194), 123(251), 126(263, 264), 128(264), 137(308), 138(308), 162, 163, 165
Mooney, J. M., **21:**21(47, 50), 22(50), 25(56), 72, 73, 219(128), 236; **23:**91(19, 25), 92(25), 105(51, 52), 135(26)
Mooney, R. L., **2:**250(16a), 304
Moore, A. C., **7:**390(208), 397
Moore, A. J. W., **4:**164(99), 189
Moore, A. L., **24:**33(132)
Moore, C. A., **13:**181(107), 185(107), 188(107), 195
Moore, C. R., **4:**169(109), 190
Moore, D. W., **3:**41(214), 58
Moore, E. J., **11:**290(118), 317
Moore, G. E., **2:**147(1), 190
Moore, H., **9:**33(117), 67
Moore, K. J., **21:**341(35), 367
Moore, L., **11:**253, 255
Moore, R., **12:**257(70), 291
Moore, R. A., **15:**67, 139; **17:**157, 159, 160, 180, 181, 182, 187, 190, 191, 192, 197, 199, 200, 201, 202, 203, 204, 205, 206, 207, 221–224
Moore, T. A., **24:**33(132)
Moore, W. J., **3:**62, 125
Moorjani, K., **16:**59, 79, 135, 137
Morancho, R., **17:**121, 143
Moravec, T. J., **12:**252(54), 253, 291; **13:**105, 135(104), 141(104), 150
Moreau, A., **17:**34, 43, 44, 45, 49, 51, 53, 75, 76
Moreau, W. M., **24:**228(22), 233(22)
Morel, B., **17:**306, 345, 347, 364, 367
Moreland, J., **16:**30–32, 129
Morey, G. W., **1:**4, 64
Morgan, D., **15:**55, 138
Morgan, D. F., **9:**22(88), 27, 67
Morgan, P. E. D., **16:**28–29, 129
Morgan, R. A., **21:**117(7), 128(7), 221(139), 222(139), 232, 237
Morgan, T. W., **7:**162, 211(84), 249
Morgan, W. L., **9:**43(189, 190, 192), 69
Morgaritondo, G., **14:**234(145), 240

Morgenthaler, F. R., **15:**32–33, 33, 137
Morgulis, N. D., **3:**65(20), 125; **12:**101(172), 162
Mori, H., **12:**274(185), 275(175, 180), 294; **13:**60(3), 77(3), 94, 104
Mori, I., **14:**19(76), 75
Mori, N., **16:**97, 140
Mori, R., **3:**15, 51(66), 54
Mori, T., **24:**34(142), 35(142)
Mori, Y., **12:**277(204), 278, 294
Moriizumi, T., **7:**191(154), 220(206), 251, 253
Morikawa, T., **17:**153, 219
Morillot, G., **11:**49(71), 98
Morimoto, H., **7:**14(32), 38
Morimoto, K., **13:**20(40), 23(40), 45(40), 57
Morimoto, M., **14:**226(88), 238
Morimoto, O., **17:**153, 220
Morimoto, Y., **13:**52(62), 58
Morin, F. G., **24:**151(8), 154(8), 155(8c, 19), 156(8, 19), 157(8c), 159(8c), 164(30), 165(30), 167(30), 168(33)
Morin, F. J., **5:**211(219), 235
Morishige, Y., **15:**113, 141
Morishita, T., **15:**168, 215
Morita, H., **17:**308, 321, 352, 364, 366, 368
Morita, M., **14:**105(54), 126
Moritani, A., **19:**51
Moriya, T., **6:**215(49), 296
Moriyama, T., **23:**166(90)
Morkoc, H., **11:**51(75), 57(75), 59(75), 98; **15:**244–245, 263
Morkoe, H., **23:**255(77)
Morningstar, O., **7:**1(3), 37
Morokoff, W., **22:**178(15), 271
Moroney, R. M., **17:**217, 224
Morosin, B., **16:**54, 134
Morozov, E. P., **7:**155(68), 249
Morozov, V. N., **13:**186(154), 190(154), 197
Morozumi, C., **7:**11(23), 37
Morris, B., **11:**249(53), 251(53), 315
Morris, C. E., **6:**36(105), 38(105), 76
Morris, D. E., **16:**4, 124
Morris, G. W., **17:**22, 74
Morris, H. B., **11:**132(89, 90), 201
Morris, R. W., **3:**92, 126
Morris, W. C., **5:**312, 322
Morris, W. F., **12:**257(69), 291
Morrish, A. H., **6:**272, 299
Morrison, C. F., **14:**167(119), 169(119), 197
Morrison, J., **5:**190, 233
Morrison, R. E., **3:**141(53), 142(53), 144(53), 163; **16:**101, 141
Morriss, R. H., **11:**114(53), 201
Morritz, F. L., **5:**152(33), 186(33), 199(33), 209(33), 230, 238(4), 261(4), 269(96), 270(4), 283, 314, 316, 319; **6:**36(113), 45, 76
Morse, P. M., **2:**77, 141; **9:**84(16), 143; **17:**44, 75
Mortenson, R. A., **5:**312, 322
Mortin, J. T., **17:**206, 223
Morton, G. A., **2:**181(63), 192; **12:**143(314), 144, 154(329), 166
Morton, H. S., **1:**207(96, 97), 230
Morton, J., **16:**51, 133
Moruzzi, J. L., **14:**91(25), 92(91), 103(3, 25), 116(25), 117(25), 118(25), 125, 127
Moruzzi, V. L., **15:**166, 177, 215–216
Morys, P., **11:**290(117), 317
Mosely, R. C., **14:**71(232), 79
Moser, A., **24:**98(85)
Moser, E. M., **17:**229, 263, 298
Moser, F., **9:**15(72), 66
Moser, J., **6:**103, 145; **22:**240(87), 274
Moskovits, M., **19:**281

Moskowitz, M. J., **16:**81, 102–105, 121, 138
Mosley, V. M., **7:**1(2), 37
Moss, M., **22:**113(28), 115
Moss, R. W., **14:**31(124), 32(124), 76
Moss, S. C., **7:**12, 13(27), 16(46), 24, 25(27), 38
Moss, T. S., **2:**297(44), 304; **3:**276, 277, 278, 293, 300; **10:**42, 67; **11:**142(113), 143(113), 202; **12:**169(8), 232
Moss, T. S., III, **16:**92, 139
Mostovetch, N., **2:**23(27), 60
Mostovskii, A. A., **12:**101, 105(176, 177), 106, 162
Motakabbir, K. A., **24:**220(87)
Motamedi, M., **17:**153, 154, 221
Motesharei, K., **24:**21(97), 23(97), 24(97)
Motizuki, K., **4:**329, 369
Motohiro, T., **22:**7(15), 77, 146(71), 173
Motomatsu, M., **24:**89(48)
Mott, L., **10:**169, 204(38), 226
Mott, N. F., **2:**64, 67, 68, 74(1), 140; **4:**197, 252, 327(11), 338(40), 365(81), 366(81), 369, 370, 371; **6:**14, 64(182), 75, 78; **7:**260(8), 306; **9:**7(36), 9, 65; **11:**226(20), 256(20), 258, 290(20, 120), 314, 317; **14:**88(18), 114(69), 120(81), 120–121(81), 125, 127; **15:**191, 217; **21:**7(24), 8(24), 11(33), 72
Mottram, A., **7:**167, 169(94), 250
Motyka, R. J., **6:**132(197), 148; **10:**207(40), 226
Moubis, J. H. A., **3:**91, 126
Mouchard, J., **13:**121(59), 148
Moulin, M., **14:**226(84), 238
Moulton, C., **3:**116, 128
Moulton, G., **5:**167(131), 232
Mountain, R. D., **22:**128(38), 172
Mountain, R. W., **13:**169(69), 184(121), 185(69), 187(69), 194, 196; **14:**232(119), 239
Mourou, G. A., **15:**245–246, 248, 251, 263; **16:**107, 141
Moustakas, T. D., **15:**279, 310; **17:**19, 20, 21, 32, 74
Movchan, B. A., **13:**62(25), 63(27), 79(25), 81, 82, 83, 92, 93(25), 104; **14:**63(200), 78, 136(16), 195; **17:**60, 64, 65, 77, 119, 143, 156, 221; **19:**194, 216; **22:**2(1), 77, 122(23), 172
Mozzi, R. L., **7:**15, 35(34), 38; **13:**218(62), 221(62), 224
Mrksich, M., **24:**14(61), 25(61, 120), 106(112), 169(34), 240(69, 71), 243(80), 245(80), 248(91), 249(91, 92, 93, 94)
Mroczkowski, S., **13:**205(18), 223
Mu, X. C., **14:**232(127), 239
Mucha, J. A., **15:**294–295, 311; **22:**211(70, 71), 212(73), 273–274
Muchmore, R. B., **1:**87(38), 121; **2:**254, 304
Mucic, R. C., **24:**36(146)
Mücke, K., **17:**356, 357, 369
Muehlhoff, H.-M., **21:**38(78, 81), 39(78), 40(78), 49(78), 73
Mueller, A., **7:**320(31), 393
Mueller, C. H., **16:**17, 126
Mueller, G. F. P., **5:**167, 231
Mueller, K., **24:**172(42), 173(42)
Mueller, M. H., **1:**204(74), 230
Muenchausen, R. E., **16:**58, 61, 75, 135, 137
Muhkerjee, A., **16:**75, 137
Mui, K., **19:**73
Mujsce, A. M., **24:**14(59), 20(59), 23(112), 27(112), 30(112)

Mukae, F., **24:**19(82), 20(82), 21(82), 27(82), 29(82)
Mukae, M., **24:**36(144)
Mukai, S., **11:**67(133), 73(123), 99, 100
Mukaida, M., **16:**41, 52, 131, 133
Mukherjee, S. P., **12:**261, 292
Mularie, W. M., **14:**166(117), 167(117), 175(132), 178(151), 183(132, 151), 184(132), 197, 198
Mulhern, P. J., **16:**28–30, 33, 129–130
Muller, A. B., **19:**41
Muller, E. K., **3:**139(40), 163
Müller, E. K., **5:**166(115), 176(115), 178, 203(115), 208(115), 232
Muller, E. K., **17:**235, 296
Müller, E. W., **1:**205(82, 84, 85), 230
Muller, G., **16:**107, 141
Muller, G. O., **3:**174, 208
Müller, H., **16:**70, 136
Muller, H. K., **9:**51(245, 246), 71
Muller, H.-R., **15:**154, 158, 164, 215
Müller, K., **4:**196(40), 197(40), 252
Müller, K. A., **16:**2, 123
Müller, K. H., **13:**119, 120, 148; **14:**60(191, 192), 61(191), 78; **17:**121, 126, 143; **18:**244, 245; **22:**5(10), 13(10), 77, 123(30), 151(30), 163(84), 164(84), 172–173
Muller, R. H., **19:**50, 52, 280, 284, 287
Muller, R. S., **5:**226, 235; **17:**210, 211, 223
Mulligan, W., **2:**392(123), 423
Mullin, J. B., **7:**170, 191(107), 250
Mullins, W. W., **4:**111(30), 135; **7:**283(82), 309; **22:**50(57), 78, 254(99), 274
Mulvey, T., **14:**23(91), 75
Mumtaz, A., **14:**71(231), 79; **17:**111, 142
Munakata, F., **16:**87, 138
Munekata, H., **12:**294
Munir, Z. A., **11:**108(24), 112(24), 200
Munoz, F. A., **7:**142(40), 248
Munteanu, O., **15:**227, 244, 261
Münz, W. D., **14:**60(193), 78, 179(154, 156), 198; **17:**88, 89, 90, 91, 97, 98, 99, 100, 108, 112, 118, 119, 121, 123, 124, 128, 129, 130, 131, 140, 141, 143
Murakami, **7:**198(174), 252
Murakami, E., **14:**89(19), 109(19), 117(19), 125
Murakami, M., **16:**14, 125
Murakami, T., **16:**25, 73, 127
Murakami, Y., **13:**64(28), 65(28), 68(28), 88(28), 91(28), 105; **15:**87, 140; **17:**153, 220
Muramatsu, M., **7:**363(157, 158, 159, 160), 396
Murarka, S. P., **12:**277(213), 279(213), 295; **13:**99(104), 107; **15:**299–300, 311
Murasaki, A., **24:**36(144)
Murasato, S., **12:**294
Murata, K., **14:**55(174), 77
Murata, O., **16:**79, 137; **17:**209, 223, 290, 299
Murawski, L., **11:**256(65), 257(65), 316
Murayama, A., **17:**308, 309, 364
Murayama, Y., **9:**53, 71; **10:**236(14), 258; **13:**79, 106; **17:**153, 219
Murbach, H. P., **1:**206, 230; **3:**220(27), 229(27), 234, 243, 271
Murdock, B., **16:**28, 128
Murduck, J., **17:**49, 58, 76
Murgia, J. E., **21:**219(128), 236

Murotani, T., **11:**51(76, 78), 52, 60, 76, 98
Murphy, D. W., **16:**3, 124
Murphy, J. A., **9:**31(115), 67
Murphy, M. J., **18:**263
Murphy, S., **24:**151(10), 154(10), 156(10), 163(10)
Murr, L. E., **4:**160(91a), 189; **9:**5(20), 65
Murray, F., **2:**379(82), 397(82), 422
Murray, L. A., **4:**261(34), 263(34), 274, 320; **5:**282(174), 318
Murray, P. T., **14:**7(23), 69(23), 73
Murray, R., **24:**206(2)
Murray, R. W., **24:**24(118, 119), 26(118, 119), 151(7, 9), 154(7, 9), 156(9)
Murry, S. J., **23:**279(22, 102)
Murthy, C. S., **22:**127(33), 172
Murthy, S., **24:**222(104)
Murti, D. K., **14:**230(103), 239
Muruyama, S., **11:**73(146), 100
Murygin, V. I., **7:**186(146), 251
Musal, H., **6:**41, 77
Musatov, Al. L., **12:**126(262), 164
Muscje, A. M., **24:**44(11), 45(11), 219(80)
Musher, J. N., **22:**178(11), 271
Musil, J., **12:**276(189), 294; **14:**14(48), 74; **17:**83, 88, 89, 90, 91, 95, 97, 98, 99, 111, 112, 114, 115, 116, 117, 118, 119, 121, 123, 124, 126, 127, 128, 129, 131, 137, 139–143; **18:**122, 124, 125, 136, 137, 151, 152, 165, 263, 265, 269, 271, 278
Musket, R. G., **14:**202(2), 236
Müsser, H. A., **6:**134, 148
Muth, D. G., **6:**4, 74
Muth, E. P., **11:**116(66), 201
Muto, S., **15:**222, 261
Muto, Y., **16:**86, 88–92, 138–139
Mutter, W., **7:**284, 286(84), 309
Mutter, W. E., **4:**257(2), 320
Mutucumarana, T. de S., **4:**200(75), 252
Myers, E., **5:**279, 318
Myers, E. R., **17:**226, 228, 232, 234, 236, 241, 248, 250, 251, 253, 255, 258, 260, 264, 291, 295–299
Myers, F. R., **22:**177(5), 193(44–46), 271–273
Myers, H. P., **9:**190(86), 257; **19:**153, 155, 168
Myers, T. A., **13:**68, 105
Myles, D. C., **24:**21(97), 23(97), 24(97)
Myoren, H., **16:**27, 98, 127, 140
Myoshi, Y., **11:**36(14), 96

N

Naber, G. T., **5:**183, 233
Naberthuis, S., **15:**171, 215
Nabitovich, I. D., **7:**27(113), 29(123), 30, 34, 35(126), 40
Nace, M., **14:**183(162, 163), 199
Nadai, J. P., **6:**6(29), 27(93), 74, 76
Nadhodkin, N. G., **22:**79
Nadier, M. P., **15:**306, 312
Naegele, J., **9:**47(220), 70
Naegele, K., **19:**281
Naemura, M., **13:**20(46), 24(46), 56(46), 58
Nag, A., **6:**188, 203
Nag, B. R., **15:**222, 261
Nagahama, Y., **2:**122(127), 144
Nagahara, L. A., **24:**98(94), 237(64)
Nagai, I., **13:**20(39), 22(39), 40(39), 57
Nagai, J., **17:**318, 352, 353, 366, 368
Nagai, K., **12:**246(19), 290; **13:**142(124), 150

Nagai, T., **12:**274(184), 294
Nagaibekov, R. B., **14:**145(65), 196
Nagakura, S., **14:**183(162), 199
Naganuma, M., **11:**44(50), 64, 65(113), 70(140), 73(140), 76, 97, 99, 100
Nagao, F., **17:**313, 314, 345, 365
Nagao, N., **13:**20(29), 23(29), 45(29), 57
Nagao, Y., **15:**171–172, 198–199, 207, 216
Nagasawa, A., **4:**149(32), 171(32), 188
Nagasawa, E., **14:**226(88), 238
Nagasawa, K., **5:**313(304), 322
Nagasawa, M., **9:**22(85), 23(85), 67
Nagasawa, Y., **17:**162, 221
Nagasima, K., **3:**229(54), 230(54), 272
Nagasima, N., **15:**300, 311
Nagata, H., **19:**250
Nagata, S., **11:**91(237), 92(237, 238), 103
Nagato, K., **15:**170, 174, 210, 215, 217
Nagatomo, T., **19:**250
Nagel, M., **2:**319, 359
Nagel, M. R., **4:**52(121), 93
Nagle, J. F., **24:**114(28)
Nagoa, N., **17:**153, 154, 160, 220
Nagpal, K. C., **16:**98, 140
Nagpal, V. P., **6:**110, 112, 132(131), 146
Naguib, H. M., **13:**117(43), 127(43), 135(43), 148; **14:**17(65), 74, 230(102), 239
Nahir, T. M., **24:**45(80)
Nahory, R. E., **7:**254; **14:**113(68), 127
Naik, I. K., **17:**228, 255, 298
Nail, N. R., **2:**376(75), 421
Nairn, J. A., **16:**279, 337
Naito, K., **7:**375, 379, 383(195), 397
Naito, M., **16:**42, 50, 131
Naito, N., **16:**28–29, 129
Naka, S., **19:**252
Nakagawa, K., **6:**208(16), 209(16), 262(16), 263(16), 264(16), 265(16), 266(16), 286(16), 295
Nakagawa, T., **14:**31(122), 76; **15:**244, 263; **24:**169(37)
Nakahara, S., **15:**147, 152, 154, 214–215; **16:**2, 3, 53, 124, 134, 251, 255, 268–269; **21:**221(137), 222(137), 237; **23:**2(2); **24:**6(24), 86(25)
Nakai, A., **16:**28–29, 129
Nakai, H., **13:**166(61), 167(64, 66), 168(61), 169(65), 180(64), 185(61, 64, 65, 66, 140), 186(149), 188(61, 64, 65, 66, 140), 189(149), 194, 196, 197
Nakai, J., **19:**51
Nakai, Y., **5:**290, 320
Nakamizo, K., **3:**232(60b), 236(60b), 272
Nakamori, T., **16:**86–87, 138
Nakamoto, K., **15:**267, 310
Nakamura, K., **1:**148(54, 55), 169(56, 55), 186; **13:**60(5), 78, 79(5), 104; **15:**152, 214; **16:**28, 33, 128
Nakamura, M., **10:**228(8), 251, 252, 258, 259
Nakamura, N., **10:**228(7), 258
Nakamura, T., **12:**135(295), 165; **15:**221, 260
Nakamura, Y., **6:**215(46), 216, 296; **15:**176, 179, 195–199, 216
Nakanisi, T., **19:**27
Nakanuma, S., **5:**268, 316
Nakase, K., **17:**353, 368
Nakase, M., **13:**180(104), 185(104), 187(104), 195
Nakashima, J., **15:**113, 141
Nakashima, S., **1:**148(55), 169(55),

Nakashima, S., (*continued*) 186; **13:**167(64), 180(64), 185(64), 188(64), 194
Nakatawi, M., **11:**76(152), 100
Nakayama, M., **4:**340(47), 370; **17:**24, 75
Nakayama, T., **4:**260(17), 320
Nakayama, Y., **16:**54, 134
Nakayima, M., **23:**166(90)
Nakhodkin, G., **3:**113(127), 127
Nakhodkin, N. G., **12:**121(243), 164
Nalk, I. K., **17:**228, 239, 240, 241, 297
Namba, S., **10:**237(20), 258; **13:**153(8), 154(8), 166(61), 167(64, 66), 168(61), 169(65), 180(64), 182(8), 183(8), 185(61, 64, 65, 66, 140), 186(149), 188(61, 64, 65, 66, 140), 189(149), 192, 194, 196, 197
Namba, Y., **6:**105, 146
Namegaya, T., **23:**228(32)
Namioka, T., **11:**23(16), 24(18), 25(18), 28(18), 29(18), 34
Namirovsky, Y., **21:**155(58), 234
Namura, T., **18:**102
Nanba, T., **17:**310, 311, 317, 365
Nanbu, K., **11:**50(72), 70(72, 141), 71(141), 72(141), 98, 100
Nandedkar, A. S., **22:**127(33), 172
Nandgeonkar, S. N., **22:**5(4), 77
Nandy, K. P., **3:**32(155), 51(155), 57
Nanev, K., **12:**87(118, 119), 161
Nankivell, J. F., **1:**191(23, 24), 228
Nanni, L. F., **4:**218(103), 219(103), 232(103), 253
Nannichi, Y., **7:**140, 141, 142, 248
Naoe, M., **14:**42–43(151), 77, 183(161), 199; **15:**152–153, 214; **17:**88, 91, 140; **18:**250–253
Narahara, C., **17:**197, 201, 222, 223
Narasawa, T., **13:**78(62), 79(62), 106
Narayan, J., **16:**10, 48, 57, 60, 65, 91, 125, 134, 136, 139; **19:**196, 205, 232, 233, 280
Narayan, S. Y., **23:**230(15), 231(15)
Narayanamurthy, V., **19:**281
Naslain, R., **17:**109, 141
Nassau, K., **16:**28–29, 129
Nastasi, M., **16:**50, 132
Nasu, H., **16:**27, 98, 127, 140
Natan, M., **14:**51(164), 77
Nath, P., **13:**60(4), 75(58), 76, 79(4), 94(89), 95, 104, 105, 106
Nathan, M. I., **13:**212(46), 217(46), 220(70), 224, 225; **15:**222, 260; **21:**240(5), 308
Nathan, R., **12:**100, 105, 109(105), 162
Nathans, R., **1:**205(81), 230
Nathanson, H. C., **12:**136(300, 301, 302, 303), 137(302), 165
Natori, A., **16:**263, 269
Natori, K., **16:**93, 139
Natta, M., **9:**167(22), 256
Naumann, R., **24:**34(141), 35(141)
Navarathna, A., **16:**214–216, 230–231
Navarro, E., **14:**105(55), 106(55), 112(97), 126, 128
Navinsek, B., **6:**102(94), 145
Navinvek, B., **14:**229(92), 239
Navratil, K., **10:**169(19), 226
Nawata, S., **4:**86(233), 96
Nawrocki, J., **24:**163(26)
Nayak, D. K., **23:**72(234)
Nayer, P. S., **21:**128(37), 233
Naylor, P. H., **9:**240(233), 261
Nazar, L., **16:**45, 63, 106, 132, 135, 141
Nazarova, R. I., **6:**37(123), 40, 76; **12:**295
Nazri, G., **15:**304–305, 312
Neal Cox, J., **15:**288, 311
Neal, W. E. J., **19:**280

Neave, J. H., **11:**42(39, 42), 97; **19:**10, 15
Nedorezov, S. S., **6:**137(209), 148
Nee, T. W., **6:**134(202), 148; **11:**293(128), 317
Neeb, H., **16:**23, 127
Needham, J. G., **2:**369(35), 420; **3:**106, 127
Neel, C. B., **2:**358(80), 361
Néel, L., **1:**280(9), 281(12), 284, 310(62), 315(62), 331, 333; **6:**210(22), 211(22, 23, 24, 25), 212(24, 25), 214(41), 215(41), 216(41), 220(63, 64), 221, 222, 225(24), 226, 229, 234(118), 241(22, 93, 139), 242(22, 23, 24, 25), 244(24, 25), 250(24, 41, 118), 253(22, 93), 254(2)
Neff, G. A., **24:**166(32)
Negami, S., **11:**229(23), 315
Negrete, G. V., **16:**81, 102–105, 121, 138
Neher, R. T., **4:**60(152), 94
Neighbor, J. E., **6:**124(187a), 148
Neikirk, D. P., **21:**193(93), 235
Neil, K. S., **12:**81, 160
Neite, B., **15:**115, 117, 141
Nelson, D. F., **15:**222, 261
Nelson, H., **7:**115(1), 116(3), 117(1), 118, 119(3), 137, 138, 139(35), 143(50), 146(53), 153(65), 186(141), 188(151), 214, 216, 217(197), 219(53), 221(213), 230(35, 234, 235), 231(236), 232(236, 236a), 243(265, 266), 244(272); **12:**135(294), 165(2), 245(273, 274), 247, 248, 249, 251, 253, 254, 255
Nelson, J., **21:**311–368, 312(4), 313(5), 323(4), 324(4, 15), 325(15, 18), 329(15), 330(27), 331(27), 332(27), 333(15), 334(5, 15, 29), 336(5, 15), 341(29), 342(29), 348(38), 351(18), 353(18, 27), 355(27, 29, 38), 356(38), 361(43), 366, 367, 368
Nelson, K. E., **2:**315(10), 330, 359, 360; **4:**60(144), 94; **10:**14(26), 43(26), 44(26), 45(26), 66
Nelson, O. L., **4:**353, 371
Nelson, R. C., **3:**169(3), 180, 208
Nelson, R. G., **13:**187(125), 196
Nelson, R. J., Jr., **11:**66(122), 99
Nelson, R. S., **14:**232(116), 239
Nemanich, R. J., **15:**277, 310
Nemchinskii, V. A., **14:**147(73), 161(107), 196, 197
Nemetz, A., **17:**310, 365
Nemiroff, M., **15:**15–16, 137
Nemirovsky, Y., **12:**282(219), 295; **14:**103(50), 124(50), 126; **21:**40(86), 74
Nemtsev, V. P., **3:**113(127), 127
Nender, C., **14:**14(50, 51), 74; **17:**111, 112, 116, 142
Neogi, P., **24:**14(62), 15(62)
Neogi, S., **24:**14(62), 15(62)
Neretina, N. A., **14:**155(87), 197
Nesbitt, E. A., **6:**209(19), 279, 295, 299; **9:**273, 296; **15:**150, 214
Nethercot, A. H., **7:**177(179), 204(179), 252
Netterfield, R. P., **13:**112(24), 120(54), 124(54, 72), 127(72, 91), 131(54), 132(54), 134(54), 135(54, 72, 105, 106, 107, 108), 136(54, 72, 105, 106, 108), 137, 138, 139(105), 140(105), 141(72, 105, 122), 143(91, 107), 144(24, 107, 133, 136), 145(138), 147, 148, 149, 150; **14:**61(194, 196), 78

Netzer, L., **24:**13(66), 15(66), 181(17), 228(21)
Neubauer, G., **24:**98(88), 221(90), 222(90)
Neuberger, H., **10:**248(41), 259
Neugebauer, C. A., **1:**58, 59, 60(141), 67, 207(103), 223(103), 227(103), 230, 324, 325, 326, 334; **2:**23(33, 35), 24, 25(33), 27(33), 28, 29(35), 30(44), 36, 54, 55(89), 56, 60, 61, 62, 371(58, 59), 383(58, 59), 384(58), 388(58, 59), 393(139), 421, 423; **3:**1(1), 7, 51(34), 53, 54, 212, 254, 255(94), 256, 257, 258(94), 259(94), 263, 264(94), 266, 270, 273; **6:**82, 143; **9:**4(6, 7), 6, 7(29), 43(194)
Neumann, G., **18:**155, 184; **22:**20(27), 77
Neumann, G. M., **7:**263(28)
Neumark, G. F., **21:**9(29), 72
Neumeier, G., **16:**92, 139
Neupert, W. M., **7:**48, 70(26), 112
Neureuther, A. R., **22:**5(4), 77, 181(19), 271
Neurgaonkar, B. R., **17:**162, 221
Neurgaonkar, R. R., **17:**261, 298
Neustadter, S. F., **4:**289(109), 322
Neverov, V. V., **14:**175(136), 198
Neviere, M., **11:**14(8, 9), 29, 30(19), 33, 34
Nevis, B. E., **14:**27(106), 75
Newbower, R. S., **6:**124(187a), 148
Newburgh, R. G., **1:**130(1la), 184
Newhouse, V. L., **1:**253(51, 52), 263(52), 263(77), 264(51, 52), 274, 275
Newkirk, J. D., **17:**43, 75
Newkirk, J. W., **1:**207(103), 223(103), 227(103), 230
Newman, N., **16:**121, 143
Newman, P. G., **21:**121(25), 233, 240(4), 241(7, 18), 245(18), 246(4, 24), 248(4), 252(4), 278(4), 308, 309
Newman, P. O., **15:**222, 260
Newman, R., **3:**137(32b), 163; **4:**315(159), 324
Newman, R. C., **1:**219(144), 231; **4:**138, 140(8), 187, 223(114), 253, 254; **6:**102(93a), 145
Newman, R. L., **5:**277, 280, 318
Newman, T. H., **16:**122, 143
Newmann, G., **22:**254(101), 255(101), 256(101), 262(101), 275
Newmann, G. M., **22:**254(101), 255(101), 256(101), 262(101), 275
Newnham, R. E., **19:**249, 272
Newton, I., **3:**2, 8, 53
Newton, J. L., **15:**286–287, 310
Ng, M-F., **16:**97, 140
Ngai, K. L., **9:**146(4a), 255; **11:**285(140), 304, 305, 317
Nghi, L. Q., **14:**8(30), 8–9(30), 17(30), 74
Nguyen, H. V., **19:**54, 94–96, 100, 101, 103, 129, 135, 152, 154, 157, 159, 160, 163, 165–167, 171, 175, 176, 181, 183, 280, 310
Nguyen, M. T., **17:**359, 369
Nguyen, N. V., **19:**56, 148, 280, 310
Nguyen, S. V., **15:**291, 311
Nguyen, V. S., **15:**291, 297–299, 311
Nguyen Van Doan, **7:**261(12, 13), 262(12, 16), 263(12, 32), 264(13), 294(13), 305(115), 306, 307, 310
Ni, W.-X., **23:**55(10)
Nicholas, J. F., **4:**142(22), 164(99), 184(22), 187, 189
Nicholas, N. C., **7:**50(32), 112
Nichols, D., **15:**53, 138

Nichols, D. R., **4:**270(87), 314(87), 322; **5:**296, 320; **12:**265(111), 292
Nichols, M. H., **4:**341, 370
Nicholson, B. J., **3:**139(40), 163; **17:**235, 296
Nicholson, R. B., **4:**225(130), 226(130), 254; **6:**228(105), 297
Nickel, J. H., **16:**4, 124
Nickerson, J., **5:**164(86), 231
Nickerson, J. W., **3:**203(82, 83), 205, 210
Nicks, L. J., **12:**287(68), 288(68), 312
Nicodemus, F. E., **10:**11(18), 65
Nicol, J., **1:**267(92), 272(100), 275
Nicolet, M.-A., **14:**66(217), 78, 104(53), 126; **17:**22, 74
Nicoll, F. H., **3:**195(58), 209; **5:**275, 277, 278, 318; **7:**171(114), 186(147), 250, 251
Nicollian, E. H., **11:**51(80, 81), 54(80), 89(227), 98, 103
Nie, H. Y., **24:**89(48)
Niebuhr, J., **6:**113(133), 146
Niedermayer, R., **3:**27(140, 141), 28(140, 141), 29(141), 30(141), 51(140, 141), 56
Niegand, J., **7:**385(201), 397
Nieh, C. W., **23:**161(64)
Niehaus, W. C., **11:**42(46), 50(46), 75(46), 76(46, 155), 97, 100; **13:**217(56), 218(59), 221(59), 224
Niehrs, H., **1:**198(50), 229; **2:**119(117), 143
Nielsen, A., **1:**324(98), 326, 334
Nielsen, J. W., **9:**275, 276, 281, 284, 288, 294, 296; **15:**7, 9, 11, 13, 136–137
Nielsen, S., **5:**147, 152, 183, 185, 199, 229, 233, 261(56), 268, 315, 316
Nieuwenhuizen, J. M., **22:**21(30), 77
Niewodniczanski, H., **3:**11(52), 51(52), 54
Nifontoff, N., **1:**218(137), 231; **2:**23(31), 60
Nigara, Y., **19:**240
Nigh, H. E., **5:**303(262), 321
Nigro, A., **16:**17, 126
Nihei, H., **18:**125
Nijs, J., **23:**2(4)
Niki, K., **24:**45(82)
Niklasson, G. A., **17:**351, 356, 367, 369; **19:**60, 129, 139
Nikolaeva, I. N., **9:**51(247), 71
Nikoshiba, N., **10:**256(57), 259
Nikulin, E. I., **1:**244(29), 274
Nill, K. W., **11:**107(19), 128, 129, 155, 175, 200, 201; **13:**219(67), 224
Nilsson, P. O., **6:**188, 203; **9:**188(75), 193(92), 194, 257, 258
Nilsson, P-A., **16:**67, 136
Nimmagadda, R., **13:**63(27), 75(59), 79(72, 74, 75, 76, 77), 80(59), 83(85); **17:**60, 77
Ninomiya, T., **12:**86(113), 99, 161
Ninua, O. A., **7:**186(148, 149), 187(148), 247(282), 251, 256
Nipotti, R., **14:**112(99), 128
Nisbino, M., **12:**277(217), 295
Nisenoff, M., **16:**101, 141
Nishi, Y., **21:**21(42), 72
Nishida, H., **9:**271, 296
Nishida, N., **24:**150(3)
Nishida, S., **15:**300, 311
Nishihara, Y., **15:**154, 156, 158, 164–165, 183, 185, 215
Nishihata, Y., **17:**24, 75
Nishikawa, S., **14:**55(174), 77
Nishimoto, C. K., **11:**100
Nishimoto, K., **6:**209(17), 295

Nishimura, M., **21:**31(62), 32(62), 73; **23:**144(35)
Nishimura, T., **10:**236, 258
Nishino, T., **9:**35(163b), 68
Nishioka, M., **17:**228, 247, 297
Nishiyama, Y., **16:**27, 98, 127, 140; **17:**317, 365
Nishizawa, J., **7:**193(159), 252; **12:**277(200), 294; **15:**300, 311; **19:**35
Niskanen, K., **16:**92, 139
Nissim, Y. I., **13:**218(63), 224
Niwa, K., **16:**34, 130; **24:**45(82)
Niwa, M., **24:**34(142), 35(142)
Niwa, T., **17:**356, 369
Nix, W. D., **17:**43, 46, 47, 49, 63, 75–77
Nixon, W. C., **2:**125, 144; **6:**103(102b), 145
Nizzoli, F., **17:**48, 49, 51, 53, 76
Nobes, M. J., **22:**113(26, 27, 30), 115
Noda, E., **17:**153, 220
Noda, J., **14:**31(121), 76; **17:**261, 298
Noda, M., **15:**279, 281, 310
Noda, T., **7:**23(79), 39
Nodvik, J. S., **1:**143(36), 163(36), 165(36), 169(36), 185
Noel, J. P., **14:**8(26), 36(26), 74
Noel, J.-P., **21:**5(7), 32(7), 33(7), 41(7), 71
Noga, K., **12:**94(148), 161
Nogami, J., **24:**86(26)
Nogami, M., **11:**68(134), 100
Nogar, N. S., **16:**58, 61, 75, 135, 137
Noge, S., **16:**111, 142
Noggle, T. S., **7:**66(54), 113
Noguchi, T., **10:**58(125), 68
Noguchi, Y., **6:**72(213), 79
Nogues, J., **16:**12, 125
Noh, D. W., **16:**91, 139
Noirhomme, B., **17:**279, 299
Nöldecke, G., **3:**11(51), 49(51), 51(51), 54
Nolder, R. L., **5:**152(33), 186(33), 199(33), 204, 209(33), 230, 234, 269(96), 316
Nöller, H. G., **1:**10, 64
Nomura, K., **14:**52(169), 77
Nomura, M., **1:**220(146), 231
Nonaka, T., **16:**96, 140
Nordberg, R., **7:**45, 48(16), 111
Nordblad, P., **16:**92, 139
Nordheim, L., **4:**351, 371; **6:**88(52), 144
Nordheim, L. W., **21:**22(53), 73
Nordine, P. C., **19:**168
Nordland, W. A., **11:**79(176), 101
Nordland, W. A., Jr., **11:**86(207), 102
Nordling, C., **7:**45, 48(16), 111
Nordman, J. E., **17:**87, 140
Noreika, A. J., **3:**100(1 45), 103(146), 118(145, 146), 128; **5:**154(45), 160(45), 161(45), 167(134), 176(209), 203(134), 222(209), 230, 232, 235, 303(263), 321; **6:**325(32), 346; **11:**171, 183, 203; **17:**153, 154, 220
Noreland, E., **11:**211(9), 314
Norika, A. J., **12:**269(150), 293
Norling, P., **16:**92, 139
Norman, J. E., **19:**148
Normand, C. E., **3:**73(46), 125
Norris, A., **7:**351, 395
Norris, M. T., **11:**67(133a), 100
Norris, P. E., **12:**264(110), 292; **16:**91, 139
Norrman, S., **19:**128, 173
Norstrbm, H., **12:**248(45), 251(45, 52), 252(52), 253(45), 254(52), 255(52), 269(52), 291
Norstrom, H., **14:**233(134), 240

North, A. M., **11:**256(57), 257(57), 315
North, J. C., **9:**271, 297
Northrop, D. A., **11:**108, 200
Norton, D., **16:**10, 48, 125
Norton, D. P., **16:**70, 136
Norton, F. J., **1:**2(6), 3, 7(12, 13), 8, 12, 64
Norton, M. G., **16:**61–62, 66, 135–136
Norton, P., **17:**347, 357, 358, 367, 369
Nos, S., **22:**134(51), 142(51), 172
Nosé, H., **1:**329(105), 334; **6:**272, 299
Nosé, S., **22:**134(52, 53), 172–173
Noskov, M. M., **2:**221, 238
Nossek, R., **6:**100, 104, 131(112, 114), 145, 146
Nosulenko, N. A., **4:**87(240), 96
Notarys, H., **15:**170, 215
Nothnick, C. E., **15:**54, 59, 138; **17:**174, 178, 222
Noto, K., **16:**88–90, 138–139
Noufi, R., **16:**97, 99, 140
Novak, F., **24:**184(35, 41), 188(41), 192(35)
Novice, M., **5:**1(7), 45
Novice, M. A., **2:**403(179), 424; **13:**61, 104
Novick, D. T., **1:**1(1), 63, 241(11), 273; **2:**393(141), 423
Novikov, E. B., **19:**2
Novikov, V. M., **9:**33(119), 68
Novoselova, A. V., **7:**31(129), 40
Nowichi, R. S., **22:**118(3), 171
Nowick, A. S., **16:**234, 266
Nowick, W., **12:**257(65), 291; **13:**99(108), 107, 111(20), 143(20), 147
Noy, A., **24:**97(83), 98(83, 84)
Noyce, R. N., **11:**140(103), 202; **21:**318(10), 366
Nozaki, T., **12:**292; **13:**217(57), 224
Nozik, A. J., **9:**58, 71; **21:**365(48, 49), 368
Nuese, C. J., **7:**193(162, 163), 205, 207, 209(183), 252, 253
Nunomora, K., **14:**65(208), 78
Nuss, M. C., **16:**101, 141
Nuttal, R., **5:**253(39), 257(39), 315
Nutting, J., **4:**225, 254
Nuzzo, R. G., **24:**2(3), 3(10, 11, 12, 13, 15), 4(15), 5(12), 10(35, 37), 12(13, 37), 13(13), 14(13, 35, 37, 60), 15(13, 37, 60), 16(13, 37), 17(10, 13), 18(11, 13, 35), 20(13), 30(13), 32(15, 37), 44(1, 2, 3, 4, 5, 6, 8, 9,12, 14), 45(6), 46(12, 14), 47(94), 55(12, 14, 94), 56(12, 14), 64(94), 65(12), 66(4, 6, 8, 9), 70(5, 6, 14), 82(7, 10), 83(10), 86(10, 40), 87(41), 88(41), 89(7, 41), 90(41), 91(41), 112(1, 3), 114(26), 125(50), 142(26), 150(2), 163(28), 181(18), 206(4), 212(54, 55, 57, 62), 213(62, 66), 217(62), 220(57, 66), 228(3, 4, 16, 18, 19), 237(16, 18), 243(80, 81), 245(80)
Nyaiesh, A. R., **14:**41(148), 76
Nyce, A. C., **7:**102(75), 113
Nye, J. F., **4:**140(19), 187
Nysen, P. A., **17:**217, 224

O

Oakamoto, H., **17:**318, 365
Oakland, L. J., **6:**224(79), 297
Oba, M., **17:**197, 223
Obasa, R., **13:**64(28), 65(28), 68(28), 88(28), 91(28), 105
Obeng, Y. S., **24:**27(125)
Oberg, P., **1:**32(76), 65; **2:**365(12), 420

Oberg, P. E., **3:**24(105), 26(126), 44, 51(105, 126), 55, 56, 59
Oberli, D. Y., **23:**255(77)
Oberly, J. J., **5:**199(174), 202(174), 234, 273, 317
Oblas, D., **22:**121(21), 154(21), 172
O'Brien, B., **3:**46(224), 59
O'Brien, D. F., **16:**92, 139
O'Bryan, H. M., **1:**127, 184; **7:**70(62), 113
Obvintseva, I. E., **17:**318, 344, 366, 367
Ochoa, E., **12:**253(66), 291
Ochs, S. A., **2:**179(61), 192
Ochsenfeld, R., **1:**247(38), 274
Ocko, B. M., **24:**3(17), 4(17), 44(19), 122(46), 180(10), 181(12), 182(23), 184(34, 36, 39, 40, 43), 185(10), 186(10), 187(34), 188(34), 190(34, 40), 192(12), 199(12, 23)
O'Connell, J., **22:**128(38), 172
Oda, E., **21:**31(62), 32(62), 73; **23:**144(35)
Oda, J., **5:**269, 316
Oda, M., **16:**25, 73, 127
Oda, S., **16:**88, 138
Oda, T., **14:**65(210), 78; **17:**153, 220
Odani, K., **12:**278(206), 294
Oden, P. I., **24:**98(94)
Odencrantz, F. K., **4:**34(64), 88(64), 92
Oding, A., **3:**255, 257, 258(96), 260, 263, 264(96), 265, 268, 273
Odom, R. W., **16:**60, 135
O'Donnell, R. W., **6:**89, 144
O'Dwyer, J. J., **6:**67, 73(216), 78, 79
Odynets, L. L., **6:**71(212), 79
Oechsner, H., **17:**130, 131, 143; **22:**6(12), 77
Oehrlein, G. S., **14:**232(125, 128), 232(129), 239; **22:**101(21), 115
Oelhafen, P., **9:**217(163), 259; **15:**166, 215
Oerate, C. F., **24:**84(21)
Oertel, G., **12:**94(149), 96(149), 161
Oeschner, H., **14:**17(64), 74
Ofer, D., **24:**21(103), 23(103, 113), 26(103), 27(103, 113), 30(113)
Offret, S., **9:**43(184, 187), 69; **14:**175(127), 198
Ogale, S. B., **16:**63, 70, 135–136
Ogata, K., **17:**83, 128, 129, 139
Ogawa, E., **16:**52, 133
Ogawa, H., **14:**52(169), 77
Ogawa, K., **16:**28, 33, 128; **24:**169(37)
Ogawa, M., **4:**68(165), 94; **13:**212(41, 44), 214(41), 215(41), 224
Ogawa, S., **1:**282(16), 331; **4:**149(32), 156(81), 159(81), 160(81), 171(32), 188, 189, 238, 239, 240, 254; **11:**44(51), 73(146), 74(147), 97, 100; **13:**143(130), 150; **15:**154, 164–165, 183, 185, 215
Ogawa, T., **22:**145(68), 173
Ogburn, F., **4:**199(62), 204(62), 227, 252
Ogilvie, R. B., **5:**189(159), 233
Ogita, N., **22:**145(68), 173
Ogletree, D. F., **24:**82(4), 89(47), 98(87, 88), 112(20), 117(20), 119(20), 128(20), 129(20), 218(78), 221(90), 222(90)
Ogmen, H., **21:**62(117), 74
Ogrin, Yu. F., **6:**137(213), 139, 140, 141, 142, 148
Oguey, H., **1:**320, 333
Ogura, S., **19:**238
Oh, B., **16:**42, 50, 117–118, 122, 131, 142–143
Oh, D. J., **22:**127(37), 172
Oh, S. J., **15:**176, 191, 216–217
O'Hanlon, J. F., **6:**37(124), 38(136a),

39, 45, 51, 76, 77; **7:**102(78), 113; **12:**276(187, 190), 294; **14:**85(7), 86(10), 86–87(10), 118(10), 125
O'Hara, M., **3:**168, 171(1), 180, 200(1), 201, 208
Ohashi, K., **15:**186, 217
Ohata, K., **13:**212(44), 217(57), 224
Ohba, Y., **21:**206(103), 235
Ohbuchi, Y., **15:**148, 214
Ohgaki, K., **17:**361, 370
Ohgihara, J., **15:**87, 140
Ohji, K., **17:**249, 297
Ohkata, R., **15:**150, 195, 214
Ohkawa, S., **13:**212(47), 224; **15:**50–51, 138
Ohkoshi, M., **15:**154–155, 161, 162, 163, 214–215
Ohlidal, I., **10:**169(19), 226
Ohnesorge, F., **24:**84(22)
Ohnishi, H., **15:**222, 261
Ohnishi, M., **10:**256(57), 259; **17:**261, 263, 298
Ohno, H., **11:**49(67), 67(67), 70, 98, 100; **17:**361, 362, 368
Ohno, M., **17:**354, 368
Ohring, M., **7:**268(59, 60), 270, 273(68), 278(59, 60), 279, 280(68), 282(68, 77), 283(68, 77), 284(59, 86), 296(59, 60), 308, 309; **9:**4(14), 5(14), 65; **16:**241–242, 244, 249, 253–257, 263, 265, 267–269
Ohsaka, T., **17:**361, 370
Ohsawa, Y., **17:**353, 368
Ohsone, H., **17:**110, 141
Ohta, K., **15:**172, 216, 244, 263
Ohta, N., **15:**185, 216
Ohta, S., **14:**29(113), 76
Ohta, Y., **15:**152, 214
Ohtsuka, N., **19:**35
Ohtsuka, T., **16:**88, 138; **17:**315, 316, 317, 329, 331, 365
Ohuchi, F. S., **14:**64–65(207), 65(207), 78
Ohuyama, N., **16:**262–263, 269
Ohya, K., **14:**19(76), 75
Oi, T., **17:**304, 357, 364, 369
Oikawa, H., **18:**221
Oikawa, M., **17:**235, 296
Oishi, K., **15:**172–173, 216
Ojha, S. M., **12:**238(11), 241(11), 247(50), 248(42, 47, 51), 251, 252, 253, 254(49, 52), 255(49, 52), 257(51), 269(51), 290
Oka, K., **17:**116, 142
Okabayashi, H., **14:**226(88), 238
Okabe, T., **7:**264(44), 307
Okada, M., **13:**71(52), 105
Okada, T., **5:**276, 277, 280, 318
Okada, Y., **15:**287, 310
Okahata, Y., **24:**10(38), 23(38), 24(38), 36(144)
Okamine, S., **15:**185, 216
Okamoto, A., **14:**12(47), 13–14(47), 14(47), 71(234), 74, 79
Okamoto, H., **11:**67(133b), 80(177), 100, 101; **12:**275(174), 294
Okamoto, J., **15:**87, 140
Okamoto, K., **13:**203, 223
Okamoto, T., **15:**112–113, 115–116, 141, 172, 216; **17:**111, 142
Okamura, T., **19:**250
O'Kane, D. F., **9:**281, 288, 291, 296
Okane, H., **13:**185(135), 187(135), 196
Okaniwa, H., **9:**47(222a), 70
Okano, H., **13:**176(91), 177(91), 180(104), 185(91, 104), 187(91, 104), 195
Okatov, M. A., **9:**33(144), 68
O'Katoy, M. A., **12:**187(83), 234

O'Keeffe, T. W., **15:**22, 25, 32, 34–35, 43, 47–58, 72–74, 78, 137–139
Oketani, S., **1:**223, 232
Okiji, A., **6:**246, 298
Okimura, H., **3:**198(69), 210
Okochi, M., **6:**272, 299
Okrent, D., **17:**91, 92, 140
Okubo, S., **16:**272, 281, 285–287, 332, 334, 337
Okuda, T., **15:**103, 113, 140
Okumura, T., **12:**289(265), 296
Okuno, Y., **7:**193(159), 252
Okura, K., **5:**276(143), 318
Okuyama, M., **16:**97, 140; **17:**290, 299; **19:**250
Okuyama, N., **16:**262–263, 269
Okuyama, S., **22:**135(59), 173
Olafson, B. D., **24:**211(47)
Olaison, R., **12:**248(45), 251(45), 253(45), 291
Olander, D. R., **16:**60, 135
Oldham, L. P., **7:**68(58), 113
Oldham, W. G., **5:**276(141), 317, 318; **7:**135(31), 248; **19:**58, 59; **22:**5(4), 77
Oleinik, O. A., **22:**91(16), 115
Olin, H., **16:**67, 136
Olin, H. J., **3:**66(24), 125
Oliner, A. A., **17:**146, 219
Oliphant, M., **14:**208(18), 237
Olive, G., **6:**37(128), 45(128), 46(128, 142a), 47(128), 48(128), 49(128); **12:**276, 294
Oliver, C. B., **7:**279(74), 309
Oliver, M. B., **12:**99, 162
Oliver, P. E., **7:**223(218), 253
Oliver, W. C., **17:**49, 62, 63, 77
Oliver, W. F., **17:**256, 257, 298
Olmstead, J., **4:**301(133), 323; **5:**297
O'Loughlin, M. J., **21:**128(33), 162(69), 233, 234
Olsen, A. L., **4:**46(81), 92
Olsen, C. D., **1:**308, 333
Olsen, G. H., **12:**125(255), 129(272), 164, 165
Olsen, J. L., **1:**241(16), 273; **6:**96, 113(140), 121, 129(75), 130, 135(75, 140), 136(140), 145, 146
Olsen, T., **6:**82(24), 144
Olson, E. C., **22:**209(66), 273
Olson, E. R., **2:**370(46, 47), 390(119), 420, 421, 423
Olson, H. L., **6:**282(196), 286(196), 299
Olson, J. M., **21:**322(14), 337(14), 367
Olson, O. H., **2:**321, 359
Olson, W. L., **16:**81, 100, 102–106, 117, 121, 138, 141–142
Olsson, E., **16:**292, 340
Omnes, F., **19:**2, 9, 29, 30, 38, 41
Omoto, O., **19:**250
Ondarcuhu, T., **24:**235(57)
Ondris, M., **1:**288(32), 289(34), 332
O'Neill, J. A., **18:**310
O'Neill, J. J., **14:**216(35), 219(57, 60), 220(60), 221(60), 222(60), 223(60), 235(152), 240
O'Neill, J. J., Jr., **9:**46(210, 211), 50(227), 54(227), 70
Ong, E., **15:**303, 311
Ong Sing Pien, **2:**125, 144
Onishi, K., **15:**147–148, 214
Onishi, M., **17:**161, 162, 221
Onishi, S., **17:**153, 154, 220
Onnagawa, H., **17:**288, 299
Ono, K., **17:**88, 91, 140; **18:**81, 82, 84
Onton, A., **7:**174(120), 176(124), 178(118e), 196(169), 203(180), 209(120), 222(124), 250, 251, 252
Onuki, K., **3:**232(60b), 236(60b), 272
Oo, K., **9:**49(224a), 70

Oohashi, T., **9:**54, 71
Oomori, T., **18:**194, 203–204, 207–208
Oona, H., **7:**66, 67(52), 113
Oosterlaken, T. G. M., **22:**267(117), 275
Opheusden, J. V., **22:**134(51), 142(51), 172
Opschoor, J., **15:**227, 244, 261
Oranovich, J., **17:**153, 154, 221
Ord, J., **6:**23, 75
Oredson, H. N., **6:**282(196), 286(196), 299
Orehotsky, J., **9:**273, 296
O'Reilly, E. P., **23:**228(30)
Orel, E. N., **9:**47(222b), 70
Orem, T. H., **4:**227, 254
Oriani, R. A., **7:**258(5), 264(41), 305, 307
Oritz, A., **17:**153, 154, 221
Orlando, T. P., **16:**28, 33, 116, 128, 142
Orlang, R., **5:**255(43), 256(44), 315
Orlinov, V. I., **14:**10(37), 51(165), 74, 77
Orlowski, R., **9:**237(224), 245(224), 261
Orme, J. E., **6:**23, 27(83), 75
Oron, M., **19:**41
Oroshnik, J., **5:**289, 320
Orowan, E., **3:**256, 273
Orr, J. C., **12:**246(20), 290
Orr, W. H., **1:**189(11), 228
Orris, E., **5:**291, 320
Ortiz, A., **12:**176(54), 177(54), 178(54), 179(54), 233
Osadin, B. A., **14:**196
Osafune, H., **3:**137(28), 141(28), 162; **4:**265(65), 314(65), 321; **5:**268, 289, 316, 320
Osaka, Y., **14:**105(54), 126; **15:**293, 311; **16:**27, 98, 127, 140
Osamura, K., **7:**198(174), 252
Osanai, H., **16:**88, 138
Osantowski, J. F., **10:**57(122), 68, 86(27), 87(27), 88(27), 89(27); **12:**10(39), 12(39), 50
Osborn, J. A., **19:**140
Osborne, J., **4:**305(145), 306(145), 323; **21:**337(33), 367
Osborne, N. R., **16:**28–30, 33, 129–130
Osbourn, G. C., **15:**222, 260; **21:**364(47), 368
Osenbach, J. W., **15:**272, 292, 311
Osgood, R. M., **11:**198(159), 203; **13:**159(35), 162(45, 46), 163(45, 46, 51), 164(45, 53), 165(58), 167(62), 169(62, 68), 171(72, 73), 174(35, 58, 72, 79, 81), 175(35, 81), 177(96, 97), 178(72), 181(111), 182(51), 183(119), 184(119), 185(35, 53)
Oshe, E. K., **6:**64(186), 78
Oshima, M., **14:**232(117), 239
Oshima, S., **6:**291(212), 300
Oshinsky, W., **5:**279, 280, 318
Osipov, V. A., **14:**164(116), 176(116), 197
Osofsky, M. S., **16:**53, 65, 78, 134, 136–137
Ossikovski, R., **19:**45
Osterberg, H., **1:**70(14), 120; **2:**247, 304; **10:**237, 258
Osterman, D. P., **16:**110, 142
Ostermayer, F. W., Jr., **13:**164(56, 57), 165(56, 57), 166(56), 167(57), 176(56, 57, 84), 178(100), 185(84), 186(56, 57, 84, 100), 188(100), 189(56, 57, 84), 194, 195
Ostling, M., **14:**233(134), 240
Ostrander, W. J., **2:**391(121), 423
Ostrovskii, L. F., **7:**262(18), 264(33, 38), 306, 307

Ostrowsky, D. B., **10:**228(4), 230(10), 238(10), 242(10, 32), 243(34), 253(4), 254(4), 258
O'Sullivan, E. J. M., **16:**5, 124
Osvald, J., **16:**249, 268
Oswald, H. R., **12:**238(2), 288(256), 289, 295
Oswald, R. G., **9:**217(162), 259
Ota, K., **15:**168, 215
Ota, Y., **11:**36, 97
Othmer, D. F., **9:**12(55), 66
Ott, J., **23:**24(97)
Ott, N. L., **17:**64, 65, 77
Ottenbacher, V., **17:**256, 257, 298
Ottermann, C., **17:**348, 356, 357, 367, 369
Otto, A., **6:**198(59), 203; **9:**202(141), 206(131), 210, 212(135, 140), 230(213, 214), 234(217), 241(235, 237), 254(251), 258, 259, 260, 261
Otto, G., **13:**31(48), 58
Ottosson, M., **16:**92, 139
Ouang, T. T., **3:**39(204), 52(204), 58
Oudar, J. L., **23:**255(78)
Oudeacoumar, **12:**169(22), 174(22), 180(22), 189(22), 194(22), 196(22), 233
Oupta, R. P., **16:**98, 140, 239, 267
Oura, M., **13:**20(38), 22(38), 34(38), 57
Ourmazd, A., **16:**42, 131
Ouyang, C., **22:**165(86), 173
Ovcharenko, V. D., **14:**161(109), 197
Overbeck, C. J., **3:**122, 128
Overcash, D. R., **6:**103(99), 122(99), 132(99), 145
Overney, G., **24:**95(76)
Overney, R. M., **24:**89(54), 92(54), 94(69, 71, 72), 95(69, 71, 72), 218(79)
Ovshinsky, S. R., **7:**32, 33(134, 135), 40
Owen, A. E., **4:**304, 305, 323
Owen, H. A., **2:**365(6), 419; **14:**175(128, 129), 198
Owen, L. W., **5:**306, 321
Owens, J. M., **15:**47–48, 48, 51–52, 64, 67–68, 75, 77, 81–83, 88, 101, 138–140
Oxley, A. E., **19:**199
Oxley, J. H., **5:**238(1), 261–263, 289(224), 316–318, 320; **9:**31(111), 67
Oxtoby, D. W., **22:**135(59), 173
Oyama, H., **16:**37–38, 130
Oyama, N., **17:**361, 370
Oychinnikov, I. M., **16:**71, 79, 136
Oye, H. A., **16:**3, 124
Ozawa, M., **13:**25(47), 27(47), 58; **14:**196
Ozawa, Y., **3:**229(55), 230(55), 234(55), 272
Ozekein, A., **17:**19, 20, 21, 32, 74
Ozeki, M., **19:**35
Öztürk, M. C., **23:**48(170)

P

Paccard, D., **6:**211(23, 24, 25), 212(24, 25), 215(42), 217(42), 219(42, 62), 220, 225(24), 242(23, 24, 25), 244(24, 25), 250(24), 254(24, 25), 255(24, 25), 256(24), 257(24), 258(24), 259(24), 260(24), 271(24), 295, 296
Pacey, C. G., **13:**120(54), 124(54), 131(54), 132(54), 134(54), 135(54, 107), 136(54), 137(54), 143(107), 144(107), 145(138), 148, 150
Pacey, D. J., **3:**24(121), 51(121), 56

Pachori, R. D., **12:**214(103a), 222(103a), 235
Packer, D. M., **1:**184(72), 186; **4:**60(143), 94
Paclaczyk, C., **11:**236(39), 237, 315
Padalka, V. G., **2:**340, 360; **14:**158(99), 163(115), 164(116), 176(116, 140), 177(146, 147), 185(115), 197, 198
Padamsee, H., **16:**106, 141
Padawer, G. E., **4:**218(103), 219(103), 232(103), 253
Padmanabhan, K. R., **16:**96, 140
Paff, R. J., **7:**173, 179(119), 186(143), 250, 251; **12:**99(162), 162
Pafomov, V. E., **9:**159, 174(20), 256
Page, C. J., **24:**166(32)
Page, D. J., **17:**198, 223
Page, G. A., **24:**70(107)
Page, T., **10:**75(8), 164
Pagel, B. R., **11:**150(121), 202
Pagel, L., **12:**256(61), 291
Paget, D., **19:**2
Pai, C. S., **18:**222, 224
Pai, C, S., **22:**211(72), 274
Pai, P., **15:**288, 311
Pai, P. G., **15:**287–288, 310
Pai, S. T., **16:**250, 268
Paiste, R. M., **17:**359, 361, 369
Pajaczkowska, A., **16:**5, 124
Pakulis, E. J., **16:**13, 125
Palatnik, A. S., **7:**22, 23(75), 39
Palatnik, L. S., **1:**209, 231; **3:**11(53), 51(53), 54, 256(101), 257, 258(104), 261, 263, 264(101, 104), 265, 266, 267(112), 271, 273; **5:**164(95), 231; **7:**21(66), 22, 23(76), 38, 39; **9:**2, 64; **17:**3, 60, 73
Pale-Grosdemange, C., **24:**10(36), 12(36), 14(36)
Palik, E. D., **3:**299(73), 301
Pallmer, M., **24:**70(106)
Palmer, E. W., **11:**21, 33
Palmer, I. C., **12:**126(261), 128(261), 132(261), 133(283), 164, 165
Palmer, M. J., **23:**47(160)
Palmetshofer, L., **11:**130(87), 201
Palumbo, G., **4:**50(97), 93
Pamplin, B. R., **12:**182(72), 183, 193(72), 234
Pan, A., **22:**121(22), 172
Pan, C. K., **17:**22, 74
Pan, D. S., **15:**236, 245–246, 259, 262–263; **21:**94(23), 111
Pan, P., **15:**291, 297–299, 311
Pan, S., **16:**52, 133
Panaiotov, I., **7:**315(2), 393
Panchishin, R. S., **6:**116(151), 147
Pandy, R. K., **11:**112, 200
Pandya, D. K., **12:**169(26, 40, 41, 42, 43), 176(26, 49, 51), 179(65), 180(65), 185(49), 186(26, 49), 187(87, 88), 189(88), 190(87, 88), 191(88), 197(26), 198(49), 199(49), 203(42), 204(43), 205(43), 207(42), 208(42), 210(42); **22:**79, 211(99), 212(99, 104, 105, 106), 213(104, 105, 106), 214(42, 108), 215(42), 217(40, 41, 42, 104), 218(99, 111), 219(106, 111, 112), 220(112), 223(106, 113), 224(120), 226(124, 125), 227(124), 229(41, 106, 111), 230(99,106), 233, 234, 235
Pang, S. W., **14:**232(119, 121), 239
Panish, M. B., **7:**145(51a), 146(54), 149(59), 154(59, 67), 155(67a), 157(69, 74), 158(75), 166(92a), 181(134), 186(142), 190(153a), 193(158), 198(171), 201(134,

Panish, M. B., (*continued*) 178), 203(134, 158, 178), 214(198), 217(198, 200a), 234(240,); **19:**281; **21:**150(54), 197(97, 243, 245), 199(97), 200(97), 201(97), 202(99, 100), 203(99), 204(99), 207(97, 99), 213(116), 214(116), 216(116), 234, 235, 236, 239(253), 248, 249, 251, 252, 253, 254, 255
Pankert, J., **16:**47, 50, 132
Pankey, T., **5:**163(65), 165(105, 106, 111), 168(65), 174, 176, 190(105), 193, 203, 209(65), 219(106), 222(106), 223, 231, 232
Pankove, J. I., **7:**169(96), 205(181), 213(181), 230(234), 232(236a), 250; **12:**133(285), 165; **13:**51(60), 52(60), 53(60), 58; **17:**304, 319, 364
Pankowska, H., **16:**98, 140
Pankratz, J. M., **9:**51(251), 52, 62(251), 71
Pannish, M B., **11:**66(121), 99
Pansack, L., **5:**289(225), 320
Panson, A. J., **16:**8, 11–12, 87, 125, 138
Pantaloni, S., **17:**361, 369
Pantano, J. V., **23:**237(33)
Panzner, M., **16:**70, 136
Paoletti, A., **16:**99, 140
Papaconstantopoulos, D. A., **17:**119, 142
Paparoditis, C., **5:**164(100), 172, 232
Papazian, S. A., **5:**274, 317
Papillon, D., **21:**96(29), 112
Papirno, R., **3:**254, 273
Papoulis, A., **10:**186, 226
Pappas, D. L., **19:**105
Pappis, J., **5:**264(65), 299(252), 316, 321
Papuchon, M., **10:**228(4), 243(34), 253(4), 254(4), 258; **21:**96(29), 112
Paquin, L., **15:**282–283, 310
Paradis, E. L., **17:**153, 220
Paranjape, B. V., **9:**159(17), 256
Paranjape, V. V., **9:**15(68), 66
Pardee, W. J., **9:**217(169), 259
Parekh, J. P., **15:**33, 37, 72–73, 101, 137–140
Parey, J. Y., **19:**8
Pariiskii, V. B., **3:**216(18), 271
Parikh, A. N., **19:**258, 282, 288, 290; **24:**3(15, 18), 4(15, 18), 32(15), 33(137), 44(8, 37), 66(8), 97(78), 98(78), 169(35, 36), 172(42), 173(42), 213(66), 220(66), 228(3, 20), 229(40)
Parikh, N. R., **16:**6, 125; **17:**291, 299
Parilia, P., **16:**97, 99, 140
Paris, B., **5:**261(58), 315
Paris, M., **17:**256, 257, 298
Parish, W., **21:**221(139), 222(139), 237
Parisi, G. I., **2:**371(63), 397(63), 421
Parisot, J. M., **3:**11(55), 12(55), 13(55), 49(55), 51(55), 54
Paritskii, L. G., **13:**162(43), 164(43), 165(43), 193
Park, B. K., **17:**271, 272, 276, 278, 298
Park, H., **12:**258(74), 291
Park, J., **21:**84(15), 111
Park, J. S., **21:**4(17), 72, 77–112, 78(1, 2), 81(13, 14), 86(14), 87(14), 88(18), 91(2, 19, 20), 95(20, 26), 97(26), 100(36), 104(43), 105(1), 108(46), 109(2), 111, 112, 155(62), 207(110), 208(112), 209(112), 210(112), 211(1); **23:**72(234), 102(43), 109(59), 146(46)12), 212(112), 213(112), 224(141), 234, 235, 236, 237

Park, J.-H., **22:**183(28), 185(28), 232(28), 270(121), 272, 275
Park, K. C., **10:**55(118), 68
Parker, C. D., **15:**244–245, 262–263
Parker, D. L., **13:**179(102), 195
Parker, E. H. C., **11:**36(24, 25), 66(127), 68(135, 136), 97, 98, 100
Parker, G. J., **22:**178(14), 271
Parker, M. A., **17:**347, 353, 367, 368
Parker, R. J., **16:**147, 151, 229
Parker, S. G., **5:**167(136), 182(136), 233, 282, 318
Parker, W., **3:**39(202), 40(202), 52(202), 58
Parkhutik, V., **14:**94(32), 103(32), 106(32), 126
Parkin, S. S. P., **16:**34–35, 130
Parkinson, D., **6:**318(31), 340(59), 341(59), 346, 347
Parks, R. D., **6:**103(102), 145
Parnieux, J. P., **11:**254, 255
Paroli, P., **15:**104, 113, 141; **16:**99, 140
Parrinello, M., **22:**142(64, 65), 143(65), 173
Parris, M., **17:**241, 287, 297
Parrish, W., **21:**117(7), 128(7), 232
Parrott, J. E., **6:**89, 91, 92, 116(164), 117(164), 132(164), 134, 144, 145, 147
Parson, K. A., **24:**163(26)
Parsons, C. A., **21:**365(49), 368
Parsons, R. R., **14:**11(41), 44–45(157), 45(158), 53(41), 54(41), 74, 77; **16:**28–30, 33, 129–130; **17:**111, 116, 142; **19:**139
Parsons, W. F., **13:**60(9), 62(9), 104
Partlow, D., **17:**206, 222
Partlow, W. D., **11:**10(4), 33
Partovi, F., **7:**364(164), 396
Pascard, H., **6:**274(183a, 184), 299
Paschen, F., **4:**60(146), 94
Pasciutti, E. R., **1:**227(185), 232
Pasco, R. W., **16:**249, 252, 268
Pascual, E., **19:**45
Pascual, R., **17:**241, 290, 297
Paserin, V., **16:**53, 133
Pashinkin, A. S., **7:**31(129), 40
Pashley, D. W., **1:**196(36, 42), 226, 227, 229, 230, 280(5), 330(113), 331, 334; **2:**13(6a), 14, 18(6a), 22(6a), 53, 54, 56, 58(83), 59(83), 60, 62; **3:**32(161), 57, 212(4, 6, 8), 213(4, 6), 216, 218(6), 219(23), 225, 243, 256(100), 257(100, 106), 258(100), 261, 263(100), 264(100), 266, 267(100), 268(111), 270, 271, 273; **4:**118, 119(45, 47), 120, 128, 131(47), 132(75), 133(44, 45, 75), 135, 136, 137, 138(1); **14:**59(197), 77
Paska, Z. F., **21:**216(119, 121), 236; **23:**213(20)
Paskin, A., **6:**137(222), 143, 148
Pasko, J. G., **11:**107(18), 157(129), 171(18), 200, 202; **21:**40(84, 88), 74
Passaglia, E., **3:**16(72), 51(72), 54; **4:**75(186), 83(206), 84(206), 95
Passerini, S., **17:**361, 369, 370
PastPnak, J., **5:**160, 230
Pastrnak, J., **19:**10
Pate, M. A., **15:**249–250, 263; **21:**334(29), 341(29), 342(29), 348(38), 355(29, 38), 356(38), 367
Pate, R., **23:**49(37)
Patel, J. R., **5:**188(158), 233, 253(38), 257(38), 315
Paterson, M. S., **4:**230(143), 254
Paterson, R., **24:**163(26)

Pathak, R. N., **21:**241(16), 309
Patil, H. R., **7:**282(81), 309
Patnaik, B., **7:**289(96), 309
Patnaik, B. K., **16:**6, 125
Patrick, W. J., **3:**171(24), 176(24), 209; **5:**164(89), 172(89), 231
Patt, R., **16:**110, 142
Pattee, H. H., **1:**33(78), 65
Pattee, H. H., Jr., **2:**125(138), 144
Patterson, A., **17:**318, 365
Patterson, D. A., **10:**84(25), 164
Patterson, G. N., **22:**189(42), 272
Patterson, J., **6:**103, 146
Patterson, J. S., **10:**293(8), 308
Patterson, R. J., **2:**370(55), 421; **3:**139(43), 163; **5:**299, 321
Patterson, R. W., **15:**22, 25, 32, 34–35, 43, 137–138
Patterson, W. E., **2:**393(131), 423; **6:**4, 7, 74
Patton, C. E., **6:**239, 240(133, 134), 298; **15:**95, 98, 107–111, 140–141
Patton, G. L., **23:**55(11)
Paul, G. L., **15:**283, 310
Paul Hed, P., **19:**210
Paul, K., **24:**239(68)
Paul, W., **1:**44(123), 45(123), 67; **5:**166(119), 178, 232; **7:**24, 26, 39; **12:**289(265), 296; **15:**277, 279, 310; **19:**72
Pauleau, Y., **14:**7(24), 8(24), 74, 98(40), 126; **17:**111, 142
Paulevé, J., **1:**281(12), 331; **6:**280(192), 299
Pauley, J. L., **7:**363(155), 396
Paulsen, J. F., **3:**106(97), 127
Paulson, W. M., **17:**36, 40, 75
Paulsson, G., **19:**2, 23, 26–29
Paulus, M. J., **15:**244, 263
Pauly, T., **1:**18(39), 64
Paunovic, M., **4:**196(36, 37, 39), 199(36, 39), 200(36, 39), 201(36), 204(36, 39), 206(36, 39), 209(36, 39), 211(36), 251, 252
Pauthenet, R., **6:**280(192), 299
Pavaskar, N. R., **12:**235
Pavlov, L. P., **3:**172(10), 174(10), 208
Pavlov, P. V., **7:**36(158), 41
Pavuna, D., **16:**26–28, 127–128
Pawel, R. E., **6:**12, 26(91), 75, 76
Pawlik, D., **15:**299, 301–302, 311
Paxman, M., **21:**312(4), 313(5), 323(4), 324(4, 15), 325(15), 329(15), 330(27), 331(27), 332(27), 333(15), 334(5, 15), 336(5, 15), 353(27), 355(27), 366, 367
Payan, R., **6:**202(80), 204
Payling, C. A., **15:**249–250, 263
Payne, D. A., **16:**99, 140; **17:**240, 241, 297; **19:**250; **24:**243(81)
Payne, E. H., **22:**101(21), 115
Paz de Araujo, C. A., **17:**253, 259, 289, 297–299
Paz-Pujalt, G. R., **16:**98, 140
Pazdzerskii, V. A., **9:**15(70), 66
Peacock, R. N., **6:**124(190), 127(190), 148
Peaker, A. R., **7:**167, 169(94), 250; **9:**33(132), 35, 37, 68
Pearlman, D., **2:**376(75), 421
Pearsall, T. P., **10:**252(68), 259; **23:**221(9), 224(9)
Pearson, A. D., **4:**310(155, 156), 324
Pearson, B. D., Jr., **10:**162(92), 166
Pearson, C., **6:**54(160), 63(160), 68, 71(205), 77, 78
Pearson, D. L., **24:**33(137), 44(37)
Pearson, G. L., **2:**149, 190; **5:**145, 229; **7:**140, 141, 142(36), 155, 166(67b), 180, 183(129), 189, 190(153), 248, 249, 251

Pearson, R. G., **24:**3(14)
Pearton, S. J., **21:**216(122, 123), 236
Pease, R. F., **24:**206(2)
Pease, R. S., **3:**67, 125
Pebere, N., **24:**163(24)
Pechenkina, R. S., **4:**271(92), 322
Pechhold, W., **11:**289(113), 317
Peck, D. D., **15:**244–245, 262–263
Peck, R. L., **12:**36(64), 38(64), 51
Peck, W. F., **16:**3, 124
Pecoraro, R. P., **4:**257(2), 320
Pedone, D., **17:**359, 369
Peekema, R. M., **6:**54(153), 77
Peers, J., **13:**186(153), 190(153), 197
Peeters, F. M., **21:**241(15, 16), 309
Pegis, R. J., **1:**70(22), 120; **5:**60(16), 86
Pei, S.-S., **21:**94(25), 111, 150(55), 155(59), 158(59), 159(59), 181(55), 234; **23:**279(22, 102), 280(105)
Peibst, A., **3:**174, 208
Peiniger, M., **16:**107, 141
Peiser, H. S., **1:**205(88), 230
Peitner, H. E., **16:**254, 269
Pelissier, A., **12:**135(293), 165
Pell, E. M., **7:**45(8), 111
Pellegrini, P. W., **23:**101(38), 102(40), 163(47), 166(91)
Pelletier, E., **19:**192, 239, 240, 280
Pelletier, J., **18:**124
Pellicori, S. F., **12:**36(66), 51
Pellin, M. J., **14:**6(21), 73
Pelloie, B., **14:**94(92, 93), 98(94–96), 101(94, 95), 111(63), 112(92, 93, 97–99), 113(93, 94), 114(93, 96), 119(64), 126–127, 127, 128
Pelouch, W. S., **21:**365(48), 368
Pelve, E., **21:**106(45), 107(45), 108(45), 112, 118(14), 128(14), 133(14, 44), 137(14), 141(14), 149(14), 150(14), 160(14), 172(14), 193(14), 209(14), 232, 233, 244(21), 249(25), 261(21), 284(25), 309
Pemble, M. E., **19:**2, 10, 15, 32
Pendergast, W. J., **2:**394(148), 423
Pendrick, V., **16:**106, 141
Penfold, A. S., **2:**69(13), 70(13), 71, 141; **14:**2(3), 14(52), 22(3), 33(3), 34(3), 39(3), 73, 74; **17:**111, 112, 116, 117, 142
Penfold, J., **24:**220(86)
Peng, J., **13:**217(55), 224
Peng, J. B., **24:**36(147), 180(11)
Pennanen, K., **24:**181(14), 184(34), 187(34), 188(34), 190(34)
Pennebaker, W. B., **3:**154(77), 164; **10:**37(67), 67; **12:**276(190), 294
Penner, T. L., **24:**12(54), 13(54), 15(54), 212(59)
Penney, R. V., **7:**263(25), 306
Penning, F. M., **3:**91, 126
Pensak, L., **1:**35(88), 66; **12:**157(334), 166
Pentyush, E. V., **17:**355, 356, 368
Penwell, R. C., **16:**282, 284, 338
People, R., **15:**222, 261; **17:**26, 75; **21:**79(6), 80(9), 111
Peralta, E., **24:**249(96)
Perdijk, H. J. R., **1:**51(126), 67
Perera, A. G. U., **21:**1–75, 4(12, 18), 6(18), 7(2, 18, 23), 8(25, 26), 9(18, 25), 10(25), 11(25), 14(37), 17(2, 25), 20(37), 22(18), 28(2, 18, 26), 29(58), 31(26, 61), 32(2, 18, 64), 35(2, 68), 36(2, 26), 37(25), 39(82), 41, 42(23, 94), 45(23), 46(23), 47(93), 48(68, 95), 49(68), 51(94), 52(94), 53(90, 91), 55(90), 56(90), 57(93), 58(93), 59(93, 94, 107,

Perera, A. G. U., (*continued*) 108), 60(109), 61(95, 107, 108, 114), 62(107, 114, 116), 64(58), 65(95, 109, 114, 121, 126), 66(126), 67(126), 68(126), 69(126), 70(126), 71, 72, 73, 74, 75, 185(83), 186(83); **23:**144(36, 37), 239(45), 242(58, 68, 90, 91, 93, 94), 243(45), 251(45), 256(64), 261(64)
Peressini, P., **16:**261, 269
Peretti, E. A., **7:**199(175a), 200(175b), 252
Perez, A., **17:**314, 365
Perez, J. M., **18:**151
Perez, R., **14:**112(99), 128; **15:**284–285, 310
Perez-Oyarzun, S., **24:**181(15)
Perez-y Jorba, M., **14:**107(59), 126
Peria, W. T., **12:**121(244), 164
Perkins, D. M., **5:**221(206), 234
Perkins, M. H., **1:**31(74), 32(74), 65; **2:**365(7, 8), 419, 420; **3:**3, 4, 5(23, 25), 51(23, 25), 53
Perneski, A. J., **9:**275, 294
Perny, G., **3:**112, 129
Perotin, M., **12:**169(21), 176(50), 179(50), 180(50), 183(50), 186(50), 187(50), 190(50), 197(50), 200(50), 232, 233
Perret, P., **23:**24(96), 30(96)
Perri, J. A., **3:**101(90), 102(90), 113(90), 126; **4:**257(1, 3, 4), 261(30); **12:**258(75), 291; **17:**273, 274, 298
Perriere, J., **14:**86(14), 87(16), 90(90), 92(26, 27), 94(92), 95(34), 98(14, 16, 26, 94–96), 99(26), 100(42–44), 101(94, 95), 102(45), 103(16), 104(26), 105(14, 16, 55), 106(55), 109(16), 111(63), 112(92, 96–99), 113(16, 94); **16:**30, 32, 76, 78, 114(96), 117(76), 118(26), 119(64), 120(16, 76, 80), 121(44), 122(34), 122–123(87), 124(90), 125, 126, 127, 128, 129–130, 137
Perrin, F. H., **4:**50(89), 75(174), 92, 95
Perrin, J., **19:**6
Perrot, M., **2:**23(28), 60
Perry, A. J., **12:**177(63), 234; **14:**135(15), 195; **17:**44, 75, 131, 143; **18:**40
Perry, D. L., **3:**13(61), 51(61), 54; **10:**168, 207, 225, 226; **12:**7, 50
Perry, J., **17:**83, 128, 139
Perry, L. E., **21:**128(33), 233
Pershan, P. S., **15:**192, 217; **24:**3(17), 4(17), 44(19), 112(23), 121(23), 180(10), 181(12, 13, 14), 184(34, 36, 39, 43), 185(10), 186(10, 44), 187(34), 188(34), 190(34), 192(12), 199(12)
Persson, K., **6:**41(138), 77
Perssons, G., **12:**294
Perstin, A. J., **24:**127(55)
Persyk, D. E., **12:**142, 165
Perthel, R., **15:**154, 158, 164, 215
Perutz, M. F., **2:**110, 143
Pery, A., **2:**334(52), 360
Pesavento, F. L., **14:**93(30), 126
Pestes, M., **12:**275(181), 294
Petch, M. J., **3:**266, 273
Petch, W. J., **17:**69, 77
Peterlinz, K. A., **24:**92(59), 93(59)
Peters, E. T., **5:**309, 322
Peters, F. G., **2:**369(40), 420; **3:**100, 112, 126, 127, 139(44), 163; **4:**267, 322; **5:**310(282), 322; **9:**51(239), 70
Peters, M. W., **22:**193(46), 273
Peters, R. C., **7:**139(35b), 248

Petersein, H., **17:**130, 131, 143
Petersen, K., **24:**228(23)
Petersen, N., **24:**96(76), 98(76)
Peterson, A., **9:**22(89), 27(89), 67
Peterson, D., **2:**370(54), 421; **3:**140, 163; **4:**265(59), 314(59, 60), 319(59), 321; **5:**287(207), 290, 302, 304, 307, 310, 311, 319
Peterson, D. R., **4:**265, 319(60), 321
Peterson, D. T., **7:**262(17), 306
Peterson, D. W., **4:**39(70), 92
Peterson, L. D., **19:**21
Peterson, L. R., **18:**16
Peterson, M. W., **21:**365(49), 368
Peterson, P. E., **12:**45(74), 51
Peterson, R. E., **9:**130(38), 144; **10:**16(29), 48(29, 98), 49(29), 55(29), 64(146), 66, 68, 69
Peterson, R. L., **16:**9, 125
Petersson, C. S., **14:**233(134), 240
Pethig, R., **11:**256(63), 257(63), 315
Petit, J., **17:**121, 143
Petit, R., **9:**223(189c), 260
Petrich, M. A., **19:**93
Petrillo, E. J., **14:**232(125), 239
Petritz, R. L., **2:**175, 191, 192; **3:**278, 279, 285, 289, 294(64), 296, 300, 301; **11:**150(121), 202
Petrocco, G., **11:**113(51), 200
Petroff, M. D., **21:**4(13), 18(13), 72
Petroff, P., **14:**232(66), 238
Petroff, P. M., **11:**85(195, 196), 102; **12:**265(114), 292; **14:**232(115), 239; **15:**221, 260; **19:**15, 32
Petrosev, V. A., **14:**175(136), 198
Petrosov, B. A., **14:**138(23), 195
Petrosov, V. A., **14:**138(27), 195; **17:**92, 95, 140
Petrov, A. I., **7:**216(197a), 253
Petrov, I., **17:**122, 143; **18:**248; **22:**240(87), 274
Petrov, I. G., **14:**51(165), 77
Petrova, R., **12:**87(119), 161
Petrucci, J. L., Jr., **18:**301, 310
Petschauer, R. J., **6:**224(81), 292(81), 297
Pettas, H. J., **12:**133(283), 165
Petti, R. D., **3:**18(83), 51(83), 55; **7:**336, 351(99), 395
Pettit, G. D., **7:**142, 171(111), 177(179), 183(138), 193(157), 204(179), 218; **10:**41(79), 67; **13:**219(72), 220(69, 72), 221(72), 224, 225
Pettit, R. B., **9:**202(117), 212(117), 258; **10:**28(49), 48, 49(99), 50(99), 63(99), 66, 68
Petty, M. C., **24:**21(99), 25(99)
Petzold, W., **3:**39(186), 52(186), 58
Pevera, A. G. U., **23:**291(64, 81, 111)
Pfann, W. G., **5:**253(38), 257(38), 315
Pfeiffer, H., **11:**112(46), 200; **15:**37–38, 137
Pfeiffer, H. G., **1:**205(91), 206(91), 230
Pfeiffer, L., **21:**130(42), 133(42), 151(42), 153(42), 154(42), 209(111), 211(42), 233, 236
Pfennig, H., **6:**130, 148
Pfund, A. H., **1:**146, 185; **4:**77, 78(191), 95; **13:**110(8), 147
Pham, L., **21:**128(37), 233
Pham Thi, M., **17:**314, 328, 365
Pharr, G. M., **17:**49, 62, 63, 77
Phelan, R. J., **3:**277(31, 32), 300
Phelps, F. P., **3:**24(123), 51(123), 56
Phil, M. A. D., **17:**92, 140
Philibert, J., **7:**258, 260(2), 305; **16:**234, 266; **22:**89(13), 115
Philip, R., **3:**1(3), 53; **4:**86(234), 96; **12:**71(70), 72, 160

Philipp, H., **12:**60(35), 101(35), 102, 103, 108, 109(35), 159, 162
Philipp, H. R., **6:**153, 202; **12:**265(113), 292
Philips, R., **1:**194(33), 229
Philips, V. A., **2:**14, 16, 17, 19, 20, 21, 60
Philips, W., **10:**253, 254, 259
Phillip, H. R., **1:**143(35), 147, 149(37), 169(35, 37), 185
Phillipp, H. R., **10:**32, 66; **12:**79, 80, 160
Phillips, A. B., **5:**211, 234
Phillips, B. F., **14:**231(108), 239
Phillips, E. A., **16:**272, 285, 332, 337
Phillips, H. R., **5:**297, 321
Phillips, J., **22:**217(79), 274
Phillips, J. C., **10:**31(54), 66; **12:**57, 58(20), 68(20), 159; **19:**73
Phillips, J. M., **16:**69, 136; **17:**209, 223, 272, 290, 293, 298
Phillips, L. S., **2:**389(116), 422
Phillips, R. C., **16:**55–56, 134
Phillips, T. E., **16:**79, 137
Phillips, V. A., **4:**154, 181, 189, 190
Phillips, W. A., **11:**241(49), 315
Phillips, W. D., **24:**230(44)
Phillpot, S. R., **17:**59, 77
Philofsky, E., **7:**290(98), 310
Philofsky, E. M., **6:**336(42), 347; **17:**41, 75
Philpott, M. R., **9:**240(234a), 261; **19:**296
Phipps, T. E., **4:**339(45), 370
Piaget, C., **12:**135(298), 165
Pianetta, P., **12:**114(206), 115(217, 220), 116(222), 128(268), 163, 164
Piccioli, N., **11:**132(94), 133, 142(94, 114), 202
Pick, H. J., **4:**199(54, 57, 58), 200(58), 202, 204, 205, 206, 207, 209, 211(57, 58), 233(58), 252
Pickard, R. M., **6:**237(124b), 298
Pickart, S. J., **1:**205(81), 230; **9:**273, 274, 296; **15:**147, 214
Picket, W. E., **17:**119, 142
Pickett, D. L., **24:**2(6), 3(6), 17(6), 181(16)
Pickett, W. E., **17:**54, 76
Pickhardt, V. Y., **11:**36(11, 12, 19, 20), 59(11), 96, 109, 116(30), 119(30), 200
Pickrell, D., **19:**105
Pictroski, C. F., **24:**94(72), 95(72)
Pidduck, A. J., **23:**24(89)
Piedmont, J. R., **9:**47(224), 70; **12:**259(82), 292
Piegari, A., **19:**192, 238, 240
Piejak, R. B., **18:**59
Piekarski, I. J., **13:**218(62), 221(62), 224
Piel, H., **16:**107, 141
Pierce, J. T., **3:**151(70), 164
Pierce, R. D., **9:**283, 288, 292, 295, 296
Pierret, R. F., **17:**172, 221
Pierson, H. O., **13:**99(99), 107; **24:**244(82)
Pike, G. E., **11:**295(136), 317
Pike, W. T., **21:**208(112), 209(112), 210(112), 211(112), 212(112), 213(112), 236
Pilkuhn, M., **7:**170(108, 109), 210(192), 214(199), 223(192), 230(199), 250, 252, 253
Pilorget, L., **14:**124(89), 127
Pimbly, J. M., **23:**91(20)
Pincherle, L., **3:**277(23), 278, 300
Pinczuk, A., **21:**155(61), 234
Pinder, A. R., **24:**9(34)
Pine, G., **16:**13, 125

Pines, B. Ia., **7:**18(52), 38
Pines, B. Ya., **6:**339(56), 347
Pines, D., **1:**126(4), 184; **2:**99, 103, 142
Pings, C. J., **7:**3(17), 37
Pinkas, E., **7:**234(243), 254
Pinkhasov, E., **14:**138(25, 26), 195
Pinnell, J. E., **5:**282, 319
Pinnis, Ya. Ya., **17:**355, 356, 368
Pinsker, Z. G., **1:**196, 198, 203, 218(134), 229, 231; **4:**223, 253
Pinto, J. F., **23:**279(22)
Pinto, R., **6:**64(181), 78
Pippard, A. B., **1:**251, 256, 258, 275; **6:**82(32), 91, 144, 145, 154, 202
Pirich, R. G., **16:**168, 177, 194, 196–201, 213, 230
Piro, O., **21:**52(102), 74
Pish, G., **3:**39(188), 52(188), 58
Pistorius, C. A., **6:**19, 75
Pistov, N., **6:**236, 298
Pitkevich, Ya. A., **17:**318, 319, 320, 366
Pitkin, E. T., **3:**73(45), 125
Pitsch, W., **3:**248, 272
Pitt, C. W., **14:**31(117), 76; **17:**153, 220
Pittelli, E., **5:**167(130), 181(130), 193(130), 199(130), 232
Pitzer, D., **17:**291, 299
Pivovonsky, M., **2:**319, 359; **4:**52(121), 93
Piwkowski, T. R., **5:**167(129), 181(129), 193(129), 199(129), 232
Pizzarello, F. A., **3:**180(36), 209; **5:**149, 229, 276, 277, 318; **15:**8, 136
Planer, G. V., **2:**389(116), 422
Plano, W. E., **21:**287(39), 310
Plante, T. D., **17:**347, 359, 360, 367, 369
Plaskett, T. S., **7:**163(88a), 225(229), 249; **9:**264, 275, 281, 288, 290, 295, 296
Platikanov, D., **7:**315(23), 393
Platonov, F. S., **6:**71, 72, 79
Platt, C. E., **16:**119, 121, 143
Platt, J. R., **3:**43, 49(218), 59
Plattner, H., **14:**8(26), 36(26), 74
Platzman, P. M., **11:**297(137), 317; **21:**240(6), 308
Plecenik, A., **16:**52, 133
Pletnev, V. A., **13:**186(154), 190(154), 197
Plew, L., **11:**66(120), 82, 99, 101
Plieth, W. J., **19:**281
Pliny the Elder, **24:**180(2)
Pliskin, W. A., **3:**135, 137, 141(52), 162, 163; **4:**257(1), 259(11, 12), 260, 261, 262(45), 263(29, 35, 39, 50a), 265(66), 266(68), 267(68, 69), 269, 270(39, 45, 68), 271, 272, 273, 274, 275, 276, 277, 278, 279, 280, 281, 282(12, 29), 283, 284, 285; **5:**290, 320; **12:**258(75), 259(80, 84), 262, 268, 291, 292, 293; **15:**287, 310; **17:**273, 274, 298
Ploog, K., **11:**42(45), 44(49), 45(45, 49, 56, 59), 59, 67(105), 82, 83(187); **12:**294
Ploog, K. H., **23:**221(10), 225(10)
Pluddemann, E. P., **24:**228(12)
Plumb, R. C., **2:**216(20), 238; **4:**83(209, 210), 96; **12:**14, 50
Plynto, A. A., **17:**93, 94, 140
Plyutto, A. A., **14:**151(83), 152–153(83), 154(83), 156(83), 158(83), 197

Poarch, S., **22:**119(13), 120(13), 161(13), 171
Poate, J. M., **12:**277(208), 279(208), 294; **16:**74, 137, 234, 266
Pockels, A., **7:**311(2), 313(18), 392, 393
Pockrand, I., **9:**223(189b), 224(190), 226(194), 244(241), 260, 261
Podlesnik, D. V., **13:**162(45, 46), 163(45, 46), 164(45), 165(58), 169(68), 171(72, 73), 174(58, 72), 178(72), 186(46, 58, 68, 72), 189(45, 46, 58, 68, 72), 190(46), 193, 194
Poehler, T. O., **3:**180(37), 209; **5:**182, 233
Poelsema, P., **24:**117(38)
Poferl, D. J., **12:**246(28), 290
Pohlack, H., **5:**60(11), 85
Pohm, A. V., **1:**308, 319, 321(80, 81, 84), 333
Pohn, A. V., **6:**224(80), 292(80), 297
Pointu, P., **13:**121(59), 148
Poirier, G. E., **24:**44(46), 45(78, 79), 46(78), 56(79), 60(78), 82(15), 86(15, 30, 31), 94(62), 112(10), 125(48), 127(52), 138(64), 206(17), 212(17), 238(67)
Poitevin, J. M., **13:**74, 105; **14:**12(44), 74; **17:**111, 112, 128, 142, 143
Poker, D. B., **16:**43, 48, 131
Pol, H. v. d., **15:**297, 311
Polak, A. J., **12:**277(212), 279(212), 280(212), 295; **14:**103(49), 126
Polanco, J. I., **11:**256(61), 257(61), 315
Polder, D., **7:**36(161), 41; **15:**27, 127, 137
Poli, G., **4:**199(65, 66, 67, 68), 200(65, 66, 67, 68), 204, 207, 209(65, 67, 68), 210, 211(68), 228, 236(66), 252
Polinsky, M., **17:**271, 272, 276, 278, 298
Polito, W. J., **17:**153, 219
Polizzotti, R. S., **24:**135(63)
Polk, D. E., **7:**24, 25(95), 39
Polkosky, J. J., **12:**85(108), 161
Polky, J. N., **10:**238(28), 258
Polla, D., **17:**208, 210, 211, 223
Polla, D. L., **17:**207, 210, 222, 228, 293, 294, 299
Pollack, P. I., **5:**267(85), 316
Pollack, S. R., **6:**36(105), 38(105), 76
Pollak, F. H., **13:**205(18), 223
Pollak, M., **11:**256(58), 257(58), 290(115), 295(136), 315, 317
Pollak, R. A., **9:**217(165, 169), 259
Pollak, T. M., **17:**237, 275, 276, 277, 280, 296
Polmear, I. J., **7:**297(113), 310
Poloniak, E. S., **9:**54(258), 56(258), 62(258), 71
Polster, H. D., **1:**70(6), 120; **5:**78(25), 81(25), 86
Polster, H. J., **3:**10(45), 51(45), 54
Polturak, E., **16:**59, 65, 68, 135
Polyak, S., **21:**66(125), 75
Pomerantz, D., **7:**137(33), 248
Pomerantz, M., **1:**329, 334; **24:**228(21)
Pommier, R., **12:**169(21), 232
Pompe, W., **16:**70, 136
Ponce, F. A., **16:**78, 137
Pongratz, S., **18:**155
Ponomarenko, I. N., **12:**105(176), 106(176), 162
Ponpon, J., **14:**232(118), 239
Ponpon, J. P., **14:**213(32), 237
Pool, P. J., **12:**132(282), 165
Poole, H. H., **3:**140(48), 163
Poole, K. M., **3:**146, 163
Poon, C.-D., **24:**151(9), 154(9), 156(9)
Pope, R. A., **9:**45(203), 69

Pope, T. P., **9:**136(50), 144
Popiano, P. L., **6:**36(109), 37(109), 44, 76
Popiel, W. J., **7:**313(15), 393
Popma, T. J. A., **9:**276, 296
Popov, O. A., **18:**124, 125, 147, 159, 160, 168, 184, 194, 204, 218
Popova, V., **14:**10(37), 74
Popovic, R. S., **13:**215, 224
Poppa, H., **1:**196(37, 39), 229; **3:**32(162), 57; **4:**165, 189; **9:**4(11), 65
Poppe, G. P. M., **19:**2
Poppe, U., **16:**5, 24–25, 28, 32, 124, 127–129
Porkolab, M., **18:**150
Porsons, R. R., **14:**48(161), 77
Portal, H. C., **15:**249, 263
Portemer, F., **17:**304, 364
Porteous, P., **7:**170(102), 174(122), 193(161), 250, 252
Porteous, R. K., **18:**83–84
Porter, A. S., **2:**48(79), 50, 62
Porter, E. F., **7:**348, 367(120, 172), 369(120, 172), 395, 396
Porter, M. D., **24:**3(11, 16), 4(16), 10(40), 12(48), 17(40), 18(40), 24(117), 26(117), 27(117), 32(16), 44(14, 15, 20, 28, 29, 30, 31, 32, 33, 34, 35), 45(15), 46(14, 85), 53(28, 33), 54(29), 55(14, 29), 56(14), 70(14, 15, 28, 29, 30, 32), 82(19), 83(20), 86(19, 32), 88(46), 90(20), 91(46), 98(82), 101(82), 105(82), 112(4, 5), 128(58, 60), 212(61), 221(92)
Porter, R. F., **11:**108(21), 200
Porteus, J. O., **4:**4(10), 12(10), 13, 17, 18, 19(10), 46(10), 90, 91; **10:**168(11), 225
Portier, E., **9:**50, 70
Portis, A. M., **3:**295, 296, 301; **16:**37, 130
Portniagin, A. I., **13:**169(70), 185(70), 187(70), 194
Portnoi, E. L., **7:**217(200), 247(281), 253, 256; **13:**162(44), 163(47), 164(44), 186(47), 189(47), 193
Porto, D. R., **14:**139(34), 195
Posner, G. H., **24:**11(44)
Postlethwaite, T. A., **24:**24(118, 119), 26(118, 119), 151(9), 154(9), 156(9)
Postma, J. P. M., **22:**134(54), 136(54), 173
Postma, M. P. M., **24:**220(85)
Poston, T. D., **15:**85, 140
Potapov, A. P., **6:**216(53, 54, 55, 56), 220(53), 269(55), 276(185), 277(54, 185), 296, 299
Potemski, R. M., **7:**221(212), 253
Potenza, J. A., **16:**60, 62, 135
Potter, D. L., **3:**32(157), 51(157, 158), 57
Potter, J. F., **5:**311(284), 322
Potter, R. C., **15:**245, 263
Potter, R. F., **4:**71, 94, 95; **5:**165(104), 174, 232
Potter, R. I., **15:**147, 180, 182, 214
Potter, W. D., **5:**309, 322
Pouch, J. J., **16:**50, 132
Poulek, V., **17:**83, 99, 119, 126, 139, 143
Pouligny, B., **17:**241, 287, 297, 299
Poulin-Dandurand, S., **15:**282, 310
Poulsen, R. G., **14:**27(107), 75
Pound, G. M., **1:**210, 231; **2:**2, 59; **4:**97(1, 3, 5, 6, 9), 98(12), 99(1 2, 14, 16), 100(5, 6), 102(3), 103, 106(3, 6), 107(3), 108(3, 26), 109(3, 6, 9, 14), 111, 112(31), 113, 114, 115, 117, 118(35, 3 7),

Pound, G. M., (*continued*)
119(3), 120, 121, 122, 123; **16:**4, 124
Povilonis, E. I., **5:**163(76), 166, 168, 196(76), 231
Powar, S. H., **12:**212(100), 235
Powell, C. F., **3:**133(20), 162; **5:**238(1), 242, 249(15), 284, 289(224), 302(191), 314, 317–326; **9:**31, 67
Powell, C. J., **9:**205(128), 214(144), 258, 259
Powell, J. M., **2:**297(45), 304
Powell, J. R., **12:**147(318), 166
Powell, R. J., **21:**21(48), 72
Powell, R. W., **6:**123(185), 147
Power, F. S., **2:**393(136), 405(136), 423
Powers, J. H., **10:**43(90), 44(90), 67
Powers, J. V., **9:**266, 295
Powers, K. D., **16:**43, 47, 131
Powers, P. E., **21:**365(48), 368
Pradad, V., **22:**129(43), 142(43), 145(43), 172
Praff, R. J., **12:**269(147), 293
Prager, W., **16:**302, 339
Prakash, M., **24:**180(11)
Pramanik, D., **22:**20(28), 77
Pranevicius, L., **13:**111(21), 134, 140(21), 141(21), 144(132), 147, 150
Prange, R. E., **6:**134(202), 148, 298
Prasad, J. J. B., **16:**244–245, 249, 257–263, 265, 267, 269
Prasad, V., **22:**117, 141(63), 152(74), 161(81, 82), 165(82, 86), 173, 194(54), 273
Prateu, C. B., **24:**94(67), 95(67)
Pratt, G. W., **3:**277(24), 300
Pratt, I. H., **2:**367(18), 371(57), 373(65), 376(65), 387(57), 388(57), 389(57), 404(181), 405(181), 420, 421, 424
Preier, H., **11:**111(38), 112(46), 114(38), 146, 147, 148, 149, 200, 202; **15:**227, 244, 261
Prem Nath, **12:**169(24, 25), 172(25), 183(25), 195(24), 196(24), 233
Prentiss, M., **24:**44(62), 229(33), 230(33, 44), 235(33), 245(83), 246(83)
Prescott, C. H., **12:**71(59), 73, 159
Presland, A. E. B., **3:**219(23), 271; **10:**37(68), 67
Pressau, J. P., **9:**50(229), 70
Pressman, L. D., **14:**57(179), 77
Preston, J. S., **3:**109(112), 124, 127, 128; **6:**106(119), 110(119), 146; **9:**9, 43(188), 47(214), 50(230), 62, 65, 69, 70
Preuss, L. E., **3:**30(149, 150), 31(150), 32(180), 38, 40, 44, 51(149, 150), 52(180, 181, 182, 183, 184), 57, 58
Pribat, D., **14:**102(45), 126
Price, A. G., **7:**324(39), 393
Price, D. J., **10:**36(65), 67
Price, F. W., **23:**124(16)
Price, G. L., **10:**37(68), 67
Price, P. B., **2:**58(91), 62; **4:**200(74), 252
Price, P. J., **6:**91, 116, 144
Price, W. C., **2:**202, 237
Pride, G. E., **4:**59(129), 94
Priest, J. R., **1:**61, 62, 63, 67, 206, 230; **2:**178(60), 192, 403(174, 175, 176, 177, 178), 404(177), 424; **3:**140(45), 148(45), 163, 220(33), 221, 229(65), 235(33), 236, 237, 238(33, 36), 242(65), 246(65), 250, 251, 271, 272; **4:**282(106), 322

Prieto, P., **16:**28, 32, 128, 129
Prigogine, I., **7:**180(131), 251
Prime, K. L., **24:**10(36), 12(36, 49), 14(36, 49), 21(94), 24(94), 181(21), 228(5)
Prince, E., **1:**205(78), 230
Pring, J. N., **5:**283, 284, 312
Pringle, J. P. S., **6:**24(84, 86), 25, 75, 76; **14:**109(60), 120(82), 126, 127
Pringsheim, P., **2:**123(132), 144
Prins, J. A., **7:**2(7), 37
Prinz, D. K., **11:**17(12), 18, 33
Prinz, E. J., **23:**24(92), 55(187, 188)
Prior, K. A., **23:**13(37)
Prior, Y., **13:**186(152), 190(152), 197
Prishivalko, A. P., **2:**221, 238
Pritchard, J. P., Jr., **3:**151(70), 164
Pritchett, W. C., **17:**49, 76
Priver, A., **6:**256(163), 260(163), 264(163), 267(163), 299
Probyn, B. A., **2:**374(69), 387(69, 108), 388(69), 389(69), 403(173), 404(173), 421, 422, 424
Prock, A., **24:**73(118, 119)
Procter, J. P., **4:**271(93), 322
Procter, S. J., **13:**214, 224
Proebster, W. E., **1:**310(61), 311, 319(76), 333
Prokhorov, A. M., **15:**140
Prokhvatilov, V. G., **17:**328, 344, 366
Prokop, G. S., **16:**240, 251, 267
Prosen, R. J., **1:**294(41), 332
Protasov, I. I., **7:**246(279), 247(281, 282), 256
Prugne, P., **3:**49(229), 59
Pruna, P., **16:**45, 132
Pruniaux, B., **12:**263(104), 264(104), 292; **13:**61, 104
Prussin, S., **4:**184(136), 190
Prutton, M., **3:**229(58), 232, 234, 272; **6:**216(59), 286(207), 296, 300
Pryde, C. A., **10:**245(35, 36), 259
Pucel, R. A., **5:**305, 321
Puchalska, I. B., **6:**228(102, 103), 236, 238(102, 103), 297
Pugacheu, A. T., **3:**256(104), 257(104), 258(104), 264(104), 273
Pugh, E. W., **1:**244(30), 274, 286, 287(27), 289(27), 332; **3:**117, 128, 214, 270; **6:**205(1), 216(1, 58), 224(83), 256(1), 257(1), 258(1), 271, 283(58, 201), 284(201), 291(83), 295, 296, 297, 300
Pugmire, R. J., **24:**153(12)
Pulfrey, D. L., **6:**36(122), 37(128), 38(122), 45(122, 128), 46(122, 128, 142a), 48(122, 128), 49(122, 128), 50(122, 128), 53, 54(147), 57, 58, 60(147), 62(147), 63, 65, 66(147), 72(215), 76, 77, 78, 79; **12:**275(186), 276(191), 294; **14:**118(77), 120(78), 127
Pulker, H., **2:**346(74), 360
Pulker, H. K., **13:**110(9), 147; **14:**135(14, 15), 195; **19:**195
Pulliam, G. R., **5:**260, 313(295, 297, 299, 300), 314(300), 315, 322; **9:**282; **12:**269(149), 293; **15:**8, 136
Pungor, A., **24:**183(33), 184(33)
Punturi, S., **24:**21(95)
Purans, J., **17:**310, 365
Purcell, J. C., **10:**123(53), 165
Purcell, J. D., **1:**184(71, 72), 186; **7:**45(10, 13), 103(13), 104(82), 105(83), 107(83, 84), 109(83), 111, 114
Purdes, A. J., **17:**153, 220
Pursch, H., **14:**139(33), 147(74), 195, 196

Pursch, M., **24:**170(40), 173(40)
Putley, E. H., **3:**276, 277, 300; **5:**209(181), 234; **17:**245, 297
Putner, T., **3:**11(55), 12(55), 13(55), 49(55), 51(55), 54, 124(162), 128
Putty, M. W., **17:**210, 223
Putvinski, T. M., **24:**12(55), 14(55, 59), 20(59), 23(112), 27(112), 30(112), 44(11, 17), 45(11, 17), 112(8), 217(68), 219(80)
Puzzer, T., **16:**26, 127
Pye, M. F., **17:**320, 366
Pylant, E., **22:**227(80), 274
Pylant, E. D., **24:**45(79), 56(79), 94(62), 238(63)
Pyle, E. A., **14:**176(137), 198
Pyne, A. E., **5:**288(217), 320

Q

Qadri, S. B., **16:**53, 134; **17:**49, 62, 77
Qian, C. T., **16:**91, 139
Qin, D., **24:**230(46), 231(49, 50), 234(46), 235(56), 241(56)
Qiu, C. X., **16:**54, 134
Quarrell, A. G., **1:**218, 231
Quarrington, J. E., **4:**60(136), 94
Quate, C. F., **24:**82(2), 86(26), 114(34)
Quattropani, A., **6:**246(144), 298
Queisser, H. J., **4:**143(23), 169(106), 184(23, 135, 137, 140), 185(23), 187, 190, 298(115), 323; **5:**249(33), 260(53), 315; **7:**135(296), 157(69), 158(75), 170(108), 247, 250; **9:**238(228), 261
Querry, M. R., **12:**46(75), 51
Quick, J. D., **15:**54–55, 138
Quincke, F., **5:**240(12), 314
Quinn, D. J., III, **1:**241(10), 272(99), 273, 275
Quinn, J., **24:**94(72), 95(72)
Quinn, J. J., **4:**328, 329, 364(17), 365(17, 20), 369; **9:**159(16), 256; **12:**62, 63, 159
Quinn, W. E., **19:**51
Quintana, G., **12:**266(138), 268(138), 293; **14:**232(66), 238
Quinto, D. T., **17:**46, 76

R

Raaijmakers, I. J., **22:**202(59), 262(108), 273, 275
Rabalais, J. W., **14:**7(23), 69(23), 73; **17:**334, 367
Rabe, K., **16:**3, 124
Rabedeau, T. A., **24:**112(23), 121(23)
Rabinovitch, K., **4:**41(78), 68(78), 92
Rabinowitch, E., **7:**331(68), 394
Rabinowitz, P., **24:**135(63)
Rabinzohn, P., **14:**122(84), 127
Rabke-Clemmer, C. E., **24:**36(145)
Rabl, A., **10:**24(40), 66
Rabolt, J. F., **15:**266, 309; **24:**15(65), 206(2)
Rabson, T. A., **17:**228, 260, 292, 298, 299
Racah, D., **16:**50, 132
Racanelli, M., **23:**60(202)
Race, H. H., **7:**348, 367, 391(121), 395
Racette, J. H., **4:**370
Racheva, T. M., **12:**204(98), 234
Radelaar, S., **22:**267(117), 275
Radeloff, C., **9:**174(32), 256
Rademacher, H., **21:**52(100), 74
Radhakrishna, S., **17:**359, 369
Radicati, F., **16:**60, 135
Radmacher, M., **24:**94(68, 70), 95(68, 70), 96(76), 98(76)
Radnoczi, G., **17:**15, 16, 17, 24, 31, 74
Rado, G., **15:**32, 135, 137

Rado, W. G., **7:**186(144a), 251
Radosz, A., **19:**180
Radparvar, M., **16:**110, 142
Radpour, F., **16:**36, 91, 130, 139
Raether, H., **4:**223, 253; **6:**197(56), 203; **7:**16(45), 38; **9:**146(4, 5), 153(9), 186(65), 195(98), 196, 197(104), 202(116, 120), 203(120, 120a, 120b), 204(5), 207(120), 211(137), 214(4, 145, 146), 215(149), 223(189a, 189b), 227(206), 231(215), 233, 237(224), 239(231)
Rafailovich, M. H., **24:**94(72), 95(72), 169(39), 170(39)
Raffel, J. I., **1:**319, 333; **13:**187(126), 196
Raffy, H., **16:**13, 30, 32, 125, 129–130
Ragay, F. W., **21:**313(6), 321(13), 341(6), 342(6), 343(40), 344(6), 351(13, 40), 355(6), 357(13), 359(13), 360(6, 40), 361(6, 40), 362(40), 365(50), 366, 368
Raghavan, K. S., **3:**255, 256, 273
Raghuram, A. C., **13:**60, 62(1), 65, 66, 68, 75(1, 30, 59), 76, 79, 80, 83(30), 87, 90, 104, 105
Ragland, R., **15:**55, 138
Rahatgi, A., **14:**232(127), 239
Rahman, A., **17:**58, 76; **22:**142(64, 65), 143(65), 173; **24:**207(22)
Rahman, T. S., **15:**37, 138
Rahmati, M., **16:**91, 139
Rahuram, A. C., **12:**275(179), 294
Raichoudhury, P., **12:**129(274), 165
Raider, S. I., **23:**47(160)
Raimes, S., **2:**99, 142
Raimondi, D. L., **9:**278, 297; **17:**153, 219
Raines, S. N., **21:**36(70), 73
Rairden, J. R., **9:**43(194), 69; **13:**61, 104
Raisch, F., **11:**45(59), 97
Raistrick, I. D., **16:**42, 47, 131; **17:**318, 326, 359, 366, 369
Raizman, A., **19:**41
Raj, R., **16:**331, 340
Rajagopalan, G., **22:**270(130), 276
Rajagopalan, S., **13:**127(86), 149
Rakhovskii, V. I., **14:**138(21, 28), 141(28, 41), 144(21, 28), 145(41, 64, 66), 155(85), 195, 196, 197
Raleigh, Lord, **7:**311, 313(1, 16, 17), 314(1), 348(17), 393
Ralston, J. D., **21:**121(17), 123(17), 152(17), 232
Ram-Mohan, L. R., **23:**279(36, 103), 280(103)
Ramachandran, S., **24:**228(10)
Ramalingam, S., **17:**95, 141, 218, 224
Ramalingham, S., **14:**167(118), 171(118), 197
Ramanarayanan, T. A., **24:**112(15, 19), 117(15), 119(19), 128(15)
Ramanlal, P., **22:**5(7), 77
Ramanowski, Z., **12:**265(109), 292
Ramans, G. M., **17:**337, 338, 349, 350, 367
Ramasubramanian, N., **6:**37(129), 77
Ramaswami, B., **22:**253(97), 254(97), 262(97)
Ramaswami, M., **22:**193(45–47), 273
Ramaswamy, V., **10:**253(48), 259
Ramberg, E., **12:**101(171), 102(171), 162
Ramberg, E. G., **2:**110(87), 143; **5:**145, 229
Rambo, A., **17:**279, 299
Ramboarina, R., **16:**292, 340
Ramdas, A. K., **21:**38(79), 73
Ramesh, R., **16:**45, 63, 112–114, 132,

Ramesh, R., (*continued*) 135–136, 142; **17:**236, 237, 270, 271, 296, 298
Ramey, R. L., **3:**51(29), 53; **5:**210, 212, 213, 234
Ramsey, J. B., **2:**214(16), 238, 284(35, 36, 39), 290(35), 297(39), 304, 334(51), 346(51), 360; **4:**9(24), 91, 282(101), 285(101), 322; **5:**167, 231; **10:**97(32), 98(32), 99(32), 100(33), 101(33), 102(33), 103(33), 104(33), 106(33), 107(34), 108(34), 109(34), 128(57), 129(64, 66), 130(64), 131(66), 141(66), 142(66), 143(66), 145(66), 146(66), 147(57, 64); **12:**2(8), 15(44), 24(49), 28(8, 54), 31(8), 40(44), 48(49), 49, 50, 51
Ramsey, J. R., **9:**130(38), 144; **10:**16(29), 48(29), 49(29), 55(29), 66
Ramsey, J. W., **12:**45(74), 51
Ramsteiner, M., **21:**121(17), 123(17), 152(17), 232
Rand, M. J., **3:**137, 162; **4:**261(33, 36), 263(33, 36), 264, 265, 314, 320; **10:**46(96), 68; **12:**269(139), 293; **13:**99, 100, 107; **15:**291, 311
Randall, J. J., **6:**6, 7(27), 26, 33, 74
Randall, J. N., **13:**182(114, 116), 183(114, 116), 195, 196
Randhawa, H., **14:**170(120, 121), 175(133), 177(142), 178(152), 179(152, 153), 180(153, 158), 181(158), 182(158, 160), 183(133, 142), 185(133), 186(153, 158, 164, 165), 187(153), 188(164), 189(164), 190(121), 191(142), 192(120, 153), 193(120), 197, 198, 199
Randhawa, H. S., **12:**169(28), 183(28), 189(28), 225(28), 233; **13:**79(71, 81), 106
Randin, J.-P., **17:**352, 358, 368, 369
Randlett, M. R., **6:**106(122a), 107, 131(122a), 146; **9:**4(13), 42(181), 65; **13:**131, 145(96), 149
Randlett, R., **9:**6(23), 65
Randon, J., **24:**163(26)
Rang, O., **1:**196(37, 38), 229; **4:**185(145), 190
Ranganath, T. R., **10:**254(53), 259
Rangarajan, L. M., **12:**85(101), 160
Rani, R., **16:**214–216, 230–231
Rank, D. H., **4:**55, 56(126), 93
Ranke, W., **12:**115(219), 163; **14:**98(39), 126
Ranyuk, A. I., **3:**124, 128
Rao, B. S., **19:**169, 180, 281, 291, 293, 294
Rao, G. B., **13:**186(148), 189(148), 197
Rao, K. V., **16:**12, 125
Rao-Sahib, T. S., **7:**159(80), 249
Rapp, G., **1:**31(73), 65
Rappe, A. K., **24:**182(25)
Raptis, Y. S., **13:**155(23), 166(23), 192
Rashba, E. I., **6:**91, 145
Rashti, M. A., **16:**53, 133
Rasigne, G., **12:**71(71), 72(71), 79(71), 160
Rasigni, G., **6:**201, 204; **9:**15(65), 66
Rasines, I., **16:**2, 124
Rasmanis, E., **2:**148(7), 190
Rasor, N. S., **3:**245(75), 268(75), 272
Rassow, J., **3:**16(74), 51(74), 55
Rastogi, A., **16:**262, 269
Rastogi, A. C., **22:**79
Rateau, M., **16:**98, 140
Rathbun, L., **11:**70(140a), 100
Rathman, D. D., **14:**232(119), 239

Ratliff, F., **21:**62(115), 63(115), 74
Ratto, J. J., **16:**28–29, 129
Ratzel, F., **16:**23, 127
Rauch, F., **17:**310, 356, 357, 365, 369
Raufuss, H., **6:**137(224), 148
Rauh, R. D., **13:**161(41), 162(41), 176(41), 186(41), 189(146), 190(41); **17:**347, 353, 359, 360, 367–369
Raupp, G. B., **22:**4(3), 77, 178(8–10), 181(22), 183(28, 29), 185(28, 29), 188(8–10), 193(8–10, 43), 202(8, 43, 58–61), 203(58), 205(43), 211(60, 68, 69), 212(68, 69), 215(75), 216(60, 61, 69, 77), 220(60, 69), 226(68), 232(28, 29, 75), 245(60, 61, 68, 77), 253(97), 254(97), 262(97), 265(111), 266(112, 113), 269(113), 270(121, 125, 127), 271–276
Rausch, D. W., **4:**118(39), 135
Rauscher, H., **23:**32(119)
Ravenscroft, M. S., **24:**21(95)
Ravi, T. S., **16:**63, 135
Rawa, R. E., **5:**290, 320
Rawdon, H. S., **4:**227(137), 254
Ray, A. K., **12:**241(229), 282, 282(229), 283, 295; **14:**89(21), 107(21), 118(21), 125
Ray, S., **15:**281, 310
Rayleigh, Lord, **1:**8(15), 64; **2:**239, 303
Raymond, L. S., **10:**37(70), 40(76), 46(97), 67, 68
Raymond, R. F., **9:**12(58), 33(58, 130), 66, 68
Razeghi, M., **15:**125, 141
Razegi, M., **19:**2, 9, 29–31, 38, 41
Read, A. A., **1:**321(80, 81, 84), 333
Read, B. E., **11:**235(36), 238(36), 315
Read, H. J., **4:**238, 239, 240, 254
Read, M. H., **2:**396(150), 423
Read, W. E., **1:**33(81), 65
Read, W. T., **11:**142, 202
Reader, P. D., **13:**112(27), 114(27), 147; **14:**23(90), 75
Readey, D. W., **12:**258(76), 259(76), 265(76), 271(76), 272(76), 291
Ready, D. W., **9:**31(113), 36(113), 37(113), 39(113), 67
Ready, J. F., **13:**155(26), 181(26), 192
Reames, J. P., **4:**6(14), 90
Reau, J. M., **11:**261(79, 80), 263(79), 270(79), 273(79), 316
Reck, J., **15:**171, 215
Recont, N. B., **10:**52(110), 68
Redaelli, G., **10:**52(113), 68
Reddaway, S. F., **9:**22(96), 23(96), 25, 26, 67
Reddington, R. W., **12:**156(332), 166
Reddy, G. B., **12:**226(124, 125), 227(124), 235
Reddy, K. V., **16:**242–245, 253–254, 257–262, 265, 267, 269
Redel, T., **16:**51, 133
Redhead, P. A., **1:**21, 26, 29(56, 65), 30(66, 67), 65
Rediker, R. H., **3:**277(31, 32), 300
Redinbo, G. F., **24:**229(41)
Redington, R. W., **5:**145(6), 229
Reeber, M. D., **1:**241(21), 245(35), 274
Reece, M. P., **14:**139(31), 195
Reed, B. S., **7:**214(196), 253
Reed, K. W., **15:**67, 77, 88, 139–140
Reed, M. A., **15:**245, 263
Reed, T. B., **9:**10(47), 66; **10:**52(105), 68
Reeder, A. A., **21:**287(37), 310
Reeder, T. M., **17:**153, 220
Rees, A. L. G., **6:**334(38), 347
Rees, D. E., **6:**36(113), 45(113), 76
Rees, H. D., **7:**240(254c), 255

Reeves, G., **16:**37–38, 130
Reeves, G. A., **16:**37, 130
Reeves, J. M., **11:**256(66), 257(66), 316
Reeves, M. E., **16:**65, 136
Regan, M. J., **24:**3(17), 184(34, 36, 39, 43), 187(34), 188(34), 190(34)
Regehr, R. W., **5:**282, 318
Regel', A. R., **5:**163(79), 171(79), 231
Regel, A. R., **6:**64(184), 78
Regenbrecht, D. E., **7:**98(73), 113
Rego, J. A., **24:**21(102), 26(102), 89(50)
Rehnberg, J. D., **5:**85(32), 86
Reiber, L., **10:**228(4), 253(4), 254(4), 258
Reible, S. A., **15:**60, 139
Reich, G., **1:**43(118), 66
Reich, R., **6:**119, 122, 128, 135(170, 178, 179), 147
Reichelt, K., **14:**66(224), 67(225), 78
Reichelt, W., **5:**167, 231
Reichert, W., **16:**24–25, 127
Reichman, B., **17:**323, 366
Reid, C. D., **4:**60(155), 94
Reid, E. E., **24:**8(27)
Reid, F. J., **5:**154(44), 158(44), 159, 230; **7:**170(101), 236(249a), 250, 254
Reidinger, F., **16:**99, 140
Reif, R., **14:**232(131, 132), 239; **23:**35(134)
Reim, W., **15:**197, 217
Reimann, R., **6:**123, 147
Reimer, J. A., **15:**278–279, 310; **19:**93
Reimer, L., **1:**188(2), 220, 228, 231; **2:**30(38), 61, 94, 131, 141, 145; **3:**32(159), 51(159), 57, 117, 128; **4:**224(126), 225, 238, 254; **6:**82(9), 143
Reinberg, A. R., **12:**266, 293
Reincke, W., **2:**30(43), 61
Reinders, W., **2:**37, 61; **3:**19(86), 51(86), 55
Reineck, S., **15:**154, 214
Reinhard, D. K., **14:**98(38), 118(38), 126
Reinhard, H. P., **1:**44(123), 45(123), 67
Reinhart, F. K., **7:**217(200a), 253; **10:**250, 259; **11:**77(157), 79(173), 80, 81, 100, 101; **16:**27, 127; **21:**121(18), 123(18), 232
Reinhoudt, D. N., **24:**23(108)
Reinke, P., **18:**91–92
Reinov, N. M., **1:**244(29), 274
Reiser, C., **13:**190(155), 191(155), 197
Reisinger, A., **17:**153, 154, 220
Reisman, A., **5:**274, 275(136), 317, 318; **12:**241(229), 282, 282(229), 295; **14:**89(20, 21), 90(20), 107(21), 118(20, 21), 123(20, 88), 124(88), 125, 127
Reiss, G., **12:**246(35), 248(35), 290
Reiss, H., **4:**151(67), 189; **17:**333, 366
Reiss, K. H., **9:**33(124), 68
Reisse, G., **12:**257(65), 291; **13:**99(108), 107, 111(20), 143(20), 147; **14:**25(99), 26(99), 75
Reiten, M., **16:**32, 129
Reitz, J. R., **15:**240, 262
Reizman, F., **3:**115, 128; **5:**167(125), 180, 196(125), 232
Reksten, G. M., **13:**159(35), 174(79, 81), 175(81), 185(35, 79, 81, 133, 134, 141), 187(35, 79, 81, 133, 134), 188(35, 79, 81, 141), 193, 194, 196
Rembetski, J. F., **22:**101(21), 115
Remeika, J. P., **9:**28(103), 67, 275, 294
Remili, N., **14:**87(16), 98(16), 103(16),

105(16), 109(16), 113(16), 120(16), 125
Remschnig, K., **16:**63, 112, 136, 142
Remy, J. C., **14:**7(24), 8(24), 74; **17:**111, 142
Ren, J. G., **16:**91, 139
Renard, C., **23:**161(60)
Renaud, P., **21:**325(17), 330(17), 332(17), 338(17), 345(36), 346(36), 350(17), 362(17), 367
Renault, A., **24:**181(12), 192(12), 199(12)
Renda, F. J., **11:**146(116), 149(116), 151(116), 202
Renk, K. F., **4:**50(103), 93; **16:**81, 138
Renninger, M. M., **2:**119, 143
Reno, C., **7:**230(234), 254
Renton, C. A., **6:**122, 147
Reny, D. J., **22:**76(74), 79
Repalov, N. S., **14:**163(115), 185(115), 197
Repinskii, S. M., **13:**185(136), 188(136), 196
Reschke, J., **14:**48(162), 55(162), 77
Reshetnikov, A. M., **7:**35(149), 40
Resnick, R., **7:**282(80), 309
Resto, O., **15:**284–285, 310
Rettner, C., **14:**19(79), 20(79), 42(79), 75
Rettner, C. T., **16:**18, 21, 126–127
Reuter, G. E. H., **6:**82(31), 144
Reuter, W., **7:**116, 118(2), 174(120), 185(139), 193(157), 209(120); **14:**8(28), 74; **22:**32(45), 78
Reutschi, M., **24:**94(71), 95(71)
Reutt-Robey, J. E., **24:**44(40)
Revell, R. S. M., **2:**109(82), 130(82), 142
Reven, L., **24:**150(4), 151(8), 154(8), 155(8b, 19), 156(8, 19), 157(8b), 158(8b), 159(8b), 161(8b), 164(30), 165(4, 30), 167(30), 168(33), 170(41), 171(41), 172(43), 173(43), 222(103)
Revesz, A. G., **4:**260(24), 301(24, 130, 131), 320, 323; **6:**47(143), 77
Rey, J. C., **22:**30(36), 77
Reynolds, D. B., **21:**5(4), 18(4), 71
Reynolds, D. C., **3:**179(26), 209; **11:**56(93), 57(93), 98
Reynolds, D. D., **24:**8(28), 9(28)
Reynolds, F. L., **9:**46(205), 70
Reynolds, F. W., **1:**1(2), 63; **3:**19, 51(87), 55; **6:**105, 131(118), 146
Reynolds, J. L., **22:**5(4), 77
Reynolds, J. M., **6:**82(26), 144
Reynolds, M. M., **2:**330(40), 860
Reynolds, S. I., **7:**348, 367, 391(111), 395
Rezek, J., **9:**6(24), 65
Rhee, S. S., **21:**81(13, 14), 84(15), 86(14), 87(14), 100(36), 111, 112
Rheinberger, P., **3:**11(50), 51(50), 54
Rhine, W. E., **16:**97, 100, 140
Rhodes, W. H., **1:**319(77), 333
Rhodin, J., **2:**109, 143
Rhodin, T. N., **1:**189(11), 218, 228, 231; **2:**2, 60; **3:**27(130, 134), 28(130), 39(1 97), 51(130, 134), 52(197), 56, 58; **4:**97(7), 100(7), 109(7), 110, 124(64), 126, 133(7), 134, 136, 150, 151, 152(45), 154, 155, 160, 188; **5:**193, 233, 260(50), 315; **9:**6, 7(30), 65; **24:**35(141), 86(34), 88(45), 89(50, 53), 90(45), 92(53), 106(111), 127(51), 212(65), 218(75), 221(89), 222(89), 228(31)
Rhyne, J. J., **9:**273, 274, 296; **15:**144, 147, 156–157, 214–215

Ribbing, C.-G., **17:**131, 143
Riben, A. R., **5:**276, 318
Riblet, H. J., **5:**60(12), 85
Ribner, A., **12:**272(167), 294
Ricco, A. J., **24:**44(55, 64, 65, 66), 46(64, 65, 66), 70(55), 92(56), 161(23), 162(23)
Ricco, B., **21:**296(50), 310
Rice, C. E., **16:**53, 96, 110, 134, 140, 142
Rice, S., **13:**181(109), 182(109), 183(109), 195
Rice, S. A., **9:**240(233), 261; **24:**184(35, 37, 41), 188(41), 192(35), 208(28, 29)
Rice, S. O., **10:**168(10), 225
Rich, G. J., **5:**261(56), 315
Rich, P., **22:**119(13), 120(13), 161(13), 171
Richard, J., **6:**201, 202, 204; **12:**272(168), 273(168), 294
Richard, P. D., **15:**288, 311
Richards, J. L., **5:**166(113, 114, 115), 176, 177, 178, 203(113, 114, 115), 208(115), 232; **6:**338(46), 347
Richardson, D., **1:**180(69), 186
Richeson, D. S., **16:**93–94, 139–140; **17:**287, 299
Richman, D., **5:**276, 318; **7:**171(115), 193(162), 244(269), 250, 252; **17:**358, 369
Richman, P., **5:**216(193), 228(193), 234
Richmond, J. C., **2:**315, 359
Richter, H., **2:**44, 62; **7:**14(30), 19(61), 20(62), 21(67, 68, 69), 24(67, 85); **14:**234(145), 240
Richter, K., **12:**286(243), 295
Richter, W., **19:**41
Rickel, D. G., **7:**66, 67(52), 113
Rickerby, D. S., **14:**191(167), 199; **17:**130, 143
Riddle, G. C., **3:**20(93), 51(93), 55
Riddoch, F. A., **21:**30(59), 73
Rideal, E. K., **7:**312(7, 11), 323(37), 330(55), 392, 393, 394
Rideout, V. L., **4:**75(187), 95; **7:**246(278), 255; **11:**87(213), 102
Ridge, M. I., **14:**44(154, 155), 56(175), 62(175), 77
Ridley, B. K., **21:**30(59), 73
Rie, K.-T., **17:**130, 143
Riecke, G., **21:**5(1), 71
Riedel, L., **6:**120, 147
Rieder, K. H., **24:**117(37)
Riedmiller, R., **3:**27(132), 28(132), 51(132), 56
Rieger, J., **14:**232(127), 239
Rieger, J. H., **17:**229, 280, 281, 282, 283, 285, 286, 290, 299
Ries, H. E., Jr., **7:**318(26, 27), 323(27), 329(26), 333(80, 81), 334(27, 83, 84, 85, 92), 393, 394
Riesenfeld, J., **3:**228, 229(53), 230, 244, 250(67), 251, 252, 272
Rietman, E. A., **16:**3, 124
Rieu, J. P., **24:**181(12), 192(12), 199(12)
Rigo, S., **12:**292; **14:**109(61), 126
Rijpers, J. C. N., **17:**111, 112, 142
Rilee, E. W., **9:**33(116a), 67
Riley, H. L., **5:**283, 319
Riley, M. E., **13:**153(9), 182(9), 183(9), 192
Rilu, E. W., **12:**177(58), 234
Rimini, E., **13:**166(60), 194
Rimmrott, M., **3:**31(152), 32(152), 51(152), 57
Rinaldi, R., **7:**331(69), 394
Rinehart, T. A., **15:**168, 215
Ring, J., **3:**15, 49(64), 51(64), 54

Ringhofer, C., **15:**249, 263; **22:**270(124), 275
Ringhofer, C. A., **22:**270(122, 123), 275
Ringsdorf, H., **24:**11(45, 46), 15(65), 18(88), 20(88), 21(45, 46, 91, 100, 102), 25(121), 26(102), 33(46), 34(46, 141)
Rios-Jara, D., **16:**4, 124
Riper, J. E., **7:**234(243), 254
Ripken, K., **9:**188(76), 257
Rippens, W., **13:**121(57), 148
Riseman, J., **2:**147(2), 178(2), 190, 365(3), 366(3), 384(96, 97), 385(97), 386(97, 101), 389(101), 419, 422; **3:**101(90), 102(90), 113(90), 118, 126, 128; **4:**257(1, 4), 261(30), 267(30), 268(30), 285(30), 288(30), 319(4), 320
Risnes, R., **6:**82(19), 92, 143, 145
Riste, T., **24:**180(10), 185(10), 186(10)
Ristolainen, E. O., **17:**119, 143
Ristow, M. L., **21:**362(44), 368
Ritchie, R. H., **2:**103, 105, 124(132a, d, f, g), 142, 144; **4:**329, 365(21), 369; **6:**193(49), 198, 203; **9:**146(1, 3), 149, 150(1), 158, 174(33), 179(54), 198, 201, 219(175), 221(183, 184, 186), 225(191), 226(191), 229(209), 237(222), 245(245), 250(250), 255, 256, 257, 258, 259
Ritchie, W. H., **15:**295–296, 311
Ritecheli, C., **16:**70, 136
Ritsko, J. J., **13:**155(28), 183(28), 184(28), 192
Ritter, D., **21:**150(54), 197(97, 98), 199(97), 200(97), 201(97), 202(99, 100), 203(99), 204(99), 207(97, 99), 213(116), 214(116), 216(116), 234, 235, 236
Ritter, E., **2:**284(38), 304, 346(71, 72, 73), 360, 402(170), 424; **4:**282, 286(102), 322; **10:**41(83), 67, 129(59, 60, 62, 63), 135(59, 62), 136(62), 165; **12:**2(6, 7), 3(13), 15(45), 41, 42, 48(76), 49, 49(13, 70, 81), 50, 51; **13:**61, 104
Ritter, H. L., **7:**11(23), 37
Ritter, K. J., **21:**191(87), 221(87), 235
Rittner, E. S., **3:**276(11), 294(62), 299, 301
Riva-Sanseverine, S., **10:**256(59), 259
Rivera, J. J., **5:**282(174), 318
Rivera, M., **1:**26(54), 65
Rizk, A. S., **12:**288(255), 295
Roas, B., **16:**55, 72–74, 107, 136–137, 141
Robbic, J. C., **12:**88, 161
Robbie, K., **22:**46(53), 49(53), 74(66), 76(66), 78
Robbins, C. R., **9:**11(49, 50, 51), 20(51), 66
Robbins, D. J., 23.24(89), 23.30(114), 23.37n(114), **23:**38(114), 39(114)
Robbins, E. J., **13:**8(13), 57
Robbins, J. L., **4:**150(51), 152, 154, 155, 188
Robbins, V. M., **21:**287(39), 310
Robert, G., **17:**320, 366
Robertazzi, R. P., **16:**117–118, 143
Roberts, A. R. V., **3:**6, 51(33), 54
Roberts, C., **21:**341(35), 348(38), 355(38), 356(38), 367
Roberts, D. A., **17:**198, 223
Roberts, E. D., **5:**287, 293, 319
Roberts, G. C., **11:**256(61), 257(61), 315
Roberts, J. F., **5:**161(58), 230, 299, 300, 321; **10:**46(96), 68; **13:**99, 100, 107
Roberts, J. S., **11:**65(118), 66(118),

Roberts, J. S., (*continued*) 70(142), 99, 100; **21:**312(4), 313(5), 323(4), 324(4, 15), 325(15), 329(15), 330(27), 331(27), 332(27), 333(15), 334(5, 15, 29), 336(5, 15), 341(29), 342(29), 348(38), 353(27), 355(27, 29, 38), 356(38), 366, 367
Roberts, N., **23:**101(33)
Roberts, R. W., **4:**9(28), 84(28), 91; **7:**327(47), 330, 334(47, 62), 335; **10:**120(51), 165
Roberts, S., **4:**84, 96; **19:**168, 173, 179
Robertson, J. M., **9:**276, 296; **15:**9, 11, 13, 136–137
Robertson, R., **19:**76, 80
Robeson, C. R., **6:**114(146), 146
Robin, S., **1:**134, 143, 147, 149(47), 168(47), 169(47), 185; **9:**178(60), 227(204), 257, 260
Robin-Kandare, S., **1:**148, 149(51), 159, 160, 185
Robins, J. L., **2:**100(61, 63), 142; **4:**126, 136
Robinson, A. L., **17:**210, 223
Robinson, C. F., **1:**42, 66
Robinson, D. H., **5:**247(24), 278, 315
Robinson, D. L., **1:**194, 229
Robinson, G., **1:**282, 284(5), 331
Robinson, G. N., **24:**14(63), 15(63), 16(63)
Robinson, G. Y., **12:**277(203), 278, 294
Robinson, I. K., **23:**2(2)
Robinson, L. B., **5:**303(296), 322; **7:**236(249a), 254
Robinson, M., **15:**70, 100, 139
Robinson, M. T., **3:**73(43), 125; **22:**32(42), 78
Robinson, McD., **9:**282, 296; **16:**106, 141
Robinson, P., **5:**302(258), 321; **17:**100, 141
Robinson, P. H., **5:**146(15), 229
Robinson, P. T., **16:**54, 134
Robinson, R. S., **14:**26(102), 75; **16:**182, 230; **22:**8(17), 77
Robinson, S., **14:**18(70), 20(70), 75
Robinson, T. S., **1:**143, 149, 163, 185; **2:**202, 237
Robrock, K- H., **16:**237, 267
Robson, A. E., **14:**147(69), 196
Roccasecca, D. D., **5:**281, 318; **7:**139(35a), 248
Rocher, A., **13:**11(20), 57
Rochet, F., **14:**98(94, 96), 101(94), 113(94), 114(96), 127, 128; **16:**30, 76, 78, 129, 137
Rocholl, P., **2:**387(109), 422
Rockett, A., **14:**65(209), 68(209), 78; **17:**131, 143
Rocztoczy, F., **3:**138(37), 163; **5:**311, 322
Rodak, H., **7:**174(121), 250
Rodbell, D. S., **1:**315, 333; **2:**31(52), 61
Roddick, R. D., **2:**315(10), 359; **4:**60(144), 94; **10:**14(26), 43(26), 44(26), 45(26), 66
Rode, D. L., **9:**284, 296
Rode, O. A., **17:**355, 356, 368
Roder, O., **2:**367(17), 420
Röder, O., **7:**170(108, 109), 250
Rodriquez, S., **21:**38(79), 73
Roe, D. W., **9:**33(128), 68
Roell, K., **15:**154, 214
Roesler, P., **3:**112(178), 129
Rogalla, H., **16:**22, 111, 117, 127, 142
Rogalski, A., **21:**40(85, 87), 74
Rogers, B. R., **22:**183(28), 185(28),

194(53), 227(53, 83, 84), 232(28, 53, 83, 84), 233(53), 235(83, 84), 238(53), 272–274
Rogers, C. T., **16:**63, 106, 111–114, 135–136, 141–142
Rogers, H. C., Jr., **2:**57(90), 62
Rogers, J., **24:**239(68)
Rogers, J. A., **24:**229(33), 230(33), 235(33)
Rogers, J. W., **10:**238(27), 258; **24:**235(55), 242(55)
Rogers, R. L., III, **12:**157(336), 166
Rogulin, V. Yu., **7:**223(217), 253
Rohatgi, A., **9:**33(135), 35, 37(135), 68
Rohde, S. L., **18:**259, 261, 262, 266, 269
Rohr, D., **17:**4, 64, 73
Rohrer, G. A., **17:**241, 287, 297
Rohrer, H., **1:**241(16), 273; **24:**21(100), 82(1, 3), 86(34), 114(33)
Rohrer, N. J., **16:**43, 50, 131–132
Rohtagi, A., **12:**180(66), 184(66), 187(66), 190(66), 234
Roizen, S., **1:**127(7), 184; **3:**46(227), 59
Roizen-Dossier, B., **3:**46(227), 59
Rojo, J. M., **14:**213(31), 237
Rol, P. K., **3:**73(57), 126
Roldan, J. M., **16:**5, 124
Rolinski, E., **12:**265(125), 293
Rollett, A. D., **16:**42, 47–48, 50, 131–132
Rollins, R. W., **4:**124(64), 126(64), 136, 150(45), 151(45), 152, 160(45), 188
Rolls, W. H., **11:**166(120), 202
Rolsten, R. F., **3:**194(55), 209
Rolston, S., **24:**230(44)
Romand, J., **1:**149(40), 185
Romankiv, L., **6:**291(208c), 300
Romano, P., **16:**17, 99, 126, 140
Romano-Moran, R., **7:**142(48), 248
Romanov, Y. U., **9:**217(164), 218(164), 259
Romanowski, Z., **12:**242(17), 243(17), 290
Romeijn, F. C., **9:**9(44), 65
Rondeel, W. G. J., **14:**155(88), 156(88), 197
Rondelez, F., **24:**169(35, 36), 181(13), 228(20)
Ronsheim, P. A., **23:**49(175)
Rook, H. L., **12:**14, 50
Rooksby, H. P., **1:**205(88), 230
Roos, J. R., **17:**110, 141
Roosendaal, H. E., **14:**15(56), 74
Roosild, S. A., **23:**87(6, 8, 11), 109(60)
Rooymans, C. J. M., **9:**288, 296
Roppel, T., **14:**98(38), 118(38), 126; **16:**123, 143
Roschen, J., **2:**376(76), 421
Rose, A., **2:**150(27, 28, 29), 191; **3:**197(63), 210; **4:**326, 369; **5:**145, 221; **12:**154(326, 330, 331), 166
Rose, C., **16:**107, 141
Rose, R., **17:**361, 370
Rose-Innes, A. C., **1:**273, 275
Rosen, S., **5:**70, 86
Rosenbaum, J., **17:**197, 199, 200, 201, 202, 203, 204, 205, 206, 207, 223, 224
Rosenberg, D., **3:**74(48), 75(53), 76(53), 78(53), 126
Rosenberg, G. V., **7:**362(154), 396
Rosenberg, R., **6:**102(95), 114(143), 132(143), 145, 146; **7:**264(46), 265(46, 51), 266(46), 267(71), 270(66, 68), 273(68, 51), 274(51, 71), 275(71), 276(51), 279(68, 76), 280(68), 282(68, 76), 283(66,

Rosenberg, R., (*continued*) 68), 284(51, 66, 68, 71), 285(108), 286(51), 290(97), 292(76); **9:**278, 282, 295; **16:**15, 126, 234, 239–240, 248, 250, 266
Rosenberg, S., **7:**315(20), 350, 393, 395
Rosenblatt, C. J., **11:**256(66), 257(66), 316
Rosenblum, M. P., **15:**278–279, 310
Rosenbluth, A. W., **24:**208(25)
Rosenbluth, M. L., **21:**128(33), 233
Rosenbluth, M. N., **24:**208(25)
Rosencher, E., **14:**109(61), 126; **21:**161(67), 170(74), 175(74), 234
Rosencwaig, A., **9:**29, 271, 288
Rosenfeld, D., **21:**40(86), 74, 155(58), 234
Rosenfeld, L., **2:**31, 32, 61
Rosenfeld, R. L., **3:**224(42a), 271
Rosenmayer, C. T., **17:**43, 47, 75
Rosenthal, P., **16:**42, 50, 131
Rosenzweig, D. N., **10:**293(8), 308
Rosette, K. H., **1:**59(144), 67
Rosetti, G. A., Jr., **19:**253
Rosier, L. L., **9:**294
Rosler, R. S., **12:**239(12), 241(12), 242, 266, 267, 268, 290, 293
Rosner, L. G., **3:**88(68), 122, 126
Ross, A., **3:**11(50), 51(50), 54
Ross, B., **16:**58, 61, 64, 72–73, 135
Ross, C. B., **24:**99(102, 103), 100(103), 106(102, 103)
Ross, D. S., **22:**91(14, 15), 115, 270(128), 276
Ross, I. M., **2:**150, 190
Ross, L., **7:**31(130), 40
Ross Macdonald, J., **11:**230(31, 32), 315
Ross, R. F., **7:**273(69), 292, 302(101), 308, 310
Rossi, D. V., **16:**42, 131
Rossi, J. A., **10:**253(69), 259
Rossignol, V., **21:**345(36), 346(36), 367
Rossing, T. D., **6:**224(79), 292(79), 297
Rossmeisl, R. A., **2:**367(14a), 405(182), 420, 424
Rossnagel, S. M., **13:**118(47), 131(47), 132(47), 134(47), 139(47), 140(47), 148; **14:**5–6(17), 33(130), 34(130, 134), 39(130), 73, 76; **17:**82, 87, 121, 139, 140; **18:**125, 180, 182, 187, 194; **22:**32(40, 41), 39(47), 45(47), 46(55), 78, 84(4, 5), 98(4), 101(4, 19, 20), 111(4, 19, 20), 112(4), 113(4), 114(4, 5, 19, 20), 115, 161(79), 173, 242(89–91), 248(89, 90), 249(89, 90), 274
Rossol, F. C., **9:**275, 276, 296
Rossteutscher, W., **7:**21, 24(67), 38
Rost, T. A., **17:**228, 260, 292, 298, 299
Rostworowski, J., **14:**48(161), 77
Rosztoczy, F. E., **7:**142(45), 152(62a), 170, 171(110), 248, 249, 250
Roth, A., **13:**164(52), 193
Roth, A. P., **17:**153, 154, 220
Roth, E. A., **5:**247(29), 274, 315, 317
Roth, H. A., **2:**373(66), 421
Roth, J., **14:**228(91), 239
Roth, L. M., **3:**281, 300
Roth, W. L., **1:**205(79), 230; **6:**215(48), 296
Rothen, A., **3:**16, 51(71), 54; **6:**186, 203; **7:**316(24), 333(74), 364(74, 162), 393, 394, 396
Rothlisberger, U., **24:**211(51)
Rothschild, M., **13:**171(74), 186(74), 190(74), 194; **17:**315, 365

Rou, S. H., **16:**14, 17, 126; **17:**229, 234, 235, 238, 239, 263, 296, 297
Rouais, J. C., **16:**254, 268
Rouard, P., **1:**70(1), 84(1), 120; **2:**249(16), 304; **4:**85(222), 86, 96; **6:**172, 201(72), 202, 202(78), 204; **9:**15(65), 66
Rougeot, H., **12:**121(236), 164
Roulet, H., **14:**98(94, 96), 101(94), 113(94), 114(96), 127, 128
Roulston, D. J., **23:**145–146(42)
Rourd, P., **12:**71(71), 72(71), 79(71), 160
Roux, D., **6:**202(80), 204
Roux-Buisson, H., **6:**277, 278(189), 299
Rovner, L. H., **12:**253(67), 291
Rovy, G., **7:**174(121), 250
Rowe, G. K., **24:**23(111), 27(111), 44(51)
Rowe, J. E., **9:**215(155), 216, 217(156, 157, 159), 259; **12:**121(238), 164
Rowe, M., **12:**85(103), 160
Rowe, T. A., **7:**27(110), 39
Rowell, J. M., **4:**370; **6:**140(228), 149
Rowlinson, J. S., **24:**184(40), 190(40), 220(82)
Rowntree, P., **24:**125(6), 127(6)
Rowntree, P. A., **24:**45(71), 46(88), 55(88), 56(88), 82(11), 83(11), 86(11), 112(6)
Rowntree, R. F., **2:**262(20), 289(20), 304
Rowntree, Y. P., **24:**199(49)
Roy, A. M., **10:**228(4), 243(34), 253(4), 254(4), 258
Roy, R., **5:**119(23), 141, 303(296), 322; **7:**34(143), 40; **9:**11(52, 54); **17:**110, 111, 119, 141
Roy, R. A., **16:**17, 23, 53, 78, 126, 133, 137; **17:**120, 121, 143, 234, 236, 296; **19:**194, 195, 230
Roybal, L L., **16:**97, 99, 140
Royds, T., **4:**60(147), 94
Royer, L. J., **14:**219(60), 220(60), 221(60), 222(60), 223(60), 238
Royer, W. A., **17:**118, 142
Royle, A., **7:**170, 191(107), 250
Rozenfel'd, I. L., **6:**64(186), 78
Rozgonyi, G. A., **7:**223(220), 253; **17:**153, 219
Rozsnyai, L. F., **24:**97(79), 98(79, 80)
Ruane, M. F., **15:**180, 186–188, 216
Rubens, S. M., **2:**365(4), 370(4), 380(4), 389(4), 390(4), 419; **3:**133(19), 162
Rubin, D., **16:**106, 141
Rubin, L. G., **1:**43(120), 66
Rubin, S., **24:**97(78), 98(78, 84)
Rubinstein, H., **1:**295(43), 322(86), 332, 333; **6:**228(100), 238(100), 294(100), 297
Rubinstein, I., **24:**33(135)
Rubinstein, M., **5:**279, 318
Ruch, R. J., **7:**329, 389(53), 394
Ruckman, M. W., **16:**51, 133
Rudee, M. L., **4:**239(166), 255; **7:**23(83), 26(99), 39
Rudenberg, H. G., **2:**370(45), 406(45), 420
Rudenko, S. I., **11:**209(4), 211(4), 314
Rudge, W. E., **24:**98(98), 106(98)
Rüdiger, O., **1:**127, 231
Rudiger, O., **4:**164(97), 171(119), 189, 190
Rudigier, H., **17:**83, 128, 130, 139
Rudisill, J. E., **3:**120(148), 128
Rudkevich, D. M., **24:**23(108)
Rudman, D. A., **16:**28, 33, 116, 128, 142
Rudner, S., **16:**100, 140

Rudnik, P. J., **17:**116, 142
Rudolph, A. C., **4:**370
Rudolph, D., **11:**23(17), 24(17), 25(17), 26(17), 27(17), 34
Ruedl, E., **1:**53, 67; **3:**70, 125
Ruehrwein, R. A., **5:**161, 230, 277, 318
Ruff, O., **4:**283(108a), 322
Rühl, W., **1:**212(122), 231; **2:**38, 41, 42(61), 61
Ruhrwein, R. E., **16:**81, 138
Ruigrok, E. W. M., **21:**365(50), 368
Rujkorakarn, R., **13:**123(69), 131(69, 99), 132(69), 133(99), 134(99), 135(99), 139(99), 140(99), 149
Rukwied, A., **7:**14(30), 38
Runtsch, B., **16:**8, 125
Runyan, W. R., **5:**264(66), 269, 296(40), 316, 320
Rup, R., **6:**141, 142(236), 149
Rupley, F. M., **22:**186(30, 31), 272
Rupp, Jr., L. W., **16:**3, 124
Rupp, L W., **11:**57(94), 99
Ruppel, W., **2:**150, 191
Ruppin, R., **9:**194(96), 258
Rupprecht, G., **9:**47(213), 70
Rupprecht, H., **7:**116, 118(2), 142(38), 183(138), 185(139), 214(199), 218, 219(38), 220(138, 139), 230(199), 247, 248, 251, 253
Rusch, D., **5:**32(23), 45
Rusek, A., **12:**265(109), 292
Rusek-Mazurek, A., **12:**242(17), 243(17), 290
Rushmeier, H. E., **24:**86(31), 138(64)
Ruska, H., **1:**189(7), 228
Ruske, W., **2:**30(39), 61; **6:**207(10, 11), 237, 273(10, 11), 295
Ruskin, R. E., **10:**78(21), 164
Russek, S. E., **16:**51, 61–62, 66, 107, 133, 135–136, 141
Russel, T. A., **3:**46(224), 59
Russell, D. A., **24:**21(99), 24(99), 25(99)
Russell, G. J., **16:**26, 127
Russell, K. C., **4:**98(12), 99(12, 14), 109(14), 134
Russell, L. A., **1:**319(77), 333
Russell, P. E., **14:**64–65(207), 65(207), 78
Russell, R. G. G., **24:**163(24)
Russell, T. P., **24:**180(10), 185(10), 186(10)
Russell, V. A., **5:**269, 317
Russer, P., **17:**217, 224
Russev, S. H., **19:**203
Russo, R. E., **16:**60, 79, 135, 137
Rustgi, O. P., **1:**124(1), 130(1), 134(1), 135(1), 143(36), 147(1), 149(1, 59), 163, 165, 169(36), 184, 185, 186; **4:**50(99), 93; **7:**45(7b), 46, 47, 48, 72(19), 111, 112
Rusu, Gh. I., **9:**33(137), 68; **12:**187(77), 234
Rusu, M., **9:**33(137), 68; **12:**187(77), 234
Ruth, R. P., **5:**275(130), 317
Ruth, V., **4:**108(27), 123, 125(27), 127(27), 135
Rutherford, S. L., **1:**19(43, 44), 65
Rutkowski, J., **21:**40(87), 74
Rutledge, R., **16:**261, 269
Ruud, J., **17:**63, 77
Ryabova, L. A., **9:**39(167, 168), 40(167, 168), 69; **12:**177(60), 234
Ryan, J. D., **9:**32(116), 33(116), 67
Ryan, J. F., **21:**30(60), 73; **23:**254(76)
Ryan, J. P., **7:**330(59), 334(95), 394
Ryan, P. J., **6:**17, 75
Ryan, R. E., **19:**9, 15, 19
Ryba, G. N., **24:**228(14)
Ryckaert, J. P., **22:**134(51), 142(51), 172; **24:**209(35), 210(42)

Ryder, E. J., **12:**64(44), 159
Rymer, T. B., **3:**226(47), 229(47, 78), 234(47), 235(47), 238(47), 244(47), 246, 247, 271, 272
Rytova, N. S., **6:**137(217), 148
Rytz, D., **19:**259
Rytz-Froidevaux, Y., **13:**155(27), 164(27), 181(27), 192
Ryuo, T., **15:**14, 137
Ryvkin, B. S., **13:**162(44), 164(44), 193
Ryvkin, S. M., **13:**162(43), 164(43), 165(43), 193
Ryzhkov, V. N., **14:**151(83), 152–153(83), 154(83), 156(83), 158(83), 197
Ryzkov, V. N., **17:**93, 94, 140
Rzhanov, A. V., **13:**185(136), 188(136), 196

S

Saba, R. G., **11:**253, 255
Sabatini, E., **24:**33(135)
Sabatini, R. L., **16:**51, 133
Sabi, B., **16:**251, 268
Sabine, G. B., **1:**146, 149(45), 185; **10:**74(7), 91(7), 164
Sablev, L. P., **14:**130(3, 4), 164(116), 166(3, 4), 176(116), 195, 197
Sacher, E., **15:**279, 281–282, 310
Sachs, R. G., **4:**339, 370
Sachse, G. W., **14:**232(114), 239
Sack, R. A., **12:**54(14), 158
Sackett, D. D., **24:**12(52), 15(52)
Sadagopan, V., **9:**281, 282, 288, 290, 295, 296, 297
Sadakata, N., **16:**88, 138
Sadeghi, N., **18:**79, 82, 85, 90
Sadhukhan, P., **1:**191(21), 228
Sadler, A. G., **14:**27(107), 75
Sadukhan, P., **2:**75(31), 76, 79(31), 93, 141
Sadwick, L. P., **15:**308, 312
Saemann-Ischenko, G., **16:**51, 72–73, 133, 136–137
Saenger, K. L., **16:**78, 137; **17:**236, 296
Saez, E. I., **24:**66(104)
Safarov, V. I., **19:**2
Sagara, T., **24:**45(82)
Sagawa, M., **16:**149, 229
Saget, P., **12:**135(298), 165
Saggan, B., **12:**72, 73(75), 160
Sagiv, J., **24:**13(66), 15(66), 181(17), 228(21)
Sagnes, I., **23:**161(60–62)
Sah, C. T., **4:**295(113), 301(113), 302(140), 304(140), 306(140), 323; **11:**140(103), 202; **17:**274, 298
Sah, C.-T., **21:**318(10), 366
Sahai, R., **12:**137(309), 165
Sai-Halasz, G. A., **11:**68(137), 70(137), 71(137), 73(137), 74(137), 79(137), 86(202, 205), 100, 102
Saifi, M. A., **16:**110, 142
Saiki, K., **11:**268(86), 269(86), 316
Saini, K. K., **17:**304, 312, 364, 365
St. Jalmes, A., **24:**181(15)
St. John, C., **6:**340(58), 347
St. Julien, D., **16:**70, 136
Sainty, W. G., **13:**112(24), 120(54), 124(54, 72), 127(72, 91), 131(54), 132(54), 134(54), 135(54, 72, 106, 107, 108), 136(54, 72, 106, 108), 137(54, 108), 141(72, 122), 143(91, 107), 144(24, 107, 136), 145(138), 147, 148, 149, 150; **14:**61(194), 78

Saito, H., **7:**235, 253; **13:**181(110), 185(139), 188(139), 195, 196
Saito, J., **15:**204, 206, 217, 221, 260
Saito, K., **11:**253, 255
Saito, M., **10:**237(21), 258
Saito, N., **6:**227(95), 297; **15:**168, 183–184, 215–216
Saitoh, K., **17:**110, 141, 228, 261, 262, 263, 298
Sakai, A., **17:**354, 368; **23:**44(151)
Sakai, S., **17:**83, 128, 129, 139
Sakai, Y., **3:**198, 210; **5:**214, 234; **9:**50(236a), 60(271), 70, 71
Sakaki, H., **11:**68(137), 70, 71, 73(137), 74, 79(137), 100; **21:**116(1), 232
Sakaley, F. E., **1:**319(77), 333
Sakamoto, Y., **17:**153, 219
Sakata, S., **13:**220(71), 221(71), 225
Sakata, T., **12:**95, 161, 162
Sakata, Y., **24:**33(134)
Saki, Y., **11:**76(149), 100
Sakomura, S., **16:**263, 269
Sakudo, N., **18:**122, 166
Sakurai, N., **19:**35
Sakurai, T., **6:**72(213), 79
Sakurai, Y., **15:**147–148, 214
Sala, C., **14:**66(217), 78
Salahuddin, M., **22:**76(73), 79
Salam, E., **10:**43(87), 44(87), 67
Salama, C. A. T., **5:**163(73), 170, 231; **12:**264(106), 292
Salama, M., **12:**246(30), 248, 290
Salamon, M. B., **17:**29, 47, 75
Salanskii, N. M., **6:**238(126, 127), 298
Salathe, R. P., **13:**155(27), 164(27), 174(82), 180(82), 181(27), 186(82, 148), 189(148), 192, 194, 197, 213(49), 224
Saleemi, A. S., **16:**279, 337
Salem, J., **14:**19(79), 20(79), 42(79), 75; **16:**18, 21, 34–35, 126–127, 130
Salem, L., **7:**320(32), 393
Sales, B. C., **19:**199, 201, 212; **24:**98(92)
Salje, E., **17:**309, 364
Sallet, V., **19:**38
Salling, C. T., **24:**99(100), 106(100)
Salmeron, M. B., **24:**44(21), 82(4), 86(28), 89(47), 90(28), 98(87, 88), 106(105, 107), 112(20), 117(20), 119(20), 128(20), 129(20), 218(78), 221(90, 91, 93, 99), 222(90, 91), 228(8)
Salome, M., **14:**96(35), 126
Salomon, R. E., **6:**36(118), 43(118), 76
Saltykov, P. A., **16:**5, 124
Salvo, H. L., **17:**153, 157, 187, 195, 220, 222
Salvo, H. L., Jr., **15:**67, 139; **17:**171, 221
Salzberg, B., **1:**70(8), 77(34), 92(8), 120, 121
Salzberg, C. D., **1:**61(147), 67; **2:**284(37), 304, 346(70), 360, 401(168); **10:**129(58), 165; **12:**26(51), 28(51), 50
Samant, M., **24:**112(22)
Samant, M. E., **24:**112(21)
Samid, I., **10:**250(44), 259
Sampere, S. M., **16:**45, 132
Sampogna, M., **12:**276(187), 294
Samson, J. A. R., **7:**45(7a), 47(21), 70(60), 111, 112, 113
Samson, S., **9:**27, 67
Samuel, D., **6:**25, 76
Samuels, . L. E., **4:**7(17), 91
Samuelson, L., **19:**2, 23, 26–29
Samukawa, S., 18.27n, **18:**100–101, 102, 104, 125, 160–162, 163, 165, 168, 179, 186

Samulski, E. T., **24:**151(9), 154(9), 156(9)
Sanchez, A., **13:**162(45, 46), 163(45, 46), 164(45), 186(46, 153), 189(45, 46), 190(46, 153), 193, 197
Sanchez, L. E., **17:**228, 239, 240, 241, 254, 255, 256, 297, 298
Sanchez-Rubio, A., **13:**186(145), 189(145), 190(145), 197
Sandburg, W. J., **12:**277(194), 294
Sander, L. M., **21:**98(33), 112; **22:**5(7), 77
Sander, W., **2:**38(63), 39(63), 44, 61, 62
Sander, X., **24:**170(40), 173(40)
Sandercock, J. R., **15:**107, 140
Sanders, D. H., **17:**94, 141
Sanders, D. M., **14:**176(137, 138), 198
Sanders, F. H. M., **14:**232(126), 239
Sanders, J. V., **4:**152(71, 72), 171, 180(71, 72), 189
Sanders, R. M., **3:**44(222), 59
Sanderson, C. C., **6:**340(58), 347
Sanderson, R. B., **4:**60(141), 94
Sandfort, R. M., **9:**265, 296
Sandhu, H. S., **4:**87(239), 96
Sandler, N. P., **5:**306, 322
Sandomirskii, V. B., **6:**137, 139, 141, 142, 148
Sandor, J. E., **3:**134, 141(53), 142(53), 143(23), 144(53), 162, 163; **5:**289(223), 320
Sands, T., **17:**271, 298
Sandstrom, A. E., **7:**45, 48(15), 111; **10:**110, 165
Sandstrom, R. L., **16:**5, 21, 124, 127
Sandy, A. R., **24:**127(56)
Sanganeria, M. K., **23:**48(170)
Sangster, R. C., **5:**265, 316
Sanjurjo, A., **16:**84, 90, 138
Sankaran, R., **12:**123(251), 130(276), 139, 164, 165
Sankur, H., **17:**153, 154, 221
Sano, M., **13:**99(100), 107; **17:**336, 346, 367
Sano, N., **17:**24, 75
Sans, C., **17:**64, 77
Sans, C. A., **14:**71(232), 79
Sansalone, F. J., **12:**246(33), 248(33), 252(33), 290; **13:**142(123), 150
Santangelo, P. G., **24:**228(14)
Santesson, L., **24:**44(42), 86(34)
Santos, F. D. D., **24:**235(57)
Sapelkin, N. P., **17:**3, 60, 73
Saperstein, W. A., **12:**128(268), 164
Sapieha, S., **11:**278(100), 316
Sarachik, M. P., **1:**248(45), 251(45), 274
Saraswat, K. C., **22:**30(36), 77, 182(25), 188(25), 193(25), 235(86), 272, 274
Sarbei, O. G., **4:**349, 354(60), 367, 370
Sard, R., **4:**238(164), 255
Sarin, V. K., **13:**79(75), 106
Saris, F. W., **13:**38(51), 58
Sarkady, K., **21:**219(127), 236
Sarkus, T. A., **17:**307, 364
Sarma, S. D., **15:**222, 261
Sarusi, G., **21:**207(105), 208(112), 209(112), 210(112), 211(112), 212(112), 213(112), 216(122, 123), 224(141), 235, 236, 237
Sasabe, H., **24:**150(3)
Sasaki, A., **13:**3(6, 7), 5(10), 11(6, 10), 18(26), 20(30, 31, 41, 42), 21(30, 31), 22(7), 23(7, 41, 42), 25(30), 27(31), 28(31), 39(31), 44(41, 53), 56, 57, 58
Sasaki, H., **15:**37, 137
Sasaki, T., **9:**34(157), 35(157), 68, 186(68), 257

Sasaki, Y., **5:**277(152), 318; **10:**236(19), 258
Saski, T., **7:**363(157, 158, 159), 396
Sassi, G., **21:**337(31), 367
Sastre, A. M., **24:**163(24)
Sasuga, E., **16:**263, 269
Satchell, J. S., **16:**52, 133
Sato, H., **4:**156(83a), 189
Sato, J., **16:**28, 33, 128
Sato, K., **5:**306, 321
Sato, M., **15:**172, 204, 206, 216–217
Sato, N., **15:**210–211, 213, 217; **17:**315, 316, 317, 329, 331, 365; **18:**91–92, 96
Sato, R., **15:**168, 183–184, 215–216
Sato, R. N., **21:**128(37), 233
Sato, T., **15:**210, 217
Sato, Y., **3:**277(28), 300; **17:**355, 369; **24:**23(109), 27(109)
Satoh, H., **17:**197, 201, 222, 223
Satoh, T., **15:**168, 215; **16:**27, 28–30, 32, 53, 57, 127, 129, 134
Satoh, Y., **15:**168, 215
Saton, M., **17:**83, 139
Satterthwaite, C. B., **1:**255(60), 265(60), 275
Satyapal, S., **21:**36(70), 73
Satz, H. G., **6:**96, 130, 145
Sauer, D. J., **23:**124(15)
Sauerbrey, G., **2:**365(10), 420; **3:**24, 51(101, 103), 55; **6:**198(60), 203; **9:**174(32), 219(177), 220(179, 181), 221(187), 227(200), 256, 259, 260
Sauerbrey, G. Z. Z., **24:**48(97)
Saul, R. H., **7:**133(22, 23), 139(35a), 166, 167(93), 223(220), 225(226), 247, 248, 249
Saunders, A. E., **9:**33(146), 68
Saunders, J., **2:**379(82), 397(82), 422
Saunders, M.-L., **21:**241(7), 308
Sauro, J., **4:**23(56), 92
Sauter, F., **9:**192(88), 257
Sautet, P., **24:**98(88), 221(90), 222(90)
Savage, A., **11:**85(195, 196), 102
Savage, C. M., **15:**206, 217
Savage, S., **15:**221, 260
Savary, A., **9:**178(60), 227(204), 257, 260
Savchenko, N. S., **6:**238(127), 298
Savehenko, V. P., **1:**205(86), 230
Savelli, M., **12:**169(21), 224(116), 232, 235
Savet-Peyman, F., **15:**55, 138–139
Savina, G. M., **6:**71(212), 79
Savitskaya, Y. S., **9:**39(168), 40(168), 69
Savornin, F., **6:**94, 145
Savoy, R., **14:**19(79), 20(79), 42(79), 75; **16:**18, 21, 34–35, 126–127, 130
Savvides, N., **14:**61(195), 66(216), 78; **17:**88, 110, 119, 123, 140, 141, 143; **18:**253, 255, 259, 261, 263, 266, 267, 272, 274, 277, 278
Sawada, Y., **14:**52(168), 77
Sawaki, T., **2:**289(34b), 304; **3:**15, 51(67), 54
Sawamuru, I., **3:**229(56), 230(56), 236(56), 272
Sawatzky, E., **9:**282, 296
Sawin, H. H., **22:**270(129), 275
Sawyer, D. L., **3:**217(19b), 271
Sawyer, R. A., **4:**50(110), 93
Sawyer, W. E., **5:**240(8), 314
Saxe, S. G., **13:**124(71), 127, 137(112), 138(112), 140, 141(112), 149, 150
Saxena, A. N., **6:**177, 202; **16:**240, 251, 267–268; **22:**20(28), 77
Saxena, R. R., **12:**54(9), 136(9), 138(9), 139(9), 158

Saxena, T. K., **17:**312, 365
Saxon, D., **1:**16(34), 64
Sayama, Y., **12:**74, 160
Sayer, M., **11:**256(66), 257(66), 316; **17:**153, 163, 220, 221, 228, 232, 241, 247, 249, 250, 290, 296, 297; **19:**250
Sayer, M. T., **14:**57–58(180), 77
Sayers, D. E., **7:**27, 31(133), 32(100), 39, 40
Sayre, C. N., **24:**19(86), 20(86), 27(86), 28(86), 30(86)
Scaglione, **19:**240
Scagnelli, H., **2:**365(2), 419
Scala, C., **19:**139
Scala, L. C., **7:**335, 369, 370, 375, 379, 381(96), 394
Scalapino, D., **21:**54(106), 59(106), 74
Scalapino, D. J., **16:**106, 141
Scanlon, J., **17:**19, 20, 21, 32, 74
Scanlon, P. J., **14:**33(129), 39(129), 68(129), 76
Scanlon, W. W., **3:**277, 278, 289(50), 300; **11:**120, 121(72), 201
Scapple, R. Y., **3:**20(94), 51(94), 55
Scardefield, J. E., **7:**203(178a), 225(229), 252, 253
Schaal, W., **14:**67(225), 78
Schaarschmidt, E., **5:**256(45), 315
Schach, M., **2:**307(1), 312(1), 358; **3:**10(48), 51(48), 54; **10:**146(78), 166; **12:**29(56), 51
Schade, H., **7:**243(266), 245(273, 274), 255; **12:**135(294), 165
Schade, K., **14:**35(137), 38–39(137), 51(138), 76
Schaefer, H., **5:**240(13), 247, 314, 315
Schaefer, H. G., **14:**72(235), 79
Schaefer, V. J., **1:**189(8), 228; **7:**324(38), 338(109), 342(111), 347(118), 350(128, 129, 130), 388(205), 393, 395, 397
Schafer, F., **7:**386(202), 397
Schafer, H., **3:**194(56), 209; **12:**269(153), 293; **15:**299, 301–302, 311
Schafer, W. A., **24:**163(26)
Schaffer, P. S., **5:**270, 317
Schaffer, W. J., **17:**287, 299
Schaffler, F., **21:**78(4), 111
Schafft, H. A., **16:**240, 251, 267 268
Schagan, P., **12:**137(306), 165
Schaible, P. M., **2:**397(158), 424; **3:**92(73), 96(73), 108(103), 109(109), 126, 127
Schairer, W., **7:**143(51), 248
Schantterly, S. E., **19:**6
Scharf, P. T., **4:**50(100), 93
Scharfe, M., **6:**36(114), 38(114), 45(114), 76
Scharli, M., **12:**270(156), 293
Schatzkes, M., **19:**135, 164
Schauer, R. F., **1:**321(80, 81, 84), 333
Schauer, W., **16:**23–24, 127
Schawlow, A. L., **1:**248(44), 251(44), 257(67), 274, 275
Scheckler, E. W., **22:**181(19), 271
Scheek, M., **2:**387(104), 422
Scheel, H. J., **5:**106(17), 141
Scheer, J., **5:**1(4), 45
Scheer, J. J., **7:**242(260), 255; **12:**54(7), 87, 89, 112, 114(204), 158, 161, 163
Scheer, M. D., **9:**46(208), 70
Scheibner, E. J., **2:**370(44), 420; **4:**85(224), 96, 301(117), 323; **9:**215(150), 259
Scheithauer, U., **17:**291, 299
Schelkunoff, S. A., **1:**92(41), 121
Schellekens, H., **14:**138(24), 195
Schellingerhout, A. J. G., **17:**77

Schemmel, R. R., **6:**142, 149
Scher, H., **11:**256(64), 257(64), 290(116), 315, 317
Scherber, W. E., **12:**266, 267, 293
Scherer, G. W., **17:**239, 297
Scherer, M., **13:**67(37), 105; **14:**48(160), 77
Schessler, H. M., **24:**70(105)
Scheuermann, M., **16:**17, 126
Scheuermann, R. J., **10:**273(7), 308
Schidlovsky, G., **7:**358, 360(148), 396
Schieber, D. J., **6:**54, 77
Schiffrin, D. J., **24:**154(15, 16)
Schilder, J., **16:**52, 133
Schiller, R., **17:**318, 319, 366
Schiller, S., **13:**67(32), 105; **14:**11(39), 34(133), 35(137), 36(138), 38–39(137), 44(39); **17:**111, 116, 142
Schilling, J., **5:**299(252), 321; **9:**214(148), 215(148, 149), 259
Schilling, M. L., **24:**12(55), 14(55, 59), 20(59)
Schilling, R. B., **13:**61, 104
Schilling, W., **16:**237, 267
Schindler, A. I., **9:**297
Schineller, E. R., **10:**237(26), 258
Schirmer, O. F., **17:**345, 349, 350, 367
Schivley, M. A., **24:**172(42), 173(42)
Schlacter, M., **3:**115, 129
Schlapp, W., **23:**254(75)
Schlayen, B., **19:**8
Schlegel, E., **2:**376(76), 421
Schlenker, C., **6:**215(42), 217(42), 219(42, 61, 62), 220, 222(61), 223, 268, 269, 296
Schlenoff, J. B., **24:**44(61), 46(89), 60(89), 150(3)
Schlesinger, H. I., **12:**257(70), 291
Schlesinger, M., **9:**6(24), 22(90), 23(90), 27(90), 65, 67
Schlesinger, T. E., **16:**26, 127; **17:**49, 62, 77; **23:**163(47)
Schlier, R. E., **5:**190(162), 233
Schlig, E. S., **2:**151(37, 37a), 176(37), 191
Schlom, D. G., **16:**53–54, 134
Schlomann, E., **15:**6, 8, 92, 136, 140
Schlosser, W. O., **11:**42(46), 50(46), 75(46), 76(46), 97
Schlossman, M. L., **24:**180(10), 181(12, 13), 185(10), 186(10), 192(12), 199(12)
Schlote, J., **22:**205(64), 273
Schlotter, P., **17:**345, 353, 367, 368
Schlötterer, H., **2:**17, 60
Schlotterer, H., **4:**149(33), 156(83), 171(33), 186, 188, 189, 190
Schlueter, J., **22:**193(51), 194(52), 226(51), 227(52), 228(52), 229(52), 273
Schlüter, M., **9:**177, 257
Schmaevskii, V. E., **5:**164(101), 232
Schmahl, G., **11:**23(17), 24(17), 25(17), 26(17), 27(17), 34
Schmalfeld, B., **9:**176(50), 178(50), 186(50), 187(50), 257
Schmatz, D. J., **14:**12(45), 74; **17:**111, 112, 142
Schmedlin, F. W., **1:**264(79, 80), 275
Schmelz, C. M., **7:**174(118b), 250
Schmid, E., **24:**236(60)
Schmid, H., **6:**214(37), 295
Schmider, P., **6:**119, 129(169), 130, 147
Schmidt, D. N., **16:**87, 138
Schmidt, F. A., **7:**262(17), 264(43), 306, 307
Schmidt, K., **14:**66(224), 78; **16:**24–25, 127
Schmidt, L. D., **22:**209(66), 273
Schmidt, M., **16:**102–105, 141

Schmidt, M. P., **19:**61, 85, 87, 93
Schmidt, M. S., **16:**81, 102–105, 121, 138
Schmidt, M. T., **16:**42, 131
Schmidt, P., **4:**283(108a), 322; **19:**2, 26, 29
Schmidt, P. F., **2:**406(188), 424; **4:**316(163), 324; **6:**6(32), 74
Schmidt, P. H., **9:**264, 267, 276, 287, 294; **12:**246(24, 33), 248(33), 252(33), 290; **13:**116(37), 117(37), 142(123), 148, 150
Schmidt, R. H., **2:**353(79), 357, 361
Schmidt, R. N., **9:**130(37), 132, 144; **10:**17(30), 18(30), 55(118), 66, 68; **12:**45, 51
Schmidt, R. V., **10:**228(5, 9), 253(48), 254(9), 258, 259
Schmidt, V., **1:**194, 229
Schmidt, W., **1:**19(41, 47), 27(47), 28(41), 43(114), 46, 47, 51(114), 64, 65, 66; **5:**6(13), 45
Schmidt-Rohr, K., **24:**151(11), 152(11), 153(11, 14)
Schmit, F., **13:**31(48), 58
Schmitt, F.-J., **24:**25(121)
Schmitt, J., **14:**99(41), 110(41), 126
Schmitt, P., **16:**79, 137
Schmitt, R. W., **1:**267, 275
Schmitz, B., **19:**142
Schmüser, P., **9:**204(124), 206(124), 207(124), 211, 258
Schnable, G. L., **12:**292
Schnackenberg, H., **2:**40, 61
Schnatterly, S. E., **9:**227(199), 228(199), 260
Schneemeyer, L. F., **16:**3, 124
Schneer, G. H., **4:**316, 324
Schneider, M., **5:**118(21), 141
Schneider, M. V., **3:**42, 59; **11:**77(159), 100
Schneider, T., **22:**134(56), 173
Schneider, T. W., **24:**27(124), 28(124), 29(124), 30(124)
Schnell, J. P., **21:**96(29), 112
Schnitzel, R., **7:**289(96), 309
Schnitzke, K., **16:**214, 230
Schnupp, P., **6:**73(219), 79
Schoder, R. B., **12:**218(110), 219(110), 235
Schoenes, J., **17:**83, 128, 131, 139, 143
Schoening, F. R. L., **4:**171(121), 186, 190
Schoer, J. K., **24:**99(103), 100(103), 106(103), 228(32)
Schofield, T. G., **16:**48, 132
Schoijet, M., **19:**252
Schölch, H. P., **16:**51, 133
Scholl, E., **23:**242(61)
Scholz, H., **7:**210(188), 252
Scholz, S. M., **19:**41
Schöneich, B., **16:**70, 136
Schonenberger, C., **24:**86(37, 38)
Schonherr, H., **24:**11(45), 21(45, 91, 102), 26(102), 88(45), 89(50), 90(45), 212(65)
Schonhorn, H., **7:**330(64), 390(210, 211), 394, 397
Schönwald, J., **9:**154(11), 256
Schoolar, R. B., **3:**299, 301; **5:**164(92), 171(92), 231; **11:**107, 108(10), 111(41), 113(52), 114(11, 55, 56, 57), 119(71), 128, 131(56), 132, 135(52), 153(124), 160, 165, 166, 186, 187, 200, 201, 202, 203
Schopper, H., **1:**70(11), 120, 224(170), 232; **6:**177, 187, 201(67), 203, 204, 314, 315, 327, 346; **7:**365(168), 396
Schossberger, F., **1:**207, 230
Schott, M., **16:**292, 334, 340

Schott, O., **2:**122(126), 123(126), 130(144), 144; **3:**39(209), 52(209), 58
Schottky, W., **4:**357(77), 371; **13:**200, 201, 223
Schowengerdt, F. D., **23:**49(37)
Schrader, G., **6:**208(15), 295
Schramm, R. J., **13:**63(27), 79(69), 105, 106
Schreiber, F., **24:**139(66)
Schreiber, G., **3:**13, 51(59), 54
Schreiber, H. U., **16:**250, 268
Schreiber, P., **7:**46(17), 112; **9:**195, 196, 197, 258
Schreiber, S. L., **24:**249(96)
Schrieffer, J. R., **1:**255(57), 256(57), 258(57, 72), 274, 275
Schrieiber, H.-U., **23:**55(185)
Schrijner, A. J., **6:**4(24), 5(24), 74
Schrijner, P., **15:**197, 217
Schroder, D. K., **12:**136(303), 165; **21:**21(44), 72
Schröder, E., **9:**195(100), 196, 197, 228(207), 258, 260
Schroder, K.-W., **22:**205(64), 273
Schroeder, A., **5:**114(19), 141
Schroeder, D. J., **9:**74, 82(6), 132(6), 143; **10:**117, 165
Schroeder, D. K., **12:**136(302), 137(302), 165; **17:**173, 222
Schroeder, H., **5:**94(8), 98(10), 101(10a), 102(13), 103(13, 14), 104(13, 14), 105(16), 114(18), 120(8), 121(8, 18), 123(26), 127(31a), 129(31b), 130(26), 132(26, 34, 35), 137(10), 139(43), 141, 313(313); **9:**74(7), 129(7), 130(7), 131(7), 322
Schroeder, H. H., **2:**258(19a), 304, 340(66), 353(66), 360; **10:**54(117), 68; **12:**3(15), 42(15), 50
Schroeder, H. Z., **1:**70(9), 84(9), 120
Schroeder, J., **17:**130, 131, 143
Schroeder, J. B., **5:**313(294), 322
Schroeder, K., **9:**273, 296
Schroeder, W. E., **11:**77(161), 100
Schroen, W., **3:**27(139, 140, 141), 28(139, 140, 141), 29(139, 141), 30(139, 141), 41(139, 140, 141), 56; **4:**298(115), 323; **6:**36(102), 76
Schubert, E. F., **21:**209(111), 236
Schubert, J., **16:**24–25, 28, 32, 121, 127–129, 143
Schuberth, G., **21:**78(4), 111
Schubin, S., **1:**131(18), 185
Schubring, N. W., **17:**229, 300
Schuegraf, K. K., **4:**257(2), 320
Schuelke, R. J., **21:**337(30), 367
Schuetze, H. J., **2:**396(149), 423
Schul, G., **7:**220(208), 253; **11:**36(3), 41(3), 47(3), 58(3), 79(3), 89(3), 96
Schuller, I. K., **16:**52, 133; **17:**24, 43, 45, 46, 48, 49, 51, 52, 53, 56, 57, 58, 74–76
Schulman, F., **24:**2(8), 3(8)
Schulman, J. H., **7:**323(37), 325, 326(40), 350, 354, 358(131), 389(40), 393, 395
Schulman, J. N., **15:**226, 261; **21:**129(41), 233
Schulte, H. J., **12:**7, 50
Schultheiss, C., **16:**23, 127
Schultz, L., **16:**55, 58, 61, 64, 72–74, 79, 107, 135–137, 141, 214, 230
Schultz, M. L., **23:**87(10)
Schultz, P. C., **12:**292
Schultz, P. G., **24:**36(147), 106(106)
Schultze-Gebhardt, F., **7:**27(101), 39

Schulz, D., **9:**186(72), 187, 257
Schulz, E., **10:**262, 308
Schulz, H., **17:**109, 110, 141
Schulz, L. G., **1:**216(129), 219(142), 231; **2:**209, 230, 237, 238, 339, 340, 360; **3:**46(226), 59; **4:**23(54), 85(223, 224), 86(229), 91, 96, 150(37, 53), 188
Schulz, M. J., **23:**101(36)
Schulz, S., **17:**357, 358, 369
Schulze, D., **17:**100, 141
Schulze, K. J., **1:**196(39), 229
Schumacher, B. W., **3:**39(210), 52(210), 58
Schumacher, E. E., **1:**5, 64
Schuppel, W., **1:**293(40), 332
Schürer, C., **12:**251(62, 63), 256(61, 62), 291
Schuster, A., **4:**55(125), 93
Schuster, K., **2:**257, 258(18), 304
Schuster, S., **3:**255, 256, 257, 258(109), 259(109), 264(109), 273
Schuster, S. E., **13:**183(118), 184(118), 196
Schvan, P., **13:**155(24), 166(24), 169(24), 178(24), 192
Schwar, M. J. R., **18:**80
Schwartz, B., **5:**147, 148(16), 196(16), 229, 254(67), 272, 297(246), 316; **12:**277(194), 294; **13:**205(19), 223
Schwartz, C. D., **4:**238(164), 255
Schwartz, D. K., **24:**163(29), 169(37), 181(13)
Schwartz, G. C., **12:**259, 263(85), 264(85), 292
Schwartz, G. P., **13:**205(19), 223
Schwartz, H., **4:**263, 321
Schwartz, J. H., **21:**61(113), 62(113), 65(113), 74
Schwartz, J. M., **2:**332(47), 360
Schwartz, L. H., **4:**228(141), 254
Schwartz, N., **2:**367(26), 368(26), 371(61), 372(61), 373(61), 374(68), 375(68), 376(68), 379(68, 85), 393(85), 397(68, 151, 152), 398(85), 406(190), 407(68, 190), 408(68, 151), 409(68, 85, 151), 420; **3:**33(163), 52(163), 57, 421, 422, 423, 424; **14:**75
Schwartz, P., **24:**44(46), 94(61), 112(10, 11), 117(11), 119(11), 139(66), 142(11), 143(11), 144(11)
Schwartz, P. V., **23:**24(92), 41(113), 55(187, 188)
Schwartz, R. F., **4:**325(2), 351(2), 369; **23:**87(7, 9)
Schwartz, R. S., **2:**413(195), 425; **4:**258(10), 319(10), 320
Schwartz, S., **7:**284, 286(83), 309
Schwartz, S. A., **23:**41(113)
Schwartz, S. S., **17:**172, 221
Schwartz, U. D., **24:**98(85)
Schwarz, H., **1:**31(75), 65; **2:**384(99), 422; **3:**5, 51(28), 53
Schwarz, J. A., **16:**249, 251–252, 268
Schwarz, R., **15:**287, 310
Schwarz, R. F., **2:**148(4), 190, 406(188), 424
Schwarz, S. A., **16:**63, 135
Schweizer, H. P., **9:**39(170, 171, 172), 40(170, 171, 172), 41(170, 171, 172); **12:**177(59, 62), 234
Schwuttke, G. H., **3:**142(54), 163; **4:**184(137), 190; **5:**191(165), 233
Sciama, D. W., **3:**277(23), 300
Scifres, D. R., **7:**234(238, 247, 248), 254
Scilla, G., **23:**24(97)

Scilla, O., **16:**17, 23, 126
Sciortino, P. F., **23:**228(31), 229(31)
Scitres, D. R., **10:**251(45), 259
Sclar, N., **21:**4(9), 7(9), 8(9), 32(9), 71
Scnable, G. L., **12:**258(77), 291
Scofield, J. H., **16:**42, 131
Scoles, C. A., **3:**33(164), 36(164), 52(164), 57
Scoles, G., **24:**45(71, 72, 73, 75), 82(11, 12), 83(11, 12), 86(11, 12, 29), 94(61), 112(6, 7, 8, 9, 10, 11, 12, 13, 14, 15), 117(7, 11, 13, 15), 119(9, 11, 13), 121(12, 14), 122(13), 123(13), 125(6, 9), 127(6), 128(15), 139(66), 142(10), 143(10), 144(10), 199(49), 206(14), 212(14), 217(68)
Scoles, G. C., **24:**44(17), 45(17)
Scoles, P., **24:**44(46)
Scott, B. A., **10:**235(11), 236(17), 258
Scott, D. M., **14:**232(123), 239
Scott, G. B., **11:**70, 100; **15:**113, 141
Scott, G. D., **1:**30(69), 65; **2:**181(62), 192, 334(55), 360; **3:**215, 270; **12:**71(68), 160; **19:**128
Scott, H. G., **3:**65(18), 125
Scott, H. L., **24:**114(28)
Scott, J., **4:**301(133), 323; **5:**290(231), 320; **19:**72
Scott, J. F., **17:**241, 253, 256, 257, 259, 287, 297–299
Scott, J. H., **5:**297, 320
Scott, L. L., **4:**50(119), 93
Scott, N. J., **4:**40(76), 92
Scott, N. W., **1:**188(6), 223(158), 228, 232; **12:**2(4), 49
Scott, Nancy J., **1:**35(88), 66
Scott, P. B., **13:**157(31), 193
Scott, W. T., **2:**74, 75(25), 81, 141
Scouler, W. J., **10:**33(60), 34(60), 66
Scovil, H. E. D., **9:**264, 265, 294
Scow, K. B., **2:**384(98), 419(98), 422; **3:**4(27), 5(27), 51(27), 53
Scribner, D. A., **21:**219(127), 236
Scrin, B., **1:**241(8, 9), 273, 275
Scrosati, B., **17:**361, 369, 370
Scurer, C., **12:**29
Scutero, M., **21:**5(1), 71
Seah, M. P., **9:**215(151), 259
Sealy, B. J., **12:**277(199), 294
Sears, G. W., **2:**2, 59; **4:**108, 120, 135, 136, 203(81), 204(82), 252; **5:**260(51), 315
Seaton, C. T., **17:**46, 76
Seavey, M. H., **1:**327, 329(103), 330(112), 334
Sebenne, C. A., **12:**114(211), 163
Secker, P. E., **14:**196
Secrist, D. R., **3:**138, 163; **4:**263(57, 58), 264, 319(57, 58), 321; **12:**260, 261, 292
Seddon, R. I., **14:**26(102), 75; **19:**195
Sedgwick, T. O., **4:**261(38), 263(38), 264(38), 321; **5:**275(137), 283, 318; **23:**24(97), 49–50(178–180)
Sedlacek, J. H. C., **13:**184(121), 196
See, Y. C., **23:**60(202)
Seeger, A., **2:**39(65), 61
Seeliger, R., **3:**63, 125
Seforest, S. E., **10:**163(95), 166
Segall, B., **19:**131, 133, 153
Seghezzi, H. D., **2:**393(140), 423
Segmüller, A., **1:**202, 229, 280(8), 290(37), 291, 292(38), 331, 332
Segmuller, A., **11:**67(130), 71, 73(130), 85(130), 99, 100; **16:**5, 64, 71–72, 78, 124, 136–137
Segmüller, A., **16:**292, 340
Segmuller, A., **17:**12, 74; **24:**228(21)
Segre, C. U., **16:**60, 135
Segré, E., **2:**66(5), 140

Segui, Y., **11:**278(97), 316
Segura, I., **16:**252, 268
Sehgal, N. K., **12:**169(40), 212(104), 213(104), 214(108), 217(40, 104), 218(111), 219(111), 229(111), 233, 235
Seidel, K. H., **14:**67(225), 78
Seidel, St., **14:**141(37), 195
Seidel, T. E., **7:**174(118b), 250; **14:**232(115), 239
Seidman, D. N., **17:**34, 43, 44, 49, 75
Seifarth, H., **17:**111, 112, 142
Seifert, H. G., **9:**219(177), 259
Seifert, K., **24:**85(24), 90(24), 99(24)
Seijka, J., **14:**104(26), 125
Seilmeier, A., **23:**254(75)
Seip, H., **7:**14(33), 38
Seiter, H., **4:**199, 200, 201, 202, 203, 204(55), 206(55), 233(55), 252
Seith, W., **6:**304(5), 307, 308(5), 314(26), 346
Seitz, F., **5:**211(198), 218, 234; **6:**73(217), 79; **12:**64(46), 159
Seitzer, D., **6:**293(225), 300
Sejourne, B., **10:**228(4), 253(4), 254(4), 258
Sekine, M., **13:**176(91), 177(91), 180(104), 185(91, 104, 135), 187(91, 104, 135), 195, 196
Sekrist, D. R., **5:**292, 320
Sekula, S. T., **16:**43, 48–49, 132
Selamoglu, N., **22:**212(73), 274
Selby, S. M., **10:**175(28), 226
Sella, C., **4:**150(43, 44), 152(43, 44), 188
Sellers, C. H., **17:**77
Sellers, H., **24:**32(131), 44(49), 64(49), 114(27), 210(46)
Sellmeier, D. J., **15:**185, 216
Sellmyer, D. J., **16:**212, 216–218, 230–231
Sello, H., **7:**265(47), 307
Selloni, A., **19:**2
Seltzer, M. S., **5:**261, 315
Selwood, P. W., **2:**31(55), 61
Selwyn, G. S., **18:**301, 305–306, 309, 310, 314
Semenov, A. S., **13:**186(154), 190(154), 197
Semenov, V. N., **12:**178(64), 234
Semenza, G., **24:**6(23), 106(109, 114)
Semiletov, S. A., **1:**218(133, 134), 231; **5:**164(98, 99), 166(116), 172, 177, 193, 232; **7:**34(146), 35(151), 40, 41
Semura, S., **13:**185(139), 188(139), 196
Sen, P. N., **19:**139
Sena, L. A., **14:**147(70), 196
Senderoff, S., **3:**222, 271
Senecal, G., **5:**164(96), 172(96), 231
Sengers, J. V., **24:**184(40), 190(40)
Senitzsky, B., **12:**60(30), 159
Senkovits, E. J., **2:**181(65), 192
Sennet, R. S., **2:**334(55), 360
Sennett, R. S., **1:**30(69), 65; **3:**215, 270; **4:**22(50), 91; **7:**352(135), 395; **12:**71(68), 160; **19:**128
Sensiper, M., **21:**117(9), 128(9), 232
Sequedo-Osorio, F., **14:**219(55), 238
Seraphim, D. P., **1:**1(1), 63, 239(7), 241(10, 11), 242(24, 25), 244(7), 245(33, 34, 36), 273, 273(104), 274; **2:**393(141), 423; **3:**132, 162, 227, 229(50), 242, 272; **4:**302(139), 306, 309, 323
Seraphin, B. O., **4:**3(8), 7(8), 8, 58(128), 90, 94; **10:**3(4), 28(48), 37(48, 69, 72, 73), 40(72, 76), 45, 46(73, 97), 47(72), 51(103), 58(4), 59(4), 62(48), 65, 66, 67, 68; **12:**42(72), 45(72), 51

Serber, **23:**177(5)
Serdyuk, A. D., **7:**264(45), 307
Sere, M., **10:**169(18), 225
Sergeeva, L. A., **3:**176(25), 209
Sergent, A. M., **11:**81(182), 101; **23:**228(31), 229(31)
Serikawa, T., **14:**12(47), 13–14(47), 14(47), 64(206), 71(234), 74, 78; **17:**111, 142
Serin, B., **6:**82(11), 143
Seriven, L. E., **7:**325(41), 393
Serre, J., **21:**13(35), 72
Serreze, H. B., **12:**176(48a), 233
Seshadri, K., **24:**172(42), 173(42), 229(40)
Sethares, J. C., **15:**6, 40, 47, 50, 64, 67, 81, 136, 138–139
Sethian, J. A., **22:**270(131), 276
Sethuraman, S., **22:**83(1), 84(1), 85(1), 92(1), 93(1), 97(1), 114
Setsune, K., **14:**20(78), 75; **16:**2, 17–19, 30, 33–34, 36, 111, 124, 126, 129–130, 142; **17:**249, 297
Setty, T. H. V., **4:**196(35), 199(64), 204(64), 206(64), 209(64), 210(64), 211(64, 91), 236, 238, 251, 252, 253, 254
Settzo, R. J., **2:**373(67), 374(67), 421
Severance, D. L., **24:**44(41)
Sevier, K., **3:**39(202), 40(202), 52(202), 58
Sevier, K. D., **9:**200(112), 258
Seward, G., **17:**347, 357, 358, 367, 369
Sewell, G., **22:**202(57), 273
Sewell, P. B., **4:**223(116), 253
Sexton, J. H., **16:**28–29, 129
Shabaik, A. H., **13:**102(116), 103(118, 119), 107
Shaeffer, C., **4:**271(90), 322
Shafeev, G. A., **13:**169(70), 185(70), 187(70), 194
Shafer, F. J., **19:**259
Shafer, M. W., **9:**11(54), 66
Shaffer, C. F., **15:**60, 139
Shaffer, L. H., **10:**14(24), 66
Shafrin, E. G., **7:**329(50, 51), 331(67), 393; **24:**2(7), 3(7)
Shah, J., **21:**193(91), 235
Shah, M. L., **10:**256(60), 259
Shah, S. I., **14:**42(150), 77; **16:**28–29, 129
Shah, S. S. P., **16:**28–29, 129
Shaha, B. M., **6:**64(181), 78
Shahidi, M., **11:**270(87, 88), 272, 316
Shaklee, K. L., **7:**254
7.(237)
Shakouri, A., **21:**290(44), 310
Shalabutov, Y. K., **9:**45(202), 69
Shalcross, F. V., **23:**124(14, 15)
Shaldervan, A. I., **22:**79
Shalev, S., **14:**177(149), 198
Shalimova, K. V., **3:**172, 174, 208; **5:**164(90), 231
Shallcross, F. V., **2:**150(23, 24, 25, 32), 153(24), 154(24), 157(40), 170, 173(40), 176(23, 58), 181(31, 58, 64), 183(24, 58), 189(67), 191, 192; **3:**171(17, 22, 40), 172, 175, 176(22), 181(40), 208, 209; **5:**164(85), 172(85), 209(85, 184), 218(85), 219(85, 200), 231, 234
Sham, L. J., **21:**97(30), 112
Shamraev, V. N., **19:**199
Shams, Q., **16:**81, 138
Shan, Z. S., **15:**185, 216
Shand, E. B., **2:**374(72), 421
Shane, S. F., **24:**12(55), 14(55)
Shanefield, D., **4:**199(71), 200(71), 209(71), 218(99), 220(99), 231(99), 252, 253
Shanfield, S., **13:**99(107), 107

Shang, T. T., **12:**277(209), 279(209), 294
Shani, A., **24:**11(41)
Shank, C. V., **10:**238(29), 251, 258; **21:**193(91), 235
Shanks, H. R., **15:**286, 310; **17:**318, 366
Shannon, J. M., **23:**102(41)
Shanthi, E., **12:**169, 176(46), 183(73), 185(30, 73), 186(26, 30, 73), 187(73), 189(29, 30), 190(30), 192(30), 196(29), 197(26, 73), 198(30, 73), 199, 200(30, 73), 228(30), 233, 234
Shao, H. H., **24:**182(22)
Shao, Z., **24:**106(110)
Shapiro, M. J., **16:**92, 139
Shapiro, S., **1:**267(92), 272(100), 275; **6:**36(103), 76
Shapkin, P. V., **13:**186(154), 190(154), 197
Shappirio, J., **14:**232(118), 239
Shaqfeh, E. S. G., **18:**71; **22:**113(31), 115, 188(37), 193(37), 272
Sharif, R. I., **11:**260, 316
Sharma, B. L., **13:**210(36), 224
Sharma, N. C., **12:**169(40, 41), 212(104), 213(104), 214, 217(40, 104), 218(107, 111), 219(111), 229(41, 107, 111), 230(107), 233, 235
Sharma, P. P., **12:**235
Sharma, R., **24:**44(22), 70(22), 169(39), 170(39), 220(84)
Sharp, D. J., **3:**138, 163
Sharp, R. W., **6:**215(43), 296
Sharpe, L. H., **7:**331(70), 394
Sharrah, P. C., **7:**29(121), 40
Shatilov, A. V., **5:**60(18)
Shatzkes, M., **6:**114(144), 132(144), 146; **16:**240, 267
Shav, J. L., **17:**304, 364
Shaw, C. C., **2:**315(9), 316(9), 318, 359; **4:**60(140), 94; **12:**6(26), 50
Shaw, C. H., **4:**50(96), 68(96), 93
Shaw, D. T., **16:**55–56, 64–65, 134, 136
Shaw, D. W., **5:**279(165), 318; **11:**76(153), 100
Shaw, J. E., **4:**60(137), 94
Shaw, J. M., **14:**235(152), 240
Shaw, T. M., **12:**238(7), 290
Shaw, W., **1:**241(22), 274
Shawabkeh, A., **17:**256, 257, 298
Shcheglov, I. M., **15:**40, 44, 138
Shcherbakova, V. Ya., **12:**204(97), 217(97), 234
She, C. Y., **13:**123(68, 69), 131(69), 132(69), 149
Shea, D. E., **5:**271, 317
Shealy, J. R., **17:**153, 154, 220
Sheard, F. W., **15:**222, 249–250, 261, 263
Shearer, J. N., **4:**55(123), 93
She'd, W., **23:**87(11)
Shedd, W. M., **23:**109(60)
Sheen, C. W., **24:**3(18), 4(18), 229(41)
Sheergar, K., **22:**55(59), 60(59), 62(59), 64(59), 78
Sheets, M. J., **11:**146(116), 149(116), 151(116), 202
Sheffield, S. M., **17:**354, 368
Shefov, A. S., **12:**71(61), 92(144), 160, 161
Sheftal, N. N., **5:**265, 273, 316
Sheftal, R. M., **6:**140(226), 141(226), 148
Shekhar, R., **24:**240(75)
Sheklein, A. V., **10:**52(109, 110), 64(145), 68, 69
Shemansky, F. A., **22:**203(62), 205(62), 273
Shen, G.-D., **23:**55(10)

Shen, J. X., **15:**185, 216
Shen, M., **12:**248(38, 40), 251(38, 40), 290
Shen, O., **17:**24, 49, 51, 53, 56, 75
Shen, V. O., **21:**103(39), 106(45), 107(39, 45), 108(45), 112, 118(14), 128(14), 133(14, 44), 137(14), 141(14), 149(14, 53), 150(14), 160(14), 161(53), 162(53), 172(14), 193(14), 209(14), 221(136), 232, 233, 234, 237, 244(21), 249(25), 261(21), 284(25), 290(45), 309, 310
Shen, Y. H., **16:**43–44, 46–47, 131
Shen, Y. R., **24:**221(99)
Sheng, K. C., **16:**43–44, 46–47, 131; **17:**7, 10, 12, 23, 27, 28, 32, 63, 64, 71, 73
Sheng, L., **16:**81, 138
Sheng, T., **22:**46(53), 49(53), 78
Sheng, T. T., **12:**265(114), 292
Sheng, Y. J., **16:**4, 124
Sheng, Z. Z., **16:**2, 9, 13, 81, 123, 125, 138
Shepard, J. W., **7:**330(59), 334(95), 394
Shepard, P. F., **21:**5(20), 6(20), 22(20), 27(20), 29(20), 45(20)
Shephead, R. G., **7:**265(49), 307
Shepherd, F. D., **21:**4(14), 31(14), 72, 196(96), 219(96, 128), 220(96), 235, 236; **23:**87(5, 6, 11), 105(51), 109(60), 144(40)
Shepherd, F. R., **14:**8(26), 36(26), 74
Shepherd, S. G., **15:**67, 139; **17:**187, 222
Shepherd, W. H., **5:**268, 316; **7:**284(85), 309; **17:**258, 298
Sheppard, C. J. R., **12:**70(56), 126(261), 128(261), 132(261), 135(245, 292), 159, 164, 165
Sheppard, E., **7:**333(82), 334(82, 91), 394
Sheppard, N., **15:**276, 310
Sher, I., **7:**349, 395
Sheridan, T. E., **22:**7(14), 77
Sheriiff, R. E., **23:**144(36, 37)
Sherman, J. C., **14:**147(71), 148(71), 196
Sherman, M. G., **23:**8(102, 103), 29(102, 103)
Sherman, R. J., **16:**42, 47, 131
Shermansky, F. A., **22:**193(43), 202(43), 205(43), 272
Sherriff, R. E., **21:**4(18), 5(2), 6(18), 7(2, 18), 8(26), 9(18), 17(2), 22(18), 27(2, 57), 28(2, 18, 26, 57), 29(26), 31(26), 32(2, 18, 57, 64), 35(2), 36(2, 26), 71, 72, 73
Sherrill, F. A., **13:**127(85), 149; **19:**252
Sherrington, D., **24:**180(10), 185(10), 186(10)
Sherwood, R. C., **1:**290, 292, 332; **9:**264, 267, 276, 287, 294
Shevchenko, O. G., **6:**124(191), 148
Shevehik, N. J., **7:**24, 26(97), 39
Shewchuk, T. J., **15:**244–245, 263
Shewmon, P. G., **16:**234, 236, 266; **22:**254(100), 275
Shi, D., **16:**4, 124
Shi, J.-X., **24:**3(18), 4(18)
Shibata, K., **10:**236(19), 258
Shibata, S., **16:**86, 138
Shibata, T., **15:**154, 156, 158, 215
Shibata, Y., **12:**296
Shibatomi, A., **15:**221–222, 260–261
Shick, L. K., **9:**276, 284, 288, 294, 296; **15:**13, 137
Shieh, H. P. D., **15:**154, 197, 204–205, 215, 217
Shifrina, R. R., **17:**318, 366

Shigeno, K., **13:**20(36), 22(36), 32(36), 57
Shigesato, Y., **17:**308, 309, 364, 365
Shih, D. K., **15:**295, 299, 311
Shih, I., **16:**54, 134
Shih, K. K., **7:**186(145), 198(172), 199(175a), 200(175b), 220(210, 211), 225(223, 230), 251, 252, 253, 254
Shih-Duan' Yin, **5:**163(79), 171(79), 231
Shikalgar, A. G., **12:**212, 235
Shiles, E., **19:**131, 133, 153, 155, 156
Shiloh, M., **12:**269(146), 293
Shilov, V. A., **14:**139(33), 195
Shilton, M. G., **17:**354, 368
Shima, R., **6:**36(113), 45(113), 76
Shimada, T., **15:**285–286, 310
Shimadzu, S., **2:**135(158), 137(158), 145
Shimakawa, Y., **16:**28–29, 129
Shimamori, T., **15:**185, 216
Shimano, A., **13:**176(86), 188(86), 189(86), 195
Shimano, T., **11:**51(76, 78), 52(78), 60(76), 76(78, 152), 98, 100
Shimanoe, T., **11:**58(100), 66(100), 99
Shimaoka, K., **1:**202(67), 229
Shimizu, H., **15:**279, 281, 310
Shimizu, I., **17:**355, 369
Shimizu, K., **5:**164(88), 172, 231
Shimizu, R., **14:**17(68), 74
Shin, J. J., **4:**120, 136, 136(51)
Shin, S. C., **15:**172, 216
Shin, S. H., **21:**40(84, 88), 74
Shinde, S. L., **16:**5, 124
Shine, C., **7:**290(99), 292(100), 294(99), 310
Shine, M. C., **16:**250, 268
Shinjo, T., **17:**2, 9, 10, 15, 22, 73, 74
Shinn, M., **17:**4, 5, 6, 13, 19, 24, 25, 49, 53, 73–76
Shinoda, M., **12:**292, 293
Shinoda, T., **15:**147, 214
Shinoda, Y., **12:**277(216), 281, 295
Shinohara, K., **16:**87, 138; **24:**119(41), 121(41)
Shinoki, F., **14:**11(43), 74; **17:**111, 142
Shinomura, Y., **7:**282(78), 309
Shinozaki, S., **4:**156(83a), 189
Shintani, Y., **14:**18(74), 19(74, 82), 20(82), 30(74), 66(220), 75, 78
Shiojiri, M., **3:**171(18, 86), 172(18), 208, 210; **12:**88(131), 96(131), 161; **17:**310, 365
Shionoya, S., **9:**22(85), 23(85), 67
Shiosaki, T., **14:**65(210), 78; **17:**151, 153, 154, 167, 219, 220, 221, 228, 232, 243, 296, 297
Shiota, I., **12:**277(200), 294
Shiota, P., **5:**291, 320
Shipp, G. S., **4:**68(165), 94
Shirai, K., **18:**91–92, 96, 159
Shirai, S., **1:**220, 231
Shiraki, Y., **11:**36, 97; **23:**10–11(38)
Shirane, G., **1:**205(76, 81), 230
Shirata, T., **19:**15
Shiri, M., **24:**163(24)
Shirikashi, J., **24:**99(104), 100(104), 101(104)
Shirley, D., **15:**177, 216
Shirley, D. A., **9:**217(165, 169), 259
Shirn, G. A., **4:**270(89), 322; **6:**54(149), 61, 77, 78
Shirose, S., **7:**191(154), 251
Shitova, E. V., **7:**36(158), 41
Shivashankar, S. A., **16:**53, 78, 133, 137
Shive, J. N., **5:**145, 229
Shizukuishi, M., **17:**355, 369
Shklyarevski, I. N., **4:**87(240), 96

Shklyarevskii, I. N., **2:**340, 360; **6:**183, 203
Shlenskii, A. A., **7:**223(217), 253
Shmerkin, I. A., **7:**216(197a), 253
Shnidman, Y., **24:**32(131), 44(49), 64(49), 114(27), 210(46)
Shockley, W., **1:**316(69), 333; **2:**149, 150(16), 161, 190; **4:**298(115), 323; **11:**140(102, 103), 142, 143, 202; **12:**64(44, 45), 159; **21:**318(10), 319(12), 366
Shockley, W. L., **3:**106, 127
Shoeing, T. T., **22:**121(20), 172
Shoemaker, D. D., **24:**228(28)
Shohata, N., **16:**27, 28–30, 57, 127, 129, 134; **17:**247, 290, 293, 299
Shokoohi, F. K., **16:**63, 135
Shombert, D. J., **5:**256(45), 266, 315, 316
Shooter, D., **3:**73(47), 100(88), 125, 126
Shor, G. I., **3:**39(208), 52(208), 58
Shore, K. A., **23:**242(61)
Short, K. T., **16:**51–53, 74, 132–134, 137
Shortt, B., **7:**240(156), 255
Shoulders, K. R., **2:**370, 411, 421; **3:**4, 5(24), 51(24), 53, 146(60), 163
Shousha, A. H. M., **6:**65(191a), 73(191a), 78
Shrednik, V. N., **1:**205(86), 230
Shrieffer, J. R., **2:**174(51), 191
Shrier, A., **4:**218(99, 103), 219(103), 220(99), 231(99), 232(103), 253
Shtrikman, H., **19:**41
Shtrikman, S., **6:**262(168a), 299
Shu, Q. S., **16:**106, 141
Shubert, R., **10:**238(28), 258
Shubin, L. D., **9:**47(224), 70; **12:**259(82), 292
Shuermeyer, F. L., **9:**45(200), 46(200), 69; **12:**259(83), 292
Shufflebotham, P. K., **15:**293, 311
Shuhara, A., **13:**134(103), 150
Shull, R. D., **16:**59, 79, 135, 137
Shul'man, S. G., **3:**20, 51(92), 55
Shultz, L. G., **10:**207(42, 43), 222(43), 226
Shum, K., **21:**330(25), 333(25), 367
Shupp, B. R., **5:**150, 155(27), 230, 277, 318
Shur, Ya. S., **6:**216(53, 55, 56), 220(53), 269(55), 276(185), 277(54, 185), 296, 299
Shurtz, R. R., **7:**154(66), 219(205), 249, 253
Shuskus, A. J., **17:**153, 220
Shuto, K., **13:**205(22), 223
Shveikin, V. I., **7:**216(197a), 253
Sibert, M. E., **12:**257, 291
Sichel, E. K., **17:**351, 367
Sickhaus, W. J., **14:**202(2), 236
Siconolfi, D. J., **15:**150, 169–170, 214
Siddall, G., **1:**224, 232; **2:**374(69), 387(69, 108), 388(69), 389(69), 403(172, 173), 404(172, 173), 405(172), 421, 422, 424; **3:**101, 110, 124, 126, 127, 132(12), 162; **4:**263(48), 267, 321, 322; **9:**43(186), 50, 51(186), 62(186), 69
Sidhu, J., **13:**154(17), 182(17), 183(17), 192
Sidhu, S. S., **1:**204, 230
Sie, C. H., **6:**292(222), 300
Sie, S. H., **13:**135(108), 136(108), 137(108), 141(122), 150; **14:**61(194), 78
Siebenmann, P. G., **6:**82(22), 144
Sieber, W., **14:**36(138), 76; **17:**111, 116, 142

Siegal, B. M., **10:**59(130), 68
Siegbahn, K., **7:**45, 48(16), 111
Siegel, B. M., **1:**189(11), 228; **3:**217(19b), 271; **4:**126, 136
Siegel, R., **10:**11(17), 13, 36(17), 65
Siegle, L., **7:**282(80), 309
Siegle, W. T., **6:**237(124a), 281(194), 282(200), 284, 285, 286(194), 298, 300
Siegrist, T., **16:**3, 124
Siejka, J., **6:**27(93), 76; **14:**86(14), 87(16), 92(26, 27), 94(92), 95(34), 98(14, 16, 26), 99(26), 100(42–44), 102(45), 103(16), 105(14, 16, 55), 106(55), 109(16), 111(63), 112(92, 97), 113(16), 117(76), 118(26), 119(64), 120(16, 76), 121(4)
Siepman, J. I., **24:**206(9, 10, 11, 12, 13), 208(29, 30, 31, 32), 209(9, 10, 11, 12, 13, 39), 210(9, 10, 43, 44), 212(9, 10, 11, 12, 13), 213(10), 214(10), 216(67), 217(12), 218(12), 219(9, 10), 221(11), 222(11)
Sieradzki, K., **17:**56, 58, 76
Sievers, A. J., **10:**60(138), 69
Sievers, A J., **16:**62, 135
Sievers, R. E., **16:**84, 90, 138
Siffert, P., **16:**76, 78, 137
Sigai, A. G., **7:**171(113, 114), 193(163), 250, 252
Sigmund, P., **14:**6(19), 8(29), 24(19), 69(19), 73, 74
Sigrist, H., **24:**18(88), 20(88), 106(111)
Sigsbee, R. A., **4:**114, 122, 127(33a), 135; **9:**43(194), 69
Sigurd, S., **14:**230(100), 239
Sihvonen, Y. T., **9:**51, 71
Siitari, D. W., **15:**168, 215
Sikina, T. V., **2:**379(81, 82), 397(81, 82), 421, 422; **3:**108(108), 127
Silberzan, P. L., **24:**169(35, 37)
Silbey, R., **24:**73(118, 119)
Silcox, J., **2:**14, 60; **9:**180(59), 202(117), 212(117), 257, 258
Silcox, N. W., **2:**397(153), 406(153), 408(191), 424; **3:**107(102), 127; **6:**54, 69(151), 70, 77
Silin, V. P., **9:**192(87), 257
Silker, T. R., **17:**197, 222
Sill, L. R., **17:**49, 51, 52, 76
Sillion, J. F., Jr., **10:**256(62), 259
Sils, V., **5:**191(165), 233, 257(47), 315
Silsbee, R. H., **3:**66, 125
Silva, L. F., **6:**230(110), 231(110, 114, 115), 232, 235(114, 115), 237(110, 114), 297
Silver, M., **2:**315(16), 334(16), 336(16), 359; **10:**130(69), 132(69), 165; **12:**8(33), 50
Silver, M. D., **4:**22, 23, 91
Silver, R. M., **16:**17, 52, 74, 126, 133
Silverman, B. D., **19:**253
Silverman, J., **21:**21(50), 22(50), 73, 219(128), 236; **23:**91(25), 92(25), 135(26)
Silverman, S. J., **3:**276, 277, 300
Silverman, S. L., **4:**274(97), 322
Silversmith, D. J., **13:**169(69), 184(121), 185(69), 187(69), 194, 196; **14:**232(119), 239
Silverster, N. R., **2:**92, 97, 141, 142
Silvestri, V. J., **5:**274, 276, 277, 279, 317, 318
Silvey, G. A., **4:**270(87), 314(87), 322; **12:**265(111), 292
Simhony, M., **5:**221(203), 234
Simmonds, P. E., **15:**249–250, 263
Simmons, G. W., **4:**223(115), 253; **24:**169(38)

Simmons, J. G., **2:**367(25), 393(25, 144), 397(144), 411(144), 420, 423; **11:**218(15), 314; **21:**5(7), 32(7), 33(7), 41(7), 71
Simnad, M. T., **1:**210(114), 231; **2:**2(2), 59; **4:**99(16), 103(16), 122(16), 123(59), 127(59), 134, 136
Simon, A., **3:**69, 125
Simon, E. S., **24:**10(36), 12(36), 14(36)
Simon, I., **1:**143, 185; **12:**5(19), 50
Simon, R. E., **7:**244(271), 255; **12:**60(33), 89(139), 113(186, 188), 135(289), 137(305), 157(336), 159, 161, 162, 165, 166
Simon, R. W., **16:**106, 119, 121, 141, 143
Simondet, F., **19:**31
Simonetti, E., **14:**220(70), 238
Simons, R. N., **16:**98, 140
Simonsohn, G., **3:**31(152), 32(152), 51(152), 57
Simpson, J., **5:**299(252), 321
Simpson, J. A., **2:**101(69, 70), 103(70), 104(70), 105(70), 106(70), 142; **7:**16(43), 38
Simpson, N. H., **4:**210(89), 253
Simpson, O., **3:**293, 300; **11:**106(2), 108(2), 199
Simpson, T. R. E., **24:**21(99), 24(99), 25(99)
Simpson, W. I., **5:**151, 153(32), 230, 269(97, 98), 281, 316–318; **11:**108(20), 121(20), 200; **12:**129(273), 165
Sinai, A. K., **22:**121(20), 172
Sinclair, G., **6:**335(39), 347
Sinclair, W. R., **2:**369(40), 390(118), 420, 423; **3:**100, 112(123), 126, 127, 139(44), 163; **4:**267, 322; **5:**310, 322; **9:**51, 70
Sinfelt, J. H., **24:**154(18)
Singer, B., **17:**227, 295
Singer, D. M., **24:**182(22)
Singer, S. J., **7:**333, 391(77), 394
Singh, A., **15:**288, 290, 311
Singh, A. D., **6:**137(222), 143, 148
Singh, A. K., **16:**53, 65, 134, 136
Singh, H. P., **9:**5(20), 65
Singh, J., **13:**143, 150
Singh, N. B., **17:**154, 155, 182, 222
Singh, P., **16:**256, 269
Singh, R., **14:**232(118), 239; **16:**10, 48, 57, 60, 91, 125, 134, 139
Singh, R. K., **16:**65, 91, 136, 139
Singh, S., **17:**355, 368; **24:**151(8), 154(8), 155(8b), 156(8), 157(8b), 158(8b), 159(8b), 161(8b), 222(103)
Singh, V. K., **22:**188(37), 193(37), 272
Singh, V. P., **12:**224(115), 235
Singhal, S., **16:**98, 140
Singhvi, R., **24:**228(6), 229(6), 249(6)
Singleton, D. E., **17:**359, 361, 369
Sinha, A. K., **12:**266, 268, 276, 277(206), 279(206), 293, 294; **16:**250, 268
Sinha, A. P. B., **12:**235
Sinha, K. P., **7:**225(227), 253
Sinha, S., **16:**91, 139
Sinha, S. K., **24:**182(23), 184(40), 190(40), 199(23)
Sinharoy, S., **17:**164, 221, 229, 236, 237, 270, 271, 272, 275, 276, 277, 278, 280, 281, 282, 283, 284, 285, 286, 290, 292, 294, 296, 298–300
Sinniah, K., **23:**29(102, 103), 38(102, 103); **24:**44(40)
Sinsheimer, R. L., **4:**50(109), 93

Siow, K.-S., **24:**228(9)
Siperko, L. M., **24:**12(48), 98(82), 101(82), 105(82)
Sirota, E. B., **24:**182(22, 23), 184(40), 190(40), 199(23)
Sirota, N. W., **7:**180(128), 251
Sirotenko, I. G., **3:**117(141), 128
Sirtori, C., **23:**239(46), 240(54), 243(65), 262(65, 80, 83, 84, 85, 87), 263(65), 265(85), 266(46), 268(46), 269(92), 271(54), 272(54, 83, 96), 273(86), 274(87), 279(83), 291(80), 298(80)
Sita, L. R., **24:**33(136), 128(57)
Sitarik, J., **6:**4, 74
Sites, J. R., **13:**118(46, 47), 123(69), 131(46, 47, 69, 99, 100, 101), 132, 133(99), 134, 135(99), 139(47, 99, 100, 101, 117), 140(47, 99, 100, 101, 140), 148, 149, 150; **14:**24(97), 75
Sith, J., **16:**52, 133
Siu-Chung Ho, **7:**263(29), 306
Sivco, D. L., **21:**196(94, 95), 207(105), 235; **23:**239(46), 240(54), 243(65), 262(65, 80, 83, 84, 85, 87), 263(65), 265(85), 266(46), 268(46), 269(92), 271(54), 272(54, 83, 96), 273(86), 274(87), 279(83), 291(80), 298(80)
Siveo, D. L., **15:**229–230, 262
Sivilotti, M. A., **21:**60(112), 65(112), 74
Sivukhim, D. V., **7:**365(169), 396
Sizelove, J., **5:**1(8), 45
Sizemore, J. T., **16:**53–54, 134
Sjöstrand, F. J., **2:**109, 143
Skaiman, J. S., **12:**168(4), 169(4), 176(4), 183(4), 190(4), 232
Skalski, J. F., **11:**189(151), 203
Skarman, J. S., **3:**197(62), 209; **5:**225(213), 235; **12:**169(9, 10), 194(9, 11), 232
Skeath, P., **12:**128(268), 164; **13:**201(4), 202(4), 203(16), 207, 208, 223
Skeath, P. R., **12:**115(212), 116(212), 117(212), 163
Skelt, E. R., **6:**45(145), 48(145), 77; **12:**282(223), 295
Skelton, E. F., **15:**157, 215; **16:**53, 134
Skerlavaj, A., **17:**131, 143
Skerritt, J. W., **2:**379(83), 392(83), 422
Skibowski, M., **9:**186(69), 187(69), 188(77), 257
Skinner, D., **17:**260, 263, 298
Sklar, M., **4:**260(25), 301(25), 320
Skocpol, W. J., **16:**42, 74, 131, 137
Skoda, R. E., **15:**149, 154, 165, 168, 171, 214
Skoglund, D. L., **16:**81, 102–105, 121, 138
Skolnick, L. P., **6:**36(112), 40, 45(142), 46, 76, 77; **7:**102(76, 77), 113
Skolnik, M. I., **15:**61, 139
Skove, M. J., **6:**103(98, 99), 122, 132(98, 99), 145
Skovlin, D. O., **12:**203(96), 234
Slack, L. H., **9:**33(116a, 135), 35(135), 37(135), 67, 68; **12:**177(58), 180(66), 184(66), 187(66), 190(66), 234
Slade, J. J., **4:**218(102), 253
Slade, M. L., **14:**234(145), 240
Slaoui, A., **14:**98(94–96), 101(94, 95), 113(94), 114(96), 127, 128
Slater, J. C., **2:**108, 142; **3:**294, 300
Slavin, W., **4:**91, 92
Slay, B. G., **3:**151(70), 164
Sleator, T., **24:**6(24), 86(25)
Slegesky, W. G., **2:**369(39), 393(39), 420

Slichter, C. P., **24:**154(18)
Sliker, T. R., **17:**198, 223
Slippy, Jr., W. A., **16:**245, 267
Sloan, D. J., **2:**393(125), 405, 406(125), 407(125), 423
Sloane, D. J., **3:**108(104), 127
Sloape, B. W., **5:**163(, 62, 63), 168(62, 63), 199, 209(63), 211, 212, 214(63), 215, 230, 234, 260, 315
Slobodin, D., **15:**287, 310
Slonczewski, J. C., **6:**231(113), 232, 235(113, 116), 280, 281(191), 297, 299; **9:**269, 271, 296
Sloneczewski, J. C., **1:**297(47a), 332
Sloope, B. W., **4:**131(71, 72), 136, 169, 171(105), 190
Slowter, E. E., **5:**306, 321
Smakula, A., **2:**240, 303
Small, D. M., **24:**141(67), 182(22)
Small, M. B., **7:**120(6), 247
Smallman, R. E., **7:**286(89), 309
Smatko, V., **16:**52, 133
Smay, T. A., **6:**293(223), 300
Smelt, J. M., **14:**21(85), 75
Smentkowski, V. S., **24:**3(19)
Smid, V., **12:**296
Smiley, V. N., **10:**256(63), 257(63), 259
Smilgys, R. V., **13:**190(162), 191(162), 197
Smirnov, A. P., **1:**244(27, 28, 29), 274
Smirnov, V. A., **12:**88(130), 161
Smit, B., **24:**207(20), 208(20, 33), 210(43, 44)
Smith, A., **1:**134, 185; **2:**418(196), 425; **4:**60(131), 94
Smith, A. B., **9:**270, 296
Smith, A. E., **2:**48(78), 50(78), 51(78), 52(78), 62; **3:**219(24), 271
Smith, A. M., **5:**255(40), 315
Smith, B. A., **7:**200(175c), 252
Smith, C., **23:**2(3)
Smith, C. G., **19:**280
Smith, Charles S., **3:**224(42), 271
Smith, Cyril S., **3:**253(91), 255(91), 257, 258(91), 259, 273
Smith, D., **19:**237–240
Smith, D. A., **17:**43, 47, 75, 121, 143
Smith, D. H., **9:**264, 267, 275, 276, 287, 294, 295, 297
Smith, D. J., **13:**141(121), 142(128), 143(128), 150
Smith, D. L., **11:**36, 59, 63, 96, 99, 109, 116(30), 119(30), 200; **12:**118(225), 131(225), 163; **13:**202(12), 223
Smith, D. O., **1:**280(7), 283, 285(24), 286(29), 289(33), 304, 306(58), 312(64), 321, 327, 331, 332, 333, 334; **3:**214, 228, 229(51), 230, 234, 247, 250(87), 255(87), 270, 272, 273; **6:**216(57), 235(119), 270, 272(57), 296, 298
Smith, D. Y., **19:**131, 133, 153, 155
Smith, E. E., **2:**369(36, 37), 420; **3:**112, 127; **4:**267(72), 322
Smith, E. L., **24:**44(31, 34, 35), 83(20), 90(20), 112(5)
Smith, F. D., **1:**98(42), 121
Smith, F. W., **12:**256(59, 60), 291; **19:**73; **23:**45–46(156)
Smith, G., **23:**102(42)
Smith, G. B., **13:**145, 150; **17:**318, 356, 365, 369
Smith, G. E., **6:**91(65), 133(65), 140(228), 144, 149
Smith, G. H., **2:**65(2), 82, 85(2), 86, 87(38), 88(38), 89, 91, 92, 93, 97, 140, 141
Smith, G. P., **7:**29(121), 40; **14:**140(35), 195
Smith, G. S., **24:**181(13)

Smith, H. A., **7:**330(58), 394
Smith, H. I., **13:**163(48, 50, 51), 182(50, 51), 183(50), 193; **17:**170, 221
Smith, H. M., **4:**9(27), 91; **12:**143(314), 144, 166
Smith, J., **13:**126(78), 149
Smith, J. P., **5:**167(136), 182(136), 233
Smith, J. R., **17:**56, 76
Smith, J. S., **15:**229, 261; **21:**116(2), 155(60), 232, 234
Smith, Jr., C. V., **15:**47–48, 48, 51–52, 64, 67, 77, 81–83, 138–140
Smith, L. E., **21:**221(137), 222(137), 237
Smith, L. W., **2:**247, 304; **10:**237, 258
Smith, M., **5:**283, 319
Smith, N. V., **9:**217(161), 259; **12:**62, 159
Smith, P., **9:**77(13), 84(13), 143
Smith, P. H., **1:**105, 121, 267(92), 272(100), 275; **2:**370(43), 420; **6:**36, 37(108), 38, 39(108), 45, 76
Smith, P. M., **16:**240, 251, 267, 268
Smith, P. S., **15:**67, 139; **17:**187, 222
Smith, R., **22:**113(30), 115
Smith, R. A., **3:**278, 300; **4:**50, 93(111, 112, 113, 114, 115), 328, 369; **11:**135(97), 202; **12:**67(51), 159; **21:**21(46), 72
Smith, R. C., **9:**47(224), 70; **12:**259(82), 292
Smith, R. S., **1:**58, 67, 280(4), 315(66), 316(68), 323, 326, 331, 333; **2:**30(50), 61; **3:**246, 272; **4:**219(104), 228(104), 253
Smith, R. T., **13:**187(125), 196
Smith, R. W., **2:**150(29), 191
Smith, S. D., **2:**297(44), 304; **3:**235(60), 236, 246, 272; **5:**60(15), 75, 76, 77, 81, 86
Smith, S. P., **24:**44(62), 245(83), 246(83)
Smith, T., **1:**83(36), 121; **24:**183(27)
Smith, T. E., **12:**263(98), 264(98), 266(138), 268(138), 292, 293; **13:**99(104), 107
Smith, W., **3:**275, 299
Smith, W. C., **17:**358, 369
Smithells, C. J., **2:**380(86), 422
Smithey, R., **16:**23, 127
Smits, J. G., **17:**153, 220
Smolenskii, G. A., **15:**140
Smolin, M. D., **7:**264(39), 307
Smolinsky, G., **10:**241(31), 258
Smoller, J., **22:**89(12), 92(12), 93(12), 115
Smoluchowski, R., **1:**220(148), 231
Smooha, Y., **16:**255, 269
Smullin, L. D., **18:**65
Smy, T., **22:**3(63), 7(48), 9(19), 10(19), 11(20, 21), 12(20), 15(23), 17(25), 18(23, 25, 26), 20(19), 22(31), 23(31–33), 24(20, 33), 25(33, 34), 26(33), 28(33), 30(37, 38, 48), 32(46, 48), 34(26), 35(26), 37(46), 38(46), 39(49), 40(48, 49), 41(49), 42(48), 44(49), 45(25, 26, 48, 49, 51, 52), 46(25, 53), 47(25), 49(53), 50(19, 56), 52(56), 53(56), 54(56, 58), 55(59), 56(56), 59(48), 60(59, 60), 61(61, 63), 62(59), 64(59), 66(61), 67(63), 68(61), 70(63), 72(63), 76(67–75), 77–79, 123(27), 172, 232(85), 274
Smy, T. J., **22:**1, 67(64), 74(64, 66), 75(64), 76(64, 66), 78
Smyth, C. P., **24:**48(98)
Smyth, D. M., **6:**54, 61, 77, 78; **17:**259, 298
Snaper, A., **13:**79, 106

Snaper, A. A., **14:**130(1, 2), 195
Sneddon, I. N., **7:**6(18), 37; **16:**279, 337
Sneed, R. J., **3:**103(175), 115, 129; **5:**167(128), 181, 193, 196(128), 197, 198, 199, 232
Snider, C. B., **7:**45, 103(13), 111
Snortland, H. J., **16:**12, 101, 125
Snouse, T. W., **3:**74(52), 75(52), 126
Snow, A. W., **17:**210, 212, 223
Snow, E. H., **4:**260(25), 292, 295(113), 301(25, 113, 124), 302, 304, 306, 311; **11:**255; **17:**274, 298
Snowden, D. P., **3:**295, 296, 301
Snyder, N. W., **2:**326, 359
Snyder, P. G., **19:**280
Snyder, W. L., **7:**142(40), 248
So, S., **19:**288
Soares, A. M., **13:**187(126), 196
Sobajima, S., **9:**47(222a), 70
Soboleva, N. A., **12:**71(61), 160
Sobolewski, R., **16:**5, 97, 99, 107, 109, 124, 140–142
Sobotka, H., **7:**315(20, 22), 330(60), 342, 348(112), 349(123, 124), 350(12, 128), 393, 394, 395
Socei, R., **7:**240(256), 255
Soffer, S. B., **6:**86, 89, 90(58), 144
Sofianek, J. C., **14:**196
Sohler, W., **9:**241(237), 261
Sohno, K., **13:**99(103), 107
Sokoewska, A., **12:**265(125), 293
Sokoewski, M., **12:**265(125), 293
Sokolov, J., **24:**94(72), 95(72), 169(39), 170(39)
Sokolova, T. P., **12:**169(39, 41), 201(39), 229(41), 233
Sokolowski, M., **12:**242(16, 17), 243(16, 17), 244(16), 265, 290, 292
Sokolwski, A., **12:**242(17), 243(17), 290
Sokowska, A., **12:**265(109), 292
Solanki, R., **17:**153, 220
Sole, M., **6:**21, 27(72), 75
Solie, L. P., **17:**170, 171, 221
Sollien, V., **6:**92, 145
Sollner, T. C. L. G., **15:**244–246, 248–251, 262–263; **23:**243(63)
Solomko, A. A., **15:**115, 118, 141
Solomon, A. L., **3:**139(41), 163; **17:**235, 296
Solomon, I., **19:**85, 93
Solomon, P. M., **21:**193(92), 235
Solomon, R., **7:**236, 241(251), 254
Solt, I., **15:**26, 137
Soltner, H., **16:**5, 28, 124, 128
Somekh, R. E., **17:**22, 74; **22:**8(16), 77, 148(73), 173
Somekh, S., **10:**228(6), 248(40), 250, 251(6), 258, 259; **11:**66(121), 79(171), 99, 101
Someya, T., **2:**122(127), 144
Sommer, A., **5:**287(192), 319
Sommer, A. H., **7:**242(259), 244(269), 255; **12:**54(5, 6, 15), 60(15), 61, 67(15), 71(62, 66), 73, 74(62, 66), 75(86), 76(66), 79, 80, 81, 82(66), 84, 85(66), 86, 87(117), 88, 92(66, 142, 143), 93, 94, 96, 99(162), 100(166), 101, 102, 103, 105(106), 113
Sommer, H., **1:**42(111), 43(113), 66
Sommer, W., **16:**53, 134
Sommerfeld, A., **1:**234, 273; **4:**351, 371
Sommerhalder, R., **1:**248(46), 251(46), 274, 292(38), 329(108), 332, 334
Sommermeyer, K., **3:**63, 125
Sommers, Jr., H. S., **7:**218(202), 253

Somoano, R., **15:**236, 259, 262
Somorjai, C. A., **3:**172, 208
Somorjai, G. A., **24:**106(109), 114(32), 144(68)
Sondag-Huethorst, J. A. M., **24:**21(105), 26(105), 27(105), 28(105), 86(37, 38)
Sondericker, J. H., Jr., **1:**204(75), 230
Sondheimer, E. H., **3:**19, 51(85), 55, 109(113), 127; **6:**82(20, 31), 86, 91, 101(20, 43), 143, 144; **9:**7(33), 8(33), 65
Sondheimer, F. H., **2:**381(88), 422
Sone, A., **24:**45(82)
Song, L., **22:**269(118–120), 275
Song, Q. X., **16:**36, 130
Song, S. N., **16:**43–44, 46–47, 131
Sonnenberg, H., **12:**113(189), 118(233), 162, 163
Sonntag, B., **7:**46(17), 48(25), 49(28), 112
Sonomura, H., **7:**171(116), 250
Soonpaa, H. H., **4:**3(6), 90; **6:**142, 149; **19:**169, 180, 281, 291
Sooryakumar, R., **21:**155(61), 234
Sopher, R. P., **7:**265(49), 307
Soprykin, L. I., **14:**177(146), 198
Sorbello, R. S., **16:**239, 267
Sorokin, N. M., **16:**71, 79, 136
Sorokin, V. K., **5:**164(95), 231
Sorrows, H. E., **2:**175(55), 192
Soshea, R. W., **4:**326(6), 346, 347(59), 349, 354(59), 367, 369, 370
Sosniak, J., **9:**50(228), 70; **14:**219(49), 232(66), 238
Sosnowski, L., **11:**106(2), 108(2), 199
Souckov, L., **5:**160, 230
Souda, R., **19:**35
Souers, P. C., **2:**384(92), 422
Soulen, R. J., Jr., **16:**53, 134
Southern, A. L., **3:**73(43), 125
Sowell, R. R., **10:**28(49), 44(92), 45(92), 48, 49(99), 50(99), 63(92, 99, 144), 64(92), 66, 67, 68, 69
Soxman, E. J., **14:**39(146), 76
Soznowski, L., **3:**294(60), 300
Spaepen, F., **15:**159, 215; **17:**35, 37, 39, 42, 63, 66, 75, 77
Spah, R. J., **16:**52, 133
Spain, R. J., **6:**225(89), 227(94), 228(103), 236, 238(103), 297
Sparks, C. J., **4:**40, 92, 144(27), 187, 230, 254; **16:**3, 124
Sparks, M., **4:**329, 369; **6:**273(178), 299; **15:**8, 136
Sparrow, E. M., **10:**57(124), 68
Spear, K. E., **17:**110, 111, 119, 141
Spear, W. E., **11:**256(60), 257(60), 315; **12:**272(164), 273, 293, 296
Spears, K. G., **18:**295–296, 298
Spears, R. A., **3:**24(99), 51(99), 55
Specchiulli, G., **16:**252, 268
Specht, E. D., **16:**3, 124
Spee, A. I. M. A., **16:**82–84, 138
Speidell, I. L., **12:**282(237), 283(237), 295
Speliotis, D. C., **1:**286(27), 287(27), 289(27), 332
Spence, W., **9:**47(219), 48(219), 70
Spencer, A. G., **17:**111, 116, 142; **18:**275
Spencer, E. G., **9:**28(103), 67, 264, 267, 276, 287, 294; **12:**246(33), 248, 252(33), 290; **13:**116(37), 117(37), 142, 148, 150
Spencer, H. E., **3:**281, 285, 300
Spencer, L. V., **2:**69, 71, 128, 141
Spencer, N. D., **24:**21(98), 22(98)
Spencer, R. A., **22:**243(92), 274
Speriosu, V. S., **17:**22, 74
Spiccia, L., **17:**318, 365
Spicer, W., **5:**1(2), 9(2), 11(17), 44, 45

Spicer, W. E., **4:**329, 344(55), 369, 370; **7:**243(265), 255; **9:**217(161), 227(203), 228(203), 259, 260; **10:**41(78), 67; **12:**54(15), 55(16), 59, 60(15, 33, 38, 43), 61, 65(23), 67(15), 69(16), 71(62), 74(62, 84, 85), 75(84), 78(85), 81, 82, 90(16), 91(16, 138), 92(143), 95(16), 96, 97, 101(16), 102, 103, 105; **13:**201(4), 202, 203(16), 207, 208, 223; **23:**91(21)

Spierings, G. A. C. M., **17:**228, 240, 256, 257, 297, 298

Spiess, H., **24:**151(11), 152(11), 153(11, 14)

Spiller, E., **10:**115, 117, 165, 168(7), 225; **13:**124(74), 126(74), 149; **17:**2, 34, 75

Spiller, G. D. T., **17:**30, 73; **19:**70

Spiller, M., **5:**103(14), 104(14), 141

Spindler, R., **17:**361, 370

Spink, J. A., **7:**325, 326, 389(40), 393

Spirito, P., **14:**232(129), 239

Spitsyn, B. V., **19:**104

Spitz, J., **9:**39(172a), 69, 282, 294; **12:**231(126), 235

Spitzer, S. M., **7:**284, 286(83), 309

Spitzer, W., **21:**21(45), 72

Spitzer, W. G., **4:**341(50, 51), 342, 343(51), 345, 346(58), 347, 349(51), 354(51), 367(51), 370; **5:**222(208), 234; **7:**146(54), 248; **12:**84(99), 160; **13:**201(5), 202, 203(5), 219(66), 223; **23:**88(16)

Spitzmüller, J., **23:**32(119)

Spivak, A. I., **3:**127

Spivak, G. V., **3:**102, 117, 118(142), 128

Spivak, J. F., **5:**184, 233

Spizzichino, A., **10:**168(13), 173(13), 176, 225

Sprague, J. L., **4:**323

Spratt, J. P., **2:**148(4), 190; **4:**325, 351(2), 369; **5:**164(87), 231; **23:**87(7, 9)

Sprecht, H., **7:**27(104), 39

Spriggs, R. S., **5:**284(196), 319

Spriggs, S., **6:**113(135), 146

Sprik, M., **24:**208(31), 211(50, 51)

Springer, K. N., **16:**42, 47, 131

Springer, R. W., **17:**4, 64, 65, 66, 73, 77

Springthorpe, A. J., **7:**126, 228, 229(16), 247

Sproul, W. D., **13:**67(35, 36), 105; **14:**6(20), 55(20, 170), 59(186), 73, 77; **17:**44, 75, 110, 116, 141, 142; **18:**269, 271, 278

Sproule, I., **16:**54, 134

Sputtered Films, Inc., **15:**152, 214

Spyridelis, J., **6:**12, 75

Squillante, M. R., **12:**176(48a), 233

Sreenivas, K., **17:**163, 221, 228, 232, 247, 296, 297; **19:**250

Sreseli, O. M., **13:**162(43), 163(49), 164(43), 165(43), 183(49), 193

Srinivasan, G., **15:**95, 98, 107–111, 140–141

Srinivasan, G. R., **22:**127(33), 172

Srinivasan, M., **12:**71(64), 72(64), 75(89), 148(320), 160, 166

Srinivasan, R., **13:**153(2, 3, 4, 10, 12, 13, 15), 154(3, 15, 18, 19), 170(71), 171(71), 172, 181(13), 182(2, 3, 4, 10, 12, 13, 15, 18, 71), 183(2, 4, 10, 12, 13, 16, 18, 19, 71), 192, 194; **14:**122–123(87), 127; **16:**55, 134

Srivastava, N. C., **15:**34, 137

Srivastava, R. D., **4:**211(90), 253

Srivastava, V. K., **7:**337, 338, 349, 353(106, 107), 354, 356, 357, 358(107), 361(151), 362(152, 153), 364(165, 170, 171), 371(178, 179, 180), 372(178, 180, 184), 373(106, 188, 189, 190, 191, 192), 374(188, 189), 376(189), 377(190), 378(191, 192), 395, 396, 397
Srivinas, D., **22:**183(28), 185(28), 232(28), 272
Sroka, R., **4:**199(52), 200(52), 204(52), 252
Ssuschkin, N. G., **2:**94(45), 141
Stach, J., **5:**303(262), 321
Stacy, W. T., **9:**288, 296; **15:**14, 137
Stafsudd, O. M., **13:**79(81), 95(90), 106
Stahl, H. A., **1:**202(69), 223, 229, 231
Stall, R., **13:**218(58), 219(64), 224
Stall, R. A., **11:**51(79), 54(79, 90), 56(90), 76(156, 156a), 88(156, 156a, 225), 98, 99, 100, 102
Stals, L. M., **17:**110, 141
Stamper, A. K., **16:**26, 127
Stancil, D. D., **15:**85, 105–106, 113, 115–117, 140–141; **16:**220, 231
Standley, C. L., **3:**108(103), 127; **14:**221(64), 222(64), 238
Standley, R. D., **10:**238, 258
Stanford, J. L., **9:**227(201, 206), 228(208), 260
Stanko, W. S., **12:**246(27), 290
Stankoff, A., **6:**274(184), 299
Stanley, C. R., **11:**67(133a), 100
Stapelbroek, M. G., **21:**4(13), 5(4), 18(4, 13), 71, 72
Staples, E. J., **17:**162, 211, 221, 223, 224, 261, 298
Staples, J. L., **13:**210(32), 212(32), 224
Stark, F., **3:**62, 125
Stark, J., **14:**196
Starke, A., **19:**195, 238, 239
Starkiewicz, J., **11:**106(2), 108(2), 199
Starkiewiez, J., **7:**210(191), 252
Starobinets, S. S., **15:**95, 140
Starostin, Yu. V., **6:**292(220), 300
Stasyuk, Z. V., **6:**116(161), 147
Statsenko, V. M., **7:**264(39), 307
Staudigl, G., **15:**299, 301–302, 311
Stavish, T., **5:**247(30), 315
Stavitskaia, T. S., **13:**32(49), 58
Stayt, J. W., **21:**117(7), 128(7), 221(131, 139), 222(131, 139), 232, 236, 237
Stayt, J. W., Jr., **21:**221(134, 135, 137), 222(135, 137), 236, 237
Stchakovsky, M., **19:**8
Stearns, F. S., **15:**64, 139
Steckelmacher, W., **1:**33(82), 65
Steeds, J. W., **22:**113(28), 115
Steekelmacher, W., **3:**11, 12(55, 56), 13(55, 56), 19(117), 24(117), 26(117), 43, 44(219), 49(55, 56), 51(55, 56, 117), 54, 56, 59
Steel, D. G., **16:**28, 33, 116, 128, 142
Steele, A. G., **21:**101(37), 112, 128(35), 179(79), 209(35), 233, 235
Steele, G. N., **7:**49, 50, 51(31), 52(35), 54(31), 56(31, 35), 59(31), 64(35), 74, 77(31), 80(31, 35), 81, 82, 84, 85(31), 86(31, 35), 87, 89(31), 90, 98, 111(35), 112
Steele, J., **23:**60(202)
Steele, S. R., **4:**349, 354, 367, 370; **5:**299(251, 252), 300, 321; **9:**31(113); **12:**258(76), 259(76), 265(76), 271(76), 272(76), 291
Steele, U., **7:**24(90), 39
Steenbeck, K., **17:**111, 112, 142
Steenbeek, K., **14:**11(40), 74

Stefan, V., **9:**33(137), 68; **12:**187(77), 234
Steffen, R., **7:**20(65), 38
Stegeman, G. I., **17:**46, 49, 51, 53, 76
Steigmeyer, E. F., **7:**174(118b), 250
Stein, A. M., **5:**266, 267, 316
Stein, B. F., **9:**282, 296
Stein, C., **23:**60(202)
Stein, G. D., **13:**8(15), 12(21), 13(22), 57
Stein, H. J., **12:**269(144), 293
Stein, K. U., **6:**239, 240(135, 136), 249, 260(165), 292(156, 216), 298, 299, 300
Stein, M., **13:**124, 142(126), 149, 150
Stein, S. J., **2:**386(101, 102, 103), 389(101, 102, 103), 422
Stein, S. M., **24:**12(55), 14(55)
Steinbeiss, E., **14:**11(40), 74; **17:**111, 112, 142
Steinberg, A. R., **18:**60
Steinberg, G. N., **18:**60
Steinberg, H. L., **3:**16(72), 51(72), 54; **4:**83(206), 84(206), 95
Steinberg, S., **24:**169(37)
Steinberger, I. T., **2:**330, 360
Steinbrüchel, C., **18:**312
Steinemann, S., **4:**238, 255
Steinfeld, J. I., **13:**190(155), 191(155), 197
Steinfelder, K., **13:**67(32), 105; **14:**11(39), 36(138), 44(39), 48(162), 55(162, 173), 74; **17:**111, 116, 142
Steinmaier, W., **3:**134(21), 137(21), 162; **4:**263(53), 264, 319(53), 321
Steinmann, P. A., **17:**130, 143
Steinmann, W., **6:**195, 203; **9:**176, 178(37), 179(37, 52, 53), 186(69); **10:**76, 77(10), 164; **12:**8(34), 50
Steinmeier, W., **5:**266, 272, 288, 316, 320
Stem, F., **11:**142(112), 202
Stem, L. A., **13:**182(116), 183(116), 196
Stemme, O., **1:**293(40), 332
Stemple, N. R., **2:**393(140), 423
Stencel, J. M., **17:**307, 364
Stenhagen, E., **7:**354, 358(141), 395
Stephanopolous, G. N., **24:**228(6), 229(6), 249(6)
Stephans, R. E., **19:**217
Stephen, J. T., **17:**291, 299
Stephen, R. E., **12:**51
Stephens, D. W., **7:**312, 313(12), 393
Stephens, P. W., **24:**151(10), 156(10), 163(10), 222(104)
Stepke, E., **17:**229, 280, 281, 282, 283, 285, 286, 290, 296, 299
Stepper, H. J., **7:**264(34), 307
Sterling, H. F., **3:**138(36), 163; **4:**270(86), 322; **5:**292, 297, 320; **12:**261, 262(92), 266, 292, 293
Stermer, R. L., **21:**5(4), 18(4), 71
Stern, E., **14:**38(143), 76
Stern, E. A., **6:**193(50), 195(55), 198, 199, 203; **7:**27, 31(133), 32(100), 39, 40; **9:**146(2), 150(2), 153(8), 174(36), 182(62), 186(62), 210(133), 220(178), 221, 225, 245(248), 247, 255, 256, 257, 258, 259, 260, 261; **15:**191–192, 217
Stern, F., **7:**213(194), 253; **15:**222, 261; **19:**296; **21:**122(27), 233, 303(52), 310
Stern, Jr., F. B., **16:**272, 285, 337
Sternberg, N., **18:**22
Sternglass, E. J., **2:**69, 71, 141
Sternling, C. V., **7:**325(41), 393
Sterret, R. R., **7:**354, 356(137), 395

Stetson, S. B., **21:**5(4), 18(4), 71
Stetter, M., **16:**51, 133
Stettmaier, K., **9:**195(101), 197(101), 258
Steudel, A., **3:**11(51), 49(51), 51(51), 54
Stevels, J. M., **6:**51(146), 77
Stevens, J., **13:**124(73), 149
Stevens, J. E., **18:**38, 45, 124, 159, 160, 167, 173, 176, 184, 186
Stevens, J. R., **17:**358, 361, 369, 370
Stevens, R. J., **14:**31(117), 76
Stevenson, A. C., **17:**212, 223
Stewart, J. E., **9:**227(197), 260
Stewart, R. M., **1:**321(80, 81, 84), 333
Stewart, R. W., **4:**268, 322
Stewart, W. C., **2:**365(6), 419; **14:**175(128, 129), 198
Sticker, R., **12:**292
Stickler, B. A., **4:**150(60), 188
Stickler, R., **4:**169(108), 190; **5:**185(150), 188, 205, 208(150), 233, 234
Stiebel, F., **7:**322(34), 393
Stiegler, J. O., **7:**66(54), 113
Stierwalt, D. S., **10:**61(143), 63(143), 69
Stiglitz, M. R., **15:**6, 55, 136, 138
Stilbans, L. S., **13:**32(49), 58
Stiles, P. J., **3:**277(26, 27), 278(27), 300
Stillinger, D. W., **5:**310(282), 322; **9:**51(239), 70
Stillinger, F. H., **24:**184(40), 190(40)
Stillman, G. E., **7:**235(249), 254; **11:**61(107), 99, 121(74), 201; **21:**37(73), 73
Stillwell, E. P., **6:**103(98, 99), 122, 129(195), 131(194), 132(98, 99), 145, 148
Stilwell, D. E., **17:**318, 366
Stilwell, G. R., **3:**19, 51(87), 55; **6:**105, 131(118), 146; **7:**18(56), 38
Stimmel, J., **17:**111, 142
Stinson, D. G., **15:**172, 216
Stirland, D. J., **3:**114, 127, 215(15), 238(15), 270, 272; **4:**150(59), 188; **5:**187(153), 233, 270(104, 105), 317
Stirling, A. J., **14:**15(57–59), 74, 219(51–54), 238
Stirling, C. J. M., **24:**14(62), 15(62), 23(107)
Stirling, H. F., **12:**271, 272(161, 162), 293
Stitzer, S. N., **15:**44, 68, 96–100, 138–140
Stix, G., **17:**288, 299
Stobie, R. W., **19:**281
Stobinski, L., **24:**53(102)
Stockbauer, R., **16:**28, 128
Stockbridge, C. D., **3:**23(98a), 24(114, 115, 119), 27, 51(98a, 114, 115, 119), 55, 56
Stockdale, G. F., **9:**12(60), 33(60, 152), 66, 68
Stocker, B. J., **3:**96, 126; **12:**128(269), 129(269), 164
Stocker, H. J., **17:**355, 368
Stoddart, C. T. H., **3:**133(16), 162
Stoebe, T. G., **4:**239, 255
Stoemenos, J., **17:**77
Stokes, J. J., **24:**151(9), 154(9), 156(9)
Stokey, W. F., **16:**272, 337
Stole, S. M., **24:**3(16), 4(16), 32(16), 212(61)
Stoll, B., **11:**289(113), 317
Stoll, E., **22:**134(56), 173
Stoll, H., **10:**248(39), 259
Stoll, I., **12:**276(189), 294
Stollman, G. M., **16:**47, 50, 132

Stone, D. E., **4:**23(55), 92
Stone, I., **6:**84(39), 103, 144
Stone, J., **9:**88, 90(17), 143
Stone, L. E., **7:**185(140), 251
Stoneham, E. B., **12:**277(196), 294
Stoner, E. C., **1:**297, 322; **16:**157, 229; **19:**140
Stoner, J. O., **7:**67(56), 113
Stoney, G. G., **3:**219, 271
Stopek, S., **5:**279, 318
Stoquert, J. P., **16:**76, 78, 137
Storey, G. G., **4:**199(57, 58, 60), 200(58, 60), 204(57, 58, 60), 205(57), 206(57, 58, 60), 207(57), 209(57, 58, 60), 211(57, 58), 233(58), 252
Storhoff, J. J., **24:**36(146)
Stork, J. M. C., **23:**55(11)
Storks, K. H., **7:**334, 359(86), 394
Storm, A. R., **17:**118, 142
Stormer, H. L., **11:**78(165), 85(165, 199, 200, 201), 101, 102; **16:**52–53, 133–134
Storp, S., **14:**230(105), 239
Story, H. S., **1:**226(177, 178), 232; **3:**220(32), 229(32), 234(32), 241, 245(32, 74), 271, 272
Storz, F., **21:**287(41), 310
Storz, F. G., **17:**354, 368
Stout, B. F., **2:**366(19), 411(19), 420
Stover, E. T., **5:**263, 316
Stowell, M. J., **3:**32(161), 57, 212(6), 213(6), 216(6), 218(6), 225(6), 243(6), 270; **4:**118, 119(45), 120(45), 132(75), 133(44, 45, 75), 135, 150(39, 41), 155(41), 156(41), 165, 167(102), 168(102), 170(102), 171(102), 172, 174, 175, 176, 188, 189, 190, 224; **5:**189(160), 233
Stowell, W. R., **13:**70, 80(51), 105
Straboni, A., **14:**87(16), 90(90), 98(16, 40), 100(44), 103(16), 105(16), 107(58), 109(16, 61), 111(63), 113(16), 119(58), 120(16), 121(44), 124(90), 125, 126, 127
Strachan, C., **7:**333, 365(98), 394
Strack, H., **13:**210(33), 213(33), 224
Stradling, R. A., **6:**82(36), 144
Strand, A. R., **16:**50, 132
Strange, M. G., **5:**85(31), 86
Stranghan, B. W., **7:**170, 191(107), 250
Stranghn, B. L., **16:**101, 107, 141
Stranski, I. N., **1:**210(116, 117), 231
Strater, K., **5:**306(269), 321
Stratton, J. A., **1:**72(30), 121; **2:**198, 237
Straumens, J., **17:**318, 319, 320, 366
Strauss, A. J., **3:**277(31, 32), 300; **11:**106(5), 107(19), 108(22, 23), 199, 200; **16:**41, 131
Strausser, Y. E., **19:**149
Strbik, V., **16:**52, 133
Street, R. A., **12:**264(103), 292; **15:**278–279, 310
Streete, J. L., **11:**162(134), 203
Streetman, B., **11:**66(120), 99
Streetman, B. G., **21:**193(93), 235
Strehlau, B., **16:**23, 74, 127, 137
Streifer, W., **10:**251(45), 259
Streit, D. C., **21:**121(19), 123(19), 232
Streitz, F. H., **17:**58, 76
Strel'nitskii, V. E., **14:**176(140), 198
Stremin, V. I., **7:**186(146), 251
Stringer, G., **4:**260(18), 301(18), 320
Stringfellow, G. B., **7:**128, 131(19), 180, 181(125), 182(136), 189(125), 193(165), 196(166), 198(125, 175), 199, 201(175), 204(136), 247, 251, 252
Strl'chenko, S. S., **12:**126(262), 164

Strom, U., **11:**256(70), 257(70), 316
Stromberg, R. R., **3:**16(72), 51(72), 54; **4:**75(186), 83(206), 84(206), 95
Strong, J., **1:**133(19), 185; **2:**240, 315(14), 316(14), 359, 803; **3:**43, 44; **10:**130(68), 165, 262, 308; **12:**5(22), 50
Strong, L., **24:**45(70), 82(9), 83(9), 86(9), 121(43), 123(43), 125(43), 199(48), 212(56), 217(56)
Strong, R., **23:**163(47, 48)
Strother, D. D., **16:**81, 102–105, 121, 138
Struchinskii, G. B., **12:**101(168), 162
Strumpfel, J., **14:**11(39), 36(138), 44(39), 48(162), 55(162, 173), 74, 76, 77
Strumpfer, J., **13:**67(32), 105
Strzelecki, L., **17:**288, 299
Stuart, P. R., **11:**23(17), 24(17), 25(17), 26(17), 27(17), 34
Stuart, R., **4:**344, 370; **23:**91(21)
Stuart, R. V., **3:**65, 72(40, 41), 125; **13:**117(41), 148; **22:**244(95), 274
Stuck, R., **12:**176(50), 179(50), 180(50), 183(50), 186(50), 187(50), 190(50), 197(50), 200(50), 233
Studer, F. J., **3:**194(54), 200(54), 209; **5:**282, 319
Studna, A. A., **19:**2, 4, 6, 9–19, 23, 26, 36, 59, 148, 150, 196–199, 201–203
Stuke, J., **7:**27(102), 39; **9:**50(234), 70
Stukin, A. D., **3:**39(208), 52(208), 58
Stukin, E. D., **3:**39(208), 58
Stumer, J. H., **3:**179(30), 209
Stünckel, D., **2:**30(49), 61
Stünkel, D., **3:**27(139, 141), 28(139, 141), 29(139, 141), 30(139, 141), 51(139, 141), 56
Stupak, R. I., **14:**164(116), 176(116), 197
Stupp, E., **12:**135(293), 165
Sturge, M. D., **9:**288, 295, 296
Sturkey, L., **1:**198(49), 229; **2:**119(115), 143
Sturm, J. C., 23.24(92), 23.30(113), 23.37n(113), **23:**41(113), 55(187, 188), 166(91)
Sturm, K., **9:**158(14c), 256; **19:**129, 131, 134, 155
Stürmpfel, J., **17:**111, 116, 142
Stutz, C. E., **15:**244, 263; **21:**121(22, 23), 123(22, 23), 194(22), 195(22), 233, 303(53), 310
Su, C. Y., **12:**115(212), 116(212, 222), 117(212), 163; **13:**201(4), 202(4), 203(16), 207(4, 16), 208(4, 16), 223
Subramanian, M. A., **16:**28–29, 129
Subramanyam, G., **16:**36, 130
Suchalkin, S. D., **23:**243(62)
Suddeth, J. A., **2:**101(69, 70), 103(70), 104(70), 105(70), 106(70), 142
Sudersena Rao, T., **17:**241, 297
Sudlow, P. D., **7:**167, 169(94), 250
Sudoh, A., **19:**128, 129, 139, 140
Sudovtsev, A. I., **1:**244(27, 28), 274
Sue, J. A., **14:**191(166), 199; **17:**83, 130, 139, 143
Suedomi, N., **17:**336, 346, 367
Suematsu, Y., **10:**236(19), 258
Suemitsu, M., **23:**30(116), 32(116)
Sugai, K., **17:**153, 154, 221
Sugano, T., **4:**343(54), 344(54), 346(54), 347(54), 348(54), 349(54), 354(54), 367(54), 370; **12:**84(98), 160, 277(198, 204, 205), 294, 295; **14:**82(2), 103(47), 107(57), 109(62),

Sugano, T., (*continued*)
113(65), 123(2), 124(2), 125, 126, 127
Sugata, S., **13:**19(27), 57, 181(110), 185(139), 188(139), 195, 196
Sugawara, F., **15:**295, 311
Sugawara, K., **7:**125(13), 247
Suge, Y., **10:**237(20), 258
Sugibuchi, K., **17:**279, 299
Sugita, Y., **6:**227, 297; **7:**125(13), 247; **9:**271, 296; **15:**185, 216
Sugiura, H., **19:**15
Sugiyama, I., **9:**47(222a), 70
Sugiyama, K., **5:**311(285), 322; **7:**186, 187(144), 235(215), 251, 253; **16:**88, 138; **17:**354, 368
Suh, S. Y., **13:**144(134), 150
Suhara, A., **16:**88, 139
Suhl, H., **15:**92, 94, 127, 140
Suhre, D. R., **17:**154, 155, 182, 222
Suhrmann, R., **2:**23(24), 40, 46, 60, 61, 62; **12:**96(159), 162
Suikovskaya, N. V., **5:**120(24), 121(24), 141
Suiter, W. B., Jr., **6:**212, 295
Suito, E., **3:**171(18, 86), 172, 208, 210
Suito, K., **2:**135(158), 137(158), 145
Suits, J. C., **1:**244(30), 274; **2:**42, 61; **6:**224(83), 271, 291(83), 297
Sukegawa, T., **12:**114(207), 135(295), 163, 165
Sukenik, C., **24:**11(42)
Sukhorukov, V. V., **12:**291
Sulich, M., **6:**214(35), 295
Sullivan, B. T., **16:**28–30, 129; **19:**139
Sullivan, D. L., **6:**228(101), 238(101), 297
Sullivan, F., **22:**128(38), 172
Sullivan, G. A., **7:**263(24), 306
Sullivan, G. J., **21:**117(8), 128(8), 221(8), 223(8), 232
Sullivan, M. V., **5:**247, 315; **7:**135(30), 248
Sumioka, A., **17:**217, 224
Sumiyama, K., **15:**176, 179, 195–199, 216
Summer, B., **11:**157(128), 202
Summitt, R., **9:**22(97, 98, 99), 23(97, 98, 99), 26, 67
Sumner, G. G., **4:**150(48), 155(48), 188
Sumomogi, T., **14:**67(226), 78
Sumski, S., **7:**158(75), 201(178), 214, 217(198), 234(240, 245), 239(253), 249, 252, 253, 254, 255
Sun, C. H., **4:**235(153), 254
Sun, F., **24:**34(139, 140), 35(139, 140)
Sun, J. Z., **16:**22, 127
Sun, L., **24:**12(50), 15(50), 18(50), 19(50), 45(54, 55, 63), 46(63), 70(55, 63), 86(36), 92(56), 99(102), 106(102)
Sun, P. H., **7:**268(59, 60), 278(59, 60), 296(59, 60), 308; **16:**241–242, 249, 255–257, 267
Sun, R. C., **22:**119(6), 171
Sun, T. I., **17:**352, 367
Sun, T. S., **11:**153, 155(127), 198(157), 199(127), 202, 203
Sunami, H., **12:**74(85), 78(85), 160; **14:**88(17), 89(19), 109(19), 117(19), 125
Sundarababu, G., **24:**18(88), 20(88), 106(111)
Sundaram, S., **22:**270(127), 276
Sundelin, R. N., **17:**153, 220
Sundgren, J. E., **13:**89, 106; **17:**7, 15, 16, 17, 24, 31, 63, 65, 73, 74, 89, 99, 110, 119, 122, 123, 124, 129, 130, 131, 136, 140, 141, 143, 144
Sundgren, J.-E., **14:**60(193), 65(209), 68(209), 78

Sundquist, B. E., **4:**164, 189
Sung, C., **22:**5(4), 77
Sungaila, Z. F., **16:**32, 129
Sunier, J. W., **11:**130(84), 131(84), 201
Sunjic, M., **9:**213(142, 143), 259
Sunseri, C., **17:**317, 365
Sunshine, S., **16:**2, 60, 124, 135
Sunshine, S. A., **16:**3, 124
Superfine, R., **24:**151(9), 154(9), 156(9)
Supertzi, E. P., **11:**10(4), 33
Suri, A. K., **13:**79(77), 102(114), 106, 107
Suris, R. A., **23:**238(42), 240(42, 53)
Surowiak, Z., **19:**250
Surridge, R., **14:**67(227), 78
Surridge, R. K., **12:**135(291), 165
Suryanarayanan, R., **16:**28–29, 98, 129, 140
Susman, S., **5:**284(196), 319
Sussi, H., **12:**288(257), 295
Susuki, T., **11:**64(114), 99
Suter, U. W., **24:**208(34)
Suto, S., **13:**185(135), 187(135), 196
Sutton, P. M., **4:**309, 324
Suurmeijer, E. P. Th. M., **14:**208(20), 237
Suyama, S., **14:**71(234), 79
Suzukawa, T., **9:**33(127), 68
Suzuki, A., **15:**172, 216
Suzuki, H., **9:**182(63), 186(63), 257; **12:**282(234), 295; **15:**300, 311; **17:**197, 201, 222, 223
Suzuki, K., **6:**227(96), 297; **18:**122, 152, 163
Suzuki, M., **12:**71, 75, 159; **16:**25, 73, 127
Suzuki, N., **17:**339, 344, 367
Suzuki, T., **6:**72(213), 79; **12:**73, 160
Suzuki, Y., **13:**143(130), 150; **17:**318, 365
Svechnikov, S. V., **12:**223(114), 229(114), 230(114), 235
Svensson, J. S. E. M., **17:**303, 315, 316, 351, 355, 356, 364, 365, 367, 369
Svinarich, K. A., **17:**49, 52, 76
Sviridov, D. V., **17:**317, 365
Sviridova, A. A., **5:**120(24), 121(24), 141
Swain, J., **11:**254, 255
Swalen, J. D., **15:**266, 309; **24:**206(2)
Swaminathan, V., **21:**221(135, 137), 222(135, 137), 236, 237
Swan, J. B., **9:**205(128), 212(140), 258, 259
Swanepoel, R., **13:**185(128), 187(128), 196
Swank, R. K., **13:**201(3), 223
Swann, C. G., **12:**261, 292
Swann, R. C. G., **5:**288, 292, 297, 298, 320, 321; **12:**266, 272(161, 162), 293
Swanson, B. I., **24:**97(78), 98(78, 84)
Swanson, C. E., Jr., **7:**264(43), 307
Swanson, M. L., **17:**291, 299
Swanston, D. M., **17:**241, 290, 299
Swarm, R. C. G., **3:**138(36), 163; **4:**270(86), 322
Swartz, J. C., **1:**17, 64; **21:**21(44), 72
Swartzendruber, L. J., **16:**59, 79, 135, 137
Sweeney, J. E., **13:**161(39), 182(39), 183(39), 193
Swendson, R. H., **24:**117(37)
Swenson, C. A., **1:**241(17), 273
Swenson, C. J., **24:**209(36), 210(36), 211(36), 220(36)
Swift, J. D., **18:**80
Swinney, H. L., **21:**50(99), 52(103), 74
Syainasundara Rao, Y., **16:**244, 257, 267

Sybert, J. R., **6:**82(27), 144
Sybertz, W., **16:**24–25, 127
Sykes, C. G., **15:**82, 140
Symons, J., **23:**2(4)
Syms, C. H. A., **12:**130(278), 165
Syrotynsky, O. I., **16:**5, 124
Szabo, A., **24:**72(117)
Sze, R. C., **13:**157(31), 193
Sze, S., **6:**67(201), 78
Sze, S. M., **4:**327(8), 330, 331(24), 332, 333(25), 334, 335(36), 337(25), 338, 339, 340, 341, 343, 344(54), 346, 347(54), 348(54), 349, 354(54), 356(75), 357, 358(75), 359, 360, 361(80), 362, 363, 364, 367; **5:**306(274), 322; **11:**141(104), 202; **12:**84, 160; **13:**209(29), 210(29), 215(29), 223; **21:**16(40), 22(52), 23(52), 24(52, 54), 30(40), 42(40), 72, 73, 314(7), 366
Szedon, J. R., **4:**316(161), 324; **5:**287(213), 320; **16:**87, 138; **17:**153, 154, 156, 197, 200, 201, 215, 220–224
Szekely, G., **5:**247(21), 314
Szentpaly, L. V., **7:**379, 385(200), 386(200, 203), 397
Szepessy, L., **12:**176(50), 179(50), 180(50), 183(50), 186(50), 190(50), 197(50), 200(50), 233
Szigeti, B., **6:**34, 51(98), 76
Szmud, D. M., **21:**365(48), 368
Szmulowicz, F., **21:**18(41), 72, 121(22, 23), 123(22, 23), 194(22), 195(22), 233, 303(53), 310
Szostak, D., **12:**118(226), 119, 163
Szostak, D. J., **12:**129(272), 165
Szundi, I., **11:**271(89), 274, 316
Szwarc, M., **3:**157(85), 164

T

Tabat, M. D., **13:**159(37), 161(40), 164(40), 184(122, 123), 185(122, 123, 124a, 130), 187(122, 123, 130), 190(123, 130), 191(130), 193, 196
Tabata, H., **16:**79, 137; **17:**209, 223, 290, 299
Tabata, O., **9:**39, 40, 69
Tabata, S., **2:**289(34b), 304
Tabor, D., **7:**334, 389(90), 394; **17:**60, 77
Tabor, H., **2:**330, 353(75), 360; **9:**129(134), 144; **10:**1, 2, 4(1), 16(1, 28), 23(1), 25, 28(28, 44, 45, 46, 47), 43(28, 46), 47, 49, 57, 65, 66
Tabor, M., **22:**139(62), 173
Tabor, W. J., **9:**271, 288, 296, 297
Tachibana, K., **22:**182(24), 272
Tachik, M., **17:**171, 221
Tachiki, M., **16:**10, 125
Tachiya, H., **12:**86(113), 99(113), 161
Tada, H., **17:**361, 362, 370
Tada, O., **14:**18(74, 75), 19(74, 77, 82), 20(75, 82), 30(74), 66(219, 220), 75, 78
Tada, T., **17:**153, 219
Tadayon, J., **7:**330(55), 394
Taft, E., **12:**60(35), 101(35), 102, 103, 108, 109(35), 159
Taft, E. A., **1:**143(65), 147, 149(37), 169(35, 37), 185; **12:**162, 265(113), 267, 292
Taft, F. A., **15:**291, 293, 299–300, 311
Taga, Y., **14:**52(168), 77; **16:**21, 127
Tagirov, R. I., **6:**216(53, 54, 55, 56), 220(53), 269(55), 276(185), 277(54, 185), 296, 299
Taglauer, E., **13:**117(40), 148;

14:16(61), 17(61), 74, 218(48), 238
Taguchi, K., **13:**143(130), 150
Tahara, S., **16:**68, 136
Tai, K. L., **6:**335(40), 347; **16:**241–242, 244, 249, 253–255, 263, 265, 267–269
Tai, Y. C., **17:**210, 223
Taiedy, F., **11:**256(68), 257(68), 279(101), 316
Tait, R. N., **22:**11(20, 21), 12(20), 22(31), 23(31–33), 24(20, 33), 25(33), 26(33), 28(33), 32(46), 37(46), 38(46), 45(52), 60(60), 61(61), 66(61), 68(61), 76(67–69, 71, 72, 75), 77–79, 227(82), 274
Tak. thashi, N., **1:**199(63), 229
Takabayashi, M., **5:**275, 276, 317
Takada, H., **17:**173, 222
Takada, S., **10:**256(57), 259; **11:**67(130a), 99; **17:**153, 154, 161, 162, 220, 221, 261, 263, 298
Takada, T., **5:**313(304), 322; **16:**51, 132, 251, 268; **17:**2, 9, 10, 15, 22, 73, 74
Takagi, H., **13:**176(86), 188(86), 189(86), 195; **15:**147, 154, 163–164, 185–186, 214, 217; **16:**3, 124
Takagi, M., **1:**197(45), 229; **7:**21(71), 39; **19:**168
Takagi, N., **9:**47(222a), 70
Takagi, S., **2:**120, 143
Takagi, T., **13:**1, 2(2, 3), 3(3, 6, 7), 5(9, 10), 8(16), 9(19), 11(6, 10), 13(22), 18(26), 20(28, 29, 30, 31, 33, 34, 35, 36, 37, 38, 39, 41, 42, 43, 44, 45, 46), 21(9, 30, 31, 33, 34), 22(7, 35, 36, 37, 38, 39), 23(7, 28, 29, 41, 42, 43); **14:**24(19, 28, 44, 45, 46), 25(30, 47), 27(31, 47), 28(31, 33), 30(34), 31(35), 32(36), 33(37), 34(38), 37(16), 38(51), 39(31), 40(39), 44, 45(29), 47(55), 48(56), 49(43, 56), 51(44), 53(45), 56, 56(46), 132(7), 133(7), 137(7, 9, 41, 42, 43); **17:**82, 83, 112(31), 116, 134, 139
Takagi, Y., **15:**287–288, 310
Takahashi, A., **15:**168, 172, 215–216
Takahashi, C., **17:**199, 201, 222, 223
Takahashi, H., **19:**128, 129, 139, 140
Takahashi, K., **5:**214, 234; **7:**191(154), 220(206), 251, 253; **11:**44(50), 64(112, 113, 114), 65(113), 70(140), 73(140), 76(149), 97, 99, 100
Takahashi, M., **1:**282, 331; **9:**271, 296; **15:**185, 216
Takahashi, S., **16:**10, 125; **17:**197, 222
Takahashi, T., **5:**311, 322; **12:**87, 95(156), 103, 104, 105, 161, 162; **13:**99, 107; **17:**153, 221, 354, 368
Takahasi, Y., **16:**96, 140
Takai, M., **13:**166(61), 167(64, 66), 168, 169(65), 180(64), 185(61, 64, 65, 66, 140), 186(149), 188(61, 64, 65, 66, 140), 189(149), 194, 196, 197
Takai, Y., **16:**92, 139
Takami, T., **24:**218(75)
Takamisawa, K., **16:**288, 338
Takano, H., **24:**94(69, 71), 95(70, 71), 218(79)
Takano, S., **17:**318, 366
Takaoka, H., **13:**2(2), 5(9), 20(28, 34, 36, 37, 43, 44), 21(9, 34), 22(36, 37), 23(28, 43), 24(28, 44), 30(34), 32(36), 33(37), 38(51), 48(56), 49(43, 56), 51(44), 56, 57
Takaoka, T., **3:**15, 51(66), 54
Takase, E., **16:**92, 139

Takaska, H., **13:**112(31), 148
Takasu, S., **5:**313(292), 322
Takayama, R., **17:**228, 232, 233, 237, 242, 243, 244, 245, 246, 247, 283, 293, 296, 297
Takayama, T., **10:**237(21), 258
Takayanagi, M., **11:**254, 255, 315
Takeda, F., **17:**153, 220
Takeda, N., **13:**172(78), 173(78), 177(78), 190(78), 191(78), 194; **15:**103, 113, 140
Takehara, K., **24:**19(82), 20(82), 21(82), 27(82), 29(82)
Takei, H., **5:**313(292), 322
Takei, W. J., **6:**317(30), 325(32), 346; **11:**171(145), 183(145), 203; **17:**229, 232, 233, 234, 237, 265, 266, 267, 268, 269, 271, 280, 281, 282, 283, 284, 285, 286, 290, 292, 296, 298, 299; **21:**185(82), 189(82), 235
Takemori, T., **7:**34(143), 40
Takemura, H., **24:**19(82), 20(82), 21(82), 27(82), 29(82)
Takenaka, H., **14:**240
Takeshita, K., **17:**87, 140
Taketoshi, K., **12:**86(113), 99(113), 161
Takeuchi, A., **13:**99(101), 107
Takeuchi, H., **17:**162, 221
Takeuchi, K., **3:**232(60b), 236(60b), 272
Takeuchi, T., **17:**217, 224
Takhonov, V. V., **15:**64, 80–81, 139
Taki, K., **15:**172, 216
Takoaka, H., **13:**116(35), 148
Talbot, M., **12:**176(48a), 233
Talebian, A., **15:**303, 312
Talisa, S., **17:**146, 148, 186, 218
Talisa, S. H., **15:**20, 62, 81, 115–117, 137, 139, 141
Talledo, A., **17:**361, 370
Tallman, R. L., **5:**199, 202, 234
Talreja, R., **16:**280, 311–313, 333, 337
Talvacchio, J., **16:**8, 17–18, 22, 52, 74, 101, 107–108, 125–127, 133, 141–142; **17:**271, 272, 276, 278, 298
Talwalker, A. T., **9:**31(115), 67
Tam, G., **23:**60(202)
Tam-Chang, S.-W., **24:**14(60), 15(60)
Tamada, H., **15:**112–113, 115–116, 141; **17:**228, 261, 262, 263, 298
Tamagawa, T., **17:**208, 223, 228, 293, 294, 299
Tamargo, M. C., **19:**2
Tamaru, K., **5:**274, 317
Tamir, T., **13:**121(62), 148
Tamm, I., **2:**123, 144
Tamm, I. G., **1:**131(18), 185
Tamm, L. K., **24:**106(110)
Tamura, K., **21:**330(26), 331(26), 340(26), 343(26), 367
Tamura, M., **7:**125(13), 247
Tamura, N., **3:**224(43), 232, 233, 271
Tan, N. X., **16:**4, 124
Tanabe, K., **14:**236(154), 240
Tanaha, K., **7:**334(88), 394
Tanaka, A., **16:**34, 130
Tanaka, F., **15:**171–172, 198–199, 207, 216
Tanaka, H., **19:**2, 16–18, 27, 32–36
Tanaka, K., **2:**135(158), 137(158), 145; **14:**31(122), 76; **15:**300, 311; **19:**76, 80
Tanaka, M., **17:**355, 357, 368, 369
Tanaka, S., **3:**137(28), 141(28), 162; **4:**265(65), 314(65), 321; **15:**171–172, 198–199, 207, 216; **16:**3, 14, 32–33, 124–125, 129–130
Tanaka, T., **9:**49(224a), 70; **11:**91(237), 92(237, 238), 103; **12:**264(105), 292

Tanaka, Y., **1:**130(11), 184; **16:**2, 123
Tanase, S., **17:**354, 368
Tanberg, R., **14:**150(78, 82), 196, 197
Tanbun-Ek, T., **21:**204(101), 205(101), 235
Tandon, J. L., **14:**66(217), 78
Tanemura, M., **22:**145(68), 173
Tanemura, S., **10:**58(125), 68
Tang, C. L., **21:**365(48), 368
Tang, H., **24:**27(124), 28(124), 29(124), 30(124), 182(23), 199(23)
Tang, M., **16:**4, 124
Tanger, C. W., **16:**123, 143
Tangherlini, F. R., **2:**230, 238, 339(60), 340(60), 360
Tangonan, G. L., **10:**255(55), 259
Taniguchi, I., **24:**45(81)
Taniguchi, N., **13:**119(49), 126(49), 148
Taniguchi, S., **1:**280(10), 331
Tannenbaum, E., **2:**370(44), 420; **4:**301(117), 323
Tannenwald, P. E., **1:**327, 329(103, 106), 330(112), 334; **15:**244–245, 262–263
Tanner, D. B., **6:**124, 125, 126, 127, 148
Tanner, H. A., **2:**265(24), 304
Tanner, K. B., **5:**88, 140
Tanno, M., **15:**14, 137
Tantraporn, W., **2:**151(33), 191; **3:**132(7), 146(7), 162; **23:**102(42)
Tao, G., **24:**19(81), 20(81, 89)
Tao, G. S. T., **16:**98, 140
Tao, L. J., **9:**273, 274, 296
Tao, T. F., **11:**113, 130(84), 131(84), 200, 201
Tao, W., **16:**91, 139
Tao, Y.-T., **24:**3(13, 15), 4(15), 11(43), 12(13), 13(13, 43), 14(13), 15(13), 16(13), 18(13), 20(13), 23(115), 30(13), 32(15, 115), 33(115), 44(8, 12, 44), 46(12), 55(12), 56(12), 65(12), 66(8), 112(3), 212(57), 213(66), 220(57, 66), 228(3, 18), 237(18)
Tarascon, J. M., **17:**236, 237, 270, 271, 296, 298
Tarascon, J.-M., **16:**13, 18, 56, 63, 125–126, 134–135
Tarbell, D. S., **24:**70(106)
Targove, J. D., **13:**122(66), 148
Tarlov, M., **24:**44(38), 45(38, 78), 46(78), 60(78), 70(38), 112(10)
Tarlov, M. J., **24:**18(77), 19(77), 44(46), 82(15), 86(15, 30, 31), 127(52), 138(64), 206(17), 212(17), 229(42)
Tarng, M. L., **9:**278, 297; **14:**17(69), 75
Tarnopol, M. S., **9:**33(149), 68; **12:**187(85), 234
Tarshis, L., **3:**254(95), 255(95), 256, 257(95), 258(95), 263(95), 264(95), 265(95), 267, 269, 273
Tarucha, S., **11:**268(86), 269(86), 316
Tarui, Y., **5:**276, 318
Tasushima, T., **15:**154, 164–165, 183, 185, 215
Tatarinova, L. I., **7:**24, 29(91), 30(125), 34(144, 145), 35(91, 147), 36(160), 39, 40, 41
Tate, J., **16:**52, 133
Tateishi, K., **11:**44(50), 70, 73(140), 97, 100
Tatham, M. C., **21:**30(60), 73; **23:**254(76)
Tatsumi, T., **16:**30, 129; **23:**44(151), 55(9)
Tatsuoka, H., **17:**324, 329, 366

Tauc, J., **10:**41(80), 67; **12:**59(26), 159; **19:**73
Tausch, F. W., **5:**279, 299(161), 318
Tavger, B. A., **6:**137(215, 216), 148
Tawada, Y., **12:**275(174), 294
Tayagi, H., **16:**2, 124
Taylor, A. G., **19:**2, 10, 15
Taylor, A. H., **4:**60(142), 94
Taylor, A. M., **2:**199(5), 237
Taylor, D. S., **22:**201(56), 253(56), 263(56), 273
Taylor, G. N., **24:**12(55), 14(55)
Taylor, G. W., **11:**218(15), 314; **17:**289, 299
Taylor, H., **5:**274(124), 274(125), 317
Taylor, H. D., **2:**240, 303
Taylor, H. E., **11:**238(41), 242, 315
Taylor, H. F., **10:**256(63), 257, 259; **21:**121(19), 123(19), 232
Taylor, J. B., **12:**114(198), 163
Taylor, K. N. R., **6:**247, 298; **15:**146, 214; **16:**26, 127
Taylor, N. J., **4:**161, 189
Taylor, P. C., **11:**256(70), 257(70), 316
Taylor, R. C., **9:**282, 297; **15:**180, 216
Taylor, R. L., **2:**393(135), 405(135), 423
Taylor, R. W., **23:**87(6)
Taylor, T. N., **24:**98(84)
Taysing-Lara, M., **21:**121(25), 233, 241(8, 9, 10, 18), 245(18), 256(8, 9, 10, 27, 29), 259(10), 260(8), 265(27), 267(29), 269(9), 308, 309
Tazoh, Y., **16:**41, 131
Tbakur, R. P. S., **16:**91, 139
Tbouless, M. D., **16:**288–289, 291–292, 334, 338–340
Te Velde, **5:**184, 233
Teague, E. C., **7:**133(24), 247
Teal, G. K., **5:**273, 317
Teale, J., **1:**20(48), 65
Teasdale-Buckell, P., **4:**85(221), 96
Tech, M., **12:**226(121), 235
Tedder, L. L., **22:**186(33), 209(33), 272
Tedrow, P. M., **19:**165
Teer, D., **13:**78(65), 106
Teer, D. G., **12:**246(30), 248, 290; **14:**132(17), 133(17), 136(17), 195
Teheuredkjian, N., **7:**333(82), 334(82, 91), 394
Teitler, S., **19:**282, 288
Tejwani, M. J., **23:**49(175)
Telesnin, R. V., **6:**292(220), 300
Telkes, M., **13:**33(50), 58
Tell, B., **14:**113(68), 127
Teller, A. H., **24:**208(25)
Teller, E., **24:**208(25)
Temkin, H., **23:**55(8)
Temmink, A., **17:**310, 332, 333, 348, 356, 357, 365–367, 369
Temple, W., **12:**248(36), 271(36), 290
TenBrinke, G., **24:**95(73)
Teng, Y., **9:**221, 225, 244(242), 260, 261
Teng, Y.-Y., **6:**199, 203
Tengvall, P., **24:**218(76)
Tennant, W. E., **21:**40(88), 74, 117(8), 128(8), 221(8, 134), 223(8), 232, 236
Tenne, R., **13:**186(152), 190(152), 197
Tenney, A. S., **15:**287, 299–300, 310–311
Teo, W.-K., **24:**228(9)
Teranishi, N., **21:**31(62), 32(62), 73; **23:**144(35)
Terao, N., **1:**223, 232
Terasaki, I., **16:**54, 134
Terashima, K., **16:**51, 133

Terashima, T., **16:**25, 51, 73, 127, 132; **17:**229, 236, 287, 290, 296
Terauchi, H., **15:**281, 310; **16:**51, 132; **17:**24, 75
Terekhova, T. S., **12:**169(38), 201(38), 203(38), 205(38), 210(38), 233
Terhune, C., **14:**57(178), 77
Terkaura, K., **15:**177, 216
Terman, L. M., **2:**167, 191
Terpstra, R. A., **17:**110, 141
Terrataz, S., **24:**44(40)
Terrill, H. M., **2:**68(9), 141
Terrill, R. H., **24:**151(9), 154(9), 156(9)
Terry, L. E., **6:**36(111), 76
Tersoff, J., **13:**202(10), 223; **22:**127(35), 172
Tertian, L., **1:**148(51), 149(51), 159(51), 160(51), 185
Terzis, A., **24:**151(9), 154(9), 156(9)
Tesar, A. A., **19:**210
Teschner, G., **14:**35(137), 38–39(137), 51(138), 55(173), 76, 77
Teshima, H., **5:**276(143), 318
Teshima, R., **9:**159(17), 256
Tesmer, J., **16:**58, 61, 75, 135
Tessler, L. R., **16:**50, 53, 132, 134
Testardi, L. R., **17:**118, 142
Tetel'baum, D. I., **12:**291
Tevebaugh, A. D., **13:**60, 104
Tewksbury, S. K., **16:**107, 141
Textor, M., **12:**288(257), 295
Thacher, P. D., **19:**259
Thacker, B. R., **21:**365(49), 368
Thakoor, A. P., **12:**169(28), 176(49, 51), 183(28), 185(49), 186(49), 187(87), 189(28), 190(87), 198(49), 199(49), 225(28), 233, 234
Thallikar, G., **22:**177(5), 271
t'Hart, W., **7:**36(161), 41
Thatcher, E. M., **7:**368(175), 396
Theeten, J. B., **13:**11(20), 57; **14:**82(4), 100(42), 103(4), 125, 126; **19:**50, 51, 59, 196, 203, 280
Theis, T. N., **14:**93(30), 126
Thekaekara, M. P., **10:**5(6, 8), 65
Thelen, A., **2:**265(24b), 290(24b), 304; **3:**124(161), 128; **5:**61(19), 65(21), 68(19), 78, 80(21), 84(27), 86, 123(28), 134(36), 141; **9:**105(19), 110, 143, 144
Theoret, A., **12:**248(39), 251(39), 290
Thermond, C. D., **12:**292
Thernquist, T., **7:**263, 302(26), 306
Theroux, P., **2:**216(20), 238
Thesis, M. S., **13:**98(96), 106
Thetcher, W. J., **14:**23(91), 75
Theuerer, H. C., **2:**367(30), 396(30), 420; **3:**87, 121(153, 154), 126, 128; **15:**150, 214
Theurer, H. C., **5:**147(21), 148, 229, 253(37, 38), 256, 257(38), 265, 266, 315, 316; **6:**335(40), 347
Thèye, M. L., **6:**100, 145, 188, 189(32), 190, 191(32), 193(47), 203
Thiele, A. A., **9:**267, 269, 271, 275, 295, 297; **15:**187, 217
Thielen, J., **4:**50(105), 93
Thiessen, P. A., **7:**374(194), 397
Thijs, P. J. A., **23:**226(11), 227(11), 228(11), 229(11), 232(11)
Thirsk, H. R., **4:**191, 233(3), 235(3), 251
Thomas, B. N., **24:**184(35, 41), 188(41), 192(35)
Thomas, C., **16:**319, 339
Thomas, C. O., **5:**266(81), 267, 316
Thomas, D. C., **15:**222, 260; **21:**240(5), 308
Thomas, D. J. D., **4:**169(114), 190;

Thomas, D. J. D., (*continued*) **7:**24(92), 35(152), 36(159), 39, 41
Thomas, G., **4:**143(23), 184(23, 140), 185(23), 187, 190, 225(127), 254; **16:**63, 135
Thomas, H., **1:**304(55), 306(55), 309(55), 312(63, 65), 313(65), 315(63), 333; **4:**345, 370; **5:**1(3), 45; **6:**104(112), 131(112), 146
Thomas, H. A., **1:**42(111), 43(113), 66
Thomas, J. E., Jr., **4:**302(138), 306, 323
Thomas, J. H., **14:**212(26), 232(124), 235(152), 237, 239, 240
Thomas, L. D., **9:**33(123), 68
Thomas, R. A., **11:**268(85), 269, 316
Thomas, R. C., **24:**44(56), 92(56), 94(64, 65), 221(88)
Thomas, R. E., **13:**155(24), 166(24), 169(24), 178(24), 192; **21:**103(38), 107(38), 112
Thomas, R. K., **24:**220(86)
Thomas, R. N., **5:**163(75), 166, 170, 196(75), 197, 199, 200, 205, 206, 227(75; **12:**136(300, 301, 302, 303), 137, 165; **21:**21(44), 72
Thommen, K., **3:**76, 126
Thompson, A. G., **7:**173(117), 250
Thompson, C. J., **6:**137(221), 148
Thompson, C. T., **1:**55(134), 67
Thompson, D., **16:**80–81, 137–138
Thompson, D. A., **6:**212, 260, 261, 262(166, 167), 263, 265, 267, 272(174), 288, 289, 290, 291(208a), 292(167, 208a, 221), 293(31), 295, 299, 300
Thompson, E. R., **4:**143, 144(24), 187, 213, 214, 219(96), 228(96), 229(96), 231, 234(96), 235(96), 242, 243, 249, 253, 255
Thompson, G. H. B., **7:**218(203a), 253
Thompson, H. C., **2:**179(61), 192
Thompson, J. B., **6:**38, 77; **14:**85(8), 125
Thompson, J. C., **7:**34(141), 40
Thompson, J. R., **16:**10, 43, 48–49, 92, 125, 132, 139
Thompson, K. T., **9:**5(22), 65
Thompson, M., **24:**45(99, 100), 49(99)
Thompson, M. E., **24:**164(31), 168(31)
Thompson, M. J., **19:**61
Thompson, M. W., **4:**190; **22:**146(70), 173
Thompson, N., **4:**159(84), 189
Thompson, T. L., **5:**285, 319
Thomson, G. P., **4:**223, 232, 236(113), 253
Thomson, J. J., **3:**19, 51(84), 55; **6:**84(40), 144; **18:**59
Thomson, M. J., **14:**99(41), 110(41), 126
Thomson, S. J., **9:**14, 66
Thornton, J. A., **13:**67(31), 68, 72(31, 41), 73(31, 41), 81, 82, 83, 105, 106; **14:**2(3, 9), 22(3, 9), 33(3, 9, 131), 34(3), 36(131), 36–37(139), 39(3, 131), 40(131), 41(131), 59(186), 63(201), 65(211–213), 68(211, 229), 70(229), 73, 76, 77, 78, 132(11), 195; **17:**23, 31, 34, 74, 87, 119, 120, 121, 140, 143, 156, 157, 221, 314, 365; **18:**246; **19:**195; **22:**2(2), 77, 119(5, 7–9), 120(18, 19), 121(18, 19), 122(18), 151(18), 154(8, 19), 171
Thornton, W. A., **3:**188(45), 203(78, 79), 205, 207(84), 209, 210
Thorp, J. S., **11:**260, 316
Thorpe, A. J. S., **15:**250, 263
Thorpe, S. C., **24:**21(99), 24(99), 25(99)

Thorpe, T. F., **17:**162, 221
Thostesen, T. O., **2:**312(5), 358; **10:**162(93), 166
Thuault, C. D., **12:**114(211), 163
Thumwood, R. F., **12:**87, 161
Thun, R. E., **1:**197(46), 199, 202(64), 211(120), 215, 218(57), 220, 221, 223, 224(56, 64, 120, 150, 151), 227(185), 229, 231, 232; **2:**147(2), 178(2), 190, 214(16), 238, 284(35, 39), 290(35), 297(39), 304, 366(3, 14, 19), 373(14), 379(83), 384(98), 390(120), 391(120), 392(83, 120), 404(14), 405(14), 411(19); **3:**4(27), 5(27), 51(27), 53, 212, 214, 219, 270; **5:**163(69), 167, 231; **17:**126, 143
Thundat, T., **24:**98(92), 237(64)
Thurgate, T., **22:**5(6), 77
Thurmond, C. D., **5:**150(29), 230, 277, 279, 294(237), 318, 320; **13:**205(19), 223
Thurnau, D. H., **4:**68(162), 94
Thurnauer, H., **2:**373(64), 421
Tian, F., **22:**240(87), 274
Tibol, G. J., **6:**36(101), 37(127), 38(101), 44, 45(101), 47, 48, 49, 77
Tickner, A. W., **6:**38(136), 51(136), 77
Tidd, E. D., **2:**374(68), 375(68), 376(68), 379(68), 397(68), 404(68), 407(68), 408(68), 421
Tideswell, N. W., **7:**35(150), 41
Tidrow, M. Z., **21:**211(114), 236, 241(9, 11, 12), 256(9, 29), 267(29), 269(9), 282(11, 36), 290(49), 300(12), 308, 309, 310
Tidswell, I. M., **24:**44(19), 112(23), 121(23)
Tiede, E., **3:**188, 209
Tiemeijer, L. F., **23:**226(11), 227(11), 228(11), 229(11), 232(11)
Tien, D. K., **11:**58(103), 80(103), 99
Tien, J., **24:**231(50), 235(56), 241(56), 249(94)
Tien, P. K., **9:**169(28), 256; **10:**177, 226, 227(3), 230, 234(3), 235(12), 236, 241, 256, 258, 259
Tiernan, R. J., **5:**163(65), 168(65), 209(65), 231
Tiersten, H. F., **17:**181, 182, 222
Tietjan, J. J., **12:**113(188), 162
Tietjen, J. J., **5:**209(183), 234, 282, 319; **7:**127(18), 132(21), 169(97), 170(105), 232(236a), 242(263), 244(272), 247, 250, 254, 255
Tietz, L. A., **16:**61–62, 66, 70, 135, 136
Tigges, C. P., **16:**123, 143
Tijburg, R. P., **15:**227, 244, 261
Tildesley, D. J., **22:**127(32), 128(32), 172; **24:**207(19), 208(19)
Tilford, S. G., **7:**91(67), 113
Tillack, T. W., **24:**106(110)
Tiller, C. O., **1:**316(68), 333; **4:**131(71, 72), 136, 169, 171(105), 190; **5:**163(62, 63), 168(62, 63), 199, 209(63), 211, 212, 214(63), 215, 230, 234, 260, 315
Tiller, W. A., **4:**184, 190; **7:**120, 121(5, 7), 247
Tillett, P. I., **7:**11, 18, 19(25), 37
Tillman, N., **24:**12(54), 13(54), 15(54, 67), 18(67), 19(67), 88(44), 94(68, 70), 95(68, 70), 128(59), 212(59, 60)
Tillmann, P., **7:**385(201), 397
Timmermans, J. H., **24:**164(31), 168(31)
Timms, C., **3:**33(164), 36(164), 52(164), 57

Timofeev, A. N., **7:**267(56, 57), 270(56, 57), 308
Timoshenko, S., **16:**280, 337
Timoshenko, S. P., **3:**223(39), 271
Timothy, J. G., **11:**24(18), 25(18), 27(18), 28(18), 29(18), 34
Timson, P., **15:**289, 311
Tinder, R. F., **9:**46(206), 70
Ting, C. Y., **14:**72(235), 79; **16:**250, 268
Tinkham, M., **1:**257(68), 258(70), 265(85, 86, 87, 89), 275; **16:**11, 70, 125
Tirabassi, A., **19:**238, 240
Tirado-Rives, **24:**209(38), 210(38)
Tischenko, V. D., **3:**65(20), 125
Tisone, G. C., **13:**174(80), 178(98), 186(80, 98, 144), 189(80, 98, 144), 194, 195, 197
Tisone, T. C., **14:**27(106), 75, 226(81), 238; **22:**119(6), 171
Title, A. M., **9:**136(50), 144
Title, R. S., **7:**26(98), 39; **10:**41(79), 67
Titov, V. A., **12:**126(262), 164
Tittel, H. O., **9:**176(49), 178(49), 186(49), 257
Titus, J. A., **16:**111, 116, 142
Tiwari, S., **13:**210(38), 224
Tjapkina, V. V., **12:**73, 160
Tkaczyk, J. E., **16:**70, 136
Tkharev, E., **14:**94(32), 103(32), 106(32), 126
Toba, H., **15:**172, 216
Tobias, D. J., **24:**207(24)
Toda, K., **17:**235, 296
Todd, B. J., **1:**2(4, 5), 5, 6, 9(17), 63
Todorova, S., **17:**7, 63, 65, 73
Toft, A. R., **10:**114, 165; **12:**10(39), 12(39), 50
Togami, Y., **15:**168, 183–184, 215–216
Togano, K., **16:**9, 125
Tohyama, S., **21:**31(62, 63), 32(62), 73; **23:**144(35)
Toigo, A., **9:**245(245), 261
Tokarsky, R. W., **10:**60(137), 69
Tokuda, J., **13:**169(65), 185(65, 140), 186(149), 188(65, 140), 189(149), 194, 196, 197
Tokuhara, S., **15:**172, 216
Tokumotu, H., **24:**89(48)
Tokunaga, T., **15:**154–155, 163, 214
Tokura, Y., **16:**2, 124
Tokushima, T., **15:**171, 216
Tokuyama, T., **3:**141(51), 163; **4:**262, 314, 321; **14:**88(17), 89(19), 109(19), 117(19), 125
Tolansky, S., **1:**158, 186, 206, 230; **2:**235(39), 238; **3:**2(14), 17(7), 36(7, 14, 169, 170), 52(7, 14, 169, 170), 53, 57; **4:**2(2), 3(7), 20, 22(47), 25(58, 59), 26(59, 62), 86(236), 90, 91, 92, 96; **7:**336, 351, 352, 353(100, 101), 395; **9:**16(73), 66; **10:**204, 226
Tolksdorf, W., **9:**276, 284, 297; **15:**11, 14, 112, 137, 141
Tolle, H. J., **15:**171, 215
Tollet, H., **16:**52, 133
Tolliver, D. L., **4:**265(61, 63), 314(61), 319(61, 63), 321; **5:**287(210), 320
Tolmasova, V. N., **12:**71(61), 160
Tolomasov, V. A., **5:**163(70), 170, 231
Tomar, M. S., **7:**361(151), 362(153), 364(151), 365(170, 171), 396
Tomar, V. S., **16:**116, 142
Tomaspolsky, Yu. Ya., **17:**327, 335, 366
Tomboulian, D. H., **7:**45(8), 46(18), 47(22), 48(26), 66(53), 70(26), 111, 112, 113

Tombs, N. C., **3:**137(32b), 163; **4:**315(159), 324; **5:**297, 321

Tomg, C. J., **16:**2, 123

Tominaga, K., **13:**20(29), 23(29), 45(29), 57; **14:**18(74, 75), 19(74, 77, 82), 20(75, 82), 30(74), 66(219; **17:**153, 154, 160, 220

Tomita, K., **5:**311(285), 322

Tomita, O., **17:**345, 367

Tomita, Y., **17:**228, 232, 233, 237, 242, 243, 244, 245, 246, 247, 283, 293, 296, 297

Tomitori, M., **24:**98(88), 221(90), 222(90)

Tomlinson, H. M., **1:**189(12), 228

Tomlinson, W. J., **10:**242(33), 245(35, 36), 258, 259

Tompa, G. S., **16:**91, 139

Tompkins, F. C., **2:**48(79), 50, 62

Tonge, L. M., **16:**87, 93–94, 138–140

Tonooka, K., **17:**318, 365

Toombs, G. A., **15:**222, 249–250, 261, 263

Toombs, P. A. B., **6:**106(123), 107, 131(123), 146

Topfer, M., **5:**287(208), 319

Toprac, A. J., **22:**178(11), 193(51), 194(52), 226(51), 227(52), 228(52), 229(52), 271, 273

Torii, Y., **18:**166

Torok, E., **17:**44, 75

Torok, E. J., **3:**44(222), 59; **6:**282(196), 286(196), 299

Torresi, R. M., **17:**326, 327, 366

Torricelli, A., **4:**199(50), 204(50), 252

Torrn, C., **16:**2, 124

Torrori, I., **22:**181(18), 271

Tostmann, H., **24:**184(39)

Toulouse, B., **11:**49(71), 98

Tour, J. M., **24:**33(137), 44(37)

Tournié, E., **23:**221(10), 225(10)

Tournier, J., **4:**162(94), 189

Tourtellotte, H. A., **4:**263, 321

Tousey, R., **1:**131(14), 136(14), 142(28), 143, 147(22, 48), 148, 149(22, 48, 50, 56), 150(22), 152(22), 153(48, 50), 170(50), 172, 173(56), 176(50), 177(50), 182(50), 184(50, 71, 72), 185, 186; **2:**285(34), 304, 319, 324(29), 334(53), 359, 360; **3:**10(47), 51(47), 54; **4:**10(33, 34), 50(101), 60(34), 68(164), 91, 93; **7:**45(7, 10, 11, 12, 13, 14), 47(14, 24), 48(14), 57, 58(24), 70(14), 103(13), 104(82), 107(84), 108

Towa, S., **12:**248(45), 251(45), 253(45), 291

Tower, J. R., **23:**124(15)

Towner, F. J., **21:**241(11), 282(11), 290(49), 309, 310

Townes, **23:**177(5)

Townes, C. H., **3:**63, 125

Townsend, P. D., **13:**126(79), 149; **14:**217(42), 237; **17:**161, 221

Toxen, A. M., **1:**241, 254(20, 56), 259(20, 74), 274; **3:**229(64), 242, 272; **6:**115, 135(151), 147

Toyama, M., **7:**158, 225(78), 249

Toyoda, K., **10:**237, 258; **13:**153(8), 154(8), 182(8), 183(8), 192; **17:**339, 344, 367

Toyosawa, K., **24:**45(81)

Track, E. K., **16:**110, 142

Tracy, C. J., **22:**227(83), 232(83), 235(83), 274

Tracy, K. M., **22:**266(115, 116), 275

Trageo, F., **24:**92(58)

Trakhtenberg, I. Sh., **7:**267(56, 57), 270(56, 57), 292(103), 308, 310

Tran, N. C., **14:**31(119), 76

Tran-Quoc, H., **19:**85, 93
Trapp, K. D. C., **21:**221(137), 222(137), 237
Traub, A. C., **3:**15, 51(65), 52(213), 54
Traub, E., **3:**40(213), 58
Trawny, E. W. L., **12:**70(56), 126(261), 128(261), 132(261), 135(292), 159, 164, 165
Traynor, S., **17:**256, 257, 298
Treble, F. C., **21:**315(8), 366
Tredgold, R. H., **5:**160(51), 230
Trefny, J., **16:**13, 125
Tregold, R. H., **24:**180(9)
Treherne, D. M., **4:**84(214), 96
Trekhov, E. S., **14:**196
Tremere, D. A., **7:**255
7.(257)
Trendelenburg, E. A., **1:**42(109), 66
Trent, H. M., **4:**23(55), 92
Tressler, R. E., **15:**288–289, 311
Tretyakov, D. N., **7:**155(68), 217(200), 246(276), 249, 253, 255
Treupel, H., **1:**214(123), 231
Treuting, R. G., **3:**243(66), 272; **4:**174, 190
Trevalion, P. A., **13:**8(13), 57
Trevor, D. J., **16:**51, 132
Triboulet, R., **13:**186(151), 190(151), 197; **19:**38
Trickett, E. A., **5:**285(202), 319
Triebwasser, S., **5:**216(196), 234
Trillat, J. J., **1:**199(63), 223, 225, 229; **4:**150(43, 44), 152(43, 44), 188
Trimm, D. L., **10:**37(68), 67
Triolo, J. J., **10:**57(122), 68, 128(56), 129(64, 65, 66), 130(56, 64), 131(66), 132(72), 134(65), 135(65), 136(65), 137(65), 138(65), 139(65), 141(66), 142(66, 77), 143(66), 145(66), 146(66), 147(64), 149(79), 150(79); **12:**28(54), 29(57, 58), 31(60), 151(64, 65, 79), 152(56, 79), 153(56), 154(79), 157(85), 158(85), 161(85), 165, 166
Tripathi, P., **6:**142(236), 149
Tripp, K. F., **9:**130(36), 131(36), 144
Tripp, T. B., **6:**21(74), 33(74), 54(149), 75, 76
Triscone, J.-M., **16:**19, 28, 126, 128
Trivelpiece, A. W., **18:**65
Trivich, D., **4:**199(56, 61), 200(56, 61), 204(56, 61), 206, 209, 242(61), 252
Trofim, V. G., **7:**186, 187(148), 246(279), 247(282), 251, 256
Trolier-McKinstry, S., **19:**249, 254, 263, 267, 272, 273, 276
Trombe, F., **10:**25(41), 66
Trommer, R., **11:**89(230), 103; **12:**294
Tromp, R. M., **14:**232(125), 239; **23:**43(74)
Trompette, J., **3:**1(3), 53
Tronin, A. Y., **19:**281
Tronstad, L., **4:**82(203), 95
Trotel, J., **10:**243(34), 258
Troue, H. H., **14:**191(166), 199; **17:**83, 130, 139, 143
Troughton, E. B., **24:**3(11, 13), 12(13), 13(13), 14(13), 15(13), 16(13), 18(11, 13), 20(13), 30(13), 44(12, 14), 46(12, 14), 55(12, 14), 56(12, 14), 65(12), 70(14), 112(3), 212(57), 220(57), 228(18), 237(18)
Trounson, E. P., **6:**195(54), 203; **9:**176(39), 256
Trousdale, W. L., **6:**246, 298
Truman, J. K., **16:**17, 126
Trumbore, F. A., **7:**148, 164, 165, 166(56), 223(221), 224(222), 248, 253

Trundle, C., **17:**237, 238, 296
Truong, K. D., **24:**46(88), 55(88), 56(88)
Truong, V.-V., **17:**355, 356, 368
Trurnit, H. J., **7:**327(46), 349, 358, 360(148), 364(163), 388(206), 391(214), 393, 395, 396, 397
Trussell, C. W., **7:**154(66), 219(205), 249, 253
Tryk, D. A., **24:**231(51), 233(51)
Tsai, B. L., **24:**228(10)
Tsai, C. C., **15:**191, 217; **19:**93
Tsai, C. S., **15:**115, 122–124, 141; **17:**174, 175, 176, 194, 222
Tsai, D. H., **22:**144(67), 173
Tsai, J. A., **16:**28, 128
Tsai, W., **22:**17(25), 18(25, 26), 34(26), 35(26), 45(25, 26, 51, 52), 46(25, 53), 47(25), 49(53), 54(58), 76(72), 77–79, 227(82), 232(85), 274
Tsakalakos, T., **17:**2, 43, 48, 49, 51, 54, 55, 56, 61, 73, 75, 76
Tsakalis, K. S., **22:**269(118–120), 275
Tsaneva, V. N., **14:**51(165), 77
Tsang, J. C., **14:**232(125), 239
Tsang, W. T., **11:**67(129), 79, 88, 89, 91, 99, 101, 102, 103; **13:**218(60), 224; **23:**228(31), 229(31)
Tsao, C. H., **16:**272, 285, 332, 337
Tsao, J. Y., **13:**155(30), 157(31), 159(31), 161(42), 165(30), 168(42); **17:**26, 28, 75
Tsarenkov, B. V., **19:**2
Tsaur, B. Y., **23:**107(58), 146(43), 165(43)
Tsaur, B.-Y., **16:**41, 131
Tsen, M., **24:**12(50), 15(50), 18(50), 19(50)
Tseng, W. F., **23:**48(162)
Tsiensuu, V. H., **7:**11(24), 37
Tsong, I. S. T., **22:**113(28), 115
Tsu, R., **11:**81(185a), 82(185b), 86(202), 101, 102; **15:**245, 252, 263; **21:**81(11), 111
Tsubouchi, K., **17:**153, 154, 221
Tsubouchi, N., **23:**105(55)
Tsubouchi, S. N., **23:**124(17)
Tsuchida, E., **17:**361, 362, 370
Tsuchikawa, H., **16:**251, 268
Tsuchimoto, J., **13:**166(61), 167(66), 168(61), 185(61, 66), 186(149), 188(61, 66), 189(149), 194, 197
Tsuei, C. C., **16:**11, 17, 41, 51, 78, 125–126, 131–132, 137
Tsuge, H., **22:**6(11), 17(11), 32(11), 77
Tsui, D. C., **9:**218(170), 259; **15:**223, 244, 249–250, 261, 263; **21:**98(34), 100(34), 112; **23:**290(109)
Tsui, E. S. M., **21:**334(29), 341(29), 342(29), 355(29), 361(43), 367, 368
Tsui, J. B., **15:**54, 138
Tsui, R. T. C., **1:**251(48), 274; **6:**102(95), 145
Tsuji, H., **15:**204, 217
Tsukada, I., **16:**54, 134
Tsukada, N., **13:**181(110), 185(139), 188(139), 195, 196
Tsukada, T., **14:**34(135), 76
Tsukahara, S., **13:**172(77), 173(77), 177(77), 190(77), 191(77), 194
Tsukamoto, K., **22:**181(18), 271
Tsukane, N., **15:**172, 216
Tsukiji, N., **23:**228(32)
Tsunashima, S., **15:**147, 154, 163–164, 171, 176, 179, 185–186, 195–199, 204, 214, 216–217
Tsunokuni, K., **18:**102
Tsuroka, K., **13:**60(5), 78(5), 79(5), 104

Tsuruoka, T., **16:**86–87, 138
Tsutsumi, M., **15:**85, 140
Tu, C. W., **21:**128(37), 233
Tu, K. N., **16:**234, 266
Tuan, H. S., **15:**33, 72–73, 101, 137, 139–140
Tubb, D. J., **3:**168, 208
Tubbs, M. R., **2:**132(153), 134(153), 145
Tubouchi, N., **23:**88(14)
Tuchkevich, V. M., **7:**246(276), 255
Tucker, A. W., **13:**185(138), 188(138), 196
Tucker, R. N., **5:**307, 322
Tucker, T. W., **5:**163(73), 170(73), 231
Tuckerman, M. E., **24:**207(23, 24)
Tufts, C., **1:**218(135), 231
Tuinstra, F., **17:**77
Tuler, F. R., **12:**287, 295
Tullos, T., **16:**240, 267
Tulock, J. J., **24:**75(122)
Tulpule, A. M., **16:**91, 139
Tuma, D. R., **18:**65
Tuma, D. T., **14:**139(34), 162(112), 163(112), 195, 197; **17:**93, 140
Tummers, L. J., **5:**163(68), 168(68), 231
Tung, C., **9:**265, 295
Tung, S. K., **4:**261(32), 263(32), 264, 320; **5:**275, 288, 317, 320
Tuomi, T., **7:**49(28), 112
Tuppen, C. G., **23:**13(47)
Tupper, K. J., **24:**221(100, 101)
Turbadar, T., **6:**163, 202; **9:**234, 260
Turban, G., **12:**272(163, 165, 166), 273, 289(266, 268), 293, 296
Turcolte, R., **12:**176(48a), 233
Turkevich, J., **12:**225(254), 227(254), 241
Turnbull, A., **4:**26(62), 92
Turnbull, A. A., **7:**242(261), 255; **12:**112(184), 118(184), 137(306), 162, 165
Turnbull, D., **1:**210(119), 231; **4:**97(2), 106, 134, 135
Turnbull, D. T., **3:**9(37), 10, 36(169), 49(37), 51(37, 169), 54, 57
Turnbull, J. C., **1:**10, 11, 12, 64
Turnbull, W. R., **5:**145, 147, 229
Turner, A. F., **1:**70(16, 18, 20), 87(16), 104(20), 120; **2:**240, 285(32), 295(43), 303, 304, 333, 340(66), 353(50, 66), 360; **3:**220(28), 234, 235(28), 271; **4:**9(27), 91; **5:**134(37), 141; **7:**62(42), 112; **9:**74(7), 79(8), 82, 91(8), 95, 106(21), 116, 129(7), 130(7), 131(7), 143, 144; **10:**29(50), 54(117), 66, 68, 115, 165; **12:**3(15), 42(15), 50
Turner, C. W., **16:**7, 12, 113, 125
Turner, D. C., **24:**98(91)
Turner, D. R., **4:**199(70), 200(70), 206(70), 252
Turner, D. T., **9:**8(37), 65
Turner, E. F., **1:**207(97), 230
Turner, E. G., **10:**253(48), 259
Turner, F., **16:**54, 134
Turner, G. L., **3:**116(133), 128, 139(40), 163
Turner, G. L'E., **5:**167(132), 203(132), 232
Turner, G. M., **22:**146(69), 173
Turner, G. W., **11:**81(185a), 101; **23:**230(14), 230–231(34), 232(17, 18), 237(33), 278(100)
Turner, J. A., **3:**19, 51(91), 55; **21:**365(49), 368
Turner, L. G., **16:**41, 43–48, 50, 52, 131–133
Turner, P. A., **6:**335(40), 347
Turner, R. G., **13:**210(30), 223

Turney, A., **11:**253, 255
Turnquist, R. D., **6:**224(81), 292(81), 297
Turos, A., **14:**230(100), 239
Turova, N. Ya., **17:**318, 366
Turski, Z., **15:**68, 139
Turyan, I., **24:**183(31), 192(31)
Tutov, E. A., **17:**344, 367
Tuttle, B., **17:**291, 299
Tuzzolino, A. J., **4:**68(163), 94
Tveten, A. B., **15:**111, 113–116, 141
Twietmeyer, H., **9:**238(230), 239(230), 261
Twose, W. D., **6:**64(182), 78
Tyagi, M. S., **21:**325(19), 367
Tyler, J. E., **15:**288, 311
Tyson, W. R., **22:**261(106), 275
Tyutikova, L. P., **5:**60(18), 86
Tzeng, Y., **16:**123, 143

U

Ubersax, F., **4:**199(48), 204(48), 252
Uchida, H. J. H., **17:**135, 144
Uchida, N., **14:**31(121), 76; **17:**261, 298
Uchida, S., **16:**2, 3, 124
Uchida, T., **22:**181(18), 271
Uchida, Y., **1:**148(54, 55), 169(54, 55), 186
Uchikawa, K., **17:**356, 369
Uchinokura, K., **16:**54, 134
Uchiyama, S., **15:**147, 154, 163–164, 171, 176, 179, 185–186, 195–199, 203–204, 214, 216–217
Udayakumar, K. R., **19:**249, 250, 253
Uddin, A., **21:**207(106), 235
Udin, H., **4:**139(12a, 13), 187
Udovichenko, Yu K., **1:**44(121), 66
Udy, M. C., **2:**387(110), 422
Uebbing, J. J., **7:**244(270), 255; **12:**76(92), 77, 81(92), 113(187), 114(200), 118(92, 231), 160, 162, 163
Ueda, A., **22:**145(68), 173
Ueda, I., **17:**228, 232, 233, 237, 242, 243, 245, 246, 247, 283, 293, 296, 297
Ueda, M., **15:**293, 311
Ueda, R. M., **11:**36(16), 96
Uehara, K., **17:**357, 369
Uesugi, N., **10:**253(49), 254(49), 259
Ufert, K. D., **17:**111, 112, 142
Ufert, K.-D., **14:**11(40), 74
Ufford, C., **9:**110, 144
Uhl, J. E., **14:**210(24, 25), 211(24), 212(24), 213(24), 220(68), 221(24, 25), 237, 238
Uhlmann, D. R., **11:**238(42), 315
Uhmann, D. R., **15:**292, 311
Ujihara, K., **10:**36(63), 67
Ukada, T., **17:**162, 221
Ulacia, F., **22:**5(5), 77
Ulenaers, M. J. E., **17:**228, 240, 256, 257, 297, 298
Ulin, P. B., **3:**276(11), 299
Ullman, F. G., **2:**400(163), 424; **6:**22, 64(188), 75, 78
Ullrich, B. M., **23:**48(162)
Ulman, A., **19:**281; **24:**2(4), 12(53, 54), 13(54), 15(54, 67), 18(67), 19(67), 23(114), 31(114), 32(131), 44(7, 22, 23, 24, 49), 45(7, 75), 47(95), 64(49), 70(22), 82(8), 86(8), 88(8, 44), 112(2, 14), 114(27), 121(14), 128(59), 150(1)
Ulmer, E. A., **7:**221(214), 253
Ulrich, R., **9:**169(28), 256; **10:**227(3), 230(3), 234(3), 236(3), 242(33), 258
Umblia, E., **5:**93(5), 138, 140

Umeda, J., **10:**228(7, 8), 251(46), 252(47), 258, 259
Umezu, N., **13:**78(62), 79(62), 106
Umezu, S., **10:**58(125), 68
Umrath, W., **6:**187, 203
Ungashe, S. B., **24:**12(55), 14(55)
Unger, P., **12:**71(65), 160
Unger, T. A., **2:**353(76), 361; **10:**43(86), 44(86), 67
Unnam, J., **16:**237, 267
Uno, F. M., **2:**367(14a), 405(182), 420, 424
Uno, T., **17:**162, 221
Unuma, H., **17:**318, 365
Unvala, B. A., **4:**169(111), 172(111), 190; **5:**163(71, 72), 168, 169, 170, 188, 196(71, 72), 197, 198, 205(72), 231
Uosaki, K., **24:**23(109), 27(109)
Uphaus, R. A., **24:**181(14)
Urabe, K., **17:**339, 344, 367
Urankar, E., **24:**23(114), 31(114), 44(23)
Urbach, F., **2:**376(75), 421; **9:**15, 66; **17:**337, 367
Urban, F. K., **19:**195
Urban, K., **16:**24–25, 28, 72–73, 127–128, 137
Urban, M., **7:**240(256), 255
Urbank, K., **12:**259(86), 292
Urbassek, H. M., **14:**15(55), 74
Uritskaya, A. A., **12:**169(34, 36), 203(34, 35), 204(35), 233
Urner-Wille, M., **15:**180, 182, 197, 216–217
Uryashara, L. D., **6:**216(56), 296
Ushio, S., **13:**20(32), 21(32), 52(62), 57, 58; **15:**295, 311
Usui, H., **13:**5(9), 13(22), 20(46), 21(9), 24(46), 56, 56(46), 57, 112(31), 116(35), 148
Usui, S., **19:**93
Utsumi, T., **14:**162(111), 197; **17:**93, 94, 140
Uusimaki, A., **16:**100, 140
Uyeda, R., **2:**106, 122(122), 142, 144; **4:**155, 171, 189, 190
Uzan, C., **13:**186(151), 190(151), 197

V

Vaglio, R., **16:**17, 126
Vahala, K. J., **23:**226(29), 228(29)
Vaidya, S., **16:**250, 268
Vaidyanathan, K. V., **11:**66(120), 99
Vail, C. R., **2:**365(6), 419; **14:**175(128, 129), 198
Vainshtein, B. K., **2:**83, 110, 141, 143
Vainshtein, V. M., **9:**51(243, 244, 247), 70, 71
Vaipolin, A. H., **7:**36(163), 41
Vaknin, D., **24:**181(14)
Vakula, S. I., **14:**176(140), 198
Valco, G. J., **16:**43, 50, 107, 131–132, 141
Valdré, U., **3:**39(205), 52(205), 58
Valdre, U., **4:**186, 190
Vale, C. R., **17:**199, 200, 202, 203, 204, 205, 206, 207, 223, 224
Valenta, L., **1:**327, 334
Valentine, K. G., **24:**164(31), 168(31)
Valentino, A. J., **9:**281, 294
Valette, S., **10:**257, 259
Valin, J., **6:**211(24), 212(24), 225(24), 242(24), 244(24), 249(157), 250(24), 254(24), 255(24), 256(24), 257(24), 258(24), 259(24), 260(24, 157), 269, 271(24, 157), 295, 298, 299
Valla, B., **17:**361, 370
Vallade, M., **24:**181(12), 192(12), 199(12)

Valleau, J. P., **24:**208(27)
Valles-Abarca, J. A., **14:**21(84), 75
Valletta, R. M., **3:**101, 102(90), 113, 126; **4:**261(30), 263(54), 264(54), 267, 268, 285, 288, 319(54), 320, 321
Valley, D. J., **2:**370(56), 421; **3:**158(86), 164
Vallin, J. T., **21:**207(106), 235
Valstyn, E. P., **6:**225, 297
Valvoda, V., **17:**83, 89, 119, 121, 126, 131, 139, 140, 143
Vampola, J. L., **21:**259(33), 309
Van Arkel, A. E., **3:**134, 162; **5:**247, 315
Van Asselt, R. L., **12:**157(336), 166
Van Bommel, A. J., **12:**122(247), 164
Van Bueren, H. G., **3:**245, 272
Van Cakenberghe, J., **3:**180, 181(38, 39), 209; **5:**184, 233; **14:**16(63), 74
Van Cakenberghe, J. L., **9:**42(179), 56(264), 58(264), 62(264), 69, 71
Van Cakenburghe, J., **2:**183(66), 192
van de Leur, R. H. M., **17:**77
Van de Ven, E. P. G. T., **12:**263(102), 264(102), 292; **14:**234(137), 240
Van de Walle, C. G., **21:**79(8), 111
van den Berg, P. M., **9:**244(243), 261
Van Den Hoek, W. G. M., **15:**297–299, 311
van der Leij, M., **12:**180(68), 234
van der Linden, P. C., **9:**33(138), 35(138), 68
Van der Merwe, J. H., **1:**218, 231; **3:**244, 272; **4:**107(22), 128(22), 135, 139, 142, 157(15), 162(15, 16, 17, 18), 174, 184, 185(143), 187, 190, 231, 254; **5:**206, 234; **7:**125(12, 14), 133(12), 247; **16:**67, 136; **17:**26, 75
van der Vaart, H., **15:**26, 137
Van der Vegte, E. W., **24:**95(73)
van der Weg, W. F., **14:**230(100), 239
Van der Ziel, J. P., **11:**79(176), 81(183), 86(208, 209, 210, 210a, 211, 212), 101, 102
van Dover, R. B., **15:**147, 150, 152, 161–162, 169–171, 185–186, 214–216; **16:**3, 96, 124, 140
Van Duzer, T., **16:**7, 12, 28, 101, 113, 125, 128, 141
Van Geel, W. C., **6:**19, 75
Van Gelder, W., **4:**316(163), 324
Van Geloer, W., **5:**288(214), 320
van Gemert, M. J. C., **11:**230(30), 315
van Gunsteren, W. F., **22:**132(48), 134(54), 136(54), 172–173; **24:**220(85)
Van Gurp, G. J., **16:**237, 261–262, 267, 269
van Hal, H. A. M., **17:**240, 297
van Harlingen, D. J., **16:**99, 140
van Heel, A. C. S., **2:**233(36), 234(36), 238; **3:**17, 18(79, 80), 51(79, 80), 55
Van Heerden, P. J., **1:**33(79), 65
Van Hove, **24:**114(32)
Van Hove, L., **12:**57, 159
Van Itterbeek, A., **2:**23(26), 60; **3:**121, 128; **6:**127, 148
Van Khai, T., **6:**292(217), 300
Van Laar, J., **5:**1(4), 45; **7:**242(260), 255; **12:**54(7), 112, 114(204, 205, 208), 129(7), 158, 163; **13:**201(2), 223
Van Leeuwen, J. M. J., **24:**184(40), 190(40)
Van Milligan, F. J., **13:**124(71), 127(71), 149
Van Oosterhout, G. W., **9:**9(44), 65

Van Overstraeten, R. J., **21:**10(30, 31), 12(30), 15(30), 32(31, 65), 72, 73
Van Poosbroeck, W., **21:**319(12), 366
van Roosbroeck, W., **11:**142(107), 143, 202
Van Rooy, T. L., **12:**114(208), 163; **13:**201(2), 223
Van Rossum, M., **21:**287(40), 310
van Silfhout, A., **19:**2, 193, 203
Van Stappen, M., **17:**110, 141
Van Uitert, L. G., **9:**264, 267, 275, 276, 287, 288, 294, 295, 297; **10:**256(62), 259
Van, V. N., **6:**106(124), 131(124), 146
Van Vechten, J. A., **7:**173(118), 250; **12:**114(196), 163; **13:**52(61), 58; **19:**73
van Veggel, F. C. J. M., **24:**23(108)
Van Winkel, P., **6:**19, 75
Van Zanten, P. G., **1:**51(126), 67
Van Zon, J. B. A. D., **17:**111, 112, 142
Van Zytveld, J. B., **6:**97, 103(82), 113, 119, 124(82), 127, 129, 130, 135(82), 145
Vancu, A., **19:**73
Vandenberg, J. M., **15:**229–230, 234, 239, 262; **21:**117(10), 128(10), 197(98), 209(10), 215(10), 232, 235
Vandenberg, M., **12:**269(145), 293
Vandenplas, P. E., **18:**146
Vander Lindea, P. C., **12:**187(78), 234
Vanderah, T. A., **16:**78, 137
Vandermeer, R. A., **7:**294(107), 310
Vanderslia, T. A., **14:**156(89), 197
Vanderslice, T. A., **1:**40, 66; **4:**9(28), 84(28), 91; **10:**120(51), 165; **14:**34(136), 76; **17:**95, 141
Vandervoort, K., **16:**4, 124
Vanderwyck, A. H. B., **11:**160(133), 203
Vankar, V. D., **12:**169(24), 195(24), 196(24), 233; **13:**79(78), 99(78), 106
Vannimenns, J., **12:**135(298), 165
Vant-Hull, L. L., **10:**25(42), 66
VanUitert, L. G., **17:**355, 368
Varadi, P. F., **1:**2(3), 63
Varaljay, N. C., **16:**76, 137
Varasi, M., **13:**131(97), 140(97), 141, 149
Varcille, A., **14:**107(58), 119(58), 126
Varga, A. J., **11:**107(13), 128(13), 130(83), 131(13, 83), 135(13), 155(13), 158(131), 190(153), 200, 201, 203
Varga, I. K., **13:**3(5), 56, 117(42), 118(42), 119, 144(42, 135), 148, 150
Varghese, R., **19:**161
Varma, B. P., **12:**68(52), 70(54), 74(82), 85(104), 86(52, 112, 115), 89(52), 90(52), 92, 97, 98, 99(52), 100, 101, 103(115), 105, 109(104, 175, 180), 110(180), 111, 145(315), 147, 148(320), 159, 160, 161, 162, 166
Varma, R. R., **12:**114(209), 163
Varnerin, L. J., **10:**256(61), 259
Varob'eva, O. V., **12:**176(52), 233
Varonides, A. C., **21:**362(45), 364(45), 368
Vasant Kumar, C. V. R., **17:**241, 290, 297
Vase, P., **16:**66–67, 136
Vasella, A., **24:**21(98), 22(98)
Vasicek, A., **2:**242(12), 304; **3:**2(15), 53; **6:**155, 202; **7:**365(167), 396
Vasil'ev, Y. B., **23:**243(62)
Vasilishcheva, I. V., **13:**186(154), 190(154), 197
Vasin, A. I., **14:**138(27), 195

Vasquez, M. V., **17:**326, 327, 366
Vassell, M. O., **15:**252, 264
Vaughan, T., **21:**121(23), 123(23), 233, 303(53), 310
Vaughan, T. B., **4:**199(57), 204, 205(57), 206(57), 207(57), 209(57), 211(57), 252
Vaughan, W. H., **10:**84(25), 164
Vaughn, J. H., **2:**367(24), 393(24), 420; **3:**103(95), 127
Vaures, A., **16:**30, 32, 129–130
Vautier, C., **6:**118(166), 147
Vavilov, V., **12:**60(37), 159
Vavra, L., **16:**249, 268
Vaynshetyn, V. M., **9:**51, 70
Veas, F., **9:**5(22), 65
Vecht, A., **3:**169(6), 172(15), 174, 179(31), 183(31, 43), 188(50); **12:**217(109), 235; **13:**43(52), 58
Vedam, K., **19:**10, 52, 54, 55, 129, 148, 196, 199, 205, 206, 213, 216, 232, 254, 263, 280, 288
Vedel, J., **12:**226(113), 235
Veeneman, D., **9:**33(138), 35(138), 68; **12:**187(78), 234
Veira, J. A., **16:**2, 124
Veith, W., **3:**179(34, 35), 198, 209
Veksler, V. I., **3:**71, 125
Velasco, G., **17:**314, 328, 365
Velde, Te., **3:**182, 209
Velicky, B., **1:**144(38), 185
Vella-Coleiro, G. P., **9:**271, 276, 296, 297
Veltrop, H., **14:**170(122), 198
Venable, D., **1:**127(5), 184
Venables, J. A., **14:**59(188), 77; **17:**30, 75; **19:**70
Venables, J. D., **6:**13(54, 55), 75; **7:**269(64), 308
Venema, A., **1:**14(25), 55, 64, 67, 241(13), 273; **2:**393(137), 423
Venevtsev, Yu. N., **17:**344, 367
Venkataraman, K., **7:**62(43), 64(45, 47), 112, 113
Venkatesan, T., **16:**55–56, 58, 60–63, 106, 110–112, 134–136, 141–142; **17:**92, 140, 229, 236, 237, 263, 264, 270, 271, 296, 298
Venkateswaran, N., **24:**235(56), 241(56)
Venskytis, F. J., **11:**10(4), 33
Venturini, E. L., **16:**54, 134
Venugopalan, M., **12:**238(5), 290
Veprek, S., **12:**238(2), 242(18), 243(18), 245(18), 269(153), 270(156); **19:**80, 93
Verbitskaya, T. N., **15:**64, 80–81, 139
Verderber, R. R., **5:**307(277), 322; **6:**64(179), 78
Vergara, W. C., **7:**50, 112
Verhaeghe, M., **10:**169(18), 225
Verhoeven, J., **16:**234, 239, 266
Verhoeven, J. D., **7:**258(1), 262(17), 305, 306
Verkin, B. I., **6:**122, 135(180), 147
Verlet, L., **22:**128(40), 132(40), 172
Verma, A. R., **4:**195, 251; **7:**337(106, 107), 338(106, 107), 349(106, 107), 353(106, 107), 354, 356, 357, 358(107), 373(106), 395
Verma, R. L., **12:**148(320), 166
Vermaak, J. S., **4:**184, 185(143), 190
Vermilyea, D. A., **1:**207(103), 223(103), 227(103), 230; **2:**406(189), 424; **4:**198, 200(74), 206, 207, 252, 253; **6:**2, 4, 6(26, 31), 7(35a), 8(35b), 11(46), 12, 14(22, 31), 15, 19, 30, 74, 76; **17:**43, 75
Verrill, J. F., **11:**21(14), 33
Vértes, A., **17:**318, 319, 366
Verwey, E. J., **9:**9, 65

Verwey, E. J. W., **6:**14, 75
Veselovskii, V. I., **6:**64(187), 78
Vesely, W. E., **3:**213(10), 214(10), 217, 270
Vest, R. W., **16:**95–96, 140
Vette, J. I., **10:**159(88), 166
Vezmar, I., **24:**151(10), 154(10), 156(10), 163(10), 222(104)
Via, G. G., **1:**199, 215, 218(57), 229; **5:**163(69), 231
Vick, G. L., **5:**255(42), 290, 315, 320
Vickers, T. M., **24:**44(61)
Vickers, V. E., **21:**21(49), 73; **23:**91(23), 144(40)
Vidal, F., **16:**2, 124
Vidal, M. A., **22:**7(13), 77
Vidimari, F., **13:**212, 224
Vieland, L. J., **7:**147(55), 180(127), 248, 251
Viennet, R., **17:**358, 369
Viggiano, J. M., **16:**117–118, 143
Vigmond, S. J., **24:**36(143)
Vigue, J. C., **9:**282, 294
Vijh, A. K., **6:**68, 79; **9:**10(48), 66
Vilato, P., **14:**95(34), 122(34), 126
Vilelea, M. F., **21:**325(17), 330(17), 332(17), 338(17), 345(36), 346(36), 350(17), 362(17), 367
Villain, J., **16:**292, 334, 340
Villani, T. S., **23:**101(33), 124(14, 15)
Vilms, J., **7:**143, 145, 146, 237, 238(49), 248; **11:**65(119), 99; **23:**91(22)
Vincent, C. A., **9:**22(91, 92), 27(91, 92), 67; **12:**180(67), 234
Vincent, G., **23:**161(62)
Vincent, R., **9:**180(59), 202(117), 212(117), 257, 258
Vincent-Geisse, J., **2:**217, 238
Vine, J., **5:**1(7), 45; **12:**136(303), 165
Vinter, B., **21:**136(46), 233
Vinzelberw, H., **16:**70, 136
Violet, A., **19:**6
Violet, C. E., **3:**213(10a), 270
Violette, K. E., **23:**48(170)
Vionter, B., **21:**98(34), 100(34), 112
Viralda, V. J., **14:**72(235), 79
Virmani, M., **22:**202(61), 216(61, 78), 241(88), 245(61), 273–274
Virshup, G. F., **21:**362(44), 368
Vis, V. A., **3:**43, 59
Viserian, I., **9:**33(156), 34(156), 68
Vispute, R. D., **16:**70, 136
Vissner, W. M., **4:**75(176), 95
Viswanathan, R., **16:**70, 136
Vitins, J., **6:**7(35), 54(35), 66(35), 74
Vitkavage, D. J., **23:**34(125)
Vits, W., **7:**153(62), 248
Vittoria, C., **15:**9, 37, 112, 129, 136–137, 141
Viverito, T. R., **9:**33(116a, 135), 35, 37(135), 67, 68; **12:**177(58), 180(66), 184(66), 187(66), 190(66), 234
Vlannes, N. P., **15:**115, 118, 141
Vlieger, J., **19:**128, 129, 139, 140, 173
Vodar, B., **2:**23(27, 29), 60; **7:**45, 111; **14:**175(127), 198
Vodar, M. B., **1:**134, 147(46), 185
Vodjdani, N., **21:**136(46), 233
Vodopynov, L. K., **5:**221(204), 234
Voegeli, O., **6:**224(84), 249(84), 271(84), 291(210), 292(210), 297, 300; **9:**271, 296
Voet, A., **24:**230(45)
Vogel, A., **7:**149, 151(58), 248
Vogel, J., **17:**83, 128, 130, 139
Vogel, J. K., **15:**137
Vogel, S. F., **9:**50(236), 51(236), 52, 70
Vogel, T. P., **12:**91(139), 161
Vogel, W., **7:**36(162), 41
Vogler, F., **3:**295(66), 301

Vogt, C., **4:**327, 369
Voigt, J. A., **16:**4, 81, 124
Voigt, R., **14:**55(173), 77
Vojnovich, T., **9:**28(105a), 67
Volante, R. P., **24:**9(31)
Vold, M. J., **22:**5(9), 77
Volger, J., **2:**175, 192; **11:**294(132), 317
Volin, E. M., **14:**175(136), 198
Volin, T. E., **7:**286(93), 309
Volkel, E., **16:**5, 124
Vollkommer, F., **16:**237, 267
Volluet, G., **15:**79–80, 139
Völmar, S., **16:**70, 136
Volmer, M., **2:**13, 60; **4:**99, 134, 150(66), 189, 192, 198, 199(47, 49), 204(47, 49), 251, 252
Voloshchuk, Ya. V., **7:**27(113), 29(123), 40
von Allmen, M., **23:**48(162)
von Allmen, P., **21:**121(18), 123(18), 232
von Ardenne, M., **13:**112(26), 114(26), 147
von Auloch, W. H., **15:**9, 13, 136
von Bassewitz, A., **6:**114, 115, 124(141), 129, 132(141), 135(141), 146
von Blanckenhagen, P., **9:**219, 259
von Borries, B., **2:**122(124), 144
Von Bosse, J., **12:**286, 295
von Dessonneck, K. E., **16:**53–54, 134
Von Engel, A., **14:**147(69), 196, 209(22), 237
von Festenberg, C., **9:**180(57), 202(116), 257, 258
von Fragstein, C., **9:**190(83), 257
von Gutfeld, R. J., **13:**183(118), 184(118), 185(127), 187(127), 190(127), 191(127), 196
von Hippel, A., **3:**63, 85(64), 125, 126, 276, 299; **6:**72, 79
von Laue, M., **1:**198(48), 229
von Neumann, J., **21:**62(119), 65(119, 123), 74, 75
von Philpsborn, H., **7:**146, 147(52), 248
von Schweidler, E., **11:**230, 315
von Willisen, F. K., **17:**304, 364
Vonnegut, B., **4:**106, 135
Vonogradova, V. V., **9:**33(154), 68
Vook, R. W., **1:**239(7), 244(7), 273; **3:**226, 271; **6:**113(136, 137), 146
Vorlicek, V., **19:**73
Vorobeva, O. B., **12:**101(168), 162
Vorob'eva, O. V., **9:**33(154), 68
Vorob'eva, O. W., **5:**94(6), 121(6), 130(6), 141
Voronoi, G. F., **7:**24(88), 39
Voskoboynikov, V. V., **13:**99(112), 100, 107
Voss, W. A. G., **9:**43(195), 44, 69
Vossen, J. L., **6:**37(130), 44, 77; **9:**42, 46(177, 210, 211), 50(177, 227), 51(177), 53(254), 54(177, 227, 257, 258), 55, 56, 57, 62(257, 258, 259, 263), 64(274), 69, 70, 71; **10:**52(107), 68, 130, 163, 165; **12:**169(6), 232, 240(13), 241(13), 290; **13:**74(56), 105; **14:**31(123), 56(176), 59(123, 185), 71(185), 76, 77, 203(9, 11), 209(23), 212(11, 23), 217(9, 11, 40), 219(60, 61), 220(60, 61), 221(60), 222(60), 223(60, 76, 78), 224(61), 225(9), 226(80), 227(9), 228(40), 229(9, 98), 230(98, 99), 231(98), 232(99), 233(23, 98), 235(151, 152), 236(151), 237, 238, 239, 240; **16:**4, 14, 124–125, 153, 229; **17:**81, 83, 84, 92, 111, 139, 140,

Vossen, J. L., (*continued*) 146, 148, 157, 162, 163, 186, 220, 221, 232, 248, 273, 274, 294, 296, 298
Voter, A. F., **22:**254(98), 261(98), 262(98), 274
Votova, E., **1:**220, 231
Vratny, F., **2:**367(29), 396(150), 420, 423; **6:**36(121), 44, 76
Vriens, L., **13:**121(57), 148
Vu, T. Q., **15:**124, 141
Vuillermoz, B., **14:**87(16), 90(90), 98(16), 103(16), 105(16), 109(16), 111(63), 113(16), 120(16), 124(90), 125, 126–127, 127
Vuillermoz, R., **14:**107(58), 119(58), 126
Vural, K., **21:**221(134), 236
Vurgaftman, I., **23:**279(103), 280(103)
Vyatskin, A. Y., **2:**71, 141
Vyskocil, J., **14:**14(48), 74; **17:**83, 95, 97, 98, 99, 111, 112, 114, 115, 116, 137, 139, 141, 142

W

Waber, J. T., **9:**12(56), 66
Wachtel, M. M., **12:**87, 88, 161
Wachtman, J. B., **16:**55, 58, 63, 134–135; **17:**236, 296
Wacrenier, J. M., **11:**236(40), 244, 315
Wada, M., **12:**95(156), 162
Wada, Y., **11:**268(86), 269(86), 316
Wadsley, A. D., **11:**255
Wagener, J. S., **1:**27(60), 65; **3:**158(86), 164
Wagenhuber, M., **11:**130(87), 201
Wager, J. F., **14:**104(52), 126
Wagers, R., **17:**184, 222
Waghorne, R., **9:**56(261), 71
Wagnar, I., **22:**52(79), 79
Wagner, A., **16:**32, 129
Wagner, A. E., **12:**187(84), 234
Wagner, B. C., **3:**40(212), 58; **15:**149, 154, 165, 168, 171, 214
Wagner, C., **9:**9, 66; **14:**114(71), 127
Wagner, C. N. J., **7:**19, 20(60a, b), 28(117, 118), 30(127), 38, 40; **13:**75(59), 80(59), 105
Wagner, D., **9:**158(14b), 256
Wagner, G., **16:**60, 69, 135–136
Wagner, G. R., **16:**107, 141
Wagner, I., **22:**254(96), 274
Wagner, J., **21:**32(66), 73, 121(17), 123(17), 152(17), 232
Wagner, J. S., **2:**370(56), 421
Wagner, J. W., **17:**49, 53, 76
Wagner, L. F., **12:**121(237), 164
Wagner, M. D., **2:**101(69, 70), 103(70), 104(70), 105(70), 106(70), 142
Wagner, P., **24:**6(23), 106(109, 114)
Wagner, R. S., **5:**268, 316
Wagner, S., **15:**287, 310; **17:**354, 368
Wagner, W., **17:**310, 356, 357, 365, 369
Wagner, W. E., **9:**33(146), 68
Wahi, P., **15:**68, 139
Waho, T., **11:**44(51), 73(146), 74(147), 97, 100; **12:**282(235, 236), 295
Waidner, F., **15:**95, 140
Wainfan, N., **4:**23(56), 92; **7:**354, 355(145), 396
Wainwright, T. E., **24:**207(21)
Wait, G. R., **3:**295(65), 301
Waits, R. K., **14:**2(4, 10), 22(4, 10), 33(4, 10), 73, 203(12), 237; **16:**14–16, 125; **17:**87, 140
Wajda, E. S., **1:**216(132), 231; **2:**147(2), 178(2), 190, 365(3), 366(3), 419; **5:**146(11), 149, 229,

246(19), 256(44), 266, 267, 273, 314–316
Wakamatsu, N., **7:**158, 225(70), 249
Wake, D. R., **23:**255(77)
Wakefield, G., **5:**309, 322
Wakiyama, T., **15:**185, 216
Wako, S., **13:**20(32), 21(32), 57
Wakramian, A. T., **4:**199(51), 204(51), 252
Waksmann, B., **6:**273, 274, 275, 299
Waldeck, D. H., **24:**3(19)
Waldo, R. A., **17:**229, 300
Waldron, H., **18:**218
Wales, J., **10:**39(75), 40(75), 41(75), 43(75), 67
Wales, R. D., **12:**258, 291
Walkenhorst, W., **3:**27(133), 28(133), 51(133), 56; **10:**128(55), 165; **12:**10(36), 50
Walker, B., **7:**162, 211(84), 249
Walker, D. C., **7:**333(81), 334(84), 394
Walker, J., **15:**222, 226, 229–230, 234, 236–239, 261–262; **21:**116(5), 117(6, 10), 123(5), 124(5), 125(5), 126(29), 128(10, 38), 129(5, 39, 40), 136(5, 39, 40), 137(47), 141(50), 146(5, 39, 40), 149(5, 39), 196(94), 209(10, 40), 215(10, 118), 232, 233, 234, 235, 236, 240
Walker, J. W., **24:**27(124), 28(124), 29(124), 30(124)
Walker, L. R., **15:**3, 26, 136–137; **17:**17, 74
Walker, M., **2:**376(76), 421
Walker, P. L., Jr., **1:**226, 232; **4:**212, 253
Walker, W. C., **1:**124(1), 130, 134, 135, 147, 149(1, 57, 59), 184, 186; **4:**50(99), 72, 93, 95; **7:**45(7a), 49(29), 111, 112
Walker, W. E., **1:**207(96), 230
Walker, W. P., **3:**253(90), 255(90), 257, 258(90), 259, 260, 263, 264(90), 265(90), 273; **4:**181(131), 190
Wallace, J. S., **16:**79, 137
Wallace, R. A., **1:**43(119), 66
Wallace, R. W., **17:**182, 222
Wallace, W. E., **16:**149, 212, 229–230
Wallenberg, L. R., **17:**15, 16, 17, 24, 31, 74
Walley, P. A., **6:**63, 78
Wallis, G., **5:**167(129, 130), 181(129, 130), 193(129, 130), 199(129, 130); **12:**95(150), 102, 103, 161
Wallmark, J. T., **5:**216(194), 228(194), 234
Wallor, R., **3:**76, 102, 126
Waloman, F., **5:**256(45), 315
Walpole, J. N., **11:**36, 96, 97, 107, 108, 109, 128, 129, 155(78), 175, 200; **13:**219(67), 224
Walsh, W. M., Jr., **12:**246(24), 290
Walters, E. M., **9:**264, 267, 276, 287, 294
Walther, A., **2:**233(36), 234(36), 238, 321, 359; **3:**17, 18(80), 55
Walther, M., **21:**290(43), 310
Walton, A. G., **4:**121(54), 136
Walton, D., **2:**2, 4(4), 59, 60; **3:**212, 270; **4:**97(7), 100(7, 19), 109(7, 19), 110, 124, 126(64), 130, 133(7), 134, 136, 150, 151, 152, 155(64, 65), 160, 188
Walzcak, M. M., **24:**3(16), 4(16), 32(16), 44(30), 70(30)
Wan, J.-C., **16:**51, 132
Wandelt, K., **14:**230(107), 239
Wang, C. C., **5:**152, 153, 230, 272, 304(265), 317, 321; **7:**133(29),

Wang, C. C., (*continued*)
247; **11:**113, 130(84), 131(84), 160(133), 200, 201, 203
Wang, C. L., **15:**124, 141
Wang, C. P., **6:**282(199), 292(199), 300
Wang, D., **22:**270(125), 276
Wang, D. I. C., **24:**228(6), 229(6), 249(6)
Wang, D. N. K., **13:**99(104), 107
Wang, E., **17:**236, 237, 270, 271, 296
Wang, F. R., **16:**41, 131
Wang, H., **24:**74(120)
Wang, J., **24:**44(17, 18), 45(17, 18), 122(46)
Wang, J. S., **17:**153, 154, 197, 198, 220–223
Wang, J. Y., **24:**181(14)
Wang, K. L., **15:**245–246, 263, 308, 312; **21:**4(17), 72, 77–112, 78(1, 2, 5), 80(10), 81(13, 14), 84(15, 16), 86(14), 87(14), 88(18), 91(2, 19, 20), 93(21), 94(21), 95(20, 26), 97(10, 26), 100(36), 104(43), 105(1), 108(46), 109(2), 111, 112, 129(41); **23:**72(234), 240(48, 51, 55), 270(48)
Wang, P., **5:**257(47), 315
Wang, R., **16:**91, 139
Wang, S., **10:**254(53), 259; **15:**92, 140
Wang, S. Z., **16:**32, 41, 130–131
Wang, S.-Q., **22:**193(51), 226(51), 227(80), 273–274
Wang, T., **16:**51, 132, 133
Wang, W., **16:**41, 131; **24:**34(140), 35(140)
Wang, W. B., **21:**330(25), 333(25), 367
Wang, W. I., **15:**222, 261
Wang, X. K., **16:**43, 131; **17:**7, 12, 73
Wang, X. M., **16:**55–56, 64, 134
Wang, X. Z., **16:**79, 137
Wang, Y., **17:**337, 367; **18:**194
Wang, Y. H., **21:**191(88), 235, 290(46), 310
Wang, Y. J., **15:**185, 216
Wang, Y. O., **16:**2, 123
Wang, Y.-Q., **24:**27(128), 31(130)
Wang, Z., **16:**111, 142
Wang, Z. L., **24:**151(10), 154(10), 156(10), 163(10), 222(104)
Wanger, R. S., **22:**121(20), 172
Wank, M. T., **13:**61, 104
Wanlass, F. M., **4:**107(23), 135
Wanuga, S., **17:**153, 219
Warabisako, T., **14:**88(17), 89(19), 109(19), 117(19), 125; **21:**330(26), 331(26), 340(26), 343(26), 367
Warburg, E., **3:**27(128), 51(128), 56; **11:**271, 292(90, 91), 316
Ward, L., **6:**188, 203; **19:**141
Ward, M. G., **22:**254(96), 274
Ward, R. N., **24:**17(74)
Wardell, J. L., **24:**8(25), 9(25)
Wardlow, W., **5:**304(264), 321
Warekois, E. P., **3:**179(28), 209
Warmack, R. J., **24:**98(92)
Warner, A. W., **3:**24(114, 115, 119), 27, 51(114, 115, 119), 56
Warner, J. D., **16:**43, 50, 76, 107, 131–132, 137, 141
Warren, B. E., **4:**221, 222(109), 230, 253; **7:**1(3), 2, 3, 15(11, 34), 35(34); **15:**157, 215
Warren, P., **23:**24(96), 30(96)
Warren, R. G., **5:**260(55), 313(295, 297), 315, 322
Warren, W. E., **24:**114(35), 115(35), 116(35)
Warshaw, S. D., **7:**70(61), 113
Wartenberg, E. W., **5:**138, 141
Wartenbury, E. W., **9:**33(118), 67

Wartic, T., **12:**257(70), 291
Wasa, K., **12:**274, 294; **14:**20(78), 26(100), 63(199), 65(100), 75, 78; **16:**2, 17–19, 30, 33–34, 36, 111, 124, 126, 129–130, 142; **17:**153, 163, 219, 220, 221, 228, 232, 248, 249, 250, 251, 264, 296, 297
Waser, J., **7:**3(17), 37
Washburn, J., **4:**140, 143, 169(106), 184(23), 185(23), 187, 190; **5:**249(33), 315
Washida, H., **17:**308, 352, 364, 368
Washwell, E. R., **11:**142(110), 202
Wasilewski, Z. R., **21:**101(37), 112, 128(35), 170(73, 75), 178(73), 179(79), 209(35), 233, 234, 235, 287(38), 310
Wasserman, S. R., **24:**11(43), 13(43), 44(19)
Wassermeier, M., **19:**15, 32
Waszczak, J. V., **16:**3, 124; **17:**17, 74
Watanabe, D., **1:**221, 231, 282(16), 331; **4:**156(81), 159(81), 160(81), 189, 238(161), 239(161), 254
Watanabe, H., **2:**100, 101(69), 122, 142, 143, 144; **9:**47(215), 70; **13:**20(40), 23(40), 45(40), 57
Watanabe, K., **1:**131(14, 17), 136(14), 185; **4:**68(161), 94; **12:**277(217), 295; **13:**119(49), 126(49), 148; **16:**86, 88–91, 138–139, 251, 268; **17:**110, 141
Watanabe, M., **2:**122(127), 132(152), 144, 145; **11:**254, 255; **17:**355, 356, 357, 368, 369
Watanabe, M. O., **21:**206(103), 235
Watanabe, W., **16:**72–73, 137
Watanabe, Y., **15:**172, 216; **18:**296–297, 298
Wataze, M., **11:**76(152), 100
Watelski, S. B., **4:**169(109), 190
Waterhouse, N., **14:**2(14), 30(14), 32(126), 59(14), 73, 76
Waterhouse, R. B., **7:**325, 326, 389(40), 393
Waterman, T., **4:**371
Watson, D. M., **21:**36(70, 71), 73
Watson, G. W., **17:**212, 224
Watson, M., **15:**206, 217
Watson, W. R., **1:**43(119), 66
Watton, R., **17:**245, 297
Watts, D. C., **11:**256(57), 257(57), 293(125), 315, 317
Watts-Tobin, R. J., **4:**197, 252
Wauk, M. T., **17:**153, 154, 220
Waxman, A., **2:**150(25), 157(40), 173(40), 191; **5:**219(200), 221(207), 234; **6:**5(25), 36(25, 103a, 103b), 74, 76
Waylonis, J., **12:**8(32), 50
Waylonis, J. E., **1:**125(2), 148(49), 149(49), 169(49), 170(49), 184, 185; **2:**226(30), 238, 334(54), 338, 339, 360; **3:**9(39), 54; **4:**84(215), 96; **6:**158, 188, 202, 203; **9:**227(195), 260; **10:**77(16, 17), 164
Wayman, C. M., **7:**11, 18(26), 38
Weakliem, H. A., **17:**359, 361, 369
Weast, R. C., **7:**63(44), 65(48, 49), 96(71), 112, 113; **9:**11(53), 12(53), 66; **12:**253(56), 254(56), 291; **22:**17(24), 77; **24:**182(24)
Weaver, A. L., **19:**2
Weaver, C., **1:**207(101), 209, 230, 231; **2:**408(192), 424; **3:**96, 126, 132(11), 162; **4:**85(225), 86(225, 235), 96; **6:**54, 77, 310(10), 311(10), 312, 313(22), 315(29), 326(33), 328(34), 332(36),

Weaver, C., (*continued*)
335(39), 346, 347; **10:**26(43), 66; **16:**234, 236, 266
Weaver, T., **24:**233(53)
Webb, A. P., **12:**238(2), 289
Webb, C., **16:**54, 134
Webb, D. C., **15:**40–42, 47, 50, 55, 138–139
Webb, J. B., **14:**38(142), 48(163), 76, 77; **17:**153, 220
Webb, M. B., **2:**23(33), 24, 25, 26(33), 27(33), 60, 371(58), 383(58), 388(58), 421; **3:**1(1), 53; **4:**200(74), 252; **9:**6, 7(29), 65
Webber, G., **15:**70, 100, 139
Webber, H. C., **13:**158(34), 169(34), 193
Webber, J. C., **14:**18(72), 75
Weber, A. H., **3:**32(178), 38, 52(178), 58
Weber, C. E., **3:**99, 126
Weber, D., **2:**325, 326, 327, 359
Weber, H. P., **10:**242(33), 245(35, 36), 258, 259
Weber, H S., **5:**284(196), 319
Weber, P. M., **24:**184(35), 192(35)
Weber, S., **5:**28(21), 45
Weber, W. H., **9:**236(219a), 261; **11:**107(14), 130(14), 200
Webster, R., **14:**147(71), 148(71), 196
Webster, R. T., **15:**63, 66, 139; **17:**187, 222
Wecker, J., **16:**214, 230
Wedler, G., **6:**116(155, 158), 147
Weed, D. S., **2:**379(83), 392(83), 422
Weeg, L. E., **3:**39(188), 52(188), 58
Weeks, M. M., **23:**101(38), 102(40), 163(47, 48), 166(91)
Weeks, R. F., **2:**315, 359; **4:**60(134), 94
Wegener, H., **1:**210(118), 231
Wegener, H. A. R., **3:**137(32b), 163; **4:**315(159), 324
Wegner, P. P., **13:**8(15), 57
Wehner, G. K., **2:**367(20, 23), 400(159), 420, 424; **3:**62, 65(17), 66(23, 24), 71(38), 72(41, 42), 74(48, 49, 50, 51), 75(53), 76(53, 55, 58), 77(42, 60), 78(53), 85(42), 102(93), 114(129), 125, 126, 127; **4:**269, 322; **7:**49, 50, 54, 55, 56(30), 112; **9:**278, 297; **13:**110(10), 111(11), 147; **14:**2(1), 17(69), 22(1), 73, 75, 175(132), 183(132), 184(132), 198, 203(8), 217(8), 228(90), 237, 239; **16:**16, 126; **22:**244(95), 274
Wehner, K., **1:**224, 232
Wei, C. Y., **23:**102(42)
Wei, G., **17:**357, 358, 369
Wei, J. S., **14:**24(95), 75
Wei, J. Y. T., **16:**4, 124
Wei, Y., **19:**281
Weibel, E., **24:**82(1)
Weidenborner, J. E., **6:**215(51), 223(51), 296
Weigmann, W., **15:**221–222, 260
Weiher, R. L., **9:**28(102, 104), 29, 30, 67
Weihs, T. P., **24:**96(75), 97(75), 98(75)
Weijsenfeld, C. H., **3:**73(44), 75(44), 76(44), 77(44), 125
Weikliem, H. A., **13:**68(45), 105
Weil, R., **4:**212, 238, 239, 240, 253, 254, 255
Weiller, B. H., **22:**178(11), 271
Weimann, G., **21:**290(43), 310; **23:**254(75)
Weimer, P. K., **2:**148(8, 9), 149(8, 9, 11), 150(8, 9, 22, 24, 25, 26),

153(8, 24), 154(24), 156(39), 157(40), 159(39), 162, 163(39), 164(9, 26), 168(22, 39), 170(49), 173(40), 176(26), 177(26), 179(9, 11, 61); **3:**131(2), 162; **5:**146, 219(200), 228(10), 229, 234; **12:**154(331), 156(333), 166
Weinard, J., **2:**331, 360
Weinbach, S. P., **24:**181(12, 15), 192(12), 199(12, 50)
Weinberg, I. J., **15:**50, 138
Weinberg, W. H., **23:**29(102, 103), 38(102, 103)
Weinberger, B. R., **16:**56–57, 134
Weinberger, H., **10:**16(28), 25(28), 28(28), 43(28), 47(28), 66
Weiner, J. A., **17:**358, 369
Weiner, M., **10:**228(4), 253(4), 254(4), 258
Weinert, R. W., **15:**64, 72, 81, 139; **17:**181, 182, 222
Weingart, J. M., **4:**82(205), 95
Weinreich, O., **1:**218(135), 231
Weinreich, O. A., **6:**37(131), 47, 77; **12:**272(167), 277, 294
Weinrich, A. R., **9:**43(189), 69
Weinstein, L. I., **2:**276, 304
Weinstein, M., **3:**174(21), 208; **5:**154(42), 230, 283, 319
Weinstein, W., **1:**70(4, 19), 71(19), 87(4), 120; **10:**262, 308
Weis, R. S., **19:**282, 288
Weisberg, L. R., **5:**209(183), 234; **6:**114, 123, 146; **7:**128, 129, 130, 131(20), 132(21), 244(272), 247, 255
Weisbuch, C., **17:**2, 8, 12, 73, 74
Weisbuch, I., **24:**181(12), 192(12), 199(12)
Weisenhorn, A. L., **24:**98(89)
Weiser, G., **19:**75
Weiser, K., **10:**41(79), 67
Weishaar, D. E., **24:**44(30, 33), 53(33), 70(30)
Weiss, E., **3:**188, 209
Weiss, G. P., **1:**280(7), 283, 285(24), 286(28, 29), 289(33), 304(54), 331, 332; **3:**214(11, 13), 228, 229(51), 230, 234, 247(11), 250(11, 87), 255(87), 270, 272, 273
Weiss, K., **2:**340, 360; **3:**198, 210
Weiss, R. J., **1:**205(77), 230; **7:**18(58), 38
Weisskopf, V. F., **4:**328(12), 369; **5:**211(220), 235
Weissler, G. L., **1:**124(1), 130(1), 134(1), 135(1), 143(36), 147(1), 149(1, 59), 163(36), 165(36), 169(36), 184; **4:**50(99), 68(165), 93
Weissmann, S., **4:**218(99, 103), 219(103), 220(99), 231(99), 232(103, 148), 253, 254
Weissmantel, C., **12:**256(61), 257(65), 291; **13:**3(4), 56, 99(108, 109), 107, 109(3), 111(3, 19, 20); **14:**24(96), 24–25(96), 25–26(96), 25(99), 26(99), 75
Weissmantel, Chr., **12:**246(34, 35), 248, 251(62, 63), 256(62), 290, 291
Weisz, Z., **15:**284–285, 310
Weiz, F., **9:**276, 284, 297
Welboum, A. D., **23:**2(3)
Welcher, F. J., **12:**202(95), 234
Weller, D., **15:**197, 217
Weller, J. F., **9:**54(256), 71
Wells, A. F., **7:**350, 358(131), 395
Wells, M., **24:**44(66), 46(66)
Wells, O. C., **7:**287(94), 309
Wells, V. A., **10:**37(69), 40(76),

Wells, V. A., (*continued*)
45(69), 46(97), 67, 68;
12:269(144), 293
Wellstood, F. C., **16:**57, 75–77, 117, 134, 137, 142
Welton, R. D., **1:**18(39), 64
Wemple, S. H., **4:**75(187), 95; **10:**256(6l, 62), 259; **13:**217(56), 224; **17:**229, 281, 282, 299
Wen, C. P., **21:**128(37), 233
Wen, L., **17:**22, 74
Wen-han, L., **17:**22, 74
Wendel, G., **9:**33(125), 68
Wendelken, I. F., **9:**203(123a), 258
Wendland, P., **3:**172, 208
Wendlandt, W. W., **10:**210, 226
Wendman, M., **16:**52, 133
Wendt, J. F., **9:**46(204), 70
Wendt, O. M., **3:**13, 51(59), 54
Wentik, D. J., **19:**2
Wentworth, E., **11:**111(47), 200
Wentz, J. L., **11:**171(145), 183(145), 203
Wentzel, G., **2:**75, 141
Wenzel, S. W., **17:**212, 213, 214, 224
Werner, C., **22:**178(15), 271
Wernick, J. H., **6:**209(19), 211(28), 295; **9:**273, 296; **17:**118, 142
Wernlund, L.-D., **16:**100, 140
Werring, N. J., **13:**43(52), 58
Wertheim, G. K., **17:**318, 366
Wertheimer, M. R., **15:**283, 310
Werthen, J. G., **21:**362(44), 368
Weschenfelder, F. W., **16:**23, 127
Wessel, P., **6:**195(54), 203; **9:**176(39), 256
Wessel, S. A., **17:**328, 366
Wessels, B. W., **16:**87, 93–94, 138–140; **17:**287, 299
West, A. R., **11:**256(59), 257(59), 315
West, B. O., **17:**318, 365
West, E. J., **9:**54(256), 71; **10:**237(25), 258
West, F. G., **1:**302(50), 332
West, K., **21:**130(42), 133(42), 151(42), 153(42), 154(42), 211(42), 233
West, K. W., **21:**209(111), 236
West, L., **23:**234(20), 235(20)
West, L. C., **15:**230, 239, 262; **21:**116(4), 117(4), 232, 240(3), 308
West, M. B., **6:**114(146), 146
Westgaard, H., **2:**369(33), 420; **3:**105(96), 127
Weston, D. G. C., **9:**22(92), 27(92), 67
Westphal, W. B., **5:**298, 321
Westra, K. L., **22:**123(27), 172
Westwood, D., **13:**140(120), 150
Westwood, D. I., **19:**41
Westwood, W. D., **13:**67(33, 34), 105; **14:**2(8, 14), 3(15), 5(16), 8(15), 9(16), 21(83), 22(8), 24(94, 95), 26(8), 27(107, 109), 29–30(114), 30(14, 115, 118), 33(129), 35(16), 36(16), 38(141, 144), 39(129, 141, 147), 41(15, 149), 42(149), 44(153, 156), 45(153, 156), 46(156, 159), 47–48(147), 51(167), 59(14), 63(204), 64(204, 205), 66(204), 67(227, 228), 68(129, 228), 69(228), 70(228), 73, 74, 75, 76, 77, 78, 217(43), 219(51–54), 237, 238; **17:**111, 142, 153, 157, 220, 221, 232, 296; **22:**8(18), 77
Wetterer, S., **24:**139(66)
Wettling, W., **15:**111, 141
Wettstein, E., **16:**169, 193, 202, 207, 230
Wever, H., **7:**264(34), 307
Wexler, A., **1:**265, 275

Wexler, S., **2:**13(9), 60
Weyl, R., **3:**39(199), 52(199), 58
Whalen, R. M., **1:**319(77), 333
Whangbo, M. H., **24:**97(78), 98(78)
Whatmore, R. W., **17:**237, 238, 296
Wheeler, A., **2:**48(78), 50(78), 51(78), 52(78), 62; **3:**219(24), 271
Wheeler, R. G., **21:**124(28), 233
Whelan, M. J., **1:**194, 227(183, 184), 229, 232; **2:**14, 60, 111(101), 113(101), 114(101), 118, 119(116), 121(101), 136(156a), 143, 145; **4:**142(21), 182(21, 133), 187, 190, 225(130), 254; **6:**228(105), 297
Whetten, N. R., **1:**11(22), 64; **12:**133(287), 165
Whetten, R. L., **24:**151(10), 154(10), 156(10), 163(10), 222(104)
Whinnery, J. R., **10:**238(30), 258
Whinney, J. R., **12:**282(238), 295
Whipple, R. T. P., **16:**236, 266
Whitaker, H. H., **12:**118(228), 163
Whitaker, J., **5:**158, 160, 230; **7:**236(249e), 254
Whitaker, J. F., **15:**245–246, 248, 251, 263; **16:**107, 141
Whitcomb, E. C., **9:**276, 294
White, A. E., **16:**51–53, 74, 132–134, 137
White, A. H., **7:**11(21), 37
White, A. M., **7:**174(122), 193(161), 250, 252
White, C. L., **2:**393(131), 423; **6:**4, 7, 74
White, C. T., **11:**304, 305(140), 317
White, D., **6:**36(110), 45(110), 48, 76
White, D. P., **14:**23(90), 75
White, E. A. D., **9:**276, 283, 284, 295; **15:**113, 141
White, E. J., **3:**51(29), 53
White, G. K., **6:**82(7), 106(125), 119, 143, 146
White, H. G., **4:**332(30), 346(58), 370; **7:**148, 164, 165, 166(56), 210, 211(190), 223(221), 224(222), 248, 252, 253
White, J. C., **13:**153(14), 161(39), 182(14, 39), 183(14, 39), 192, 193
White, J. F., **16:**250, 268
White, J. P., **3:**140(49, 50), 163
White, J. R., **15:**293, 311
White, J. U., **4:**50(120), 93
White, L. C., **2:**370(45), 406(45), 420
White, P., **1:**37(102), 38, 66; **2:**132, 145, 403(171), 424; **3:**24(118), 27(118), 51(118), 56, 113(15), 146, 147, 151(72), 162, 163, 164; **4:**282(105), 310, 322, 324
White, R., **15:**26, 137
White, R. G., **7:**225, 226, 227, 228(231), 254
White, R. M., **17:**211, 212, 213, 214, 217, 223, 224
White, R. W., **17:**210, 223
White, W. B., **5:**303(296), 322
White, W. C., **3:**36, 57; **4:**4(12), 90
White, W. H., **5:**149(24), 229, 246(19), 273(19), 314
Whitehead, S., **3:**160(89), 164
Whitehouse, C. R., **15:**234–235, 262; **21:**117(12, 13), 128(36), 152(13), 153(13), 232, 233
Whitham, G. B., **4:**207, 253
Whitlock, W. S., **6:**51(145a), 77
Whitmell, D. S., **12:**248(46), 250, 251(46), 253(46), 254, 291
Whitmore, R. E., **6:**37(130), 44, 77
Whitnell, D. S., **17:**130, 143
Whittingham, M. S., **17:**305, 320, 364, 366

Whittle, K. M., **5:**255(42), 290, 315, 320
Whitton, J. L., **6:**24, 75
Whyman, R., **24:**154(15)
Wicher, P. J., **17:**212, 213, 224
Wickramasekara, L., **16:**164, 168, 177, 190–191, 193–202, 206–211, 213–214, 230
Wicks, G., **13:**203(15), 223
Wickstrom, R. A., **12:**136(302), 137(302), 165
Widdington, R., **2:**68(8), 141
Widdowson, E. E., **7:**349, 395
Wider, H., **3:**27(142), 28(142), 51(142), 56
Widing, K. G., **7:**45, 103(13), 107(84), 111, 114
Widmer, H., **4:**169(116), 190; **5:**163(74), 166, 170, 196(74), 197, 231
Widmer, R., **9:**51, 52, 70; **14:**226(79), 238; **17:**153, 159, 219
Widom, B., **24:**184(40), 190(40), 220(82)
Widrig, C. A., **24:**3(16), 4(16), 32(16), 44(20, 28), 53(28), 70(28)
Widrow, B., **21:**62(120), 74
Wiech, U., **16:**9, 125
Wieck, A. D., **21:**96(27), 111, 155(64), 234
Wied, O. J., **4:**323
Wieder, H. H., **5:**184(146), 185, 216, 217, 218, 233, 234; **7:**201(175d), 252; **11:**50(73), 57(97), 89(226), 98, 99, 102; **13:**205(21), 223
Wiegel, R., **17:**217, 224
Wiegert, R. F., **17:**171, 221
Wiegmann, W., **4:**332(29), 370; **7:**210, 211(190), 225, 226, 227, 228(231), 252, 254; **11:**53(89), 54(89), 55(89), 78(165), 79(174, 176), 81(178, 179), 84(190, 192, 193, 193a), 85(165, 195, 196, 199, 200, 201), 86(179, 207), 98, 101, 102; **17:**14, 36, 38, 74, 75; **21:**155(61), 234, 240(6), 308
Wielders, P. G. E., **14:**160(106), 197
Wielunski, L. S., **17:**318, 365
Wiemer, P. K., **3:**181(41), 197(41), 209
Wiener, O., **3:**35(168), 52(168), 57
Wierer, P. G., **13:**121(63), 148
Wiesmann, H., **16:**51, 133
Wiess, E., **3:**188, 209
Wigen, P. E., **6:**273(178), 299; **15:**95, 140
Wigger, N., **24:**8(29)
Wignall, G., **24:**151(9), 154(9), 156(9)
Wigner, E. P., **15:**228, 252, 261, 264; **23:**292(114)
Wijers, C., **19:**2
Wijn, A. P. J., **9:**275, 297
Wijn, H. P. J., **11:**294(133), 317
Wiklund, P., **14:**233(134), 240
Wilber, W. D., **15:**111, 141
Wilbur, J., **24:**229(37), 234(37), 235(37)
Wilbur, J. L., **24:**44(58), 98(83), 228(25), 230(44), 241(79)
Wilcock, J., **4:**199(53, 54), 202, 204(53, 54), 206(54), 252
Wilcock, J. D., **22:**143(66), 173
Wilcox, D. A., **12:**143, 166
Wilcox, J. Z., **15:**55, 138
Wilcox, P. S., **6:**53(147), 54(147), 57(147), 58(147), 60(147), 62(147), 63, 65(147), 66(147), 77; **14:**2(14), 30(14), 32(126), 59(14), 73, 76
Wild, D., **6:**221(70), 240, 296, 298
Wild, W. J., **16:**57, 60, 134
Wilde, D. K., **16:**42, 47, 131
Wilems, R. E., **9:**250(250), 261

Wilkens, B., **16:**56, 63, 112, 134–135, 142; **17:**236, 237, 270, 271, 296, 298
Wilkens, B. J., **16:**71, 136; **19:**9, 15, 19; **23:**41(113)
Wilkes, E., **11:**107(12, 13), 109(29), 115(12), 121(12), 128(13), 131(13), 135(13), 155(13), 200
Wilkie, E. L., **15:**222, 261
Wilkins, B. I., **16:**42, 45, 131
Wilkins, C. W., **7:**170, 191(106), 250
Wilkinson, P. G., **1:**180(68), 186
Wilkinson, P. H., **1:**216, 231
Wilkinson, R. R., **6:**6(27), 7(27), 26, 33(27), 74
Will, H. A., **12:**246(27), 290
Willander, M., **23:**55(10), 146(45)
Willardson, R. K., **5:**222, 235
Wille, D. A., **10:**236(18), 258
Wille, H., **9:**179(53), 257
Willems, D. A., **15:**68, 139
Willems, R. E., **6:**198, 203
Willens, R. H., **2:**42(71, 72), 61, 62; **7:**28(115), 40
Williams, A., **13:**208(25), 223
Williams, A. R., **15:**166, 177, 215, 216
Williams, B. E., **3:**109(112), 127; **6:**106(119), 110(119), 146; **9:**9(41), 43(9), 62(9), 65
Williams, B. F., **7:**242(263), 244(269, 271), 255; **12:**113(186, 188), 118(228, 229), 120(229), 126(229), 127(229), 135(289), 162, 163, 165
Williams, C. K., **14:**89(20), 90(20), 118(20), 123(20), 125
Williams, C. M., **9:**297
Williams, D., **11:**36, 96, 97
Williams, D. A., **10:**61(141), 69
Williams, D. E., **24:**210(45), 213(45), 214(45)
Williams, D. F., **14:**48(163), 77; **17:**153, 154, 220
Williams, E. J., **2:**75, 107, 141, 142
Williams, E. W., **7:**142(42, 46, 47), 147(47), 157(70), 174(122), 193(161), 248, 249, 250, 252
Williams, F., **7:**162(87), 249
Williams, F. E., **3:**179(33), 180, 209
Williams, F. V., **5:**161, 230, 277, 318
Williams, G., **6:**54, 77; **11:**235(36), 238(36), 254, 255, 256(57), 257(57), 293(125, 129), 315, 317
Williams, G. M., **21:**40(88), 74, 117(8), 128(8), 221(8, 134), 223(8), 232, 236, 251(26), 309
Williams, H. J., **1:**290, 292, 316, 332, 333; **6:**279(190), 299
Williams, J. C., **2:**375(73), 421; **3:**112, 127; **4:**267(77), 322
Williams, M., **21:**287(41), 310
Williams, M. W., **9:**221(183), 227(205), 228(205), 229(205), 237(222), 260, 261
Williams, P., **23:**48(162)
Williams, R., **4:**341(52), 370; **16:**123, 143; **21:**22(51), 73; **23:**88(18)
Williams, R. E., **11:**76(153), 100
Williams, R. H., **12:**114(209), 163; **19:**41
Williams, R. S., **13:**205(23), 223
Williams, V. A., **9:**51, 71
Williams, V. L., **10:**124(53a), 165
Williams, W. S., **1:**37(96), 66
Williamson, D. L., **15:**284–285, 310
Williamson, R., **12:**248(46), 250, 251(46), 253(46), 254, 291
Williamson, R. C., **17:**147, 167, 170, 219, 221
Williamson, W. J., **5:**164(97), 172(97), 231
Willich, P., **15:**112, 141, 197, 217

Willicut, R. J., **24:**19(87), 20(87), 27(87), 28(87), 30(87, 129), 33(129), 34(129), 44(26, 39), 45(26), 46(26), 70(26, 39)
Willis, W. R., **3:**73(43), 125
Willmott, D. J., **14:**32(126), 76
Willner, I., **24:**19(81, 83), 20(81, 83, 89)
Wills, J. M. M., **4:**126(66a), 136
Wills, L. A., **17:**287, 299
Willson, C. G., **13:**159(36), 160(36), 182(115), 183(115), 193, 195
Wilman, H., **1:**206, 226(181), 230, 232; **3:**218, 220(27), 229(27), 234, 243; **11:**107(9), 200; **12:**73, 160
Wilmot, D. W., **10:**237(26), 258
Wilmsen, C. W., **14:**104(52), 126, 234(144), 240
Wilmshurst, J. K., **2:**202, 237
Wilsdorf, H. G. F., **3:**256, 269, 273
Wilsey, N. D., **15:**9, 129, 136
Wilson, A. H., **4:**328, 369
Wilson, A. J. C., **1:**5(88), 20, 230
Wilson, B. A., **4:**60(133), 94; **21:**192(90), 235
Wilson, D. A., **15:**8, 105, 136, 140
Wilson, D. J., **23:**130(23)
Wilson, D. K., **4:**301, 323
Wilson, E. B., **19:**299
Wilson, E. B., Jr., **15:**266, 309
Wilson, H., **12:**262(94), 292
Wilson, H. L., **2:**151(35), 176(35), 191; **5:**216(195), 234
Wilson, J. R., **6:**71, 79
Wilson, J. W. L., **15:**154, 215
Wilson, M. N., Jr., **2:**331(45), 332(45), 360
Wilson, R. G., **23:**234(37)
Wilson, R. H., **6:**82, 143
Wilson, R. J., **24:**85(27)
Wilson, S. R., **13:**130(94), 132(102), 133(94), 134(94), 135(94), 149; **22:**270(127), 276
Wilson, T. E., **24:**36(147)
Wilson, W. L., **24:**12(55), 14(55)
Wilson, W. R., **14:**196
Windischmann, H., **14:**69(230), 78
Window, B., **14:**66(216), 78; **17:**88, 123, 140; **18:**253, 255, 259, 261, 263, 266, 267, 271, 272, 274, 278
Windte, V., **16:**23–24, 127
Winegarner, R. M., **10:**51(104), 68
Winkler, G., **15:**137
Winograd, N., **14:**230(101, 104), 239
Winslow, D., **17:**184, 222
Winslow, D. K., **12:**269(148), 293; **13:**61, 104; **17:**153, 219
Winslow, E. H., **1:**205(91), 206(91), 230
Winslow, K. D., **17:**153, 154, 220
Winston, R., **10:**19, 24, 66
Winter, T. C., **7:**56(38), 112
Winterbottom, A. B., **4:**82(204), 95
Winters, H. F., **9:**278, 297; **12:**248(41), 290; **14:**6(19), 7–8(25), 8(27), 24(19), 69(19, 25), 73, 74, 234(136, 138), 235(150), 240
Winters, J. H., **16:**107, 141
Winterton, S. S., **22:**76(70), 78
Winterton, W., **22:**50(56), 52(56), 53(56), 54(56), 56(56), 78
Wintle, H. J., **11:**209(3), 278(3, 99, 100), 314, 316
Winz, E., **12:**288, 295
Wirick, M. P., **19:**238
Wirth, M. J., **24:**228(7)
Wirtz, K., **4:**271(90), 322
Wirz, P., **13:**67(37), 105; **14:**48(160), 77
Wise, K. D., **17:**210, 223
Wise, M. L., 23.35n(138), **23:**41(145)

Witanachchi, S., **16:**55–56, 64, 134
Withers, R. S., **15:**60, 139
Witt, A. T., **5:**285(201), 319
Witt, F., **3:**226, 271
Witt, M., **5:**287(208), 319
Witteborn, F. C., **3:**74(52), 75(52), 126
Wittekoek, S., **9:**276, 296
Wittels, M. C., **13:**127(85), 149; **19:**252
Witter, K., **15:**112, 141, 171, 215
Wittingham, M. S., **13:**97(94), 106
Wittke, J. P., **10:**257(66), 259
Wittmer, M., **14:**59(183), 77
Wittry, D. B., **7:**159(80), 249
Wittwer, T. Y., **23:**124(16)
Wittwer, V., **17:**345, 350, 367
Wixwat, W., **17:**358, 369
Woeckel, E., **9:**221(187), 260
Woelke, C. D., **9:**51(249), 71
Woertz, B. B., **10:**23(37), 66
Wohlers, M. D., **17:**170, 221
Wohlfarth, E. P., **1:**297, 332; **16:**157, 229
Wohltjen, H., **17:**210, 212, 223
Wojciechowski, P. H., **16:**287, 291, 328, 334, 336, 338–339
Wojtowicz, P. J., **17:**319, 320, 366
Wolf, B. M., **17:**209, 223
Wolf, D., **17:**59, 76, 77
Wolf, E., **1:**72(32), 121; **5:**53(8), 57, 62, 85; **10:**171(25), 226; **24:**221(99)
Wolf, G. K., **17:**82, 130, 139, 143
Wolf, H., **24:**11(46), 18(88), 20(88), 21(46, 102), 26(102), 33(46), 34(46,141), 35(141), 88(45), 89(50), 90(45), 106(111), 127(51), 212(65), 218(75), 228(31)
Wolf, I. W., **3:**133(17), 162
Wolf, K. L., **7:**312(9), 392
Wolf, M. O., **24:**20(90), 44(36), 45(36)
Wolf, P., **1:**321(83), 327, 333, 334
Wolf, R. M., **17:**292, 299
Wolf, S. A., **16:**53, 134
Wolf, S. G., **24:**180(11)
Wolf, T., **16:**9, 125
Wolfe, C. M., **7:**235(249), 254; **11:**61(107), 99, 121(74), 201; **21:**37(73), 73
Wolfe, E., **19:**281, 287, 288
Wolfe, J. C., **16:**54, 134
Wolfe, J. E., **3:**40(212), 58
Wolfe, R., **9:**271, 297
Wolfe, W. L., **4:**75(188), 95; **10:**11(18), 65
Wolff, G. A., **3:**174(21), 179(29), 208, 209; **5:**285(202), 319
Wolff, P. A., **6:**211(28), 295; **11:**297(137), 317; **12:**60, 84(36), 159
Wölfing, C., **6:**116(158), 147
Wolfram, T., **15:**32, 37, 135–136, 137
Wolfson, R., **13:**99(107), 107
Wolfstirn, K. B., **7:**149(59), 152(62a), 154(59), 162, 164(82), 248, 249
Woll, C., **24:**45(77), 65(77), 86(27, 35), 92(57), 237(65), 238(65)
Wolley, E. D., **5:**199(174), 202(174), 234
Wollman, E. W., **24:**19(84, 85), 20(84, 85), 28(127), 228(29)
Wollmann, K., **5:**150(28), 230
Wolsky, S. P., **1:**43(120), 66; **3:**27(137, 138), 28(137, 138), 51(1 37, 138), 56, 62, 65, 73(47), 100, 125, 126; **5:**167(129, 130), 181, 193, 199, 232
Wolter, A. R., **4:**23(52), 91
Wolter, H., **1:**71, 120; **3:**2(16), 53; **6:**171, 201(70), 202, 204; **9:**74, 143

Wolter, J. H., **21:**313(6), 321(13), 341(6), 342(6), 343(40), 344(6), 351(13, 40), 355(6), 357(13), 359(13), 360(6, 40), 361(6, 40), 362(40), 365(50), 366, 368
Wong, C. A., **16:**96, 140; **17:**229, 300
Wong, J., **12:**258(79), 292; **15:**287, 310
Wong, K. K., **17:**306, 345, 347, 357, 358, 364, 367, 369
Wong, Y. M., **21:**221(135), 222(135), 236
Wonsidler, D. R., **12:**293
Woo, J. C. S., **23:**72(234)
Woo, K., **24:**24(117), 26(117), 27(117)
Wood, B. M., **15:**250, 263
Wood, C. E. C., **11:**41(36), 42(36), 45, 48(36), 49(67, 69), 50, 51(74, 77, 79), 52(85, 87), 54(79, 90), 56(90), 57(77, 98), 58(77), 59, 60, 61, 62, 63(106), 65(77, 118), 66(87, 118), 70(140a), 76(96, 154, 156, 156a), 77, 78(166), 81(185), 82, 83(166, 188a), 88(69, 77, 156, 156a, 220, 225), 97, 98, 99, 100, 101, 102
Wood, D. M., **19:**135
Wood, G., **15:**286–287, 310
Wood, G. C., **6:**54(160), 61, 63(160), 68, 71(205), 77, 78
Wood, M. I., **14:**191(167), 199
Wood, R. F., **13:**166(59), 181(106), 194, 195
Wood, R. W., **4:**121(56), 136; **7:**43(1, 2), 111
Wood, W. A., **4:**227, 254
Woodall, J. M., **7:**116, 118(2), 142(38), 183(138), 185(139), 193(157), 218, 219(38), 220(138, 139), 221(212), 225(230), 246(278, 280), 248, 251, 253, 254, 255, 256; **10:**58(126), 68; **13:**176(85, 87), 185(85, 87), 187(85), 188(87), 189(87); **14:**229(96), 239
Woodard, D. W., **11:**51(74), 98
Woodbridge, K., **21:**341(35), 367
Woodcock, J., **11:**51(77), 57(77), 58(77), 59(77), 60(77), 65(77), 77(77), 88(77), 98
Woodcock, L. V., **22:**134(50), 172
Woodhouse, J. B., **7:**234(248), 254
Woodhouse, J. D., **13:**182(114), 183(114), 195
Woodruff, W. H., **4:**83(211), 96
Woods, C. E. C., **13:**203(14, 15), 218(58), 219, 223, 224
Woods, J., **3:**179(27), 209
Woods, J. F., **2:**175(55), 192, 392(123), 423; **3:**295, 301; **7:**163(88b), 182(137), 249, 251
Woods, R., **24:**228(11)
Woods, R. C., **18:**82
Woods, S. B., **6:**82(7), 106(125), 119, 143, 146
Woodward, J. T., **24:**163(29)
Woodward, P. M., **3:**277(23), 300
Woodyard, J. R., **3:**70, 125
Wooley, J. C., **7:**173(117), 174(118a), 177(118c), 200(175c), 250, 252
Wooley, M. C., **2:**371(63), 397(63), 421
Woolf, D. A., **19:**41
Woollam, J. A., **13:**79(80), 106; **16:**80–81, 137–138; **19:**280
Wooten, F., **4:**344(55), 370; **10:**32(57), 33(57), 66; **12:**59(24), 85(110), 91(138), 109, 111, 112(179), 159, 161; **17:**350, 367; **19:**132; **23:**91(21)
Wooten, F. O., **1:**54, 67
Worcester, G. R. C., **24:**180(3)

Worden, D. G., **6:**104, 105, 131(110), 146
Work, R. N., **11:**253, 255
Worledge, P. L., **6:**36(110), 45(110), 48, 76
Worley, R. D., **3:**32(157), 51(157), 57
Wormeester, H., **19:**2
Worobey, W., **6:**82(11), 143
Worthing, A. G., **4:**60(154), 71(170), 94, 95
Worthing, F. L., **4:**302, 323; **5:**297(246), 321
Worthington, T. K., **16:**110, 142
Wortman, R., **1:**21(52), 65
Wortmann, A., **13:**212(40), 224
Wosik, J., **16:**54, 134
Wouk, M. T., **12:**269(148), 293
Wranglen, G., **4:**199(59), 200(59, 73), 202, 204(59), 252
Wright, A. C., **16:**88, 94, 138, 140
Wright, D. A., **9:**22(88, 96), 23(96), 25, 26, 27, 67
Wright, D. C., **15:**171, 216
Wright, D. R., **7:**174(122), 250
Wright, G. T., **2:**148(6), 150(6), 165(46), 190, 191
Wright, K. H. R., **3:**226(47), 229(47), 234(47), 235(47), 238(47), 244(47), 246(47), 271
Wright, P., **3:**2(17), 40(17), 53
Wright, P. D., **11:**66(122), 99
Wright, S., **24:**229(42)
Wright, S. L., **15:**227, 244, 261; **21:**193(92), 235
Wrighton, M. S., **24:**19(84, 85), 20(84, 85), 21(103), 23(103, 113), 26(103), 27(103, 113), 28(127), 30(113), 97(79), 98(79, 80), 228(29), 240(74)
Wrinch, D., **7:**350, 395
Wrobel, A. M., **15:**291, 311
Wrobel, J. S., **11:**121(73), 146(117), 148(117), 149(117), 169(142), 201, 202, 203
Wroe, H., **14:**175(130), 196, 198
Wronikowski, M., **12:**242(17), 243(17), 290
Wtorow, E. N., **2:**94(45), 141
Wu, A. Y., **17:**234, 251, 252, 296, 297
Wu, C., **16:**123, 143
Wu, C. S., **12:**169(12, 13), 194(13), 196(12), 232; **14:**232(123), 239; **21:**128(37), 233
Wu, C. T., **14:**65(214), 78
Wu, H. J., **15:**47–48, 48, 138
Wu, I. C., **21:**5(3), 18(3), 36(72), 71, 73
Wu, K., **16:**41, 131, 252, 268
Wu, L. S., **19:**250
Wu, M. C., **23:**228(31), 229(31)
Wu, M. K., **16:**2, 123
Wu, P. T., **16:**25, 36, 127, 130
Wu, Q.-H., **19:**281
Wu, S. Y., **17:**228, 229, 239, 240, 241, 254, 255, 256, 261, 266, 267, 268, 269, 275, 292, 297, 298
Wu, T. Y., **7:**189, 190(153), 251
Wu, T.-B., **17:**54, 76
Wu, X., **16:**42, 131; **24:**181(12), 182(23), 184(40), 190(40), 192(12), 199(12, 23)
Wu, X. D., **16:**55–56, 58, 60–63, 75, 106, 110–111, 134–135, 141–142; **17:**236, 237, 270, 271, 296
Wu, Y., **22:**134(55), 173
Wu, Z., **17:**241, 249, 250, 290, 297; **19:**250
Wucher, A., **14:**17(64), 74; **22:**32(45), 78
Wudl, F., **17:**354, 368
Wuensch, B. J., **17:**29, 47, 75
Wuhl, H., **16:**8, 125

Wulff, J., **4:**139(12a, 13), 187
Wurfel, P., **19:**253
Wurlingham, H., **17:**184, 222
Wyatt, P. W., **13:**187(126), 196
Wyder, P., **6:**82(8), 96, 115, 121, 124(148), 135(148, 173), 143, 145, 147
Wyekoff, R. W. G., **9:**22(100), 28(100), 30(100), 67
Wyman, J., Jr., **7:**348, 367(120, 172), 369(120, 172), 395, 396
Wyndrum, R. W., **6:**54(155), 77
Wynne, K. J., **24:**206(12)
Wyrobisch, W., **9:**217(168), 259
Wyrwich, H., **2:**75(27), 141
Wysocki, C. A., **2:**369(39), 393(39), 420

X

Xanthopoulos, N., **24:**18(88), 20(88), 106(111)
Xi, X. X., **16:**23–24, 106, 127, 141
Xia, M., **22:**163(83), 173
Xia, Y., **24:**44(60), 229(33, 34, 36, 39), 230(33, 46, 48), 231(49, 50), 232(39), 233(52), 234(46, 54), 235(33, 39, 52, 56, 59), 240(48, 69, 71, 77), 241(56, 77, 78), 242(54), 243(52, 80), 245(80), 246(34, 36, 86), 248(91), 249(91)
Xianglei, M., **16:**4, 124
Xiao, D. G., **16:**41, 131
Xiao, X., **23:**55(187, 188), 166(91); **24:**221(93), 228(8)
Xiao, X. D., **24:**106(107), 221(99)
Xiaoxian, L., **16:**4, 124
Xie, L. M., **16:**91, 139
Xiong, G. C., **16:**32, 130
Xu, C., **24:**44(54, 65), 46(65), 86(36), 161(23), 162(23)
Xu, D.-X., **23:**55(10)
Xu, G., **17:**318, 319, 366
Xu, J. M., **23:**241(52)
Xu, R., **17:**228, 229, 260, 287, 298, 299
Xu, S., **24:**94(63), 98(86), 106(108)
Xu, Y., **17:**228, 229, 260, 287, 298, 299
Xue, J. M., **16:**41, 131

Y

Ya, M., **21:**296(50), 310
Yablonovitch, E., **23:**48(161), 226(25, 26)
Yabumoto, N., **14:**232(117), 239
Yacha, T., **14:**64(206), 78
Yada, K., **2:**109(84), 143
Yadav, M. S., **12:**226(121), 235
Yagil, Y., **19:**161
Yaguchi, H., **23:**10–11(38)
Yagudaev, A. M., **14:**155(85), 197
Yakovlev, V., **19:**38
Yakovleva, A. U., **1:**148(52), 149(52), 185
Yakoyama, M., **12:**278, 294
Yaloz, H., **14:**177(149), 198
Yamada, A., **17:**228, 261, 262, 263, 298
Yamada, E., **5:**303(260), 321
Yamada, I., **11:**81(180), 86(180), 101, 102; **13:**1(1), 2(2), 3(6, 7), 5(9, 10), 9(19), 11(6, 10), 13(22), 18(26), 20(29, 30, 31, 33, 39, 41, 42, 46), 21(9, 30, 31, 33), 22(7, 39), 23(7, 29, 41, 42), 24(19, 46), 25(30, 47), 27(31, 47), 28(31, 33), 38(51), 39; **15:**114–115, 141, 229, 261; **16:**40(39), 44(41, 53), 45(29), 48(56), 56, 56(46), 57, 58, 116(34, 35), 138(115), 139(115), 143(131), 145(115), 148, 150,

250, 268(31); **19:**281, 310; **21:**103(40), 112, 116(2), 133(45), 155(60), 220(130), 232, 233, 234, 236, 290(44), 310
Yamada, K., **9:**182(63), 186(63), 257
Yamada, S., **17:**313, 324, 329, 339, 344, 365–367
Yamada, T., **15:**152, 214; **17:**228, 232, 243, 279, 296, 297, 299
Yamada, Y., **3:**39(211), 40(211), 52(211), 58; **4:**50(95), 93
Yamafuji, K., **11:**239(44), 247(44), 254, 255, 315
Yamaguchi, H., **19:**12; **24:**45(81)
Yamaguchi, J., **7:**375, 379, 383(195), 397
Yamaguchi, S., **9:**186(64), 257; **15:**185, 216; **19:**128, 139
Yamaguchi, T., **9:**190(84), 257; **16:**88, 138; **19:**128, 129, 139, 140
Yamaguchi, Y., **15:**154, 164–165, 183, 185, 215; **17:**173, 222
Yamaguehi, M., **12:**277(216), 281, 295
Yamaka, E., **19:**250; **23:**166(90)
Yamakage, Y., **13:**164(54), 188(54), 190(54, 159), 191(54, 159), 193, 197
Yamamoto, A., **13:**176(83), 186(83), 188(83), 194
Yamamoto, H., **16:**149, 229
Yamamoto, K., **16:**22, 25, 51, 73, 127, 132; **17:**229, 236, 287, 290, 296
Yamamoto, M., **5:**303, 321; **16:**41, 52, 131, 133
Yamamoto, O., **17:**354, 368
Yamamoto, R., **17:**361, 370
Yamamoto, Y., **3:**258(110), 259, 273; **14:**65(210), 78; **17:**15, 16, 17, 24, 31, 74, 153, 219, 220
Yamanaka, K., **17:**318, 353, 354, 365, 368
Yamanaka, M., **16:**87, 138
Yamanaka, N., **23:**228(32)
Yamanaka, S., **9:**54, 71; **14:**183(161–163), 199
Yamananaka, K., **17:**352, 368
Yamane, H., **16:**84, 86, 88–92, 111, 116, 138–139
Yamane, Y., **9:**33(127), 68
Yamanuchi, **17:**197, 223
Yamasaki, J., **15:**197, 217
Yamasaki, K., **14:**113(65), 127
Yamashina, T., **12:**277(217), 295; **17:**111, 142
Yamashini, K., **13:**134(103), 150
Yamashita, T., **16:**111, 116, 142
Yamauchi, N., **17:**279, 299
Yamauchi, Y., **19:**12, 19
Yamazaki, O., **16:**17–19, 126; **17:**163, 221, 228, 248, 297
Yamazaki, T., **23:**47(158)
Yamda, I., **17:**153, 154, 160, 220
Yameda, K., **17:**153, 220
Yamin, M., **4:**302, 306, 309, 323, 324
Yamoda, I., **13:**112(31), 148
Yamoka, H., **15:**168, 215
Yan, G., **6:**37(142b), 47(142b), 77
Yan, J., **16:**36, 130
Yan, S. L., **16:**36, 130
Yang, A. C., **23:**87(5, 6, 8, 11), 109(60), 144(40)
Yang, A. C. M., **16:**292, 334, 340
Yang, B., **19:**195, 196, 216
Yang, B. Y., **19:**52, 103
Yang, C. L., **15:**236, 259, 262
Yang, E. S., **16:**42, 131; **21:**84(17), 111
Yang, H. C., **24:**12(52), 15(52), 44(65, 66), 46(65, 66), 161(23), 162(23)
Yang, H. Q., **17:**7, 12, 73
Yang, J., **24:**34(141), 35(141), 106(110)

Yang, K. Y., **16:**52, 111, 115, 117, 122, 133, 142
Yang, L., **4:**99(16), 103(16), 122(16), 123(59), 127(59), 134, 136, 234(152), 254
Yang, Ling, **1:**210(114), 231
Yang, M., **24:**48(99, 100), 49(99)
Yang, R. Q., **23:**241(52), 279(102, 103), 280(103, 105)
Yang, S. J., **12:**75(87), 160
Yang, W. M. C., **17:**2, 36, 39, 40, 43, 48, 49, 51, 73, 75
Yang, X. M., **16:**36, 130; **24:**231(51), 233(51)
Yang, Y. N., **21:**5(20), 6(20), 22(20), 27(20), 29(20), 45(20), 72
Yang, Z. P., **24:**22(106)
Yano, M., **11:**68, 100
Yano, O., **11:**268(86), 269, 316
Yano, S., **13:**176(83), 186(83), 188(83), 194
Yanovskaya, M. I., **17:**318, 344, 366, 367
Yao, I., **15:**100, 140
Yao, J. Y., **21:**207(107), 235
Yao, T., **11:**36(13, 14), 96
Yaozhong, R., **16:**4, 124
Yapsir, A. A., **18:**96, 97, 98
Yaqub, M., **6:**123, 129(182, 183, 184), 131(182, 183, 184), 147
Yarar, B., **16:**13, 125
Yarbrough, W. A., **19:**61, 105, 106
Yaremenko, V. G., **16:**71, 79, 136
Yariv, A., **10:**228(7), 247(38), 248(39, 40), 250(44), 251(46), 258, 259
Yariv, Y., **15:**229, 261
Yarwood, J., **4:**9(22), 91
Yashiro, A., **24:**23(109), 27(109)
Yashiro, K., **15:**50–51, 138
Yasouka, H., **17:**22, 74
Yasuda, H., **13:**68(47), 105
Yasui, I., **17:**310, 311, 317, 365
Yasukouchi, K., **24:**45(81)
Yasunaga, H., **16:**262–263, 269
Yates, J. T., Jr., **23:**27(100), 29(102, 103)
Yau, S.-L., **24:**12(52), 15(52)
Yavorskaya, T. V., **3:**102(92), 127
Yazawa, Y., **21:**330(26), 331(26), 340(26), 343(26), 367
Yazu, S., **16:**32–33, 37–38, 129–130
Ye, C., **17:**208, 223, 228, 293, 294, 299
Ye, C. Q., **16:**91, 139
Ye, Q., **21:**81(13), 111
Ye, S., **24:**23(109), 27(109)
Ye, Z. Y., **16:**41, 131
Yeagley, H. L., **6:**314, 327, 346
Yee, D. S., **14:**65(214), 78; **16:**17, 23, 51, 53, 78, 126, 132–133, 137
Yee, E. M., **12:**135(299), 165
Yee, K. K., **13:**68, 69(49), 75, 80(49), 105
Yee, W. M., **23:**242(61)
Yeganeh, M. S., **24:**135(63)
Yeh, G. S., **4:**126(66a), 136
Yeh, J. T. C., **13:**153(6), 181(6), 182(6), 183(6), 192
Yeh, J.-J., **16:**21, 28–29, 127, 129
Yeh, P., **11:**81(180), 86, 101, 102; **19:**258, 281, 282, 287, 288, 290, 310
Yellott, J. I., **5:**130, 141
Yelon, A., **6:**206(2), 209(2), 211(23, 24, 25), 212(24, 25), 215(42), 217(42), 219(42), 220(42), 224, 225(24), 226, 230(109), 233(109), 234(109, 118), 235(109), 242(23, 24, 25), 243(140), 244(24, 25, 140), 245(140), 249; **15:**250(24, 108, 118), 254(24, 25), 255(24, 25), 256(24), 257(24, 160), 258(24),

259(24), 260(24, 160), 271(24, 84), 277(140), 279, 280, 281–282, 281(193), 282(193), 283(201), 284(201), 285(203), 286(203), 291(210), 292(210), 295, 296, 297, 298, 299, 300, 310
Yen, H. W., **10:**228(7), 258
Yen, K. H., **17:**153, 154, 220
Yeo, D., **5:**284, 319
Yeom, G. Y., **17:**87, 140
Yep, T. O., **7:**191, 192(155), 251; **12:**271, 293
Yeryomenko, A. A., **13:**169(70), 185(70), 187(70), 194
Yeshurun, Y., **16:**90, 139
Yeung, K. F., **11:**151(122), 158(131), 164(137), 166(140), 202, 203
Yi, G., **17:**249, 250, 297; **19:**250
Yilmaz, S., **17:**229, 263, 264, 298
Yim, W. M., **7:**171(114), 250; **12:**269(147), 293
Yin, D. L., **16:**32, 41, 130, 131
Yin, S., **12:**169(16, 18), 183(16, 18), 232
Ying, Q. Y., **16:**55–56, 64–65, 134, 136
Ying, S. C., **9:**159(16), 256
Yoder, J. R., **5:**85(32), 86
Yoder, M., **13:**212(43), 224
Yohino, T., **17:**318, 366
Yokoto, K., **12:**176(48), 224(48), 233
Yokoyama, N., **13:**212(47), 224; **15:**222, 261
Yokoyama, S., **13:**164(54), 188(54), 190(54, 159), 191(54, 159), 193, 197
Yolken, H. T., **4:**37, 38, 92
Yon, E., **4:**305(146), 309, 323
Yonts, O. C., **3:**73(46), 125; **14:**218(47), 238
Yoo, W. J., **18:**302, 310–312
Yoon, S.-G., **19:**250
Yorimoto, Y., **17:**354, 368
York, D. B., **2:**402(169), 404(169), 405(169), 424; **4:**282(107), 322
York, P. K., **23:**230(15), 231(15)
Yorozu, T., **15:**170, 174, 210, 215, 217
Yoshi, S., **14:**232(117), 239
Yoshida, K., **6:**246, 298; **17:**354, 368
Yoshida, M., **5:**306(273), 321; **16:**14, 125
Yoshida, S., **7:**282(78), 309; **9:**190(84), 257; **11:**67(130a), 99; **17:**153, 154, 219, 220; **19:**128, 139, 140
Yoshida, T., **3:**52(187), 58; **16:**51, 133
Yoshihara, A., **15:**185, 216
Yoshihara, H., **12:**274(185), 275(175, 180), 294; **13:**60(3), 77(3), 94, 104
Yoshiike, N., **17:**309, 324, 325, 327, 328, 329, 365, 366
Yoshikawa, A., **9:**60(271), 71
Yoshimatsu, Y., **12:**176(48), 224(48), 233
Yoshimura, T., **17:**348, 353, 355, 356, 357, 367–369
Yoshinaga, H., **4:**50(95), 93
Yoshino, J., **16:**93, 139
Yoshioka, H., **2:**106, 142
Yoshitake, T., **16:**27–30, 32, 53, 57, 127, 129, 134
Yoshiwa, T., **10:**58(125), 68
Yoshizawa, S-I., **16:**93, 139
Yoshizumi, S., **16:**111, 115, 117, 122, 142
Yotsumoto, H., **2:**110(89), 143
Youdelis, W., **7:**33(138), 40
Young, A. H., **3:**194(53), 209; **5:**282(178), 319
Young, D., **15:**115, 122–124, 141
Young, D. J., **16:**234, 266
Young, D. R., **4:**302(138), 306, 323

Young, E. T., **21:**5(1), 71
Young, F., **4:**153, 189
Young, I. G., **2:**384(93), 422; **6:**116(154), 147
Young, J. E., **7:**330, 334, 389(54), 394
Young, J. F., **15:**250, 263
Young, J. R., **1:**11(22), 64; **2:**71, 141
Young, J. T., **24:**33(133), 211(52)
Young, L., **2:**369(42), 405(42), 406(42), 420; **4:**9(20); **5:**65, 78, 86, 163(73), 170(73), 231; **6:**2(3, 4), 3(3, 10), 4, 5, 6(30, 34), 7, 8(34, 37), 9, 10, 11(42a), 12(52a), 13(3, 52a, 56), 15(63, 64), 16, 17, 18, 19(23, 34), 20, 22, 23, 26(3), 27, 30, 33, 35(37), 36(122), 37(128), 38(122), 45(122, 128); **12:**275(186), 276(191); **14:**118(77), 120(78), 127; **15:**8, 136
Young, M. G., **16:**42, 131
Youngren, G. K., **24:**72(114)
Youtz, J. P., **6:**310, 335, 346; **17:**2, 73
Yu, D., **14:**8(28), 74
Yu, E. T., **23:**281(108)
Yu, H., **24:**206(2)
Yu, H.-Z., **24:**27(128), 31(130)
Yu, L. S., **21:**103(42), 112, 190(85, 86), 191(88), 235, 298(51), 310
Yu, M. L., **19:**35; **23:**34(125)
Yu, P. C., **15:**304–305, 312
Yu, X., **17:**332, 333, 366
Yu, Y. P., **19:**10
Yu, Z., **15:**295–296, 311
Yuan, C. W., **16:**52, 133
Yuan, H., **24:**24(117), 26(117), 27(117)
Yuan, H. X., **21:**8(25), 9(25), 10(25), 11(25), 14(37), 17(25), 20(37), 37(25), 72
Yuan, X.-Y., **17:**24, 49, 51, 53, 56, 75
Yuasa, T., **14:**18(75), 20(75), 66(219), 75, 78
Yue, H., **15:**15–16, 137
Yueqiang, S., **16:**66–67, 136
Yuh, P. F., **21:**161(68), 234; **23:**240(48, 51, 55), 270(48)
Yuheng, Z., **16:**4, 124
Yunogami, T., **18:**105
Yurshin, B. Y., **13:**185(137), 188(137), 196
Yurshina, N. I., **13:**185(136, 137), 188(136, 137), 196
Yutani, N., **23:**105(55), 124(17)
Yuuki, A., **22:**182(24), 272
Yvon, K., **16:**2, 124

Z

Zacek, F., **12:**276(189), 294
Zachar, F., **16:**249, 268
Zacharias, P., **9:**188, 194(95), 198(110), 202(118), 218(110), 257, 258
Zachariasen, W. H., **1:**205(90), 230; **15:**268, 310
Zachariou, A., **21:**345(37), 367
Zaghloul, A. R. M., **19:**288
Zah, C. E., **23:**226(29), 228(29)
Zaharchuk, G., **16:**121, 143
Zahl, P. A., **7:**368(175, 176), 396
Zahn, D. R. T., **19:**41
Zahn, U. V., **1:**44(123), 45(123), 67
Zahurak, S. M., **16:**3, 124
Zainiger, K. H., **6:**5(25), 36(25, 103a), 74, 76
Zaininger, K. H., **4:**260(24), 301(24, 130, 131), 320, 323; **5:**152(36), 153(36), 230, 304(265), 321; **12:**264(110), 292
Zak, J., **24:**24(117), 26(117), 27(117)
Zakharov, V. E., **15:**95, 140

Zakrzewski, Z., **18:**65
Zallen, R., **19:**102, 135
Zalm, P., **12:**87, 89, 161
Zalm, P. C., **14:**15(56), 74; **22:**110(25), 115
Zalph, B. L., **12:**258(74), 291
Zalucki, Z., **14:**141(37), 156(91), 195, 197
Zama, H., **16:**88, 138
Zamel, J. N., **11:**107, 108(10), 109, 113, 114(15), 115, 135(52), 200, 201
Zamerowski, T. J., **12:**129(272), 165
Zamgvil, Z., **17:**153, 219
Zander, W., **16:**121, 143
Zandin, M., **21:**40(88), 74
Zang, D. Y., **17:**174, 194, 222
Zang, Y., **16:**121, 143
Zangwill, A., **24:**182(26)
Zanoni, R., **17:**46, 49, 51, 53, 76
Zanovick, R. L., **5:**274, 317
Zappella, P. I., **11:**192(154, 156), 203
Zaromb, S., **9:**19, 20, 66
Zarudiansky, A., **17:**361, 370
Zarzychi, J., **7:**35(153), 41
Zasadzinski, J. A. N., **24:**98(95), 169(37)
Zaslavskii, Yu. S., **3:**39(208), 52(208), 58
Zaslavsky, A., **15:**249–250, 263
Zaugg, F., **24:**106(109)
Zavaritski, N. V., **1:**254(55), 274
Zavaritskii, N. V., **6:**123, 148
Zavaritsky, N. V., **16:**111, 142
Zavaritsky, V. N., **16:**111, 142
Zavracky, P. M., **9:**116(26), 144
Zawadzki, P. A., **16:**91, 139
Zawodzinski, T. A., **24:**97(78), 98(78, 84)
Zazulin, V. S., **3:**24(116), 51(116), 56
Zdanis, R., **6:**121, 147
Zdanuk, E. J., **1:**43(120), 66; **3:**27(138), 28(138), 51(138), 56, 65(19), 73(47), 100(88), 125, 126
Zega, B., **3:**5, 51(30), 53; **13:**78, 79(63), 106
Zegarski, B. R., **24:**44(3, 5), 55(94), 64(94), 70(5), 114(26), 121(44), 212(55)
Zehden, W., **3:**10(44), 51(44), 54
Zehender, E., **1:**210(113), 231; **3:**40(213), 58
Zehnder, D., **13:**98(97), 106
Zehner, R. W., **24:**33(136), 127(56), 128(57)
Zeidenbergs, G., **5:**272(114), 317
Zeira, E. C., **13:**98(96), 106
Zeitler, E., **1:**191(20), 228; **2:**97, 142
Zeitman, S. A., **5:**176(209), 222(209), 235; **17:**153, 220
Zeitsov, N. S., **12:**71(63), 160
Zektser, M. P., **14:**145(64), 160(105), 196, 197
Zeldov, E., **16:**110, 142
Zeldovich, J. B., **4:**98(10), 99(10), 134
Zelez, J., **17:**351, 367
Zeller, H. R., **17:**310, 314, 315, 365
Zellers, E. T., **17:**212, 213, 224
Zellmer, D. L., **9:**35(165), 36(165), 37(165), 69; **12:**190(91), 234
Zelmer, D. L., **12:**190(90), 234
Zemany, P. D., **1:**205(91), 206(91), 230; **3:**39(194, 195), 52(194, 195), 58
Zemel, J. N., **2:**157(42), 168(42), 191; **3:**228, 272, 299(73), 301; **5:**164(92, 93), 171, 216, 231; **12:**218(110), 219(110), 235
Zener, C., **7:**44(3), 111
Zenner, R. L. D., **14:**180(159), 199
Zeppenfeld, K., **9:**202(116), 258
Zeppenfeld, P., **24:**98(99), 106(99)

Zerlaut, G. A., **10:**133, 166
Zernike, F., **2:**110(90), 143; **7:**2(7), 37; **10:**236, 258
Zettler, R. A., **5:**306, 322
Zhang, B., **13:**127(87), 149
Zhang, B.-R., **15:**295–296, 311
Zhang, J., **16:**93–94, 111, 115, 117, 122, 139–440, 142; **17:**315, 347, 365; **19:**10, 15; **21:**128(37), 233
Zhang, J. M., **16:**93, 139
Zhang, K., **16:**83–84, 88, 94, 138, 140
Zhang, Q. M., **16:**36, 130
Zhang, S.-Y., **17:**24, 49, 51, 53, 56, 75
Zhang, X. K., **16:**91, 139
Zhang, Y.-H., **23:**234(20, 37), 235(20)
Zhang, Y-J., **16:**51, 132
Zhang, Z., **24:**33(133), 211(52)
Zhang, Z. H., **16:**96, 140
Zhao, H., **15:**229, 249–251, 261, 263
Zhao, J., **16:**87, 93–94, 138–140
Zhao, J.-W., **24:**27(128), 31(130)
Zhao, W., **24:**169(39), 170(39)
Zhao, X.-M., **24:**44(60), 229(33, 36), 230(33), 235(33), 240(77), 241(77, 79), 246(36)
Zhao, Z. R., **16:**212, 230
Zhaojia, C., **16:**4, 124
Zheleznyakov, V. V., **18:**136
Zheng, D. C., **17:**22, 74
Zheng, G. G., **16:**41, 131
Zheng, J. P., **16:**55–56, 64–65, 134, 136
Zheng, J. Q., **16:**41, 131
Zheng, L., **17:**218, 224
Zheng, Z. T., **16:**26, 127
Zhidong, Y., **16:**4, 124
Zhiqiang, M., **16:**4, 124
Zhong, C.-J., **24:**44(29, 32), 54(29), 55(29), 70(29, 32)
Zhong, Q., **17:**318, 328, 366
Zhotikov, V. G., **15:**95, 140
Zhou, J. M., **16:**262, 269
Zhou, Q. F., **19:**250
Zhou, W. K., **16:**91, 139
Zhou, X., **21:**207(108), 235
Zhou, X. D., **16:**36, 130
Zhu, R., **19:**281
Zhukhovitskii, A. A., **16:**234, 266
Zhuralev, N. N., **12:**88(130), 161
Zi-qin, W., **17:**22, 74
Ziborov, E. S., **6:**292(220), 300
Ziegler, J. F., **10:**58(126), 68; **14:**228(91), 229(96), 239
Ziegler, J. M., **13:**153(9, 11), 182(9, 11), 183(9, 11), 192
Zil'berman, P. E., **15:**64, 80–81, 139
Zilko, J. L., **21:**221(137), 222(137), 237
Ziman, J. M., **2:**105(72), 142; **6:**89, 144
Zimmerer, G., **7:**49(28), 112
Zimmerman, E., **5:**154(41), 155(41), 156(41), 228(41), 230
Zingaro, R. A., **12:**203(96), 234
Zinn, W., **6:**210, 295
Zintl, E., **12:**88(132), 161
Zipperian, T. E., **15:**222, 260; **16:**123, 143
Zirkle, T. E., **22:**215(76), 270(127), 274, 276
Zisk, M., **16:**84, 90, 138
Zisman, W. A., **7:**329(49, 50, 51, 52), 331(66, 67), 367(173), 393, 394, 396; **24:**2(6, 7, 8), 3(6, 7, 8, 9), 17(6, 9), 181(16)
Zitelli, L. T., **17:**153, 219
Zlotnicka, A., **23:**60(202)
Zng, L. M., **24:**98(96)
Zobel, F. G. R., **6:**2(4), 4(19), 11(42a), 21, 22, 27, 30, 33, 50, 63(4), 64(4), 73, 74
Zocher, H., **7:**322(34), 393
Zorin, E. I., **12:**291

Zorll, U., **6:**339(52, 55), 341(60), 347
Zou, C., **24:**23(113), 27(113), 30(113), 240(74)
Zozulya, G. V., **6:**339(56), 347
Zschauer, K. H., **7:**149, 151(58), 157(71), 236(250a), 248, 249, 254
Zscheile, H., **12:**251(62), 256(62), 291
Zubkov, V. I., **15:**37–38, 138
Zubler, E. G., **2:**393(145), 423
Zuccaro, D. E., **10:**159(90), 166
Zucker, M., **1:**241(8, 9), 273
Zuckermann, M. J., **15:**185, 216
Züger, O., **24:**221(89), 222(89)
Zuleeg, R., **2:**151(34, 38), 165(34), 181(65), 183, 191, 192; **7:**241(258), 255; **17:**259, 298; **19:**250
Zunger, A., **13:**99(106), 107
Zunick, M. J., **9:**33(129), 68
Zur, A., **13:**202, 223
Zurheide, M., **9:**186(72), 187, 257
Zussman, A., **21:**128(30), 129(30), 130(30), 131(30), 132(30, 43), 133(43), 135(30), 136(30), 137(30, 43), 138(30), 139(30), 140(30), 141(30), 142(30), 143(30), 144(30), 145(30, 52), 146(30), 147(30, 52), 148(30), 149(30, 43,52), 150(30, 43, 52), 160(30), 213(52), 219(43), 221(135), 222(135), 233, 236
Zwanzig, R., **11:**293(128), 317; **24:**72(113, 115)
Zwanzig, R. W., **7:**320(33), 393
Zweben, C., **16:**333, 339
Zwingman, R., **15:**154, 215
Zworykin, V. K., **2:**181(63), 192; **5:**145, 229; **12:**101(171), 102(171), 154(328, 329), 162, 166
Zydzik, G., **14:**57(179), 77
Zydzik, G. J., **9:**275, 297; **17:**355, 368
Zysset, B., **13:**174(82), 180(82), 186(82), 194

Subject Index

A

AA *see* Arachidic acid
Abb refractometer, **2:**209–210; **4:**55
Abels-Brewster angle, **4:**77–78
Abels method, optical constants, **2:**210–213
Ablation
  laser-induced etching, **13:**153–154
  materials other than organic polymers, **13:**181
  pulsed laser deposition, **17:**236
  threshold laser fluence, **13:**172–173
Abrasion, ion-beam processing, **13:**125
Absorbers
  black chrome or black nickel, **10:**47–49
  coatings
    electroplated, **10:**47
    oxidation, **10:**42–43
  tandem, **10:**38–54
Absorbing films
  oxide layers and, **5:**124–126
  for sunglasses, **8:**92–93
  wave fields and, **5:**7
Absorptance, **2:**307–310
  in admittance notation, **1:**95
  intensity absorptance, **1:**77
  solar, of evaporated coatings, **2:**305–361
    determination, **2:**313–333
    measurement, **2:**330–333
    thermal emittance and, **2:**312–313
Absorption
  at "pure" ECR, **18:**131–132
  at second harmonic resonance, **18:**189–190
  at upper hybrid resonance, **18:**139–142
  eigenoscillation and, **5:**6–7
  gray, **5:**125
  of left hand polarized wave, **18:**127, 129, 136–138, 167, 187–190
  level of, **18:**178
  microwave power, **18:**124, 126, 131, 150, 162, 172
    in electron cyclotron resonance, **18:**126–128, 188–190
  non-linear, **18:**130, 148, 150–151
  photocurrent, **5:**36
  radiation, in superconducting film, **1:**265–267
  of right hand polarized wave, **18:**126, 129, 136–138, 167–168, 170, 188–190
  surface, **18:**162, 194

Absorption (*continued*)
of whistler wave, **18:**129, 134–136, 151, 184, 188–190
Absorption band profile, porphyrin-doped Langmuir-Blodgett films, **20:**301
Absorption coefficient, **1:**141; **15:**232
amorphous silicon films, **13:**42–43
as function of photon energy, **13:**52
for negative electron affinity materials, **12:**70
in optical absorption, **12:**55
photons, **5:**2
Absorption energy, of electron, **2:**122–123
Absorption peak, in discontinuous film, **8:**143
Absorption spectra, of ionic crystals, influence of structural defects, **1:**227–228
Absorption spectroscopy, **14:**15
Absorption zone, microwave power of, **18:**126, 150, 162–163, 176, 185, 194, 206
AC bridge measurements, in dielectric polarization, **6:**54–56
Acceleration
of electrons, **18:**148
of ions, **18:**159
Accelerometers, topologically structured films and, **8:**314–316
Acceptors, in amorphous solids, **8:**178–183
Accommodation of misfit
between nuclei and substrate, **4:**161–163
between overgrowth and substrate, **4:**139–149
AC conduction, theory, **11:**206
AC conductivity
in dielectric response, **11:**217
frequency dependence, **11:**290
for various materials, **11:**257
vs. DC conductivity, **11:**258
Accumulation barrier, in nonmetallic films, **8:**199
Accumulation layer, **2:**157
Achromatic quarter-wave plate, **19:**199, 254
Acoustic performance, **17:**159
coupling coefficient, **17:**159, 163
coupling factor of ZnO vs. angle of tilt, **17:**191
quality factor, **17:**188, 189, 192, 194, 195, 196
Acoustic sensors, **17:**210
comparison sensitivities, **17:**216
FBAR sensor, **17:**215
Lamb wave sensor, **17:**212
SAW resonator sensors, **17:**211
Acoustic wave devices, **17:**247, 248, 261
Acoustic waves, **18:**151
Activated dissociation reduction reaction, **13:**99
Activated reactive evaporation, **13:**60–62, 64–74, 76–79, 94–95, 95–97, 97–98, 98–101, 101–102, 102–103, *see also* Reactive evaporation
Activation, of II-VI compound films, **3:**165–210
Activation distance, Frenkel defects and, **6:**18–19
Activation energy
of diffusion, **2:**12
for thin films, **8:**104, 109
Activators, simultaneous evaporation of, and host crystals, **3:**179–180

Adatom energy, sputter deposition, **22:**150–152, 170, 171
Adatom migration effect, **13:**3
Adatoms, incorporation on silicon surface, **23:**15–18
Addition process, in optical element correction, **10:**263–265
Adherent silver mirrors, **12:**36–39
Adhesion, **17:**106, 130, 133
cancer cells to silane films, **20:**136
cathodic arc plasma deposition, **14:**173–174
fracture mechanics definition, **16:**276
interfacial adhesion, **16:**267
ion-beam processing, **13:**125
of sputtered films, **3:**96–98
testing, **16:**268
thin film, **1:**206–207
Adhesion strength, **13:**25, 55; **16:**276
Adhesive failure, **16:**276, 278, 314, 330–331
Adhesive forces, by atomic force microscopy, **24:**96, 105, 106
Admittance
charts, **1:**102–111
concept, **1:**92–96
in interference filter theory, **9:**75–77
reduced, **1:**93
Admittance matching, **9:**74
Adsorption, **24:**3, 45–77
Aerosil, **5:**118
AES *see* Auger electron spectroscopy
AFM *see* Atomic force microscopy
Age theory, Fermi, **5:**9
Agglomerated film, **8:**102–118, 139
Aggregate critical size, **2:**3–4
Aging, **15:**169, 172–174
effect on reflectance, **1:**165–167
aluminum, **1:**152–154, 157–158
germanium, **1:**159–161
gold, **1:**168
platinum, **1:**168
parameters, **1:**127–130
Air mass
solar insolation and, **10:**6
terrestrial solar spectral irradiance and, **10:**8–9, 21
Airy summation, for rough layer, **10:**177–182
Alcohol monolayer, headgroup orientation, **20:**169–170
Algorithms
constant temperature control algorithm, **22:**134–135
extended system method, **22:**138–142, 161
generalized Langevin equation, **22:**135–138, 141–142, 170
Gear 4 algorithm, **22:**132
leap-frog algorithm, **22:**132
molecular dynamics, **22:**128–131, 132, 170
shock-tracking algorithm, **22:**97–98, 113
SIMBAD model, **22:**9–114
SIMSPUD model, **22:**6–8
Verlet-type algorithm, **22:**132, 169
Aliphatic alcohol monolayer, X-ray structure, **20:**149–153
Aliphatic chain, annealing, **24:**65–69
Alkali antimonides, **12:**84–112
band structures, **12:**105
bi- and trialkali, **12:**86–87
cathode preparation in, **12:**85–87
chemical composition and crystal structure, **12:**87–90
electrical properties, **12:**94–101
high-temperature instability, **12:**91
monoalkali, **12:**85–86
optical properties, **12:**101–112

Alkali antimonides, (*continued*)
photocathode superficial oxidation and, **12:**92
photoelectron spectroscopy, **12:**108
photoemission properties, **12:**90–94
as secondary emitters, **12:**101
Alkali halides
IV-VI layers in, **11:**113–115
as parting agents, **7:**65–66
Alkali metal film, resistivity studies in, **6:**104–105
Alkanethiol film, structure on liquid mercury, **24:**180–201
Alkanethiol/gold monolayers, **24:**2
aliphatic chain annealing, **24:**65–69
atomic force microscopy, **24:**82–83, 114
depressions and domains, **24:**86
formation
atomic force microscopy, **24:**83, 89–93
quartz crystal microbalance, **24:**45–69
of steps, **24:**62–63
thermodynamics, **24:**45–69, 83, 89–93, 150
gold/thiol interface, chemistry, **24:**52–55
motional dynamics, **24:**70–77
quartz crystal microbalance, **24:**46, 48–52, 65–66, 68
scanning tunneling microscopy, **24:**92
structure
by atomic force microscopy, **24:**82–86
by differential studies, **24:**112, 114, 121–130
Alkanethiols
chain dynamics, **24:**156–159
containing ringed structures, **24:**21–33
films on liquid mercury, X-ray studies, **24:**180–201
on gold nanoparticles, **24:**154–163, 175, 222
with internal aromatic functional groups, **24:**30–33
omega-terminated, **24:**9–11
pyrrole-terminated, **24:**30
synthesis, **24:**6–11
terminating in groups commensurate with hydrocarbon lattice, **24:**12–20
Alkyl disulfides, asymmetric, **24:**20–21, 89, 90
Alkyl-lithium reagents, surface modification, **20:**96–99
Alkylphosphonates, on metal oxides, **24:**163–169, 175
Alkylsiloxane SAMs, NMR spectroscopy, **24:**169–175
Alloyed ohmic contacts, to GaAs, **13:**209–219
alloyed AuGeNi contact, **13:**215–216
alternative contact methods, **13:**218–219
band model, **13:**215
characterization, **13:**213–214
contact resistivity *see* Contact resistivity
fabrication, **13:**210–213
Ga outdiffusion, **13:**217
models, **13:**215–217
process parameters, **13:**211
sheet resistance, **13:**214
tunneling probability, **13:**218–219
Alloying effects, in spray pyrolytic process, **12:**181–187

Alloy superconductors, critical temperatures, **1:**245–246
Aluminosilicate glass film, pyrolytic, **4:**314
Aluminum
  chemical vapor deposition, **5:**307–308
  evaporated *see* Evaporated aluminum
  films with coatings, **2:**345–353
  grain boundary diffusion, electromigration and, **7:**294
  low-temperature resistivity, **6:**119–120
  nonabsorbing surface films on, **12:**24
  optical functions
    bulk aluminum, **19:**131–33
    thin films, **19:**150–83
  reflectance, **10:**75, 118; **12:**9, 17–19
    as function of incidence angle, **12:**21
      vacuum ultraviolet, **1:**150–159
      aging effect, **1:**152–154, 157–158
      oxidation effect, **1:**135, 178–179
      on platinum, **1:**178–179
      purity effect, **1:**150–151
  as reflective coating, **12:**8–10
  single-crystal wire studies, **6:**119
  sputter deposition, **22:**25–26, 30–33
    high temperature, **22:**49–52, 56–58
    ionized magnetron sputter deposition, **22:**242–243, 244, 248–252
Aluminum/aluminum nitride superlattice, **17:**64
Aluminum/aluminum oxide coatings
  evaporation of with electron gun, **10:**147–149
  in satellite temperature control, **10:**128
Aluminum-antimony film, evaporation, **5:**175–176
Aluminum arsenide, in liquid phase epitaxy, **7:**171–172
Aluminum arsenide/gallium arsenide system, in liquid phase epitaxy, **7:**183–188
Aluminum-chromium film, **6:**313; **7:**292
Aluminum-copper film, **16:**250
  electromigration in, **7:**287–293, 297
  surface diffusion, **22:**253–265
Aluminum-copper system, superlattice, **17:**33, 60, 70
Aluminum evaporation, on arsenic-rich GaAs surfaces, **11:**87
Aluminum evaporator, **12:**10–14
Aluminum film, *see also* Aluminum thin film
  apparent voids in, **7:**53
  contact resistivity, as function of annealing temperature, **13:**28–29
  deposition, in evaporator, **12:**12–14
  evaporated *see* Evaporated aluminum film
  evaporation technique, **7:**59–60
  as filters, **7:**47
  fuchsin as parting agent for, **7:**62
  isotransmittance curves for, **9:**118
  Parlodion films and, **7:**70
  pinholes in, **7:**109–110
  purity, **12:**14
  SEM images, **13:**30
  silicon monoxide coating for, **7:**89
  size-dependent resistivity in, **6:**113–115

Aluminum film, (*continued*)
X-ray diffraction intensity at different acceleration voltages, **13:**29
Aluminum filters, *see also* Aluminum film
mounting of, **7:**99
pinhole transmittance for, **7:**88
spurious pinholes in, **7:**52
Aluminum fluoride, microstructure and optical functions, **19:**222, 227, 229
Aluminum-gold film, **6:**301, 315–326
Aluminum-magnesium thin film, electromigration in, **7:**295–298
Aluminum-manganese superlattice, **17:**33
Aluminum mirrors, **12:**11, 35–36, *see also* Evaporated aluminum mirrors
aged, **10:**89
optical constants and reflectance, **12:**17–19
reflectance enhancement in, **8:**44–45
Aluminum nitride, **14:**8, 11, 18, 20, 25–26, 42, 57, 65, 67–70
chemical vapor deposition, **5:**302–304
semiconductor films, **5:**160
Aluminum nitride film, plasma chemical vapor deposition for, **12:**269–270
Aluminum nitride/niobium nitride superlattice, **17:**33, 49, 58
Aluminum oxide, **12:**10; **14:**4, 8, 9, 15, 18, 31, 35, 39, 40, 47–48, 50, 57, 67, 70; **16:**275; **17:**64; **19:**223, 235, 237, 241
chemical vapor deposition, **5:**259–262
electrical conductivity, **11:**261
electron gun evaporation, **4:**283–285
ion-beam processing, **13:**140–141
microhardness, versus base temperature, **13:**93
microstructure and optical functions, **19:**223, 227, 235, 238, 241
Aluminum oxide coatings, **10:**85–91, 149–154
on aluminum, **2:**352–353
deposition, **12:**28
reduced reflectance of at higher incidence angles, **12:**34
Aluminum oxide film, **8:**37
infrared spectra, **15:**303–304
band assignments, **15:**303
reflection spectrum, **15:**303
transmission spectrum, **15:**303
as overcoatings, **12:**25
plasma chemical vapor deposition for, **12:**264–265
Aluminum oxide/molybdenum coatings (AMA), **10:**55
Aluminum oxide/molybdenum/sapphire sandwich, **8:**292
Aluminum oxide window, **18:**166, 167
Aluminum phosphide, in liquid phase epitaxy, **7:**171–172
Aluminum phosphide/gallium phosphide system, **7:**201
Aluminum/silicon dioxide coatings
evaporation of with electron gun, **10:**141–147
flight experiments with, **10:**138–139
in satellite temperature control, **10:**128, 133–140
Aluminum-silicon thin film, **16:**251

Aluminum-silver coatings
aluminum and silver coatings for, **10:**149–154
with double layers of aluminum oxide and silver dioxide, **10:**149–154
Aluminum-silver couples, **6:**328–333
Aluminum-surfaced diffraction gratings, coating for, **11:**13
Aluminum/Teflon sheets, **10:**162
Aluminum thin film, **16:**249; **19:**127–87, *see also* Aluminum film
copper diffusion in, **16:**253
copper effective charge in, **16:**242, 248
electromigration in, **7:**287–288
failure in, **7:**266–267
nucleation and growth, **19:**150–62
optical functions of continuous films, **19:**162–68
optical functions of particle films, **19:**168–83
particles as nanocrystallite clusters, **19:**179
Aluminum-vanadium compounds, as semiconductor films, **5:**158
Aluminum wires, temperature-dependent resistivity, **6:**130
Alzak, **10:**159–160
AMA (aluminum oxide-molybdenum) coatings, **10:**55
alpha-Amino acids
crystal structure, **20:**176
hydrogen bonding, **20:**178
interaction energy, **20:**179
p-Aminobenzoic acid
intercalation of Cu2 +, **20:**44
orientation in films, **20:**52
p-Aminobenzoic acid film
contact angle for organic solvents, **20:**47
defect ratio, **20:**47–48
Amorphous boron film, **18:**159
Amorphous carbon, high-field effects in, **8:**220
Amorphous films, **10:**235–238; **14:**60, *see also* Thin films
defects in, **9:**289
local atomic order in, **7:**1–37
for magnetic bubbles, **9:**277–279
types, **7:**1–2
Amorphous germanium, Poole-Frenkel conduction in, **8:**213
Amorphous metals, radial distribution function measurements for, **7:**18–28
Amorphous selenium, mobility measurements in, **8:**231
Amorphous silicon
density of localized states in, **8:**236
hydrogenated *see* Hydrogenated amorphous silicon
mobility measurements in, **8:**230
Amorphous silicon film, **18:**159
infrared spectra, **15:**277–283
air exposure and, **15:**282
band assignments, **15:**279
crystallinity and, **15:**281
deposition gas and, **15:**278
light induced changes in, **15:**283
structural defects in, **15:**279
substrate bias and, **15:**279, 281
temperature and, **15:**279
thickness dependence, **15:**282
Amorphous solids, *see also* Disordered nonmetallic films
avalanche multiplication in, **8:**211
defined, **8:**170–171
disorder in, **8:**171–172
donors, acceptors, and traps in, **8:**178–183

Amorphous solids, (*continued*)
donor-trap hopping conduction in, **8:**207–208
electrical properties, **8:**173–174
electron spin resonance in, **8:**239–240
empty state in, **8:**179
energy level structure, **8:**174–183
field enhanced emission from donors in, **8:**199
field ionization in, **8:**208
forbidden gap and, **8:**177, 240
hopping in, **8:**176–177, 207–208, 212–217
mobility measurements in, **8:**229–232
occupied state in, **8:**179
photoconduction in, **8:**235–239
physical properties, **8:**170–171
Poole's law and, **8:**224–226
recombination in, **8:**233
space charge effects in, **8:**210–211
space-charge-limited conduction and, **8:**228–229
structure, **8:**171
$T^{\frac{1}{4}}$ law of conductivity in, **8:**189
thermally stimulated currents in, **8:**233–235
Amorphous state, indeterminate nature of, **8:**173
Amphiphile
electron acceptor, **20:**244, 251–253
electron donor, **20:**244, 251–253
ferrocenes as, **20:**244, 247, 249–252
in molecular diode, **20:**244–245
Amphiphilic alcohol film, **20:**145–181
Amphiphilic dyes, **20:**296
Amphiphilic mesophases, **20:**368–369
Amphiphilic monolayer
nucleation template, **20:**146
phases, **20:**367–371
Amphiphilic porphyrin, **20:**71, 73
Amphiphilic porphyrin film, **20:**69–82
Amphiphilic receptors, **20:**225
Amphiphilic triads, **20:**246–249
Amplifiers, semiconductor films as, **5:**145
Amplitude
diffracted, **2:**113–114
phase diagram, **2:**120–121
reflected
double layer coating, **2:**249
transparent medium with transparent layer, **2:**243
Amplitude reflection, in interference filter theory, **9:**75–77
Anatase structure, in oxide coatings, **5:**114
Angle of principle incidence, **2:**201
Angle shift, for 5-layer bandpass filter, **10:**224
Angular distribution
ion, **18:**73, *see also* Energy distribution
of scattered flux, **10:**213–216
Anisotropic materials, **19:**258, 279–312
anisotropic surface layer, **19:**2–41
anisotropic thin films, **19:**309–312
Brewster's angle for anisotropic substrates, **19:**304
simulation of optical function tensors, **19:**283–312
simulation of SE spectra, **19:**295–312
spectro-ellipsometry, **19:**279–312
Anisotropic samples
analysis methods, **19:**281
anthracene on glass, **19:**292
approximate ellipsometric models, **19:**281
biaxial, **19:**281

bismuth tellurium sulfide crystal, **19:**292
examples, **19:**281
metals, thin-wire resistivity, **6:**121–122
surfaces and thin films, **19:**281, 296
uniaxial, **19:**281, 304
Anisotropy, magnetic *see* Magnetic anisotropy
Anisotropy constant, **6:**225
Anisotropy energy, **16:**151
domain wall, **1:**314
magnetic, **1:**314; **15:**129
cubic, **15:**130
uniaxial, **15:**131
per unit volume, **1:**297
of strained polycrystalline film, **1:**283–284
total, **1:**298
Anisotropy field, **16:**150
magnetic, **15:**5, 26
defined, **15:**130
observed in ferrites, **15:**9
uniaxial, **15:**131
Anistropic films, *see also* Very thin films
general expressions for, **6:**177–178
intensifies for, **6:**178
phases in, **6:**179
polarization in, **6:**178–181
total reflection in, **6:**180–181
very thin, **6:**181–182
Annealing, **4:**186
aliphatic chain, **24:**65–69
crystal size and, **3:**217–218
films condensed at low temperature, **2:**38–40
in situ SE studies, ferroelectric films, **19:**264–70
of island films, **2:**27–28
magnetic, **1:**221–223
nickel films, **1:**294
permalloy films, **1:**289–294
Anode, **14:**170
Anodic oxide films, **6:**1–73
cathodic breakdown voltage for, **6:**70–71
dc electric conduction in, **6:**63–66
defined, **6:**2
dielectric breakdown, **6:**66–73
dielectric properties, **6:**51–73
experimental breakdown in, **6:**68–69
gaseous anodization and, **6:**36–51
overpotential for, **6:**21
photostimulated growth, **6:**13
polarization, **6:**52–54
Anodization, **2:**369–370
constant current formation in, **6:**5–8
constant field transient and, **6:**10
constant voltage formation in, **6:**10–12
current efficiency in, **6:**6
gaseous *see* Gaseous anodization
general features, **6:**3–13
ionic conductivity and, **6:**8–10
overfield in, **6:**3
plasma, **18:**96
preparative techniques in, **6:**5
substrates in, **6:**3–5
transients and, **6:**8–10
wet, **6:**44
Anodization constant, **6:**5, 45
Antenna, **16:**108; **18:**126, 153–154
Antenna molecules, **20:**254–256
and energy transfer, **20:**254
spectra, **20:**255
structure, **20:**247
Anthracene clusters, **13:**54, 55
Antiferromagnetic films, **6:**213–223, 268

Antimonides, alkali *see* Alkali antimonides
Antimony
doping, **23:**14–15
in films for MBE fluxes, **11:**73–74
radial distribution function measurements, **7:**19–20
Antimony film
diffraction patterns, **13:**13–14
as filters, **7:**49
gases, effect on, **1:**53–54
resistivity, **6:**141
specular reflection in, **6:**132
Antimony oxide film, **8:**40
Antireflection coatings, **2:**239–304; **5:**48, 60
dielectric films as, **5:**20; **8:**42–43
inhomogeneous films as, **8:**85–88
oxide layers for, **5:**123–124
quarter wave thickness, **2:**244ff.
for semiconductor oxide films, **9:**19–20
stepdown design, **2:**253
theory and design, **2:**242–284
Application Technology Satellite, **10:**159
Arachidic acid
lattice constant, **20:**31
monolayers, **20:**16–18
Arc, **14:**32; **17:**39, 43, 47
Arc discharge, **17:**92
Arc evaporation, **13:**79; **17:**94–96
Argon
chemical vapor deposition in, **5:**313
film growth in, **5:**295
Argon plasma, **18:**162–163, 173–178, 181, 203, 205, 209–213
Argon sputtering, **13:**118
Aromatic rings, in terminal group of alkanethiols, **24:**21–33
Aromatic thiols, **24:**33
Arrhenius plots, of Pb barrier PbTe photodiodes, **11:**160–161
Arsenic, radial distribution function measurements for, **7:**19–20
Arsenic chloride, in semiconductor films, **5:**151
Arsenic oxide film, **8:**40
Arsenic-phosphorus alloys, smaller band gapin, **12:**122–123
Artificial intelligence, neural networks, **21:**60, 61
Aspect ratio, etching rates, **13:**170
Asymmetrical quantum-well infrared photodetectors, **21:**116
n-doped, **21:**161–167
p-doped, **21:**167–169
Asymmetric disulfide, SAMS containing, **24:**20–21, 89, 90
ATM filters, **7:**83–86, 99
Atom
reflected, **14:**7, 8, 27, 51, 65–69
sputtered, **14:**4, 15–21
target, **14:**6
Atom clusters, **14:**15–21
Atomic force microscopy (AFM), **24:**82, 114
high critical temperature superconductors, **16:**54
nanofabrication, **24:**83, 98–102, 103, 104, 106
nanomechanical properties, **24:**83, 104, 105
adhesive and frictional forces, **24:**96–98
elastic compliance, **24:**93–96
structure characterization, **24:**83–89
Atomic movements, during oxide formation, **14:**107–114
Atomic peening, **22:**121
Atomic stoichiometric matrix, **22:**185

Atomic weight, refractive index, **10:**41–42
Atomistic modeling, **22:**177
ATR method, **9:**233–234, 238
Attenuated total reflectance, **15:**273, 283, 289, 309, *see also* Internal reflection spectroscopy
Attenuation, magnetostatic *see* Propagation loss
Attenuation length, **5:**9–11
  measurements, **4:**345–350
Auger analysis, **10:**125–126
Auger electron spectroscopy (AES), **11:**44–45; **17:**22
  high critical temperature superconductors, **16:**5, 63, 69, 71
  ion-beam processing, **13:**122
  of plasma grown oxides, **12:**277
Auger recombination, **3:**278
Augmented double-cavity filter, **9:**127–128
Augmented spaces, design of filters with, **9:**106–110
Autoradiography, **16:**248–249
Avalanche breakdown, band shape and, **12:**60
Avalanche diodes, **7:**241
Avalanche multiplication, in amorphous solids, **8:**211
Average electron energy, **13:**72
Azbel-Gurzhi curve or model, **6:**99, 127
Azobenzenethiol monolayers, terminal groups, **24:**88, 90

B

Background
  limitations of PbS films, **3:**284–285
  radiation effects in PbSe films, **3:**291
Background limited performance, **21:**27, 257–258
Background noise limited detection, **11:**124–126, 158
Back-sputtering, semiconductor films, **5:**181
Back surface reflectors, quantum-well solar cells, **21:**338–340
Backward volume waves *see* Volume waves, magnetostatic
Bacteriorhodopsin
  photocell, **20:**287–290
  photocycle, **20:**280–281
  photoelectric behavior, **20:**279–292
Baffles, **1:**13–15
Baines curve, in thin-film resistivity, **6:**99
Baked films, **5:**101–104
Ballistic transport and reaction model, **22:**193–202
  diffuse re-emission, **22:**194–197
  gas/solid interactions, **22:**194
  governing equations, **22:**201–202
  LPCVD, **22:**202–203
  nonlinear overrelaxation, **22:**202
  specular transport, **22:**197–200
  surface species, **22:**200–201
Band-bending
  heterojunction model, **12:**119–120
  surface dipole model, **12:**120
  surface escape probability and, **12:**127
  vs. lattice constant for III-V compounds, **12:**131

Band edge
  conduction, **15:**221–222
  valance, **15:**221
Band filling, quantum efficiency and, **11:**139
Bandgap, **20:**11–12, 21–23, 29
  of hydrogenated amorphous silicon, **19:**73–75
  indirect, **7:**207
Bandgap energy
  cadmium sulfide film, **20:**24–26
  in liquid phase epitaxy, **7:**175–177, 185–186
  negative vs. positive electron affinity in, **12:**113
  vs. surface escape probability, **12:**127
  zinc sulfide film, **20:**25
Band model, metal-GaAs contact, **13:**215
Bandpass filter, **16:**101, 105, 106; **21:**258, 275–282
  all-dielectric, **10:**199–203
  attributes, **9:**93–94
  characteristics, **5:**52
  choice of metals for, **9:**95–102
  design
    bandpass filter attributes and, **9:**93–94
    in metal dielectric interference filters, **9:**93–102
  narrow *see* Narrow-bandpass filter
  scattering characteristics, **10:**200
  thickness of layers in, **9:**102
  transmittance, **9:**122–123
  transmittance curve, **5:**66
Band-reject filters, **16:**105, 106
Band shape, effective mass ratio and, **12:**60
Bandwidth, **15:**238
  broadening, in multiple casting filters, **9:**126
  magnetostatic, **15:**5, 28, 36
  spectrometer, **4:**55–57
Barium chloride, as parting agent, **7:**66
Barium copper oxides, **14:**20
Barium ferrite, properties, **15:**9
Barium fluoride substrates, IV-VI semiconductor layers and, **11:**115–122, 198
Barium magnesium fluoride film, **17:**280–286
  annealing treatments, film structure, **17:**281
  crystal structure, ferroelectricity, **17:**279
  C-V plot, MIS structure, **17:**285–286
  epitaxial orientations, multiple, **17:**281–282
  hysteresis data, **17:**282–283
  memory window, **17:**285
  molecular beam epitaxy, **17:**280
  programming voltages, **17:**286
  properties, **17:**279–280
  switching voltage, polarization, **17:**284–285
Barium titanate, **14:**18, 19, 31
Barium titanate film, **17:**228–229, 287
Barker coder-decoder, magnetostatic, **15:**48
Barrier height
  of heterojunction internal photoemission detectors, **23:**145–148
  of platinum silicide on silicon-germanium detector, **23:**155–159
Barrier-modulation theory, **3:**293–294
BAW device structures, **17:**179
  acousto-optic tunable filters, **17:**182

delay lines, **17:**183
lateral field resonators, **17:**181, 182
mosaic transducers, **17:**184, 185
propagation loss, **17:**180
resonators, **17:**181
transducers, **17:**150, 151
Beams, elastic theory, **3:**222–223
Beam splitters, **5:**48, 131
Beckmann's formalism, **10:**173
Beck's force field, **24:**213
Bell System Technical Journal, **10:**228
Bending moment light valves, **8:**326–330
Berg-Barrett X-ray topography, **5:**191–192
Beryllium
in MBE doping, **11:**65–66
radial distribution function measurements, **7:**21–22
Beryllium film
absorption coefficient, **7:**46
evaporation methods for, **7:**58
as filters, **7:**45–46
massive soluble substrates for, **7:**73–74
Beryllium oxide film, **8:**40
infrared reflectivities, **13:**48–49
ionized cluster beam, **13:**46–49
sound velocity of phonon, **13:**48
surface-state density, **13:**49
uniaxially oriented growth, **13:**47–48
X-ray diffraction pattern, **13:**46–47
Bialkali antimonide
electrical properties, **12:**97
as mutlialkali photocathode, **12:**86
optical properties, **12:**103
Bias, **14:**132
Bias-assisted photocathode, energy band diagram for, **12:**139
Biasing, **18:**90
dc, **18:**91–92
rf, **18:**92–95
Bias sputtering, **3:**92–96
SIMBAD model, **22:**30–38
Biaxial anisotropy, **6:**281–282
BIB detectors *see* Blocked-impurity-band detectors
BICMOS process, heterojunction bipolar transistor fabrication, **23:**60, 62–64, 68, 69
Bilinearity factor, in convolvers, **15:**101
Bimetallic film filters, **7:**73
Binary compounds, in liquid phase epitaxy, **7:**139–172
Binary oxides, oxygen vacancy and, **9:**10
Binding energy, adsorbed atoms, **2:**12–14
Binding modulation, spatial, **20:**224
Binding site, depth versus binding character, **20:**213–214
Biological oligomers, SAMS and, **24:**35–36, 150
Biomimetic systems, specificity, **20:**210
Biomineralization, **20:**11, 29
Biomolecular interaction
association rate, **20:**217–218
bond strength, **20:**219
concentration effect, **20:**214–215
force measurement, **20:**207–222
ligand effect, **20:**214–215
membrane fluidity, **20:**212–213
receptor blocking, **20:**215–219
specificity, **20:**210
steric inhibition, **20:**213–214
steric interaction, **20:**210–212
surface mobility, **20:**210–212
temperature effects, **20:**212–213
Biosensors, **20:**129, 135–137, 207

Biotin-streptavidin film
bond strength, **20:**219
layer failure scheme, **20:**219
receptor concentration, **20:**214
Biotin surface, description, **20:**208–209
Bipolar junction transistors, **23:**51–54
heterojunction bipolar transistor compared with, **23:**52–54
operation, **23:**51–52
Biprism interference microscopy, **1:**191, 194
Birefringence, **4:**47; **15:**103
Bismuth
radial distribution function measurements for, **7:**21
specular reflection in, **6:**132
thickness-dependence oscillations in, **6:**141
Bismuth crystals, electrical size effects in, **6:**116–117
Bismuth film
condensed at low temperature, **2:**44–46
specular reflection in, **6:**132
Bismuth oxide film, **8:**39
Bismuth-silver-oxygen-cesium photocathode, **12:**84
Bismuth-strontium-calcium-copper-oxide (Bi-Sr-Ca-Cu-O) superconductor, **16:**2, 4
Bismuth-strontium-calcium-copper-oxide (Bi-Sr-Ca-Cu-O) thin film, **16:**28–34, 51–54, 78–80, 92–94, 98–100
electrodeposition, **16:**99
flash evaporation, **16:**54
in situ or low-temperature process, **16:**29–32, 54, 93
laser ablation, **16:**76, 78–80
layered deposition, **16:**32–34, 54, 79, 93
lead doping, **16:**4, 32–34, 79, 93, 98–100
low oxygen pressure annealing, **16:**4, 32, 34
melt growth, **16:**99
MOCVD, **16:**92–94
MOD, **16:**98, 99
110 K phase formation, **16:**30, 32–34, 78, 93, 94, 99, 100
plasma spray, **16:**99
post-annealed, **16:**28, 29, 33, 34, 52–54, 76, 78–80, 98–100
screen printing, **16:**100
solution growth, **16:**99
spray pyrolysis, **16:**99
sputter-deposition, **16:**28–34
vacuum evaporation, **16:**52–54
Bismuth-substituted YIG
Bragg cell in, **15:**123
magnetooptic tensor in, **15:**105
utility in devices, **15:**112
Bismuth telluride film
orientation in, **5:**195
structure, **5:**206
Bismuth tellurium sulfide crystal, **19:**291
Bismuth titanate film, **17:**264–278
display structure, contrast ratio, **17:**267–270
electro-optic switching, **17:**264–265, 267, 269, 270
epitaxial buffer, YSZ, **17:**271
hysteresis, **17:**267–268
lamellar habit, **17:**265
metal-ferroelectric-semiconductor transistor, **17:**275–278
properties and applications, **17:**228–229

pulsed laser deposition, epitaxial sandwich structures, **17:**271
single-domain film, **17:**266
spontaneous polarization, **17:**264
stoichiometric range, **17:**265
switching times, **17:**270
twin structures, **17:**267
Bistability, **15:**247, 249–251
Bitter pattern, in multilayer films, **6:**230–231, 248
Blackbody, **10:**11, 12
Blackbody absorber, effectiveness, **10:**18–19
Blackbody radiation, cutoff wavelength, **10:**16–17
Black chrome, as absorber surface, **10:**47–49
Black nickel, as absorber surface, **10:**47–48
Blazed concave holographic grating, efficiency maps, **11:**24
BLIP *see* Background limited performance
Bloch lines, in magnetic domain dynamics, **9:**269–271
Bloch wall, **1:**312, 315
in multilayer films, **6:**229, 240
Bloch wave functions, **12:**59
Blocked-impurity-band detectors, **21:**4
germanium-based, **21:**36
silicon-based, **6:**19; **21:**17
Blocking agent, **20:**215–219
addition procedure, **20:**190
n-octadecyl mereaptan, **20:**184, 189–202
n-octadecylsilane, **20:**190–202
Blocking filters, in metal-dielectric filter production, **9:**140–142
BNL detection *see* Background noise limited detection
Boats
electrically heated, **1:**126
refractory metal, **3:**31–32
Bohm velocity, **18:**8, 17
Bohr magneton, **15:**126
Bolometer, resistive, as infrared detector, **23:**125–126
Boltzmann relation, **18:**18
Boltzmann transport equation, **6:**84, 98, 190
Bombardment
ion, **17:**82, 83, 99, 133
microstructure, **17:**121, 127
particle, **17:**81, 82
subsurface effects, **17:**82
surface effects, **17:**82
Bond percolation, in cermet films, **8:**123
Bond type
polarizability, **10:**42
sensitivity to disorder in, **8:**183
Borides, **17:**101–104
Born approximation, for screened coulomb field, **2:**64
Boron
chemical vapor deposition, **5:**309
contamination from glass, **1:**8
doping, **23:**12–14
radial distribution function measurements, **7:**22
removal of, from molecular bean epitaxy system, **23:**10
Boron compounds, in semiconductor films, **5:**161
Boron film, as filters, **7:**47
Boron nitride, **14:**67
chemical deposition, **5:**298–302
cubic
activated reactive evaporation applications, **13:**98–101
ir spectra, **13:**99
lattice spacing, **13:**101

Boron nitride, (*continued*)
physical properties, **13:**101
polycrystalline films, **13:**100–101
transmittance and reflectance spectra, **13:**100
Boron nitride coatings, **18:**122
Boron nitride film
electron reflection diagram, **5:**301
plasma chemical vapor deposition, **12:**271–272
Boron oxide film, **8:**40
Boron trichloride, dissociation, **12:**257
Borophosphosilicate glass, infrared spectra, **15:**301
band assignments, **15:**299
composition by, **15:**301–302
Borosilicate glass, infrared spectra, **15:**300
band assignments, **15:**299
boron content by, **15:**300
Bose-Einstein function, **8:**245
Böttcher formula, **8:**61–65
Boundary conditions
between doubly positioned nuclei, **4:**173–174
magnetostatic, **15:**28
described by surface permeability, **15:**132
molecular dynamics, **22:**142–144
spins with exchange, **15:**135
Boundary motion, Hamilton-Jacobi equation, **22:**84–85, 98–99
Bound carriers, **11:**226
Bound charges, vs. free, **11:**313–314
Bound-to-bound quantum-well infrared photodetectors, **21:**123–126, 131, 184–187
Bound-to-bound transition, **15:**230–233; **21:**118
Bound-to-continuum quantum-well infrared photodetectors
miniband detectors, **21:**187–190
n-doped, **21:**126–155
p-doped, **21:**156–160
Bound-to-continuum state, **21:**118
Bound-to-free transition, **15:**231–233
Bound-to-miniband detectors, **21:**190–191
Bound-to-quasicontinuum quantum-well infrared photodetectors, **21:**224
Bragg cell, magnetooptic, **15:**111, 120, 123
Bragg diffraction, **15:**118, 120
Bragg reflection
of magnetostatic waves by surface array, **15:**77
of planar wave, **1:**196–197
Bragg rod profile, **24:**116, 135
Breakdown voltage, in anodic oxide films, **6:**70–72
Breit-Wigner resonance, **15:**227
Brewster's angle, **2:**201, 202; **6:**175; **19:**301–302, 304
measurement, **4:**76–78
Brillouin scattering
by spin waves, **15:**95, 103, 107, 109
in magnetic media, **15:**102
Brillouin zone, **8:**174, 178
Broad-beam ion source, **13:**114–115
Bruggeman effective medium theory, **19:**41–42, 59–61, 72, 107, 110, 153, 156, 162, 177, 206, 257
BTRM *see* Ballistic transport and reaction model
Bubble etch chamber, **8:**266
Buckbee-Mears fine mesh screens, **7:**78
Buffer layer
Bi-Sr-Ca-Cu-O thin film, **16:**79, 98, 99

epitaxial, **16:**27, 69, 98, 121
Tl-Ba-Ca-Cu-O thin film, **16:**37–38
Y-Ba-Cu-O thin film, **16:**26–28, 42, 45, 63–65, 69, 70, 88, 91, 97, 121
Built-up molecular films, **7:**311–392, *see also* Films; Monolayers; Multilayers; Thin films
conduction in, **7:**379–386
electron optical studies, **7:**358–360
future applications, **7:**386–392
insoluble monolayers and, **7:**313–323
optical properties, **7:**360–365
precautions, **7:**337–342
structural properties, **7:**350–360
thickness, **7:**350–354
tunneling in, **7:**379–386
types, **7:**343–346
X-ray diffraction studies, **7:**354–358
X, Y, and Z forms, **7:**343–346
Bulk diffusion, SIMBAD model, **22:**53–54
Bulk metals, optical properties, **6:**152
Bunching, **4:**209
Burgers vector, **7:**125; **17:**66
Burge and Smith scattering calculations, **2:**82–87
Buried collector technique, **5:**225

C

Cadmium arachidate film, **20:**296–312
absorption threshold, **20:**16
doped with porphin, **20:**302–304
Cadmium oxide, electrical and optical properties, **9:**30
Cadmium oxide film, sputtering, **9:**1
Cadmium selenide film
properties, **5:**172
solution growth process, **12:**205
thickness vs. time, **12:**207
Cadmium selenide thin-film transistor, **2:**176–177
Cadmium stearate film
lattice constant, **20:**31
transfer ratios, **20:**322–323
water evaporation rate, **20:**320
X-ray diffraction, **20:**323–325
Cadmium sulfide, **14:**39
evaporation, **2:**181–183
gold ohmic contacts to, **2:**170
Hall mobility, **2:**173–176
particle size, **20:**25
thin-film transistor
depletion type, **2:**154
drain current, **2:**156, 157
gain vs frequency, **2:**154–155
gate capacitance, **2:**155–156
operating characteristics, **2:**153–157
switching waveforms for, **2:**156
Cadmium sulfide film
character, **20:**19
doping with indium and gallium, **12:**181
electrochemical apparatus, **20:**17
evaporation, **5:**171–172
exciton diameter, **20:**21
particle size, **20:**16, 21
properties, **12:**193–194
spray deposited, **12:**169, 172–173
spray pyrolysis, **12:**174–176, 180
transmission spectra, **12:**194
zinc addition to, **12:**196
Cadmium tin oxide, **14:**44
Cadmium wires, resistivity size effects in, **6:**122
Calix–4-resorcinarene, **24:**23
Cameras, infrared, platinum silicide based, **23:**105–106

Cancer cells, adhesion to silane film, **20:**136
Candidate acoustic materials, **17:**152
Capacitors, thin film, **2:**398–409
CAPD *see* Cathodic arc plasma deposition
Carbide cutting tools, wear, **13:**102–103
Carbides
 microhardness, **13:**80
 properties, **13:**91–92; **17:**101–104
Carbon
 diamondlike, ion-beam processing, **13:**141–142
 i-C, ion-beam processing, **13:**141–142
 as MBE impurity, **11:**57–58
 radial distribution function measurements, **7:**22–23
 removal of, from molecular beam epitaxy system, **23:**9
Carbon dioxide, laser emission line for, **12:**32–34
Carbon film
 dc and rf deposition systems, **12:**250
 as filters, **7:**47
 infrared spectra, **15:**305–307
Carbon impurities, by infrared spectroscopy, **15:**308
Carbon monoxide, reduction in Langmuir-Blodgett films, **20:**270–271
Carnot efficiency, converter efficiency and, **10:**14
Carrier diffusion, laser-induced etching, **13:**164–165
Carrier generation and control, etching rates, **13:**173–176
Carrier-to-noise ratio, **15:**209–213
Carrier transport, of negative electron affinity materials, **12:**123–125
Cascade process, quantum efficiency, **23:**247–252
Case hardening, by plasma carburizing, **12:**285
Catalysts
 for electroless metallization, **20:**118
 ice nucleation, **20:**145
Cathode, **5:**1–2, 12–14, 15–43, 36–39, *see also* Photocathode; Reflective interference cathode
 border, **14:**167
 erosion rate, **14:**155
 large area, **14:**167
 material, **14:**171
 spot, **14:**138
 voltage, **14:**141
Cathode-ray tube fields, semiconductor films in, **5:**145
Cathodic arc, **14:**138–147, 165
Cathodic arc plasma deposition (CAPD), **14:**130, 164–174
Cathodoluminescence, **3:**200–202; **7:**209
Cerenkov radiation, **2:**123
Cerium fluoride, microstructure and optical function, **19:**221–222, 237
Cerium oxide, **14:**61
 ion-beam processing, **13:**141
Cermet films, **8:**122–127
 activation energy, **8:**104, 109
 bond percolation model, **8:**123
 conduction mechanism, **8:**113
 conductivity, **8:**102–129
 defined, **8:**100, 122
 electrical properties, **8:**102–139, 152–160

electron transfer, **8:**110–112
electron tunneling, **8:**110–111
Cesiated silver film, photoemission from, **12:**81
Cesium, role of in multialkali photocathodes, **12:**97–98
Cesium activation, from silver oxide films, **12:**74–75
Cesium antimonide
absorption constant, **12:**102
crystal structure, **12:**88
electrical behavior, **12:**96
energy distribution spectra, **12:**108
Cesium-antimony cathodes
absorption and photocurrent in, **5:**38–39
attenuation length in, **5:**2
conventional and adapted, **5:**40
optical properties, **5:**14
quantum yield in, **5:**13
Cesium-bismuth compounds, energy distribution spectra, **12:**108
Cesium dioxide film, **8:**38
Cesium fluoride film, **8:**54
Cesium monolayer coverage, Fermi level position and, **12:**118
Cesium oxide film
absorption and photocurrent, **5:**37
reflectance-photocurrent relations in, **5:**19
Cesium/oxygen activation, electron affinity lowering by, **12:**114–123
Chalcogen deficiency, n-type conductivity and, **11:**107
Chalcogenide film, **12:**168–169, 201, 206, 217, 218–222, 230, *see also* Chemical solution deposition; Solution-grown films; Spray pyrolysis
Chalcogenide glass, in electronic switching devices, **7:**34
Charge carriers, DC conductivity, **11:**231–232
Charged particle, effect on film quality, **13:**20
Charge hopping, frequency dependence in, **11:**291, *see also* Hopping electronic systems
Charge injection, in solids, **11:**278–279
Charge neutrality, **15:**227, 239, 257–258
Charge oscillations, plasmas as, **9:**145–147
Charge separation units, **20:**243–246
Charge storage mode infrared detectors, **21:**38–40
Charge transfer, collisional, **18:**15, 72, 82
Charging currents, logarithmic representation, **11:**276
Chebyshev polynomials, **5:**59–60, 66
Chemical contamination, of absorbed coatings, **10:**123
Chemical deposition, **12:**226–227, 228, *see also* Chemical solution deposition
activated, of films, **2:**370
Chemical etching, in transparent conducting films, **9:**61
Chemical solution deposition, **12:**167–232, *see also* Solution growth process
complexing agents in, **12:**203, 208
electrical and optical properties of films in, **12:**192–201
impurity and dopant effects in, **12:**211–212
large-area applications, **12:**223–230

Chemical solution deposition, (*continued*)
  for multicomponent films, **12:**212–213
  oxide films and, **12:**213–214
  solution growth process and, **12:**201–211
  structural properties of films in, **12:**187–191
  substrate in, **12:**210–211
  transparent conducting oxides and, **12:**200
  transport properties in, **12:**217–223
Chemical sputtering *see* reactive sputter etching
Chemical transport processes, in chemical vapor deposition, **5:**245–247
Chemical vapor deposition (CVD), **4:**113–114; **9:**31–41; **14:**130; **17:**3, 154, 237–238; **24:**243
  advantages, **5:**239
  of alumina, **5:**259–262
  of aluminum, **5:**307–308
  of aluminum nitride, **5:**302–304
  of aluminum pentoxide, **5:**302–304
  of boron, **5:**309
  of boron nitride, **5:**298–302
  chemical beam deposition, PbTiO3, **17:**238
  chemical processes in, **5:**241–247
  chemical transport in, **5:**245–257
  chloride hydrolysis in, **9:**31–38
  of chromium, **5:**309–310
  cleaning in, **5:**254
  closed and open systems in, **5:**248–249
  closed space vapor transport (CVST), **17:**154
  conducting compounds, **5:**310–311
  control systems in, **5:**252
  of copper, **5:**308
  decomposition in, **5:**242–243
  defined, **5:**238
  deposits formed by, **5:**257–264
  of dielectric thin films, **5:**285–306
  disproportionation reactions in, **5:**246, 272–273
  doping techniques in, **5:**255–257, 277
  epitaxial layer growth of GexSi1-x by, **23:**21–51
    kinetics, **23:**37–43
    low-temperature, **23:**22–23, 27–37
    minimum temperature for, **23:**44
    systems for, **23:**23–27
    transition abruptness in, **23:**43–44
  examples, **5:**264–314
  of ferrimagnetic oxides, **5:**312–314
  generation and injection of vapors in, **5:**249–253
  of germanium, **5:**273–276
  of gold, **5:**308
  halide disproportionation in, **5:**272–273
  halide pyrolysis in, **5:**274
  halogens in, **5:**253
  heterojunction bipolar transistors grown by, **23:**55–56
    process for, **23:**59–64
  hydride pyrolysis in, **5:**270
  hydrogen as transport in, **5:**285
  low-pressure CVD, **22:**202–210
  mass transfer in, **5:**266
  metalorganic CVD, PbTiO3, **17:**237–238
  metals and conductive materials in, **5:**306–314
  MOCVD, **17:**154
  of molybdenum, **5:**306–307

in negative electron affinity photocathode fabrication, **12:**129
of niobium pentoxide, **5:**304
nonthermal activation in, **5:**311–313
nucleation and growth in, **5:**260
plasma *see* Plasma chemical vapor deposition
plasma-enhanced CVD, **22:**210–215
polymerization in, **5:**245
programmed rate CVD, **22:**265–269
pyrolysis in, **9:**38–41
pyrolytic, **12:**238, *see also* Spray pyrolysis
of pyrolytic carbon, **5:**262
of pyrolytic graphite, **5:**258–262
reactions in, **5:**241, 253–255
reduction in, **5:**243–244
remote microwave PECVD, **22:**215–226
of semiconductor films, **5:**263–314, *see also* Semiconductor films
semiconductors deposited by, **10:**40
setups for, **5:**248–257
silane pyrolysis in, **5:**271
silicon dioxide
LPCVD, **22:**202–210
PECVD, **22:**210–215
reaction mechanism, **22:**186–188
remote microwave PECVD, **22:**215–226
of silicon nitride, **5:**283–298
of silver carbide, **5:**283–285
of silver dioxide, **5:**286–293
SIMBAD model, **22:**29–30
simultaneous deposition and sputter etch, **22:**241–252
substrate heating in, **5:**254–255
surface diffusion, **22:**252–265
surface impedance method in, **5:**252
surface preparation for, **23:**44–50
HF-last clean in, **23:**47–49
thermal desorption cleaning in, **23:**45–47
thermal desorption in hydrogen ambient in, **23:**49–50
of tantalum pentoxide, **5:**304
of thin films, **5:**237–314
of tungsten, **5:**309
tungsten, programmed rate CVD, **22:**265–269
ultra-high vacuum, *see also* Ultra-high vacuum chemical vapor deposition (UHV/CVD)
of vanadium, **5:**306–307
vs. physical vapor deposition, **5:**238, 257
Child law *see* Sheath, Child law
Chlorides, hydrolysis of in film deposition, **9:**31–38
Chlorobenzene-*cis*-decalin system, loss peak for, **11:**246
Chopper, light, for vacuum ultraviolet reflectance, **1:**135–136
Chromium, **16:**284, 328
chemical vapor deposition, **5:**309–310
film resistors, **2:**384–386
Chromium film
as filters, **7:**48
optical parameters, **6:**186
rf sputtering, **13:**145
Chromium-iron system, **17:**22
Chromium oxide, **14:**16, 51, 70
Chromium-rhenium-ruthenium system, **17:**34
Chromium-silicon monoxide cermet film, **8:**124–125, 160
Chromium-silicon monoxide resistors, **2:**390–392

Chromium steel, plasma nitriding, **12:**288
Chromonyx, **10:**49
Chromophore, dynamics, **24:**70–77
Circuits, thin-film, **2:**409–419
Circular collectors, efficiency, **11:**181–182
Circular polarizer, **18:**169
Circulator, **18:**184, 210
Cladded magnetic films, **16:**222
Cladding, **24:**246
Clausius-Mossotti Lorentz-Lorenz theory, **8:**55, *see also* Lorentz-Lorenz theory
Close-packed monolayers, **7:**326–327, *see also* Monolayers
  molecule ''overturning'' in, **7:**329–330
Cluster
  diameter distribution, **13:**12–13
  diffraction patterns, **13:**13–14
  interatomic distances, **13:**13–15
  kinetic energy, **13:**8–9
  phase, **13:**13
  size, **13:**8–12
  structure, **13:**12–15
  velocity, **13:**11–12
Cluster-beam formation, **13:**6, 8
Coalescence, of nuclei, **4:**165–175
Coating receiver, **10:**22–23
Coatings, **5:**89–90, 91–92, 94–99, 123–126, 126–132, *see also* Oxide layers
  antireflection
    double-layer, **2:**249–265
    experimental techniques, **2:**284–303
    filming materials for, **2:**284–285
    four layer, **2:**272–275
    on germanium, **2:**263–265
    for glass in visible and near IR, **2:**287–293
    for IR optical materials, **2:**293–303
    with low visible and high IR reflectance, **2:**353–357
    multilayer, **2:**275–280, 280–284
    single-layer, **2:**243–249
    for studying erosion effects in space, **2:**357–358
    triple layer, **2:**265–272
  bulk, **17:**107
  corrosion resistance, **14:**190
  decorative, **14:**180, 191, 193
  erosion resistance, **14:**190
  gradient coatings, **17:**109
  hard coatings, **14:**186; **17:**100, 110
  macroparticle coatings, **14:**174
  multilayer coatings, **17:**109
  reflectance *see* Reflectance coatings
  reflecting, vacuum ultraviolet, **1:**123–186
  thin film, **14:**186
  tool life, **14:**189
  uniformity, **17:**137
  upper layer, **17:**108
  wear resistance, **14:**186
  wear-resistant, **17:**100, 109
Coaxial antenna, **18:**153–154
Cobalt, phase transition, **1:**220–221
Cobalt film, as filter, **7:**48
Cobalt oxide, **14:**16
Cobalt thin film, **16:**262
Coercivity, **15:**183, 186–187, 206
Coevaporated homogeneous films, **8:**51–96
Coevaporation process, for thin films, **8:**59–64
Coherence, **13:**158; **15:**222
Coherence length, **16:**74, 110, 118
Coherence range, **1:**256

Cohesion modulus, **16:**273
Cohesive failure, **16:**276, 314, 330
Coil or flux transformer, **16:**75–78, 117–118
Cold cathode electron emitters, in negative electron affinity devices, **7:**244–246
Cole-Cole diagram, **11:**222, 226–228
Collector configurations, in photothermal solar energy conversion, **10:**22–25
Collimated sputtering, **22:**162–165
  SIMBAD model, **22:**39–45
  titanium nitride, **22:**227–229
Collisional excitation, **18:**14, 73
Collision cross section, **14:**34
Collisions
  charge-exchange, **14:**8
  electron-ion, **18:**149
  electron-neutral, **18:**126, 128, 130–132, 146–151, 203
  hard-sphere, **3:**68
  high-energy ion, **3:**67–68
  inelastic, **18:**129
  molecular, **14:**6
  weakly screened Coulomb, **3:**68
Color filters, oxide layers for, **5:**134–136
Columnar microstructure, **15:**161–163
Commensurate growth, **23:**3
Communications, long-range, high-data-rate, multiple-quantum-well detectors in, **23:**210–211
Compensation temperature, **15:**180–182
Compensator *see* Achromatic quarter-wave plate
Complex dielectric constant, **6:**153
Complex dielectric susceptibility, formal expression for, **11:**288
Complex energy method, energy and lifetime from, **23:**291–298
Complex impedance diagrams, **11:**218–220, 244
Complex impedance plot, representation of, **11:**222
Complex permittivity
  diagrams for triglycine sulfate, **11:**237
  frequency dependence, **11:**220
Composite films, **10:**60, *see also* Thin films
Composite structures, **10:**64; **16:**266–268
Composition
  change, **14:**19, 20
  chemical, **17:**107, 111
  phase, **17:**107, 118
  stoichiometric, **14:**12, 24–26, 37, 67
Composition modulation, **17:**11–14, 18–22
  amplitude, **17:**9, 10, 14, 35
  chemical order, **17:**8, 15
  coherency strains, **17:**9, 14, 24–29, 40, 41, 55
  diffusion, **17:**14
  Fourier coefficients, **17:**12, 14
  Fourier components, **17:**12, 17
  interdiffusion, **17:**19, 23, 24, 28, 35–42, 56
    amplification factor, **17:**35
    effective diffusion coefficient, **17:**36–41
    free energy vs. composition, **17:**35–41
    gradient energy coefficient, **17:**35–41
    immiscible systems, **17:**37–40

Composition modulation, (*continued*)
interdiffusion coefficient, **17:**35–41
linearized equations, **17:**35–41
miscible systems, **17:**37, 38
nonlinear equations, **17:**41–42
regular solution model, **17:**36
spinoidal decomposition, **17:**39
interface roughness, **17:**8, 10, 19, 23, 30–34
interface width, **17:**16
intermixing, **17:**8, 14
interplanar spacing modulation, **17:**11, 12
lattice-matched, **17:**13
lattice-mismatched, **17:**13
sinusoidal, **17:**8, 9, 12
small-amplitude, **17:**35, 36
square wave, **17:**8, 9, 12, 14, 15
stability, **17:**22
structural order, **17:**8
trapezoidal, **17:**15
wavelength, **17:**10, 14
Compound dielectric films, **8:**20–22
Compound-forming systems, **6:**305–309
Compounds
formation, **14:**4
layer, **14:**5, 6
radial distribution function measurements for, **7:**28–37
stable, **14:**4
Compressive receiver, **15:**57–58
Compressive stamping, **24:**233
Computers
high speed switches, **16:**101, 123
simulations
stress field in coating, **16:**312
thin-film fracture, **16:**282–283
thin film calculation
analog, **1:**118–120
digital, **1:**116–118
thin-film elements, **2:**184–186
Condensation
films, **1:**208–210
deposition rate, **1:**212–213
substrate temperature, **1:**209–210, 211–212
substrate-mediated, **20:**356–358
Condensation coefficient, nucleation and, **4:**107–108
Condensation energy, **14:**59
Condensation flux, temperature and, **5:**175
Condensation rate, **3:**214–215
Condensed films, molecular structure, **7:**320
Conductance, thermal, between film and low temperature bath, **1:**253
Conducting compounds, chemical vapor deposition, **5:**310–311
Conduction, *see also* Electrical conduction
high-field, **8:**195–211
low-field, **8:**185–195
through traps, **8:**206–207
Conduction band offset, size, **23:**5
Conductivity *see* Electrical conductivity
Conductors
sputtered film, **3:**108–110
thin-film, **2:**410
Cone formation, **14:**228–230
Conservation of energy, **18:**17, 67
Conservation of particles, **18:**19
Constant field transient
anodization and, **6:**10
theory, **6:**16–18
Constant rate CVD, **22:**265
Constant temperature control algorithm, **22:**134–135

extended system method, **22:**137–142, 161
generalized Langevin equation, **22:**135–138, 141–142, 170
Constant voltage formation, anodization and, **6:**10–12
Constriction-type contact films, **8:**118
Contact angle
organic solvents-gold, **20:**192
organic solvents on *p*-aminobenzoic acid, **20:**47
water-*p*-aminobenzoic acid, **20:**59
water on alpha-amino octanoic acid, **20:**177
water-diamond, **20:**124–125
water-ethylene diamine surface, **20:**132
water-gold, **20:**192
water-ligand-coated gold, **20:**189
water-phenyltrichlorosilane, **20:**114
water-polypropylene, **20:**92
Contact angles, simulations, **24:**220–221
Contact layer, **15:**227
Contact masking, laser-induced etching, **13:**159–161
Contact phenomena, in semiconductor films, **5:**219–222
Contact potential difference, on multilayer films, **20:**260–265
Contact resistivity, **13:**209–210
device geometries used to measure, **13:**213
as function of annealing temperature, **13:**28–29
as function of *n*-GaAs doping, **13:**210–211
space-charge density and, **13:**222
true, **13:**216
tunneling, **13:**221
Contamination
gas, in sputtering, **3:**86–87
particle, plasma, **18:**289, 291, 314
process-inherent, **18:**290
Continuous thin films, strain coefficient of resistance in, **8:**133, *see also* Thin films
Continuum states, **15:**223, 231–232, 254
Continuum thermal spike model, **13:**119
Contrast amplitude, **1:**189
Contrast phase, **1:**189, 191
Converter
lifetime, operating environment and, **10:**26–27
surface
energy flow at, **10:**13–19
overall requirements for, **10:**25–27
Convolution
in compressive receiver, **15:**57
defined, **15:**100
Convolver, magnetostatic, **15:**100
bilinearity factor, **15:**101
Copper
adsorption of thiols onto, **24:**3
chemical vapor deposition in, **5:**308
diffusion in tellurium and selenium, **6:**340–342
sputter deposition, **22:**154, 156
Copper-aluminum alloys, electromigration in, **7:**287–293
Copper-aluminum film, surface diffusion, **22:**253–265
Copper film
adhesion strength as function of acceleration voltage, **13:**25
Parlodion films and, **7:**70
Copper-gold multilayer film, **6:**310

Copper ion complex, formation, **20:**183–205
Copper-iron superlattice, **17:**3, 60, 70
Copper mirror coatings, **12:**17
Copper-nickel-iron superlattice, **17:**33, 49, 51
Copper-nickel superlattice, **17:**17, 27, 33, 49–51, 54, 56, 58, 60–62, 80
Copper-niobium superlattice, **17:**33, 41, 43, 44, 49, 51
Copper oxide, **14:**9
Copper-palladium superlattice, **17:**33, 41, 43, 44, 49, 51
Copper sulfide, **14:**37
Copper-titanium superlattice, **17:**22
Copyrolysis, in spray pyrolytic process, **12:**183–187
Core excitation, **2:**124
Correction mask, for optical elements, **10:**265
Correlated states, in many-body interactions, **11:**304
Correlation, **15:**48, *see also* Convolution
Correlation junctions, in dielectric response, **11:**293–294
Corresponding state theory, **20:**371–383
  experimental verification, **20:**374–378
  refinement, **20:**378–382
Corrosion prevention, hexadecanethiolate SAM, **24:**240
Corrosion resistance, **15:**171–174
Cotton-Mouton effect, **15:**103
  constants for, in YIG and Bi:YIG, **15:**106
Coulomb friction element, **16:**296
Coulombic potential, of donor center, **8:**200
Coulomb interactions, **15:**226–227, 239–243, 250
Coupling aperture, **18:**167–169, 178–179
Coupling effects, **6:**247–290
  angular dependence and, **6:**252–253
  by grating, **9:**221
  domain structures and, **6:**274–275
  dynamic reversal behavior and, **6:**269–272
  magnetostatic waves to current, **15:**40
  in multilayer films, **6:**247–290
  quasistatic reversal behavior and, **6:**251–261
  resonance phenomena as, **6:**272–274
Coupling energy, measurement, **6:**284–285
CPMAS *see* Cross-polarization and magic-angle spinning
Crack density, **16:**266, 281, 285–286, 314, 320, 324
Crack extension force, **16:**272
Crack gap, **16:**318
Crack onset strain *see* First cracking strain
Crack resistance force, **16:**272
Creep, **3:**258, 259–261
Creme-Cote, as parting agent, **7:**67
Critical angle, **1:**165; **2:**227–228
Critical current density, **16:**8–11
  anisotropy, **16:**8–10, 72, 73
  criteria for determination, **16:**11, 35, 36, 47, 48, 68, 70, 80, 90
  epitaxial thin film, **16:**9, 23
  flux-pinning, **16:**10, 22, 23, 49, 50, 68, 73, 90

grain alignment, **16:**9, 10, 23, 90, 119
magnetic field, **16:**9, 10, 23, 32, 48–50, 52, 71–74, 80, 81, 88–90, 110
radiation damage, **16:**10, 74
two-step characteristic, **16:**9, 10
weak links, **16:**9, 50, 71, 110, 119
Critical film thickness, **16:**284, 286–287, 329
Critical magnetic field, **1:**257–260
in terms of free energy, **1:**249
variation with thickness, **1:**251
Critical plasma density, **18:**126–128, 131–141, 146–147, 150–151, 179, 185–186, 190–191
Critical pumping speed, **17:**99, 115
Critical self-current, **1:**254
Critical shear stress, **16:**286
Critical super-current, **1:**260–265
Critical temperature
superconductors
1:240ff.
alloying effect, **1:**241
alloys, **1:**245–246
deposited at low temperature, **1:**243–244
film edge effects, **1:**251–252
stress effects, **1:**241–243
structural effects, **1:**241–243
Critical tensile force, **16:**276–277
Crooke's dark space, **3:**81–82
Cross-polarization and magic-angle spinning (CPMAS), **24:**151–152
Cross-sectional scanning tunnel microscopy (XSTM), **23:**281
Cross-section transmission electron microscopy (XTEM), **19:**232
Cross-slip, **4:**147–149
''Crowded'' characteristic, thin-film transistors, **2:**170–172
Cryolite
as dielectric film material, **8:**30
as substrate, **10:**207
Crystalline materials
avalanche multiplication, **8:**211
band structure, **8:**174
energy gap, **8:**175
forbidden gap, **8:**177, 240
hopping, **8:**212–217
semiconductors, **8:**178, 198
Crystallinity, interaction energy, **20:**155
Crystallization, **15:**167
Crystals
feather and plate, **3:**183
growth
by screw dislocation step advancement, **4:**193–195
ferrites, **15:**7, 8
from vapor, **4:**192–195
morphology
lead selenide, **20:**32
lead sulfide-arachidic acid film, **20:**30
oscillator, **3:**21–27
size determination from electron diffraction, **1:**198
Crystal structure, **14:**8, 65–66; **17:**15, 17, 22, 23, 31, 33, 126
alpha-amino acids, **20:**176
alcohol monolayers, **20:**148, 152
alkyl alcohols, **20:**151
amorphous, **17:**15
crystalline, **17:**15
epitaxial, **17:**16, 17, 24
CSA powder patterns, **24:**12
Curie law, **11:**225
Curie point, **9:**268

Curie temperature, **15:**180–182
alloy composition, dependence on, **15:**181
de Gennes factor dependence on, **15:**182
in multilayer films, **6:**293
Curie-von Schweidler law, **11:**230, 232, 275, 295, 302
Current density, **14:**6, 36, 141
Current distribution, in radiating microstrip, **15:**40, 43
Current flow, ionic, in glass films, **4:**293–300
Current measurement, **14:**139, 145, 151
Current-voltage characteristics
double barrier diode, **15:**247
multiquantum well, **15:**236–237
resonant tunneling, **15:**253
Cutoff
below cutoff, cavity/chamber, **18:**190–192, 197, 200–201
cavity/chamber, **18:**154, 167–168
frequency, **18:**139
plasma density
left hand polarized wave, **18:**137–138, 187, 190–191
right hand polarized wave, **18:**140
Cutoff filters, ultraviolet, **5:**137
Cutoff wavelength, concept, **10:**16–17
Cutting tools, wear, activated reactive evaporation applications, **13:**102–103
CVD *see* Chemical vapor deposition

D

Damage, **18:**10, 71, 96
atomic displacement, **18:**96
charging, **18:**91
plasma uniformity and, **18:**99
rf bias and, **18:**102
contamination, **18:**98
laser-induced etching, **13:**178–181
radiation, **18:**104
Dark current density, internal photoemission devices, **23:**92
Dark current emission, of negative electron affinity materials, **12:**135, 139
Dark current multiple-quantum-well detectors, **23:**189–190
Dark current noise, *n*-doped quantum-well infrared photodetectors, **21:**137–138
Dark currents, **15:**236–237, 259
n-doped quantum-well infrared photodetectors, **21:**132–133
quantum-well solar cells, **21:**316–318, 349–355
Type II HIWIP detectors, **21:**23
Dark current thick barrier quantum-well infrared photodetectors, **21:**240, 257–260
Dark discharge, boundary and initial conditions applicable to, **11:**281
Darken's equation, **6:**309
Dark mirror coatings, low visible reflectance, **12:**45
Dark mirrors, in three-cavity interference filter, **9:**74
Dark resistance, cadmium sulfide particulate film, **20:**26–27
Dark space, **14:**27–29, 34
DC conductivity
charge carriers and, **11:**231–232
frequency dependence in, **11:**271
theory, **11:**206
vs. AC conductivity, **11:**258

DC electronic conduction, dielectric properties and, **6:**63–66
Deb device, **17:**355
De Broglie wavelength, **6:**89–90, 132, 139
Debye equation, in diffraction theory, **7:**3–5
Debye length, **18:**20
Debye mechanisms, **11:**223
Debye peaks, distribution, **11:**288–289
Debye relaxation time, **11:**224, 299
Debye response
  broadened, **11:**236–237
  dipole relaxation and, **11:**293
  ideal, **11:**219, 234
  loss peak frequency and, **11:**236
  polarization decay and, **11:**224
  pure, **11:**227
Debye system, ideal, **11:**225
Debye temperature, **6:**97, 154
Decohesion
  modes, **16:**283
  number, **16:**282, 284
Decomposition process, in chemical vapor deposition, **5:**242–243
Decoupling layer, **5:**84
Defects, **15:**221; **17:**10, *see also* Pinhole
  in films, **4:**149–150, 180–182
  formation, **4:**165–175
  models, **3:**245–246
Defocusing effect, of thick samples, **4:**45–46
De Haas-van Alphen effect, **8:**174
Delay lines, **16:**106–108
  application
    in compressive receiver, **15:**57
    in phased-array radar, **15:**61
    in various systems, **15:**67
  magnetostatic, **15:**71
    approaches, **15:**64
    constant delay, **15:**72
    dispersive delay, **15:**57, 74
    ferrite films, **15:**75
    ferroelectric control, **15:**81
    negative dispersion in, **15:**75
    phase error in, **15:**60, 75
    propagation loss, **15:**6, 77
    reflective array, **15:**77
  in YIG film evaluation, **15:**20
Delta doped potential well infrared detectors, **21:**32–33
Delta function, in dielectric response, **11:**210
Demagnetization energy, **16:**158
Demagnetization factors, **16:**156
Demagnetizing field
  effect on ferromagnetic resonance, **15:**129
  uniformity in thin films, **15:**7
Dendrites, **3:**183–187
Density
  critical vapor stream, **1:**209
  energy, magnetization ripple, **1:**296
  ion current, **18:**22
  plasma, determining, **18:**22
  super current, **1:**257, 261–262
  thin film, **1:**212; **3:**34–35
  thin-film circuits, **2:**413
Density of states, **1:**267–268; **15:**223, 225, 227
  intersubband transition, **15:**243
  workfunction calculations, **21:**12–14
DEPICT process simulator, **22:**5
Depletion layer, **2:**157; **15:**247
Deposition, **14:**59–72; **18:**8, *see also* Chemical vapor deposition; Sputter deposition
  alloy, **14:**182
  angle of incidence and, **1:**286–289
  compound, **14:**183

Deposition, (*continued*)
evaporation, **17:**81
arc, **17:**91–100
electron beam, **17:**91
film morphology, **14:**173
Hamilton-Jacobi equation, **22:**83, 84–97, 113
homogeneity, **17:**117, 137
ion assisted, **14:**61
low-pressure processes, **22:**175–176
chemistry, **22:**181–188
EVOLVE simulations, **22:**180–269
kinetics, **22:**179–180
transport, **22:**177–179, 188–193, 193–202
macroparticle, **14:**185
mathematical models, **22:**81–84, 113–114
metal and elemental film, **14:**175
microdroplet, **14:**161
modeling *see* Modeling
plasma optic system, **14:**163
process, **17:**131–137
process parameters, **15:**148–155
Ar pressure, **15:**149, 151, 165, 171
sputtering, **3:**92–96
sputter yield, **15:**148, 164
structure dependence, **1:**211–212
substrate bias, **15:**149, 154, 163
substrate temperature, **15:**154
rate, **14:**6, 8, 35, 42, 45–48, 51, 69; **17:**117, 123, 128, 138
aluminum reflectance and, **1:**150–151
hydrogen effect, **14:**38
monitoring devices, **3:**3–8
sputtering, **3:**84–86; **14:**5, 6, 8, 34
reactive, **14:**176
shock-tracking algorithm, **22:**97–98, 113
SIMBAD model, **22:**5–6, 9–15, 20–76
simulation, **22:**122–123
simultaneous deposition and sputter etch, **22:**241–252
surface diffusion, **22:**252–265
system, **14:**8
techniques, **2:**365–370; **15:**148
electron beam evaporation, **15:**148
thermal evaporation, **15:**148
vapor
chemical, **17:**107
physical, **17:**80, 107
Depressions, in SAMS, **24:**86
Destructive readout memory schemes, **6:**291
Detection, vacuum ultraviolet radiation, **1:**131–133
photoelectric, **1:**124–125
Detectivity, **15:**238
of a background-limited IR photoconductor, **3:**278–281
n-doped quantum-well infrared photodetectors, **21:**149–155
spectral
of PbSe films, **3:**291
of PbTe films, **3:**287
Detectors, *see also* Infrared detectors; Multiple-quantum-well (MQW) detectors
BNL (background noise limited), **11:**124–128, 148–149, 158–159, 172–173
JNL (Johnson noise limited), **11:**124–128, 148–149, 158–159, 172–173
lead salt, **3:**275–301

noise, in thin-film IV-VI semiconductor photodiodes, **11:**149–152
noise equivalent power, **21:**27
vacuum ultraviolet, **4:**68–69
Deuterium oxide, nucleation, **20:**159
Dewald's theory, in ionic conduction process, **6:**14–15
DFB (distributed feedback) lasers, **10:**251–252
DHLS *see* Double heterojunction lasers
Diagnostics, plasma
Doppler-shift spectroscopy, **18:**77, 79, 82
electrostatic energy analysis, **18:**76, 81, 90
Faraday cup, **18:**100
Langmuir probes, **18:**80
laser-induced fluorescence (LIF), **18:**78
magnetic pickup coil, **18:**38, 45
Diamond, **17:**101, 103, 111
self-assembled films on, **20:**124–125
Diamond film, **18:**122, 159
growth studies by RTSE, **19:**105
analysis of microstructural evolution, **19:**108
contamination by tungsten and its minimization, **19:**108
nucleation and growth using heater-filament method, **19:**104
Raman spectroscopy, **19:**105, 111
substrate treatment and annealing, **19:**105
true substrate temperature evaluated by RTSE, **19:**108–09
Diamond-like carbon, coatings, films, **18:**122, 159
Diatron, **1:**20, 40
Diborane, **23:**34, 41
Dichlorosilane, **23:**35–37, 41–43
Dichroic beam splitters, **5:**133
Dielectric breakdown, Fröhlich's theory of, **8:**211
Dielectric characteristics, effect of gases on, **1:**60–63
Dielectric constant, **13:**48
complex, **6:**153
island-structure films, **8:**149–150
and joint density of states, **12:**57–58
Dielectric data, methods of presentation, **11:**217–223
Dielectric films, *see also* Dielectric materials; Dielectric mixtures; Discontinuous films; Thin films
adhesion in, **8:**10–11
aftertreatment in, **8:**25–27
as antireflective coatings, **5:**16, 20; **8:**23, 42–43
applications, **8:**42–45
bismuth oxide, **8:**39
cesium dioxide, **8:**38
chemical stability, **8:**11 14
chemical vapor deposition, **5:**285–306
composition, **8:**6
compound film production in, **8:**20–22
condensation phenomena, **8:**25–27
crucible and evaporation techniques in, **8:**19
damage from high energy radiation, **8:**4
electron beam evaporation, **8:**19–20
evaporated *see* Evaporated dielectric films
film materials and thin film properties, **8:**28–29

Dielectric films, (*continued*)
- film structure, **8:**25–27
- grain size, **8:**5
- hardness, **8:**9–11
- humidity test, **8:**12
- indium oxide, **8:**39
- inhomogeneous *see* Inhomogeneous films
- inorganic, **3:**133–134
- intrinsic stress, **3:**234–241
- ion plating and, **8:**24
- laser beam evaporation, **8:**20
- laser damage to, **8:**4, 14–17
- as laser mirrors, **8:**43–44
- loss in, **8:**4
- magnesium fluoride, **8:**30–32
- magnesium oxide, **8:**37
- materials, **8:**27–42
- mechanical properties, **8:**9–11
- metal-film combinations, **8:**44–45
- in multilayer stacks, **8:**3
- for optical applications, **8:**1–45
- organic, **8:**41–42
- oxides as, **8:**35–40
- packing density, **8:**5, 8–9
- peeling or staining, **8:**12
- preparation, **8:**18–27
- rare earth oxides as, **8:**37–38
- rate monitors and mass spectrometers in, **8:**22–23
- as reflector stacks, **8:**43–44
- refractive index, **8:**5–8
- requirements for, **8:**2–18
- resistance to high-energy radiation, **8:**14–17
- Scotch Tape test for, **8:**10–11
- semiconductor materials and, **8:**40–41
- silicon dioxide, **8:**36
- sputtering, **8:**23–24
- standing waves, **5:**6
- substrate temperature and, **8:**22
- sulfides and selenides as, **8:**34–35
- temperature stability, **8:**17–18
- thorium dioxide, **8:**37
- thorium fluoride, **8:**32–33
- three-temperature method, **8:**21–22
- tin oxide, **8:**39
- titanium dioxide, **8:**38
- transparency, **8:**2–5, 22
- ultraviolet irradiation, **8:**15–16
- water vapor absorption in, **8:**5–6
- wave field and, **5:**4–5
- zirconium dioxide, **8:**37

Dielectric friction, **20:**343
Dielectric function, determination for metals, **9:**234–238
Dielectric induction, defined, **11:**210
Dielectric interference filters, **10:**168
Dielectric lattice, **11:**232–233, 259
Dielectric lattice response, **11:**267–270
Dielectric loss
- empirical classification, **11:**226–235
- frequency dependence on, **11:**269
- hollandite, **11:**263
- polyethylene, **11:**268–269
- stearic acid films, **11:**265–267
- temperature dependence in, **11:**268–269
- triglycine sulfate, **11:**262

Dielectric loss peaks, **11:**229
- ceramics, **11:**260
- general analysis, **11:**251–256
- poly-*n*-butyl methacrylate, **11:**240
- polydian carbonate, **11:**238
- polyethylene, **11:**243
- polyethylene terephthalate, **11:**247
- polymeric solids, **11:**250
- polymethyl methacrylate, **11:**248
- polyvinyl acetate, **11:**239

Dielectric loss spectrum, for silica glass, **11:**252
Dielectric materials, **8:**27–42; **13:**128–142, *see also* Dielectric films
  cryolite as, **8:**30
  fluorides and other halogenides as, **8:**30–34
  time domain response and, **11:**272–277
  types and properties, **8:**28–29
Dielectric matrix, **11:**232–233
Dielectric medium, equations for, **8:**53
Dielectric mirrors, Smith chart and, **10:**196
Dielectric mixtures
  refractive index, **8:**52–67
  structure, **8:**53
Dielectric overcoatings, **15:**171
Dielectric permittivity
  defined, **11:**213
  of free space, **11:**210
  high-frequency, **11:**216
Dielectric polarization
  disorder in, **11:**306–307
  types, **6:**51
Dielectric properties, **11:**253–255
  of anodic oxide films, **6:**51–73
  dc electronic conduction and, **6:**63–66
  dielectric breakdown and, **6:**66–73
  polarization processes and, **6:**52–63
Dielectric relaxation
  energy relations in, **11:**311–312
  time-dependence in, **11:**309–311
Dielectric response
  charge carriers and, **11:**285
  correlation function approach, **11:**293–294
  defined, **11:**206–207, 209–210
  diffusive models, **11:**292–293
  frequency dependence, **11:**207, 245
  Ising model, **11:**302–303
  many-body interpretation, **11:**296–312
  Maxwell-Wagner effects, **11:**294–295
  Ngai's infrared divergence model, **11:**303–304
  in optical region, **11:**207–208
  screened hopping model, **11:**296–301
  superposition principle, **11:**211
  temperature dependence, **11:**233–235
  time dependence, **11:**206–207
  two-level systems, **11:**305–312
  types, **11:**284–286
  universal *see* Universal dielectric response
Dielectric response function, **11:**210
Dielectric susceptibility
  defined, **11:**213
  formal expression for, **11:**288
  glancing incidence reflection electron type, **11:**42–44
Dielectric window, **18:**11, 13
Differential expansion coefficient, film-substrate, **3:**227–228
Differential transmission probability, **22:**191
Diffraction, **1:**191–194; **24:**112–114, *see also* Electron diffraction analysis; Magnetooptic diffraction; X-ray diffraction
  Debye equation, **7:**3–5
  neutron, **1:**203–205
  order and phase behavior, **24:**130–144
  pattern types, **1:**198
  probes, **24:**114–117
  recording, continuous, **1:**199–202
  structure, **24:**121–130

Diffraction, (*continued*)
theory, **7:**3–15
Diffraction gratings
aluminum coatings for, **11:**3, 13
aluminum-gold interdiffusion in, **11:**14–17
blazed concave holographic, **11:**23
characteristics, **11:**13–14
in conical diffraction mountings, **11:**28–30
echelle, **11:**3
efficiency
in conical diffraction mountings, **11:**28–30
improvements, **11:**13–14
measurement, **11:**17–33
uniformity, **11:**17
vs. wavelength, **11:**25–29
wavelength and, **11:**25–29
efficiency maps, **11:**18–25
grating anomalies in, **11:**19–21
groove efficiency curve, **11:**26
holographic, **11:**9–11, 23
intermetallic diffusion, **11:**14–17
master conventional, **11:**11
nonuniform surface, **11:**19–20
overcoating, **11:**12
reflectance coatings, **11:**1–33
replica conventional, **11:**11–12
roughness, **11:**8
ruling, **11:**2–3, 19
ruling errors, **11:**6
stray light measurements, **11:**30–33
target pattern, **11:**21
wavelength vs. efficiency, **11:**25–29
Diffraction microscopy, **1:**191–192
Diffraction patterns, by type of film, **7:**1–2
Diffraction probes, **24:**114–117
Diffuse interphase, **20:**100
Diffuse re-emission, BTRM, **22:**194–197
Diffusion
ambipolar, **18:**8, 17–18, *see also* Transport
equations for, **6:**345
experimental methods in, **6:**310–315
lateral, **6:**338–343
in metallic films, **6:**301–346
miscibility and, **6:**302–304
in miscible systems, **6:**335–338
solid state, **1:**224–226
under external force, **16:**237
Diffusion coefficient, **13:**19
determination, **6:**336
Fab'-lipids, **20:**228
lateral versus chain entanglement, **20:**338–339
Diffusion equation solution, **16:**236
Diffusion flux, in size space, **4:**98–99
Dignam effective field, **6:**16–18
Diketopiperazine, selective crystallization, **24:**247
Dingle calculation, for thin wires, **6:**92–93, 129
Dingle theory, **6:**191
Diode-laser mixing technique, for multiple-quantum-well detectors, **23:**198–199
Diodes
AC, **14:**28
DC, **14:**27
RF, **14:**28
systems, **14:**26–31
thin-film field-effect, **2:**183–184
topologically structured thin films and, **8:**301
Dipolar interactions, **15:**185
Dipolar materials
beta peaks, **11:**246–251

dielectric properties of, **11:**253–255
Dipping, **5:**90; **7:**118
Direct evaporation, **13:**62–63
Direct exchange coupling, in multilayer films, **6:**208–213
Director field, **20:**369
Direct overwrite techniques, **15:**204–207
Direct tunneling, **21:**249–250
Direct writing, laser-induced etching, **13:**161
using thermally activated chemical process, **13:**167–168
Discharge
arc, **17:**92
cathode spots, **17:**92
distributed, **17:**92
capacitive, **18:**5, 90
limitations, **18:**9
typical parameters, **18:**9
design
examples, **18:**23, 51, 57
principles, **18:**13
double-site-sustained, **17:**89
emission, optical, **17:**116
heating, **18:**23
ohmic, **18:**24, 54
stochastic, **18:**24, 54
wave-particle, **18:**25
high efficiency, **18:**10, 13
remote, **18:**12
low pressure, **18:**13
magnetically enhanced, **18:**10
magnetron, **14:**2, 35, 39; **17:**86
MERIE, **18:**10
RF, **14:**2, 39, 44, 51
sputtering, **13:**110; **14:**2
triode, **18:**10
Discharging currents
logarithmic representation, **11:**276
time domain response, **11:**230–231
Discontinuous films, *see also* Dielectric films; Thin films
absorption peak, **8:**143
classes, **8:**117–118
conduction, **8:**113–115, 121–122
electrical properties, **8:**102–139, 152–160
filamentarylike, **8:**121, 134
future studies, **8:**164
gauge factor, **8:**139
optical properties, **8:**139–152, 161–164
scattering efficiency, **8:**148
sheet resistance, **8:**159–160
strain coefficient of resistance, **8:**132–139, 158
temperature coefficient of resistance, **8:**129–132, 156–158
Discrete microwave devices, and topologically structured films, **8:**299–310
Dislocations, **1:**226–228; **2:**19–20
formation, **4:**169–171
interfacial, stress due to, **3:**244–245
misfit, **4:**142–143, 182–186
during diffusion across interface between crystals, **4:**184–186
generation, **4:**146–147
pinning, **3:**268–269
Disorder
in amorphous material, **8:**171–172
diffraction studies, **24:**131–132
NMR studies, alkylphosphonate monolayers, **24:**168–169
structural, **2:**1–62
introduced during film growth, **2:**14–22
Disordered nonmetallic film, electrical conduction in, **8:**169–246

Dispersion
anisotropy, **1:**302–303
of easy axis direction in Permalloy films, **1:**302–303
Displacement, of thiols, **24:**99–102
Disposable emitter mandrel process, **23:**60, 62–64
Disproportionation reactions
in chemical vapor deposition, **5:**246, 261
halide, **5:**272, 275
iodide, **5:**273
Dissado-Hill analysis, of two-level dielectric response systems, **11:**305–312
Dissipation factor, capacitor, **2:**400
Dissociation, **14:**7, 8, 25, 69
Distributed feedback lasers, **10:**251–252
Distributed parameter circuitry, **2:**416
Distribution of relaxation times, dielectric susceptibility and, **11:**288
DNA, immobilization on SAMs, **24:**106
Docosanoic acid film
phase diagram, **20:**368
on water and polymers, **20:**377
Domain formation
alloy composition, relation to, **15:**189
domain wall width and size and, **15:**189
high density recording and, **15:**202
stability, **15:**187–189
Domain structures
in multilayer films, **6:**274–278
in SAMS, **24:**86
Domain wall
coercive force, **1:**315–316
cross-tie, **1:**312ff.
devices, **1:**321–322
energy, **1:**310–316
jagged structure, **6:**237
magnetostatic coupling and, **6:**227–240
pinning, **15:**185
pole density, **6:**228
reversal property, **6:**237–238
ripple in, **6:**277
structural changes in, **6:**276–278
structure and energy, **6:**228
transition thickness, **6:**232
velocity, **1:**316
width, **1:**315
Donor center
coulombic potential and, **8:**200
ionized, **8:**204
Donors
in amorphous solids, **8:**178–183
field-enhanced emission for, **8:**199–207
Donor-trap hopping conduction, **8:**207–208, 213–215, *see also* Hopping
Dopants
in chemical solution deposition, **12:**211–212
N-type, **11:**57–63
Doped oxides, in semiconductor oxide films, **9:**10–12
Doping, **14:**3; **21:**8, 12, 19
gradient, **14:**29, 32, 38
of MBEs, **11:**52–66
multicomponent, in spray pyrolysis, **12:**181–187
nanocrystalline particulate films, **20:**22
optimum level, **9:**11
quantum wells, **21:**79, 95–101, 152–153
techniques, **5:**255–257, 277

Doppler broadened resonance, **18:**132–134, 146, 173, 177
Doppler effect
  for ECR resonance, **18:**37
  for optical measurements, **18:**77
Doppler shifted resonance, **18:**132–134, 136, 155, 162, 169–178
Double barrier diode, **15:**244–247, 250–251
  for oscillators and fast logic, **15:**245
Double heterojunction lasers
  GaAlAs heterojunctions in, **11:**78–79
  minority traps and, **11:**53
Double-positioning, **4:**156
Doughnut-shaped plasma, **18:**174
Dowanon-EB solvent, **10:**124
Drain current saturation, in thin-film transistors, **2:**161–162
Drift mobility, **16:**238
  in amorphous solids, **8:**231
  in nonmetallic films, **8:**188
Drift velocity, **16:**242, 245
DRO memory schemes, **6:**291–292
DRT *see* Distribution of relaxation times
Drude equations, **6:**153, 167, 177, 189–190, 197
Drude mirrors, **10:**51–52
Drude theory, **8:**55–57, 69–71; **17:**350
Druke-Zener type reflector, **10:**35
DT *see* Direct tunneling
Duoplasmatron, ion sources, **13:**114
Du Pont, **16:**296–297
Duran glass, heat-resistant, **5:**133
Dust particle prevention, **4:**6
Dynamic model, **16:**293
Dynode materials, for photomultipliers, **12:**141–144
Dynode systems, focusing designs for, **12:**142

E

Earth Observing System (NASA), **21:**223–224
Ebanol treatments, of absorber coatings, **10:**44
e-beam lithography, **24:**229, 230
e-charge carrier superconductors, **16:**2, 78
Echelle gratings, coating thickness for, **11:**3, *see also* Diffraction gratings
ECR *see* Electron cyclotron resonance discharge
ECR deposition *see* Electron cyclotron resonance deposition
Eddy-current damping, **3:**20–21
Edge displacement method, **16:**247–248
EELA *see* Electron energy loss analysis
Effective distribution coefficient, for dopants, **7:**154–156
Effective mass, deviation from, **15:**225
Effective medium theory (EMT), **17:**351; **19:**59, 99, 153, 162, 177, 206
  Bruggeman effective medium theory, **19:**41–43, 59, 107, 110, 153, 156, 162, 177, 206, 257
  generalized Maxwell-Garnett effective medium theory, **19:**139–40, 177
  Lorentz-Lorenz theory, **19:**156
  Maxwell-Garnett theory, **19:**43, 127–28
Effective phase thickness, **1:**78

Effective refractive index, **1:**78
Effective thickness, **15:**274
Effective-work-function model, **13:**202–209
  Fermi-level pinning, **13:**207
  principles, **13:**203–204
  special cases, **13:**206–207
Efficiency maps, **11:**18–25
Ehrenfest's theorem, **23:**250
Eicosanoic acid film, on water and SiO, **20:**375–376
Eigen oscillation, damped, **5:**6
Eigenvalue function, **15:**223
Elastic compliance, by atomic force microscopy, **24:**93–96, 104, 105
Elastic limit, **16:**288
Elastic modulus, **3:**257–259
Elastic properties, **17:**2, 26–29, 42–46, 48–59, 66–70
  anomalies, **17:**56–59
  biaxial modulus, **17:**43, 48, 49, 56
  elastic constants, **17:**45, 46, 49, 53
  shear modulus, **17:**66–70
  supermodulus effect, **17:**2, 48–51, 54–56
  theories, **17:**54–59
    embedded-atom method, **17:**56, 58
    Fermi surface perturbation, **17:**54, 57–58
    lattice expansion, **17:**56–58
    molecular dynamics, **17:**56, 58
    nonlinear elasticity, **17:**55, 56
  Young's modulus, **17:**41, 43, 46, 48, 49
Elastic scattering, collisional, **18:**15, 72
Electrical conduction
  bulk, **6:**82–84
  in disordered nonmetallic films, **8:**169–246
  experimental results in, **8:**212–240
  high-field, low-temperature, **8:**227–228
  hopping and, **8:**185–199, 212–217
  inhomogeneity and internal fields in, **8:**183–185
  low-field, by hopping carriers, **8:**185–195
Electrical conductivity, **8:**185–195; **20:**26–27
  Debye temperature and, **6:**97
  experimental measurements in, **6:**101–136
  formula, **6:**83
  Fuchs-Dingle model, **6:**93–96
  of island film, **2:**24–25
  light transmission and, **9:**2
  measurements, specimen preparation, **6:**101–103
  in metals, **6:**82–83
  nonlinear, of polymer films, **3:**160–162
  output, of thin-film transistor, **2:**161
  quantum size effects in, **6:**137–143
  in semimetal films, **6:**116–118
  single-crystalline anisotropies, **6:**90–93
  single-sample measurements in, **6:**99–101
  size-dependent, **6:**81–843
  small-angle scattering models, **6:**96–99
  specimen size and, **6:**81–143
  specularity parameter, **6:**89–90
  $T^{\frac{1}{4}}$ law, **8:**189
  temperature dependence in, **6:**93–100
  of thin films, **6:**84, 103–118
  of thin wires, **6:**87–89, 118–123

ZnSb amorphous films, **13:**35
Electrical contacts, **5:**219–222; **16:**7, 8, 73
Electrical double layer, in lipid membrane, **20:**249–253
Electrically conducting surface films, temperature control coatings with, **10:**163
Electrical properties
of alkali antimonides, **12:**94–101
of spray pyrolysis films, **12:**192–201
Electrical resistance, **1:**237, 238
in gold-aluminum films, **6:**319–320, 323
island films, **2:**22–30
metal film
effect of glow discharge, **3:**107
measurement, **3:**18–20
temperature coefficient, **2:**29–30, 379, 381
Electrical resistivity *see* Resistivity
Electrical size effects
quantum size effects and, **6:**137–143
specular reflection and, **6:**131–134
Electric breakdown, in anodic oxide films, **6:**67–68
Electric field
ambipolar, **18:**159, 200
amplitudes, **1:**76, 80
perpendicular to static magnetic field, **18:**126–127, 139, 148–151
refraction and, **5:**2–3
threshold, **18:**127, 149–151
Electric field intensity, **1:**73–76
Electric transition dipole moment vector, **19:**295
azimuthal angle, **19:**299
tilt angle, **19:**299
Electrochemical generation, silver film, **20:**15
Electrochromic device
applications, **17:**303
general design, **17:**303
liquid electrolyte, **17:**352, 353
polymer electrolyte, **17:**358, 359, 361
solid inorganic bulk-type ion conductor, **17:**354
solid inorganic thin-film ion conductor, **17:**355, 356, 357
Electrochromic film, materials for, **17:**303
Electrochromism
absorptance modulated, **17:**345
reflectance modulated, **17:**345
Electrode, associated character of film-covered gold, **20:**191–193
Electrode line, **20:**332–333
Electrodeposition, **4:**191–255
Bi-Sr-Ca-Cu-O thin film, **16:**99
impurity effects, **4:**210–212
morphology and, **4:**199–217
nucleation rate, **4:**198
process parameters, **16:**95, 96
screw dislocation mechanism, **4:**203–204
structure studies, **4:**226–250
substrate metal and, **4:**212–217
temperature effects, **4:**210–212
thickness effects, **4:**210–212
on thin metal films, **4:**243–248
two-dimensional surface nucleation, **4:**198–199
vapor deposition, differs from, **4:**195–196
Y-Ba-Cu-O thin film, **16:**97
Electroless metallization, **20:**117–123
Electroless plating, in gold film processing, **9:**60

Electroluminescence, **3:**202–207; **7:**178
Electroluminescent diodes, in liquid phase epitaxy, **7:**205–235
Electroluminescent junctions, **7:**206, 212
Electromagnetic theory, of multilayered samples, **19:**282, 283, 289, 290
Electromigration
  alloying effects, **7:**287–305
  in aluminum-chromium thin films, **7:**292–293
  in aluminum-copper thin films, **7:**287–292, 297–298
  in aluminum-magnesium thin films, **7:**295–298
  copper-aluminum alloying effects, **7:**287–294
  discontinuous flux across interfaces, **7:**277
  grain boundary, **7:**266–268
  lattice results, **7:**258–264
  oxygen-silicon effects, **7:**296
  surface effects, **7:**284–287
  in thin films, **7:**257–305
  three-grain junction, **7:**271
  triple-point damage mode, **7:**274
  void nucleation and growth, **7:**278–283
Electromigration activation energy
  aluminum-copper film, **16:**250
  aluminum film, **16:**250
  aluminum-silicon film, **16:**251
  copper in silver film, **16:**261
  ICB deposited aluminum film, **16:**250
  silver film, **16:**259
Electromigration drift velocity, **16:**242, 245
Electromigration effective charge, **16:**237–239
  cobalt thin film, **16:**261
  copper in aluminum, **16:**253
  gold thin film, **16:**254
  indium thin film, **16:**258
  tin thin film, **16:**257
  under pulsed condition, **16:**252
Electron acceptor amphiphiles, **20:**244, 251–253
Electron affinity, **12:**66, 70, 112–113, 114–123, *see also* Negative electron affinity
Electron beam evaporation, **9:**45–46
  in dielectric film preparation, **8:**19–20
  in silica and alumina coating deposition, **12:**28
Electron-beam melting, **2:**72
Electron beams
  heating, **1:**127
  lithography, **8:**260
  zone heating, **5:**182
Electron bombardment, film formation, **3:**146–150
Electron current, **14:**27, 34
Electron cyclotron resonance
  definition, **18:**126–129
  ''pure,'' **18:**126, 128, 131–132
Electron cyclotron resonance deposition, silicon dioxide, **22:**241, 243, 244, 246, 247, 251
Electron cyclotron resonance discharge, **18:**12, 25
  cavity, **18:**31, 40
  configurations, **18:**26
  design example, **18:**23
  distributed ECR, **18:**31, 39
  heating, **18:**31
  microwave systems, **18:**26

Electron-defect scattering, **6:**99, 126
Electron density, **18:**125–126, 128, 130, 132–140, 145–147, 159, 167, 171–190, 194–195, 200–202, 205–208
Electron diffraction analysis, **1:**196–202; **2:**109–121
  diffraction pattern photography, **7:**15–16
  dynamical theory, **2:**115–121
    relativistic, **2:**134–135
  electrodeposit structure, **4:**223–224, 232–238
  electronic recording technique, **7:**16–17
  kinematical theory, **2:**110–115
  line broadening, **2:**98–99
  of local atomic order, **7:**1–37
  low energy, **1:**203
  low temperature, **1:**202
  radial distribution function and, **7:**5–11
  techniques, **7:**12–17
Electron donor amphiphiles, **20:**244, 251–253
Electron-electron interaction, **15:**22, *see also*, Coulomb interaction
Electron-electron scattering
  energy loss in, **12:**59–60, 110
  secondary carriers in, **12:**60
Electron energy, average, **13:**72
Electron energy loss analysis, in plasma chemical vapor deposition, **12:**257
Electron energy losses, in nonradiative surface plasma excitation, **9:**199–219
Electron energy spectral filter, spike-doped, **23:**99–100
Electron escape, surface barrier and, **12:**66–67
Electron field microscopy, **1:**205
Electron gun
  in aluminum film deposition, **12:**2, 14, 28
  aluminum/silicon dioxide coating with, **10:**141–147
  in production of evaporation coatings, **10:**147–149
Electron-hole pairs, generation, **12:**60
Electron hopping
  in mixed monolayers, **20:**344–346
  osmium *tris*-diphenylphenanthroline film, **20:**340–346
  solvent effects, **20:**343–346
Electronic density of states, **15:**176–177
  magnetooptic effects, relation to, **15:**192
  optical conductivity and absorption, relation to, **15:**190
Electronic imaging, topologically structured films in, **8:**272–294
Electronic warfare, **15:**52
Electron mean free path, **6:**102; **19:**133–39, 162–68, 176–78, 185–86
Electron microscopy, **1:**188–196; **2:**109–121; **5:**188–189
  bright and dark field, **1:**189–191
  diffraction, **1:**191–194
  electrodeposit structure, **4:**224–226, 238–250
    copper, **4:**238–239
    nickel, **4:**239–243
  sample and replica preparation, **1:**188–189
  stereo, **1:**191

Electron-phonon interaction
pair production and, **12:**60
probability, **6:**97
Electron-phonon scattering, resistivity and, **6:**131
Electrons
absorption, practical applications, **2:**72–73
beam interaction with thin films, **2:**63–145
changes in materials by, **2:**129–134
characteristic energy losses in solids, **2:**99–105
collision frequency, **18:**126–128
de Broglie wavelength, **6:**89–90
Fermi energy and, **6:**139
resistivity size effects and, **6:**132
drift (oscillation) velocity, **18:**127, 149
effective collision frequency, **18:**128, 133, 146–151
energy distribution function, **18:**124, 127–128, 132–133, 146–151
energy distribution tail (fast/hot electrons), **18:**148, 150–151, 190–193, 196–198, 201–205
excitation, **9:**219–223
nonradiative surface plasmons by, **9:**199–219
radiative surface plasmons by, **9:**173
heating time, **18:**148–149
interactions, nonlocalized nature, **2:**107–109
mean free path, **18:**149, 237
at low temperature, **1:**236
Houston theory, **1:**235
mobility in semiconductor films, **5:**210–211
motion
in electromagnetic field, **18:**240
in magnetic field, **18:**240
path, **14:**31, 33
reflection of at surface, **9:**213–214
scattering, comparison of theory and experiment, **2:**92–98
temperature, **18:**14, 19, 126, 128, 132–133, 146–151, 159
thermal velocity, **18:**127, 131–133, 135, 146–151
transmission and reflection, **9:**214
tunneling in thin film structures, **4:**325–371
wavelength, **1:**196
Electron spin polarization, **6:**246
Electron spin resonance, in amorphous solids, **8:**239–240
Electron states, stratification, **6:**137
Electron transfer
free energy difference, **20:**244–245
photoinduced
in Langmuir-Blodgett film, **20:**249
rate, **20:**243–245
rate in monolayer, **20:**244–245
Electron transport, in photoemission, **12:**59–66
Electron traps
DLTS spectra, **11:**54–55
in MBE, **11:**52–56
Electron tunneling, **4:**325–371
between islands, **2:**25, 383
between two superconductors, **1:**268
in built up films, **7:**379
through thin insulating films, **1:**267–271
through transparent surface barriers, **8:**288
Electron wind effect, **7:**261

Electro-optics
birefringence shift, PLZT, **17**:249–251
bismuth titanate films, **17**:264–270
coefficients, **17**:260
frequency doubling, PLZT, **17**:251–252
$LiNbO_3$, $KNbO_3$, KTN films, **17**:260–264
linear, quadratic effects, **17**:249–250
magnetron sputtered PLZT films, **17**:250
PZT, PLZT films, **17**:248–252
second harmonic generation, PLZT, **17**:252
sol-gel PZT films, **17**:249–250
Ellipsometric parameters, **6**:165–170; **19**:286, 289
Ellipsometry, **4**:79–84
in gaseous anodization, **6**:46–48
in situ, **6**:46
in ionic conductance measurements, **6**:27–29
Langmuir film of palmitic acid, **19**:291–293
in optical studies, **6**:157
Ellipticity, **15**:202; **19**:285
Embedded atom method, **22**:127
Embedding techniques, **3**:188–194
Emission *see* Infrared emission; Thermionic emission
Emission spectroscopy, **14**:15, 16; **15**:276
Emittance, **2**:310–312
calorimetric, **2**:327–330
radiometric, **2**:325–327
spectral, **4**:10
thermal
determination, **2**:313–333
of evaporated coatings, **2**:305–361
Emitted material, **14**:150, 158
Emitter characteristics, **4**:356–360
Empty state, in amorphous solids, **8**:179
Emulsion, photographic, blackening by electrons, **2**:123
Energy
band structure of PbTe, **3**:277
domain wall, **1**:314
loss per electron-ion pair created, **18**:15–17
optical thin films, **1**:76–77, 80
of sputtered particles, **3**:71–72
Energy diagram
amphiphilic triad film, **20**:248
lipid membrane, **20**:266
molecular photodiode, **20**:244–245
photocurrent via triad molecule, **20**:266
photodiode in lipid film, **20**:244–245
Energy distribution
electron, **18**:14
ion, **18**:71, 89
control of, **18**:90, 92
measurements of, **18**:76, 81, 90
transverse, **18**:87
variation with frequency, **18**:90
Maxwellian, **18**:14, 72
Energy gap
crystals, **8**:175
semiconductor, **3**:277
superconductor, **1**:265–271
magnetic field dependence, **1**:269–271
temperature dependence, **1**:269, 270
Energy levels, quantum well, superlattice, **15**:222–226

Energy levels, quantum well, superlattice, (*continued*)
excited, virtual, **15:**222–224, 232, 242
Energy transfer
antenna molecules and, **20:**254
hole-burning spectroscopy, **20:**299–300
Entropy of mixing, **4:**102
Epi-base, after extrinsic base and emitter opening process, **23:**60, 61
Epitaxial films, **5:**152–154; **14:**27, 65
substrate orientation in, **5:**261
vapor growth apparatus for, **5:**158–159
Epitaxial gold, **24:**6
Epitaxial layer growth, **15:**221
by MBE
maximum temperature, **23:**18–21
minimum temperature, **23:**18, 19
chemical vapor deposition, **5:**282–283; **23:**21–51
dislocations and, **5:**205
of GexSi1-x, **23:**1–74
promotion, **5:**268–269
of semiconductors, **5:**151, 196–203
of silicon carbide, **5:**284
Epitaxial layers
infrared detectors
delta-doped platinum silicide on silicon detectors, **23:**141–143
epitaxial silicides, **23:**161–162
SiGe/Si heterojunction internal photoemission detectors, **23:**144–153
Si-homojunction detectors, **23:**143–144
silicide/SiGe Schottky detectors, **23:**153–161
thin films, on silicon for infrared signal detection, **23:**115, 168
Epitaxy, **1:**216–220, 289
aliphatic alcohols on ice, **20:**152–155
Dixit formulation, **1:**219
germanium, **1:**218
and heterogeneous nucleation, **4:**128–134
contact angle and, **4:**130–131
temperature and, **4:**129–130
theory, **4:**129–131
lead selenide, **20:**33–37
lead sulfide, **20:**31–32
liquid phase *see* Liquid phase epitaxy
molecular beam *see* Molecular beam epitaxy
water on alpha-amino acids, **20:**175
Epoxy cements, in metal film filters, **7:**85
Equilibrium critical thickness, **23:**3
Equilibrium temperature, for metal film filters, **7:**94–98
Equivalence concept, **1:**91–92
Equivalent circuit, **15:**49, 50
Equivalent layer spacers, MDM filter design and, **9:**110–112
Equivalent screen background input (ESBI), for image intensifiers, **12:**150–151
Equivalent thickness
defined, **10:**128
for multilayer systems, **5:**64–67
Escape cone/cap, **23:**90–91
Escape probability, *n*-doped quantum-well infrared photodetectors, **21:**146–149
Essential work of fracture, **16:**273
Ester-alcohol monolayer, structure, **20:**169–170

Etching, **14:**18
anisotropic, **18:**8, 73, 87
glass film evaluation using, **4:**272–273
isotropic, **18:**8
model, **18:**70
patterned hydrophobic SAMs as ultrathin resists, **24:**240–244
rate, **13:**168–176
carrier generation and control, **13:**173–176
geometry, **13:**169–170
laser-induced etch enhancement vs. laser-power density, **13:**174–175
normalized photoinduced rate enhancement vs. laser-power density, **13:**174–175
temporal characteristics of light source, **13:**168–169
wave-guiding effects, **13:**170–171
wavelength dependence, **13:**171–173
reactive ion etching, **22:**60
SIMBAD model, **22:**60–76
simultaneous deposition and sputter etch, **22:**241–252
of thin films, **8:**261–266
three-dimensional, **8:**262
wet etching, **22:**25
Etch pits, **24:**86
Ethyl eicosanoate film, phase diagram, **20:**373
Eutectic systems, miscibility and, **6:**304
Evanescent zone, **18:**136
Evaporated aluminum, reflectance as function of incidence angle for, **12:**21
Evaporated aluminum film
purity, **12:**14
silicon monoxide effects on UV irradiation in, **12:**27–28
Evaporated aluminum mirrors
germanium and silicon monoxide coatings for, **12:**42–43
hafnium dioxide and yttria coatings for, **12:**35
infrared reflectance data, **12:**29–30
magnesium fluoride coating for, **12:**48–49
with overcoatings, **12:**25–32
Evaporated films, **10:**71–163, 93–110, 261–308, *see also* Thin films
dielectric films, water absorption in, **12:**46–49
for satellite temperature control, **10:**127–164
Evaporated rhodium film, reflectance-enhancing coatings for, **12:**39–42
Evaporated rhodium mirrors, overcoated vs. bare film, **12:**41
Evaporated silver film, *see also* Silver; Silver mirror coatings
high reflectance and visibility, **12:**24
minimum polarization introduced by, **12:**24
reflectance of vs. incidence angle for, **12:**20
Evaporation, **13:**212; **17:**81, 91
arc, cathodic, **17:**91, 92, 96
chamber, vacuum ultraviolet monochromator, **1:**136–138
electron beam, **17:**91
high-voltage, **17:**91
low-voltage, **17:**91
electron bombardment sources for, **5:**169

Evaporation, (*continued*)
filaments, electron emission and, **9**:45
flash, **5**:166, 176–179
laser beam, **17**:90
methods, **1**:126–127
multisource, **5**:173–176
sources, **2**:286
techniques, **2**:365–366
thermal, **17**:91
ultra-high vacuum, **1**:1–38
uniform thickness formation, **3**:42–45
in vacuum deposition process, **5**:162–173
Evaporation masking, in thin film deposition, **8**:254–261
Evaporation rate, **14**:157
Evaporative and ablative techniques, **17**:235–237
activated reactive evaporation, **17**:236
epitaxial films, BTO, **17**:237
evaporative components, **17**:235
flash evaporated, BaTio3, **17**:235
fluence, **17**:236
pulsed laser deposition, **17**:236
Evaporator, aluminum *see* Aluminum evaporator
EVOLVE simulation, **22**:180–181
low-pressure chemical vapor deposition, **22**:202–210
physical vapor deposition, **22**:226–241
plasma-enhanced chemical vapor deposition, **22**:210–215
programmed rate CVD, **22**:265–269
remote microwave PECVD, **22**:215–226
simultaneous deposition and sputter etch, **22**:241–252
surface diffusion, **22**:252–265
EXAFS, **17**:22
Exchange anistropy
hysteresis loop in, **6**:218
in multilayer films, **6**:213–223, 268
Exchange constant, **15**:25, 126, 134
Exchange-dominated modes, **15**:135
Exchange field
derived, **15**:126
dominance at short wavelength, **15**:3, 135
neglect in magnetostatics, **15**:26
Exchange interaction, transition metal, dependence on, **15**:180
Excitation hot-electron spectroscopy
optically stimulated hot-electron spectroscopy, **21**:252–255
thermally stimulated hot-electron spectroscopy, **21**:246–251, 253–254
Expanded monolayers, **7**:321–322, *see also* Monolayers
Extended system method, **22**:138–142, 161
External diode efficiency, increase in, **7**:212
External reflection, **15**:272
Extinction contours, **2**:112
Extreme ultraviolet, **7**:43–111
Extreme ultraviolet astronomy, **10**:71–163
coatings for, **10**:73–127
gold, **10**:93–94
iridium, **10**:94–97
multilayer coatings, **10**:114–117
osmium, **10**:97–100
platinum, **10**:91–92
rhenium, **10**:100–103
rhodium, **10**:110–114
ruthenium, **10**:108–110
tungsten, **10**:103–107

two-layer coatings, **10:**118–122
contamination of reflecting surfaces, **10:**122–127
reflectance vs. thickness in coatings, **10:**93–107
reflecting filters for, **10:**114–116
transmitting filters for, **10:**117
for wavelengths larger than 1000, **10:**75–91
for wavelengths shorter than 1000, **10:**91–114
Extrinsic infrared photoconductors, **21:**4

F

Fab'-lipid, diffusion coefficient, **20:**228
Fab'-lipid films
diffusion, **20:**228–230
gas-fluid transition model, **20:**230–233
preparation, **20:**226
Fabrication, **24:**103, 104
with atomic force microscopy, **24:**83, 98–102, 103, 104, 106
e-beam lithography, **24:**229–230
focused ion beam lithography, **24:**229, 230
microcontact printing, **24:**228–251
micromachining, **24:**229, 230
microtransfer molding, **24:**240
"nanografting," **24:**101–102
neutral metastable atom lithography, **24:**229, 230
scanning probe lithography, **24:**229, 230
with scanning tunneling microscopy, **24:**99
UV photopatterning, **24:**228–229
Fabry-Perot filters, **8:**44–45
Fabry-Perot formula, **5:**75
Fabry-Perot interferometer, **9:**74, 88–90
radiant transmittance, **9:**88
transmittance of symmetrical multilayer in, **9:**90–92
Faceting, **14:**226–227
Faraday effect, **1:**329–330
coefficients in YIG and Bi:YIG, **15:**106
in collinear mode conversion, **15:**113
contrasted with Cotton-Mouton effect, **15:**103
dominance, **15:**106, 112
in transverse mode conversion, **15:**122
Faraday rotation, MnBi films, **13:**31
Farey fractions, **21:**52
Far infrared (FIR) emission
detection, **21:**35–37
from multiple-quantum-well structures, **23:**243–247
Fatty acids, phase diagrams, **20:**372–372
FBAR *see* Film bulk acoustic resonator
Feco fringe system, **2:**235
Feco interferometric method, **4:**31–37
absolute phase change measurement, **4:**88–90
accuracy, **4:**36–37
alignment, **4:**33–34
coating of plates, **4:**31–32
interferogram analysis, **4:**34–36
measurement of surface roughness, **4:**20–21
Feedback
control, **14:**53, 55
time constant, **14:**55

Feedback lasers, double heterostructure distributed, **10:**228
Feed-throughs, **1:**36–37
Fellgett's advantage, **15:**271
FEMFET *see* Ferroelectric memory field effect transistor
FEP Teflon type A, **10:**154–155
Fermi age theory, **4:**344; **5:**9; **12:**64
Fermi-Dirac distribution function, **8:**105, 186–187
Fermi electron, scattering with, **23:**91–92
Fermi energy, **15:**227
collector, emitter, **15:**252
electron de Broglie wavelength, **6:**139
in hopping carriers, **8:**186–193, 212–217
Fermi function, **8:**245
Fermi level, in electron escape, **12:**67
Fermi-level pinning, **13:**201–209, 221–222
effective-work-function model, **13:**207
heterojunctions, **13:**219
position, **12:**115–117, 118
Fermi momentum, **6:**246
Fermi surface
in electrical conductivity, **6:**83–84, 93
galvanometric measurements and, **6:**101
in polyvalent metal films, **6:**112
quantum size effects and, **6:**137
specular reflection and, **6:**91
spherical and ellipsoidal, **6:**92
Ferrell mode, for radiative surface plasmons, **9:**159–163
Ferrimagnetic oxides, chemical vapor deposition, **5:**313–314
Ferrite film, by sputtering, **3:**120
Ferrites
magnetic properties compared, **15:**9
magnetooptic properties compared, **15:**106
opposing spins on sublattices, **15:**12
types, **15:**8–9
YIG *see* Yttrium iron garnet
Ferrocene amphiphilics, **20:**244, 247, 249–252
Ferrocene terminated SAMs, **24:**23, 27–29
Ferroelectric films, **17:**150; **19:**249–76
$Bi_4Ti_3O_{12}$, **17:**209
deposition, **17:**230–242
inhomogeneities in films, **19:**250–75
$LiNbO_3$, **17:**161
$LiTaO_3$, **17:**161
material properties and applications, **17:**227–230
microstructure-electrical properties correlation, **19:**272–75
PZT, **17:**162
SE studies on ferroelectric films
$BaTiO_3$ film on $SrTiO_3$, **19:**263
(Pb, La)TiO_3 film on sapphire, **19:**256–61
$PbTiO_3$ film on $SrTiO_3$, **19:**260–62
(Pb, Zr)TiO_3 film on Pt coated Si, **19:**269–70
(Pb, Zr)TiO_3 film on sapphire, **19:**264–69
size effect mechanisms, **19:**250–53, 259
Ferroelectric memories
aging, **17:**257–259
barium magnesium fluoride, **17:**285–286
bismuth titanate, **17:**272–278

capacitor type, PZT, **17:**253–255
fatigue, **17:**257–259
memory circuit elements, **17:**254
PZT films for memories, **17:**252–259
select transistor, **17:**255
status, **17:**294
switching speed, PZT, **17:**256–257
Ferroelectric memory field effect transistor (FEMFET), **17:**253, 272
BMF gate dielectric, **17:**285, 286
channel conductance, **17:**275, 277
$LiNbO_3$ gate dielectric, **17:**260–261, 292–293
memory window, **17:**285
MFST with BTO gate dielectric, **17:**275–277
non-destructive readout, **17:**272
programming voltage, **17:**277–278, 286
SiO_2 buffer layer effect, **17:**276–278
switching mechanisms, **17:**275–276
switching time, **17:**27
Ferromagnetic-antiferromagnetic films, **6:**268
Ferromagnetic bridges, in multilayer films, **6:**242
Ferromagnetic films, *see also* Multilayer magnetic films; Very thin films
domain wall coupling in, **6:**227–240
high-order and helical anisotropies in, **6:**278–290
multilayer, **6:**210–211
quasiwalls in, **6:**234–235
uncoupled, **6:**216
Ferromagnetic resonance
Larmor frequency, **15:**5, 27
derived, **15:**126
increased by transverse demagnetizing field, **15:**129
linewidth
measurement, **15:**19
minimum observed in YIG, **15:**16
observed values in ferrites, **15:**9
related to damping, **15:**128
related to propagation loss, **15:**6
premature decline at high power, **15:**92
Ferromagnetism, spin model, **15:**125
Fiber-matrix composite, **16:**281, 306, 328
Fibronectin, adsorption, **24:**248, 249
Field-effect transistors (FET), **8:**305–307
doping profile in, **11:**75–76
high-power, **11:**75
low-noise, **11:**75–76
MBE substrate and, **11:**49–51
power layers, **11:**76
Schottky barrier and, **8:**306–307
thin films as, **8:**305
Field electron microscopy, **1:**205
Field-emission imaging, point arrays for, **8:**287–294
Field-enhanced emission, from donors, **8:**199–207
Field ionization, low-temperature, **8:**208–210
Figures of merit, **21:**256–257
Filamentarylike thin films, **8:**121, 134
Filament evaporation, electron emission and, **9:**45
Filament sources, **3:**30–31
Film bulk acoustic resonator (FBAR), **17:**196
device structure, **17:**198, 199
equivalent circuit, **17:**200
figure of merit (FOM), **17:**200
filter approaches, **17:**201

Film bulk acoustic resonator (FBAR), (*continued*)
ladder filter, **17:**201, 204
monolithic crystal filter, **17:**202
stacked FBAR filter, **17:**202, 205
resonator, **17:**196–198
Film deposition, **9:**30–60; **14:**2; **17:**3–8, 153, 230–241, *see also* Chemical vapor deposition; Physical vapor deposition; Sputtering
amorphous substrates and, **9:**42
comparison of methods, **17:**231
electrodeposition, **17:**3
electron-beam evaporation (EB), **19:**215
epitaxial, **17:**10, 23
evaporation, **17:**3, 4, 7; **19:**148
film thickness control in, **9:**133
in filter production, **9:**132–133
geometry
fixed substrate, **17:**4
movable substrate, **17:**7
rate distributions, **17:**4, 6
ion-assisted deposition, **17:**34; **19:**195, 239–44
ion-beam sputtering, **19:**240
ionized cluster beam, **19:**195
magnetron sputtering, **19:**83, 148
metal organic chemical vapor deposition, **19:**263
molecular beam epitaxy, **19:**50
multi-ion-beam reactive sputtering, **19:**254, 264
nucleation mechanisms, **17:**22, 23, 27, 30–34
critical thickness, **17:**27, 31
Frank-van der Merwe (layer-by-layer), **17:**30, 32
Stranski-Krastanov, **17:**31
surface energies, **17:**30
Volmer-Weber (three-dimensional island), **17:**30
nucleation-modifying layers in, **9:**42–43
oxide target sputtering, **9:**53–59
plasma enhanced chemical vapor deposition (PECVD), **19:**50, 62–63, 77–79
properties related to, **9:**63
rate monitoring, **17:**8
reactive evaporation, **9:**47–49
reactive low-voltage ion plating, **19:**195
reactive sputtering, **9:**49–53
sheet resistivity and, **9:**62
sol-gel technique, **19:**254, 260–61
steps, **13:**66
surface diffusion, **17:**23, 31
types, **9:**59–60
Film formation, **4:**8–12; **13:**15–20; **14:**137
by generation, growth and coalescence of nuclei, **4:**150–186
charged particle effect, **13:**20
ion-assisted, **13:**1–2
kinetic energy effect, **13:**15–18
mass deposited, **13:**16–17
migration deposits, **13:**19
migration effect, **13:**18–19
modes, **4:**138
nucleation processes, **4:**97–136
optimum value of kinetic energy, **13:**2–3
Film-forming chemicals
N-(2-aminoethyl)(3-aminopropyl)-trimethoxysilane, **20:**131
amphiphilic porphyrin, **20:**70
arachidic acid, **20:**30
behenic acid, **20:**263
benzyltrichlorosilane, **20:**115

cadmium arachidate, **20:**296
cadmium stearate, **20:**319–323
p-(chloromethyl)-phenyltrichlorosilane, **20:**123
cholesterol, **20:**324
L-alpha-dimyristo-phosphatidylethanolamine, **20:**355–361
dioctadecyldimethylammomium bromide, **20:**20
dioctadecylferrocene, **20:**339
dipalmitoylphosphatidylcholine, **20:**322
docosanoic acid, **20:**368, 372
ethyl eicosanoate, **20:**372
ethyl stearate, **20:**325
heneicosanoic acid film, **20:**376–377
lanthanum stearate, **20:**325–326
(3-mercaptopropyl)-trimethoxysilane, **20:**129
nickel hexadecylcyclams, **20:**267–269
18-nonadecenyltrichlorosilane, **20:**45
N-octadecylferrocenecarboxamide, **20:**336
n-octadecyl mercaptan, **20:**185
n-octadecyltrichlorosilane, **20:**46
oleic acid, **20:**324
osmium *tris*-diphenylphenanthroline film, **20:**340–346
perfluorolauric acid, **20:**263
phenethyltrichlorosilane, **20:**115
phosphatidylethanolamine, **20:**324
poly(l-naphthol), **20:**185
stearyl alcohol, **20:**324
2,2'-thio-*bis*-ethyl acetoacetate, **20:**185
3,3'-thio-*bis*-propyl acetoacetate, **20:**186
2-(trimethoxysilyl)ethyl–2-pyridine, **20:**120
trimethoxysilylpropyl-diethylenetriamine, **20:**136
undecyloxypropionic acid, partially fluorinated, **20:**258
Film growth, **14:**59–63
photostimulated, of anodic oxide films, **6:**13
Film lifting, comparison of methods, **20:**321–328
Film morphology, **17:**23, 157
columnar growth, **17:**157, 160
columnar structure, **17:**23
grain size, **17:**23
porosity, **17:**23, 31
structure zone model, **17:**157
surface texture, **17:**23
Film optics, inhomogeneous films in, **8:**83–84
Film orientation, **17:**154, 156
C-axis orientation, **17:**156, 158, 159
effect of ion bombardment, **17:**157, 158
RHEED, **17:**158–159
X-ray 0–20, scan, **17:**155
Films, **5:**89–90, 90–91, 92, 96, 237–314, *see also* Film deposition; Film formation; Film morphology; Film optics; Film orientation; Metal films; Monolayers; Multilayer magnetic films; Multilayers; Semiconductor films; Thin films
above 950, **1:**172–177
amorphous *see* Amorphous film
application, **14:**3, 59
at low temperature, **1:**233–275

Films, (*continued*)
normal state, **1**:234–240
superconducting state, **1**:255–273
of Bi, Ga and Sb condensed at low temperature, **2**:44–46
bombardment, **14**:51, 60, 65
boundaries, scattering at, **9**:8
coalescence, with nucleation-modifying layers, **9**:45
composition, **22**:176
SIMBAD modeling, **22**:45–49
in spray pyrolytic process, **12**:179–181
titanium-tungsten films, **22**:229–241
condensation, **1**:208–210
in contact, electronic interaction between, **1**:272–273
device applications, **7**:386–387
dielectric breakdown, **7**:372–379
dielectric properties, **7**:367–372
electrical properties, **7**:365–387
ellipsometric studies, **7**:363–364
environment effect on, **1**:129–130
epitaxial *see* Epitaxial films
evaporated metal, optical properties, **2**:333–358
future potential, **7**:391–392
highly disordered, prepared at low temperature, **2**:37–46
inverted, **1**:304ff.
low-temperature, **1**:233–275
mixture films, scattering and, **10**:280–285
multilayer
computer calculation, **1**:116–120
theory for normal incidence, **1**:73–78
nucleation and growth, **1**:207–215
opaque
optical constants, **1**:142–144
true quantum yield, **12**:63
optical *see* Optical films
optical absorption studies, **7**:363
optical properties, **7**:361–363
optical reflection and transmission studies, **7**:365
organic films, **8**:41–42
ionized cluster beam, **13**:54–56
reactivity, **1**:129
in region (500–950A), **1**:177–179
residual gas effects, **1**:51–63
magnetic, **1**:58–60
superconducting, **1**:55–58
semiconductor oxide *see* Semiconductor oxide film
size-dependent electrical conduction, **6**:81–143
solution grown *see* Solution-grown films
stress, **16**:183
surface deformation by, **10**:274–276
structure, **1**:129; **3**:214–219; **14**:63–65, *see also* Structure
alkanethiol films on liquid mercury, **24**:180–201
in chemical solution deposition, **12**:187–191
optical properties and, **6**:199–202
thin *see* Thin films
transparent conducting *see* Transparent conducting films
vacuum ultraviolet reflectance, multilayer, **1**:171–179
very small grain size, **2**:38–41
''Voll-formierte'' films, **6**:11
Film thickness, **4**:21–41; **16**:311
composite survivability after film failure, **16**:309–311
distribution, **3**:33–50

mechanical methods, **3:**33
optical methods, **3:**34–38
production, **3:**40–50
radiographic methods, **3:**38–40
Feco method, **4:**31–37
Fizeau method, **4:**24–31
fracture and, **16:**275, 280, 282, 284, 307–309, 324, 326–329
measurement, **4:**21–24
Dyson method, **2:**234–235
electrical methods, **1:**236–240
multiple beam interferometry, **2:**235–236
optical, **2:**260, 262–263
X-ray methods, **4:**23–24
monitoring devices, **3:**8–32
crystal oscillator, **3:**21–27
electrical methods, **3:**18–21
gravimetry, **3:**27–30
optical methods, **3:**8–18
X-ray methods, **3:**32
oxide films, **4:**37–41
quartz crystal oscillator method, **4:**23
Filterbanks, magnetostatic, **15:**88
Filters, *see also* Metal film filters
all-dielectric, **5:**78, 135
bandwidth determination, **5:**81
channelized receiver, **15:**54
circular variable, **5:**84
color-conversion, **5:**135
equivalent-layer systems, **5:**78
high-frequency-pass, **5:**80
infrared hot-electron transistors, **21:**258–259, 275–282
lighting or optical, **5:**134–138
long- and short-wavelength-pass, **5:**78–80
magnetostatic, **15:**87, 88
metal films as, **7:**45–49
narrow-band-pass, **5:**80–84
one-M *see* One-M filter
preparation of metal films for, **7:**53–77
production techniques, **9:**132–142
cementing and addition of blocking filters in, **9:**140–142
and deposition techniques for metals, **9:**132–134
optical monitoring in, **9:**136–140
reflection or transmission monitoring in, **9:**134–136
short-wavelength-pass, **5:**51–52
technologies, **15:**55
types, **5:**51–52
unbacked metal films as, **7:**43–111
FIR *see* Far infrared (FIR) emission
First cracking strain, **16:**281, 307–310, 318, 320, 324, 326–328
Fizeau fringes, **2:**235–236
Fizeau interferometer, **4:**24–31; **10:**204
alignment, **4:**27
coating of flats, **4:**26
producing sharp fringes, **4:**24–27
surface roughness
measurement, **4:**19–20
sensitivity to, **4:**29–31
Flash evaporation, of IV-VI pseudobinary alloys, **11:**112–113
Flat plate collector, **10:**23
Floating potential, **14:**208–209; **18:**186, 192–193, 198–199, 202–206
Fluid receiver, **10:**23
Fluorescein, as parting agent, **7:**64
Fluorescence, electron-induced, **2:**123
Fluorescence spectroscopy, **14:**16
Fluorides
as dielectric film materials, **8:**34

Fluorides (*continued*)
high critical temperature superconductor production
Bi-Sr-Ca-Cu-O thin film, **16:**29, 53
low oxygen partial pressure annealing, **16:**43
Y-Ba-Cu-O thin film, **16:**42–51, 60
Fluorinated thiol SAMs, structure, **24:**88, 129
Fluoropolymers
alkylation, **20:**96–97
fluorine removal, **20:**99–100
indirect synthesis modification, **20:**100
oxidation, **20:**100–102
surface modification, **20:**95–104
Flux
diffusion, in size space, **4:**98–99
directional effect, **14:**59
distributions, **22:**191–193
electron, **14:**29
gas, **14:**43
impingement, **4:**101, 109
incoming flux, **22:**101–102
ion flux, **17:**99, 113, 121, 128, 137; **18:**9, 11; **22:**107–110, 111
low-pressure deposition, **22:**189–193, 196, 197–200
mathematical models, **22:**107–110, 111
metal vapor, **17:**111
neutral flux, **22:**107–110, 111
outgoing flux, **22:**102, 103
ratio, **14:**5, 24
reactive gas, **17:**111
sputtered, **14:**17, 21, 27, 44
Flux density change, light, due to refraction, **1:**79
Flux transformer, **16:**75–78, 117, 118
FMR *see* Ferromagnetic resonance
Focal-plane arrays, **21:**257
Focused ion beam lithography, **24:**229, 230
Focused stray light
in diffraction gratings, **11:**31–32
in vacuum ultraviolet, **11:**32
Focusing concentrator, **10:**24
Foil light valves, **8:**319–333
Forbidden angle of emission, **23:**251–252
Forbidden gap
in amorphous solids, **8:**240–241
hole concept and, **8:**177–178, 185
localized carrier movement in, **8:**198
mobility vs energy in, **8:**192–193
Force fields, **24:**209–211
Force measurement
biomolecular interaction, **20:**207–222
in monolayer, **20:**208
technique, **20:**208
Force profile
and association rate, **20:**217–218
biotin-inactive streptavidin, **20:**216
biotin-streptavidin, **20:**210
temperature effect on biotin-streptavidin attraction, **20:**212
Ford circles, **21:**52
Formvar substrates, **6:**341
FORTRAN IV, **10:**188
Forward volume waves *see* Volume waves, magnetostatic
Fourier coefficients, **5:**60
Four-mirror polarizer, **12:**33–34
Fowler emission coefficient, **23:**91–92
Fowler emission efficiency, platinum silicide arrays, **23:**102–104

Fowler-Nordheim field emission, **8:**227, 291
Fractals, **14:**64
Fracture, **3:**261–263
  bulge test method, **16:**327
  tensile fracture, **16:**273, 311
Frame storage targets, for television, **8:**282–287
Free carriers, **11:**225–226
Free charges, vs. bound, **11:**313–314
Free-electron metal, plasma scattering in, **12:**62–63
Free electron movement, in crystalline semiconductors, **8:**175–178
Free electron theory, ultraviolet and, **7:**44
Free energy
  adsorbed atom aggregate, **2:**4–5
  of formation, nuclei, **4:**101–102
  lipid membranes, **20:**249–252
  minimum, configurations, **4:**157–158
  total, of substrate system, **4:**102–103, 111
Free radicals, recombination on a surface, **3:**157–158
Free spins, boundary condition, **15:**135
Freezing temperature
  alcohol-covered water, **20:**149–153, 171
  carboxylic acid-covered water, **20:**149–150, 171
  versus aliphatic chain characteristics, **20:**149
  water
    on alpha-aminooctanoic acid, **20:**177
    under amide alcohols, **20:**165
    under ester alcohols, **20:**165
    under mixed alcohol layers, **20:**160–163, 165–168, 172–174
Frenkel defects, **6:**9, 15–19
Frequency
  electron cyclotron, **18:**126–130, 131–133, 135–137, 150–151
  electron gyration, **18:**27
  electron plasma, **18:**34
  ion cyclotron, **18:**129–130
  lateral shift, **11:**233–235
  lower hybrid, **18:**41
  microwave, **18:**126, 128, 130–140, 143, 145–147, 151
  plasma, **18:**130–131, 133, 136, 139, 147, 150, 238
  resonant, in helical resonator, **18:**63
  upper hybrid, **18:**40
Frequency cutoff *see* Cutoff
Frequency dependency, dielectric response to, **11:**207–208, 269
Frequency dispersion, **11:**270–272
Frequency shift, quartz crystal microbalance, **24:**49
Frequency spectrum, dielectric response and, **11:**214
Fresnel amplitude transmission coefficient, **10:**177
Fresnel coefficient, **5:**4, 24; **19:**285
Fresnel equations, **1:**142–143; **5:**2
Fresnel reflectance
  matrix, **19:**286
  for single surface, **2:**240–241
Friction, dielectric *see* Dielectric friction
Frictional forces, by atomic force microscopy, **24:**97–98, 104, 105
Fringes, equal thickness, **2:**112
Front-surface mirrors, **12:**1–49, *see*

Front-surface mirrors, (*continued*) *also* Mirror coatings; Reflection-type mirrors
FSL *see* Focused stray light
Fuchs (Fuchs-Dingle) theory
  for aluminum and indium films, **6:**113
  failure, **6:**136
  free electron density and, **6:**134
  for polyvalent metallic films, **6:**112
  for thin films and wires, **6:**93–96, 98
Fuchsin, as parting agent, **7:**61–63
Fuchs-Sondheimer theory of resistivity, **2:**381

G

Gadolinium alloys
  binary phase diagram, **15:**146
  coercivity, **15:**175, 183, 186–187
  compensation temperature, **15:**180–182
  crystallization, **15:**167
  Curie temperature, **15:**180–182
  dipolar interaction in, **15:**185
  domain wall pinning in, **15:**185
  Kerr rotation, **15:**195–196, 199, 203
  magnetic anisotropy, **15:**174–176, 179–180
  magnetization, **15:**174–176, 179–180
  microstructure, **15:**161–166
  radial distribution function, **15:**160
Gadolinium iron film, ionized cluster beam, **13:**31–32
Gadolinium iron garnet, **5:**313–314
Gain
  photoconductive, **3:**197
  vs. frequency for CdS thin-film transistor, **2:**153–157
Gain-bandwidth product for thin-film transistor, **2:**164
Gallium
  radial distribution function measurements, **7:**21
  wires, resistivity size effects, **6:**123
Gallium-aluminum-arsenide alloys, in MBE, **11:**70–72
Gallium antimonide, in liquid phase epitaxy, **7:**169–170
Gallium arsenate film
  chemical vapor deposition, **5:**276–283
  evaporation, **5:**172–173
  optical absorption measurements, **5:**223
  semiconductor, **5:**150–151, 155
  texture variation, **5:**194
Gallium arsenide, **13:**188–189, 201, 201–209, 205, 219–221, 222; **15:**304
  atomic layer epitaxy, **19:**35–38
  crystal stoichiometry, **7:**140–142
  field effect transistors, **8:**307; **11:**75
  forward-biased p-n junction, **12:**135
  growth oscillations under CBE conditions, **19:**26–27
  heteropolar nature, **12:**128
  high-purity preparation, **7:**235–239
  homojunctions, infrared detectors, **21:**37
  impurity incorporation theory and, **7:**147–159
  in situ control of V/III ratio, **19:**27–29
  lattice mismatch with GaP, **12:**103–131
  in liquid phase epitaxy, **7:**117, 120–121, 126–137, 140–159
  luminescent efficiency in, **7:**155
  major dopants and, **7:**142–147

in MBE doping, **11:**52–69
minority carrier diffusion length, **7:**158
in monocrystalline films, **10:**245–253
negative resistance in, **7:**239–240
plasma oxidation, **12:**277–282
RA study of growth
under atmospheric pressure AP MOCVD, **19:**32–38
under chemical beam epitaxy, **19:**22–29
under MBE conditions, **19:**12–19
under migrationenhanced atomic layer epitaxy, **19:**19–22
under vacuum chemical epitaxy, **19:**22–29
surface reconstructions, **19:**32–34
ternary compounds and, **7:**184
waveguides, laterally bounded, **10:**248–249
zinc concentration in, **7:**155
Gallium arsenide/aluminum gallium arsenide
quantum-well infrared photodetectors, **21:**240, 241, 300–301
asymmetrical, **21:**116, 160–169
indirect band-gap, **21:**191–195
superlattice miniband, **21:**117, 184–191
VLWIR, **21:**224
quantum-well solar cells, **21:**313, 335–340, 347, 354, 360, 365
carrier escape efficiency, **21:**331–333
spectral response, **21:**323–324, 333
Gallium arsenide/aluminum indium phosphide
quantum-well infrared photodetectors, **21:**206–207
quantum-well solar cells, **21:**337, 347
Gallium arsenide-antimony alloys, in MBE, **11:**72–73
Gallium arsenide film
Ga: As ratio in, **12:**280
plasma-grown, **12:**277–282
arsenic deficiency in, **12:**280
composition and electrical characteristics, **12:**279–280
Gallium arsenide/gallium antimony system, **7:**196
Gallium arsenide/gallium indium phosphide, quantum-well infrared photodetectors, **21:**204–205
Gallium arsenide/gallium phosphide system, **7:**196–197
Gallium arsenide/indium gallium arsenide, **21:**313
quantum-well infrared photodetectors, **21:**207–213, 267
quantum-well solar cells, **21:**331–332, 335, 341–344, 347, 360
Gallium arsenide-phosphorus alloys, in MBE, **11:**72–73
Gallium nitride, in liquid phase epitaxy, **7:**169
Gallium nitride film
absorption coefficient, **13:**52
fractured edges, **13:**51–52
ionized cluster beam, **13:**51–53
Gallium oxide film, **8:**40
Gallium phosphate, chemical vapor deposition, **5:**276–283
Gallium phosphide
distribution coefficients of dopants in, **7:**165–169

Gallium phosphide (*continued*)
dopant effects in, **7:**161–164
dynodes, photomultiplier applications, **12:**141–144
green emitting diodes, **7:**225
lattice mismatch with GaAs, **12:**130–131
light-emitting diode, **7:**225–227
in liquid phase epitaxy, **7:**159–169
MOSFET, **13:**205
red-emitting diodes, **7:**223
Gallium phosphide film
ionized cluster beam, **13:**45
laser-induced etching, **13:**188–189
Gallium telluride, in liquid phase epitaxy, **7:**138
Garnet films, **10:**256
defects, **9:**289–293
growth, **5:**313–314
liquid phase epitaxy, **9:**283–284, 287–288
single-crystal, **9:**281–284
sputtering, **9:**282
Garnets
chemical vapor deposition, **9:**282
for magnetic bubbles, **9:**275–276
Gas
captured by condensed film, **1:**128
control, **14:**165
discharges, **3:**80–84
distributed ring, **18:**211, 224
entrapment, sputter deposition, **22:**152–156, 170
flow, **14:**3–6; **18:**212–215, 217, 219–226
hydrogen gas, **14:**2
igniter, **14:**170
molecules striking a surface per unit time, **1:**51
non-reactive, **14:**4
permeation through a solid, **1:**2–3
glass, **1:**7–8
metal, **1:**11
reactive, **14:**4–7
residual, **14:**2
effect on film properties, **4:**9–10
in vacuum evaporators, **1:**38–63, 51–63, 211, 279
scattering, **14:**135
Gaseous anodization, **6:**36–51
ellipsometry in, **6:**46
experimental technique, **6:**43–45
ionic current in, **6:**50
oxide growth mechanism, **6:**49–51
plasma production in, **6:**37–43
vs. wet anodization, **6:**44
Gas-film interface reaction, **1:**223
Gaskets, Viton A, **1:**35
Gas-phase crystallites, **24:**222
Gas-phase transport *see* Vapor transport
Gas/solid interactions, BTRM, **22:**194
Gas source MBE growth systems, **23:**23
Gate capacitance, effect of surface states on, **2:**167–168
Gating-type image tube, **12:**149–150
Gauge factor, for discontinuous films, **8:**139, 159–160
Gear 4 algorithm, **22:**132
Gel films, transformability, **5:**89
Generalized Langevin equation, **22:**135–138, 141–142, 170
Generation-recombination noise technique, photoelectric, **23:**196–198
Geometrical resonances, **18:**129, 145–146
Gerdy-Goddard dimer (GGD) model, **24:**211, 213, 214
Gerlach theory, **17:**350
Germane, **23:**33

Germanium
  abrupt transitions in, **23:**10, *12*
  amorphous, **8:**213
  band structure, **12:**57
  chemical vapor deposition, **5:**273–276
  coatings on, **2:**263–265
  as dopant, **7:**142; **11:**58–59, 65
  double layer films on, **2:**258, 299, 301, 302
  homojunctions, **21:**4, 5, 35–36
  incorporation, **23:**10
  joint density of state spectra, **12:**58
  laser-induced etching, **13:**188
  optical constants, **1:**162, 165
  on platinum, reflectance, **1:**179
  radial distribution function measurements, **7:**24–27
  single layer films, **2:**247–249
  surface segregation, **23:**10, 11
  transmittance, **8:**87; **12:**251–252
  vacuum ultraviolet reflectance, **1:**159–165
Germanium coatings, for metal-dielectric mirrors, **12:**42–43
Germanium film
  deposition, **5:**147
  experimental parameters, **5:**213
  as filter, **7:**48
  ionized acceptor concentration, **5:**212
  sputtering, **5:**181
Germanium oxide
  optical constants, **1:**163, 165
  reflectance, **1:**161–162, 164
Germanium oxide film, **8:**40
Germanium-silicon alloys, **23:**2
Gettering, **14:**14
GexSi1-x
  channel MOSFET for, **23:**68–73
  doping, **23:**12–15
  epitaxial layer growth, **23:**1–74
    by chemical vapor disposition, **23:**21–51
    kinetics, **23:**37–43
    low-temperature, **23:**22–23, 27–37
  high electron mobility transistor, **23:**70
  molecular beam epitaxy growth, **23:**6–21
GGD *see* Gerdy-Goddard dimer
Ginsberg-Landau theory, **1:**256–257
GIXD *see* Grazing-incidence X-ray diffraction
Glancing incidence reflection electron diffraction, **11:**42–44
Glass
  composition, **1:**4–5
  corrosion-sensitive, **8:**40
  formation, **5:**98; **15:**146
  heat-resistant, **5:**133
  outgassing, **1:**9–11
  spectral transmittance, **5:**99
  sun-shielding, **5:**127–129
  vacuum characteristics, **1:**4–9
Glass films, **4:**257–324
  application to semiconductor devices, **4:**316–320
  boiling water and humidity tests, **4:**277–279
  borosilicate, **4:**309
  deposited, **4:**309–316
  electrical properties, **4:**289–316
  formed during diffusion, **4:**305–309
  frit fusion, **4:**317–318, 319
  from sedimented glass powders, **4:**266–267
  phosphosilicate, **4:**262, 305–309
  preparation, **4:**259–270
    pyrolytic decomposition, **4:**263–266, 318

Glass films, (*continued*)
radio-frequency sputtered, **4:**268–269, 318–319
reactively sputtered, **4:**267–268
space-charge buildup in, **4:**310–311
vacuum evaporated, **4:**263
properties, **4:**270–289
and RF sputtered, structural differences, **4:**274–277
Global rate expressions, low-pressure deposition processes, **22:**185
Glow-discharge cleaning, substrates, **13:**125
Glow-discharge nitriding, **12:**286–289
Glow-discharge sputtering, **3:**80–87
cathode-to-anode spacing, **3:**85–86
magnetic field effect, **3:**90–92
optimum pressure, **3:**84–85
thermionically supported, **3:**88
Glyptal, in filter fabrication, **7:**81
Goddard Space Flight Center, **10:**158–159
Gold, **16:**327–328
adsorption of alkanethiol SAMs onto, **24:**45–77
alkanethiols on gold nanoparticles, **24:**154–167, 175, 222
atomic force microscopy studies, **24:**7, 211, 222
diffusion
in copper, **6:**337
in silver, **6:**336
evaporation, **12:**15
growth
on silver, **4:**142–143
on sodium chloride, **4:**176–180
(111) nuclei, rapid growth, **4:**164–165
polymeric monolayers attached to, **24:**34
preparation, **24:**4–6, 117–118
reflectance, in vacuum ultraviolet, **1:**165–168
Gold aluminate, formation, **6:**301
Gold-aluminum diffusion couple, **6:**321–324
Gold-aluminum film, **6:**313
phase diagram, **6:**318
phase-forming systems, **6:**315–326
reflectivity changes, **6:**316
resistance changes, **6:**319–320
stability, **6:**325
Gold-black deposits, in spectrally selective surfaces, **10:**59
Gold bump contacts, in deep-etched targets, **8:**282
Gold bump structures, electroplated, **8:**279
Gold cathodes, **5:**40–42
Gold coatings
for extreme ultraviolet astronomy, **10:**93–94
groove density and, **11:**5
holes or voids in, **11:**5–7
Gold-copper multilayer film, **6:**310
Gold electrode
blocking component, **20:**190–192
surface treatment, **20:**188–189
Gold film, **6:**194
absorption and photocurrent in, **5:**41–42
chemical vapor deposition, **5:**308
electrical properties, **8:**102–104
ion-beam processing, **13:**144–145
isotransmittance, **9:**114
nucleation, **9:**43
optical parameters, **6:**185, 192, 194
optical properties, **6:**199–200
oxidation experiments, **6:**243–244

packing density, as function of acceleration voltage, **13:**25–26
Parlodion films and, **7:**70
photoemission, **5:**40
reflectance gratings, **11:**3–4
resistance-strain coefficient vs sheet resistance, **8:**159
resistivity, **6:**106–108
as function of thickness, **13:**25–27
room temperature resistivity, **6:**109
temperature coefficient, **6:**110
specular reflection, **6:**106
sputtered, **3:**109–110
structural disorder tensile properties, **2:**54–58
thickness dependence, **6:**108
thin films, **16:**253–255
Gold-lead films, reflectivity in, **6:**326–327
Gold mirror coatings
optical constants for, **12:**19
reflectance, **12:**17, 19
Gold-nickel system, diffusion in, **6:**337
Gold oxides, **14:**16
Gold-silicon monoxide film, **8:**124–125, 160
Gold-sulfur interaction, **24:**2, 3
Gold-thiol interface
chemistry, **24:**52–55
structure, **24:**135–137
Goniometer-reflectometer, in reflectance measurements, **12:**4
Grain boundary, **17:**34
critical current, **16:**9, 10
grain alignment and Jc, **16:**9, 10, 11, 23, 90
Josephson junction, **16:**9, 50, 121
in lattice migration, **7:**265
ledge defects and, **7:**300
mobile defect density in, **7:**301
weak link, **16:**9, 45, 48–50, 53, 67, 71, 97, 110, 119, 121
Grain boundary diffusion, **16:**236
Grain boundary diffusion coefficient
aluminum thin films, **16:**249
copper thin films, **16:**253
gold thin films, **16:**254
indium thin films, **16:**258, 259
tin thin films, **16:**256, 257
Grain boundary grooving, **16:**247
Grain size
dependence on condensation rate, **1:**212–213
tensile strength and, **3:**266–267
Graphite, pyrolytic, in vapor deposition process, **5:**258–259
Grating equation, Littrow mounting, **4:**56
Gratings, **15:**234–235, *see also* Diffraction gratings
anomalies, in diffraction gratings, **11:**21
coupling by, **9:**221
Gravimetry, **3:**27–30
Grazing angle reflection, **15:**272
Grazing-incidence X-ray diffraction (GIXD), **24:**186, 192, 197–198
Green's function, **21:**13
Griffith's energy criterion, **16:**276, 280, 282, 284, 323
GROFILMS model, **22:**30, 70, 72, 74–76
Groove efficiency curve, for diffraction gratings, **11:**26

Ground plane, and magnetostatic propagation, **15:**31, 35, 72–76
Group II-VI compounds
activation of films, **3:**174–175
applications, **3:**166
chemical deposition, **3:**194–196
future potential, **3:**207–208
physical properties, **3:**196–207
preparation of evaporated, **3:**168–174
spectral response, **3:**197–198
Group III-V compounds
films, properties and applications, **7:**115–247
in MBE doping, **11:**66–69
Group IV-VI compounds
layers
in alkali halides, **11:**113–115
in insulating substrates, **11:**113–122
vacuum deposition on, **11:**108–113
semiconductor photodiodes, **11:**127–157
maximum blackbody photon flux, **11:**127
semiconductors
on BaF_2 and SrF_2 substrates, **11:**115–122
epitaxial growth, **11:**108
growth of on alkali halide substrates, **11:**107
infrared detectors and, **11:**106
quantum efficiencies, **11:**132–140
static dielectric constants, **11:**168
surface effects, **11:**153–157
thin films, **11:**105–128
surface effects, **11:**153–157
Growth
by molecular beam epitaxy, **23:**6–21
of multiple-quantum-well structures, **23:**184–185
commensurate, **23:**3
epitaxial layer
by MBE, maximum temperature for, **23:**18–21
of GexSi1-x, **23:**1–74
incommensurate, **23:**3
of lattice-matched multilayered structures by LP-MOCVD, **19:**29–31
of lattice-mismatched III-V and II-VI structures, **19:**38–43
lead growth on (111) surface of silver, **4:**138–139
of thin films, **1:**207–215; **3:**212–214
Growth layer, **4:**204–209
Growth rate
silicon dioxide film, **18:**217–226
silicon nitride film, **18:**209, 213–215, 217
Guided light modes, **9:**168–170
excitation, **9:**241–242
Guided light waves, roughness factor in, **9:**254
Gunn devices, millimeter-wave, **11:**77
Gurzhi model, of thin-film resistivity, **6:**99, 127
Gyromagnetic ratio, **15:**27, 126

H

Hafnium carbide, spectral reflectance, **10:**29–30
Hafnium dioxide, for aluminum mirror coatings, **12:**35
Hafnium fluoride, microstructural and optical function, **19:**222, 229

Hafnium oxide, microstructure and optical function, **19:**223, 225, 239
Half-frequency spin waves
  generated at high powers, **15:**92
  limitation on resonator power, **15:**81
  optical observation, **15:**95
  threshold for, **15:**94–95
  used in *S/N* enhancer, **15:**96
Halide disproportionation, in chemical vapor deposition, **5:**275
Halide reduction
  in chemical vapor deposition, **5:**273–274
  in semiconductor films, **5:**147–148
Hall coefficient, temperature dependence, **11:**122
Hall effect, **8:**175, 188, 240
  in PbSe films, **3:**296–297
  in PbS films, **3:**295
  in PbTe films, **3:**296
  in semiconductor films, **5:**209–219, 222–223
Hall mobility, **6:**140–141; **8:**174, 187
  CdS films, **2:**173–176
  temperature dependence, **11:**122
Halogenides, as dielectric film materials, **8:**30–34
Hamilton-Jacobi equation, **22:**83
  boundary motion, **22:**84–85, 98–99
  characteristics, **22:**87–89
  entropy conditions, **22:**91–97
  spurious solution, **22:**85–87
  viscosity solutions, **22:**89–91
Hamming's method, **5:**61
Hard coatings, **17:**100
  composition
    chemical, **17:**111
    phase, **17:**118
  microstructure, **17:**119, 121–129
  properties, **17:**100
    electrical, **17:**129–130
    mechanical, **17:**129
    optical, **17:**129, 131
Hard films, in multilayer structures, **6:**209–210
Hard materials, **17:**101
  covalent, **17:**103–104
  ionic, **17:**104
  metallic, **17:**102, 104
  physical properties, **17:**101–105
Hardness, **17:**3, 46, 47, 59–71
  ceramic/ceramic, **17:**63, 64
  metal/metal, **17:**60–63
  theory, **17:**65–71
Harmonic resonance
  higher, **18:**129
  second (SHR), **18:**142–145, 189, 190, 209–216
  upper (UHR), **18:**129, 139–141
Hartree-Fock, **15:**226, 240
Hartree potential, **21:**98
Hautman-Klein united-atom (HKUA) model, **24:**209–210, 211, 213–216, 221
HBAR *see* High-overtone bulk acoustic resonators
HBT *see* Heterojunction bipolar transistor
Head group, **24:**112, 123, 135
  interfacial tension difference, **20:**373–375
  quenching rate and, **20:**252–253
Head group-substrate interaction, silver substrate, **24:**127–128
Heat of formation, intermediate states and atomization, **13:**118
Heat mirrors, chemical deposition process for, **12:**228
HEED *see* High-energy electron diffraction

Helical anisotropy, in multilayer films, **6:**288–290
Helical resonator discharge, **18:**12, 60, 62
Helicon discharge, **18:**12, 40
antenna, **18:**42, 44
configurations, **18:**41
coupling, **18:**46
density measurements, **18:**49
design example, **18:**51
matching network, **18:**42
modes, **18:**42
Helium
chemical vapor deposition, **5:**313
diffraction studies, **24:**112–144
film growth in, **5:**295
in high-power devices, **5:**225
liquid, pumping, **1:**21
permeation through glass, **1:**8
Helmholtz double layer, **17:**321
Helmholtz pair, **18:**157–158, 162–163, 170–172, 211
HEMT *see* High electron mobility transistor
Henry's law, **6:**309
Herpin index, **8:**88
Heteroepitaxial films, **5:**186, 228, *see also* Epitaxial films
Heteroepitaxy
in growth plane, **7:**123–124
lattice mismatch in, **7:**126–135
Heterogeneous SAMs, structure, **24:**89, 90
Heterojunction bipolar transistor (HBT), **23:**51–68
CVD-grown, **23:**55–56, 59–64
fabrication
BICMOS process, **23:**60, 62–64, 68, 69
disposable emitter mandrel process, **23:**60, 62–64
future prospects, **23:**68
history, **23:**54–56
MBE-grown, **23:**55, 56–59
operation, **23:**51–54
Heterojunction contacts, **13:**219–221
Heterojunction internal photoemission (HIP) detectors, **23:**144–153
barrier height, **23:**145–148
quantum efficiency, **23:**148–153
Heterojunction laser diodes, **7:**230
Heterojunction MOSFET, **23:**70, 71–73
Heterojunctions, **5:**152; **21:**2, 81–84, 90–101, 117–123, 241–246, *see also* Quantum-well infrared photodetectors
GexSi1-x, **23:**3–5
in light-emitting diode, **7:**221–222
MBE in, **11:**84–86
Heterostructures, or multilayer, **16:**28, 63, 76–78, 111–114, *see also* Superlattice
Hexadecanethiolate SAM, corrosion prevention, **24:**240
Hexadecane thiols, **24:**212–213, 221
Hexaferrites, for magnetic bubbles, **9:**275
Hexamethyldisiloxane, **10:**241
High critical temperature superconducting thin films, **16:**2–5
applications, **16:**100–119
antenna, **16:**108
band-pass filters, **16:**101, 105, 106
band-reject filters, **16:**105, 106
coil or flux transformer, **16:**75–78, 117, 118
computer switches, high speed, **16:**101, 123
delay lines, **16:**106–108
hybrid circuits, **16:**108

Josephson junctions, **16:**101, 110–115
microwave components, **16:**39, 95, 101–108
optical sensors, **16:**101
oscillators, **16:**101, 109, 110
passive interconnections, **16:**95, 101, 108–109
radiation detectors, **16:**101, 109, 110
resonators and oscillators, **16:**101–105
SQUIDs, **16:**76, 101, 114–123
transistors, **16:**123
characterization, **16:**5–14
electrical resistivity, **16:**7, 8, 12, 13
direct current (dc) resistivity, **16:**7, 8, 12
microwave surface resistance, **16:**12, 13, 38, 39, 48, 67, 75, 100, 103, 104, 106, 108
normal-state resistivity, **16:**7, 8, 24, 25, 65, 70, 71, 73, 74, 101
laser ablation, **16:**55–80
MOCVD, **16:**81–94
MOD, **16:**95–100
sputtering and ion-beam deposition, **16:**14–39
vacuum evaporation, **16:**39–54
High-density phase, **24:**93
High-density theory, workfunction calculations, **21:**10–12
High-detectivity infrared hot-electron transistors, **21:**260–269
High electron mobility transistor (HEMT), **23:**70
High-energy electron diffraction (HEED), **11:**42
HEED pattern, ZnS: Mn films, **13:**44
High-energy radiation, dielectric films and, **8:**4, 14–17
High-energy surfaces, **24:**13
Higher harmonic resonance, **18:**129
High-field conduction
low-temperature, **8:**226–228
through traps, **8:**206–207
High-overtone bulk acoustic resonators (HBAR), **17:**188
effect of rf bias on shear mode transduction, **17:**191, 192
factors influencing *Q*, **17:**194, 195, 196
longitudinal-mode HBAR, **17:**188
shear-mode HBAR, **17:**190
temperature compensation, **17:**192, 194
tunable resonator, **17:**191, 193
High-pass filters, **21:**258
High-power devices, **15:**95, *see also* Magnetostatic devices
High-sensitivity photocathodes, chemical analysis, **12:**99
High-speed steel tools, wear, **13:**103
High temperature deposition, SIMBAD modeling, **22:**49–53
High-vacuum evaporator, in reflecting coating deposition, **12:**8
High-yield photolithography, in thin film deposition, **8:**255–261
HIP *see* Homojunction internal photoemission detectors
HKUA model *see* Hautman-Klein united-atom model
Hole-burning spectroscopy
apparatus, **20:**296–299
energy transfer, **20:**299–300
frequency shift from environment, **20:**306–308

Hole-burning spectroscopy (*continued*)
hole shape in magnetic field, **20:**308–310
hole width, **20:**299–306
Langmuir-Blodgett films, **20:**293–314
shifts in magnetic field, **20:**308–311
Stark effect, **20:**308–311
Hole concept, in semiconductors, **8:**178–179
Hole transport, minibands, **21:**88–91
Hollandites
dielectric loss, **11:**261–264
strong dispersion in, **11:**270
Hollow cathode, ion sources, **13:**112–114
Holographic etching, **13:**161–162
Holographic gratings, **11:**9–11
blazed concave, **11:**23
energy distribution, **11:**23
supporting blank and, **11:**10
Homoepitaxial films, growth in, **5:**196–197
Homoepitaxy, doping variation defects in, **7:**125–126
Homogeneous films, **5:**58; **8:**51–96
conditions for, **5:**88–89
metallic, **6:**173–174
formulas, **6:**155–170
Homogeneous solar cells, **21:**325–327
Homojunction internal photoemission detectors, **21:**3–7
gallium arsenide, **21:**37
germanium, **21:**4, 5, 34–36
indium gallium arsenide, **21:**36–37
silicon
delta doped potential well detectors, **21:**32–33
Type I HIWIP detectors, **21:**14–19
Type II HIWIP detectors, **21:**19–31
Type III detectors, **21:**31–32
uncooled avalanche injection detectors, **21:**33–34
workfunction theory, **21:**7–14
Homojunction laser diodes, **7:**228–230
Homojunctions, **21:**2, 3ff., 4, 5, 30, 35–36, 37, 40–51, 50–59, 59–68, 239–240, *see also* Homojunction internal photoemission detectors
Hopf bifurcation, **21:**69
Hopping
in amorphous solids, **8:**176–177
at Fermi level, **8:**212–217
donor-trap, **8:**207–208
in elemental materials, **8:**216
hole concept and, **8:**178–179
ohmic and nonohmic contacts, **8:**196–199
transport and, **8:**245
trap density, **8:**215
Hopping-charge carriers, **11:**259
Hopping electronic systems, **11:**290–291
Hopping model, screened, **11:**296–302
Hopping transition, energy loss and storage in, **11:**299
Horizontal Bloch line, in magnetic domain dynamics, **9:**269–270
Horn antenna, **18:**153
Hot-electron phototransistors
excitation hot-electron spectroscope, **21:**246–255
infrared hot-electron transistors, **21:**255–289
quantum-well infrared photodetectors

intersubband transitions, **21:**241–246
multicolor, **21:**289–294
Hot-electrons
collection, **4:**330
generation, **4:**357–359
transport, **4:**325–371
Hot-electron spectroscopy *see* Excitation hot-electron spectroscopy
Hot-well vacuum deposition technique, **11:**111–112
Houston theory, electron mean free path, **1:**235
HPFS, **21:**258
Hund's rule, **15:**144
Husstad-Lothe model, of thin-film conductivity, **6:**98–99
Hybrid circuits, **16:**108
Hydride pyrolysis
in chemical vapor deposition, **5:**270, 274–275
in semiconductor films, **5:**148–149
Hydride reactants, layer growth of GexSi1-x,, **23:**37–41
Hydrocarbon gas, in plasma carburizing, **12:**284–285
Hydrocarbon packing structure, **24:**133–134
Hydrocarbon tilt structure, **24:**132–134
Hydrogen
GexSi1-x growth and, **23:**27–30, 41
hydrogen content, **18:**209, 213, 215
as transport agent, **5:**285
Hydrogenated amorphous silicon, **19:**61–84
infrared spectra, **15:**283–287
band assignments, **15:**284, 286
hydrogen incorporation and, **15:**283
oxidation and, **15:**285
silicon-fluorine alloys and, **15:**285
ion-beam processing, **13:**142–143
optical function, **19:**70–73
silicon-germanium alloys and, **15:**286
silicon-tin alloys and, **15:**284
synthesis by activated reactive evaporation, **13:**96–97
Hydrogenated amorphous silicon-carbon alloys, **19:**85
hydrogen dilution of reactive gases, **19:**88–93
optical gaps from real time observations, **19:**85–88
Hydrogen bond, **20:**45–64
in alpha-amino acids, **20:**178
in biotin-streptavidin surface, **20:**210
in film matrices, **20:**42–68
in ice nucleation, **20:**164
in monolayer, **20:**164
substrate/film interaction, **20:**377
templates, **20:**42
Hydrogen bonded films
p-aminobenzoic acid intercalation, **20:**48–62
defects, **20:**47–48
infrared characterization, **20:**46–48
matrices, **20:**42
Hydrogen discharge lamp, **1:**130
Hydrogen sulfide
on silver film, **12:**37–38
in spray pyrolysis, **12:**202
Hydrogen termination molecular beam epitaxy system, **23:**9
Hydrophobic interaction, in biotin-blocked streptavidin attraction, **20:**216–217
Hydroquinone, **24:**23

Hysteresis, **15:**249–251
high field rotational, **1:**304
pressure, **14:**3–6, 12–14, 53
reactive sputtering, **17:**111, 114–116
thin-film transistors, **2:**169
voltage, **14:**36
Hysteresis loops, **1:**301; **16:**148, 159
in exchange anisotropy, **6:**218–220
in multilayer films, **6:**256, 266–267

I

Ice, structure, **20:**147–148
Ice nucleation, **20:**149–158
alcohol chain length, **20:**155–157
on amino acid, **20:**174–177
by organic solvents, **20:**170–174
catalysts, **20:**145
factors, **20:**149
film solvent and, **20:**172–174
headgroup orientation, **20:**164
templates, **20:**145–181
under immiscible alcohols, **20:**162–163
under miscible alcohols, **20:**160–162
under modified amide-alcohol layers, **20:**164–169
ICP *see* Inductive discharge
Ideal contacts, **13:**200
Ideal Debye response, **11:**234
Ideal Debye system, **11:**275
Ideality factor, **21:**348
Ideal photodiode, quantum efficiency, **11:**125
Ideal quantum efficiency, **21:**16
IHET *see* Infrared hot-electron transistors
ILEED, **9:**215
Image charge, in silicon, **4:**296–297
Image converters, **12:**150
Image intensifiers
characteristics, **12:**150–152
electrostatic focusing, **12:**147–148
magnetic focusing, **12:**146–147
mechanical transfer function for, **12:**151–152
microchannel plate type, **12:**148–149
photoemissive materials for, **12:**145–149
proximity focusing, **12:**145–146
Image isocon, **12:**156–157
Image orthicon, **12:**156
Image projection, laser-induced etching, **13:**159
Image sensor, **20:**288–290
Image tubes
gating tubes as, **12:**149–150
image converters and, **12:**150
image intensifiers and, **12:**150–152
photoemissive materials for, **12:**145–157
signal-generating tubes and, **12:**152–157
Imaging arrays, quantum-well infrared photodetectors, **21:**219–228
Imaging sensors, internal emission, **23:**87, 88
Immiscible systems, **17:**14
IMPATT (impact ionization avalanche transit time) diodes, **8:**295, 301–302; **11:**77
Impedance, polarized plane wave, **1:**92
Imperfection, oriented structural, **1:**286–289
IMS deposition *see* Ionized magnetron sputter deposition
Incidence angle
polarization values for, **12:**23
reflectance as function, **12:**18–23

Incident energy, **22:**150
Incoming flux, **22:**101–102
Incommensurate growth, **23:**3
Inconel, plasma nitriding, **12:**288
Indices condition
  four-layer coatings, **2:**272, 274
  infrared transparent materials, high index, **2:**273–274
Indirect band-gap quantum-well infrared photodetectors, **21:**191–195
Indirect gap semiconductor, **15:**236
Indium, low-temperature electrical resistivity, **6:**120
Indium antimonide film, dendritic, **5:**185, 216
Indium arsenate film
  doping level, **5:**217
  electron mobility in, **5:**210
Indium arsenide
  in liquid phase epitaxy, **7:**170
  M-n-InAs contact, band diagram, **13:**220
  M-n-InAs *n*-GaAs contact, band diagram, **13:**220
  M-n-InAs-*n*-GaInAs-*n*-GaAs contact, band diagram, **13:**220–221
Indium arsenide-gallium arsenide system, in liquid phase epitaxy, **7:**189–191
Indium arsenide-indium antimony system, in liquid phase epitaxy, **7:**198–201
Indium arsenide-indium phosphide system, in liquid phase epitaxy, **7:**198–201
Indium film
  critical temperature with thickness, **1:**241–242
  effect of gases on, **1:**52–53
  as filters, **7:**48
  size-dependent resistivity, **6:**113, 115
  thin films, **16:**258–259
    diffusion coefficient, **16:**258, 259
    electromigration effective charge, **16:**259
Indium-gallium-arsenic alloys, smaller band gap, **12:**122–123
Indium-gallium-arsenic-phosphorus quaternary systems, **12:**123
Indium-gallium arsenide alloy cathodes, thermalized electron distribution in, **12:**126
Indium gallium arsenide/gallium arsenide
  quantum-well infrared photodetectors, **21:**207–213, 267
  quantum-well solar cells, **21:**331–332, 335, 341–344, 347, 360
Indium gallium arsenide homojunctions, infrared detectors, **21:**36–37
Indium gallium arsenide/indium aluminum arsenide, quantum-well infrared photodetectors, **21:**196
Indium gallium arsenide/indium phosphide
  quantum-well infrared photodetectors, **21:**196–201
    p-doped, **21:**213–215
  quantum-well solar cells, spectral response, **21:**344–347
Indium gallium arsenide phosphide/indium phosphide, quantum-well infrared photodetectors, **21:**201–204

Indium oxide, electrical and optical properties, **9:**28–30
Indium oxide coatings, **10:**52
Indium oxide film, **8:**39
  from pyrolysis of indium acetylacetonate, **9:**39
  properties, **9:**41
  reactively sputtered, **9:**51
Indium phosphide
  in liquid phase epitaxy, **7:**170–171
  spatial frequency response of gratings, **13:**164–165, 165, 167
Indium phosphide-gallium phosphide system, in liquid phase epitaxy, **7:**193–196
Indium tin oxide, **14:**44, 52, 55–56, 62
  coatings, **10:**52
  ion-beam processing, **13:**143–144
Induced transmission filter, **9:**74
Induction heating, **1:**127
Inductive discharge, **18:**12, 52
  coil coupling, **18:**55
  configurations, **18:**52
  design example, **18:**57
  efficiency, **18:**58
  low density, **18:**58
  power absorption, **18:**54
  regimes, **18:**54
Inelastic electron scattering, cross-section, **2:**77–79
Inert gas atoms, implanted, **6:**24–26
Inflection coefficients, for inhomogeneous media, **8:**89
Infrared absorption, **15:**229–230
  description of bands, **15:**267
  oxides, **13:**122
  source, **15:**266
  symmetry, **15:**268
Infrared detection, **15:**220, 229–230
  Group IV-VI compound semiconductors, **11:**106–107
  photoconductive, **15:**230–231, 236, 259
  photovoltaic, **15:**230–231, 259
  thin film epitaxial layers on silicon, **23:**115–168
Infrared detectors, **10:**168; **21:**3–7, 68–71; **23:**120–127
  epitaxial devices, **23:**141–162
  extrinsic detectors, **21:**4
  FIR, **21:**35–37
  heterojunction detectors
    infrared hot-electron transistors, **21:**241–308
    quantum-well infrared photodetectors, **21:**101–110, 114–231, 240–300
  homojunction detectors
    charge storage mode detectors, **21:**38–40
    HIP detectors, **21:**5–37
    mercury cadmium telluride homojunctions, **21:**40–41, 239–240
    pulse mode detectors, **21:**49–50, 70
  intrinsic detectors, **21:**4, 114
  LWIR, **21:**221–223
  MWIR, **21:**191–195
  silicon-based, **23:**133–135
  types, **21:**4
  VLWIR, **21:**223–228
Infrared (IR) emission
  far *see* Far infrared (FIR) emission
  interband-type, **23:**220–238
  intersubband
    with sequential resonant tunneling, **23:**244–247
    transition processes, **23:**238–262
  long wavelength, **23:**256–274

middle wavelength, **23:**262–274
quantum-well devices for, **23:**217–298
Infrared emitter, **15:**220
Infrared filter, **5:**53
Infrared hot-electron transistors, **21:**241, 255–256, 307–308
dark currents, **21:**257–258
figures of merit, **21:**256–257
filtering, **21:**258–259, 275–282
functions, **21:**258–260
high-detectivity, **21:**260–269
high-temperature operation, **21:**275–282
intersubband transitions, **21:**241–246
long-wavelength, **21:**282–289
multicolor, **21:**300–306
small noise equivalent temperature difference, **21:**269–275
Infrared imaging arrays, internal emission, platinum silicide, **23:**83–111
Infrared layer, thickness measurement, **20:**46
Infrared optical materials, coatings for, **2:**293–303
Infrared photodiodes, properties, **11:**122
Infrared photoelectron tunneling spectroscopy, **21:**290–291
Infrared radiation, **15:**234
Infrared reflectance data, for SiO-coated aluminum mirrors, **12:**29–30
Infrared reflectivities, BeO films, **13:**48–49
Infrared signals, thermal, requirements imposed by, **23:**84–86
Infrared spectra, **23:**117–120
of amorphous solids, **15:**268
of crystals, **15:**268
Infrared spectroscopy, glass films, **4:**270–271
Infrared staring sensors
dynamic range, **23:**84–85
freedom from excess low-frequency noise, **23:**86
photoresponse, uniformity, **23:**85–86
pre–1975, **23:**86–87
Infrared transmission, **15:**272
in superconducting films, **1:**265–267
Inhomogeneity, **19:**193–95, 250–51
in dielectric films, **8:**7
internal fields and, **8:**183–185
measurement in thin films, **8:**81–83
phase separation in, **8:**184
transparent films, **19:**193–95, 250–51
optical characterization, **19:**191, 249
optical functions of ferroelectric films, **19:**261, 268
optical functions of fluoride films, **19:**236
optical functions of oxide films, **19:**236
Inhomogeneous films, **6:**173, 175; **8:**51–96
antireflection coatings and, **8:**85–88
applications, **8:**83–93
approximate methods, **8:**77–81
Maxwell equations, **8:**73
multilayer film as, **8:**88–92
reflectance, **8:**84, 87–88
reflection and transmission coefficients, **8:**72–81
refractive index, **8:**81–83
specific index variations, **8:**81
for sunglasses, **8:**92–93

Inhomogeneous films, (*continued*)
TE and TM waves, **8:**73–76
Inorganic film deposition, **12:**162–232, *see also* Chemical solution deposition; Spray pyrolysis
In-plane anisotropy, **16:**154, 155, 161, 166, 181, 183, 189, 198, 200
In situ process, **16:**18–28, 19, 29–32, 51, 52, 54, 61–78, 89–92, 93, 112, 113, *see also* Low-temperature process
Insoluble monolayers, formation and spreading, **7:**315–317, *see also* Monolayers
Instron, **16:**296, 304, 319
Insulated-gate thin-film transistors, **2:**147–192
circuit applications, **2:**187–190
description of structure, **2:**151–153
drain characteristic, **2:**159–165
enhancement type, **2:**153–154
fabrication, **2:**178–183
physical processes, **2:**157–178
Insulating films
crystal film, **10:**253–258
gas-phase deposition, **3:**131–164
Insulators
sputtered thin film, **3:**110–113
sputtering, **3:**100–102
Insulator-semiconductor systems, MBE and, **11:**89
Integral interference filter, in thin-film photodiodes, **11:**188–189
Integrated circuits, tuning devices for, **8:**316–319
Integrated devices structures, **17:**292–294
FEMFET, $LiNbO_3$ gate, **17:**292–293
hydrogen-terminated Si surface, **17:**293
MgAl2O4 buffer layer, **17:**293
polysilicon membrane, **17:**293
pyroelectric detector, **17:**293–294
Integrated electronics, issues, **17:**289–294
Integrated optical systems, MBE and, **11:**80
Integrated optics
defined, **10:**228
non-single-crystal films for, **10:**235–238
thin films for, **10:**227–258
Integrating sphere, **10:**210
Intensity absorptance, **1:**77
Interaction energy
crystallinity, **20:**155
water-alpha-amino acids, **20:**179
Interaction phenomena, in multilayer films, **6:**208–250
Interband, **15:**238–239
Interband transitions, **6:**154
Intercalation, **20:**48–49
p-aminobenzoic acid, **20:**44
copper, **20:**44
lateral transport in, **20:**49–50
matrix stability, **20:**51
p-nitroaniline, **20:**54
reversal conditions, **20:**49
Interdiffusion coefficient, **6:**303
Interdigital transducer, magnetostatic, **15:**43
Interface and stability effects, **17:**291–292
electrode effects, **17:**291
La0,5Sr0,5CoO_3 films, **17:**292
lattice matched oxides, **17:**292
refractory compound buffer, **17:**291
superconducting YBCO films, **17:**292
Interface structure, **4:**106–107
Interfacial adhesion, **16:**267
Interfacial force microscopy, **24:**221

Interference, **9:**16–18
Interference cathodes, **5:**15–43
object, **5:**43–44
yield and spectral response, **5:**43–44
Interference coatings, for spectrally selective surfaces, **10:**54–57
Interference effects, absorbing coatings and, **5:**124–126
Interference films, application, **5:**47–48
Interference filters
metal dielectric *see* Metal-dielectric interference filters
multilayer *see* Multi-layer interference filters
multistage, **9:**74
Interference fringes, **3:**17–18, 35–38
Interference function, **24:**116
Interference microscopy, **4:**271–272
Interference photocathodes, **5:**1–44
Interference techniques, laser-induced etching, **13:**161–163
Interferogram, **15:**270
Interferometer
basic equation, **4:**28, 35–36
optical parameters of metallic films, **6:**187
Interisland separations, trapping centers for, **9:**6
Intermetallic compounds, **6:**305–307, 318
Intermetallic diffusion, prevention, **11:**16
Internal emission, **23:**88–100
field, **23:**93–100
in photodetectors, **23:**88–92
themionic, **23:**92–93
model, **23:**93–100
process, **23:**88–100
tunneling and, **23:**93–96
Internal emission imaging sensors, **23:**87, 88
Internal field, **11:**313
Internal photoemission, **23:**88–92
Internal photoemission detectors, **21:**4
Internal quantum efficiency, **21:**21–22
Internal reflection spectroscopy, **15:**273
Internal workfunction, **21:**4–6
Interphase, **20:**89–91
Interplanetary Explorer, **10:**159, 161–163
Interpulse time intervals, **21:**3
nonlinear dynamics, **21:**50–59
spiketrains, **21:**41, 44–47
Intersubband infrared emission, **15:**229
device emitting, radiative transitions in, **23:**239, 240–243
first 5m, graded-gap injection structures, **23:**263–266
from superlattices, **23:**238–239
with sequential resonant tunneling, **23:**244–247
Intersubband photoconductors, **21:**4
Intersubband transitions
quantum wells, **21:**78–81, 117–123, 241–246
p-type delta-doped Si, **21:**95–101
p-type SiGe/Si, **21:**90–95
Intervalence band transition, quantum wells, **21:**79
Interwell photon-assisted tunneling, **23:**240
Intraband, **15:**222, 229, 238–39
Intrinsic infrared detectors, **21:**4, 114
Inversion layer, **2:**157
Ion-assisted deposition, **18:**243–249
effect of ion energy, **18:**244
effect of ion-to-vapor-flux, **18:**245

Ion-assisted deposition, (*continued*)
microstructural changes, **18:**246
Ion beam, **14:**22–26, 51
self-sputtering, **17:**94
Ion-beam deposition, **2:**412; **16:**14, 16, 17, 27, *see also* Sputtering
Bi-Sr-Ca-Cu-O thin film, **16:**29, 32
in plasma chemical vapor deposition, **12:**246–248
sequential, **16:**17
Y-Ba-Cu-O thin film, **16:**17, 18, 26
Ion-beam processing, **13:**110–145
Ion bombarding energy, **18:**7, 9, 11, 19, 71, 81
Ion bombardment, **14:**135; **17:**82, 121, 136; **18:**157, 198
back-sputtering, **14:**135
condensing atom, ratio, **17:**90, 99, 125
microstructure, **17:**121, 127
sputter deposition, **22:**152–156, 170
sputtering, **17:**113
step coverage and, **22:**167–169
surface effects, **17:**82, 83, 84
Ion cleaning, **17:**133–136
Ion current, **14:**161
Ion emission, **14:**158, 160
Ion energy, **14:**4, 34; **18:**192, 198
average, **18:**267
control, **18:**90
defect formation, **17:**122
per deposited atom, **18:**267
surface effects, **17:**83
threshold, **18:**71
transverse, **18:**87
Ion etching (plating), **8:**24, 265–266
Ion-exchanged films, **10:**237
Ion field microscopy, **1:**205
Ion flux, **14:**5, 7, 158; **17:**99, 113, 121, 128, 137; **22:**107–110, 111
Ion guns, electromagnetic, **6:**43
Ionic binding constant, monolayers, **20:**197
Ionic conduction process, **6:**34
activation distance, **6:**18–19
amorphous oxide, **6:**19
bulk or interface control, **6:**33–34
classical model limitations, **6:**34–36
classical work, **6:**13–14
constant field transients, **6:**16–18
Dewald's theory, **6:**14–15
Dignam effective field, **6:**16–18
electrolyte-labeled marker layers, **6:**26
electronic effect, **6:**36
Frenkel defects, **6:**15–16
ion mobility, **6:**24–27, 29–30
nonlinear dependence of log J on E, **6:**19–24, 32–33
normal mode treatment, **6:**34–36
site parameters, **6:**19
slow polarization, **6:**16–18
Tafel slope anomaly, **6:**14–15
theories, **6:**13–36
transients in, **6:**15–16
transition state in, **6:**22
two-layer model, **6:**27–29, 30–34
variable instantaneous values in, **6:**8–10
Ionic conductors, **11:**292
Ionic current
instantaneous, **6:**9
in oxide formation, **6:**50
Ionic networks, **20:**72–76
Ionic polarization, **11:**223
Ionic replacement, reversible, **20:**195
Ion implantation, **14:**232
Ion-implanted thin films, **10:**237
amorphous films, **10:**237
Ionization
additional, magnetron, **17:**87
balance, **18:**139, 180, 182, 202

collisional, **14:**33; **18:**14, 19, 73, 86
efficiency, **18:**177, 180, 187
energy, **14:**34
enhanced, sputtering, **17:**88
Penning, **14:**31
two-step, **18:**190
vapor, **17:**97, 133
zone, **18:**163, 194
Ionization-gauge deposition rate monitor, **3:**3–5
Ionized cluster beam, **13:**3–12, 17, 18, 21–56, 115–116, *see also* Cluster; Film formation
Ionized impurity scattering, **17:**350
Ionized magnetron sputter deposition, **22:**101, 110–113, 165–169, 242–243, 244, 248–252
Ion milling, **13:**127–128
Ion mobility, **14:**27, 39
in ionic conduction process, **6:**24–27, 29–30
Ion plating, **14:**44, 131; **17:**85, 128
magnetron sputtering, **17:**85, 87
plasma chemical vapor deposition, **12:**248–249
reactively sputtered films and, **9:**53
Ion polishing, **13:**126
Ion probe current, **18:**134, 138, 140–144, 161, 171–175, 181–182
Ion range, **14:**7
Ions
cyclotron resonance, **18:**129
gaseous, **17:**99, 134
metallic, **17:**99, 134
negative, **14:**18, 19, 20
size, monolayer binding and, **20:**198–199
sources, **13:**112–116
Ion sheath, **3:**84; **14:**28
Ion-to-deposited-atom ratio, **18:**266
Ion transport, **18:**73
IPET *see* Infrared photoelectron tunneling spectroscopy
Iridium coatings, in extreme ultraviolet astronomy, **10:**94–97
Iron film, as filters, **7:**48
Iron-manganese superlattice, **17:**33
Iron oxide, **14:**2, 29, 51
Iron oxide film, **13:**49–51
Iron silicide film, ionized cluster beam, **13:**34–37
Iron-vanadium superlattice, **17:**33
Irtran–2, antireflection coatings for, **2:**297–298, 301
Ising model, in dielectric response, **11:**302–303
Island structure, model, **9:**6
Island-structured films, **8:**110–113
annealing, **2:**27–28
dielectric constants, **8:**149–150
electrical resistance, **2:**22–30
energy diagram, **8:**103
optical properties, **8:**139–140
properties, **2:**22–37
strain coefficient of resistance, **8:**135
superparamagnetism of ferromagnetic metals, **2:**30–37
as transport conductors, **9:**6
Isoelectronic doping, in MBE, **11:**66
Isotropic electron mean free path, **6:**84
Isotropic metallic films, formulas for, **6:**155–170
Isotropic metals, thin-wire experiments with, **6:**119–121
Iterative system corrections, in optical element correction, **10:**300–304

J

Jacobean elliptic function, **5:**66
Jacquinots's advantage, **15:**271
Jagged domain wall structure, **6:**237
JFET *see* Junction field effect transistor
Johnson noise, in thin-film IV-VI semiconduct or photodiodes, **11:**150–151, 165
Johnson noise limited detector, **11:**124–128, 148–149, 158–159, 172–173
Joint density of states, in optical absorption, **12:**57
Josephson junction, **16:**101, 110–115
  bicrystal grain boundary, **16:**119–122
  edge junction, **16:**110, 118, 119
  grain boundary, **16:**9, 50, 121
  heteroepitaxial multilayer, **16:**111–114
  step-edge junction, **16:**110, 119
Junction capacitance, reduction of, **11:**168–171
Junction field effect transistor (JFET), **11:**190

K

Kapton, **10:**154–156
  mechanical behavior, **16:**296–306
Kaufman ion source, **13:**114–115
Kelvin model, **16:**294–295
Kelvin resistor structure, **13:**214
Kerr effect, **1:**329–330
Kerr rotation, **15:**202–203
Kerr technique, **6:**288–289
Kinematic model, **16:**296
Kinetic energy, **14:**7, 15
Kinetics, **14:**4
  growth, **14:**59
  low-pressure deposition, **22:**179–180
    LPCVD, **22:**203–205
  of plasma anodization, **14:**97–107
    compound semiconductor, **14:**103–104
    metal, silicide, **14:**104–107
    SiO, **14:**98
  reaction, **14:**10–15
Kirchhoff's law, **10:**13
Knox–60, as parting agent, **7:**69
KPR photoresist guides, **10:**242
Kramers-Kronig relations, **6:**183; **11:**214, 232, 245, 267, 270
Kramers-Kronig transformation, **19:**296, 297

L

Laboratory instruments
  contamination, **10:**125
  unbacked filters, **7:**90–92
Lamella, absorption in, **5:**7, 15
Laminin, **24:**249
LAMs *see* Langmuir monolayers
Landau damping, **18:**50, 131, 135–137, 139, 146, 192, 197
Langley Research Center, **10:**133
Langmuir adsorption isotherm, **24:**57
Langmuir-Blodgett films, **24:**180
  adhesion promoters, **20:**109
  amide alcohols, **20:**164–169
  amphiphilic alcohols, **20:**145–181
  artificial photosynthesis, **20:**267–271
  cadmium arachidate, **20:**296–312
  catalytic amphiphilic nickel cyclams, **20:**267–271
  chain length, **20:**155–157
  controllable factors, **20:**351–355

dye-doped, spectroscopy, **20:**293–314
dyes for, **20:**296
effect on spectral broadening, **20:**302
electrical field control, **20:**358–361
electron hopping in, **20:**339–346
ester alcohols, **20:**164–169
horizontal lifting mechanisms, **20:**317–18, 326–328
and ice nucleation, **20:**145–181
ionic networks, **20:**72–78
lead sulfide crystals in, **20:**30
low temperature dynamics, **20:**300–306
mechanical relaxation, **24:**221–222
multilayered, **20:**267–271, 349
odd/even carbon chain, **20:**157–158
palmitic acid on water surface, **19:**291–92
polymeric, **20:**69–82
purple membrane, **20:**283–284
Stark effect, **20:**297
stripe deposition, **20:**356–362
structure control, **20:**349–364
structure variety, **20:**350 351
substrate/monolayer interactions, **20:**355–361
transfer ratios, **20:**322–323
water evaporation rate from cadmium stearate, **20:**320
X and Y lifting methods, **20:**317–318
Langmuir-Blodgett techniques, recognition patterns, **20:**223–236
Langmuir monolayers (LAMs), **24:**180–183, 198–200
Langmuir probe, **3:**83–84; **18:**138, 161–163, 171–177
Langmuir wetting configuration, **20:**350
Lanthanum fluoride film, microstructure and optical function, **19:**222, 224, 238
Lanthanum hexaboride, as spectrally selective surface, **10:**30
Large-area thin-film devices, chemical solution deposition processes in, **12:**223–230
Laser ablation, **16:**55–81
Bi-Sr-Ca-Cu-O thin film, **16:**76, 78–80
elemental velocity distribution, **16:**56–67
forward or direct stoichiometric component, **16:**58, 60, 78
particulates, **16:**59, 70, 75, 77, 122, 5516.57
plume species, **16:**55, 56, 58
process parameters, **16:**55–59, 65, 67–69
rare earth-based thin film, **16:**59
selective resputtering during, **16:**63
thermal non-stoichiometric component, **16:**58, 78
Tl-Ba-Ca-Cu-O thin film, **16:**56, 80, 81, 102–104, 106, 117, 121
Y-Ba-Cu-O thin film, **16:**56–78, 106, 108, 111–114, 117, 118, 122
Laser beam evaporation, in dielectric film preparation, **8:**20
Laser diode emission, efficiency requirements, **7:**211–218
Laser diodes
homojunction and heterojunction types, **7:**229–230
liquid phase epitaxy and, **7:**228–235
Laser-induced etching
applications

Laser-induced etching (*continued*)
metals, **13:**183–184
other inorganic materials, **13:**190–191
polymers, **13:**181–183
semiconductors, **13:**185–190
damage, **13:**178–181
etch depth versus
applied voltage, **13:**179
sheet resistance, **13:**176–177
flat-band voltage shifts versus injection charge density, **13:**180
groove broadening as function of percentage of light transmitted, **13:**160
lower-power, **13:**181
mechanisms, **13:**152–156
ablation, **13:**153–154
photochemical processes, **13:**155–156
thermal processes, **13:**155
process-induced defects, **13:**178
reactant generation by photochemical means, **13:**184
resolution *see* Resolution
selectivity, **13:**176–179
Laser mirrors, **8:**43–44
Lasers
dielectric films, damage from, **8:**4, 14–17
mirrors for, **10:**167–225
principal wavelengths, **13:**157–158
Laser spot scanning, focusing optics, **13:**161–162
Laser structures, in aluminum gallium arsenide system, **7:**216
Lateral collection photodiodes, **11:**177–185
in bulk crystals, **11:**183–184
collector geometries, **11:**178
with integral interference filter, **11:**188
potential barrier in, **11:**178
Lattice, thermal oscillation, **1:**235–236
Lattice constant, **20:**31
Lattice defects, **1:**235; **17:**122
generation, **13:**180
Lattice disorder, optical properties and, **4:**11
Lattice effects, in electromigration in thin films, **7:**258–264
Lattice electromigration
experimental results in, **7:**262–264
grain boundary in, **7:**265
Lattice-matched heterojunction, **13:**219
Lattice matching
in epitaxial fabrication, **23:**225–226
YIG to substrate, **15:**15
Lattice misfit dislocation network, **7:**130–132
Lattice mismatch, **17:**17, 31, 33, 40, 41
in heteroepitaxy, **7:**126
in liquid phase epitaxy, **7:**122–135
Lattice parameter, **13:**80; **17:**126
stress-induced changes, **3:**225–226
Lattice relaxation, **17:**23, 24–29, 41, 56
critical thickness, **17:**27–29
kinetic limitations, **17:**28
Matthews theory, **17:**24–29
Lattice sites
in polarization process, **6:**62
in YIG, **15:**13–14
Lattice spacing, cubic boron nitride, **13:**101
Layered magnetic films, **16:**215
Layer thickness, **14:**7, 9; **20:**46
LB *see* Langmuir-Blodgett films
Lead

bonding to thin films, **2:**412
as dopant in MBE, **11:**60–61
electrical resistivity, **6:**120
Lead barrier thin-film PbTe devices, **11:**130–132, 135–136
Lead chalcogenides, **11:**62, 116
Lead film, as filter, **7:**49
Lead fluoride film, stress in, **8:**13
Lead germanate film, **17:**228, 288
Lead-gold film, **6:**326–327
(Lead, lanthanum) titanate film (PLT), microstructure and optical functions, **19:**258–59
(Lead, lanthanum) zirconate titanate film (PLZT), microstructure and optical functions, **19:**259–60
Lead oxide film, **8:**40
Lead salts, single crystal, **3:**276–278
Lead selenide
crystal morphology, **20:**32
epitaxy, **20:**33–37
particulate film, **20:**22–23, 33–37
Lead selenide film, **3:**287–291
optical and electrical properties, **12:**219–221
Lead sulfide
crystal structure, **20:**31–32
epitaxy under films, **20:**31
ion-beam processing, **13:**144
for tandem absorber combinations, **10:**61
Lead sulfide-arachidic acid film, crystal morphology, **20:**30
Lead sulfide film, **3:**281–285
Lead telluride
Auger lifetimes, **11:**143, 146
detectivity for, **11:**146
diffusion-limited behavior, **11:**158–159
on mica substrates, **11:**153
parlodian replica of layer, **11:**117–120
RLL quantum efficiency, **11:**137–138
Lead telluride devices
Arrhenius slopes, **11:**160–161
heat treatment for, **11:**155
Lead telluride film, **3:**286–287
Lead telluride photodiode array, configuration, **11:**195–196
Lead telluride phototransistors, **11:**189–191
Lead telluride thin-film detectors, **11:**130
Lead-tellurium film, ionized cluster beams, **13:**32–33
Lead titanate film (PT), **17:**228, 242–259
microstructure and optical functions, **19:**260–62
sputtered film, **3:**110–111
Lead zirconate titanate, **14:**24
(Lead, zirconium) titanate film
microstructure and optical function
on Pt-coated Si, **19:**269–70
on sapphire, **19:**264–68
Leap-frog algorithm, **22:**132
LEDs *see* Light-emitting diodes
LEED *see* Low-energy electron deposition; Low-energy electron diffraction
Left-hand polarized waves
absorption, propagation, reflection, **18:**136–139, 176, 186–187, 190
definition, **18:**127
transformation into plasma waves, **18:**137–139, 186–187, 190
Lennard-Jones potential, **22:**126–127, 132

Lenz theory, elastic electron scattering, **2:**75–77
Ligand
  octadecylimidazole, **20:**73
  organosilane, **20:**121
  planar alignment by, **20:**72–76
Ligating films, **20:**120–123
Light
  excitation of surface plasmons by, **9:**180–199
  modulation for information displays, **8:**319–333
  plasma radiation excited by, **6:**197
  scattering by magnetic waves *see* Brillouin scattering
  vacuum ultraviolet, **1:**130–131
Light absorption
  by silver films, **8:**161
  in transparent conducting films, **9:**13–16
Light coupling methods, quantum-well infrared photodetectors, **21:**215–219
Light emission
  and coupling by statistical surface roughness, **9:**219–221
  via roughness, **9:**219–223
Light-emitting diodes (LEDs)
  graded-gap injection structures, **23:**265–266
  heterojunction in, **7:**221–222
  high-intensity, **11:**78
  in liquid phase epitaxy, **7:**218–228
  minority traps in, **11:**53
Light guiding, in thin films, **10:**231–235
Lighting filters, **5:**134–138
Light resonator, **5:**2–8
Light sensor, sandwich structure, **20:**287–290
Light valves, **8:**319–333
  bending-moment type, **8:**326–333
  Schlieren, **8:**324–326
  tension-aided membrane type, **8:**320–324
Light waves, excitation of surface plasmons by, **9:**223–229
Limiter
  frequency selective application in systems, **15:**69
    magnetic film design, **15:**98–100
Limiting efficiency, quantum-well solar cells, **21:**357–362
Linear regression analysis (LRA), **19:**58, 62, 64, 97, 146, 153–56, 169, 205
Line-averaged plasma density, **18:**174, 176–177, 184–185, 187–189, 206
Line electrode, **20:**332–333
Line pairing, **24:**237
Line shape measurements, **24:**153–154
Linewidth, ferromagnetic
  observed in ferrites, **15:**9
  phenomenological theory, **15:**128
  related to propagation loss, **15:**6
Liouville transformation, **5:**57
Lipid complex films
  biotin-streptavidin, **20:**225–234
  crystallization in, **20:**233
  diffusion coefficient, **20:**228
  Fab'-lipid, **20:**226–234
Lipid membranes
  by Langmuir-Blodgett technique, **20:**241–249
  charge separation units in, **20:**242–249
  contact potential difference, **20:**260–265
  electrical double layer, **20:**249–253
  electron transfer kinetics, **20:**249–254

energy diagram, **20:**266
ferrocene derivatives in, **20:**244, 247, 249–253
free energy difference, **20:**249–252
mixed monolayer, **20:**247–249
molecular photodiode, **20:**243
multilayered systems, **20:**243–246
nanodomains by phase separation, **20:**256–267
phase separation by atomic force microscopy, **20:**265–267
photodiode energy diagram, **20:**244–245
reaction centers, **20:**257
scanning surface-potential microscopy, **20:**260–265
simulation of transfer process, **20:**254–256
spacer molecules, **20:**243
triad systems, **20:**246–249
Liquid-phase epitaxy, **10:**248
aluminum arsenide, **7:**171–172
avalanche diodes, **7:**24
bandgap energy, **7:**175–177
binary compounds, **7:**169–172
cathodoluminescence, **7:**209
dopant temperature dependence, **7:**154–156
electroluminescence and cathodoluminescence, **7:**178
electroluminescent diodes, **7:**205–235
film growth by, **9:**287–288
gallium antimonide, **7:**169–170
gallium arsenide, **7:**140–159
gallium nitride, **7:**169
gallium phosphide, **7:**159–169
growth apparatus, **7:**117–119
heteroepitaxy, **7:**123
heterojunctions in, **7:**128
high-frequency devices, **7:**241–242
homoepitaxy, **7:**125–126
indium arsenide, **7:**170
indium phosphide, **7:**170–171
kinetics, **7:**119–122
laser diode emission, **7:**211–218
lattice mismatch effects, **7:**122–135
light-emitting diodes and, **7:**218–228
microwave devices, **7:**235–242
negative electron affinity, **7:**242–246; **12:**129
nucleation, **7:**119
in PbTe substrates, **11:**171
photoemissive surfaces, **7:**242–244
properties and applications, **7:**115–247
solar cells, **7:**246
substrates, **7:**135–139
ternary systems, **7:**172–205
YIG films, **15:**10, 11, 13, 15, 17, 18
Lithium, polymer modifying agents, **20:**96
Lithium ferrite, properties, **15:**9
Lithium film, reflectivity, **9:**236
Lithium fluoride coated aluminum, reflectance, **1:**173
Lithium fluoride coatings, for extreme ultraviolet astronomy, **10:**75–91, 117
Lithium niobate film, **10:**256; **17:**260–263
Lithium niobate guides, **10:**253
Lithium tantalate film, **17:**260–263
Lithium tantalate guides, **10:**253
Lithography
electron beam, **8:**260; **24:**229, 230
film requirements, **20:**110–112
nanometer, **20:**267–268
patterned films, **20:**109–141
Local anisotropy, **15:**144

Local atomic order, electron diffraction analysis, **7:**1–37
Localized carriers, **11:**226
Localized corrosion resistance, activated reactive evaporation applications, **13:**94–95
Local stress function, **22:**144–145, 151
Logic elements, **1:**319–321
London theory of superconductivity, **1:**248ff
Longitudinal optical phonon intersubband lifetime, **21:**30
Long wave infrared (LWIR) arrays, **21:**221–223
Long wave infrared (LWIR) band, **23:**120
Long wave infrared (LWIR) detection, material considerations, **23:**127–135
Long-wavelength infrared hot-electron transistors, **21:**282–289
Lorentz-Lorenz theory, **8:**55–57, 61, 69–71, 91
Lorentz microscopy, **6:**228, 233, 248, 278
Low-energy electron deposition (LEED), reconstruction of surface atoms, **12:**115
Low-energy electron diffraction (LEED), **9:**215–216; **11:**45–46
Low oxygen partial pressure annealing
  Bi-Sr-Ca-Cu-O thin film, **16:**4, 32, 34
  fluoride route, **16:**43
  110 K phase, **16:**4, 33
  Y-Ba-Cu-O thin film, **16:**45
Low-pressure deposition processes, **22:**175–176
  ballistic transport and reaction model, **22:**193–202
  chemistry, **22:**181–183
    global rate expressions, **22:**185
    kinetics, **22:**179–180
    reaction mechanisms, **22:**186–188
    stoichiometry, **22:**181, 183–185
  EVOLVE simulation, **22:**180–181
    chemical vapor deposition, **22:**202–210
    plasma-enhanced chemical vapor deposition, **22:**210–215
    programmed rate CVD, **22:**265–269
    remote microwave PECVD, **22:**215–226
    simultaneous deposition and sputter etch, **22:**241–252
    surface diffusion, **22:**252–265
  physical vapor deposition, **22:**226–241
  transport, **22:**177–179
    BTRM, **22:**193–202
    coordinate systems, **22:**188–189
    flux, **22:**189–191
    flux distributions, **22:**191–193
    single particle distribution function, **22:**189
Low-pressure/low-throw sputtering, sputter deposition, **22:**165–169
Low-pressure plasma deposition process, **13:**78
Low-temperature epitaxial growth of GexSi1-x, **23:**22–23
  surface reactions in, **23:**27–37
    of diborane, **23:**34
    of dichlorosilane, **23:**35–37
    of germane, **23:**33
    of hydrogen, **23:**27–30

of phosphine, **23:**34–35
of silane, **23:**30–33
Low-temperature field ionization, **8:**208–210
Low temperature process, **16:**18–28, 29–32, 52, 54, 61–78, 89–92, 93, 112, 113
LPE *see* Liquid-phase epitaxy
LRA *see* Linear regression analysis
LWIR *see* Long wave infrared
Lyman-alpha live thickness monitor, **10:**78–82

M

Macroparticle, **17:**94–96, 108
elimination, **17:**94
emission, **17:**94
Magnesium, in MBE doping, **11:**64–65
Magnesium film, as filter, **7:**47
Magnesium fluoride, **14:**18
as dielectric film material, **8:**30–32
ion-beam processing, **13:**137–139
mechanical properties, **13:**138–139
refractance, **10:**121
Magnesium fluoride coated aluminum, **1:**172–173, 174
Magnesium fluoride coatings, **12:**48–49
on aluminum, **2:**346
for extreme ultraviolet astronomy, **10:**75–91, 117
on glass, **2:**287ff.
Magnesium fluoride spacer, in filter production, **9:**138
Magnesium oxide, microstructure and optical function, **19:**217
Magnet, permanent, **18:**153–155, 163–165
Magnetic anisotropy, **1:**278–297; **9:**284–285; **15:**183
amorphous materials and, **9:**285–286
angle of incidence, **1:**286–289
biaxial, **6:**281–282
crystal orientation dependence, **1:**281
differential optical absorption and, **1:**287
dispersion, **1:**302–303
"exchange," **1:**285, 294–297; **6:**213–223
in garnets, **9:**286–288
heat treatment induced, **1:**280–281
helical, **6:**288–290
high-order, **6:**278–290
imperfection, **1:**284–285
impurity, **1:**284–285
magnetic bubble films, **9:**284–288
magnetostatic averaging, **1:**294–297
multiaxial, **6:**279–280
in multilayer films, **6:**212–213, 225
Nel-Taniguchi, **1:**280–282
pair ordering, **15:**183–185
of strained polycrystalline film, **1:**283–284
Magnetic bubble films, **9:**263–294, *see also* Magnetic bubbles
anisotropy, **9:**284–288
defects, **9:**288–293
in information storage industry, **9:**293–294
inhomogeneity, **9:**288–290
materials selection, **9:**265–266
static properties, **9:**266–269
Magnetic bubble materials, growth, **9:**276–284
Magnetic bubbles
amorphous film preparation for, **9:**277–280

Magnetic bubbles (*continued*)
anisotropy energy required for, **9:**268
defined, **9:**263
information stored in, **9:**264
magnetic theory and, **9:**266–272
materials, **9:**272–276
amorphous, **9:**273–275
crystalline, **9:**275–276
performance and cost factors, **9:**265–266
single-crystal garnet films for, **9:**281–284
size, **9:**263–264
velocity, **9:**269
Magnetic coils, **18:**126, 152–154, 157–165, 170–172, 210–211
Magnetic domain configurations, MnBi films, **13:**30
Magnetic excitation, instability, **15:**93
Magnetic field, **14:**31, 33
absorption, **18:**160–162, 171, 173
axial, **18:**17
breakdown, **18:**170–172
in closed superconducting loop, **1:**271–272
divergent, **18:**157–165, 179
"high-field-side," **18:**134–136, 151, 159–163, 177–178, 186–187
magnetron, **17:**88
mirror, **18:**142, 147, 155, 160, 180
mirror-like, **18:**158–160, 163
multidipole, **18:**12, 42, 52
and multilayer matrix, **5:**53–54
multipolar, **17:**88, 90
resonant, **18:**126–129, 131, 134, 136–140, 171, 173, 194
uniform, **18:**133, 157, 159, 165, 177, 180, 211
vacuum window, **18:**159–162, 171–172, 188, 203
Magnetic field intensity, **1:**73–76
Magnetic films, *see also* Ferromagnetic films; Metallic films; Multilayer magnetic films
amorphous, **16:**184
directly crystallized, **16:**189
GdFe films, **13:**31–32
ionized cluster beam deposition and epitaxy, **13:**29–32
MnBi films, **13:**29–31
multistage, **6:**205–209
subsequently crystallized, **16:**188
Magnetic free energy, supplied by magnetic field, **1:**249
Magnetic materials, Curie or Nel point, **9:**268
Magnetic moments, **15:**176
electronic density of states, relation to, **15:**176
of ferromagnetic films, **2:**30–37
transition metal alloying, effect, **15:**177–189
Magnetic penetration depth, **16:**12, 39, 101
Magnetic stripes, vs. magnetic bubbles, **9:**267
Magnetic theory, **9:**266–272
domain dynamics in, **9:**269–272
Magnetic tuning, **15:**5, 88, 116
Magnetic units, **16:**144
Magnetization
array of superparamagnetic particles, **2:**31ff.
control by ionic substitution, **15:**13
Curie and compensation temperatures, **15:**180–182
equations of motion, **15:**126
exchange interaction, **15:**174, 180

ferrites, observed values, **15:**9
films in normal and superconducting state, **1:**249–250
function of film thickness, **1:**322–327
mean field parameters, **15:**174
in multilayer films, **6:**261–269
permeabilities resulting from, **15:**3, 27, 127
reversal process, **1:**297–317
reversal processes, **1:**297–317; **15:**186
by domain wall motion, **1:**316–317
failings of idealized model, **1:**302–304
ripple in Permalloy films, **1:**295–297
temperature dependence, **1:**323–325; **15:**14
transition metal magnetic moment, **15:**176
Magnetization angle, **6:**212
Magnetoelasticity, **1:**282–284
Magnetooptic devices, **15:**116–117, 119, 121
Magnetooptic diffraction
anisotropic, **15:**118, 120
Bragg, **15:**118, 120
isotropic, **15:**119
Magnetooptic effects, **15:**192–200, 202
electronic density of states and, **15:**192
ellipticity, **15:**202
Kerr rotation, **15:**202–203
non-magnetic metal substitution and, **15:**197–200
rare earth substitution and, **15:**196–197
transition metal substitution and, **15:**195–196
Magnetooptic interactions, **15:**102
Magnetooptic medium design, **15:**201
corrosion resistance, **15:**207
direct overwrite, **15:**204–207
signal-to-noise ratio for readout, **15:**202–204, 207–208
storage capacity, **15:**201–202
Magnetooptic tensor, **15:**103
components observed in ferrites, **15:**106
dominance of Faraday rotation, **15:**106, 112
related to mode conversion, **15:**106, 114
Magnetoresistance
resistivity ratio and, **6:**140
size effects, in thin films, **6:**82
Magnetostatic coupling
of domain walls, **6:**227–240
in multistage films, **6:**223–227
Magnetostatic devices, **15:**48, 59, 71, 81, 85, 86, 96, 98, 100
Magnetostatic modes
double ferrite layer, **15:**37, 38, 39, 133
multiple ferrite layers, **15:**37
single ferrite layer, **15:**26
bandwidths, **15:**5, 28, 36
dipole-exchange, **15:**136
exchange-dominated, **15:**135
ground plane effects, **15:**35
shock waves, **15:**45
surface waves, **15:**5, 28, 131
thickness resonances, **15:**25, 29, 134
volume waves, **15:**5, 28, 134
width modes, **15:**32, 34
Magnetostatic potential, **15:**26
Magnetron, **18:**165

Magnetron, (*continued*)
baffled, **14:**44–47
conventional, **17:**86, 99
cylindrical, **14:**33, 36, 40; **17:**87
DC, **14:**32–39
double-ended, **14:**41, 42
electric field, **17:**88
enhanced ionization, **17:**87, 88
magnetic field, **17:**86, 88
modified, **14:**43–52
multipolar, **17:**88, 90
planar, **14:**2, 33; **17:**87
RF, **14:**39–43
sputter-gun, **14:**48–49
systems, **14:**32–52
unbalanced, **17:**88, 89, 99, 123
Majority carrier devices, **11:**75–77
Manganese, in MBE doping, **11:**65
Manganese-bismuth films, **13:**30, 31
Many-body effects, quantum wells, **21:**95, 97
Many-body phenomena, in dielectric responses, **11:**296–312
Mar-Klein all-atom (MKAA) model, **24:**210, 211, 213–215
Masking
conformal masks, **2:**411
mechanical masks, **2:**410–411
negative preparation, for optical elements, **10:**266–269
precision masking of evaporated patterns, **2:**178–181
Massive soluble substrates, in metal film filters, **7:**73–74
Mass spectrometer, **1:**38–47; **8:**22–23
cycloidal, **1:**41–42
data interpretation, **1:**45–47
differential electrochemical, **20:**270–271
electrical filter, **1:**44–45
magnetic analyzer, **1:**39–41
omegatron, **1:**42–44
Mass spectroscopy, **14:**16
Master conventional gratings, **11:**11–12, *see also* Diffraction gratings
Matching
in ECR, **18:**27
in helical resonator, **18:**64
in helicon, **18:**42
in inductive discharge, **18:**53, 58
in surface wave discharge, **18:**69
Matching stacks, **9:**113–114, 119
Materials processing, **17:**289–290
epitaxial buffer layer, **17:**290
processing temperature, **17:**289
rad-hard FEMFET fabrication, **17:**290
Mathematical models, **22:**81–84
collisionless transport, **22:**98–100
Hamilton-Jacobi equation, **22:**83, 84–97, 113
ionized magnetron sputter metal deposition, **22:**101, 110–113
ion and neutral fluxes, **22:**107–110, 111
shock-tracking algorithm, **22:**97–98
two-energy model, **22:**100–107
Matrices
hydrogen-bonded multilayers, **20:**42
hydrogen-bonded silane, **20:**42–68
monolayer, **20:**12
multilayer, **20:**45–48
sealing components, **20:**184
stability to intercalation, **20:**51
Matrix notation, thin film optics, **1:**87–92
Matthiessen's rule, **1:**237; **2:**381, 382; **6:**113, 124–131
Maxwell-Boltzmann distribution function, **8:**105
Maxwell-Garnett absorption

in photocathode materials, **12:**71–72, 79
for thin silver films, **12:**71–72
Maxwell-Garnett theory, **8:**146, 149, 152; **10:**60–61
Maxwell model, **16:**294, 295
Maxwell-Wagner mechanism, **11:**272
Maxwell-Wagner-Sillars effect, **11:**294
Maxwell-Wagner systems, **11:**294
MBE *see* Molecular beam epitaxy
MCP *see* Microchannel plate image intensifier
MDMDMDM filters, **9:**124, 143
MDMDM filters, **9:**113
MDM filters, **9:**88–90, 113
augmented, **9:**125–127
cementing, **9:**140–142
design, **9:**102–104
equivalent layer spacers and, **9:**110–112
maximum transmittance, **9:**109
monitoring, **9:**138
optical constants vs. wavelength, **9:**104
phase shift in, **9:**103
reflectance and transmittance, **9:**137
spectral transmittance, **9:**107–108
wedged, **9:**105–106
MDM structure, in reflection filters, **9:**130
Mean free path, **14:**18, 29
collisional, **18:**14, 15
elastic electron scattering, **2:**75
electron-electron, **4:**328–329
electron-optical-phonon, **4:**331–333
hot-electron and electrical conductivity, **4:**327–328
hot-electron-plasma interactions, **4:**328–329
Mean free time, between collisions, **11:**226
Mechanical properties, **13:**125; **24:**83, 104, 105
atomic force microscopy, **24:**93–98
carbides, **13:**91–92
magnesium flouride, **13:**138–139
nitrides, **13:**87, 89
refractory compound deposits, **13:**87, 89–94
silicon dioxide, **13:**135
simulations, **24:**221–222
titanium dioxide, **13:**133
titanium nitride, **13:**87, 89
Mechanical property measurements, **17:**42–47
elastic moduli and constants, **17:**42–46, 49
acoustic phonon dispersion, **17:**45
biaxial modulus, **17:**43, 49
Brillouin scattering, **17:**43, 45, 46, 49, 51, 52
bulge test, **17:**43, 49, 51
cantilever beam, **17:**43, 49
flexural modulus, **17:**43, 49
nanoindentation, **17:**43, 46, 47, 49
pulse-echo method, **17:**46, 49
Raman scattering, **17:**43
shear modulus, **17:**45
ultrasonic techniques, **17:**44–46, 49
uniaxial tension, **17:**43, 49
Young's modulus, **17:**43, 46, 49
hardness and yield strength, **17:**46, 47
Knoop test, **17:**46
nanoindentation, **17:**46, 47
uniaxial tension, **17:**47
Vickers test, **17:**46

Mechanical property measurements, (*continued*)
stress-strain curves, **17:**43, 44, 47
Mechanical relaxation, self-assembled
Mechanical relaxation, (*continued*)
monolayers (SAMs), **24:**221–222
Mechanical transfer function, for image converters, **12:**151
Mechanomigration, **16:**237
Median time to failure, **16:**239–240
Arrhenius plot, **16:**240
dependence on gross defects, **16:**240
Medium-energy electron diffraction, **11:**42
Medium noise, **15:**210
Meissner effect, **1:**247
Membrane light modulation, **8:**320–324
Membrane light modulator (MLM) device, **8:**320–324
Membranes *see* Lipid membranes; Purple membrane
Memory elements, **1:**317–319
2-(10-Mercaptodecyloxy)–3,6,7,10,11-pentapentyloxytriphenylene SAMS, terminal groups, **24:**88, 91
(4-Mercaptophenyl)phthalimide, force fields for, **24:**211
Mercaptosilane film, **20:**128–130
Mercury cadmium telluride (HCT)
homojunctions, **21:**40–41, 239–240
as LWIR material, **23:**128–133
Mesh screens, for metal film support, **7:**78–85
Mesophase
amphiphilic, **20:**368–390
definition, **20:**369
Metal, sputtered, **14:**4
Metal-barrier devices, surface effects and nature, **11:**153–157
Metal cathode, **21:**6
Metal-containing SAMs, **24:**24
Metal deficiency, p-type conductivity and, **11:**107
Metal-dielectric film combinations, **8:**44–45
Metal-dielectric interference filters, **9:**73–143
admittance and amplitude reflection in, **9:**75–77
bandpass filter design and, **9:**93–102
characteristic matrix, **9:**77–78
design
multiple cavity and other designs, **9:**124–128
one-M filter design, **9:**112–114
Fabry-Perot type, **9:**74–75, 88–90
future developments, **9:**142–143
MDM type *see* MDM filter
net flux ratio, **9:**79–86, 87–88
assembly of layers, **9:**82–87
maximum value, **9:**86–87
production, **9:**132–142
radiant absorptance, **9:**81
radiant reflectance and transmittance, **9:**78–79
reflection filters as, **9:**128–132
single-cavity type, **9:**102–112
theory, **9:**75–93
Metal-dielectric-metal interference filters *see* MDM filters
Metal-dielectric mirrors, germanium coatings for, **12:**42–43
Metal film filters, *see also* Metal films; Unbacked filters
aging and storage, **7:**86–89
bimetallic, **7:**70–73
cleaning, **7:**89–90
equilibrium temperature, **7:**94–98

fabrication, **7:**77–86, 81–85
massive soluble substrates, **7:**73–74
plastic films and, **7:**69–73
thin film substrates, **7:**69–73
Metal films, *see also* Cermet films; Metal-dielectric interference filters; Metal film filters; Multilayer magnetic films; Thin films; Very thin films
adhesion, **20:**123–124
of alkali and noble metals, **6:**104–112
bulk resistivity ratio, **9:**8
carrier scattering sites, **9:**7
conductivity, **8:**152–154
contact electrodes, **13:**27–28
contact resistivity, **13:**28–29
continuous, **9:**7–9
corrosion, **7:**60–61
defects, **9:**7
dielectric constant, **6:**201–202
diffusion in, **6:**301–346
ellipsometric parameters, **6:**165–170
evaporation methods, **7:**57–60
experimental diffusion in, **6:**310–315
film structure, **6:**199–202
film support and filter fabrication, **7:**77–86
filter fabrication, **7:**81–85
fine mesh screen for, **7:**78–80
floating of, **7:**75–76
fuchsin as parting agent for, **7:**61–63
Group IV-V, **13:**80
impurities, **9:**8
ionized cluster beam deposition and epitaxy, **13:**25–29
laser-induced etching, **13:**183–184
miscible systems and, **6:**302–304, 335–338
ohmic contacts, **13:**27
optical properties, **6:**151–202; **9:**13–14
parting agents for, **7:**60–69, 74
photometric measurements, **6:**162, 182–188
pinholes, **7:**56, 78–80, 83
polyvalent, **6:**112
preparation, **7:**53–77
reflectance, **6:**156
removal from substrates, **7:**74–77
resistivities, **6:**104–112; **9:**8
separate thickness measurements, **6:**187–188
sheet resistance, **8:**100
size effect, **9:**7
specific optical properties, **6:**188–202
stripping technique, **7:**56
substrate and, **7:**53–57
surface wear, **13:**102
theoretical models, **8:**102
thin homogeneous and isotropic, **6:**155–170
transmittance in, **6:**156
transparent conductivity *see* Transparent conducting films
uses, **6:**152
very thin, **6:**170–177
Zapon films and, **7:**70
Metallic front surface mirrors, reflectance of without overcoatings, **12:**16–21
Metallic layers, physical properties, **5:**89–90
Metallic wires, Nordheim expression for, **6:**129
Metallized halftone correction mask, **10:**271

Metalorganic chemical vapor deposition (MOCVD), **16:**81–94, 109, 116
Bi-Sr-Ca-Cu-O thin film, **16:**92–94
hot-wall reactor, **16:**84–88, 92, 93
low-pressure MOCVD (LPMOCVD), **16:**85–94
mass transport, **16:**82–85
precursors, **16:**81–84, 87, 90
volatility and stability, **16:**82–84, 87, 90
process parameters, **16:**81–85
Tl-Ba-Ca-Cu-O thin film, **16:**94
Y-Ba-Cu-O thin film, **16:**85–92, 111, 116
Metalorganic deposition (MOD), **16:**95–100
Bi-Sr-Ca-Cu-O thin film, **16:**98–99
particulates, **16:**95
precursors, **16:**95–97
process parameters, **16:**94–96
Tl-Ba-Ca-Cu-O thin film, **16:**100
Y-Ba-Cu-O thin film, **16:**96, 97
Metal oxide coatings, organogenic, **5:**120–121
Metal oxide semiconductor field effect transistors (MOSFETs), **11:**77–78; **12:**237
channel, for GexSi1-x system, **23:**68–73
GaAs plasma oxidation for, **12:**277
high-dielectric-constant films, **11:**89
Metal oxide semiconductors, **20:**112, *see also* Semiconductor oxide film
capacitor, **20:**127
transitor, **20:**112, 128
Metals
chemical vapor deposition, **5:**306–309
free-electron theory, **1:**234–235
highly reflecting, **6:**159–160
moderately reflecting, **6:**160–165
optical properties, **6:**152–155; **12:**18
radial distribution function measurement, **7:**18–37
transmitting windows, **7:**45–49
vacuum characteristics, **1:**9–12
Metal semiconductor systems, MBE, **11:**86–88
Metal systems
compound-forming, **6:**305–309
miscible systems and, **6:**302–304
Metal transition
amorphous, **17:**121
borides, **17:**101–104
carbides, **17:**101–104
nitrides, **17:**101–104, 109, 119
oxides, **17:**101, 104, 108
Metal wires, Dingle calculation for, **6:**92
Metastable critical thickness, **23:**3
Methyl glycidyl methacrylate copolymer, **10:**242
Michelson interferometer, **15:**269
Microbalances, **3:**27–30
Microchannel plate, image intensifier, **12:**148–149
Microcrystalline films, atomic order in, **7:**10–11
Microcrystalline silicon, **19:**93
optical function, **19:**94–104
optical and microstructural analysis, **19:**94
size effect in microcrystallites, **19:**99
Microfractures, **13:**180–181
Microhardness, **3:**265–266; **17:**127, 130
base temperature and, **13:**93
carbon-to-metal ratio and, **13:**90

deposition temperature and, **13:**91, 93
Group IV and V metals, **13:**80
TiC, **13:**80, 83, 90–91
TiN, **13:**89
Ti(N, C), **13:**91–92
Y2O3, **13:**93
Micromachining, **24:**229, 230
Micromechanical devices, in signal modulation, **8:**310
Micromolding in capillaries (MIMIC), **24:**246
Microscopy, *see under* individual types of microscopy
Microstain, **17:**126
Microstrip transducers, **15:**3, 27, *see also* Radiation resistance
Microstructure, **15:**161–166; **17:**119; **22:**2–5, 125–126, 170
aluminum oxide, **13:**141
correlation
process, **17:**129
properties, **17:**129
ion-beam processing, **13:**123–124
magnesium fluoride, **13:**138
refractory compound deposits, **13:**81–85
silicon dioxide, **13:**135
tantalum oxide, **13:**140
titanium dioxide, **13:**132–133
titanium nitride films, **17:**121–126
under ion bombardment, **17:**121
zirconium oxide, **13:**136–137
Microtransfer molding, **24:**246
Microwave
breakdown, **18:**170–172
modes, **18:**129, 152–155, 164, 176, 178–179, 186
power, absorbed, forward, reflected, **18:**126–130, 143–144, 147–151, 167–169, 172–174, 176–181, 184–193, 203, 207, 210
window, **18:**126, 137, 152, 154, 159–160, 165–169, 210–211
Microwave components, **16:**39, 95, 101–108
antenna, **16:**108
band-pass filters, **16:**101, 105, 106
band-reject filter, **16:**105, 106
delay line, **16:**106–108
oscillator, **16:**101, 103
quality factor *Q*, **16:**12, 13, 101–106
resonator, **16:**101–105
Microwave devices
discrete, **8:**299–310
in liquid-phase epitaxy, **7:**235–242
Microwave discharge, emission spectra, **13:**69
Microwave rectification technique, bandwith of multiple-quantum-well detectors, **23:**199–200
Microwave surface resistance, **16:**12, 13, 38, 39, 48, 67, 75, 100, 103, 104, 106, 108
Mid-wave infrared (MWIR) band, **23:**119–120
Migration, **13:**18–19
average distance transversed, **13:**19
diffusion coefficient, **13:**19
Millimeter-wave Gunn devices, **11:**77
MIMIC *see* Micromolding in capillaries
Minibands, superlattice *see* Superlattice minibands
Minimum-energy trajectory method, **6:**259
Minority traps, MBE and, **11:**53
Mirror coatings
aluminum, **12:**17–19
copper, **12:**17

Mirror coatings (*continued*)
gold, **12:**17
metal-dielectric, **12:**42–45
optical constants and calculated reflectance, **12:**19
platinum, **12:**17
with protective layers and reflectance-enhancing surface films, **12:**22–42
reflectance measurements for, **12:**3–8, 19
rhodium, **12:**15–19, 39–42
silver, **12:**17–19
Mirrors
for lasers, **10:**167
scattering by, **10:**167–225
Miscible systems, **6:**302–304
diffusion coefficients for, **6:**336
diffusion measurements in, **6:**335–338
Misfit dislocations, **17:**17, 19, 24–29, 41
Burgers vectors, **17:**24–29
Misfit lattice, **17:**106
MIS structures
4.293ff.
band diagram, **17:**273–274
BMF gate dielectric, **17:**286
BTO gate dielectric, **17:**275, 276
C-V behavior, **17:**273–275
flatband voltage, shift, **17:**273–275
polymer gate dielectric, **17:**279
tunnel injection, **17:**274, 275
MKAA model *see* Mar-Klein all-atom model
MLM device *see* Membrane light modulator device
Mobility
amorphous solids, **8:**229–232
atom, **14:**59, 64
electron, **14:**28, 39
in forbidden gap, **8:**192–194
ion *see* Ion mobility
localized, **8:**188–192
in nonmetallic films, **8:**188–190
polymer surface chains, **20:**85–87
thin-film semiconductors, **2:**172–176
MOCVD *see* Metalorganic chemical vapor deposition
MOD *see* Metalorganic deposition
Mode, *see also* Wave
helicon, **18:**42
absorption, **18:**50
damping, **18:**50
measurements, **18:**45
standing, **18:**49
slow wave, **18:**61, 65
in helical resonator, **18:**62
surface, **18:**65
Mode conversion, optical, in magnetic media, **15:**103
collinear, **15:**113
from Faraday rotation, **15:**106, 114
transverse, **15:**122
Modeling, **22:**3–6, 122–123, *see also* Simulations
anisotropic medium, **19:**295
arc evaporation, reactive, **17:**112
atomistic modeling, **22:**177
DEPICT model, **22:**5
EVOLVE simulation, **22:**180–181
low-pressure chemical vapor deposition, **22:**202–210
physical vapor deposition, **22:**226–241
plasma-enhanced chemical vapor deposition, **22:**210–215
programmed rate CVD, **22:**265–269
remote microwave PECVD, **22:**215–226

simultaneous deposition and sputter etch, **22:**241–252
surface diffusion, **22:**252–265
GROFILMS model, **22:**30, 70, 72, 74–76
mathematical *see* Mathematical models
molecular dynamics *see* Molecular dynamics modeling
SAMPLE model, **22:**5
SHADE model, **22:**84, 111
SIMBAD deposition model, **22:**5–6, 9–15, 20–29, 76
bias sputtering, **22:**30–38
bulk deposition, **22:**53–54
chemical vapor deposition, **22:**29–30, 36
collimated sputtering, **22:**39–45
etching, **22:**60–76
film composition, **22:**45–49
high-temperature deposition, **22:**49–53
interpolated 3-D SIMBAD model, **22:**55–59
SIMSPUD vapor transport model, **22:**6–8, 14, 15–20, 40, 45, 76
SPEEDIE model, **22:**5
sputter deposition, **22:**6, 7
sputtering, reactive, **17:**112–118
structure, **17:**119–121
Mode locking, **21:**53, 54
Moderately reflecting films, optical properties, **6:**160–165
Modulation transfer function, calculation, **12:**125
Moir fringe patterns, **1:**194–196; **4:**226
of nuclei misalignment, **4:**156–157
Moir magnification, **1:**195
Molecular area
alkyl alcohols, **20:**150, 164
amphiphilic porphyrin, **20:**71
arachidic acid, **20:**31
ester alcohols, **20:**169
in mixed monolayers, **20:**75
Molecular devices, **20:**256, 265–267
N-octadecylferrocenecarboxamide, **20:**336
osmium *tris*-diphenylphenanthroline, **20:**340, 342
p-aminobenzoic acid, **20:**56
Molecular beam epitaxy (MBE), **10:**248; **11:**75–86
advantages, **11:**95–96
buffer layers, **11:**49–52
carbon as impurity, **11:**57–58
computer-controlled, **11:**41
diagram, **11:**37
doping, **11:**52–66
beryllium dopant, **11:**65–66
gallium arsenide dopant, **11:**52–69
germanium as atmospheric dopant, **11:**58–59, 65
Group III-V binaries other than GaAs, **11:**66–69
Group IV elements as dopants, **11:**57–61
Group VI elements as dopants, **11:**61–63
intentional doping control, **11:**53
isoelectronic doping, **11:**66
lead as dopant, **11:**60–61
magnesium as dopant, **11:**64–65
manganese dopant, **11:**65
N-type dopants, **11:**57–63
P-type dopants, **11:**63–66
silicon as dopant, **11:**58
tin as dopant, **11:**59–60
electron traps, **11:**52–56
epitaxial layer growth, **23:**18–21

Molecular beam epitaxy (MBE), (*continued*)
FETs and, **11:**49–51
growth by, **23:**6–21
growth stages in, **11:**48–52
growth systems, gas source, **23:**2
heterojunction bipolar transistors, **23:**55, 56–59
heterojunction structures and, **11:**84–86
hydrogen background pressure, **11:**58
impurities, **11:**95
in situ assessment techniques, **11:**40–48
insulator-semiconductor systems, **11:**89
liquid nitrogen-cooled shroud, **11:**38
majority carrier devices in, **11:**75–77, 109
metal semiconductor systems, **11:**86–88
minority carrier devices and, **11:**78–81
multiple-quantum-well structures, **23:**184–185
in multistage structures, **11:**82–84
periodic structures, **11:**86
problems and applications, **11:**35–96
projections, **11:**92–95
quadrupole mass spectrometry, **11:**41
quaternary III-V layer growth in, **11:**74–75
reflection electron diffraction, **11:**42–44
research, **11:**93
in selected-area epitaxy, **11:**89–92
source assemblies, **11:**38
substrate preparation, **11:**48–49
surface analytical facilities, **11:**40
surface cleaning and preparation, **23:**9–10
system design in, **11:**36–40
technological developments, **11:**86–92
ternary III-V compound alloys in, **11:**69–75
vacuum deposition on IV-VI layers, **11:**108–110
Molecular beam evaporation, of IV-VI semiconductors, **11:**110
Molecular devices, **20:**256, 265–267
Molecular dynamics, self-assembled monolayers of thiols on gold, **24:**207–212
Molecular dynamics modeling, **22:**123–125, 169–171
fundamentals
average film stress, **22:**144–145
constant temperature control algorithms, **22:**134–142, 169–170
discretization and integration, **22:**132–134
local stress function, **22:**145
periodic boundary conditions, **22:**142–144, 170
potential function, **22:**126–128
sorting algorithms, **22:**128–131
sputter deposition, **22:**125–126
adatom energy, **22:**150–152
collimated sputtering, **22:**162–165
deposition, **22:**149–160
ion bombardment and gas entrapment, **22:**152–156
low pressure/low-throw sputtering, **22:**165–169

step coverage, submicron liner, **22:**161–169
transport, **22:**146–148
voltage bias, **22:**156–160
Molecular films, built-up *see* Built-up molecular films
Molecular photodiode, **20:**244–245
Molecular reaction flux, steady-state values as function of pressure, **13:**169–170
Molecular rearrangement, in biotin film, **20:**211–212
Molecular sieve, **1:**14, 15
Molecular simulation, **24:**206–222
Molecular spectroscopy, high-resolution, multiple-quantum-well detectors, **23:**207–210
Molecular structure, condensed films and, **7:**320
Molybdenum
chemical vapor deposition, **5:**306–307
nitriding rate, **12:**288
Molybdenum hexacarbonyl, reduction, **5:**307
Molybdenum-nickel, **17:**33, 49, 53, 56–58
Molybdenum-tantalum, **17:**33, 49
Molybdenum-vanadium, **17:**16, 17, 33
Momentum
of arriving atoms, **3:**7–8
conservation in internal emission, **23:**89–91
theory of sputtering, **3:**63–70
Monitors, **14:**51–56
deposition rate, **1:**30–32
optical, **14:**29, 53, 55
Monoalkali antimonides, **12:**85–86
Monobasic trisodium phosphate, as parting agent, **7:**66
Monocrystalline films, **10:**245–258, *see also* Thin films
gallium arsenide family and, **10:**246–253
insulating crystal films in, **10:**253–258
Monokinetic range, of electrons, **2:**68–71
Monolayer assemblies
electron transfer, **20:**239–241
reaction centers, **20:**239–241
Monolayer growth mechanism, **4:**196–199
Monolayers, **5:**126–130; **20:**12, 12–14, 22, 149–153, 183–205, 184, 197, 198–199, 208, 209–221, 258–261, 331–346, 367–384, *see also* Built-up molecular films
of alkylphosphonic acids on metal oxides, **24:**163–169, 175
amphiphilic *see* Amphiphilic monolayers
build-up, **7:**337–343
close-packed, **7:**326–327
expanded or "liquid expanded," **7:**321–322
homogeneity and, **7:**322
insoluble, **7:**313–323
interferometric techniques, **7:**336
of long-chain compounds, **7:**319
metric thickness, **7:**335
organic, alkanethiol film structure, **24:**180–201
organic monolayers, **24:**181–201
polymeric, **24:**34
properties
deposited on solids, **7:**329–337
simulations, **24:**211–218
radioactive tracer technique, **7:**334

Monolayers, (*continued*)
self-assembled *see* Self-assembled monolayers (SAMs)
stability, **7:**317–323
transfer of on solids, **7:**323–329
X and Y types, **7:**340–341
Monomolecular films, on liquid surfaces, **7:**311, *see also* Monolayers
Monopartite ruling, for diffraction gratings, **11:**19
Monte Carlo techniques
attenuation length, **4:**344
defined, **24:**207–208
self-assembled monolayers of thiols on gold, **24:**207–212
Morse function, **6:**22
MOSFETS *see* Metal oxide semiconductor field effect transistors
Mossbauer spectroscopy, **17:**22
Motional dynamics, alkanethiol/gold SAMs, **24:**70–77
Motion sensor, **20:**288–290
Moving thin film foils, device functions achieved with, **8:**311
MOWIP *see* Multiquantum-well infrared photodetectors
MQW detectors *see* Multiple-quantum-well (MQW) detectors
MSSW *see* Surface waves, magnetostatic
MTF *see* Mechanical transfer function
Multialkali antimonides, **12:**107, 112
Multialkali cathodes, **5:**12
Multialkali films
first-order RIC and, **5:**26–27
optimum thickness, **5:**14
reflectance and absorption, **5:**25, 28–29
RIC principle and, **5:**30
sensitivity, **5:**44
transmittance and reflectance, **5:**25
Multialkali photocathodes
band-bending at surface, **12:**100
cesium role in, **12:**97–99
sensitivity, **12:**92
Multiaxial anisotropy, **6:**279–280
Multibandgap solar cells, **21:**321–323
Multicolor infrared hot-electron transistors, **21:**300–306
Multicolor quantum-well infrared photodetectors, **21:**289–300
Multicomponent materials, sputtering, **3:**98–100
Multilayer coatings
for extreme ultraviolet astronomy, **10:**114–117
high energy radiation damage to, **8:**15
surface scattering in, **10:**176–188
Multilayer fatty acid soap films, **7:**349
Multilayer films, *see also* Built-up molecular films; Films; Monolayers; Multilayer Langmuir films; Multilayer magnetic films
build-up, **7:**327–343
contact potential and, **7:**365–366
general properties, **7:**346–350
inhomogeneous films as, **8:**88–92
substances used for, **7:**350
Multilayer interference filters
absorbing, **5:**50
advantages, **5:**47–48
applications, **9:**74
band pass, **5:**48
change points in, **5:**69
circular variable type, **5:**84–85

coating materials for, **5**:50–51
construction, **5**:48
decoupling layer, **5**:84
design, **5**:47–58, 59–60
effective-interface method, **5**:75–78
equivalent indices, **5**:63–64
equivalent layers, **5**:60–68
filter types, **5**:51
infrared, **5**:53
long-wavelength, **5**:48
mathematical theorems and computations, **5**:61–62, 69–70
optical description of designs, **5**:72–74
refining methods, **5**:68
short-wavelength, **5**:48
substrates and media for, **5**:49–50
symmetrical, **5**:61
theoretical basis, **5**:53–59
Multilayer Langmuir films, **7**:391–392
Multilayer magnetic films, **6**:205–294
advantages, **6**:206
anisotropy, **6**:212, 225
exchange anisotropy, **6**:213–223, 268–269
applications, **6**:291–294
Bitter pattern, **6**:230–231, 248
Bloch walls, **6**:229–230, 240
coercive force and wall mobility in, **6**:237–238
constant magnetization, **6**:251–261
coupling effects, **6**:247–290
coupling strength and angular dependence, **6**:252–253
demagnetization effects, **6**:236
direct coupling, **6**:226
direct exchange coupling, **6**:208–213
domain walls and structures, **6**:227–230, 274–278
DRO memory schemes and, **6**:291–292
dynamic reversal behavior, **6**:269–272
effective mean values, **6**:263
exchange coupling, **6**:245
ferromagnetic, **6**:210–211, 242
ferromagnetic-antiferromagnetic coupling in, **6**:212–223
flux closure, **6**:238
hard and soft layers, **6**:209
high-order and helical anisotropies, **6**:278–290
hysteresis loops, **6**:256, 266–267
hysteresis processes, **6**:278
indirect coupling, **6**:241–247
interaction phenomena, **6**:208–250
magnetization variable, **6**:261–269
magnetostatic coupling, **6**:223–227
minimum-energy trajectory method, **6**:259
NDRO memory schemes and, **6**:292–293
Nel walls, **6**:229, 231–232, 234, 238
''orange peel'' coupling, **6**:225–226, 241
oxidation experiments, **6**:243
pinhole coupling, **6**:226, 241–247
quasistatic reversal behavior, **6**:251–269
quasiwalls, **6**:234–235
Rayleigh region model, **6**:222
resonance phenomena, **6**:272–274
RKKY interaction, **6**:245–247
sequence dependent switching, **6**:265–266
stray-field coupling, **6**:224, 248
structure properties, **6**:265–269
wall motion phenomena, **6**:240
Multilayer matrix, **5**:53–56
Multilayers, **24**:17–18, 193–197

Multilayers, (*continued*)
homogeneous, **5:**53, 58
inhomogeneous, **5:**65
irregular, **5:**65, 134
nonabsorbing, **5:**60
optical thickness in, **5:**57
periodic structures in, **5:**59
reflectance and transmittance in, **5:**57–58
reversing order, **5:**56
selectively reflecting, **5:**132–133
symmetrical, **5:**53
Multilayer stacks, of dielectric film, **8:**3
Multilayer structures, MBE in, **11:**82–84
Multimode cavity, **18:**154
Multiple antenna, **18:**154
Multiple cavity filters, **9:**124–128
augmented double-cavity, **9:**127–128
transmittance and absorbance curves for, **9:**128
Multiple layers, **14:**51
Multiple-quantum-well (MQW) detectors, **23:**174–215, 174–245
absolute responsivity, **23:**190–194
applications, **23:**204–211
dark current for, **23:**189–190
design, **23:**178–185
direct, **23:**174–175
electrical bandwidth, **23:**196–204
by diode-laser mixing technique, **23:**198–199
by microwave rectification technique, **23:**199–200
by photoelectron generation-recombination noise technique, **23:**196–198
design, **23:**214–215
time constants governing, **23:**200
energy levels, **23:**178–184
fabrication, molecular beam epitaxy, fabrication, **23:**188–189
grown by molecular beam epitaxy, **23:**184–185
intersubband absorption, **23:**185–188
packaging, **23:**188–189
heterodyne, **23:**175–178
for high-resolution molecular spectroscopy, **23:**207–210
improvements in, **23:**211–215
heterodyne sensitivity, **23:**201–204
instrumental resolution, **23:**205–207
instrumental sensitivity, **23:**205–207
intersubband absorption, **23:**185–188
in long-range, high-data-rate communications, **23:**210–211
optical-heterodyne experiments and, **23:**196–204
photoconductive gain in, **23:**194–195
quantum efficiency, external, **23:**195
spectral responsivity, **23:**190–194
Multiple-quantum-well infrared photodetectors, **21:**4, *see also* Quantum-well infrared photodetectors
Multiple-quantum wells, (MQW); **21:**240; **23:**228–238
Multiple-quantum-well (MQW) structures, **21:**333
far infrared emission from, **23:**243–247
hot-electron phototransistors, **21:**239–308

quantum-well solar cells, **21:**364–366
Multiple scattering technique, **21:**12
MWIR detectors, **21:**191–195

N

Nanocrystalline particulate films
absorption coefficient, **20:**23
absorption spectra, **20:**23–25
bandgap energy, **20:**24–26
cadmium sulfide, **20:**25–26
characterization, **20:**16–17
chemical generation, **20:**14
copper sulfide doping, **20:**22
crystal structure of silver, **20:**28
electrochemical generation, **20:**15
growth mechanism, **20:**19–20
lead selenide, **20:**22–23, 33–37
lead sulfide, **20:**23
microscopic characterization, **20:**18
mixed sulfides, **20:**22
semiconductor, **20:**23–26
silver, **20:**15, 26–29
silver resistivity, **20:**29
size quantization, **20:**21–23
template, **20:**12
zinc sulfide, **20:**23
Nanodomains
analysis by microscopy, **20:**257–265
by phase separation, **20:**256–267
Nanofabrication *see* Fabrication
"Nanografting," **24:**101–102
Nanomechanical properties, **24:**83, 93–98, 104, 105
Narrow-band-pass filter, transmittance, **5:**77, 82–83
Narrow plasma mode, **18:**160, 161, 176, 178, 188–190, 208
NASA, **10:**133, 158
NEA materials *see* Negative electron affinity materials
Near infrared spectral region, **23:**117
Néel wall **1:**312ff.
in multilayer films, **6:**229, 231–232, 234, 238
Nel point, magnetic material, **9:**268
Nel-Taniguchi anistotropy, **1:**280–282
Nel temperature, **6:**213, 293
Negative differential, conductance, resistance, **15:**236–246
Negative electron affinity, **12:**114–115, 121
Negative electron affinity devices
cathodes, quantum efficiency, **12:**157–158
GaP dynodes, in photomultiplier tubes, **12:**135
in liquid phase epitaxy, **7:**242–246
photocathodes, fabrication, **12:**129–133
Negative electron affinity materials, **12:**112–140
absorption coefficient, **12:**70
activation techniques, **12:**128–129
carrier transport and, **12:**123–125
cold cathodes and, **12:**134–135
dark current, **12:**135
diffusion length and doping, **12:**125–126
electron equilibrium, **12:**65
escape depth vs. diffusion depth, **12:**133
photoemission, **12:**62
in photoemitters, **12:**136–140
secondary emission, **12:**133–134
stability, **12:**135–136
surface escape probability, **12:**126–128
Negative temperature coefficient, for resistivity, **6:**142

Neodymium-based superconductors, **16:**2, 78
NEP *see* Noise equivalent power
Net flux ratio
  for assembly of layers, **9:**82–86
  maximum value, **9:**86–87
  for metal-dielectric interference filters, **9:**79–80
  properties, **9:**87–88
  for single layer in metal-dielectric interference filters, **9:**81–82
Neural networks, **21:**60–68
Neuron emulation, semiconductors, **21:**59–68
Neutral flux, **22:**107–110, 111
Neutral metastable atom lithography, **24:**229, 230
Neutron activated tracer scanning, **16:**244
Neutron activation sectioning, **16:**242
Neutron diffraction, **17:**22
Ngai infrared divergence model, **11:**303–304
Nichrome film resistors, **2:**386–389
Nickel
  crystal growth in, **5:**313
  deposited on copper, surface structure, **4:**212–214
  sputter deposition, voltage bias, **22:**156–159
Nickel cyclams, as amphiphilic catalysts, **20:**268–271
Nickel film
  magnetic annealing, **1:**294
  outgassing characteristics, **1:**3
  saturation magnetization, **1:**58–59
Nickel-iron magnetic film, **1:**277–234
  growth process, **1:**279–280
  studies using, **1:**322–331
Nickel-iron-manganese alloy films, **6:**215–216
Nickel iron oxide, **14:**15, 27
Nickel oxide film, infrared spectra, **15:**304–305
Nickel-palladium, **17:**33
Nickel-platinum, **17:**49, 53, 56, 57
Nickel-titanium, **17:**49, 53, 56, 57
Nickel-titanium carbide, **17:**65
Niobium, nitriding rate for, **12:**288
Niobium nitride, **14:**55
Niobium pentoxide, **14:**17
  chemical vapor deposition, **5:**304–306
Niobium pentoxide film, **6:**12–13
Niobium substrates, ionic conduction in, **6:**3
Niobium-tantalum, **17:**33
Niobium-vanadium, **17:**33
Nitride layer, **14:**7
Nitrides, **17:**101–104, 109, 119
  mechanical properties, **13:**87, 89
Nitriding, glow-discharge, **12:**286–289
Nitrogen, **18:**138, 140–144, 177, 180, 185–186, 189–193, 195–202, 204, 206–207, 209–217
  aluminum reflectance and, **1:**154–156
  film growth in, **5:**295
  nitriding process and, **12:**286
Nitrous oxide ionization chamber, **10:**78–79
NLORM *see* Nonlinear overrelaxation
NMR spectroscopy, **17:**22; **24:**150–151
  alkanethiols on gold nanoparticles, **24:**154–163, 175
  alkylphosphonates on metal oxides, **24:**163–169, 175
  alkylsiloxane SAMs, **24:**169–175
  techniques

cross-polarization and magic-angle spinning (CPMAS), **24:**151–152
line shape measurements, **24:**153–154
relaxation parameters, **24:**152–153
NMR studies, **24:**150–176
order and disorder, **24:**131–132
organosulfur compounds used to form, **24:**12–36
patterning, **24:**228–229
phthalocyanine-based, **24:**24–25
quartz crystal microbalance, **24:**46, 48–52, 65–66, 68
redox-active, **24:**26
scanning tunneling microscopy, **24:**82, 92, 99, 114
silver substrate, **24:**127–128
simulations, **24:**205–222
stability, **24:**112
structure, **24:**150–151
atomic force microscopy, **24:**83–89
differential studies, **24:**112, 114, 121–130
terminal groups, **24:**13–17, 88–89, 159
3D-SAMs, **24:**154
two-component patterned SAMs as templates, **24:**244–250
uses, **24:**228
viologens, **24:**29–30
X-ray-induced damage, **24:**119–121
Noise, **3:**285; **15:**230, 238
Noise currents, in semiconductor films, **5:**225
Noise equivalent power, **21:**27
Noise equivalent temperature difference, **21:**219–221, 256, 269–275
Non-Debye loss peak, **11:**288
Non-Debye response, **11:**234–235
Nondestructive readout memory schemes, **6:**292–293
Nonhomogeneous films, plasma resonance absorption in, **9:**190
Nonidentical stacks, scattering in, **10:**219–224
Nonlinear magnetic effects, **15:**92
devices using, **15:**95
frequency doubling in, **15:**101
half-frequency generation and subsidiary absorption, **15:**92
Nonlinear microwave absorption processes, **18:**148–152, 184–187
Nonlinear overrelaxation, BTRM, **22:**202
Nonmetallic films
accumulation barrier, **8:**199
avalanche multiplication, **8:**211
disordered, **8:**169–246
forbidden gap, **8:**192–193
high-field conduction, **8:**195–211
low-temperature field ionization, **8:**208–210
mobility, **8:**188–190
space charge effects, **8:**210–211
Nonparabolicity, **15:**25
Nonradiative decay processes, **23:**254–255
Nonradiative surface plasmons, *see also* Surface plasmons
angular dependence of differential excitation, **9:**204
asymmetric layer system, **9:**207–208
asymmetry of surface loss intensity and, **9:**210–213
coating with plasma film, **9:**209

Nonradiative surface plasmons, (*continued*)
coupling by grating, **9:**221–223
and coupling dispersion of symmetric layer system, **9:**206–207
dispersion, **9:**208–209
dispersion curve, **9:**226
electron energy losses and, **9:**199–219
excitation, **9:**199–255
by electrons, **9:**215–223
by evanescent light waves, **9:**230–255
of guided light modes, **9:**241–242
excitation probability of surface losses for, **9:**200–201, 215
light scattering and, **9:**245–248
nonnormal incidence, **9:**209–210
and optical constants or dielectric functions of metals, **9:**234–238
photoelectron emission, **9:**240–241
probability for exciting surface losses in, **9:**204–206
reflection of electrons at surface, **9:**213–215
retardation, **9:**202
roughness effects, **9:**242–244
small spheres and, **9:**218–219
surface coatings and, **9:**205
Nonuniformity, imaging arrays, **21:**219–228
Nordheim expression, for thin wires, **6:**129
n-p-n structure, **21:**2, 32
NRDO memory schemes, in multistage films, **6:**292–293
NTT/Matsuo source, **18:**152–153, 155, 157–162, 167–168, 178–180, 183, 188, 189, 194, 209, 217–218, 221–222
N-type dopants, in MBE, **11:**57–63
Nuclear magnetic resonance *see* NMR
Nucleation, **1:**207–215; **2:**2–7; **3:**212–214; **20:**157–158, 159, 164, 169–170, 170–174, 174–177, *see also* Ice nucleation
angle of incidence, **3:**214
barrier size
nature of film and substrate, **2:**7–10
substrate temperature and deposition rate, **2:**10–11
binding energy to substrate, **2:**11–14
frequency, **2:**6
lattice misfits and, **7:**124
preferred, **3:**213, 214
structural consequences, **2:**7–14
surface diffusion, **2:**11–14
in thin film formation, **4:**97–136, 150–186
at surface steps, **4:**152
data analysis, **4:**122–124
deposition rate and, **4:**154
electrodeposits, **4:**192–217
epitaxial, theory, **4:**129–131
experimental results, **4:**114–127
heterogeneous, **4:**98–114
imperfections and impurities, **4:**120–121
rate, **4:**151–152
direct measurement, **4:**124–126
substrate perfection and purity, **4:**152–154
substrate temperature, **4:**155
Nucleation modifying layers, **9:**42, 45
Nuclei
alignment, effect of contaminants, **4:**161

coalescence, **4:**165–175
critical, **2:**4ff.
formation and growth, **1:**210
free energy of formation, **4:**101–102
growth, **4:**163–165
misalignment, **4:**156–157
orientation, **4:**156–161
perfection, **4:**155
shape, **4:**100–101, 161
Nucleus radius, critical, **4:**198
Numbers-modulation theory, **3:**293

O

Occupied state, in amorphous solids, **8:**179
1-Octadecanethiol/gold monolayer
aliphatic chain annealing, **24:**67–68
thermodynamics of formation, **24:**63–65
N-Octadecylferrocenecarboxamide, **20:**333–339
Octadecylphosphonic acid (ODPA), on metal oxides, **24:**163–164
Octadecylsiloxane, monolayers on silica, NMR, **24:**170–175
Octadecyltrichlorosiloxane (OTS), **24:**170
Ohmic contacts
formation at low annealing temperature, **13:**27
in hopping conduction, **8:**196–199
preparation, **11:**88
in semiconductor films, **5:**220–221
Oil molecules, monolayers, **7:**314
Oligo(ethylene glycol), **24:**248–249
Oligonucleotides, organosulfur analogs of biological oligomers, **24:**35–36
Omegatron, **1:**42–44
One-M filter
admittance vs. wavelength, **9:**120–121
configuration, **9:**116
defined, **9:**112
design, **9:**112–124
examples, **9:**117–124
matching stacks and, **9:**113–114, 119
simplest type, **9:**113
spectral absorbance, **9:**122
spectral transmittance, **9:**123
110 K phase, **16:**2, 4
Bi-Sr-Ca-Cu-O thin film, **16:**30, 32–34, 78, 93, 99, 100
lead addition, **16:**4, 30, 32–34, 99, 100
low oxygen partial pressure annealing, **16:**4, 33
One-way glass, **5:**126
OPD *see* Optical path difference
Open-circuit voltage, quantum-well solar cells, **21:**355–357, 365
Optical absorption, **15:**192
joint density of states, **12:**57
in photoemission, **12:**55–59
in semiconductor films, **5:**222–223
Optical absorption coefficient, **12:**55
FeO*x* films, **13:**50
Optical anisotropy, **19:**279, 280
in films, **19:**168–69, 173–74, 178, 257
Optical coatings, **14:**22
activated reactive evaporation applications, **13:**101–102
Optical conductivity, **15:**192
Optical constants
ion-beam processing, **13:**121
metals, **9:**234–238
thin films, **2:**193–238
Abels method, **2:**210–213
critical angle method, **2:**227–228

Optical constants (*continued*)
metals, **2**:339–340
polarimetry, **2**:218–227
reflection computations using, **2**:322–325
spectrophotometry, **2**:213–228
in vacuum ultraviolet, **1**:141–145, 169–171
from measurement of electron energy losses, **1**:145
gold, **1**:167
opaque films, **1**:142–144
semitransparent films
on absorbing substrate, **1**:145
on transparent substrate, **1**:144–145
Optical density, thin film, **3**:34–35
Optical elements
autopositive preparation in, **10**:269
correction capabilities, **10**:305–308
correction procedure, **10**:291–294
durability criteria, **10**:285
figuring experiments for, **10**:286–291
future applications, **10**:305–308
mask preparation, **10**:266–267, 269–270
Polaroid negative, **10**:293
Polaroid negative of, **10**:293
polishing, **10**:263
process error consideration, **10**:294–296
stress-free mixtures, **10**:278–280
stress variation, **10**:276–278
test correction, **10**:297–299
thickness distribution, **10**:296–297
Optical films, **1**:69–121
Optical functions
aluminum in bulk, **19**:131–33
aluminum fluoride, **19**:236
aluminum oxide, **19**:236, 238, 241
aluminum thin films, **19**:150–83
during nucleation and growth, **19**:150–62
liquid aluminum film, **19**:167–68
particle films, **19**:168–83
anisotropic material-tensors-simulation, **19**:298–300
cerium fluoride, **19**:236, 237
hafnium amorphous silicon, **19**:70–73
hafnium oxide, **19**:236, 239
hydrogenated amorphous silicon, **19**:70–73
hydrogenated fluoride, **19**:236
lanthanum fluoride, **19**:236, 238
(lead, lanthanum)titanium oxide film, **19**:261
lead titanium oxide film, **19**:260
magnesium oxide, **19**:217
microcrystalline silicon, **19**:94–104
scandium fluoride, **19**:236, 243
scandium oxide, **19**:236, 239
thorium oxide, **19**:236
transparent fluoride films, **19**:236–42
transparent oxide films, **19**:236–42
vitreous silica, **19**:214
yttrium oxide, **19**:236, 240
zirconium oxide, **19**:236, 240
Optical gap
hydrogenated amorphous silicon, **19**:73–75
hydrogenated amorphous silicon-carbon alloys, **19**:85–88
Optically stimulated hot-electron spectroscopy, **21**:252–255
Optical mass, defined, **6**:154
Optical measurements, coordinate system, **2**:195–197
Optical methods, for semiconductor film analysis, **5**:187–188

Optical monitoring, in filter production, **9:**136–140
Optical parameters
  angle of incidence, **6:**182–185
  interferometry, **6:**186–187
  polarimetry, **6:**185–187
Optical path difference map, **10:**265–267, 293
Optical performance, deterioration in dielectric films, **8:**2–4
Optical phase relations, **1:**77–78, 80
Optical plasma resonances, in thin films, **6:**193–197
Optical projection printer, **8:**259
Optical properties
  absorption band in, **6:**201
  alkali-antimonide photocathodes, **12:**101–112
  aluminum oxide, **13:**140
  bulk metals, **6:**152–155
  discontinuous films, **8:**139
  ellipsometric measurements in, **6:**157
  film structure and, **6:**199–202
  ion-beam processing, **13:**120–121
  local field and, **6:**201
  magnesium fluoride, **13:**137
  metallic films, **6:**151–202
  normal incidence in, **6:**158–159
  silicon dioxide, **13:**133–134
  tantalum pentoxide, **13:**139
  temperature dependence, **10:**34–35
  titanium dioxide, **13:**130–131
  zirconium dioxide, **13:**135–136
Optical response functions, **19:**287, 298
Optical sensor, **16:**101, 109, 110
Optical transitions between subbands, **23:**238–262, *see also* Intersubband transitions
Optical transmission, thin films, **2:**194–209
Optical waveguides, **10:**168; **14:**24, 47–48
Optics
  thin films
    computational methods, **1:**84–120
    concepts, **1:**71–83
    graphical methods, **1:**101–115
    matrix notation, **1:**87–92
    precision measurement, **4:**1–96
    vector diagrams, **1:**101–102
Optoelectronics, **21:**63, 240
"Orange peel" coupling, in multilayer films, **6:**225–226, 241
Orbiting Solar Observatory, **10:**158, 160
Order
  AFM studies, **24:**83–89
  diffraction studies, **24:**131–132
  IR studies, **24:**66–69
  NMR studies, alkylphosphonate monolayers, **24:**168–169, 175
  in SAMs, **24:**83
  simulations, **24:**212–214
Ordering, directional, iron atom pairs, **1:**280–282
Organic guides
  laterally bounded, **10:**242
  plasma polymerization type, **10:**241
  solution deposition techniques, **10:**240–241
Organic solutes, oxide layers deposited from, **5:**87–140, *see also* Oxide layers
Organosilane films
  monomers, **20:**113
  photochemical properties, **20:**115–117

Organosilane films (*continued*)
two-step patterning, **20:**122–123
ultraviolet photochemistry, **20:**112–117
Organosilane ligands, **20:**121
Organosilanes
structure of aromatic hydrocarbon, **20:**113
structure of ligands with nitrogen, **20:**121
Organosulfur compounds, **24:**2
analogs of biological oligomers, **24:**35–36
polymeric, **24:**33–35
used to form SAMs, **24:**12–38
Orientation
nuclei, **4:**156–161
at surface steps, **4:**160–161
changes in, **4:**175–180
deposition rate and, **4:**160
substrate temperature and, **4:**158–160
preferred, **17:**126–129
refractory compound deposits, **13:**85, 87
titanium carbide, **13:**71–72
Orthoferrites, for magnetic bubbles, **9:**275
Orthogonality, **15:**254
Oscillators, **16:**101, 103
magnetostatic, **15:**85, 86
OSHES *see* Optically stimulated hot-electron spectroscopy
Osmium coatings, in extreme ultraviolet astronomy, **10:**97–100
Osmium *tris*-diphenylphenanthroline, film electron exchange rate, **20:**342–344
Outgassing, vacuum materials, **1:**2
glass, **1:**5–6
metal, **1:**9–11
Outgoing flux, **22:**102, 103
Overcoatings, **1:**171–179
aluminum oxide as, **12:**25
diffraction gratings, **1:**180–184
evaporated aluminum mirrors with, **12:**25–32
silicon monoxide for, **12:**25–26
Overfield, defined, **6:**3
Overlap matrix, **15:**254
Overpotential
analogy with, **6:**3
anodic oxide films, **6:**21
Oxidation, **15:**168–174
of absorption coatings, **10:**42–43
double bonds to acid groups, **20:**45
overlayers on corrosion resistance, **15:**171–174
silicon, **4:**260–262
Oxidative coupling
by diffusion, **20:**78–79
by reactants from substrate, **20:**79–81
polymer networks, **20:**76
Oxide coatings, ortho ester, **5:**95
Oxide films
semiconductor *see* Semiconductor oxide film
solution growth technique, **12:**213–214
Oxide layers
as antireflection coatings, **5:**123–124
applications, **5:**123–139
baked, **5:**96, 101–105
characteristics, **5:**94–105
coating
apparatus and techniques, **5:**90–94
materials, **5:**94–99
"disturbed structures," **5:**109

glass substrates, **5:**101
high-index, **5:**105
high-temperature, **5:**140
high-valence, **5:**122
insoluble, **5:**140
for lighting filters, **5:**134–138
multiple layers, **5:**99–101, 130
for organic solutions, **5:**87–140
organogenic, **5:**120–121
"priming," **5:**101
as semiconductor coating, **5:**138
silica, **5:**115–122
special, **5:**105–122
titania, **5:**105–115
trouble zones in, **5:**93
Oxides, **17:**101, 104, 108, 135, *see also* specific oxides
binary, **9:**9–10
as dielectric films, **8:**35–40
doped, **9:**10–12
film thickness, **4:**37–41
capacitance measurement, **4:**39–40
x-ray measurement, **4:**40–41
growth
in gaseous anodization, **6:**49–51
germanium reflectance and, **1:**163–165
in vacuum, **4:**37
heats of formation, **13:**118
infrared absorption, **13:**122
mechanical properties, **13:**92–94
in semiconductor oxides, **9:**9–12
as substrates in semiconductor film, **5:**154
target sputtering, **9:**53–59
hollow-cathode source in, **9:**50
Oxidized aluminum, reflecting properties, **11:**13
Oxygen, **18:**173, 176–177, 187–188, 207, 216–226
aluminum reflectance and, **1:**154–156
in high Tc superconductors, **16:**4
Bi-Sr-Ca-Cu-O superconductors, **16:**3, 4, 27, 65, 66
c-axis parameter, **16:**3, 21, 27, 64, 86
N2O, **16:**22, 61, 68, 72, 79
ozone, **16:**24, 51, 54
Y-Ba-Cu-O bulk or thin film, **16:**3, 21, 27, 64–66, 69, 78, 86, 89, 90
magnetic anisotropy and, **1:**285
removal from molecular beam epitaxy system, **23:**9
Oxygen isotopes, ionic conductance and, **6:**24–26
Oxygen plasma, **14:**83–88
AC discharge, **14:**86–88
DC discharge, **14:**85–86
general characteristics, **14:**83–85
Oxygen plasma-solid interaction, **14:**88–96
at anodic bias, **14:**90–94
at cathodic bias, **14:**94–96
at floating potential, **14:**88–90
Oxygen vacancy, in crystal, **9:**10

P

PABA *see* *p*-Aminobenzoic acid
Packing density
dielectric films, **8:**8–9
as function of acceleration voltage, **13:**25–26
Paint, **20:**221
Pair ordering, **15:**164–165
magnetic anisotropy and, **15:**185
sputter yield, dependence on, **15:**164

Pair production, in electron transport, **12:**60–61, 110
Palladium silicide, internal emission imaging sensors using, **23:**87
Palladium-silver glaze, **2:**392–393
PAPVD, **13:**60, 67–68
Parallel asynchronous processing, **21:**62–68
Parallel-plate-type reactor, **12:**267–268
Parallel strip transducer, **15:**47, 51
Parametrons, **1:**321
Parlodion coating, **7:**56–57
  plugging, **7:**111
  in space applications, **7:**107
  spurious, **7:**52
  substrate cleaning and, **7:**56
  transmittance, **7:**51
  in unbacked metal filters, **7:**49–53
Parlodion films, **7:**56
  problems using, **7:**70–71
  as thin film substitute, **7:**70
Partially miscible systems, **6:**304
Partial yield, **14:**6
Particle bombardment, **17:**82
  uniformity, **17:**117, 137
Particle cloud, plasma, **18:**292, 294, 307
Particle contamination, plasma, **18:**289, 314
Particle distribution
  energy, **17:**93
  mass, **17:**93
  spatial, **17:**93, 137
  velocity, **17:**93
Particles
  macroparticle, **17:**93–96
  in thin-film processing plasma
    charging, **18:**292, 312
    cloud, **18:**292, 294, 307
    concentration, **18:**296, 297, 313, 315
    deposition, **18:**294
    etching, **18:**309
    forces on particle, **18:**313, 315
    formation, **18:**291
    in-situ detection, **18:**291, 295, 299, 301, 309
    laser light scattering, **18:**291, 295, 301, 309
    modeling, **18:**308
    nucleation, **18:**294, 298, 300, 315
    size, **18:**292, 299, 300, 304
    sputtering, **18:**301
    structure, **18:**305, 306, 310
    trapping, of particle, **18:**306
Parting agents
  alkali halides and related compounds, **7:**65–66
  barium chloride, **7:**66
  chemical activity, **7:**65
  Creme-Cote as, **7:**67
  fluorescein, **7:**64
  fuchsin, **7:**61–63
  Knox–60 as, **7:**69
  for metal films, **7:**60–69
  monobasic trisodium phosphate, **7:**66
  PCB–100, **7:**69
  phenolphtalein, **7:**64
  polyvinyl alcohol, **7:**68
  polyvinyl pyrrolidone, **7:**68–69
  sodium chloride, **7:**66
  sodium metaphosphate, **7:**63–64
  sugar-aerosol solution, **7:**67–68
  Teepol 610 as, **7:**67
  Victawet as, **7:**66–67
Passivation, electrical requirements, **4:**289–292
Passive interconnections, **16:**95, 101, 108

Patterning, of SAMs, **24:**228–229
Patterns
generation, **2:**410–412; **20:**224–225
selective removal, **2:**411–412
PCB–100, as parting agent, **7:**69
PCTFE *see* Poly(chloro-trifluoroethylene)
Peak detectivity, **21:**149, 160
Pease's theory of sputtering, **3:**67–69
PEC reactions, etch rates, **13:**176
PECVD, **22:**182
Pendulum chain, waves of, **5:**5–6
Penetration depth, superconductor, **1:**257–260
Peptides, organosulfur analogs of biological oligomers, **24:**35–36
Percolation threshold, thin aluminum films, **19:**161, 186
Periodic boundary, sputter deposition, **22:**142–144, 149
Permalloy films, **1:**277–334
by sputtering, **3:**117–120
cracking, **16:**282, 305, 324–325
magnetic annealing, **1:**289–294
magnetization ripple, **1:**295–297
Permanent magnet, **18:**153–155, 163–165
Permeability tensor, **15:**3, 27
altered by anisotropy, **15:**131
components displayed, **15:**4
derived, **15:**127
in magnetostatic waves, **15:**27
Permeation studies
of gas through solids, **1:**2–3, 7–8, 11
helium through glass, **1:**8
permeation rate, **1:**7
velocity of hydrogen, **1:**12
ion-beam processing, **13:**124
Perpendicular anisotropy, **16:**158, 196, 203, 210, 212
Phase-adapting film, **5:**15
Phase behavior, diffraction studies, **24:**130, 138–141
Phase boundary, superconductor-normal state, **1:**247
Phase change
absolute, on reflection measurement, **4:**84–90
Feco interferometric method, **4:**88–90
error sources, **4:**86–88
methods, **4:**85–86
internal stress due to, **3:**243–244
nonabsorbing film on absorbing substrate, **4:**38
photocurrent and, **5:**36
Phase contrast, **1:**189, 191
Phased-array radar, **15:**61, 63
Phase diagram
aliphatic chain molecules, **20:**367–383
docosanoic acid, **20:**368
ethyl eicosanoate, **20:**373
fatty acid films, **20:**372–373
Phase separation, by scanning surface-potential microscopy, **20:**260–265
Phase transfer catalyst, **20:**100
Phase transitions, **1:**220–223
Phase velocity, **18:**126, 130–132, 134–136, 139, 146, 179
Phenolphthalein, as parting agent, **7:**64–65
Phenyltrichlorosilane film, mass spectra, **20:**114–117
Phillips process, **3:**137
Phonoelectric effect, multilayer homojunction system, **21:**30
Phonons, defined, **6:**153
Phosphine, GexSi1-x growth and, **23:**34–35, 41

Phosphor, sodium salicylate, **4:**68–69
Phosphor films, transparent, **3:**200
Phosphorus nitride film, plasma chemical vapor deposition, **12:**270–271
Phosphosilicate glass, **4:**262, 305–309
 infrared spectra, **15:**299
  band assignments, **15:**293
  carbon incorporation and, **15:**294
  fluorinated silicon nitride and, **15:**295
  hydrogen incorporation and, **15:**290–291
  nitrogen content determined by, **15:**293
 in multilevel passivation, **4:**313–314
 sodium migration through, **4:**307–309
Phosponic acids, **24:**163
Photoablation, **13:**153
 metals, **13:**184
 polymers, **13:**171, 182
Photocathodes
 alkali antimonide *see* Alkali antimonides
 fibromultiplier use and, **12:**141
 high-sensitivity, **12:**99
 interference, **5:**1–44
 multialkali *see* Multialkali photocathodes
 phase- and amplitude-adapting films and, **5:**30–31
 photoemission, **12:**90–94
 silver-oxygen-cesium *see* Silver-oxygen-cesium photocathode
 spectral response, incident light and, **12:**92–93
 thermionic emission from, **12:**100
Photochemical reactions
 gas-phase, **13:**155–156, 171, 177
  gallium arsenide, **13:**188
  metals, **13:**184
  semiconductors, **13:**185–186
  silicon, **13:**187
 laser-induced etching, **13:**155–156
 liquid-phase
  metals, **13:**184
  semiconductors, **13:**185–186
  silicon, **13:**187
 (3-mercaptopropyl)-trimethoxysilane, **20:**129
 solid-phase, **13:**156, 173, 176–177
  gallium arsenide, **13:**188–189
  semiconductors, **13:**185–186
  silicon, **13:**187
Photoconduction, **3:**197–200
 in amorphous solids, **8:**235–239
 in lead salt films, **3:**291–299
 in semiconductor films, **5:**223–225
Photoconductive gain
 multiple-quantum-well detectors, **23:**194–195
 n-doped quantum-well infrared photodetectors, **21:**138–143
 p-doped quantum-well infrared photodetectors, **21:**160
Photoconductors, **23:**122–123
 fabricated from mercury cadmium telluride, **23:**129–131
 infrared detectors, **21:**4
 speed of response, **3:**197–198
 types, **11:**106
Photocurrent, **21:**314–316
 absorption and, **5:**36
 amplification, **11:**189
 attenuation length and, **5:**11
 energy diagram for triad molecule, **20:**266
 function of purple membrane layers, **20:**287
 generation mechanism, **20:**285–287

phase change and, **5**:36
rectification
by biorhodopsin films, **20**:284–286
mechanism, **20**:285
in RIC'S, **5**:26–27
versus amphiphilic triad shape and orientation, **20**:247–249
Photodetectors, chemical deposition process, **12**:228–230
Photodiodes, **11**:124, 125, 130, 168–172, 172–177, 177–185, 186–189, 279–281, *see also* Thin-film IV-VI semiconductor photodiodes
from mercury cadmium telluride, **23**:131–133
molecular, **20**:244–245
Photoelectric conversion efficiency, **20**:249–254
Photoelectric film
optical properties, **5**:26
reflectance, **5**:24
Photoelectric generation-recombination noise techniquc, bandwith of multiple-quantum-well detectors, **23**:196–198
Photoelectrochemical cell
multiquantum-well structures, **21**:365
purple membrane, **20**:283–285
Photoelectrochemical etching, **13**:156, 185–186
Photoelectrochemical solar cells, chemical deposition process for, **12**:226
Photoelectronic analysis, of semiconductor films, **5**:224
Photoelectrons
attenuation length, **5**:2, 9–10
bulk generation, **12**:55
emission, **5**:8–12
production rate, **5**:10
registration of plasma resonance by, **9**:188–190
Photoemission
electron-electron scattering, **12**:60–62
electron escape from surface, **12**:66–70
electron transport, **12**:59–66
experiments, **4**:341–350
internal, **23**:88–92
as detection mechanism in Schottky barrier detectors, **23**:135–136
mechanism, **12**:54–70
negative electron affinity materials, **12**:62
nonradiative surface plasmons, **9**:240–241
optical absorption, **12**:55–59
as ''surface'' phenomenon, **12**:54
Photoemission quantum efficiency, **12**:64, 69, 112
Photoemissive detectors, **21**:14
Photoemissive film, **5**:2
Photoemissive materials, **12**:53–158
Ag-O-Cs(S–1) photocathode, **12**:70–84
alkali antimonides, **12**:84–112
applications, **12**:140–157
for image tubes, **12**:145–157
photoemission mechanisms and, **12**:54–70
in photomultipliers, **12**:140–144
quantum efficiency improvement in, **12**:112
resistance peak in, **12**:95
spectral response, **12**:80

Photoemissive properties, of alkali antimonides, **12:**90–94
Photoemissive surfaces, **7:**242–244
Photoemitters
alkali antimonide type, **12:**59
quantum efficiency, **12:**64, 69, 112
transferred-electron type, **12:**136–137
Photoetching, of polymers, **13:**182
Photoexcitation
broad-band, **15:**231–232, 234
narrow-band, **15:**230, 233
Photoexcited coherent tunneling, **21:**294–298
Photogenerated carriers, transport, **12:**123–125
Photolithography
high-yield, **8:**255–261
oxide holes formed in, **8:**273
Photoluminescence, silicon-carbon-hydrogen films, **12:**274–275
Photometry
film thickness measurement by, **3:**9–16
metallic films, **6:**182–185
Photomultipliers
dynode materials, **12:**141–144
photoemissive materials in, **12:**140–144
uses, **12:**141
windowless, **1:**131–132
Photomultiplier tubes
with glass-bonded GaAs cathode, **12:**132
negative electron affinity-GaP dynodes in, **12:**135
Photon detectors, **21:**4; **23:**121, 122–124
Photon emission, electron-induced, **2:**123–125
Photon energy, **24:**121
Photon flux, quantum efficiency of infrared detectors and, **23:**121
Photons
absorption, **5:**2
photocathode and, **5:**1
quantum yield, **5:**13
Photon wave function, in superlattice, **23:**252–256
Photopolymers, **3:**151–155
Photorelaxation, boundary and initial conditions, **11:**281
Photoresist, **3:**151; **11:**9–10
Photoresponse, per absorbed photon, **4:**342–343
Photosynthesis
artificial
Langmuir-Blodgett films for, **20:**271
reaction centers in, **20:**231–243
multielectron transfer reaction, **20:**269
Photothermal conversion
absorbers, **10:**26
conversion efficiency, **10:**13–14
thermal efficiency, **10:**15
Photothermal solar energy conversion, **10:**4–27
collection configurations, **10:**22
and energy balance at converter surface, **10:**13–19
figure of merit for real surfaces in, **10:**19–22
goals, **10:**1–2
spectrally selective surfaces, **10:**1–64
thermal radiation and, **10:**11–13
Phototransistors, **11:**189–191, *see also* Hot-electron phototransistors
Photovoltage responsibility, **21:**16

Photovoltaic detectors, **23:**122
Photovoltaic effect, infrared detectors, **21:**4
Photovoltaic energy conversion, for quantum-well structures, **21:**312–366
Photovoltaics, principles, **21:**312–323
Phthalocyanine-based SAMs, **24:**24–25
Physical vapor deposition (PVD), **17:**153; **22:**182, 226–229
  evaporation, **17:**153
  laser ablation, **17:**154
  MBE, **17:**153
  sputtering-magnetron, **17:**154
Piezolelectric films, **17:**150
  aluminum nitride, **17:**153
  lithium tantalate, **17:**161
  lithium tetraborate, **17:**162
  tantalum pentoxide, **17:**162
  zinc oxide, **17:**153
Pinched-off photodiodes, **11:**172–177
Pinch-off voltage, thin-film transistor, **2:**160, **2:**166ff.
Pinhole coupling, in multilayer films, **6:**226
Pinhole fogging maps, **7:**83, 109
Pinholes
  determination, **20:**189
  double, **7:**79, 110
  in fine mesh screen supports, **7:**78–80
  fogging due to, **7:**108–109
  from laboratory instruments, **7:**90
  location of, **7:**83
Pinned spins, boundary condition, **15:**135
Piranha solution, **24:**51
Planar guides, **10:**239–240, 255
Planar stress, **15:**163
  argon incorporation and, **15:**164
  bias voltage and, **15:**163
  magnetic anisotropy and, **15:**185
Planck function, **10:**11
Plasma
  arc discharge, **17:**93
  confinement, magnetic, **17:**86, 88, 90
  emission, optical, **17:**116
  etching *see* Reactive sputter etching
  frequency *see* Frequency
  high-frequency induced, **3:**88–90
  magnetron, **17:**86
  modes, **18:**160–163, 184–189, 205–211
  polymerization, **14:**217
  potential, **14:**208–209; **18:**159, 175, 180, 192–193, 195–196, 200–202, 204–206
  resonance, **18:**131, 134–137, 147, 150–151, 185–187, 190
  surface roughness and, **6:**198–199
  thin-film processing, particle, **18:**289, 290
  velocity, **14:**152–153
Plasma-assisted physical vapor deposition, **13:**60, 67–68
Plasma boundary, very thin films in, **9:**238–240
Plasma carburizing, **12:**283–285
Plasma chemical transport, film deposition by, **12:**243
Plasma chemical vapor deposition, **12:**238–245, 246–275, 275–283, 283–285, 286–289, *see also* Chemical solution deposition
Plasma density *see* Electron density
Plasma deposition reactor potentials, **12:**241
Plasma deposition system, Roster-Engle design, **12:**242

Plasma electron-beam gun, **13:**78
Plasma-enhanced chemical vapor deposition, silicon dioxide
Plasma-enhanced chemical vapor deposition, (*continued*)
from TEOS, **22:**210–215, 215–226
Plasma etching, **12:**237
laser-enhanced, silicon, **13:**187–188
Plasma films
radiative plasma oscillations and, **9:**168
resonant light emission, **9:**176
Plasma oxidation
of GaAs devices, **12:**277–282
in magnetic field, **12:**278
mechanism, **14:**114–123
Cabrera-Mott theory, **14:**115–116
comparison to experimental data, **14:**117–123
mobile defect in oxide, **14:**116–117
Wagner theory, **14:**114–115
in oxide-layer production on GaAs, **12:**237
rf oxidation and, **12:**282–283
for silicon dioxide films, **12:**282
sputter etching, **14:**212–215
for thin insulating films, **12:**275–277
Plasma production
anodizing apparatus for, **6:**39
electromagnetic ion guns in, **6:**43
in gaseous anodization, **6:**36–43
high-frequency plasmas and, **6:**42–43
low-pressure glow discharge in, **6:**37–42
in thin films, **6:**195–197
Plasma radiation, **6:**195–197
theory, **9:**173–176
Plasma resonance
optical, **6:**193–199
in reflection at thin films, **9:**188
registration by photoelectrons, **9:**188–190
in small spheres, **9:**198–199
Plasma resonance absorption, **9:**180–186
by optical excitation of volume plasmons, **9:**191–192
fine structure, **9:**191
in nonhomogeneous films, **9:**190
in thin films, **6:**196
Plasma resonance emission, **9:**195–198
Plasma slab
asymmetric layer system and, **9:**154–157
symmetric layer system and, **9:**149
Plasma spray, **16:**96–99
Plasma-substrate interactions, **13:**73–74
Plasma transport, **18:**69
Plasma wavelength, **6:**154
Plasmons, **9:**164–166
dispersion in thin films, **6:**195
nonradiative surface, **6:**163
optical excitation and radiative delay, **6:**198
scattering, in free-electron metal, **12:**62–63
surface *see* Surface plasmons
Plastic films
removal, **7:**71–72
in X-ray astronomy, **7:**69
Plasticity, **3:**259–261
Plated-beam technology, **8:**295–299
Plate glass
coated, **5:**128
heat-resistant, **5:**133
partially reflecting, **5:**126–132
spectral transmittance, **5:**99

Platinum
evaporation, **1:**167
growth on gold, **4:**143–144
reflectance, **12:**15
reflectance in vacuum ultraviolet, **1:**165–168
Platinum coatings
for extreme ultraviolet astronomy, **10:**91–92
mirror coatings, **12:**17
Platinum oxide, **14:**27, 31
Platinum silicide
formation, long-term annealing, **23:**101
infrared cameras based on, **23:**105–106
internal emission imaging sensors using, **23:**88
internal emission infrared imaging arrays, **23:**83–111
on silicone (PtSi/Si) detectors, delta-doped, **23:**141–143
on silicon-germanium (PtSi/SiGe/Si) infrared detectors, **23:**155–161
spectral response, **23:**100–102
Platinum silicide detector arrays, **23:**100–106
Fowler emission
efficiency, **23:**102–104
excess low-frequency noise, **23:**105
parameters for, **23:**105
response uniformity, **23:**104–105
future, **23:**106–110
Schottky barrier infrared detectors, **23:**123–124
barrier height, **23:**139–140
detection mechanism, **23:**135–139
emission coefficient, **23:**138–139
expression of Y for Shottky diodes, **23:**136–137
summary, **23:**135–140
PLT *see* (Lead, lanthanum) titanate film
PLZT *see* (Lead, lanthanum) zirconate titanate film
PM *see* Purple membrane
PMMA
nonuniform hole profiles, **13:**171–172
uv photoetching characteristics, **13:**154
p-n doping, of semiconductor films, **5:**177
p-n junction, **21:**4, 32, 313–314
dielectric loss, **11:**244
electron distribution, **7:**215
from mercury cadmium telluride, **23:**131–133
in semiconductor films, **5:**151
Point defects, generation, **13:**178–180
Polarimetric measurements, **4:**69–74
film thickness, **3:**16–17
metallic films, **6:**185–187
optical constants, **2:**218–227
Polarizability, bond type, **10:**42
Polarization, **15:**234
charge carriers and, **11:**281–283
dielectric, **11:**306–307
electric field vector, **1:**79, 81
ionic, **4:**293–300
mechanisms, **11:**223–226
monochromator, in reflectance measurement, **1:**133–134
reflection measurements, **4:**66–67
space-charge, in glass-over-oxide structures, **4:**311–313
Polarization increment, **11:**215
Polarization interferometer, optical constants by, **2:**231–235

Polarization process
ac bridge and step-response measurements, **6:**54–58
in anisotropic films, **6:**179–182
atomic and electronic, **6:**51
dielectric properties and, **6:**52–63
electron spin and, **6:**246
film history and, **6:**52–54
''lattice sites,'' **6:**62
models, **6:**60–63
space charge vs. step response in, **6:**58–60
in very thin films, **6:**171–177
Polarized light
elliptical, **19:**285
linear, **19:**284
Polarized waves, **6:**155
Polarizer
four-mirror, **12:**33–34
reflection type, **12:**33–34
three-mirror, **12:**33
vacuum ultraviolet, **1:**124
Polarizing angle, **2:**201
Polaron absorption, **17:**345
Polar tail groups, **24:**17–20
Polishing techniques, bowl-feed, **4:**3–5
Poly(chlorotrifluoroethylene)
kinetics of modification, **20:**97–99
mechanism of alkylation, **20:**96–99
surface modification, **20:**95–100
Polycrystalline films, **10:**235–238
gold, **6:**106, 110
Polycrystalline permalloy, in multilayer film, **6:**273–274
Polycrystalline silica, plasma-chemical transport deposition, **12:**243
Polydian carbonate, dielectric loss peaks in, **11:**238
Poly(dimethylsiloxane) (PDMS)
microcontact patterning, **24:**229
microcontact printing, **24:**231–234, 250
Polyethylene, chlorination types, **20:**102
Polymer films, **3:**144–162
by gaseous discharge, **3:**155–157
MIS transistor, **17:**279
polyvinylidine fluoride (trifluorethylene), **17:**279
Polymeric organosulfur compounds, **24:**33–35
Polymerization
in chemical vapor deposition, **5:**245
of DC–704 silicone oil, **3:**146–147
photolytic, **3:**151–155
Polymer networks, **20:**69–82
coupling reaction, **20:**76
forced planar alignment, **20:**72–76
formation requirements, **20:**69–70
Langmuir-Blodgett technique, **20:**69–82
oxidative coupling, **20:**76
perpendicular alignment, **20:**71–72
Polymers
dewetting, microcontact printing, **24:**246
laser-induced etching, **13:**181–183
Polymer surface
alkyl-lithium reagents, **20:**96–99
carbonyl modification, **20:**103–104
chain mobility, **20:**85–87
chemical modification, **20:**89–104
types, **20:**91
chlorination, **20:**102
definition, **20:**85–87
environmental effect, **20:**86–87
interphase modification, **20:**89–90
modification of fluoropolymer, **20:**100–102
modification methods, **20:**87–88
modification of

poly(chlorotrifluoroethane), **20:**95–100
modified, characteristics, **20:**90–91
polypropylene, hydroxylation, **20:**91–95
types, **20:**87
variation, **20:**85–87
Polyphenylsiloxane guides, **10:**242
Polypropylene hydroxylation, **20:**91–95
characterization of surface, **20:**92–95
hydroboration, **20:**94
rate, **20:**91–93
side reactions, **20:**94–95
Polysilicic acid ester, **5:**97
Polyvalent metal films
electrical size effects, **6:**112–113
resistivity size effects, **6:**134–136
Polyvinyl acetate, dielectric loss peaks, **11:**239
Polyvinyl alcohol, as parting agent, **7:**68
Poly(vinylidene fluoride)
removal of fluorine, **20:**99–100
surface modification, **20:**101–103
Polyvinyl pyrrolidone, as parting agent, **7:**68–69
Poole-Frenkel constant, **8:**227
Poole-Frenkel effect, **6:**23; **8:**199–207, 217–224
Poole's law, **8:**190, 224–227
Population inversion, tunneling and, **23:**240–243
Porphyrin films, **20:**70–82
coupling prior to formation, **20:**79–81
covalent linking, **20:**76–81
diffusion polymerized, **20:**78–79
electron spin resonance, **20:**71
ionic linking, **20:**72–76
linear dichroism, **20:**71
mixed porphyrins, **20:**71
structure of Langmuir-Blodgett films, **20:**77–79
Position index method, **22:**129, 169
Positioning, double, **4:**223–224
Position sensitive detector, diagram, **20:**260, 265
Positive electron affinity, vs. negative electron affinity, in band-gap energy, **12:**113
Postdeposition bombardment, ion-beam processing, **13:**127–128
Potassium antimonide
crystal structure, **12:**88
optical absorption, **12:**103
Potassium cesium antimonide, superficial oxidation, **12:**92
Potassium film, plasma resonance emission from, **9:**197
Potassium foil, absorption of *p*-polarized light by, **9:**193
Potassium niobate/tantalate (KTN) films, **17:**263–264
defect structures, **17:**263
Potassium thin film, in ultraviolet region, **7:**43
Potential
distributed, **18:**16, 84
distribution in space, in tunnel, and flat-bed reactors, **12:**240
floating, **18:**16, 91
plasma, **18:**8, 86, 91
measurements, **18:**80, 81
surface, **18:**101
Potential barrier shape, **4:**333–335
Potential energy, molecular dynamics, **22:**126–128
Power
absorption, **18:**36, 50, 54, 67

Power (*continued*)
balance, **18:**17, 67
dissipation in thin-film circuits, **2:**416
minimum, for discharge, **18:**37, 68
Power tower, **10:**25
Poynting vector, **9:**79–80
PPS *see* Polyphenylsiloxane guides
Praesodymium oxide, **14:**16
Preferred orientation
refractory compound deposits, **13:**85, 87
TiC, **13:**71–72
Prepolymers, **24:**244
Pressure
in evaporator, aluminum reflectance and, **1:**150–151
high pressure, sputtering rate and, **3:**96
initial, **14:**4
low pressure, sputtering and, **3:**87–92
partial pressure, **14:**5, 11, 27
with mass spectrometer, **1:**45–47
reduction, **14:**4
rise during evaporation, **1:**140–141
vapor pressure, gas at low temperature, **1:**20–21
Primary-beam deposition, **13:**145
Primary-beam ion source, **13:**115–116
Primary cracks, **16:**320–322
Prism coupler, **10:**233
Prism method, in light excitation of non-radiative surface plasmons, **9:**233–234
Process modeling *see* Modeling
Process temperature, **18:**209, 211, 215, 216, 219, 223–224
Programmed rate CVD, **22:**265–269
Propagation loss
due to ground plane, **15:**77
in ferrite, **15:**6
Proportional limit, **16:**288
Proteins, immobilization on SAMs, **24:**106, 248–249
Proton pumping, **20:**280, 286
Proximity masking, laser-induced etching, **13:**159–161
Pseudobinary alloys, band crossings in, **11:**106
Pseudo-Brewster's angle, **19:**301
Pseudo-dielectric function, **19:**115
Pseudo domain walls, in multilayer films, **6:**219
PTCS *see* Phenyltrichlorosilane
p-type material, in electron affinity reduction, **12:**70
Pulse compression, **15:**67
Pulsed-plasma deposition system, **12:**244
Pulse mode infrared detectors, **21:**49–50, 70
Pulse separation, **15:**68
Pulsing, spiketrain, **21:**41–48
Pumping, **17:**115
magnetic, instabilities in response, **15:**93–95
sputtering, **3:**106–107
Pumps
cryogenic, **14:**164
diffusion, **14:**164
turbomolecular, **14:**164
vacuum, **1:**12–23
cryogenic, **1:**20–23
diffusion, **1:**13–15
evapor-ion, **1:**16–18
getter-ion, **1:**15–20
sputter-ion, **1:**18–20
Pump speed, **14:**4, 5, 11–14
Purification of II-VI compound materials, **3:**168–169
Purity

aluminum reflectance and, **1:**150–151
evaporated material, **4:**10–11
evaporation parameters and, **1:**127–129
Purple membrane (PM), **20:**280–290
photoelectrochemical cell, **20:**283–285
quantum efficiency, **20:**284–285
on SnO2, **20:**283–285
''Purple plague,'' development, **6:**301
PVD *see* Physical vapor deposition
p-wave, **6:**155
Pyramidal growth forms, **4:**200–204
1-Pyrenebutanethiol, motional dynamics, **24:**70–77
Pyroelectric detectors, **17:**242–247; **23:**126–127
detectivity, **17:**245, 247
epitaxial PbTiO3, **17:**242
figure of merit, **17:**242
frequency dependence, **17:**247
IR detector arrays, **17:**245–247
lead germanate, **17:**228, 288
morphotropic boundary, PZT, **17:**243
phase diagram, PZT, **17:**243
PLT films, **17:**243, 245–247
pyroelectric coefficient, **17:**244, 288
PZT films, **17:**243–244
responsivity, **17:**245
SBN films, **17:**287–288
Pyrohydrolysis, in chemical vapor deposition, **5:**313
Pyrolitic carbon, **5:**263
Pyrolitic graphite, **5:**258–259, 262
Pyrolysis
film deposition, **3:**133ff.
in plastic film backing removal, **7:**71
reaction, **3:**137–138
Pyrolytic chemical vapor deposition, **9:**38–41; **12:**238, *see also* Spray pyrolysis
for silicon carbide films, **12:**272
Pyrolytic deposition, **2:**370
Pyrrole-terminated alkanethiols, **24:**30

Q

QCM *see* Quartz crystal microbalance
Q state, **20:**21
Quadrupole mass spectrometry, **11:**41
Quantum cascade laser
development, **23:**266–274
higher-gain, **23:**279–280
Quantum effect, purple membrane, **20:**283–285
Quantum efficiency, **15:**233
of Ag-O-Cs photocathodes, **12:**84
band filling and, **11:**139
of Cascade process, **23:**247–252
of heterojunction internal photoemission detectors, **23:**148–153
external, **23:**152–153
internal, **23:**148–152
of infrared detectors, **23:**121
of internal emission photodetectors, **23:**89, 91–92
internal quantum efficiency, **21:**21–22
of IV-VI semiconductors, **11:**132–140
of multiple-quantum-well detectors
effective-heterodyne, **23:**177
external, **23:**195, 211–214
n-doped quantum-well infrared photodetectors, **21:**144–146
of negative electron affinity cathodes, **12:**157–158
in photoemission, **12:**64, 69, 112

Quantum efficiency, (*continued*)
of platinum silicide, **23:**160–161, 165–166
quantum-well solar cell, **21:**350–353
Quantum size effects
electrical conductivity, **6:**137–143
energy bands in, **6:**138
experimental measurements, **6:**140–143
resonant tunneling, **21:**81–84
superlattice minibands, **21:**85–90
Quantum-well devices
for infrared emission, **23:**217–298
application potential for, **23:**274–282
at 1.5 to 3.0μm, **23:**228–231
between 3.0 and 4.5μm, **23:**231–236
limitations, **23:**274–282
midwave, **23:**228–238, 236–238
recent trends in, **23:**274–282
Quantum-well infrared photodetectors, **21:**4, 114–117, 228–231
asymmetrical, **21:**116, 160–169
bound-to-bound state, **21:**123–126, 131, 184–187
bound-to-continuum state, **21:**126–160, 187–190
design, **21:**101–104
imaging arrays, **21:**219–228
indirect band-gap quantum-well infrared photodetectors, **21:**191–195
infrared hot-electron transistors, **21:**241–308
intersubband transitions, **21:**90–101, 117–123, 241–246
light coupling methods, **21:**215–219
materials
gallium arsenide/aluminum gallium arsenide, **21:**116, 117, 160–169, 184–195, 224, 240, 241, 300–301
gallium arsenide/aluminum indium phosphide, **21:**206–207
gallium arsenide/gallium indium phosphide, **21:**204–205
indium gallium arsenide/gallium arsenide, **21:**207–213, 267
indium gallium arsenide/indium aluminum arsenide, **21:**196
indium gallium arsenide/indium phosphide, **21:**196–201, 213–215
indium gallium arsenide phosphide/indium phosphide, **21:**201–204
multicolor quantum-well infrared photodetectors, **21:**289–300
multiquantum-well infrared photodetectors, **21:**4
n-doped
absorption spectra, **21:**129–132
asymmetrical barriers, **21:**161–163
asymmetrical quantum wells, **21:**163–167
bound-to-continuum state, **21:**125–155, 155–160
dark current, **21:**132–133
dark current noise, **21:**137–138
detectivity, **21:**149–155
escape probability, **21:**146–149
gallium arsenide/gallium aluminum arsenide, **21:**240–241
gallium arsenide/gallium indium phosphide, **21:**204–205

indium gallium arsenide/gallium arsenide, **21:**207–213
indium gallium arsenide/indium aluminum arsenide, **21:**196
indium gallium arsenide/indium phosphide, **21:**196–201
indium gallium arsenide phosphide/indium phosphide, **21:**201–204
photoconductive gain, **21:**138–143
quantum efficiency, **21:**144–146
responsivity, **21:**133–137
single, **21:**170–178, 179
structure, **21:**126–129
well capture probability, **21:**143–144
n-doped gallium arsenide/aluminum indium phosphide, **21:**206–207
p-doped
asymmetric, **21:**167–170
indium gallium arsenide/indium phosphide, **21:**213–215
peak detectivity, **21:**160
photoconductive gain, **21:**160
single, **21:**180–184
photoresponse, **21:**104–110
single quantum-well infrared photodetectors, **21:**170–184
superlattice miniband quantum-well infrared photodetectors, **21:**184–191
thick barriers, **21:**240, 248–251, 298–300
Quantum wells, **21:**77–79, 110, 114–116, *see also* Quantum-well infrared photodetectors
asymmetrical, **21:**78–81, 163–167
energy level, **15:**222, 226
excited, virtual, **15:**222–224, 232, 242
in interband-type, quantum wells in, **23:**220–238
intersubband transitions, **21:**90–101, 117–123, 241–246
many-body effects, **21:**95, 97
primary functions, **23:**220
quantum size effects
resonant tunneling, **21:**81–84
superlattice minibands, **21:**85–90
size effects, **23:**221–226
strain effects, **23:**226–228
Quantum-well solar cells, **21:**312 313, 322–323
applications, **21:**362–367
limiting efficiency, **21:**357–362
materials
gallium arsenide/aluminum gallium arsenide, **21:**313, 323–324, 331–340, 347, 354, 360, 365
gallium arsenide/aluminum indium phosphide, **21:**337, 347
indium gallium arsenide/gallium arsenide, **21:**331–332, 335, 341–344, 347, 360
indium gallium arsenide/indium phosphide, **21:**344–347
multiquantum well structures
in doped regions, **21:**365–366
for improved transport, **21:**364–365
in photoelectrochemical cells, **21:**365
photovoltaics, **21:**322–323
dark currents, **21:**316–318, 349–355
limits to efficiency, **21:**318–321

Quantum-well solar cells, (*continued*)
multibandgap solar cells, **21:**321–323
open-circuit voltage, **21:**355–357, 365
photocurrent, **21:**314–316
p-n junction solar cell, **21:**313–314
radiative limit, **21:**360
spectral response, **21:**323–327
transport properties, **21:**364–365
voltage behavior, **21:**347–357
Quantum-well structures, **23:**173–215, *see also* Multiple-quantum-well (MQW) detectors
multiquantum-well structures, **21:**333
hot-electron phototransistors, **21:**239–308
quantum-well solar cells, **21:**364–366
photovoltaic energy conversion, **21:**312–266
Quantum wires, **20:**22
Quantum yield
for copper sample, **12:**63–64
of photons and conventional cathodes, **5:**12–14
Quartz crystal microbalance (QCM), **20:**77
equation, **20:**318
in situ adsorption studies, **24:**46, 48–52
Quartz crystal oscillator, **12:**73
Quasiballistic transport, **21:**253–255
Quasistatic reversal behavior, in multilayer films, **6:**251–269
Quasi-steady-state method, **16:**248
Quasiwalls, in multilayer ferromagnetic films, **6:**234–235
QWIPS *see* Quantum-well infrared photodetectors
QWSC *see* Quantum-well solar cells

R

RA *see* Reflectance anisotropy
Radar, phased-array *see* Phased-array radar
Radial distribution function, **15:**156–159
amorphous metals and, **7:**18–28
calculation, **7:**5–12
for compounds, **7:**28–37
measurements in, **7:**17–37
for semiconductors, **7:**29–37
Radiant absorptance, for metal-dielectric interference filters, **9:**81
Radiant energy flow ratio, **6:**176
Radiant reflectance and transmittance, for metal-dielectric interference filters, **9:**78–79
Radiant transmittance, defined, **9:**78–79
Radiation, infrared, *see also* Infrared radiation
Radiation damage
in dielectric films, **8:**4, 14–17
due to electrons, **2:**134
X-rays, **18:**105
Radiation detectors, **16:**101, 109, 110
Radiation reactance, magnetostaic, **15:**47
Radiation resistance, magnetostatic wave emission
by complex transducera
equivalent circuit, **15:**50
weak coupling theory, **15:**47
by microstrip transducers, **15:**40
coupling constant, **15:**41

equivalent circuit, **15:**49
self-constant theory, **15:**43
Radiative decay time, **23:**250
Radiative limit, quantum-well solar cells, **21:**360
Radiative surface plasmons
electron excitation, **9:**173–180
excitation, **9:**171–199
Ferrell mode for, **9:**159–163
interaction with nonradiative plasmons, **9:**164–166
light and, **9:**180–199
photoelectron current produced by, **9:**189
plasma films and, **9:**168
plasma radiation and, **9:**193
roughness and, **9:**164–166, 248–250
Radiative transitions, in intersubband emitting devices, **23:**239, 240–243
Radioautographic studies, of monolayer transfers, **7:**323–329
Radiotracers, in evaporant, **3:**38
Random arc, **17:**95, 96, 99
Raoult's law, **2:**387; **6:**309
Rapid ion-beam sputtering, titania, **13:**131
Rapid thermal annealing (RTA), **16:**42, 51, 60, 91, 94, 96
Rare earth-based superconducting thin film, **16:**19, 41, 51, 52, 111
Rare earth-based superconductors, **16:**2
Rare earth oxide films, **8:**37–38
RAS *see* Reflectance anisotropy
Rate monitors, dielectric films and, **8:**22–23
Rayleigh criterion, **20:**111
Rayleigh region model, for multilayer films, **6:**222
Ray propagation, in thin films, **10:**231, *see also* Thin films
RBS *see* Rutherford backscattering
RDF *see* Radial distribution function
RDS *see* Reflectance anisotropy
Reactant diffusion, laser-induced etching, **13:**163–164
Reaction center
in artificial photosynthesis, **20:**231–243
in lipid membranes, **20:**257
photosynthetic, **20:**239–240, 280
single molecule, **20:**265–267
Reactions
electrochemical, **1:**226
solid state, **1:**224–226
surface, **14:**7
Reactive evaporation, **12:**26; **13:**63–64, *see also* Activated reactive evaporation
film deposition, **9:**47–49
ionized reactive gas, **13:**61–62
rate-controlling steps, **13:**65
reaction kinetics, **13:**65
reactions in synthesis of oxides, carbides, and nitrides, **13:**65
without ionization of reactive gas, **13:**60–61
Reactive ion etching, **22:**60
Reactive ion plating, **13:**78–79
Reactive sputter etching, **14:**233–236
Reactive sputtering, **9:**49–53; **13:**67
Reactivity template, **20:**133
Real time ellipsometry, at single wavelength, **19:**50–51
Real time spectroscopic ellipsometry (RTSE), **19:**54–109, 148–87
aluminum films, **19:**147–87
optical functions from real time observations, **19:**162–68, 168–83

Real time spectroscopic ellipsometry (RTSE), (*continued*)
optical functions in nucleation and growth stages, **19:**150–62
data collection and interpretation, **19:**56
diamond, **19:**104
analysis of microstructural evolution, **19:**108
substrate treatment and annealing, **19:**105
true substrate temperature evaluated by RTSE, **19:**109
hydrogenated amorphous silicon, **19:**61
microstructural evolution, **19:**62
optical functions from real time observations, **19:**70
process-property relationship, **19:**75
hydrogenated amorphous silicon-carbon alloys, **19:**85
hydrogen dilution of reactive gases, **19:**88–93
optical gaps from real time observations, **19:**85–88
instrumentation, **19:**55, 69, 96, 108, 147
microcrystalline silicon, **19:**93
optical and microstructural analysis, **19:**94
size effects in microcrystallites, **19:**99
nucleation and growth of films, **19:**50–121, 127–87
aluminum film, **19:**147
hydrogenated amorphous silicon, **19:**61
Rearrangement, molecular, **20:**211–212
Rear-view mirror, **17:**356
Receiver
broad band, **15:**52
channelized, **15:**54
filter technologies for, **15:**55
magnetostatic filterbank for, **15:**88
compressive, **15:**57
delay lines for, **15:**57
allowed phase error, **15:**58
magnetostatic, **15:**59
Receptor, amphiphilic, **20:**225
Recognition patterns, Langmuir-Blodgett films, **20:**223–236
Recombination processes, etching rates, **13:**174–176
Recording characteristics, **15:**209–213
carrier-to-noise ratio, **15:**209
cyclability, **15:**212
medium noise, **15:**210
modulation transfer function, **15:**203, 211
signal-to-noise ratio, **15:**209
Recording process, **15:**200
Recrystallization
Group II-VI compounds, **3:**165–210
by heating in presence of organometallic compounds, **3:**183–187
induced, **3:**180–182
methods, **3:**179–196
thin films by electrons, **2:**132–133
RED *see* Reflection electron diffraction
Redeposition in sputter etching, **14:**224–226
Redox-active SAMs, **24:**20, 26
Red phosphorus, chemical transport, **12:**245
Reduction, in chemical vapor deposition, **5:**243–244

Reevaporation of adsorbed atoms, **2:**2–3
Reflectance, **2:**196, 307–310
absorbing film
on absorbing substrate, **1:**87
on non-absorbing substrate, **1:**86
in admittance notation, **1:**94
aluminum in vacuum ultraviolet, **1:**150–159
angle of incidence, **10:**143–145; **12:**18–23
angular dependence, 2.340ff.
circle diagrams, **1:**111–115
complex
admittance, **1:**103
amplitude, **1:**84–87
computation using optical constants, **2:**322–325
defined, **6:**156; **12:**3
double layer coatings
2.249ff.
inside layer twice outside layer, **2:**254–257
layer thickness not related integrally, **2:**257–265
two layers of equal optical thickness, **2:**250–254
error due to beam polarization, **1:**133
germanium in vacuum ultraviolet, **1:**159–169
effect of oxide growth, **1:**163–165
incident angle dependence, **1:**161–162, 164
glass surface, decrease with magnesium fluoride film, **1:**175
gold in vacuum ultraviolet, **1:**165–168
incident light polarization states and, **1:**82
intensity, **1:**77
magnesium fluoride on glass, **2:**245–247
measurement, **2:**314–318
for mirror coatings, **12:**3–8
in normal incidence, **6:**183
in vacuum ultraviolet, **1:**130–141
measuring instruments, **12:**4–5
multialkali films, **5:**28–29
multiple combination films, **1:**99
normal incidence, air-absorbing medium boundary, **1:**142
optical constants and, **12:**19
parallel vs. average components, **12:**23
photoelectric film, **5:**24
platinum, vacuum ultraviolet, **1:**165–168
quarter-half coatings on glass, **2:**256–257
single coating on Ge, **2:**247–249
specular
correction for surface roughness, **4:**17–18
surface roughness and, **4:**13–16
standard, **12:**4
substrate second surface effect, **1:**82–83
thin films, **2:**203–209, 336–338
absorbing film, **2:**207–209
transparent film, **2:**203–207
true, **1:**134–141
tungsten, vacuum ultraviolet, **1:**169
vacuum ultraviolet, **1:**145–169
at non-normal incidence, **1:**133
combination films, **1:**171–179
very thin films, **6:**170–177
vs. thickness, in evaporated film coatings, **10:**93–110
vs. wave number, **8:**86

Reflectance, (*continued*)
wavelength dependence of gold and platinum, **1:**167–168
Reflectance anisotropy (RA), **19:**2–43
applications, **19:**11–43
GaAs and AlAs under MBE conditions, **19:**12–19
GaAs under CBE and VCE growth conditions, **19:**22–29
Group III-V semiconductors, **19:**11–43
to migration-enhanced atomic layer epitaxy (MEALE), **19:**19–22
experimental, **19:**8–10
growth of GaAs under atmospheric pressure AP MOCVD, **19:**32–38
atomic layer epitaxy, **19:**35–38
surface reconstructions, **19:**32–34
growth of lattice-matched multilayered structures by LP MOCVD, **19:**29–31
growth of lattice-mismatched III-V and II-VI structures, **19:**38–43
theory, **19:**3–8
Reflectance anisotropy spectroscopy (RAS) *see* Reflectance anisotropy
Reflectance coatings, **11:**1–33, *see also* Dopants; Gold coatings
Reflectance difference spectroscopy (RDS) *see* Reflectance anisotropy
Reflectance-enhancing surface films, **12:**22–42
Reflectance-filter
film combinations, **1:**175–177
overcoats for diffraction gratings, **1:**181–183
spectral transmittance, **9:**131
Reflectance-increasing film combinations, **1:**172–175
Reflectance-increasing overcoatings, diffraction gratings, **1:**181–183
Reflecting coatings, deposition, **12:**8–16
Reflecting metals, optical properties, **6:**159–162
Reflection
attenuated total, **4:**74–76
measurement
absolute phase change on, **4:**84–90
non-normal incidence, **4:**69–78
normal incidence, **4:**57–69
unpolarized light, **4:**72–74
using polarized light, **4:**70–72
quantum-mechanical, **4:**338–341
by thin films, **9:**16–18
at absorbing medium, **2:**197–203
at interface between isotropic transparent media, **2:**194–197
theory, **2:**194–209
total, **2:**227
Reflection coefficient, for inhomogeneous films, **8:**72–81
Reflection electron diffraction
applications, **11:**42
glancing incidence, **11:**42–44
patterns, **11:**43
scanning high energy, **11:**44–48
Reflection filters, **9:**128–132
antireflection coating for, **9:**130
Reflection loss limit, **11:**135–137
Reflection loss limit quantum efficiency
for lead telluride, **11:**137–138

for strip *p-n* junctions, **11:**178–179
Reflection mode secondary emission, **12:**133
Reflection monitoring, in filter production, **9:**134–135
Reflection-type mirrors, metal-dielectric mirrors as, **12:**42–45
Reflection-type polarizers, **12:**32–34
Reflective filters, for extreme ultraviolet astronomy, **10:**114–116
Reflective interference cathode (RIC), **5:**21–30
  arrangement, **5:**22
  optical properties, **5:**21–25
  photoelectric properties, **5:**26–30
  principle, **5:**21
  tuning, **5:**28–30
  wave field and, **5:**23
Reflective layers, selective, **5:**132–134, *see also* Antireflection coatings
Reflective scattering, **10:**58
Reflectivity
  defined, **12:**3
  in optical absorption, **12:**55
Reflectometer, **10:**208–210
  Coblentz, **2:**317
  designs for vacuum ultraviolet, **1:**134
  double-reflection spherical mirror, **4:**61–65
  Gier-Dunkle blackbody, **2:**317–318
  in reflectance measurements, **12:**4–5
  single-reflection movable detector, **4:**65–69
  Strong-type, **2:**315, 316; **12:**5–6
  types, **4:**59–60
Reflector/absorber combinations, **10:**51–54
Refractive index, **5:**2–3; **12:**18, *see also* Optical function
  atomic weight and, **10:**41–42
  complex, **1:**141
  dielectric materials, **8:**52–67
  experimental results, **8:**58–67
  microwave, **18:**126, 131, 135–136, 139
  semiconductor oxide films, **9:**20
  silica film, **18:**218–221, 224–225
  silicon nitride film, **18:**209, 213, 215, 217
  wavelength and, **8:**84
Refractory compound deposits, **13:**62
  mechanical properties, **13:**87, 89–94
  microstructure, **13:**81–85
  preferred orientation, **13:**85, 87
  structural morphology evolution, **13:**82–83
  structural zones in condensates, **13:**81–82
Relative yield, **14:**6
Relaxation
  crystallise boundary, **3:**247–249
  time dependence, **11:**309–311
Relaxation parameters, **24:**152–153
Relaxation time
  distribution, **11:**288–289
  magnetic annealing, **1:**292–293
Remnant flux density, **16:**149
Remote microwave PECVD, silicon dioxide from TEOS, **22:**215–226
Replica conventional gratings, **11:**11–12
Replica techniques, **1:**188–189; **4:**224–225
Resistance *see* Electrical resistance
Resistance-heated evaporation source, **13:**76

Resistive bolometer, as infrared detector, **23:**125–126
Resistivity, **14:**3, 44, 48; **17:**130
  anisotropy, **1:**239
  at low temperature, **1:**234–236
  of chromium, **2:**384–385
  film thickness and, **2:**381
  Fuchs expression in, **6:**98
  Matthiessen's rule for, **6:**113, 124–131
  negative temperature coefficient, **2:**383–384; **6:**142
  quantum size effects and, **6:**139
  sheet, **2:**378
  of silver in particulate film, **20:**29
  size effects, *see also* Electrical conductivity; Electrical size effects
    Azbel-Gurzhi model, **6:**127
    in polycrystalline and crystalline wires, **6:**132
    for polyvalent metals, **6:**134–136
  specific, **1:**212–213; **2:**391
  stress effects, **1:**238–239
  superconducting state upper limit, **1:**271–272
  temperature-dependent, **6:**97
  temperature variation, **1:**236–240
  thickness and, **13:**25–27
Resistivity ratio, **6:**110
Resistometric method, **16:**245–247
Resistors
  fired-glass, **2:**392–393
  sputtered film, **3:**108–110
  thin-film, **2:**378–398
Resolution, **13:**157–168
  magnetic mass spectrometer, **1:**40
  megatron, **1:**43
  optical limitations, **13:**157–163
    coherence, **13:**158
    pattern-formation technique, **13:**159–163
    wavelength, **13:**157–158
  process limitations, **13:**163–168
    carrier diffusion, **13:**164–165
    reactant diffusion, **13:**163–164
    thermal effects, **13:**165–168
  temperature profile along beam scanning direction, **13:**166, 168
Resonance
  ferromagnetic *see* Ferromagnetic resonance
  in multilayer films, **6:**272–274
  premature decline, **15:**92
  in TRIC systems, **5:**31
Resonance field, vs. temperature, **6:**275
Resonant cavity, **18:**130, 143, 154, 155, 167–168, 178–179
Resonant frequency, quartz crystal microbalance, **24:**48, 50
Resonant-gate transistors, **8:**318
Resonant light emission, of thin plasma films, **9:**176
Resonant scattering, **10:**58–59
Resonant tunneling, **15:**220, 224, 227–228, 252; **21:**81–84; **23:**291
  instability, **15:**248
  interband, **23:**241
  kinematic correction, **15:**252
  quantum wells, **21:**81–84
  sequential, **15:**236; **23:**244–247, 286–291
  switching time, **15:**246–248
  time delay, **15:**228
Resonators, **16:**101–105
  light waves and, **5:**2–8
  magnetostatic, **15:**81
    disc, **15:**85

for non-destructive film evaluation, **15:**20
power limitations, **15:**81
Q attainable in ferrites, **15:**9
reflective arrays, **15:**82
temperature coefficient, **15:**84
Responsivity, **15:**233, 238; **21:**21, 124
asymmetric quantum wells, **21:**165–167
infrared detectors, **23:**121
n-doped quantum-well infrared photodetectors, **21:**133–137
single quantum-well infrared photodetectors
doped, **21:**173, 175
p-doped, **21:**181–182
undoped, **21:**178
Resputtering, **13:**119; **14:**8, 25, 31, 45, 51, 62, 71
Retarding field method, **13:**11
Reversible-cathode materials, solid electrolyte batteries, **13:**97–98
rf bias, **18:**182–183, 210–211, 218, 221–223
rf sputtering, **13:**111, 145
Rhenium coatings, in extreme ultraviolet astronomy, **10:**100–103
Rhenium film, rf sputtering in preparation, **9:**45
Rhenium trioxide, spectral reflectance and, **10:**33–34
Rhodium, nonabsorbing surface films and, **12:**24–25
Rhodium coatings, in extreme ultraviolet astronomy, **10:**110–114
Rhodium mirror coatings
optical constants, **12:**19
reflectance, **12:**15–19, 39–42
RIC *see* Reflective interference cathode
Richardson-Dushmann equation, **21:**23, 43
Richardson equation, **9:**6
Ridge-type deposits, **4:**209–210
Right-hand polarized wave, **18:**126–129, 131, 134–140, 151, 159–165, 169, 172–178, 186–190
RKKY interaction, in multilayer films, **6:**245–247
RLL *see* Reflection loss limit
Rocket spectrographs, unbacked aluminum films, **7:**45
Rock-salt-structured IV-VI semiconductors, on fluorite-structured substrates, **11:**121
Roster-Engle plasma deposition system, **12:**242
Rotating disk method, **13:**11–12
Rotation matrix, **19:**300
Rotator phase, **24:**214–216
Rough layer, Airy summation for, **10:**177–182
Roughness
in leakage of guided light waves, **9:**254
light emission via, **9:**219–223
light scattering by, **9:**245–248, 250–255
in nonradiative surface plasmons, **9:**242–244
Rough stacks, properties, **10:**170–173
RTA *see* Rapid thermal annealing
RTSE *see* Real time spectroscopic ellipsometry
Rubidium antimonide
optical absorption coefficient, **12:**105
structure, **12:**88

Runge-Kutta-Gill method, **5:**61
Ruthenium coatings, in extreme ultraviolet astronomy, **10:**107–110
Rutherford backscattering (RBS), **16:**6, 17, 39, 42, 47, 75; **17:**22; **19:**232
  channelling, **17:**22
  ion-beam processing, **13:**122
  RBS channeling yield, **16:**7, 8, 23, 25, 27, 61, 63
  silicon epitaxial film, **13:**38

S

SA films *see* Self-assembled films
SAGFET (self-aligned GaAs field-effect transistor), **8:**309
SAMPLE process simulator, **22:**5
SAMs *see* Self-assembled monolayers
Sandormirskii model, of quantum size effects, **6:**138
Santa Barbara Research Center, **12:**201
Sapphire
  epitaxial layer growth on, **12:**130
  heteroepitaxial films, **5:**285
  substrates of, **5:**272
Satellites, temperature coatings for, **8:**44–45; **10:**72–73, 127–164
Saturation flux density, **16:**149
Sauerbrey equation, **24:**48
Savart plate, **2:**232–234
SAW device structures, **17:**172
  acoustic charge transport device, **17:**175, 177
  ACT transversal filter, **17:**177
  heterojunction ACT, **17:**178
  MOSFET filter, **17:**172, 173
  reflection dot array, **17:**171
  SAW transduction techniques, **17:**150, 151
  transducers, **17:**171, 172
Scaling law, **13:**54
Scandium fluoride film, microstructure and optical functions, **19:**222, 230
Scandium oxide film, microstructure and optical functions, **19:**223, 226, 239, 241
Scanning electron microscopy (SEM), **17:**125; **19:**232
  high critical temperature superconductors, **16:**7, 25, 27, 38, 39, 68, 69, 94
Scanning Fizeau interferometer, **10:**204, 206–207
Scanning force microscopy (SFM), **24:**82
Scanning high-energy reflection electron diffraction, **11:**44–48
Scanning probe lithography, **24:**229, 230
Scanning tunneling microscopy (STM), **20:**16–18; **24:**82, 114
  arachidic acid films, **20:**16–18
  high critical temperature superconductors, **16:**51, 68
  nanofabrication with, **24:**99
  self-assembly mechanism, **24:**92
  structural measurements, **24:**82, 127
Scatter, **13:**121
Scattered flux, angular distribution, **10:**213–216
Scattering, **10:**58, 58–59, 167–225, 170–176
  by silver film, **9:**252–253
  elastic, **18:**14
  electron-phonon, **4:**330–338

of electrons
by single atoms, **2:**73–92
plural and multiple, **2:**79–82
proximity of atoms and, **2:**98–109
thermal diffusion, **2:**105–107
hot-electron, processes in metals, **4:**327–329
for lasers
autocorrelation function, **10:**211–213
experimental sample fabrication, **10:**203–207
graphical applications and design examples, **10:**188–203
identical films, **10:**216–219
measurement, **10:**207–211
measuring instruments, **10:**207–211
nonidentical stacks, **10:**219–224
Smith chart applications, **10:**191–197
theoretical models, **10:**203–224
of light in spectrometers, **4:**46, 49–51
on rough silver surfaces, **9:**250
in semiconductor oxide films, **9:**21–22
Scattering efficiency, discontinuous films and, **8:**148
Scattering matrix
analytic applications, **10:**188–191
in surface scattering, **10:**185–188
Scattering parameter, **6:**190
Scatter levels, **10:**283
Schlieren light valve, **8:**324–326
Schottky barrier
internal photoemission over, **23:**87–92
series capacitor and, **11:**219
spike-doped, **23:**95–99
tunneling through, **23:**93–95
Schottky barrier detectors, **23:**123–124
platinum silicide, **23:**123–124
silicide/SiGe, **23:**153–161
summary, **23:**135–140
Schottky barrier field-effect transistor (FET), **8:**305–306
Schottky barrier GaAs transistors, **8:**307
Schottky barrier height, **13:**200, 203; **23:**139–140
Al-(GaAs-AlAs) interface, **13:**203
Au-III-V, **13:**206
valence band and, **13:**202
Schottky barrier hyperabrupt varactors, **11:**77
Schottky defects, **6:**9, 15
Schottky effect, vs. Poole-Frenkel effect, **8:**222
Schottky emission, **4:**354–367
Schottky process, **6:**73
Schrödinger equation, **21:**97–98; **23:**178
Schulz and Tangherlini method, optical constants, **2:**230–231
Screened hopping model
on dielectric surface, **11:**296–302
loss peak in, **11:**301–302
Screen printing, **16:**96, 98, 100
SE *see* Spectroscopic ellipsometer
Seals, demountable vacuum, **1:**32–35
Sebastian test, **16:**305
Secondary cracks, **16:**320–322
Secondary electron coefficient, **14:**11, 29, 35
Secondary-electron emission
in alkali antimonide photocathodes, **12:**101
for negative electron affinity materials, **12:**132–134
Secondary ion mass spectrometry, **11:**45, 49; **17:**22

Second harmonic analysis, silane films, **20:**57
Second harmonic resonance, **18:**142–144, 189, 209, 212–213, 216
Seebeck coefficient, **13:**32, 34–37
Selected-area epitaxy
  MBE and, **11:**89–92
  shadow technique for, **11:**91
Selective surfaces *see* Spectrally selective surfaces
Selectivity, laser-induced etching, **13:**176–179
Selenides, as dielectric films, **8:**34–35
Selenium, **16:**278, 312
  copper diffusion in, **6:**340–342
  radial distribution function, **7:**27–28
  silver diffusion in, **6:**339
Self-aligned gallium arsenide FET (SAGFET), **8:**309
Self-assembled films
  amphiphilic alcohols, **20:**146–149
  biological patterning, **20:**128–130
  cell adhesion, **20:**135–138
  chemically controlled, **20:**42–45
  complexation, **20:**196–199
  coordination geometry, **20:**197–199
  defect removal, **20:**193–194
  on diamond, **20:**124–125
  electroless metallization, **20:**117–123
  on functionalized diamond, **20:**124–125
  on gold, **20:**183–205
  ionic binding selectivity, **20:**197–200
  ionic competition, **20:**194–196
  ligand complexation, **20:**119–123
  ligation reversal, **20:**195–196
  organosilane, **20:**112–141
  patterned, **20:**109–141
  pattern transfer, **20:**125–128
  pinhole effect, **20:**200–202
  preparation method for blocking agent, **20:**190
  protein, **20:**129–130
  third level, **20:**41–68
  wetting control, **20:**130–133
Self-assembled monolayers (SAMs), **24:**44–45
  aliphatic chain annealing, **24:**65–69
  alkylsiloxane SAMs, **24:**169–175
  atomic force microscopy, **24:**82–106
  biological oligomers and, **24:**35–36, 150
  characteristics, **24:**228
  on colloidal gold particles, **24:**222
  depressions and domains, **24:**86
  diffraction studies, **24:**112–144
  dynamics, **24:**70–77
  fabrication, **24:**83, 98–102, 103, 104, 106
    microcontact printing, **24:**228–251
  ferrocene terminated, **24:**23, 27–29
  force fields, simulations, **24:**209–211
  formation, **24:**2
    atomic force microscopy, **24:**83, 89–93
    quality crystal microbalance, **24:**45–69
    steps, **24:**62–63
    thermodynamics, **24:**45–69, 83, 89–93
  functionalized, **24:**159
  gold-thiol interface, chemistry, **24:**52–55
  history, **24:**2–4
  hydrophobic, **24:**240–244
  immobilized proteins and DNA on, **24:**106

infrared studies, **24:**66–69
mechanical properties, **24:**83, 104, 105
adhesive and frictional forces, **24:**96–98
elastic compliance, **24:**93–96
relaxation, **24:**221–222
metal-containing, **24:**24
mixed SAMs, **24:**89, 90, 218
patterned two-component, **24:**244
molecular level studies, **24:**112–144
reaction kinetics, alkanethiol SAMs, **24:**45–69, 83, 89–93
3D-SAMs, **24:**154
Self-filtered narrow-band photodiodes, **11:**186–188
Self-sputtering yield, **22:**39–45, 100
SEM *see* Scanning electron microscopy
Semiconducting diodes, photovoltaic conversion, **21:**313–323
Semiconductor cathodes, **5:**1–2; **21:**6
Semiconductor devices
topologically structured films and, **8:**294–310
vs. power triode, **8:**300
Semiconductor films, **5:**143–229, 260, 264–314, *see also* Chemical vapor deposition
deposition
by liquid phase epitaxy, **7:**115–243
on inert substrate, **2:**181–183
electrical resistivity, **6:**137–138
materials, **8:**40–41
Semiconductor-on-metal tandem absorbers, **10:**45–46, 62
Semiconductor oxide film, **9:**2, 14–15
antireflection coatings, **9:**19–20
binary oxides in, **9:**9–10
doped oxides in, **9:**10–12
electrical conduction, **9:**9–12
electrical and optical properties, **9:**22–30
film side absorption in, **9:**17
impurities, **9:**12
refractive index variation in, **9:**20
scattering and surface morphology, **9:**20–22
substrate and, **9:**12, 17
Semiconductors, **21:**2–3, 8, 68–71, 240, *see also* Homojunction internal photoemission detectors; Quantum-well infrared photodetectors
anodic oxide film growth on, **6:**2
charge transport in, **8:**178
evaporation, **5:**162–173
gallium arsenide, **7:**239–240
Group II-VI, laser-induced etching, **13:**190
Group III-V
laser-induced etching, **13:**189–190
oxide reactions, **13:**204–205
interpulse time series analysis, **21:**50–59
ion-beam processing, **13:**142–144
ionized cluster beam deposition and epitaxy, **13:**37–45
laser-induced etching, **13:**185–190
low energy surface plasmons in, **9:**238
neuron emulation, **21:**59–68
photochemical reactions, **13:**185–186
quantum wells, **21:**77–101
radial distribution function measurements, **7:**29–37
as solar absorbers, **10:**38–40
solid solutions, **5:**178
spiketrains, **21:**3, 5, 40–51

Semiconductors, (*continued*)
sputtered, **3:**113–117
topologically structured thin films, **8:**251–331
transparency in, **8:**4
Semiconductor waveguides, types II-VI, **10:**257–258
Semimetal films
electrical size effects in, **6:**7–138
specular reflection in, **6:**132
Semitransparent films, interference effects, **10:**96–97
Sensors, **20:**288–290
Sequence dependent switching, in multilayer films, **6:**265–266
Sequential resonant tunneling, **23:**286–291
Series frequency, quartz crystal microbalance, **24:**49
Settling time, **15:**248
SFI *see* Scanning Fizeau interferometer
SFM *see* Scanning force microscopy
SHADE model, **22:**84
Shear lag, **16:**–286, 281, 312–318, 325–326, 329–331
Shear stress, interfacial, **16:**279, 312, 318, 326
Sheath, **18:**6, 18
Child law, **18:**8, 22, 94
distributed, **18:**21
magnetically enhanced, **18:**10
thickness, **18:**20
voltage, **18:**20
SHEED *see* Scanning high-energy reflection electron diffraction
Sheet resistance, **8:**159–160; **13:**214
Shockley-Read recombination mechanisms, **11:**147
Shock-tracking algorithm, **22:**97–98
Shock waves, magnetic, **15:**45
Short range order, **15:**156–161
coordination numbers and distances, **15:**158
dense random packing and, **15:**159
Laves phase and, **15:**157
normalized coefficient, **15:**159
radial distribution function, **15:**156
X-ray scattering and, **15:**156
Short wave infrared (SWIR) band, **23:**118
Shutter, rotating, **3:**45–46
Signal generating tubes, **12:**152–157
Signal-to-noise enhancer, **15:**45
application in systems, **15:**69
frequency selectivity, **15:**98
threshold power, **15:**97
Signal-to-noise ratio, detector, **3:**279
Silane
decomposition, **4:**263–264
epitaxial growth of GexSi1-x, **23:**30–33
gas, **18:**209–217
matrices, **20:**41–68
nitridation, **5:**294–295
pyrolysis, **5:**271
Silane film
copper ion intercalation, **20:**48–62
infrared spectroscopy, **20:**50–53
intercalation, **20:**51–54, 57
molecular alignment, **20:**50–59
p-nitroaniline intercalation, **20:**54
preparation, **20:**45–47
second harmonic analysis, **20:**57
silane orientation, **20:**57–58
thiol-terminated, **20:**128–130
ultraviolet spectroscopy, **20:**50–53
Silane-*p*-aminobenzoic acid film
contact angle, **20:**59
second harmonic generation, **20:**57–59

stoichiometry, **20:**54–57
wettability, **20:**59
Silica *see* Silicon dioxide
Silicic acid ester, **5:**97, 113
Silicides, epitaxial, detectors involving, **23:**161–162
Silicide-silicon germanium Schottky detectors, **23:**153–161
metal/SiGe/Si band diagrams for, **23:**153–155
Silicon
adatom incorporation on surface, **23:**15–18
coated with SiO, transmittance, **2:**295
depositions, **13:**17
diffraction patterns, **13:**15–16
as dopant in MBE, **11:**58
double-layer coatings, **2:**300
epitaxial film, **13:**37–40
infrared detectors involving, **23:**141–162
ion-beam processing, **13:**143
ion microanalysts, **13:**37–38
negative-carrier concentrations, temperature dependence, **13:**39
open-circuit photovoltage versus wavelength characteristics, **13:**40–41
Rutherford backscattering analysis, **13:**38
in frame storage targets for television, **8:**289
growth rates, **23:**38, 39
infrared detectors based on, **23:**133–135
growth/fabrication, **23:**162–167
isotropic etching, **8:**289
laser-induced etching, **13:**185, 187–188
photoconductive operation, **8:**273
point arrays in, **8:**290
radial distribution function, **7:**24–27
spectral reflectance, **10:**53
substrates, **5:**296
surface steps on, **23:**15–16
thin film epitaxial layers on, **23:**115, 168
Silicon amorphous film
absorption coefficients, **13:**42–43
acceleration voltage, **13:**41–42
crystallinity, **13:**40–41
hydrogen pressures, **13:**40, 42
ion-beam processing, **13:**142–143
ionized cluster beam, **13:**40–43
photoconductivity and dark conductivity, **13:**43–44
photovoltaic applications, **13:**95–97
Silicon carbide
chemical vapor deposition, **5:**283–285
low-temperature depositions, **5:**291
nonthermal activation methods, **5:**291
optical band gap, **12:**272
in semiconductor films, **5:**156–157
Silicon carbide film
ionized cluster beam, **13:**53
pyrolytic decomposition, **12:**272
Silicon-carbon-hydrogen film, photoluminescence in, **12:**274
Silicon diode, *p-n* junction, *C-V* characteristics, **13:**39–40
Silicon diode arrays
cavity etching, **8:**265
gold-plated bumps on, **8:**279
polycrystalline silicon landing pads on, **8:**280
Silicon dioxide, **16:**326
on aluminum, **2:**351–352

Silicon dioxide, (*continued*)
chemical vapor deposition, **5:**286–293
as dielectric, **5:**286–287
ion-beam processing, **13:**133–135
composition, **13:**134
mechanical properties, **13:**135
microstructure, **13:**135
optical properties, **13:**133–134
laser-induced etching, **13:**190–191
low-pressure deposition
electron cyclotron resonance deposition, **22:**241, 243, 244, 246, 247, 251
LPCVD, **22:**202–210
PECVD, **22:**210–215
reaction mechanism, **22:**186–188
remote microwave PECVD, **22:**215–226
open structure, **5:**293
in oxide layers, **5:**115–122
phosphorus-doped, **5:**289
pyrogenic, **5:**118
in satellite temperature control, **10:**128, 134
sputtered films, **3:**112–113
storage trap, **8:**286–287
Silicon dioxide coatings, **12:**225
Silicon dioxide film, **5:**115–122; **8:**36
etch rate, **4:**286–289
evaporated, **4:**282–285
oxygen deficiency in, **4:**285–286
pyrolytic, **4:**314
refractive index, **4:**286–289; **5:**117
RF sputtered, **4:**277
sodium in, **4:**301–305
thermally activated chemical vapor deposition, **12:**258–268
ultraviolet transmittance in, **5:**118
Silicon esters, pyrolytic decomposition, **3:**134–136
Silicon film
autodoping, **5:**215
epitaxial growth in, **5:**200, 269
as filters, **7:**48
homoepitaxial growth in, **5:**207
substrate temperature vs. growth in, **5:**202
vapor deposition, **5:**147
Silicon germanium/silicon
heterojunction internal photoemission detectors, **23:**144–153
barrier height, **23:**145–148
quantum efficiency, **23:**148–153
quantum wells, **21:**81–84
Silicon halides, nitridation, **5:**297
Silicon heterojunctions, quantum wells
infrared detectors, **21:**101–110
intersubband transitions, **21:**90–101
material, **21:**79–81
quantum size effects, **21:**81–84
Silicon homojunction detectors, for infrared radiation, **23:**143–144
Silicon homojunctions
infrared detectors, **21:**2–5, 36
charge storage mode detectors, **21:**38–40
delta doped potential well detectors, **21:**32–33
Type I HIWIP detectors, **21:**14–19
Type II HIWIP detectors, **21:**19–31
Type III detectors, **21:**31–32
uncooled avalanche injection detectors, **21:**33–34
spiketrains, **21:**3, 5, 41
generation, **21:**41–48
infrared detection, **21:**48–50

Silicon monoxide
on aluminum, **2:**346–351
deposition, **2:**401–402
dielectric for thin film capacitors, **2:**401–405
evaporation, **12:**25–26
in satellite temperature control, **10:**129, 133–140
structure, **2:**401–402
Silicon monoxide coatings
adsorptance in ultraviolet region, **12:**26
deposition, **12:**28
infrared reflectance data, **12:**29–30
for metal-dielectric mirrors, **12:**44–45
oxidation state increase in, **12:**28
reflectance vs. wavelength data for, **12:**32
Silicon monoxide film, **1:**60–63
evaporated, **4:**282–285
gettering by, **1:**62–63
as overcoatings, **12:**25–26
photoconduction in, **8:**238
reactive evaporation, **12:**26
thermally stimulated current in, **8:**237
Silicon nitride, **14:**12, 13, 24–26, 31, 49, 56, 67
amorphous, **4:**314–316
chemical vapor deposition, **5:**293–298
deposition, **12:**243
pyrolytic and sputtered, **4:**269–270
Silicon nitride film
deposition rate *see* Growth rate of film
hydrogen content, **12:**269
nitrogen as reducing agent in, **12:**269
parallel-plate-type reactor and, **12:**267–268
as passivating layers, **12:**266
plasma chemical vapor deposition, **12:**265–272
pyrolytic dissociation and, **12:**265–266
thickness, **18:**209, 212–213
uniformity, **18:**177, 212–213
Silicon oxide, **14:**8, 9, 31, 48, 56, 57, 71
as protective layer for mirror coatings, **12:**2
Silicon oxide coatings, in silver film delamination, **12:**38
Silicon oxide film
by pyrolysis, **3:**133–137
infrared spectra, **15:**287–291
band assignments, **15:**288
bandshape analysis, **15:**288
cerium oxide inclusion and, **15:**290
deposition parameters and, **15:**287
moisture content and, **15:**289–290
stress relaxation and, **15:**288–289
plasma grown, **12:**282–283
thickness, **18:**220–223
uniformity, **18:**177, 218, 220–223, 225
Silicon oxynitride, **14:**51
Silicon oxynitride film, infrared spectra, **15:**295–299
band assignments in, **15:**299
deposition temperature and, **15:**295
nitrogen incorporation and, **15:**295
oxygen incorporation and, **15:**298
Silicon slice, transmittance curves for, **12:**253

Siloxane SAMs, displacement, **24:**99–100
Silsbee effect, **1:**254
Silver, **16:**327
  adsorption of thiols onto, **24:**3
  diffusion in selenium, **6:**339
  evaporation, **12:**15–16
  incidence angle, **12:**20
  infrared reflectance, **10:**152–153
  nonabsorbing surface films and, **12:**23
  photoemission from, **12:**81
  reflectance, **12:**20–23
  rhodium and, **12:**22
Silver-aluminum couples, **6:**328–333
Silver coatings, in satellite temperature control, **10:**128
Silver-copper systems, miscibility, **6:**304
Silver film
  absorption, **12:**71
  crystal structure, **20:**28
  dispersion relations, **9:**156
  electrochemical generation, **20:**15
  evaporated, **12:**24
  growth, **12:**71
  hydrogen sulfide and, **12:**37
  light absorption by, **8:**161
  light intensity scattering by, **9:**253
  Maxwell-Garnett absorption, **12:**71–72
  number of photons on, **9:**177
  Parlodion films and, **7:**70
  plasma resonance emission from, **9:**195–196
  radiation pattern for, **9:**252
  reflection minima for, **9:**237
  reflectivity, **9:**235; **20:**16
  resistivity, **20:**29
  for silver-oxygen-cesium photocathode, **12:**71
  size-dependent resistivity, **6:**110–111
  transfer efficiency, **20:**16
  transmittance spectra, **9:**194
Silver-gold systems, miscibility, **6:**302
Silver ions, defect removal in mercaptan film, **20:**193–194
Silver mirror coatings, optical constants and reflectance, **12:**17–19
Silver mirrors
  adherent, **12:**36–39
  infrared reflectance and polarization, **12:**30–31
Silver oxide
  formation, **12:**73
  as insulator, **12:**73
  silver layer deposition on, **12:**73–74
Silver oxide film, cesium-vapor treatment, **12:**74
Silver-oxygen cathodes, high-sensitivity, **12:**72
Silver-oxygen-cesium, Maxwell-Garnett absorption, **12:**79
Silver-oxygen-cesium photocathode, **12:**70–84
  discovery, **12:**54, 70
  electrical properties, **12:**76
  light absorption, **12:**78
  optical properties, **12:**78–79
  photoemission from, **12:**80–84
  preparation, **12:**71–75
  properties, **12:**76–84
  quantum efficiency, **12:**84
  thermionic emission, **12:**76–77
Silver sulfide tarnish, growth, **8:**163
Silver/Teflon sheets, **10:**162
SIMBAD deposition model, **22:**5–6, 76
  algorithms, **22:**9–14
  applications

bias sputtering, **22:**30–38
bulk deposition, **22:**53–54
chemical vapor deposition, **22:**29–30, 36
collimated sputtering, **22:**39–45
etching, **22:**60–76
film composition, **22:**45–49
high-temperature deposition, **22:**49–53
interpolated 3-D SIMBAD model, **22:**55–59
flow chart, **22:**14
program structure and user interface, **22:**14–15
verification, **22:**20–29
SIMS *see* Secondary ion mass spectrometry
SIMSPUD vapor transport model, **22:**76
algorithms, **22:**6–8
applications
collimated sputtering, **22:**40
film composition, **22:**45
flow chart, **22:**14
verification, **22:**15–20
Simulations, *see also* Modeling
anisotropic effects in ellipsometry, **19:**295
orientation of crystal principle axis, **19:**305
spectroscopy, **19:**305–08
variable ambient medium experiments, **19:**308
variable angle experiments, **19:**301
variable thickness measurements, **19:**310
force fields, **24:**209
Beck's force field, **24:**213
complex SAMs, **24:**211
Gerdy-Goddard dimer (GDD) model, **24:**211, 213, 214
Hautman-Klein united-atom (HKUA) model, **24:**209–210, 211, 213–216, 221
Mar-Klein all-atom (MKAA) model, **24:**210, 211, 213–215
of SE spectra of anisotropic materials, **19:**295–312
multiple-interface, ambient-film-substrate systems, **19:**309
optical function tensors, **19:**283–312
for single-interface ambient-substrate system, **19:**301, 302, 305, 308
thiol SAMs on gold, **24:**206–222
Single bandgap solar cells, **21:**319–322, 347
Single-cavity filter design
augmented spacers and, **9:**106–110
equivalent layer spacers and, **9:**110–112
MDM filter and, **9:**102–104
wedged filters and, **9:**105–106
Single-crystal films
evaporated, **4:**137–190
monolayer growth, **4:**138–150
Single-crystalline anisotropies, **6:**90–93
Single interface, scattering from, **10:**170–176
Single particle distribution function, **22:**189
Single quantum well, **21:**331–332
Single quantum-well infrared photodetectors, **21:**170–184
directly doped well *n*-type, **21:**170–176
p-doped well, **21:**180–184

Single quantum-well infrared photodetectors, (*continued*)
  undoped tunneling filled well, **21:**177–180
Single-site anisotropy, **15:**145
Size effects, *see also* Electrical conductivity; Quantum size effects
  classical theory of, **6:**82–101
  in electrical conductivity, **6:**81–143
Skin depth
  collisional, **18:**55
  collisionless, **18:**55
Skin effect
  measurement, **6:**133
  resistivity and, **6:**136
  in thin films, **6:**189–193
SKYLAB, **10:**125
Slitwidth, spectrometer spectral, **4:**55–57
Slow states, **2:**168
Small spheres
  fast-electron experiments on, **9:**218
  plasma resonance effects in, **9:**198–199
Smart window, **17:**303
Smith chart
  amplitude reflectance in, **10:**197
  dielectric mirrors and, **10:**196–197
  scattering application to, **10:**191–197
  showing constant amplitude reflectance and phase change on reflectance, **10:**192–195
Smith charts, **1:**102–111
SMST structures, experimental measurements, **4:**360–366
Sodium antimonide
  absorption coefficient, **12:**104
  band gap and electron affinity, **12:**96
  crystal structure, **12:**88
Sodium chloride, as parting agent, **7:**66
Sodium metaphosphate, as parting agent, **7:**63–64
Sodium-potassium antimonide structure, **12:**89
Sodium selenosulfate, synthesis, **12:**203
Sodium thin films, in ultraviolet region, **7:**43
Soft films, in multilayer structures, **6:**209–210
Soft lithography, **24:**229, 246
Solar absorbers
  semiconductors as, **10:**39–42
  tandem, **10:**38–54
Solar cells, **21:**313–314, 319–322, 321–323, 325–327, *see also* Quantum-well solar cells
  coatings, **2:**306
  I-V characteristics, **13:**28
  in liquid phase epitaxy, **7:**246
  photoelectrochemical, **12:**226
Solar energy conversion, photothermal, **10:**1–64, *see also* Photothermal solar energy conversion
Solar films, spray-deposited, **12:**224–225
Solar flux
  components, **10:**6–7
  incident to absorber surface, **10:**17
  standard, **10:**6
Solar heat collectors, intermaterial diffusion in, **11:**16
Solar insulation, **10:**5–11
Solar spectral irradiance, **10:**8–9
Sol-gel deposition, **17:**238–241
  annealing temperatures, **17:**241
  flow diagram for growth process, **17:**240

$KNbO_3$, KNSBN, **17:**241
$LiNbO_3$, optical waveguides, **17:**2421
precursor, for PZT, **17:**239–240
thicknesses, **17:**239
Solids
charge injection into, **11:**278–279
optical engineering, **10:**31
Soluble substrates, in metal film filters, **7:**73–74
Solution-grown chalcogenide films, **12:**218–219, 230
Solution-grown films, **12:**214–223
Solution growth chemistry, **12:**202–206
Solution growth process
complexing agents, **12:**203, 208
film growth and deposition parameters, **12:**206–211
ions and, **12:**203, 208
for multicomponent films, **12:**212–213
in organic film deposition, **12:**201–211
oxide films, **12:**213
pH value, **12:**209
Solutions, formation of solid layers from, **5:**88–94
Sommerfeld electrical conductivity, **1:**235
Source-drain contacts effect, thin film transistors, **2:**165–170
Space charge
buildup in thin glass films, **4:**310–311
nonmetallic films, **8:**210–211
polarization in glass-over-oxide structures, **4:**311–313
Space-charge-limited conduction, **8:**228–229
Spacer molecules
effect on quenching rate, **20:**252–253
lipid membrane, **20:**243, 246
polyimide Langmuir-Blodgett film, **20:**246
Space technology
evaporated films, **10:**71–163
filters, **7:**101
pinholes, **7:**107–108
unbacked, **7:**92–111
Spatial modulation, in molecular binding, **20:**224
Spectra, infrared transmission and reflection, **4:**273–274
Spectral distribution, of extraterrestrial sun blackbody, **2:**314
Spectral flux, defined, **10:**13
Spectral hole
as label, **20:**306–312
width, model for variation, **20:**305–306
Spectrally selective absorbers, thin metal films and, **10:**36–37
Spectrally selective surfaces
benefits and characteristics, **10:**2
composite films as, **10:**60
defined, **10:**2
gold-black deposits in, **10:**59
of hafnium carbide, **10:**29
interference coatings and, **10:**54–57
intrinsic materials for, **10:**29–38
materials, **10:**29–32
metals as reflectors and absorbers in, **10:**32
new technologies, **10:**4
photothermal solar energy conversion, **10:**1–64
physical processes and materials systems, **10:**28–64
requirements, **10:**25–27
surface morphology in, **10:**57–64

Spectrally selective surfaces (*continued*)
  tandem absorbers and, **10:**38–54
Spectral radiance distribution, of blackbody, **10:**11
Spectral response
  definition, **21:**315
  pulse mode infrared detectors, **21:**49–50
  quantum-well solar cell, **21:**323–327
  Type I HIWIP detectors, **21:**15–17
Spectral selectivity *see* Spectrally selective surfaces
Spectral solar insulation, for various air masses, **10:**6
Spectro-ellipsometry, **19:**279
Spectrometer
  dispersive infrared, **15:**269
  Fourier transform infrared, **15:**269
  mass spectrometer *see* Mass spectrometer
  photometric nonlinearity, **4:**47–49
  sample and shutter emission, **4:**51–52
  scattered light, **4:**49–51
  spectral slitwidth, **4:**55–57
  wavelength calibration error, **4:**52–55
Spectrophotometer, **5:**85
  for optical constants, **2:**213–218
  schematic diagram, **12:**7
Spectroscopic ellipsometer (SE)
  achromatic quarter-wave plate, **19:**198, 254
  data analysis, **19:**205–07
  film characterization, **19:**191, 249
  nonlinearity of detector, **19:**202–205
  optical functions
    aluminum in bulk, **19:**131–33
    aluminum films in disordered state, **19:**168–83
    aluminum thin films, **19:**150–83
    hydrogenated amorphous silicon, **19:**70–75
    inhomogeneous transparent films, **19:**192
    transparent fluoride films, **19:**234
    transparent oxide films, **19:**234
  photodiode array detector, **19:**54
  rotating analyzer system, **19:**196, 205
  rotating polarizer system, **19:**55
Spectroscopy, **21:**307, *see also* Spectrometer
Specular amplitude reflectance, **10:**174
Specularity parameter, **6:**89–90
Specular reflection
  in gold films, **6:**106
  in other thin films, **6:**131–134
Specular scattering coefficient, **10:**173–174
Specular transport, BTRM, **22:**197–200
SPEEDIE process simulator, **22:**5
Sperimagnetism, in non S-state rare earth alloys, **15:**144
Spherical particles, Maxwell-Garnett theory, **8:**146
Spherulites, **3:**187
Spiketrains, **21:**3, 5, 41–50
Spin
  alignment within single film domain, **1:**295–297
  wave resonance, **1:**327–329
Spinel, epitaxial layer growth on, **12:**130
Spinel films, **5:**313–314
Spin waves
  at half applied frequency, **15:**81, 92
  exchange-dominated, **15:**135

in infinite medium, **15:**3, 127
related to volume waves, **15:**29
Spirals, growth, **4:**202
Spot
autograph, **14:**139
fast moving, **14:**141
fast registry, **14:**139
motion, **14:**147
retrograde, **14:**147
size, **14:**139
slow moving, **14:**141
splitting, **14:**145
velocity, **14:**148
Spray hydrolysis, in chemical vapor deposition, **9:**32
Spraying processes, in oxide coatings, **5:**93
Spray pyrolysis, **12:**168–169, 169–171, 172–174, 174–177, 192–201, 224–225
Spray pyrolytic process, **12:**169–177
characteristics, **12:**178–181
copyrolysis in, **12:**183–187
film composition in, **12:**179–181
growth rate in, **12:**178
multicomponent doping and alloying effects in, **12:**181–187
physical parameters of oxide films in, **12:**184–185
spraying solutions used in, **12:**186
stannite, adamantine, and chalcopyrite compounds in, **12:**182
substrate effects in, **12:**178–179
Spread films, stability of, **7:**317–323
Spurious pinholes, in aluminum filters, **7:**52
Sputter, mechanisms, **14:**2
Sputter deposition, **15:**148, 150–152; **17:**81, 83; **22:**119–120
adatom energy, **22:**150–152, 170, 171
aluminum, **22:**25–26, 30–33, 49–53
high-temperature sputter deposition, **22:**49–52, 56–58
ionized magnetron sputter deposition, **22:**242–243, 244, 248–252
bias sputtering, **22:**30–38
collimated sputtering, **22:**39–45, 162–165
computer automated ion beam system, **17:**234–235
copper, **22:**154, 156
discrepancy between theory and experiments, **22:**120–122
ECR plasma sputtering, **17:**234
electrode configurations, **17:**232
electron cyclotron resonance deposition, **22:**241, 243, 244, 246, 247, 251
film stoichiometry, **17:**233
gas entrapment and, **22:**152–156, 170
ion bombardment and, **22:**152–156, 170
ion bombardment, **17:**234
ionized magnetron sputter deposition, **22:**101, 110–113, 165–169, 242–243, 244, 248–252
low-pressure/low-throw sputtering, **22:**165–169
mathematical model, **22:**101, 110–113
molecular dynamics modeling, **22:**125–126
adatom energy, **22:**150–152
collimated sputtering, **22:**162–165
deposition, **22:**149–160

Sputter deposition, (*continued*)
ion bombardment and gas entrapment, **22:**152–156
low pressure/low-throw sputtering, **22:**165–169
step coverage, submicron liner, **22:**161–169
transport, **22:**146–148
voltage bias, **22:**156–160
nickel, voltage bias, **22:**156–159
periodic boundary, **22:**142–144, 149
phase formation, **17:**233
self-sputtering yield, **22:**39–45, 100
silicon dioxide, **22:**241, 243, 244, 246, 247, 251
SIMBAD model, **22:**49–53
simulation, **22:**6, 7
simultaneous CVD and PVD, **22:**241–252
tantalum, **22:**154, 155
techniques, **15:**148
facing target, **15:**152
getter, **15:**150
ion beam, **15:**148
RF diode, **15:**148
RF diode with triode, **15:**152
rnagnetron, **15:**151–152
titanium, **22:**26, 29, 34, 35, 40–43, 61, 67, 69–73, 162–165
titanium nitride, **22:**46, 48, 49, 226–229
titanium-tungsten alloy, **22:**45–48, 229–241
tungsten, **22:**23–27, 30–33, 37, 38, 55, 61–66, 68, 69
Sputtered glass films, **4:**267–269, 279–282
Sputter etching
backscattering, **14:**219–224
chemical effects, **14:**230–231
contaminant removal, **14:**217–219
contamination
from chamber surfaces, **14:**208–210
from shields, **14:**210–213
crystallographic damage, **14:**232–233
electrode configuration, **14:**203–208
instrumentation, voltage measurement, **14:**205–208
radiation damage, **14:**233
recontamination, **14:**217–219
sublimation, **14:**232
techniques
DC, **14:**203–204
planar magnetron, **14:**204
RF, **14:**203–208
of thin films, **8:**265
Sputtering, **16:**14–40; **17:**3–6, 81, 83; **18:**236
altered layer, **16:**14
argon, **13:**118
Bi-Sr-Ca-Cu-O thin film, **16:**28–34
cathode, **3:**61–129; **14:**2
application to thin film deposition, **3:**80–100
experimental data, **3:**72–80
mechanisms, **3:**67–72
particles, **3:**70–71
ratio, **3:**69
reactive, **3:**100–101
technology, **3:**102–108
theories, **3:**66–70
chemical, **14:**8
of dielectric films, **8:**23–24
diode, **17:**85; **18:**237
direct current (dc), **16:**14–17, 19, 24–26, 28, 32, 36–38; **17:**85
discharge, **13:**110
electron cyclotron resonance plasma sputtering, **16:**30
electron source in, **9:**45

film deposition, **16:**146
of garnet films, **9:**282
high magnetic field magnetron, **16:**15, 16, 25, 26
inverted cylindrical magnetron, **16:**23
for magnetic bubbles, **9:**266
magnetron, **16:**15, 19, 25–28, 36, 38; **17:**4, 5, 85, 86; **18:**239, 241, 243
benefits, **18:**242
E x *b* drift, **18:**241
ionization, enhanced, **17:**87
ion plating, **17:**85, 87
plasma, **17:**86
reactive, **17:**96, 111–118
shadowing in, **18:**241
unbalanced, **17:**88, 89, 99, 123
negative oxygen ions, **16:**15, 16, 21, 24
nucleation and, **8:**99
off-axis, **16:**21, 22, 27, 34
of oxide targets, **9:**53–59
positive ions in, **9:**46
preferential resputtering during, **16:**15, 21, 26, 29
process parameters, **16:**14–16
radio frequency (rf), **13:**111, 145; **14:**2; **16:**14–16, 19, 21, 26–28, 30, 32–34, 36–38; **17:**85; **18:**238
rare earth-based thin film, **16:**19
reactive, **1:**223–224; **9:**49–53; **13:**67; **16:**16; **17:**10
semiconductor films, **5:**167, 179–182
silica, **13:**134
spherical target, **16:**16
substrate chemical balance, **17:**113
tantalum, **2:**393–397
films, **8:**100–102
target chemical balance, **17:**112
thermalized, **16:**171
of thin films, **2:**366–369
Tl-Ba-Ca-Cu-O thin film, **16:**34–40
triode, **16:**33; **18:**238
two-target, **17:**88, 91
unbalanced magnetron, **18:**259
Y-Ba-Cu-O thin film, **16:**17–28, 47
yield, **13:**117; **17:**83, 85, 114
Sputtering targets, **15:**148–154
alloy, **15:**148
gap type, **15:**152
mosaic, **15:**154
Sputter yield, **14:**4, 6–10, 24; **15:**148
effective, **3:**93
variation with ion energy, **3:**64–66
SQUID, **16:**36, 76, 101, 114–123
SQW *see* Single quantum well
S-state and non *S*-state RE alloys, **15:**185
Stability condition, **15:**249
Stabilization, impurity, of disordered film structure, **2:**41–44
Stacked-wafer architecture, **21:**65
Stacking faults, **2:**15–17
formation, **4:**166–169
semiconductor films, **5:**204
Standard linear model, **16:**294–295, 299–301
Standing-up phase, **24:**93
Stannic oxide, **14:**16, 17
doped films, **2:**389–390
resistors, **2:**389–390
Stannic oxide crystals, resistivity and carrier concentration, **9:**27
Stannic oxide film, **8:**39
conductivity vs. temperature, **9:**28
donor activation energy, **9:**25
doped, **9:**36, 37
electrical and optical properties, **9:**22–26, 34

Stannic oxide film, (*continued*)
Hall mobility, **9:**24
indium oxide and, **9:**39–40
preparation
hydrolysis, **9:**34, 36
pyrolysis, **9:**41
reactively sputtered, **9:**51
refractive index, **9:**25
undoped, **9:**34
Stannous oxide, electrical and optical properties, **9:**22–26
Staring-mode operation, *see also* Infrared staring sensors
of imaging sensors, **23:**84
Stark shift, **15:**231; **20:**297, 308–311
Stearic acid film, capacitance and dielectric loss in, **11:**265–267
Steel
glow-discharge nitriding, **12:**286–289
plasma carburizing, **12:**283–285
Stefan-Boltzmann concept, **10:**12, 132
Step coverage, **22:**176, 205–206
ion bombardment and, **22:**167–169
pressure and distance and, **22:**166–167
submicron liner, **22:**161–169
Step response
in polarization, **6:**54–56
space-charge contribution to, **6:**58–60
Sterred arc, **14:**167; **17:**95, 96
Sticking coefficient, **14:**10, 11, 17, 25, 30, 45, 55–56
"Sticking factors," **22:**179–180, 182, 204
STM *see* Scanning tunneling microscopy
Stochastic heating, **18:**146–148
Stoddard solvent, **10:**124
Stoichiometry
ion-beam processing, **13:**121–123
low-pressure deposition, **22:**181, 183–185
magnesium fluoride, **13:**138
silicon dioxide, **13:**134
sputtered films, **17:**112, 117
of superconducting thin film, **16:**15–20, 58–60, 78
compensated target for, **16:**19, 21, 27, 29, 32, 33, 58, 79
forward or direct component in laser ablation, **16:**59, 60, 78
high field magnetron gun, **16:**15, 16, 25, 26
laser-ablation characteristics, **16:**58, 60
oxygen pressure, **16:**16, 18–20, 22, 24, 32, 56, 59
spherical target, **16:**17
substrate temperature, **16:**18–20, 22, 24
total pressure, **16:**16, 23
titanium dioxide, **13:**131
Stoke's parameters, **2:**221
Stopping power, **2:**67–73
Storage capacity, **15:**201
Strain, **16:**290
equilibrium, of an epitaxially aligned film, **4:**139–140
generation, **4:**174–175
Strain coefficient of resistance
for discontinuous films, **8:**132–139, 158–159
transverse, **8:**136
Strain hardening, **16:**288
Stray-field coupling, in multilayer films, **6:**248
Stray light measurements, for diffraction gratings, **11:**30–33

Streptavidin, **20:**208–209; **24:**25
Stress, **14:**5, 65–70; **16:**290
anisotropic, **3:**250–253
gas contamination, **3:**234
hard coatings, **17:**126, 130
internal in evaporated films, **3:**219–253
bending-plate methods, **3:**219–225
thermal effects, **3:**225
X-ray and electron diffraction methods, **3:**225–226
intrinsic, **3:**228–241
ion-beam processing, **13:**125
measurement, **10:**273–274
oxide layers as cause, **3:**244
silicon dioxide, **13:**135
in silicon oxide films, **2:**403–404
substrate temperature dependence, **3:**232–233
thermal, **3:**227–228
in superconducting films, **3:**241–243
thickness dependence, **3:**228–233
on thin film, **1:**206–207, 226–228; **22:**123, 144–145, 150, 170
magnetic, **1:**282–283
voltage bias, **22:**156–160
titanium dioxide, **13:**133
variation in optical elements, **10:**276–278
Stresscoat, **16:**266, 275, 279–281, 326
Stress free mixtures, in optical element corrections, **10:**278–280
Stress-strain, thin film, **3:**257
Strip-coating cleaning, for metal films, **7:**61
Stripe collectors, collection efficiency, **11:**180
Striped phase, **24:**142–143
Strong-Gaviola method, for optical elements, **10:**264–265
Strong low-frequency dispersion, in ionic conductors, **11:**270–272
Strong-type reflectometer, **12:**5–6
Strontium fluoride, IV-VI semiconductor layers on, **11:**115–122
Strontium titanate, **14:**65
Structural anisotropy, **15:**161–166
argon incorporation, **15:**165
columnar microstructure, **15:**161–163
pair ordering, **15:**164–165
planar stress, **15:**163
Structural constraints, in thin films, **8:**267–272
Structural morphology, evolution, **13:**82–83
Structural properties, **15:**155
oxidation and aging, **15:**168–171
short range order, **15:**156–161
stability and crystallization, **15:**166
structural anisotropy, **15:**161–166
Structural stability, **15:**166
activation energy for crystallization, **15:**167
structural relaxation, **15:**168
Structural studies
atomic force microscopy, **24:**82–83, 86, 115
alkanethiols, **24:**83–86
fluorinated thiols, **24:**88
heterogeneous SAMS, **24:**89
terminal groups, **24:**88–89, 90
thiodiethyl aklyl decanoate, **24:**87
diffraction studies, **24:**114, 121–130
hydrocarbon tilt structure, **24:**132–134
interface structure, **24:**135–137

Structural studies (*continued*)
submonolayer phases, **24:**142–144
films on liquid mercury, **24:**180–201
NMR, **24:**150–154
alkanethiols on gold nanoparticles, **24:**154–163, 175
alkylphosphonates on metal oxides, **24:**163–169, 175
alkylsiloxane SAMs, **24:**169–175
simulations, **24:**206–222
submonolayer phases, **24:**142–144
X-ray study, **24:**180–201
Structure
high critical temperature superconductors, **16:**2
thin films, **1:**187–232
changes in, **1:**220–226
comparison with bulk, **1:**212–214
dependence of physical properties on, **1:**214–215
methods and instruments, **1:**188–207
optical methods, **1:**206
substrate influence, **1:**210–211
Structure zone model, **19:**194, 195
Sublimation
apparatus for, **5:**168
of semiconductor films, **5:**162–173
Submicron liner, step coverage, **22:**161–169
Submicron structures, growth techniques for, **11:**94
Submonolayer phases, structure, **24:**142–144
Subsidiary absorption, **15:**92
Substrate/coating interface, **17:**106, 133
Substrates, **2:**370–377
aluminum, **16:**284, 324
bias, **14:**61, 64, 66, 71
biasing, **16:**30, 65
bombardment, **13:**73–74
ion-beam processing, **13:**110–111, 125–127
boron nitride, **5:**299
chemical balance, **17:**112–113
cleaned, by high energy ion bombardment, **14:**171
copper, **16:**326
effect, **14:**172
on evaporated II-VI compounds, **3:**172–174
gaseous etching, **5:**247
glass, **5:**111; **16:**282
heating, **14:**68; **17:**132
for high critical temperature superconducting thin film, **16:**5
consil, **16:**38
copper oxide, **16:**29
corundum, **16:**17, 18, 28, 29, 51, 52, 60, 63, 64, 75, 78, 81, 86–88, 90, 94, 98, 103, 121
gallium arsenide, **16:**5, 25, 28
lanthanum aluminate, **16:**5, 22, 36, 48–50, 63, 67, 75, 76, 81, 103, 106, 107, 121
lanthanum gallate, **16:**5, 63, 71, 72, 108, 110
magnesium oxide, **16:**5, 17, 18, 22–24, 28–30, 32–34, 36, 38, 41, 42, 51, 52, 54, 60, 61, 66–68, 76, 78–81, 85, 87, 90, 92, 93, 96–100, 103, 107, 111, 117, 118
potassium tantalate, **16:**5, 43, 48, 49
sapphire, **16:**107

silicon, **16:**5, 17, 21, 25–28, 42, 45, 50, 52, 60, 63, 65, 68–70, 78, 79, 91, 98
strontium titanate, **16:**5, 17–19, 21–26, 28, 29, 34, 36, 37, 41, 42, 47–52, 54, 60, 61, 63–69, 71–73, 75, 76, 78–81, 86–91, 96, 97, 110, 112, 113, 117, 118, 120–122
yttrium stabilized zirconia, **16:**5, 17, 18, 22, 23, 28, 29, 34–36, 38, 41, 51, 54, 61, 65, 66, 68, 70, 71, 75, 76, 78, 80, 86, 88, 91, 92, 94, 96, 97, 100
holder, **14:**50
imperfections, effect on nucleation, **4:**111–113
iron alloy, **5:**299
kapton, **16:**278, 282, 288, 296–306, 319–325
microfinish, **2:**371–373
motion, **14:**50–52
mylar, **16:**278, 312
orientation, **5:**261
platinum, **16:**312
potential in sputtering, **3:**92–96
preferential, **14:**17, 19
preparation, **2:**285; **4:**2–8; **14:**171
crystalline, **4:**7–8
glass, **4:**3–7
processes, **14:**6–9
rotating, **3:**46–50
sensitization, **3:**40
silicon, **5:**296
sodium content and, **2:**374
stainless steel, **16:**284, 324
systems, **14:**22–52
temperature, **5:**198–203; **17:**19, 31, 33, 119–121, 133
dependence of stress, **3:**232–233
during sputtering, **3:**103
high, **4:**108–109
low, **4:**109–111
temperature control, **14:**172
thermal expansion, **17:**104
wafer holder, **18:**153, 163, 164, 170, 182, 184, 209, 211, 221–224
for YIG growth, **15:**22
Sugar-aerosol solution, as parting agent, **7:**67–68
Sulfides, as dielectric films, **8:**34–35
Sunglasses, absorbing films for, **8:**92–93
Sun-shielding glass, **5:**127–129
Superconductive and magnetostatic delay lines, **15:**6, 60
Superconductivity
effect of gases, **1:**55–58
theory, **1:**255–257
Superconductors
activated reactive evaporation applications, **13:**95
critical field, **1:**257–260
energy gap, **1:**265–271
magnetic field transition, oxygen effect, **1:**55–57
penetration depth, **1:**248
transition, **1:**240–255
in magnetic field, **1:**246–255
in zero field, **1:**240–246
Superlattice miniband quantum-well infrared photodetectors, **21:**184
bound-to-bound miniband detectors, **21:**184–187
bound-to-continuum miniband detectors, **21:**187–190
bound-to-miniband detectors, **21:**190–191
Superlattice minibands, **15:**222, 300
quantum wells, **21:**85–89

Superlattices, **16:**28, 36, 54, *see also* Heterostructure
  energy level, **15:**222–225
  functions, **23:**220
    photon wave function, **23:**252–256
  gratings, thermal emission induced by, **23:**243–244
  intersubband emission, **23:**238–239
Superparamagnetism, **2:**30–37
Superposition notation, **1:**96–100
Supersaturation, critical, **4:**121–127
Supported organic layers, **24:**181
Supramolecular assemblies, **20:**1–6, 42
Surface
  acoustic and magnetostatic waves, **15:**6, 60
  analytic features, molecular beam epitaxy, **11:**40
  atomic mobility, of vapor atoms, **1:**209
  cleaning, **17:**106–133
  coverage, **14:**10, 11
  escape probability, **12:**106–128
  etching, **17:**133
    ions
      gas, **17:**134
      metallic, **17:**134
  films, **17:**129
    arc evaporated, **17:**96, 129
    sputtered, **17:**125, 129
  film strength and, **3:**267–268
  layers, internal stress due to, **3:**244
  morphology, **17:**96, 108, 125, 129
  particle bombardment, **17:**82
  permeability, **15:**132
  reactivity, alkylphosphonate monolayers and multilayers, **24:**165–166
  roughness, **4:**12–31; **19:**150, 153–56, 210–13, 221–27, 233
    correction of reflectance measurement for, **4:**17–18
    coupling by, **9:**219–221, 227–229
    Fizeau thickness measurement and, **4:**29–31
    Gaussian, **4:**13–17
    ion-beam processing, **13:**121, 126
    measurement, **4:**18–21
    non-Gaussian, **4:**17
    specular reflection and, **4:**13–16
  scattering, **10:**176–188
  state effects on thin-film transistor performance, **2:**165–170
  steps
    nucleation at, **4:**152
    orientation of nuclei at, **4:**160–161
  substrate characteristics, **2:**371–374
  tension, **3:**246–247
  voltammogramic analysis, **20:**188, 191, 194–196, 199, 201
Surface area, of evaporated metal films, **2:**47–52
Surface barriers
  electron affinity and, **12:**66
  electron tunneling through, **8:**288
Surface bonding, alkylphosphonate monolayers and multilayers, **24:**164–165
Surface charge oscillations, **9:**145
Surface density, intercalants, **20:**56
Surface diffusion, **22:**252–253
  aluminum-copper film, **22:**253–265
Surface diffusion coefficient, cobalt thin films, **16:**262
Surface losses
  asymmetry in intensity, **9:**210–213
  probability and position, **9:**200–204

total probability for exciting, **9:**204–206
Surface mirrors, vacuum systems for film deposition, **12:**2, *see also* Front surface mirrors; Mirror coatings
Surface plasma, surface roughness and, **6:**198–199
Surface plasma oscillations, **9:**145–255, *see also* Surface plasmons
plasma resonance emission and, **9:**195–198
Surface plasmons, **9:**145–171
as charge oscillations, **9:**146–147
coupling of by grating, **9:**224–227
coupling by statistical surface roughness, **9:**227–229
dispersion relations, **9:**157–158
electron energy losses and, **9:**180
excitation of by light, **9:**180–199, 223–229, 233–255
existence of, **9:**145
guided light modes compared with, **9:**168–171
low energy, **9:**238
nonradiative, **9:**149–159, 164–166
nonradiative surface plasmon excitation and, **9:**199–255
radiative *see* Radiative surface plasmons
reflectance deficit, **9:**228–229
spectroscopy, **24:**25
in spheres and on voids, **9:**166–167
surface fields, **9:**169
Surface potential, substrate, for vapor atoms, **1:**209
Surface segregation, germanium, **23:**10, 11
Surface species, BTRM, **22:**200–201
Surface-state density, **13:**49
Surface wave discharge, **18:**65, 67
Surface waves, magnetostatic anisotropy effects in, **15:**131
bandwidth, **15:**6, 28, 36
dispersion, **15:**28
in double ferrite layers, **15:**38
ground plane effect, **15:**36
width modes, **15:**35
Susceptors, graphite, **5:**255
SWIR *see* Short-wave infrared
Switching
domain wall, **1:**301–302
high-speed, **1:**308–310
idealized single domain, **1:**297–302
critical curves, **1:**299–301
mode, partial rotation, **1:**306–307, 308
quasi-static, **1:**304–307
superconducting transition, hysteresis in, **1:**252–255
Symmetrical multilayer, transmittance, **9:**90–92
Symmetry
anisotropy field, **15:**26, 129
magnetostatic propagation, **15:**31

T

Tafel line
curvature, **6:**23
gaseous anodization and, **6:**50–51
in ionic conductance, **6:**19–20
Tafel slope anomaly, **6:**14–15, 21
Tail groups, polar, **24:**17–20
Tandem absorbers, **10:**38–54
examples, **10:**42–44
optical properties, **10:**34
reflectance of semiconductor layer in, **10:**61

Tandem absorbers, (*continued*)
with transparent electrically conducting surface films, **10:**163
Tantalum
anodic oxide film growth and, **6:**6–7
resistors, **14:**2
sputter deposition, **22:**154, 155
sputtering, **2:**393–397
substrates, **6:**3
and tantalum nitride resistors, **2:**393–398
Tantalum aluminum oxide, **14:**52
Tantalum film, **8:**100
gauge factor vs sheet resistance in, **8:**160
low-density, **8:**127–129
temperature coefficient of resistance, **8:**158
Tantalum film, sputtered, **3:**108–109
Tantalum nitride, **13:**68; **14:**2
Tantalum oxide, **14:**17, 18, 24, 27, 55
Tantalum oxide, anodic, for capacitors, **2:**405–409
Tantalum pentoxide
chemical vapor deposition, **5:**304
ion-beam processing, **13:**139–140
Tantalum pentoxide film, **6:**12–13
Tantalum/tantalum carbide, **17:**64
Tantalum-vanadium, **17:**33
Target, **14:**56–58
bonding, **14:**57
compound, **14:**17, 18, 56
cooling, **14:**57
heating, in oxide target sputtering, **9:**54
insulating, **14:**2
magnetron, **14:**39
metal, **14:**17, 28, 56
multiple, **14:**51–52
powder, **14:**57–58
for superconducting thin film
compensated, **16:**19, 21, 27, 29, 32, 33, 58, 79
composite, **16:**16, 29, 34, 36, 80, 1716.22
metallic, **16:**17, 38
multiple, **16:**17, 28, 32, 33, 36
ring, **16:**29
spherical, **16:**16
surface, **14:**139
Tauc's method of determining the band gap of amorphous films, **19:**73, 86
TBEA *see* 2,2'-Thio*bis*ethyl acetoacetate
TBPA *see* 3,3'-Thio*bis*propyl acetoacetate
TCP *see* Inductive discharge
TCR *see* Temperature coefficient of resistance
Teepol 610, as parting agent, **7:**67
Telecommunications equipment, activated reactive evaporation applications, **13:**94
Television
camera tubes, **12:**152–157
frame storage targets, **8:**282–287
vidicon imaging targets, **8:**272–282
Tellurium film
copper diffusion in, **6:**340–341
as filter, **7:**49
vacuum sublimation, **5:**171
Tellurium ions, in spray pyrolysis, **12:**203
Tellurium oxide film, **8:**40
Tellurium p-type thin-film transistors, **2:**177–178
TEM *see* Transmission electron microscopy

Temperature
crystal frequency and, **3:**24
GexSi1-x epitaxial growth at, **23:**22–23, 27–37, 44
ion, **18:**71, 89
objects in space, controlling temperature, **2:**305–306
source, germanium reflectance and, **1:**159–160
substrate, **14:**29, 59, 64, 65, 69
aluminum reflectance and, **1:**152–153
target, **14:**57–58
Temperature coefficient of resistance (TCR), for discontinuous films, **8:**129–130, 156–158
Temperature dependence
in thin films and wires, **6:**93–100, 124–131
of Z^*, **16:**264–265
Temperature stability, of dielectric films, **8:**17–18
Template
amphiphilic film nucleation, **20:**146
hydrogen bonded, **20:**42
ice nucleation, **20:**145–181
nanocrystalline particulate films, **20:**12
reactivity, **20:**133
Tensile properties
films, **3:**253–270
models for thin film strength, **3:**266–270
results, **3:**256–266
strength, **3:**263–265
techniques, **3:**253–256
thickness dependence, **3:**269–270
thin metal films, **2:**53–59
metal whiskers, **2:**58–59
Tensile strength, **16:**290
TEOS (tetraethoxysilane), **18:**223–226
pyrolytic decomposition, **3:**134–138
vapors, chemical vapor deposition, **12:**260–261
TEOS (tetraethyl orthosilicate) method, **5:**289–290, 292, 302
Terbium alloys
coercivity, **15:**175, 183, 186–187, 206
compensation temperature, **15:**180–182
Curie temperature, **15:**180–182
direct overwrite techniques, **15:**204
domain formation, **15:**187–189
domain wall pinning in, **15:**185
electronic density of states, **15:**176–177
Kerr rotation, **15:**184, 196, 199, 203
magnetic anisotropy, **15:**183–185
magnetization, **15:**174–180
medium design, **15:**201–208
medium noise, **15:**209–211
optical conductivity and absorption, **15:**193–194
oxidation, **15:**169 171
radial distribution function, **15:**156–159
recording characteristics, **15:**209–213
short range order, **15:**159
sperimagnetism, **15:**144
structural relaxation, **15:**168
wall coereivity, **15:**185–187
Terminal groups, **24:**88–89
alkanethiols on gold nanoparticles, **24:**159–164
ferrocenes, **24:**23, 27–29
producing high energy surfaces, **24:**13–15
producing low-energy surfaces, **24:**17

Terminal groups, (*continued*)
  producing surfaces with intermediate energies, **24:**16–17
Ternary III-V alloys, in MBE, **11:**69–75
Ternary systems or compounds
  bandgap variations, **7:**173
  liquid phase epitaxy, **7:**172–205
  phase diagrams, **7:**174–205
  properties, **7:**183–205
  quasi-regular solution, **7:**180–182
Terpyridine, **24:**23
Terrestrial solar peak, **10:**6
Terrestrial solar spectral irradiance, **10:**8–9
Terrestrial spectral flux, vs. extraterrestrial insulation, **10:**5
Tetraethyl orthosilicate method *see* TEOS
Tetraethyl silicate, **5:**289
Tetramethyllead, photolysis, **5:**312
TE waves, for inhomogeneous films, **8:**73–76
Texture, **17:**126–130
  fiber, **1:**215–216
Textured film growth, **16:**140, 173, 180, 183, 190
Textured interface, **15:**235
Texturization, **1:**215–216
TFT devices *see* Thin-film transistors
Thallium-barium-calcium-copper-oxide superconductor, **16:**2, 4
Thallium-barium-calcium-copper-oxide thin film
  laser ablation, **16:**56, 80, 81, 102–104, 106, 117, 121
  MOCVD, **16:**94
  MOD, **16:**100
  post-incorporation of Tl, **16:**37, 38, 54, 81, 94
  sputter deposition, **16:**34–40
  vacuum evaporation, **16:**54
Thallium wires, resistivity size-effects in, **6:**123
Thermal conductance, of substrates, **2:**375–377
Thermal control coatings, in-flight tests, **10:**158–163
Thermal desorption cleaning
  for epitaxial growth, **23:**45–47, 49–50
  for molecular beam epitaxy system, **23:**9
Thermal detectors, of infrared radiation, **23:**121, 124–127
Thermal emission, field-assisted, **8:**217–224
Thermal emittance, **2:**305–361
Thermal equilibrium, **17:**107
Thermal expansion, substrate, **2:**28–29
Thermal expansion mismatch, **16:**270
Thermal growth, of thin films, **2:**369–370
Thermal imaging, high-speed, **21:**260
Thermalization
  atom, **14:**21, 27, 62
  in internal photoemission, **23:**91
Thermally stimulated currents, in amorphous solids, **8:**233–235
Thermally stimulated hot-electron spectroscopy (TSHES), **21:**246–251, 253–254
Thermal misfit, **17:**106
Thermal processes, laser-induced etching, **13:**155
Thermal radiation, in photothermal solar energy conversion, **10:**11–13

Thermal transport, size effects in, **6:**82
Thermionic emission, **4:**354–367; **9:**6; **15:**236–238
Thermionic thermal detectors, **23:**127
Thermodynamics
alkanethiol/gold SAM formation, **24:**45–69, 83, 89–93, 150
first law of thermodynamics, **16:**270
Thermoelectric films, ionized cluster beam deposition and epitaxy, **13:**32–37
Thermomigration, **16:**237
Thick barrier quantum-well infrared photodetectors, **21:**240, 248–251, 298–300
Thick films, **16:**183
strain coefficient of resistance, **8:**133
Thickness, optical and physical, **5:**57
Thickness dependence
aluminum reflectance, **1:**158
critical supercurrent, **1:**264
germanium reflectance, **1:**163
gold films, **6:**108–109
magnetization, **1:**322–327
penetration depth, **1:**257–260
Thickness resonances, **15:**25, 29
exchange effects in, **15:**134
Thin-film devices, unconventional, **11:**168–199
Thin-film etching, **8:**261
Thin-film IV-VI photodiodes, **11:**128–157
history, **11:**128
noise, **11:**150
response, **11:**163
temperature-dependent properties, **11:**158
Thin-film IV-VI semiconductor photodiodes
for 3–5 ?m operation, **11:**157–165
for 8–12 ?m operation, **11:**166–168
detector noise, **11:**149–153
flip-chip mounting, **11:**192
with integral interference filter, **11:**188
properties, **11:**167
resistance estimates, **11:**140–149
surface effects in, **11:**153–157
Thin film lateral-collection photodiodes, **11:**181–182
Thin-film PbTe detectors, **11:**130
Thin-film PbTe photodiodes
configuration, **11:**195–197
temperature-dependent noise properties, **11:**151
thermal image from, **11:**197
Thin-film photodiode arrays, **11:**191–199, *see also* Thin-film IV-VI devices
configuration, **11:**194–197
schematic diagram, **11:**193
Thin-film processing, plasma, particle, **18:**289, 290
Thin-film resistivity, **6:**85–90
experiments, **6:**103–118
Fuchs model, **6:**98
Husstad-Lothe model, **6:**98
temperature dependence in, **6:**100, 124–131
Thin films, **22:**118–119, *see also* Cermet films; Dielectric films; Discontinuous films; Electromigration; Filamentarylike thin films; Homogeneous films; Inhomogeneous films; Island-structured films; Metal films; Multilayer magnetic films; Thin wires; Very thin films
activation energy, **8:**104, 109

Thin films, (*continued*)
amorphous and polycrystalline, **10:**238
anomalous skin effect in, **6:**189–193
apparent dielectric constant, **6:**201
Böttcher formula, **8:**61–65
classes, **8:**117–118
coevaporation, **8:**59–64
cold-rolling process, **6:**102
colors, **3:**8–9
conductivity, **6:**103–118
correction of optical elements in, **10:**272–286
coupling, **10:**234–235
cryogenic by sputtering, **3:**120–122
defined, **8:**141
degradation, **7:**264–287
deposition, **8:**253–254
diffusion process, **10:**242
electrical conductivity size effects, **6:**81–143
electrical properties of pyrolytic, **3:**140–141
electromigration, **7:**257–305
electron microscopy, **7:**333–334
electron tunneling, **8:**111
epitaxial, by sputtering, **3:**114–116
evaporation and sputtering, **9:**42–47
field distribution, **10:**232–234
in field-effect transistors, **8:**305–310
growth in spray pyrolytic process, **12:**206
guiding and coupling in, **10:**230–235
history, **10:**228–229
hybrid vs. total integration, **10:**229–230
inhomogeneities in, **8:**81–83
in integrated circuits, **8:**316–319
integrated optics, **10:**227–258
intrinsic stresses, **8:**268–269
ion etching, **8:**265–266
ion-exchanged and ion-implanted, **10:**237
light absorption, **9:**15
light guide, **10:**231–235
magnetic, by sputtering, **3:**117–120
mechanical motion, **8:**310–330
mechanical properties, **3:**211–273
monocrystalling films and, **10:**245–258
nonradiative surface plasmons, **9:**149–159
non-single-crystal films, **10:**235–238
nonuniform grain structure, **7:**270
nucleation and growth, **8:**99
nucleation-modifying layers in, **9:**9
optical, by sputtering, **3:**122–124
as optical interference coatings, **8:**51
optical plasma resonances, **6:**193–199
organic, **10:**238–245
oriented, **3:**218
pattern definition, **8:**254–261
on plasma boundary, **9:**238–240
plasma chemical vapor deposition, **12:**237–287
plasma films, **9:**176
plasma radiation, **6:**195
plasma resonance, **9:**188
plasma resonance absorption, **6:**196
plasmon dispersion, **6:**195
polycrystalline, **3:**214–218
polycrystalline and amorphous, **10:**235–238
preparation, **2:**333–336
quantum effects, **9:**15
radial distribution function measurements in, **7:**17–37

radiative surface plasmons on, **9:**159–163
reflection, refraction, and interference in, **9:**16–19
refractive index, **8:**51–52
resonance, **10:**232
single crystal, **3:**218–219
specimen preparation, **6:**101–103
specular reflection, **6:**131–134
sputtered, applications, **3:**108–214
sputter etchings, **8:**265
sputtering, **3:**80–100; **8:**99
stress measurement in, **10:**273–274
structure, **3:**212–219; **8:**261–267, 267–272
temperature dependence in, **6:**124–131
temperature difference in, **6:**93–99
tensile properties, **3:**253–270
as thin film substrates, **7:**72
transitions between conduction mechanisms in, **8:**121–122
transparent conducting, **3:**124
of uniform thickness, **3:**1–59
vacuum deposited, **10:**236–237
as waveguides, **10:**231
wave propagation in, **8:**52

Thin-film substrates, in metal film filters, **7:**69–73

Thin-film transistors (TFT), **2:**187; **5:**171, 216, 228

Thin metal films
aircraft industry, **9:**1
deposition process, **9:**3
discontinuous, **9:**4–7
electrical conduction, **9:**4–9
history, **9:**1
island films, **9:**6
island size and interisland spacing, **9:**4
oxidation state, **9:**2
postoxidation, **9:**47
properties, **9:**3
silver films, **12:**71, 72, *see also* Silver film
spectrally selective absorbers, **10:**36
''thinness,'' **9:**5–6

Thin surface layers, surface-to-volume ratio, **10:**3

Thin wires
Dingle theory expression, **6:**129
electrical conductivity, **6:**87–93
experiments in electrical conductivity, **6:**118–123
resistivity, **6:**88, 121–123
size effects in, **6:**103

2,2'-Thio*bis*ethyl acetoacetate, **20:**184, 186

3,3'-Thio*bis*propyl acetoacetate as receptor, **20:**184

Thiodiethyl alkyl decanoate SAMs, structure, **24:**87

Thiols
aromatic, **24:**33
displacement, **24:**99–102

Thiol SAMs, **24:**206–222, *see also* Alkanethiol/gold monolayers

Thompson energy distribution, **22:**146

Thomson-Whiddington law, **2:**64

Thorium dioxide film, **8:**37

Thorium fluoride film, **8:**32–33

Thorium oxide film, microstructure and optical function, **19:**223, 226, 236

Three-dimensional waveguide structure, preparation, **11:**81

Three-layer symmetrical systems, **9:**112–116

Three-mirror polarizer, **12:**33–34

Three-port transducer model, **15:**41

Threshold

Threshold (*continued*)
electric field, **18:**149–150
microwave power, **18:**176–177
plasma density, **18:**176–177
Threshold energy, sputtering, **3:**64–65
TIC *see* Transmissive interference cathode
Tight binding theory, workfunction calculations, **21:**9
Tilt structure, **24:**132–134
Time constant(s)
of PbTe films, **3:**287
photoconduction, **3:**197–198
Time domain, **11:**212, 230–231, 272–277
Tin
as dopant in MBE, **11:**59–60
resistivity size effects, **6:**122, 141
surface-state resonances in, **6:**133
Ti(N, C)
grain morphology, defect distribution, and texture, **13:**92
microhardness, **13:**91–92
Tin film
critical temperature change with stress, **1:**242–243
as filters, **7:**48
thin film, **16:**255–257
Tin/NbN, **17:**10, 31–33, 49, 63, 64, 71
Tin oxide *see* Stannic oxide
Tipping furnace, in liquid phase epitaxy, **7:**117–118
Titanium
liners, **22:**166–169, 171
nitriding rate, **12:**288
sputter deposition, **22:**26, 29, 34, 35, 40–43, 61, 67, 69–73
collimated deposition, **22:**162–165
Titanium aluminum nitride, **14:**60
Titanium carbide, **14:**67
activated reactive evaporation, **13:**68–69
coatings, **13:**70–72
lattice parameter, **13:**80
low-temperature deposit, annealing effect, **13:**84–85
microhardness, **13:**80, 83
deposition temperature and, **13:**91
vs. carbon-to-metal ratio, **13:**90
microstructure, **13:**83–86
preferred orientation, **13:**87–88, 90
Titanium carbide film, **5:**311
plasma CVD, **12:**275
properties, **12:**193
in spray pyrolytic process, **12:**190
Titanium carbide/titanium boride, **17:**64
Titanium dioxide
chemical vapor deposition, **5:**304–306
ion-beam processing, **13:**129–133
composition, **13:**131–132
crystalline phases, **13:**129
mechanical properties, **13:**133
microstructure, **13:**132–133
optical properties, **13:**130–131
rapid, **13:**131
Titanium dioxide film, **8:**38–39
Titanium film, as filters, **7:**48
Titanium/lithium niobate couple, **10:**255
Titanium nitride, **14:**3, 5–9, 14, 22, 36, 55, 66, 68
ion-beam processing, **13:**144
mechanical properties, **13:**87, 89
microhardness, **13:**89
[N/Ti] ratio, **13:**81
synthesis, **13:**69

Titanium nitride film, **22:**46, 48, 49, 226–229
Titanium nitride/NbVn, **17:**24, 25, 33, 63, 64, 65, 71
Titanium oxide, **14:**6, 35, 43, 46, 48, 61
Titanium oxide film, **5:**105–115
diffraction pattern, **5:**106
glass substrates, **5:**111
optical constants, **5:**115
refractive index, **5:**112–114
spectral characteristics, **5:**127
spectral transmittance, **5:**110
structural and optical characteristics, **5:**111
Titanium sulfide, **13:**97–98
Titanium-tungsten alloy, sputter deposition, **22:**45–48, 229–241
TM *see* Transmission mode
TM waves, in inhomogeneous films, **8:**73–76
Tokamaks, fusion, **12:**237
Topologically structured films
in discrete microwave devices, **8:**299–310
electronic imaging and, **8:**272–294
formative techniques for, **8:**253–272
in individual and hybrid semiconductor devices, **8:**294–270
in integrated circuits, **8:**316–333
intrinsic stresses, **8:**268–269
mechanical motion, **8:**310–330
pattern definition, **8:**254–261
plated-beam technology, **8:**295–299
in semiconductor device operation, **8:**251–331
structure constraints, **8:**267–272
structure generation, **8:**261–267
Torsion balances, **3:**28
Total integrated scatter, **13:**121
Totally reflective interference cathode (TRIC), **5:**30–43
absorption and photocurrent, **5:**34–35, 38–39
arrangement, **5:**31
optical model, **5:**32–36
principles, **5:**30–32
reflectance and transmittance, **5:**34–35
S–11 film and, **5:**39–40
Total radiative decay rate, **23:**249–250
Tracer method, **16:**241–244
Tracer scanning, **16:**244
Transconductance, thin-film transistors, **2:**161–163
Transducers, high-frequency, **5:**171
Transfer process
efficiency in silver films, **20:**16
vertical dip film, **20:**319–321
Y-horizontal dip film, **20:**321–326
Transfer rates, **20:**323–326
Transfer ratios, cadmium stearate, **20:**322–323
Transferred-electron type photoemitter, **12:**136–137, *see also* Photoemission; Photoemitters
Transients, in ionicconduction process, **6:**15–16
Transistors, **16:**123
high electron mobility, **23:**70
Transition broadening, **1:**244–245
by thickness variation, **1:**251
Transition coefficient, thin films, **8:**106
Transition layer, on solid surface, **2:**222–223
Transition-metal oxides, **10:**33, 37
Transition metals, vacuum ultraviolet reflectance, **10:**120
Transition radiation, **9:**174

Transition state, **6:**22
Transmission
measurement
optical, **4:**41–57
commercial spectrometers, **4:**41–42
instrumental sources of error, **4:**47–57
systematic errors, **4:**44–47
substrates for, **15:**272
Transmission coefficient, inhomogeneous films, **8:**72–81
Transmission electron microscopy (TEM), **12:**214; **17:**9, 10, 17–20, 122–125
cross-sectional, **17:**18, 30–32, 34
high critical temperature superconductors, **16:**7, 17, 37, 43, 44, 74, 90
high-resolution, **17:**19
plan-view, **17:**19, 24, 25
satellite spots, **17:**19
Transmission mode, secondary emission, **12:**103
Transmission mode photocathode quantum efficiency, calculation, **12:**124–125
Transmission mode silicon photocathode, **12:**132
Transmissive interference cathode (TIC), **5:**15–21
photocurrent enhancement, **5:**20
principle, **5:**15–16
properties, **5:**17–21
second-order, **5:**18
Transmittance, **2:**197; **14:**44
absorbing film
chart determination, **1:**107
on non-absorbing substrate, **1:**86
admittance notation, **1:**94–95
complex amplitude, **1:**84–87
defined, **6:**156
incident light polarization and, **1:**82
indium arsenide, **2:**296–298
infrared optical materials, **2:**293–295
intensity, **1:**77
measurement in normal incidence, **6:**183
multiple combination films, **1:**99
silicon carbide films, **13:**53
silicon coated with SiO, **2:**295
substrate second surface and, **1:**82–83
thin film, on nonabsorbing substrate, **4:**42–44
total fields at oblique incidence, **1:**81
very thin films, **6:**170–177
Transmitting filters, for extreme ultraviolet astronomy, **10:**117
Transmitting windows, of various metals, **7:**45–49
Transparency
of dielectric films, **8:**2–5
electrical conductivity and, **9:**2
Transparent conducting coatings, **12:**228
activated reactive evaporation applications, **13:**95
Transparent conducting films, **9:**1–64
applications, **9:**60–63
chemical etching, **9:**61
film deposition, **9:**30–60
high pressure regime and, **9:**50
light absorption, **9:**13–16
light loss, **9:**13
oxide target sputtering and, **9:**53–59
reactive evaporation, **9:**47–49
thin metal films as, **9:**45

transparence, **9:**13–22, 60
Transparent conducting oxides, properties, **12:**200
Transport function, **5:**9
Transport properties
optically stimulated hot-electron spectroscopy, **21:**252–255
quantum wells, **21:**85–89
quantum-well solar cells, **21:**364–365
quasiballistic transport, **21:**253–255
Trapping centers, for interisland separators, **9:**6
Traps, electron, **11:**52–56
Trap-tunneling mechanism, MBE, **11:**53
Trenching, **14:**226
Trialkali antimonides, **12:**186–187
Trialkali cathodes, electrical properties, **12:**97
TRIC *see* Totally reflective interference cathode
Trichlorophenyl silane, **10:**242
Trichlorotrifluoroethane, **10:**124
Triethylaluminum, **5:**290, 302
Triglycine sulfate
complex permittivity diagrams, **11:**237
dielectric loss, **11:**262
Triisobutylaluminum, **5:**307
Triisopropyl borate, **5:**290, 302
Triodes, **4:**366–367; **14:**31–32, 38
Tripartite ruling, for diffraction gratings, **11:**19
Triple point
damage mode in thin films, **7:**274
depletion of solute at, **7:**304
Trisodium phosphate, monobasic, as parting agent, **7:**66
Trombe-Meinel cusp collector, **10:**24
True strain, **16:**290
True stress, **16:**290
TSHES *see* Thermally stimulated hot-electron spectroscopy
Tsu-Esaki formula, **15:**252–254, 257
TTMSB, **18:**223–226
Tuner (three-stub, plunger) microwave, **18:**152–154, 168–170, 177, 210
Tungsten
chemical vapor deposition, **5:**309
programmed rate CVD, **22:**265–269
sputter deposition, **22:**23–27, 30–33, 37, 38, 61–66, 68, 69
Tungsten coatings, in extreme ultraviolet astronomy, **10:**103–107
Tungsten oxide
optical properties, as-prepared films, **17:**337
reflectance, **17:**345, 350
semiconductor bandgap
crystallized structure, **17:**335
disordered structure, **17:**335
transmittance, **17:**345
Tungsten oxide bronze, **17:**306
Tungsten oxide crystal
color, **17:**306
ion intercalation, **17:**304
reflectance, **17:**30
structure, **17:**304
Tungsten oxide film
characterization, **17:**315
chemical vapor deposited, **17:**315
coloration efficiency, **17:**346
coloration method, **17:**338, 344, 345
depth profiling by SIMS, **17:**326
dip coated, **17:**317
electron paramagnetic resonance, **17:**330, 348

Tungsten oxide film (*continued*)
evaporation-produced, **17:**307, 310, 312
infrared absorption spectroscopy, **17:**328
ion intercalated, **17:**318, 319, 321, 322, 325, 326
microbalance, **17:**326
plasma enhanced chemical vapor deposited, **17:**315
radial distribution function, **17:**310, 314, 317, 326
Raman spectra, **17:**307, 314, 315, 317
Raman spectroscopy, **17:**329
sol-gel-produced, **17:**317
spray pyrolyzed, **17:**315
sputter-deposited, **17:**313, 314
X-ray photoelectron spectroscopy, **17:**330, 335
Tungsten silicide, deposition, stoichiometry, **22:**183
Tungsten-titanium alloy, sputter deposition, **22:**45–48, 229–241
Tunnel-emission triode, thin film, **4:**350–354
base transport, **4:**353–354
emitter characteristics, **4:**351–353
Tunneling
between allowed states, **9:**7
hot-electron phototransistors, **21:**249–251
infrared photoelectron tunneling spectroscopy, **21:**290–291
interwell photon-assisted, **23:**240
phonon assisted, **15:**236–237
photoexcited coherent tunneling, **21:**294–298
resonant *see* Resonant tunneling
thermally assisted tunneling, **21:**249
Tunneling equation, **8:**105
Tunneling filled well, **21:**177–180
Tunneling path length, in discontinuous film, **8:**114–117
Tunneling probability, **13:**218–219; **21:**24–25
Tunneling transition coefficient, **8:**114
Tunnel reactor potentials, **12:**241
Twin boundaries, **2:**17–19
Twin formation, **4:**171–173
Twinning effects, in semiconductor films, **5:**204
Two-dimensional electron gas, **15:**227
Two-dimensional polymers *see* Polymer networks
Two-energy model, **22:**100–107
Two-layer model, in ionic conductance studies, **6:**30–34
Two-level system, Dissado-Hill model, **11:**305–312
Type I HIWIP detectors, **21:**114–19
Type II HIWIP detectors, **21:**19–31
Type III HIP detectors, **21:**31–32

U

Ultimate tensile stress, **16:**290
Ultra-high vacuum chemical vapor deposition (UHV/CVD), **23:**24, 25
heterojunction bipolar transistors grown by, **23:**55
collector doping, **23:**67–68
performance, **23:**67–68
process, **23:**59–64
profile design, **23:**64–68
Ultra thin films, **16:**262–263
Ultrathin resists, patterned hydrophobic SAMs as, **24:**240–244, 250

Ultraviolet cutoff filters, **5:**137–138
Ultraviolet filters, *see also* Metal films; Unbacked filters
  reflection monitoring system for, **9:**135
  spectral transmittance, **9:**111–112
Ultraviolet photopatterning, **24:**228–229
Ultraviolet radiation, *see also* Extreme ultraviolet; Vacuum ultraviolet
  aging and, **10:**134
  aluminum and, **2:**348–349
  dielectric films, **8:**15–16
  free electron theory and, **7:**44
  high transparency in, **7:**43–44
  infrared reflectance and, **10:**136
  pinhole transmittance and, **7:**51
  silicon monoxide coating
    absorptance in, **12:**26
    effects on, **12:**27–28
Umklapp process, **6:**99, 131; **12:**60
Unbacked filters, **7:**43–111
  diffraction from screen, **7:**106
  equilibrium temperature, **7:**94
  in laboratory instruments, **7:**90–92
  long-term degradation, **7:**101–103
  mechanical damage, **7:**92–93
  optical effects, **7:**103–107
  pinholes, **7:**49–53, 107
  resolution loss, **7:**105
  space applications, **7:**83–84, 92–111
  supporting screen, **7:**103–107
  thermal damage, **7:**93–98
  in XUV spectroscopy, **7:**91
Unbalanced magnetron sputtering, **18:**255, 271
  applications, **18:**272–281
    corrosion protection, **18:**275–277
    electronic and optical uses, **18:**274–275
    elemental thin films, **18:**272–274
    wear and abrasion resistance, **18:**277–281
  commercial applications, **18:**281–283
  future trends, **18:**283–285
  initial development, **18:**255
  ion bombardment in, **18:**265–271
  magnetic fields in, **18:**259–265
  precursors, **18:**249–255
  principles, **18:**259–261
  type I, **18:**255
  type II, **18:**255
Uncoated silver mirrors, IR reflectance and polarization data for, **12:**330–331
Uncooled avalanche injection infrared detectors, **21:**33–34
Uniaxial magnetic material, **16:**151
Uniformity, **15:**238
  deposition, **14:**24, 49
  plasma, **18:**99
    magnetically enhanced, **18:**10
    relation to damage, **18:**101
  process, **18:**13
Unipolar field-effect transistors, **2:**150
Unitarity, **15:**256
United States Naval Research Laboratory, **7:**46; **10:**133
Universal dielectric response, **11:**205–314, 229–230
  current interpretations, **11:**287–295
  defined, **11:**229–230
  experimental evidence, **11:**283–286
  historical background, **11:**287–288
  local field in, **11:**287
  without loss peaks in carrier systems, **11:**256–257
Universal power law, **11:**232
Upper harmonic resonance, **18:**129, 139–141

Upper hybrid resonance, **18:**139–142, 151
Ural Polytechnic Institute, **12:**201
UV *see* Ultraviolet radiation

V

Vacuum-deposited films, **10:**236–237
Vacuum deposition, **10:**287
  hot-wall technique, **11:**111–112
  in multilayer interference filter construction, **5:**48–49
Vacuum evaporation, **16:**27, 39–54, 41–53, 92, 105
  Bi-Sr-Ca-Cu-O thin film, **16:**52–54
  co-deposition of co-evaporation, **16:**41, 42, 45, 47–50, 52–54
  e-gun, **16:**27, 40–43, 52–54
  flash evaporation, **16:**54
  molecular beam epitaxy, **16:**51, 52, 54
  process parameters, **16:**39–41
  rare earth-based thin film, **16:**41, 51, 52
  rate and thickness monitoring, **16:**40, 41
  sequential, **16:**41–43, 45, 51, 54, 108
  Tl-Ba-Ca-Cu-O thin film, **16:**54
  vapor source, **16:**39
  Y-Ba-Cu-O thin film, **16:**41–52, 92, 108
Vacuum field, **18:**127, 149–151
Vacuum gauges, **1:**26–30
  filament, chemical changes produced by, **1:**27–28
  ionization
    conductance effect in, **1:**26
    deposition rate monitor, **1:**31–32
    residual currents in, **1:**28–29
Vacuum pumps, **1:**12–23
Vacuum systems, **1:**23–26
  coating deposition, **11:**2
  materials, **1:**2–12
  residual gases in, **1:**38–63
Vacuum ultraviolet
  focused stray light in, **11:**32
  optical constants, **1:**141–145
  reflectance
    aluminum, **1:**150–159
    germanium, **1:**159–165
    gold and platinum, **1:**165–168
    tungsten, **1:**169
  reflectance coatings for diffraction gratings in, **11:**1–33
  reflectance determination, **1:**134–141
  reflecting coatings, **1:**123–186
  spectroscopy in, **11:**17
Vacuum ultraviolet monochromator, **11:**17
Vacuum ultraviolet spectroscopy, coatings for, **12:**2
Valence band offset, **23:**5
Valence polarization, **11:**223
Valves, all-metal vacuum, **1:**35–37
Vamfo, **4:**271–272
Vanadium films, epitaxial, **5:**306
van der Waals interactions, **24:**113
Vapor
  ionization, **17:**97
  metallic, **17:**81, 91
Vapor atoms, atomic surface mobility, **1:**209
Vapor deposition, *see also* Chemical vapor deposition; Plasma chemical vapor deposition
  halide reactions in, **5:**147–148
  hydride pyrolysis in, **5:**148–149
  physical vs. chemical, **5:**238, 257
  refractory compound films and, **5:**156–161

of semiconductor films, **5:**146–185
transfer reactions in, **5:**149–151
Vapor-liquid-solid method, for semiconductor films, **5:**185
Vapor phase epitaxy (VPE), **7:**116, 133
Vapor plating, **3:**133
Vapor stream, **14:**150
Vapor transport
ballistic transport and reaction model
diffuse re-emission, **22:**194–197
gas/solid interactions, **22:**194
governing equations, **22:**201–202
LPCVD, **22:**202–203
nonlinear overrelaxation, **22:**202
specular transport, **22:**197–200
surface species, **22:**200–201
collisionless, **22:**98–100
low-pressure deposition, **22:**177–179
coordinate systems, **22:**188–189
flux, **22:**189–193, 196, 197–200
flux distribution, **22:**191–193
single particle distribution function, **22:**189
mathematical model, **22:**98–100
molecular dynamics, **22:**146–148
SIMSPUD model, **22:**76
algorithms, **22:**6–8
applications, **22:**40, 45
flow chart, **22:**14
verification, **22:**15–20
Varactors, Schottky barrier hyperabrupt, **11:**77
Variable angle spectroscopic ellipsometry (VASE), **19:**301
Velocity, domain wall, **1:**316
Verlet-type algorithm, **22:**132, 169
Vertical Bloch lines, in magnetic domain dynamics, **9:**270–271
Vertical dip Langmuir-Blodgett film, transfer process, **20:**319–321
Very thin films, *see also* Thin films; Thin wires
ellipticity of light in, **6:**177
of fixed wavelength, **6:**171
homogeneous, **6:**171
inhomogeneous, **6:**173
polarization in, **6:**171–177
radiant energy flow ratio, **6:**176
reflectance and transmittance in, **6:**170–177
thickness limits, **6:**189
''Voll-formierte'' films, **6:**11
Vibrational sensors, thin films and, **8:**314–316
Victawet, as parting agent, **7:**66–67
Vidicon imaging targets, **8:**272–282
block diagram, **8:**274
Vinyltrimethyl silane, **10:**241
Viologens, **24:**29–30
Virtual leak due to substrates, **14:**216
Viscoelastic behavior, by atomic force microscopy, **24:**93, 94
Viscoelastic deformation, **16:**292
Viscosity solutions, Hamilton-Jacobi equation, **22:**89–91
Visibility factor, **22:**191
Vision, neural networks, **21:**62, 65–66
Vitreous silica, **19:**207–15
optical function, **19:**213, 214
surface roughness, **19:**210–13
VLWIR arrays, **21:**223–228
Void growth, in electromigration, **7:**279–283
Void nucleation, in electromigration, **7:**279–283
Voids
film, **14:**60, 63–64, 66

Voids (*continued*)
target, **14:**56
Voigt model, **16:**294–295
Voltage
quantum-well solar cells, **21:**347–357
self bias, **14:**28, 40, 44, 51
substrate bias, **14:**28
target, **14:**29, 34, 44, 49
Voltage bias, film stresses, **22:**156–160
Volume fraction, **16:**311
Volume plasma oscillations, **9:**146
Volume requirement, **16:**311
Volume waves, magnetostatic
bandwidth, **15:**5, 28
dispersion, **15:**28, 29
double ferrite layers, **15:**39
generation, **15:**42
ground plane effects, **15:**35, 72, 75–76
spin waves and, **15:**29
thickness modes, **15:**25, 29, 134
width modes, **15:**33
VPE *see* Vapor phase epitaxy
VUV *see* Vacuum ultraviolet

W

Walk-off
ground plane effects, **15:**31
spin waves, **15:**128
Wall coercivity, **15:**185–187
domain formation and, **15:**187
magnetic anisotropy and, **15:**185
magnetization reversal process and, **15:**186
Warburg mechanism, **11:**271, 292
Water, *see also* Ice; Ice nucleation
absorption, in evaporated dielectric films, **12:**46–49
Wave equation
classical, **6:**191
single-particle interaction, **2:**74–75
Wave field
absorbing film and, **5:**7
dielectric film and, **5:**4–5
in photoelectric film, **5:**23
in reflective interference cathode, **5:**22
refractive index and, **5:**2–3
Wave-guiding effects, etching rates, **13:**170–171
Wavelength
calibration errors, **4:**52–55
diffraction grating efficiency and, **11:**25–29
filter types and, **5:**51–52
maximum groove efficiency and, **11:**26
multialkali film and, **5:**27
RIC reflectance and, **5:**25
Wavelength dependence, etching rates, **13:**171–173
Wavenumber, **8:**86; **18:**126, 131–133, 135, 139, 146
Wave velocity, kinematic, **4:**208
WC/Co superlattices, **17:**19–21, 32, 33
Weak-link behavior, **16:**9, 45, 48–50, 52, 67, 71, 97, 110, 111, 123
Wear
cutting tools, **13:**102–103
metal surfaces, **13:**102
Wedged filters, design, **9:**105–106
Well capture probability, *n*-doped quantum-well infrared photodetectors, **21:**143–144
Wentzel electron scattering formula, **2:**75
Wentzel-Kramers-Brillouin approximation, **21:**133
Wentzel-Kramers-Brilouin (WKB) wave expansion, **18:**34–35

Wet etching, **22:**25; **24:**241, 243, 250
Wettability, silane-*p*-aminobenzoic acid film, **20:**59
Wetting configuration, Langmuir, **20:**350
Wetting control, **20:**130–133
Whiskers, single-crystal, **6:**102–103
Whistler wave, **18:**1, 34–136, 160–163, 176–177, 185–190, 196–198, 207–208, 209, 213
Wideline separation (WISE) NMR experiment, **24:**153–154
Wien displacement law, **10:**12
Wigner-Eisenbud time delay, **15:**228, 246, 248
Wigner function, **15:**252–253
Windshield deicing, transparent electrical heaters for, **9:**1
Wobbling diffusion constant, **24:**73
Wolter's relation, **6:**182–183
Work functions, **13:**207–208; **21:**4–6, 7–14
Work hardening, **16:**288

X

X films, **7:**343–346
XIPS *see* X-ray photoelectron spectroscopy
X-ray astronomy, plastic films in, **7:**69
X-ray diffraction, (XRD); **1:**205; **16:**6, 7, 17, 24, 26, 30, 35, 39, 63, 68, 70, 72, 74, 77, 80, 86, 89–92, 94, 100; **17:**9–13, 126; **24:**112–144
X-ray diffraction (XRD)
  alcohol-covered water ice, **20:**154–155
  aluminum films, intensity at different acceleration voltages, **13:**29
  atomic scattering factor, **17:**11, 12, 14
  Bragg reflections, **17:**15
  built-up molecular films, **7:**354–358
  cadmium stearate films, **20:**323–325
  electrodeposit structure, measurement, **4:**218–232
  film thickness monitoring devices, **3:**32
  grazing angle scan, **16:**7, 72
  high critical temperature superconductors, **16:**6, 7, 17, 24, 26, 30, 36, 39, 64, 67, 70, 72, 74, 77, 80, 81, 86, 89, 90, 92–94, 100
  intensity at different acceleration voltages, **13:**29
  ion-beam processing, **13:**123
  low-angle, **17:**16
  molybdenum-vanadium, **17:**16
  numerical calculations, **17:**15
  pattern, BeO films, **13:**46–47
  rocking curve or mosaic distribution width, **16:**7, 8, 23, 24, 27, 45, 67, 68, 71, 80, 81, 89
  rocking curves, **17:**15
  semiconductor film analysis, **5:**189–192
  small-angle scattering, **17:**15, 19, 21, 37
  superlattice reflections (satellite
X-ray diffraction (XRD) (*continued*)
    peaks), **17:**11–13, 16, 19, 36–39
X-ray diffraction pattern, anthracene films, **13:**55
X-ray-induced damage, self-assembled monolayers (SAMs), **24:**119–121
X-ray microanalyzer, electron probe, **2:**125
X-ray microscope, **2:**125

X-ray mirrors, **17:**2
X-ray photoelectron spectroscopy (XIPS)
  high critical temperature superconductors, **16:**6, 13, 45, 92
  ionbeam processing, **13:**122
X-ray radical distribution function, GdFe films, **13:**32
X-ray rocking curve studies, **5:**155
X-rays
  fluorescence, **3:**38–39
  line profile analysis, **4:**218–222
  reflectivity, alkanethiols, **24:**188–192, 193–197
  thickness measurement, metal film, **4:**23–24, 40–41
X-ray spectroscopy, **1:**205–206
XRD *see* X-ray diffraction
XSTM *see* Cross-sectional scanning tunnel microscopy
XTEM *see* Cross-section transmission electron microscope

Y

Y-horizontal dip Langmuir-Blodgett films, transfer process, **20:**321–326
Yield ratio, **14:**6
Yield strength, **16:**288; **17:**59–71
  coherency stress effect, **17:**68
  dislocation formation effect, **17:**68
  dislocation source effect, **17:**68–70
  Hall-Petch relationship, **17:**69, 70
  interface line energy effect, **17:**66–68
  metal/ceramic, **17:**64, 65
  metal/metal, **17:**60–63
  theory, **17:**65–71
    large periods, **17:**68–70
    small periods, **17:**66–68
YIG *see* Yttrium iron garnet
Young's modulus, **5:**102; **24:**95
Yttrium barium copper oxide, **14:**42, 51, 62, 65
Yttrium-barium-copper-oxide thin film
  activated or plasma assisted deposition, **16:**51, 52, 61, 62, 64, 65, 68, 72, 75, 77, 91
  buffer layer, **16:**26–28, 42, 45, 63–65, 69, 70, 88, 91, 97
  degradation, **16:**28, 52
  electrodepostion, **16:**97
  epitaxial and uniaxial, **16:**23, 24, 27, 43, 48, 49, 61, 64, 66, 67, 71, 72, 74, 109, 111, 112, 117–122
  fluoride route, **16:**42–51, 60
  in situ or low-temperature process, **16:**18–28, 51, 52, 61–78, 89–92, 112, 113
  laser ablation, **16:**56–78, 106, 108, 111–114, 117, 118, 122
  low oxygen partial pressure annealing, **16:**43
  MOCVD, **16:**85–92, 111, 116
  MOD, **16:**96, 97
  plasma spray, **16:**97, 98
  post-annealed, **16:**17, 18, 26, 28, 41–51, 53, 54, 57–61, 65, 86, 88, 91, 96–98
  screen printing, **16:**98
  spray pyrolysis, **16:**97
  sputter depostion, **16:**17–28, 48
  triaxial, **16:**43, 44, 46–48, 50
  unaligned, **16:**43–47
  uniaxial, **16:**43–47, 49
  vacuum evaporation, **16:**41–52, 92, 108
Yttrium-barium-copper-oxide (Y-Ba-Cu-O) superconductor, **16:**2

c-Axis parameter, **16:**3, 26, 64, 74, 89
oxygen concentration, **16:**3, 21, 64
structure, **16:**3
thermodynamic stability, **16:**19–21
Yttrium film, **7:**343–346
Yttrium iron garnet (YIG)
bismuth-substituted, **15:**105, 112, 123
control of magnetization, **15:**13
growth, **5:**313–314
ionic substitution
bismuth, **15:**106
lanthanum, **15:**13–14, 16
rare earth, **15:**13–14
samarium, **15:**18
scandium, **15:**15–16, 18
linewidth, least, **15:**16
magnetic properties, **15:**9
magnetooptic properties, **15:**106
propagation loss, least, **15:**6
Yttrium oxide
ion-beam processing, **13:**141
load at fracture versus deposition temperature, **13:**94
microhardness, versus deposition temperature, **13:**93
Yttrium oxide-coated aluminium mirrors, reflectance, **12:**35
Yttrium oxide film, microstructure and optical functions, **19:**223, 231, 236, 240

Z

Zapon (cellulose acetate) film, **7:**70
Zeldovich nonequilibrium factor, **4:**99
Z films, **7:**343–346
Zinc, **16:**312
resistivity size effects, **6:**122, 132
Zinc antimonide film, ionized cluster beam, **13:**33–35
Zinc compounds, chemical vapor deposition, **5:**282
Zinc oxide, **14:**18–21, 30, 32, 42, 44–46, 48, 61, 64–65, 66, 70
Zinc oxide film
electron current effect on crystallinity, **13:**20
ionized cluster beam, **13:**45–46
spray pyrolysis, **12:**190
Zinc selenide film, properties, **12:**194
Zinc sulfide
bandgap energy, **20:**25
reflectance, **1:**176
as substrate, **10:**207
Zinc sulfide coatings, **12:**48
Zinc sulfide film, electron transparent parameters, **12:**197
Zinc sulfide/magnesium fluoride, film reflectance, **1:**175–176
Zinc sulfide/manganese film, ionized cluster beam, **13:**43–45
Zirconium oxide
ion-beam processing, **13:**120, 135–137
microhardness, versus base temperature, **13:**93
Zirconium oxide film, **8:**37
microstructure and optical functions, **19:**223, 227, 236, 240
Zone diagram, **14:**63–64
Zone I, **17:**120, 121, 123, 126
Zone II, **17:**120
Zone III, **17:**120
Zone T, **17:**120, 121, 123, 126

Appendix
Tables of Contents Volumes 1–24

Physics of Thin Films Volume 1

Edited by
Georg Hass

CONTRIBUTORS v
PREFACE vii
ARTICLES PLANNED FOR VOLUME 2 xi

Ultra-High Vacuum Evaporators and Residual Gas Analysis

Hollis L. Caswell

I. Introduction 1
II. Selected Topics in the Design of Ultra-High Vacuum Evaporators 2
III. Residual Gas Analysis in Vacuum Evaporators 38
IV. Significance of Residual Gases in Determining Film Characteristics 51
References 63

Theory and Calculations of Optical Thin Films

Peter H. Berning

I. Introduction 69
II. Fundamental Concepts in Thin Film Optics from the Standpoint of Electromagnetic Theory 71
III. Survey of Computational Methods and Techniques in Thin Film Optics 84
References 120

Preparation and Measurement of Reflecting Coatings for the Vacuum Ultraviolet

Robert P. Madden

I. Introduction 123
II. Experimental Techniques 126
III. Vacuum Ultraviolet Reflectance of Vacuum Deposited Films 145
IV. Optical Constants of Evaporated Films in the Vacuum Ultraviolet 169
V. Vacuum Ultraviolet Reflectance of Combinations of Evaporated Films 171
VI. Overcoating of Diffraction Gratings for the Vacuum Ultraviolet 180
References 184

Structure of Thin Films

Rudolf E. Thun

I. Introduction 187
II. Methods and Instruments 188
III. Nucleation and Growth of Thin Films 207
IV. Textures and Epitaxy 215
V. Structure Changes in Films 220
VI. Stresses and Dislocations 226
References 228

Low Temperature Films

William B. Ittner, III

I. Introduction 233
II. The Normal State at Low Temperature 234
III. The Transition between the Normal and the Superconducting State 240
IV. The Superconducting State at Low Temperature 255
References 273

Magnetic Films of Nickel-Iron

Emerson W. Pugh

I. Introduction 277
II. Origin of Magnetic Anisotropies 278
III. Magnetization Reversal Processes 297
IV. Device Applications 317
V. Fundamental Studies Using Ni-Fe Films 322
References 331
AUTHOR INDEX 335
SUBJECT INDEX 345

Physics of Thin Films Volume 2

Edited by
Georg Hass and Rudolf E. Thun

CONTRIBUTORS TO VOLUME 2 v
PREFACE vii
CONTENTS OF VOLUME 1 xi
ARTICLES PLANNED FOR VOLUME 3 xii

Structural Disorder Phenomena in Thin Metal Films

C. A. Neugebauer

I. Introduction 1
II. Structural Disorder Introduced during Thin Film Formation 2
III. Properties of Island Films 22
IV. Highly Disordered Films Prepared at Low Temperatures 37
V. The Surface Area of Evaporated Metal Films 47
VI. The Tensile Properties of Thin Metal Films 53
References 59

Interaction of Electron Beams with Thin Films

C. J. Calbick

I. Historical Introduction 63
II. Some General Considerations 65
III. Stopping Power of Matter for Electrons 67
IV. Theory of Scattering of Medium-Fast Electrons by Single Atoms 73
V. Comparison of Theory and Experiment 92
VI. Effects Due to Proximity of Atoms 98
VII. Electron Microscopy and Diffraction 109
VIII. Physical Effects of Energy Absorption 122
IX. Electron Energies Greater than $U = 0{,}2$ (~100 kev) 134
Appendix: Symbols and Notations 139
References 140

The Insulated-Gate Thin-Film Transistor

Paul K. Weimer

I. Introduction 147
II. Historical Background of the Insulated-Gate Thin-Film Transistor 149
III. Description of the TFT Structure 151
IV. Operating Characteristics of the Cds TFT 153
V. Physical Processes in the Insulated-Gated TFT 157
VI. Fabrication of Thin-Film Transistors 178
VII. Special Forms of Thin-Film Transistors 183
VIII. Thin-Film Circuit Applications of the TFT 187
References 190

Measurement f Optical Constants of Thin Films

O. S. Heavens

I. Introduction 193
II. Theory of Reflection and Transmission by Thin Films 194
III. Principal Methods for Determining Optical Constants 209
IV. Special Methods of Determining Optical Constants and Film Thicknesses 229
V. Choice of Method 236
References 237

Antireflection Coatings for Optical and Infrared Optical Materials

J. Thomas Cox and Georg Hass

I. Introduction 239
II. General Considerations 240
III. Theory and Design of Antireflection Coatings 242
IV. Experimental Techniques and Practical Coatings 284
References 303

Solar Absorptance and Thermal Emittance of Evaporated Coatings

Louis F. Drummeter, Jr. and Georg Hass

I. Introduction 305
II. Parameters of Interest and Their Significance 307
III. Determination of Solar Absorptance and Thermal Emittance 313
IV. The Optical Properties of Evaporated Metal Films with and without Surface Coatings. . 333
References 358

Thin-Film Components and Circuits

N. Schwartz and R. W. Berry

I. Introduction 363
II. Thin-Film Deposition Techniques 365
III. Substrates 370
IV. Resistors 378
V. Capacitors 398
VI. Thin-Film Circuits 409
References 419
AUTHOR INDEX 427
SUBJECT INDEX 437

Physics of Thin Films Volume 3

Edited by
Georg Hass and Rudolf E. Thun

CONTRIBUTORS TO VOLUME 3 v
PREFACE vii
CONTENTS OF PREVIOUS VOLUMES xi
ARTICLES PLANNED FOR FUTURE VOLUMES xiii

Film-Thickness and Deposition-Rate Monitoring Devices and Techniques for Producing Films of Uniform Thickness

Klaus H. Behrndt

I. Introduction 1
II. Rate-Monitoring Devices 3
III. Thickness Monitoring 8
IV. Measurement of Thickness Distribution 33
V. Uniform Thickness Distribution 40
VI. Comparison of Methods 50
References 53

The Deposition of Thin Films by Cathode Sputtering

Leon I. Maissel

I. Introduction 61
II. Sputtering Mechanisms 62
III. Experimental Data on Sputtering 72
IV. The Application of Sputtering to the Deposition of Thin Films 80
V. The Sputtering of Insulators 100
VI. Sputtering Technology 102
VII. Application of Sputtered Films 108
References 124

Gas-Phase Deposition of Insulating Films

L. V. Gregor

I. Introduction 131
II. Inorganic Dielectric Films 133
III. Polymer Films 144
References 162

Methods of Activating and Recrystallizing Thin Films of IIVI Compounds

A. Vecht

I. Introduction 165
II. The Preparation of Evaporated IIVI Films 168
III. Activation and Recrystallization 174
IV. Methods of Activation and Recrystallization 179
V. Physical Properties of Activated and Recrystallized Films of IIVI Compounds 196
VI. The Future Potential of IIVI Films 207
References 208

The Mechanical Properties of Thin Condensed Films

R. W. Hoffman

I. Introduction 211
II. Structural Consideration 212
III. Internal Stresses 219
IV. Tensile Properties 253
References 270

Lead Salt Detectors

D. E. Bode

I. Introduction 275
II. Single-Crystal Lead Salts 276
III. Detectivity of a Background-Limited Infrared Photoconductor 278
IV. Lead Sulfide Films 281
V. Lead Telluride Films 286
VI. Lead Selenide Films 288
VII. Mechanism of Photoconductivity in Lead Salt Films 291
References 299
AUTHOR INDEX 303
SUBJECT INDEX 315

Physics of Thin Films Volume 4

Edited by
Georg Hass and Rudolf E. Thun

CONTRIBUTORS TO VOLUME 4 v
PREFACE vii
CONTENTS OF PREVIOUS VOLUMES xi
ARTICLES PLANNED FOR FUTURE VOLUMES xiii

Precision Measurements in Thin Film Optics

H. E. Bennett and Jean M. Bennett

I. Introduction 1
II. Substrate Preparation 2
III. Film Preparation 8
IV. Surface Roughness 12
V. Film Thickness 21
VI. Transmission Measurements 41
VII. Normal Incidence Reflection Measurements 57
VIII. Nonnormal Incidence Reflection Measurements 69
IX. Ellipsometric Measurements 79
X. Absolute Phase Change on Reflection Measurements 84
References 90

Nucleation Processes in Thin Film Formation

J. P. Hirth and K. L. Moazed

I. Introduction 97
II. Heterogeneous Nucleation Theory 89
III. Experimental Results 114
IV. Heterogeneous Nucleation and Epitaxy 128
References 134

Evaporated Single-Crystal Films

J. W. Matthews

I. Introduction 137
II. Modes of Film Growth 138
III. Monolayer Growth of Single-Crystal Films 138
IV. Film Formation by the Generation, Growth, and Coalescence of Nuclei 150
V. Annealing of Thin Films 186
References 187

The Growth and Structure of Electrodeposits

Kenneth R. Lawless

I. Introduction 191
II. The Nucleation and Growth of Electrodeposits 192
III. The Structure of Electrodeposits 217
References 251

Thin Glass Films

W. A. Pliskin, D. R. Kerry, and J. A. Perri

I. Introduction 257
II. Preparation 259
III. Evaluation of Physical and Chemical Properties 270
IV. Electrical Properties of Glass Films 289
V. Applications of Glass Films to Semiconductor Devices 316
References 320

Hot-Electron Transport and Electron Tunneling in Thin Film Structures

C. R. Crowell and S. M. Sze

I. Introduction 325
II. Hot-Electron Scattering Processes in Metals 327
III. Collection of Hot Electrons 330
IV. Photoemission Experiments 341
V. Tunnel Emission 350
VI. Schottky (Thermionic) Emission 354
VII. Discussion and Summary 368
References 369
AUTHOR INDEX 373
SUBJECT INDEX 386

Physics of Thin Films Volume 5

Edited by
Georg Hass and Rudolf E. Thun

CONTRIBUTORS TO VOLUME 5 v
PREFACE vii
CONTENTS OF PREVIOUS VOLUMES xi
ARTICLES PLANNED FOR FUTURE VOLUMES xiii

Interference Photocathodes

D. Kossel, K. Deutscher, and K. Hirschberg

I. Introduction 1
II. Light Waves and Resonators 2
III. The Emission of Photoelectrons 8
IV. Conventional Cathodes of High Quantum Yield 12
V. Interference Cathodes 15
VI. Summary 43
References 44

Design of Multilayer Interference Filters

Alfred Thelen

I. Introduction 47
II. Theoretical Basis 53
III. Design Methods 59
IV. Equivalent Layers 60
V. Refining Methods 68
VI. Method of Effective Interfaces 75
VII. Long-Wavelength-Short-Wavelength-Pass Filters 78
VIII. Narrow-Band-Pass Filters 80
IX. Circular Variable Filters 84
References 85

Oxide Layers Deposited from Organic Solutions

H. Schroeder

I. Introduction 87
II. Formation of Solid Layers from Solutions 88
III. General Characteristics of Oxide Layers Obtained from Solutions 94
IV. Special Oxide Layers 105
V. Applications 123
References 140

The Preparation and Properties of Semiconductor Films

M. H. Francombe and J. E. Johnson

I. Introduction 143
II. Chemical Vapor Deposition 146
III. Vacuum Deposition 161
IV. Structural and Chemical Characterization 185
V. Electrical and Optical Characterization 208
VI. Some Important Problem Areas 226
References 229

The Preparation of Films by Chemical Vapor Desposition

W. M. Feist, S. R. Steele, and D. W. Readey

I. Introduction 237
II. Chemical Processes Used in Chemical Vapor Deposition 241
III. Setups for Chemical Vapor Deposition 248
IV. Morphology of Deposits Formed by Chemical Vapor Deposition 257
V. Examples of Films Prepared by Chemical Vapor Deposition 264
References 314
AUTHOR INDEX 323
SUBJECT INDEX 334

Physics of Thin Films Volume 6

Edited by
Maurice H. Francombe and Richard W. Hoffman

CONTRIBUTORS TO VOLUME 6 vii
PREFACE ix
CONTENTS OF PREVIOUS VOLUMES xi
ARTICLES PLANNED FOR FUTURE VOLUMES xiv

Anodic Oxide Films

C. J. Dell'Oca, D. L. Pulfrey, and L. Young

I. Introduction 2
II. Introductory Survey of Some General Features of Anodization 3
III. Theories of the Ionic Conduction Process 13
IV. Gaseous Anodization 36
V. Dielectric Properties 51
References 73

Size-Dependent Electrical Conduction in Thin Metal Films and Wires

D. C. Larson

I. Introduction 81
II. Theory of Classical Size Effects 82
III. Experimental Measurements 101
IV. Quantum Size Effects 137
References 143

Optical Properties of Metallic Films

F. Abelès

I. Introduction 151
II. Optical Properties of Metals 152
III. Formulas for Thin Homogeneous and Isotropic Metallic Films 155
IV. Very Thin Films 170
V. Anisotropic Films 177

VI. Measurement of the Optical Parameters of Metallic Films 182
VII. Specific Optical Properties of Thin Metallic Films.............................. 188
References 202

Interactions in Multilayer Magnetic Films

Arthur Yelon

I. Introduction.. 205
II. Interaction Phenomena.. 208
III. Effects of Coupling.. 250
IV. Applications of Multilayer Films... 291
V. Conclusions.. 294
References 295

Diffusion in Metallic Films

C. Weaver

I. Introduction.. 301
II. Metal Systems 302
III. Experimental Methods 310
IV. Experimental Results 315
V. Conclusions.. 343
References 346
AUTHOR INDEX.. 349
SUBJECT INDEX 362

Physics of Thin Films Volume 7

Edited by
Georg Hass, Maurice H. Francombe, and Richard W. Hoffman

CONTRIBUTORS TO VOLUME 7 vii
PREFACE ix
CONTENTS OF PREVIOUS VOLUMES xi
ARTICLES PLANNED FOR FUTURE VOLUMES xiv

Electron Diffraction Analysis of the Local Atomic Order in Amorphous Films

D. B. Dove

I. Introduction 1
II. Diffraction Theory 3
III. Experimental Techniques 15
IV. Review of rdf Measurements 17
References 37

The Preparation and Use of Unbacked Metal Films as Filters in the Extreme Ultraviolet

W. R. Hunter

I. Introduction 43
II. General Considerations 45
III. Preparation of Films 53
IV. Film Support and Filter Fabrication 77
V. Aging, Storage, and Cleaning of Filters 86
VI. Use of Unbacked Filters in Laboratory and Space Applications 90
References 111

Properties and Applications of IIIV Compound Films Deposited by Liquid Phase Epitaxy

H. Kressel and H. Nelson

I. Introduction 115
II. General Considerations 117
III. Properties of Binary Compounds 139
IV. Ternary Systems 172
V. Electroluminescent Diodes 205
VI. Microwave Devices 235
VII. Negative Electron Affinity Devices 242
VIII. Miscellaneous Devices 246
References 247

Electromigration in Thin Films

F. M. d'Heurle and R. Rosenberg

I. Introduction 257
II. Lattice Effects 258
III. Degradation in Thin Films 264
IV. Alloying Effects 287
References 305

Built-Up Molecular Films and Their Applications

V. K. Srivastava

I. Introduction 311
II. Insoluble Monolayers and Their Characteristics 313
III. Transfer of Monolayers on Solids: Building Up of Multilayers 323
IV. Structural Properties of Built-Up Molecular Films 350
V. Optical Properties of Films 360
VI. Electrical Properties of Films 365
VII. Applications of Built-Up Molecular Films 387
VIII. Future Potential of Films 391
References 392
AUTHOR INDEX 399
SUBJECT INDEX 415

Physics of Thin Films Volume 8

Edited by
Georg Hass, Maurice H. Francombe, and Richard W. Hoffman

CONTRIBUTORS TO VOLUME 8 vii
PREFACE ix
CONTENTS OF PREVIOUS VOLUMES xi
ARTICLES PLANNED FOR FUTURE VOLUMES xv

Dielectric Film Materials for Optical Applications

Elmar Ritter

I. Introduction 1
II. Requirements for Dielectric Films 2
III. Preparation of Dielectric Films 18
IV. Survey on Dielectric Materials 27
V. Applications of Dielectric Single Films and Film Combinations 42
References 45

Inhomogeneous and Coevaporated Homogeneous Films for Optical Applications

R. Jacobsson

I. Introduction 51
II. The Refractive Index of Dielectric Mixtures 52
III. Generation of Refractive Index Profiles with Continuous Index Variation 67
IV. Calculation of the Reflection and Transmission Coefficients of Inhomogeneous Films . . . 72
V. Measurement of Inhomogeneities in Thin Films 81
VI. Applications of Inhomogeneous Films 83
Appendix 94
References 97

Discontinuous and Cermet Films

Z. H. Meksin

I. Introduction 99
II. Theoretical Models 102
III. Experimental Studies 152
IV. Suggestions for Future Studies 164
References 165

Electrical Conduction in Disordered Nonmetallic Films

A. K. Jonscher and R. M. Hill

I. Introduction 170
II. Energy Level Structure of Amorphous Solids 174
III. Inhomogeneity and Internal Fields 183
IV. Low-Field Conduction by Hopping Carriers 185
V. High-Field Conduction 195
VI. Experimental Results 212
VII. Conclusions 240
Appendix I 242
Appendix II 245
References 246

Topologically Structured Thin Films in Semiconductor Device Operation

H. C. Nathanson and J. Guldberg

I. Introduction 251
II. Formative Techniques for Topologically Structured Thin Films 253
III. Role of Topologically Structured Thin Films and Foils in Electronic Imaging 272
IV. Role of Topologically Structured Thin Films in Individual and Hybrid Semiconductor Devices 294
V. Function through Mechanical Motion of Thin Films 310
VI. Summary 330
References 331
SUBJECT INDEX 335

Physics of Thin Films Volume 9

Edited by
Georg Hass, Maurice H. Francombe, and Richard W. Hoffman

CONTRIBUTORS TO VOLUME 9 vii
PREFACE ix
CONTENTS OF PREVIOUS VOLUMES xi
ARTICLES PLANNED FOR FUTURE VOLUMES xiii

Transparent Conducting Films

J. L. Vossen

I. Introduction 1
II. Physics of Conductivity and Transparency 4
III. Processes for Film Deposition 30
IV. Application of Materials and Processes 60
V. Conclusion 64
References 64

Metal-Dielectric Interference Filters

Georg Hass, Maurice H. Francombe, and Richard W. Hoffman

I. Introduction 74
II. Basic Theory 75
III. General Considerations in Bandpass Filter Design 93
IV. Single-Cavity Filter Design 102
V. One-M Filter Design 112
VI. Multiple Cavity and Other Designs 124
VII. Reflection Filters 128
VIII. The Production of Filters 132
IX. Future Developments 142
References 143

Surface Plasma Oscillations and Their Applications

H. Raether

I. General Considerations on Surface Plasmons 145
II. Excitation of Radioactive Surface Plasmons 171
III. Excitation of Nonradiative Surface Plasmons 199
References 255

Magnetic Bubble Films

P. Chaudhari, J. J. Cuomo, R. J. Gambino, and E. A. Giess

I. Introduction 263
II. General Magnetics 266
III. Materials for Magnetic Bubbles 272
IV. Growth of Magnetic Bubble Materials 276
V. Anisotropy in Magnetic Bubble Films 284
VI. Defects in Films 288
VII. Summary and Conclusions 293
References 294
AUTHOR INDEX 299
SUBJECT INDEX 310

Physics of Thin Films Volume 10

Edited by
Georg Hass and Maurice H. Francombe

CONTRIBUTORS TO VOLUME 10 vii
PREFACE ix
ARTICLES PLANNED FOR FUTURE VOLUMES xi

Spectrally Selective Surfaces for Photothermal Solar Energy Conversion

R. E. Hahn and B. O. Seraphin

I. Introduction 1
II. Fundamentals of Photothermal Solar Energy Conversion 4
III. Physical Processes and Materials Systems for Spectrally Selective Surfaces 28
References 65

The Use of Evaporated Films for Space Applications—Extreme Ultraviolet Astronomy and Temperature Control of Satellites

G. Hass and W. R. Hunter

I. Introduction 72
II. Coatings for Extreme Ultraviolet Astronomy 73
III. Evaporated Surface Coatings for Controlling the Temperature of Satellites 127
References 164

Scattering by All-Dielectric Multilayer Bandpass Filters and Mirrors for Lasers

Jay M. Eastman

I. Introduction 167
II. Scattering from a Single Interface 170
III. Theory of Surface Scattering in Multilayer Interference Coatings 176
IV. Graphical Applications and Design Examples 188
V. Experimental Verification of the Theoretical Methods 203
VI. Summary 224
References 225

Thin Films for Integrated Optics

D. B. Ostrowsky and C. Vanneste

I. Introduction . . . 227
II. General Background . . . 228
III. Guiding and Coupling . . . 230
IV. Non-Single-Crystal Films for Integrated Optics . . . 235
V. Monocrystalline Films . . . 245
References . . . 258

Correction of Optical Elements by the Addition of Evaporated Films

J. R. Kurdock and R. R. Austin

I. Introduction . . . 261
II. The Addition Process Concept . . . 263
III. Generation of the Correction Mask . . . 265
IV. Selection of Thin Film Materials . . . 272
V. The Figuring Experiments . . . 286
VI. Capabilities and Future Application . . . 305
References . . . 308
AUTHOR INDEX . . . 309
SUBJECT INDEX . . . 317
CONTENTS OF PREVIOUS VOLUMES . . . 323

Physics of Thin Films Volume 11

Edited by
Georg Hass and Maurice H. Francombe

CONTRIBUTORS TO VOLUME 11 vii
PREFACE ix
ARTICLES PLANNED FOR FUTURE VOLUMES xi
CONTENTS OF PREVIOUS VOLUMES xiii

Preparation and Testing of Reflectance Coatings for Diffraction Gratings in the Extreme Ultraviolet

W. R. Hunter and G. Hass

I. Introduction 1
II. Coatings for Grating Ruling 2
III. Techniques for Coating Gratings 9
IV. Measuring Grating Efficiencies 17
References 33

Progress, Problems, and Applications of Molecular-Beam Epitaxy

Colin E. C. Wood

I. Introduction 36
II. Some Considerations in System Design 36
III. *In Situ* Assessment Techniques 40
IV. Initial Stages of Growth 48
V. Doping 52
VI. IIIV Binaries Other Than GaAs 66
VII. Ternary IIIV Alloys 69
VIII. Device Applications 75
IX. Technological Developments 86
X. Projections 92
XI. Conclusions 95
References 96

Thin-Film IVVI Semiconductor Photodiodes

H. Holloway

I. Introduction ... 106
II. Techniques for Vacuum Deposition of IVVI Layers ... 108
III. Properties of IVVI Layers on Insulating Substrates ... 113
IV. Some Properties of IR Photodiodes ... 122
V. Thin-Film IVVI Photodiodes ... 128
VI. Thin-Film Photodiodes for 3–5 μm Operation ... 157
VII. Thin-Film Photodiodes for 8–12 μm Operation ... 166
VIII. Unconventional Thin-Film Devices ... 168
IX. Thin-Film Photodiode Arrays ... 191
X. Conclusions ... 198
References ... 199

The Universal Dielectric Response: A Review of Data and Their New Interpretation

A. K. Jonscher

I. Introduction ... 206
II. Definitions and Basic Relations ... 209
III. Methods of Presentation of Dielectric Data ... 217
IV. Basic Mechanisms of Polarization ... 223
V. Empirical Classification of Loss Characteristics ... 226
VI. Survey of Experimental Data ... 235
VII. Currently Accepted Interpretations ... 287
VIII. The Many-Body Interpretation ... 296
IX. Concluding Comments ... 312
References ... 314
Note Added in Proof ... 317
AUTHOR INDEX ... 319
SUBJECT INDEX ... 329

Physics of Thin Films Volume 12

Edited by
Georg Hass, Maurice H. Francombe, and John L. Vossen

CONTRIBUTORS TO VOLUME 12 ... vii
PREFACE ... ix
ARTICLES PLANNED FOR FUTURE VOLUMES ... xi
CONTENTS OF PREVIOUS VOLUMES ... xiii

Reflectance and Preparation of Front Surface Mirrors for Use at Various Angles of Incidence from the Ultraviolet to the Far Infrared

G. Hass, J. B. Heaney, and W. R. Hunter

I. Introduction ... 2
II. Reflectance Measurements and Preparation of Mirror Coatings ... 3
III. Reflectance of Metallic Front Surface Mirrors without Overcoating ... 16
IV. Mirror Coatings with Protective Layers and Reflectance-Enhancing Surface Films ... 22
V. Metal-Dielectric Mirrors for Use as Reflection-Type Filters ... 42
VI. Water Absorption in Evaporated Dielectric Films ... 46
References ... 49

Photoemissive Materials

C. Ghosh

I. Introduction ... 54
II. The Mechanism of Photoemission ... 54
III. The AgOCs (S1) Photocathode ... 70
IV. The Alkali Antimonides ... 84
V. Negative Electron Affinity Materials ... 112
VI. Applications of Photoemissive Materials ... 140
VII. Conclusions ... 157
References ... 158

Chemical Solution Deposition of Inorganic Films

K. L. Chopra, R. C. Kainthla, D. K. Pandya, and A. P. Thakoor

I. Introduction 168
II. Spray Pyrolytic Process 169
III. Characteristic Features of the Spray Pyrolytic Process 178
IV. Multicomponent Doping and Alloying Effects 181
V. Structural Properties 187
VI. Electrical and Optical Properties 192
VII. Solution Growth Process 201
VIII. Impurity and Dopant Effects 211
IX. Multicomponent Films 212
X. Oxide Films 213
XI. Structure 214
XII. Transport Properties 217
XIII. Some Large-Area Application 223
XIV. Concluding Remarks 230
References 232

Plasma-Enhanced Chemical Vapor Deposition of Thin Films

S. M. Ojha

I. Introduction 237
II. Deposition Techniques and Systems 238
III. Preparation and Properties of Films 246
IV. Plasma Oxidation 275
V. Plasma Carburizing 283
VI. Glow-Discharge Nitriding 286
VII. Conclusions 289
References 289
AUTHOR INDEX 297
SUBJECT INDEX 312

Physics of Thin Films Volume 13

Edited by
Maurice H. Francombe and John L. Vossen

CONTRIBUTORS TO VOLUME 13 vii
PREFACE ix
EDITORS' NOTE xi

Ionised Cluster Beam Deposition and Epitaxy

Toshinori Takagi

I. Introduction 1
II. Equipment for Ionized Cluster Beam Deposition and Epitaxy 4
III. Film-Formation Mechanism 15
IV. Film Deposition and Epitaxial Growth 20
V. Conclusions 56
References 56

The Activated Reactive Evaporation Process

R. F. Bunshah and C. Deshpandey

I. Introduction 60
II. Processes for the Deposition of Refractory Compounds 62
III. Direct Evaporation 62
IV. Reactive Evaporation Processes 63
V. Activated Reactive Evaporation and the Role of Plasma 64
VI. Implementation of the Activated Reactive Evaporation Process 75
VII. Compound Synthesized by the ARE Process and the Effect of Process Variables 79
VIII. Microstructure, Preferred Orientation, and Mechanical Properties of Refractory Compound Deposits 81
IX. Applications of the ARE Process 94
X. Summary 103
References 104

Ion-Beam Processing of Optical Thin Films

Ursula J. Gibson

I. Introduction 109
II. Ion Sources 112
III. Ion-Surface Interactions 116
IV. Film Properties and Analysis Method 120
V. Substrate Bombardment 125
VI. Postdeposition Bombardment 127
VII. Ion-Beam Deposition of Optical Films 128
VIII. Conclusion 145
References 147

Laser-Induced Etching

Carol I. H. Ashby

I. Introduction 151
II. Mechanisms of Laser-Induced Etching 152
III. Methods of Laser-Induced Etching 156
IV. Applications of Laser-Induced Etching 181
V. Concluding Remarks 191
References 192

Contacts to GaAs Devices

J. M. Woodall, N. Braslau, and J. L. Freeouf

I. Introduction 199
II. "Ideal" Contacts 200
III. The GaAs Surface and Interface—Fermi-Level Pinning 201
IV. Alloyed Ohmic Contacts to GaAs 209
V. Heterojunction Contacts 219
VI. Recent Results 221
VII. Summary 222
References 223
AUTHOR INDEX 227
SUBJECT INDEX 241

Physics of Thin Films Volume 14

Edited by
Maurice H. Francombe and John L. Vossen

Contributors.......... vii
Preface.......... ix

Reactive Sputtering

W. D. Westwood

I. Introduction 1
II. Basic Mechanisms 3
III. Sputtering System 22
IV. Films and Applications 58
IV. Summary 72
Acknowledgements.......... 73
References.......... 73

Plasma Oxidation

J. Siejka and J. Perriere

I. Introduction 82
II. Properties of Oxygen Plasmas 83
III. Physical-Chemical Processes During Low-Temperature Plasma Oxidation 88
IV. Kinetics of Oxide Growth under Anodic Bias 97
V. Atomic Movements during Plasma-Oxide Growth 107
VI. Mechanisms of Plasma Oxidation 114
VII. Applications of Plasma Oxidation and Properties of the Oxide Formed 123
VIII. Conclusions 124
References 125

The Cathodic Arc Plasma Deposition of Thin Films

Philip C. Johnson

I. Introduction 130
II. Ion Plating.......... 131

III. The Cathodic Arc 138
IV. Emitted Material 150
V. Cathodic Arc Deposition 164
VI. Thin Film Deposition 175
VII. Applications 184
VIII. Conclusion 194
References 195

The Preparation of Substrates for Film Deposition Using Glow-Discharge Techniques

J. L. Vossen

I. Introduction 201
II. Sputter-Etching Configurations and Techniques 203
III. Physical Sputtering Effects 217
IV. Reactive Sputter (Ion) Etching 233
V. Conclusion 236
References 236
Author Index 241
Subject Index 255

Physics of Thin Films Volume 15

Edited by
Maurice H. Francombe and John L. Vossen

Contributors ix
Preface xi

Magnetostatic Waves

J. D. Adam, M. R. Daniel, P. R. Emtage, and S. H. Talisa

I. Introduction 2
II. Magnetic Materials for Magnetostatic Waves 7
A. Magnetic Materials 7
B. Growth of YIG Films 10
C. YIG Film Evaluation 18
III. Magnetostatic Propagation and Generation 25
A. Propagation in an Isolated Ferrite Layer 25
B. More Complex Systems 35
C. Generation of Magnetostatic Waves 40
IV. Applications of Magnetostatic Wave Devices 52
A. Broad-Band Receiver Applications 52
B. Magnetostatic Wave Delay Lines for Phased-Array Antennas 61
C. Simple Delay-Line Applications Using Magnetostatic Waves 67
D. Nonlinear Devices 68
V. Linear Devices 70
A. Wide-Band Devices—Delay Lines 71
B. Resonators, Oscillators, and Filters 81
VI. Nonlinear Devices 92
A. Coupling to Half-Frequency Spin Waves 92
B. High-Power Devices 95
VII. Magneto-optic Interactions and Devices 102
A. Scattering: General Considerations 102
B. Application of Magnetostatic Wave Interactions with Light 111
VIII. Conclusions 124
Acknowledgments 125
Appendix A. Infinite Magnetic Medium 125
Appendix B. Magnetic Anisotropy and Surface Waves 129
Appendix C. Calculation from Surface Permeability 132

Appendix D. Thickness Modes and Boundary Conditions 134
References 136

Thin-Film Rare Earth—Transition Metal Alloys for Magnetooptic Recording

Brent S. Krusor and G. A. N. Connell

I. Introduction 143
II. Materials Preparation and Properties 147
A. Preparation of Transition-Metal Rich Alloys 148
B. Structural Properties of Transition-Metal Rich Alloys 155
C. Oxidation and Aging Properties 168
III. Magnetic and Magnetooptic Properties 174
A. Magnetization, Magnetic Moment, and Exchange 174
B. Coercivity, Magnetic Anisotropy, and Domain Formation 183
C. Magnetooptic Properties 190
IV. Application to Magnetooptic Recording 200
A. Magnetooptic Medium Design 201
B. Recording Characteristics and Medium Noise 209
V. Summary 213
References 214

New Quantum Structures

D. D. Coon and K. M. S. V. Bandara

I. Introduction and Background 219
II. Electronic States in Quantum Wells and Superlattices 221
A. Energy Levels in Quantum Wells and Superlattices 222
B. Corrections to Standard Models 225
C. Density of States Considerations 227
D. WignerEisenbud Time Delay 228
III. Intersubband Transitions, Infrared Absorption, and Infrared Detection 229
A. Dark-Current Mechanisms 236
B. Noise, Detectivity, Responsivity, and Uniformity 238
C. Intraband vs. Interband Transitions 238
D. Exchange Interactions 239
IV. Resonant Tunneling and Device Applications 244
A. Application to Fast Logic and High-Frequency Oscillators 245
B. Intrinsic Bistability 249
C. Theoretical Approaches to Resonant Tunneling 252
V. Trends and Possibilities 258
Acknowledgments 260
References 260

Fourier Transform Infrared Analysis of Thin Films

D. M. Back

I. Introduction . . . 265
II. Theory of Infrared Absorption . . . 266
A. Source of Infrared Absorption . . . 266
B. Description of Infrared Absorption Bands . . . 267
C. Symmetry in Infrared Absorption . . . 268
D. Infrared Spectra of Amorphous Solids . . . 268
III. Instrumentation . . . 269
A. Fourier Transform Spectrometers . . . 269
B. FTIR vs. Dispersive Infrared Spectrometers . . . 271
C. Techniques of Infrared Analysis . . . 271
IV. Applications to Thin Film Analysis—Specific Systems . . . 277
A. Amorphous Silicon Films . . . 277
B. Miscellaneous a-Si:H Alloys . . . 283
C. Silicon Oxide Films . . . 287
D. Silicon Nitride Films . . . 291
E. Silicon Oxynitride Films . . . 295
F. Silicate Glass Films . . . 299
G. Aluminum Oxide Films . . . 303
H. Nickel Oxide Films . . . 304
I. Carbon Films . . . 305
J. Detection of Carbon Impurities . . . 308
V. Summary . . . 309
References . . . 309
Author Index . . . 313
Subject Index . . . 325

Physics of Thin Films Volume 16

Edited by
Maurice H. Francombe and John L. Vossen

Contributors ix
Preface xi

High-T_c Superconducting Thin Films

Neelkanth G. Dhere

I. Introduction 2
A. Characterization of High-T_c Superconducting Thin Films 5
II. Sputtering and Ion-Beam Deposition 14
A. YBa_2–Cu_3–O_{7-x} and Other Rare-Earth-Based Thin Films 17
B. Bi–Sr–Ca–Cu–O Thin Films 28
C. Tl–Ba–Ca–Cu–O Thin Films 34
III. Vacuum Evaporation 39
A. YBa_2–Cu_3–O_{7-x} and Other Rare-Earth-Based Thin Films 41
B. Bi–Sr–Ca–Cu–O Thin Films 52
C. Tl–Ba–Ca–Cu–O Thin Films 54
IV. Laser Ablation 55
A. YBa_2–Cu_3–O_{7-x} and Other Rare-Earth-Based Thin Films 59
B. Bi–Sr–Ca–Cu–O Thin Films 78
C. Tl–Ba–Ca–Cu–O Thin Films 80
V. Metalorganic Chemical Vapor Deposition (MOCVD) 81
A. YBa_2–Cu_3–O_{7-x} and Other Rare-Earth-Based Thin Films 85
B. Bi–Sr–Ca–Cu–O Thin Films 92
C. Tl–Ba–Ca–Cu–O Thin Films 94
VI. Metalorganic Deposition (MOD) and Other Processes 95
A. YBa_2–Cu_3–O_{7-x} and Other Rare-Earth-Based Thin Films 96
B. Bi–Sr–Ca–Cu–O Thin Films 98
C. Tl–Ba–Ca–Cu–O Thin Films 100
VII. Applications 101
A. Microwave Components 101
B. Passive Interconnections in Microelectronic Systems 108
C. Optical Sensors and Radiation Detectors 109
D. Josephson Junction 110
E. SQUIDs 114

F. High-Speed Computer Switches and Transistors 123
References 123

Permanent Magnet Thin Films: A Review of Film Synthesis and Properties

Fred J. Cadieu

I. Introduction and Scope of Review 146
A. Permanent Magnet Properties of Interest 147
B. Historical Perspective 149
C. Magnetic Units, Cgs and Mks 150
D. Film Preparation Methods 152
II. Properties of Permanent Magnet Films 154
A. Ideal Magnetization versus Field Dependence 154
B. Magnetization Behavior for Randomly Oriented Grains 166
C. Films Crystallized from Amorphous Deposits 168
D. Textured Directly Crystallized Films 169
E. Thermalization and Selective Thermalization Effects 177
F. Film Stress Effects and Film Thickness Limitations 189
III. Specific Permanent Magnet Film Systems 189
A. As-Grown "Amorphous" or Microcrystalline Films 189
B. Sm-Co-Based Films 192
C. Sm (Co, Fe, Cu, Zr) Directly Crystallized Films, Sm 11 to 19 at %, Fe/Co 0.4 203
D. Sm (Co, Fe, Cu, Zr) Directly Crystallized Films, Sm 11 to 19 at %, Fe Enriched to 30 at % 205
E. High-Coercive-Force Nd-Fe-B Films 206
F. Sm-Fe Third-Element Hard Magnet Films 213
G. Binary $SmFe_{12}$ - and Sm $(Fe, T)_{12}$ - Type Films 215
H. Pr-Containing Pr_2Fe_{14} B Films 217
IV. Permanent Magnet Film Devices and Arrays 217
A. Applications in Integrated Electromagnetic Circuits 217
B. Film Magnet Arrays and Device Geometries 220
C. Concluding Remarks 227
Acknowledgments 229
References 229

Lateral Diffusion and Electromigration in Metallic Thin Films

K. V. Reddy

I. Introduction 234
II. Formal Background 235
A. Grain Boundary Diffusion 236
B. Diffusion under External Force 237
C. Electromigration Effective Charge 237
D. Theoretical Estimation of Wind Force 239
III. Techniques for Measurements 239
A. Median Time-to-Failure 239
B. Tracer Method 241

C. Resistometric Method ... 245
D. Edge Displacement Method ... 247
E. Quasi-Steady-State Method ... 248
F. Other Direct Methods ... 248
IV. Electromigration Studies ... 249
A. Aluminum Thin Films ... 249
B. Gold Thin Films ... 253
C. Tin Thin Films ... 255
D. Indium Thin Films ... 258
E. Other Metallic Films ... 259
F. Ultrathin Films ... 262
V. Interpretation of the Results ... 263
A. Temperature Dependence of Z^* ... 264
Acknowledgments ... 266
References ... 266

Fracture and Cracking Phenomena in Thin Films Adhering to High-Elongation Substrates

Paul H. Wojciechowski and Michael S. Mendolia

I. Introduction ... 272
II. Background ... 275
A. The System ... 275
B. Previous Investigations ... 276
III. Behaviour of High-Elongation Materials ... 293
A. Introduction ... 293
B. Definitions and Background ... 294
C. Modeling of High-Elongation Materials ... 297
D. Kapton: A Case Study ... 302
IV. On the Multiple Fracture of the Low-Elongation Component in Thin-Film Composites ... 311
A. Talreja's Four Cases ... 311
B. Thickness Requirement ... 314
C. Load Transfer between Film Segment and Substrate: Shear Lag ... 316
D. Shear Lag Analysis ... 319
E. Permally on Kapton: A Case Study ... 324
V. Closure ... 333
Appendix ... 336
References ... 337
Author Index ... 341
Subject Index ... 361

Physics of Thin Films Volume 17

Edited by
Maurice H. Francombe and John L. Vossen

Contributors. ix
Preface. xi

Deposition and Mechanical Properties of Superlattice Thin Films

Scott A. Barnett

I. Introduction 2
II. Deposition Techniques 3
III. Characterization 8
A. X-Ray Diffraction 10
B. Transmission Electron Microscopy 17
C. Other Techniques 22
IV. Deposition Mechanisms, Structure, and Stability 22
A. Crystal Structure and Lattice Relaxation 23
B. Nucleation and Layer Morphology 30
C. Interdiffusion 35
D. Summary 42
V. Mechanical Property Measurements 42
A. Elastic Moduli and Constants 42
B. Hardness and Yield Strength 46
VI. Elastic Properties 47
A. Experimental Results 48
B. Theoretical Predictions 54
VII. Mechanical Strength and Hardness 59
A. Experimental Results 60
B. Theoretical Predictions 65
C. Comparison of Experiment and Theory 70
VIII. Conclusions 71
Acknowledgments 73
References 73

Hard Coatings Prepared by Sputtering and Arc Evaporation

Jindrich Musil, Jiří Vyskočil, and Stanislav Kadlec

I. Introduction 80
II. Deposition Techniques Based on Sputtering and Evaporation 81
A. Effect of Particle Bombardment in Film Deposition Process 82
B. Sputtering 83
C. Evaporation 91
D. Comparison of Reactive Magnetron Sputtering and Arc Evaporation Techniques 96
III. Hard Coatings 100
A. Physical Properties of Hard Materials 101
B. Substrate/Coating Interface 106
C. Bulk of the Coating Material 107
D. Upper Layer of the Coating 108
E. Multilayer and Gradient Coatings 109
IV. Growth and Properties of Hard Coatings 110
A. Chemical Composition 111
B. Phase Composition 118
C. Structure Zone Models 119
D. Microstructure of TiN Films Grown under Ion Bombardment 121
E. Crystalline Structure and Stress 126
F. Preferred Orientation 128
G. Surface Morphology 129
H. Correlation Process/Microstructure/Properties 129
V. Deposition Process 131
A. Substrate Heating 132
B. Surface Etching and Interface Formation 133
C. Film Deposition 136
D. Coating Uniformity 137
VI. Concluding Remarks 138
References 139

Thin Films in Microwave Acoustics

S. V. Krishnaswamy, B. R. McAvoy, and Maurice H. Francombe

I. Introduction 146
II. Thin Film Materials—Growth and Characterization 148
A. General Requirements 148
B. Piezoelectric and Ferroelectric Films 150
C. Other Film Materials 163
III. Surface Wave Device Structures 165
A. Basic Device Principles and Technology Survey 165
B. SAW Attenuation in Thin Ferromagnetic Films 171
C. Integrated Device Structures 172
D. SAWACT 174
IV. Bulk Wave Device Structures 179
A. General Background 179

B. Delay Lines . . . 183
C. High-Overtone Bulk Acoustic Resonators (HBAR) . . . 188
D. Film Bulk-Wave Acoustic Resonator (FBAR) . . . 196
V. Emerging Technology . . . 206
A. Thin-Film Materials Issues . . . 208
B. Acoustic Sensors . . . 210
C. Other Emerging Applications . . . 217
VI. Conclusions . . . 218
References . . . 219

Ferroelectric Films for Integrated Electronics

Maurice H. Francombe

I. Introduction and Background . . . 225
II. Ferroelectric Film Materials . . . 227
A. Properties and Applications of Interest . . . 227
B. Film DepositionGeneral . . . 230
III. Growth and Applications Examples . . . 241
A. $PbTiO_3$ and Solid Solutions . . . 242
B. $LiNbO_3$, $LiTaO_3$, $KNbO_3$, and $KTaO_3$ . . . 260
C. Bismuth Titanate, $Bi_4Ti_3O_{12}$ . . . 264
D. Polymers and Inorganic Fluorides . . . 278
E. Other Materials . . . 286
IV. Integrated Electronics Issues . . . 289
A. Materials Processing . . . 289
B. Interface and Stability Effects . . . 291
C. Integrated Device Structures . . . 292
V. Conclusions . . . 294
Acknowledgments . . . 295
References . . . 295

Electrochromic Tungsten-Oxide-Based Thin Films: Physics, Chemistry, and Technology

Claes-Göran Granqvist

I. Introduction . . . 302
II. Comments on W Oxide Bulk Crystals . . . 304
III. As-deposited Films: Structure and Composition . . . 307
A. Evaporated Films . . . 307
B. Sputter-deposited Films . . . 313
C. Films Made by Chemical and Electrochemical Techniques . . . 315
IV. Ion Intercalated Films: Electrochemical Characterization . . . 318
A. Diffusion Constants . . . 318
B. Electromotive Force . . . 319
C. Chronoamperometry . . . 319
D. Cyclic Voltammetry . . . 322

E. Impedance Spectrometry 325
F. Beam Deflectometry 326
V. Ion Intercalated Films: Physical Characterization 326
A. Microbalance Studies 326
B. Depth Profiling of Intercalated Species 327
C. Electron- and X-ray Extinction 327
D. Vibrational Spectroscopy 328
E. Magnetic Resonance 330
F. X-ray Photoelectron Spectroscopy 330
VI. Optical Properties 335
A. Semiconductor Bandgaps 335
B. As-deposited Films: Luminous and Near-infrared Properties 337
C. Survey of Coloration Methods 339
D. Ion Intercalated Films: Luminous and Near-infrared Properties 345
E. Theoretical Models 348
VII. Device Data 351
A. Liquid Electrolyte 351
B. Solid Inorganic Bulk-type Ion Conductor 353
C. Solid Inorganic Thin Film Ion Conductor 355
D. Polymer Electrolyte 358
VIII. Summary and Conclusions 362
References 364
Author Index 371
Subject Index 389

Physics of Thin Films Volume 18

Edited by
Maurice H. Francombe and John L. Vossen

Contributors. ix
Preface. ix

Design of High-Density Plasma Sources for Materials Processing

Michael A. Lieberman and Richard A. Gottscho

I. Introduction 2
A. Capacitively Coupled Radio Frequency Discharge Sources 5
B. Limitation of Capacitively Coupled Radio Frequency Discharges. 9
C. Overview of High-Efficiency Sources 10
II. Principles of Low-Pressure, High-Efficiency Source Design 13
A. Unified Analysis of Source Operation. 19
B. Discharge Heating. 23
III. Electron Cyclotron Resonance (ECR) Discharges 25
A. Source Configurations. 26
B. Electron Heating 31
C. Resonant Wave Absorption. 34
IV. Helicon Discharges 40
A. Helicon Configurations 41
B. Helicon Modes 42
C. Antenna Coupling 46
D. Helicon Mode Absorption 50
V. Inductive Discharges. 52
A. Inductive Source Configurations 52
B. Power Absorption and Operating Regimes 54
C. Source Operation and Coupling 56
D. Low-Density Operation and Source Efficiency. 58
VI. Helical Resonator Discharges 60
VII. Surface Wave Discharges 65
VIII. Plasma Transport 69
A. The Ion Energy Distribution Function. 71
B. Methods for Measuring Ion Energy Distribution Functions 76
C. Methods for Measuring Plasma Potentials. 80
D. Measurements for Energy Distributions and Potentials 81

E. Ion Energy Control 90
IX. Device Damage 96
A. Atomic Diplacement Damage 96
B. Contamination 98
C. Charging 98
D. Radiation 104
X. Summary and Remaining Questions 195
XI. Symbol Definitions 108
Acknowledgments 112
References 112

Electron Cyclotron Resonance Plasma Sources and Their Use in Plasma-Assisted Chemical Vapor Deposition of Thin Films

Oleg A. Popov

I. Introduction 122
II. ECR Fundamentals and Microwave Power Absorption 126
A. Principles of Electron Cyclotron Resonance 126
B. Microwave Power Absorption in ECR Plasmas 129
III. ECR Plasma Sources: Designs and Characteristics 152
A. Basic Types of ECR Plasma Sources 152
B. Hitachi/NTT ECR Plasma Source 157
C. ECR Plasma Source Characteristics and Parameters 183
IV. ECR Plasma Sources in PA CVD of Thin Films 208
A. Silicon Nitride Film ECR CVD 209
B. Silicon Oxide Films 216
C. ECR Plasma Planarization with Silicon Dioxide Films 221
V. Conclusion 226
Acknowledgments 227
References 227

Unbalanced Magnetron Sputtering

Suzanne L. Rohde

I. Introduction 235
II. Motivation for Unbalanced Magnetron Sputtering 236
A. Sputtering in General 236
B. Electron Motion in a Magnetic Field 240
C. Magnetron Sputtering 241
D. Ion-Assisted Deposition 243
III. Development of Unbalanced Magnetron Based Techniques 249
A. Precursors to the Unbalanced Magnetron 249
B. Unbalanced Magnetron Sputtering 255
IV. Principles of Unbalanced Magnetron Sputtering 259
V. Applications of UBM Sputtering 272
A. Elemental Thin Films 272

B. Films for Electronic and Optical Applications ... 274
C. Films for Corrosion Protection ... 275
D. Films for Wear and Abrasion Resistance ... 277
VI. Commercial Applications for UBM Sputtering ... 281
VII. Potential Future of UBM Sputtering ... 283
Acknowledgments ... 285
References ... 285

The Formation of Particles in Thin-Film Processing Plasmas

Christoph Steinbrchel

I. Introduction ... 289
II. General Phenomena ... 291
III. Particles in Deposition Plasmas ... 294
IV. Particles in Sputtering Plasmas ... 301
V. Particles in Reactive Ion Etching Plasmas ... 309
VI. Modeling of Particles in Plasmas ... 312
VII. Particle Contamination and Equipment Design ... 314
VIII. Conclusions ... 314
Acknowledgment ... 317
References ... 317
Author Index ... 319
Subject Index ... 323

Physics of Thin Films Volume 19

Edited by
K. Vedam

Contributors ix
Preface xi

In Situ Studies of Crystalline Semiconductor Surfaces by Reflectance Anisotropy

B. Drévillon and V. Yakovlev

I. Introduction 2
II. The Reflectance Anisotropy Technique 3
- A. Reflectance Anisotropy and the Other Characterisations Based on Polarized Light 3
- B. RA Configuration Based on the Use of a Photoelastic Modulator 6
- C. Comparison With Other RA Configurations 9
- D. Nature of the RA Under Crystal Growth Conditions 10

III. Applications of RA to the Growth of IIIV Semiconductors 11
- A. The Various Deposition Techniques 11
- B. Reflectance Anisotropy of GaAs and AlAs under MBE Conditions 12
- C. Applications of RA to Migration-Enhanced Atomic Layer Epitaxy (MEALE) 19
- D. Reflectance Anisotropy of GaAs Under Chemical Beam and Vacuum Chemical Epitaxy Growth Conditions 22
- E. Growth of Lattice-Matched Multilayered Structures by LP-MOCVD 29
- F. Reflectance Anisotropy of GaAs Under Atmospheric Pressure (AP) MOCVD Growth Environments 32

IV. Growth of Lattice-Mismatched IIIV and IIVI Structures 38
V. Summary and Conclusions 43
References 45

Real-Time Spectroscopic Ellipsometry Studies of the Nucleation, Growth, and Optical Functions of Thin Films, Part I: Tetrahedrally Bonded Materials

Robert W. Collins, Ilsin An, Hien V. Nguyen, Youming Li, and Yiwei Lu

I. Introduction 50
- A. Real-Time Ellipsometry at a Single Photon Energy 50
- B. Real-Time Spectroscopic Ellipsometry 52

II. Techniques of Real-Time Spectroscopic Ellipsometry . . . 54
A. Instrumentation . . . 55
B. Data Collection and Interpretation . . . 56
III. Studies of Tetrahedrally Bonded Thin Films . . . 61
A. Hydrogenated Amorphous Silicon . . . 61
B. Hydrogenated Amorphous Silicon-Carbon Alloys . . . 85
C. Microcrystalline Silicon . . . 93
D. Diamond . . . 105
IV. Summary . . . 121
Acknowledgments . . . 121
References . . . 122

Real-Time Spectroscopic Ellipsometry Studies of the Nucleation, Growth, and Optical Functions of Thin Films, Part II: Aluminum

Hien V. Nguyen, Ilsin An, and Robert Collins

I. Introduction . . . 127
II. Theoretical Background . . . 130
A. The Optical Functions of Bulk Aluminum . . . 131
B. Size Effects of the Optical Functions . . . 133
C. The Generalized Effective Medium Theory . . . 139
III. Experimental Apparatus for Real-Time Monitoring . . . 147
IV. Results and Discussion . . . 149
A. Overview of the Optical Functions of Aluminum Thin Films in the Nucleation and Growth Stages . . . 150
B. Analysis of the Optical Functions of Continuous Films . . . 162
C. Analysis of the Optical Functions of Particle Films . . . 168
V. Summary . . . 183
Acknowledgments . . . 187
References . . . 187

Optical Characterization of Inhomogeneous Transparent Films on Transparent Substrates by Spectroscopic Ellipsometry

P. Chindaudom and K. Vedam

I. Introduction . . . 191
II. Inhomogeneity in Thin Films . . . 193
III. Experimental . . . 196
A. Spectroscopic Ellipsometry . . . 196
B. RAE with a Compensator . . . 198
C. Detector Non-Linearity . . . 202
D. Data Analysis . . . 205
E. Measurements on Vitreous Silica as a Test of Accuracy of our Modified SE System . 207
IV. Experimental Results on Transparent Films . . . 215
A. Particulars of Films Studied . . . 215
B. Preliminary Results on MgO and LaF_3 Films . . . 216

C. Results on Transparent Films on Vitreous Silica Substrates 220
V. Discussion 231
A. General 231
B. Effect on Surface Roughness 233
C. Optical Functions of the Dielectric Film Materials 234
IV. Summary and Conclusion 243
References 244

Characterization of Ferroelectric Films by Spectroscopic Ellipsometry

S. Trolier-McKinstry, P. Chindaudom, K. Vedam, and R. E. Newnham

I. Introduction 249
II. Description of the SE System 254
III. Experimental Procedure 256
IV. Results and Discussion 258
A. Epitaxial and Oriented Films on Single Crystal Substrates 258
B. *In-situ* Annealing Studies on Ferroelectric Films 264
IV. Relation Between Film Microstructure and Electrical Properties 272
VI. Conclusions 275
Acknowledgments 276
References 276

Effects of Optical Anisotropy on Spectro-Ellipsometric Data for Thin Films and Surfaces

Atul N. Parkh and David Allara

I. Introduction 279
II. The Generalized Approach 283
A. Definition of the Fundamental Electromagnetic Problem 283
B. Implementation of the Yeh 4×4 Transfer Matrix Method 290
III. Comparison with Available Data and Previous Methods 291
IV. Simulation of Anisotropic Effects 295
A. Construction of a Model Anisotropic Medium from an Ensemble of Oriented Oscillators 295
B. Single Interface Ambient-Substrate Systems 295
C. Multiple Interface, Ambient—Film—Substrate Systems 309
Acknowledgment 312
References 313
Author Index 315
Subject Index 323

Thin Films Volume 20

Edited by
Abraham Ulman

Preface ... xv
Contributors ... xvii

Introduction

Supramolecular Assemblies: Vision and Strategy

Hans Kuhn and Abraham Ulman

1.1 Introduction ... 1
1.2 Efforts to Achieve Well-Defined Organised Molecular Assemblies ... 4
1.3 Perspective in Supramolecular Engineering ... 6
References ... 7

Advanced Materials

Oriented Growth of Nanocrystalline Particulate Films at Monolayers: A Colloid Chemical Approach to Advanced Materials

Janor H. Fendler

2.1 Introduction ... 11
2.2 Experimental Methodologies and Characterizations ... 12
2.3 Growth of Nanocrystalline Particulate Films ... 19
2.4 Size Quantization and Bandgap Engineering ... 21
2.5 Semiconductor Particulate Films ... 23
2.6 Silver Particulate Films ... 26
2.7 Organizational Imperative: Molecular Recognition and Epitaxial Growth ... 29
2.8 Conclusions and Future Directions ... 37
Acknowledgments ... 37
References ... 38

Third-Level Self-Assembly and Beyond: Polar Hybrid Superlattices via Postassembly Intercalation into Noncentrosymmetric Multilayer Matrices of Hydrogen Bonded Silanes

Rivka Maoz, Ruth Yam, Garry Berkovic, and Jacob Sagiv

3.1 Introduction 41
3.2 The Multilayer Matrix 45
3.3 Intercalated Superlattices 48
3.4 Conclusion 62
Acknowledgments 65
References and Notes 65

Building Two-Dimensional Polymers by the Langmuir-Blodgett Technique

Serge Palacin, Florence Porteu, and Annie Ruaudel-Teixier

4.1 Introduction 69
4.2 General Properties of Amphiphilic Porphyrins in LB Films 69
4.3 Ionic Network 72
4.4 Two-Dimensional Covalent Polymer 76
4.5 Conclusion 81
Acknowledgments 82
References 82

Surface Modification

Polymer Surface Modification

Thomas J. McCarthy, Timothy G. Bee, Joan V. Brennan, Elisa M. Cross, Anthony J. Dias, Nicole L. Franchina, Kang-Wook Lee, and Molly S. Shoichet

5.1 Introduction: Polymer Surfaces and Dynamic Interfaces 85
5.2 Polymer Surface Modification 87
5.3 Chemical Modification 89
5.4 Hydroxylation of Polypropylene 91
5.5 Functionalization of Poly(chlorotrifluoroethylene) 95
5.6 Dehydrofluorination of Poly(vinylidene fluoride) 99
5.7 Indirect Syntheses of Carboxylic Acid-Functionalized Fluoropolymers 100
5.8 Gas Phase Chlorination of Polyethylene 102
5.9 Carbonyl Surface Chemistry of Poly(ether ether ketone) 103
5.10 Conclusion 104
Acknowledgments 104
References 104

Lithographically Patterned Self-Assembled Films

Jeffrey M. Calvert

6.1 Introduction 109
6.2 Lithographic Considerations for Ultrathin Films as Imaging and Resist Layers 110
6.3 Deep-UV Photochemistry of Aromatic Hydrocarbon Organosilane Films 112
6.4 Selective Electroless Metallization of Patterned SA Films 117
6.5 Pattern Transfer and Microelectronic Device Fabrication by Means of Selectively Metallized SA Films 125
6.6 Deep-UV Photochemistry of Thiol-Terminal Organosilane Films and Patterned Proteins 128
6.7 Mixed Monolayers, Coplanar Molecular Assemblies, and Controlled Wetting 130
6.8 Patterned Amine Surfaces as Reactivity Templates 133
6.9 Selective Cell Adhesion and Outgrowth on Patterned SA Films 135
6.10 Conclusions and Future Areas of Research 138
Acknowledgments 139
References 139

Recognition at Surfaces

Langmuir Films of Amphiphilic Alcohols and Surfaces of Polar Crystals as Templates for Ice Nucleation

Ronit Popovitz-Biro, Jaroslaw Majewski, Jinn-Lung Wang, Kristian Kjaer, Jens Als-Nielsen, Leslie Leiserowitz, and Meir Lahav

7.1 Introduction 145
7.2 Self-Aggregates of Amphiphilic Alcohols as Templates for Ice Nucleation 146
7.3 Nucleation of Ice Under Monolayers of Aliphatic Alcohols 149
7.4 Nucleation of Ice Under Mixed Monolayers of Aliphatic Alcohols 160
7.5 Nucleation of Ice Under Alcohol Monolayers Bearing Amide or Ester Groups in the Hydrocarbon Chain 164
7.6 Induced Nucleation of Ice by Amphiphilic Alcohols Aggregated at the Oil/Water Interface 170
7.7 Induced Nucleation of Ice on Surfaces of Polar Crystals of α-Amino Acids 174
7.8 Conclusion and Outlook 177
Acknowledgments 179
References and Notes 180

Ion-Selective Monolayer Membranes Based on Self-Assembling Tetradentate Ligand Monolayers on Gold Electrodes: Nature of the Ionic Selectivity

Suzi Steinberg, Yitzhak Tor, Abraham Shanzer, and Israel Rubinstein

8.1 Introduction 183
8.2 Experiments 185
8.3 Results and Discussion 188
8.4 Conclusions 202
Acknowledgments 203
References and Notes 203

Specific Recognition at Functionalized Interfaces: Direct Force Measurements of Biomolecular Interactions

Deborah Leckband and Jacob Israelachvili

9.1 Introduction 207
9.2 Results 209
9.3 Discussion 220
9.4 Conclusion and Implications for Future Bioengineering Materials Technology 221
Acknowledgments 221
References 221

Formation of Recognition Patterns by Langmuir-Blodgett Techniques

Tilman Schwinn, Sven-Peter Heyn, Martin Egger, and Herman E. Gaub

10.1 Introduction 223
10.2 Materials and Methods 226
10.3 Conclusions 234
Acknowledgments 235
References 235

Electronic Properties

Photoinduced Electron Transfer in Monolayer Assemblies and Its Application to Artificial Photosynthesis and Molecular Devices

Masamichi Fujihira

11.1 Introduction 239
11.2 Construction of Charge Separation Units with LB Films 243
11.3 Photoelectric Conversion Efficiencies in Terms of the Electron Transfer Kinetics 249
11.4 Simulation of the Primary Process by the Use of a Mixed Monolayer with Triad and Antenna Molecules 254
11.5 Construction of Nanodomains in Monolayers by the Use of Phase Separation 256
11.6 Artificial Photosynthesis With Multilayered LB Films 267
11.7 Conclusions 271
Acknowledgments 272
References 272

Photoelectric Behavior of Bacteriorhodopsin Thin Films at the Solid/Liquid Interface

Tsutomu Miyasaka

12.1 Introduction 279
12.2 Photoelectric Response Caused by bR Photocycle 280
12.3 Photoelectric Behavior of bR at Solid/Liquid Interfaces 283
12.4 Device Application for Optical Sensing and Image Processing 287
12.5 Conclusions 290
Acknowledgments 291
References 291

Hole-Burning Spectroscopy of Dye-Doped Langmuir-Blodgett Films

Michel Orrit and Jacky Bernard

13.1 Introduction 293
13.2 Experimental Setup 296
13.3 The Hole Width and Its Temperature Dependence 299
13.4 The Hole as a Spectral Label 306
13.5 Conclusion and Prospects 312
Acknowledgments 313
References 313

Dynamic Processes

Evaluation of a Transfer Process for Langmuir-Blodgett Films by Means of a Quartz-Crystal Microbalance

Yoshio Okahatu, Katsuhiko Ariga, and Kentaro Tanaka

14.1 Introduction . . . 317
14.2 Quartz-Crystal Microbalances . . . 318
14.3 Vertical Dipping Depositions of LB Films on a QCM Plate . . . 319
14.4 Horizontal Lifting Depositions of LB Films on a QCM . . . 321
14.5 Conclusions . . . 328
References . . . 329

Translational Diffusion and Electron Hopping in Monolayers at the Air/Water Interface

Marcin Majda

15.1 Electrochemistry on the Water Surface . . . 331
15.2 Translational Diffusion in Octadecyl Ferrocene Monolayers . . . 334
15.3 Electron Hopping in Langmuir Monolayers . . . 339
15.4 Conclusions . . . 346
Acknowledgments . . . 346
References . . . 346

On-Line Structure Control of Langmuir-Blodgett Films

Hans Riegler and Karl Spratte

16.1 Introduction . . . 349
16.2 Structural and Compositional Variety of Organic Multilayers . . . 350
16.3 Schematic of the Preparation of Langmuir-Blodgett Films . . . 351
16.4 Structure Control During the Transfer Due to Substrate/Monolayer Interactions . . . 355
16.5 Conclusions . . . 361
Acknowledgments . . . 364
References . . . 364

Structure

Phase Diagrams and Chain Order in Monolayers of Aliphatic Chains

Ian R. Peterson

17.1 Phases of Amphilic Monolayers . . . 367
17.2 The Principle of Corresponding States . . . 371
17.3 Unambiguous Tests on the Principle . . . 375
17.4 Refinement of the Principle . . . 378
17.5 Conclusions . . . 382
References . . . 383
Subject Index . . . 385

Thin Films Volume 21

Edited by
Maurice H. Francombe and John L. Vossen

Contributors ix
Preface xi

Physics and Novel Device Applications of Semiconductor Homojunctions

A. G. U. Perera

I. Introduction and Background 2
II. Homojunction Internal Photoemission IR Detectors 4
A. Workfunction Variation with Concentration 7
B. Type I HIWIP Detectors $N_d < N_c$ $(E_F < E_c^{n+})$ 14
C. Type II HIWIP Detectors: $N_c < N_d < N_0$ $(E_c^{n+} < E_F < E_c^{i})$ 19
D. Type III HIP Detector: $N_d > N_0$ $(E_F > E_c^{i})$ 31
E. Delta Doped Potential Well IR Detectors 32
F. Uncooled Avalanche Injection IR Detector 33
G. Other Material Systems 34
III. Other Homojunction IR Detectors 37
A. Charge Storage Mode IR Detectors 38
B. HgCdTe Homojunctions 40
IV. Spontaneous Spiketrain Generation in Si Homojunctions 41
A. Device Physics of Spiketrain Generation 41
B. Infrared Detection without Amplifiers 48
C. Interpulse Time Series Analysis (Nonlinear Dynamics) 50
D. Neuron Emulation with Semiconductors 59
V. Summary and Future 68
References 71

Progress of SiGe/Si Quantum Wells for Infrared Detection

R. P. Gamani Karunasiri, J. S. Park, and K. L. Wang

I. Introduction . . . 77
II. Material Considerations . . . 79
III. Quantum Size Effects . . . 81
A. Resonant Tunneling . . . 81
B. Superlattice Minibands . . . 85
IV. Intersubband Transition in *p*-Type SiGe/Si Quantum Wells . . . 90
V. Intersubband Transition in *p*-Type (-Doped Si Quantum Wells . . . 95
VI. Detector Design . . . 101
VII. Detector Photoresponse . . . 104
VIII. Summary . . . 110
References . . . 111

Recent Developments in Quantum-Well Infrared Photodetectors

S. D. Gunapala and K. M. S. Bandara

I. Introduction . . . 114
II. Intersubband Absorption . . . 117
III. Bound-to-Bound State QWIPs . . . 123
IV. Bound-to-Continuum State QWIPs . . . 126
A. *n*-Doped QWIPs . . . 126
B. *p*-Doped QWIPs . . . 155
V. Asymmetrical GaAs/Al_xGa_{1-x} As QWIPs . . . 160
A. *n*-Doped QWIPs . . . 161
B. *p*-Doped QWIPs . . . 167
VI. Single QWIPs . . . 170
A. Directly Doped Well *n*-Type . . . 170
B. Undoped Tunneling Filled Well . . . 177
C. *p*-Doped Well . . . 180
VII. Superlattice Miniband QWIPs . . . 184
A. Bound-to-Bound Miniband Detectors . . . 184
B. Bound-to-Continuum Miniband Detectors . . . 187
C. Bound-to-Miniband Detectors . . . 190
VIII. Indirect Band-Gap QWIPs . . . 191
IX. QWIPs with Other Material Systems . . . 196
A. *n*-Doped QWIPs . . . 196
B. *p*-Doped QWIPs . . . 213
X. Light Coupling Methods . . . 215
XI. Imaging Arrays . . . 219
A. Effect of Nonuniformity . . . 219
B. Array Performance . . . 221
XII. Summary . . . 229
References . . . 232

Multiquantum-Well Structures for Hot-Electron Phototransistors

K. K. Choi

I. Introduction 239
II. Intersubband Transitions in QWIP Structures 241
A. Absorption Wavelength 242
B. Absorption Strength 244
III. Excitation Hot-Electron Spectroscopy 246
A. Thermally Stimulated Hot-Electron Spectroscopy 246
B. Optically Stimulated Hot-Electron Spectroscopy 252
IV. Infrared Hot-Electron Transistors 255
A. Figures of Merit 256
B. Dark-Current Requirement 257
C. Functions of an IHET 258
D. High-Detectivity IHETs 260
E. IHETs for Small NEΔT 269
F. IHET for High-Temperature Operation 275
G. Long-Wavelength IHET 282
V. Multicolor QWIP and IHET 289
A. Multicolor QWIP 291
B. Photoexcited Coherent Tunneling 294
C. Multicolor QWIP with Thick Barriers 298
D. Multicolor IHET 300
VI. Conclusion 307
References 308

Quantum-Well Structures for Photovoltaic Energy Conversion

Jenny Nelson

I. Introduction 312
II. Principles of Photovoltaics 313
A. The *pn* Junction Solar Cell 313
B. Photocurrent 314
C. Dark Current 316
D. Limits to Efficiency 318
E. Multibandgap Cells 321
F. The Quantum-Well Solar Cell 322
III. QWSC Spectral Response 323
A. Calculation of Spectral Response of Homogenous Solar Cell 325
B. Calculation of QWSC Spectral Response 327
C. Optimization of the Spectral Response 334
D. Spectral Response of GaAs/$In_xGa_{1-x}As$ QWSCs 341
E. Spectral Response in InP/$In_xGa_{1-x}As$ QWSCs 344
IV. QWSC Voltage Behavior 247
A. Current Gain vs. Voltage Loss 347
B. Theory of Solar-Cell Dark Currents 349
C. Theory of QWSC Dark Current 352

D. Analysis of QWSC Dark Currents 353
E. Analysis of V_{OC} 355
V. Limiting Efficiency of the QWSC. 357
A. Approaches to the Calculation of Limiting Efficiency 357
B. Experimental Study of Radiative Limit 360
C. The Question of Limiting Efficiency 361
VI. Future Work and Other Novel Approaches 362
A. Practical Applications for the QWSC. 362
B. Other Novel Approaches 364
References 366
Author Index 369
Subject Index. 381

Thin Films Volume 22

Edited by
Steve Rossnagel and Abraham Ulman

Contributors. ix
Preface. xi

Thin Film Microstructure and Process Simulation Using SIMBAD

Michael J. Brett, Steven K. Dew, and Tom J. Smy

I. Introduction 2
A. Thin Film Microstructure and Its Implications 2
B. Modeling Thin Film Microstructure 3
II. Description of the Transport and Deposition Algorithms 6
A. The SIMSPUD Vapor Transport Model 6
B. The SIMBAD Deposition Model 9
C. SIMBAD Program Structure and User Interface 14
III. Model Verification 15
A. Verification of SIMSPUD 15
B. Verification of SIMBAD 20
IV. Examples of Applications of the SIMBAD/SIMSPUD Models 29
A. Refractory Metal Chemical Vapor Deposition 29
B. Bias Sputtering 30
C. Collimated Sputtering 39
D. Simulation of Film Composition 45
E. High-Temperature Deposition and Bulk Diffusion 49
F. Interpolated Three-Dimensional SIMBAD Model 55
G. Etching Processes 61
V. New Developments in Film Growth Simulation 67
VI. Conclusions 76
References 77

Mathematical Methods for Thin Film Deposition Simulations

S. Hamaguchi

I. Introduction 81
II. The HamiltonJacobi Type Equation 84
A. Boundary Motion 84
B. A Spurious Solution 85
C. Characteristics and Slope Discontinuities 87
D. Viscosity Solutions 89
E. Entropy Conditions 91
F. Geometric Interpretation of Entropy Conditions 95
III. The Shock-Tracking Algorithm 97
IV. Deposition and Etch Rates 98
A. Collisionless Transport 98
B. The Two-Energy Model 100
C. Ion and Neutral Fluxes 107
V. Sample Simulations 110
VI. Summary 113
References 114

A Process Model for Sputter Deposition of Thin Films Using Molecular Dynamics

C.-C. Fang, V. Prasad, R. V. Joshi, F. Jones, and J. J. Hsieh

I. Introduction 118
A. Introductory Remarks 118
B. Sputter Deposition of Thin Films 119
C. Discrepancy between Theory and Experiments 120
D. Simulation of Thin Film Deposition 122
E. Molecular Dynamics Modeling 123
F. Modeling of Sputtering from the Target 125
II. Fundamentals of Molecular Dynamics 126
A. Potential Function 126
B. Sorting Algorithms 128
C. Discretization and Integration Scheme 132
D. Constant Temperature Control Algorithms 134
E. Periodic Boundary Conditions: Fixed versus Moving 142
F. Average Film Stress 144
G. Local Stress Function 145
III. Transport of Sputtered Atoms from Target to Substrate 146
IV. Deposition of Plane Substrates: Microstructure and Stresses 149
A. Effect of Adatom Energy 150
B. Combined Effect of Ion Bombardment and Gas Entrapment on Ni Film 152
C. Effect of Voltage Bias on Film Stresses 156
V. Step Coverage of a Submicron Liner 161
A. Collimated Sputtering 162
B. Low-Pressure/Low-Throw Sputtering 165

VI. Conclusions . . . 169
References . . . 171

Feature Scale Transport and Reaction during Low-Pressure Deposition Processes

Timothy S. Cale and Vadali Mahadev

I. Introduction . . . 176
A. Transport during Deposition Processes . . . 177
B. Chemical Reaction Kinetics . . . 179
C. Transport and Reaction Simulations . . . 180
II. Chemistry . . . 181
A. Reaction Stoichiometry . . . 183
B. Global Rate Expressions . . . 185
C. Reaction Mechanisms . . . 186
III. Low-Pressure Transport . . . 188
A. Coordinate Systems . . . 188
B. Single Particle Distribution Function . . . 189
C. Fluxes . . . 189
D. Flux Distributions . . . 191
IV. Ballistic Transport and Reaction Model . . . 193
A. Model Statements . . . 193
B. Gas/Solid Interactions . . . 194
C. Diffuse Re-emission . . . 194
D. Specular Transport . . . 197
E. Surface Species . . . 200
F. Solution of Governing Equations . . . 201
V. LPCVD of Silicon Dioxide from TEOS . . . 202
A. BTRM for LPCVD . . . 202
B. Kinetics Models . . . 203
C. Results . . . 205
D. Simulation Using Sandia's Mechanism . . . 209
VI. Plasma-Enhanced Chemical Vapor Deposition of Silicon Dioxide from TEOS . . . 210
A. Experimental . . . 211
B. Model Development . . . 212
C. Results . . . 213
VII. Remote Microwave PECVD of Silicon Dioxide from TEOS . . . 215
A. Experimental . . . 216
B. Model Development . . . 216
C. Results . . . 219
VIII. Uncollimated PVD of TiN . . . 226
A. Experimental . . . 226
B. Uncollimated PVD of TiN . . . 226
C. Collimated PVD of TiN . . . 227
IX. Composition of PVD TitaniumTungsten Films . . . 229
A. Composition Profiles . . . 232
B. First Experiments . . . 232

C. Model Development. 233
D. Second Experiments 235
E. Simulations. 238
X. Simultaneous Deposition and Sputter Etch 241
A. IMS Deposition of Aluminium 242
B. ECR Deposition of SiO_2 243
C. Simulations. 243
XI. Curvature-Driven Surface Diffusion 252
A. Experiment. 253
B. Model Development. 253
XII. Programmed Rate CVD of Tungsten 265
A. Programmed Rate CVD. 266
B. Experimental 266
XIII. Conclusions 269
References 271
Author Index 277
Subject Index. 285

Thin Films Volume 23

Edited by
Maurice H. Francombe and John L. Vossen

Contributors. ix
Preface. xi

Ge_xSi_{1-x} Epitaxial Layer Growth and Application to Integrated Circuits

David W. Greve

I. Introduction 2
II. Ge_xSi_{1-x} HeterojunctionsGeneral Considerations 3
III. Growth by Molecular Beam Epitaxy 6
A. MBE Systems 7
B. Surface Cleaning and Preparation 9
C. Germanium Incorporation and Abruptness 10
D. p-type Doping 12
E. n-type Doping 14
F. Incorporation of Adatoms and Growth Temperature Limits 15
G. Gas Sources 21
IV. Growth by Chemical Vapor Deposition 21
A. Low-Temperature Growth 22
B. Growth Systems 23
C. Surface Reactions 27
D. Kinetics of Layer GrowthHydride Reactants. 37
E. Kinetics of Layer GrowthDichlorosilane 41
F. Transition Abruptness 43
G. Minimum Growth Temperature 44
H. Surface Preparation. 44
I. Overview/Summary 50
V. Application to Heterojunction Bipolar Transistors 51
A. Operation of the HBT 51
B. Early Reports of HBTs. 54
C. MBE-Grown HBT Process. 56
D. UHV/CVD-Grown HBT Process 59
E. Profile Design for the UHV/CVD HBT. 64
F. HBT Future Prospects 68
VI. The Ge_xSi_{1-x}Channel MOSFET 68

VII. Conclusions and Future Prospects 73
Acknowledgments 74
References 74

Platinum Silicide Internal Emission Infrared Imaging Arrays

Freeman D. Shepherd

I. Introduction 84
A. Staring-Mode Operation 84
B. Requirements Imposed by Thermal Infrared Signals 84
C. Early Efforts Leading to Current PtSi IR Camera Technology 86
II. The Internal Emission Process 88
A. Internal Photoemission 88
B. Thermionic Emission (Dark Current) 92
C. Internal Field Emission 93
III. State of the Art of Platinum Silicide Detectors and Arrays 100
A. PtSi Spectral Response 100
B. Fowler Emission Efficiency 102
C. Array Response Uniformity 104
D. Excess Low-Frequency Noise 104
E. Array Parameters 104
F. Infrared Cameras 105
IV. Future of Platinum Silicide Detector and Sensor Development 106
A. Array Size 106
B. Pixel Dimensions 106
C. Optical Absorption in the Silicide Electrode 106
D. Detector Fill Factors 107
E. Industrial vs. Military Sensor Requirements 107
F. Improvement of Emission Efficiency 107
G. Extension of Cutoff Wavelength 108
H. General Observations 110
V. Summary 111
References 111

Thin Film Epitaxial Layers on Silicon for the Detection of Infrared Signals

Paul W. Pellegrini and Jorge R. Jiminez

I. Introduction 116
II. Infrared Bands, Detectors, and Materials 117
A. Infrared Spectral Bands 117
B. Detectors of Infrared Radiation 120
C. Material Considerations for LWIR Detection 127
D. Summary of PtSi Detector Basics 135
III. Group-IV Epitaxial Devices for Infrared Detectors 141
A. Delta-Doped PtSi/Si Detectors 141
B. Si-Homojunction Detectors 143
C. SiGe/Si Heterojunction Internal Photoemission (HIP) Detectors 144

D. Silicide/SiGe Schottky Detectors . . . 153
E. Detectors Involving Epitaxial Silicides . . . 161
IV. Growth and Fabrication of Si-Based Infrared Detector Structures . . . 162
A. Doping and Temperature . . . 162
B. Surface Preparation and Cleaning . . . 163
C. Uniformity . . . 164
D. Silicide/SiGe Fabrication . . . 165
V. Conclusion . . . 168
References . . . 168

III-V Quantum-Well Structures for High-Speed Electronics

E. R. Brown and K. A. McIntosh

I. Background on Quantum-Well Infrared Detectors . . . 174
A. Direct Detection . . . 174
B. Heterodyine Detection . . . 175
II. Quantum-Well Detector Design and Intersubband Absorption . . . 178
A. Quantum-Well Energy Levels . . . 178
B. Epitaxial Growth . . . 184
C. Intersubband Absorption Measurement Techniques . . . 185
D. Intersubband Absorption Results . . . 187
III. MQW Detector Fabrication and DC Response Characteristics . . . 188
A. Fabrication and Packaging . . . 188
B. Dark Current . . . 189
C. Spectral and Absolute Responsivity . . . 190
D. Photoconductive Gain . . . 194
E. External Quantum Efficiency . . . 195
IV. Electrical Bandwidth and Optical-Heterodyne Experiments . . . 196
A. Photoelectron Generation-Recombination Noise Technique . . . 196
B. Diode-Laser Mixing Technique . . . 198
C. Microwave Rectification Technique . . . 199
D. Discussion of Bandwidth and Lifetime . . . 200
E. Heterodyne Sensitivity Measurement Technique . . . 201
F. Heterodyne Sensitivity Results . . . 202
V. Applications . . . 204
A. Instrumental Resolution and Sensitivity . . . 205
B. High-Resolution Molecular Spectroscopy . . . 207
C. Long-Range, High-Data-Rate Communications . . . 210
VI. Improvements in MQW Heterodyne Detectors . . . 211
A. Enhancement of External Quantum Efficiency . . . 211
B. Design of Detectors Having Lifetime-Limited Electrical Bandwidth . . . 214
References . . . 216

Quantum-Well Devices for Infrared Emission

A. G. U. Perera, J.-W. Choe and M. H. Francombe

I. Introduction and Background 218
II. Quantum Wells in Interband-Type IR Sources 220
A. Role of Size and Strain Effects 220
B. MQW Heterostructures for Midwave IR 228
III. Intersubband Transition Processes for IR Emission 238
A. Background and Summary 238
B. Radiative Transitions and Population Inversion 240
C. FIR Emission from MQW Structures 243
D. Quantum Efficiency of the Cascade Process 247
E. Photon Wave Function in the Superlattice 252
F. Early LWIR Emission Results 256
IV. Graded-Gap Injection Structures for MWIR and LWIR Emission 262
A. First 5-um Intersubband IR Emission 263
B. Development of Quantum Cascade Laser 266
V. Recent Trends, Limitations, and Application Potential 274
References 282
VI. Appendix 286
A. Sequential Resonant Tunneling 286
B. Physics of Resonant Tunneling 291
C. Energy and Lifetime from Complex Energy Method 291
Author Index 299
Subject Index 305

Thin Films Volume 24

Edited by
Abraham Ulman

Contibutors xi

The Synthesis of Organothiols and Their Assembly into Monolayers on Gold

Paul E. Laibinis, Bentley J. Palmer, Seok-Won Lee, and G. Kane Jennings

1.1 Introduction 2
1.1.1 Overview 2
1.1.2 Background on Self-Assembled Monolayers 2
1.2 Preparation of Gold Surfaces 4
1.3 Synthesis of Alkanethiols 6
1.3.1 Synthetic Methods for the Preparation of Alkanethiols 8
1.3.2 Precursors for the Synthesis of ω-Terminated Alkanethiols 9
1.4 Organosulfur Compounds Used to Form Self-Assembled Monolayers 12
1.4.1 Alkanethiols Terminating in Groups Commensurate with the Hydrocarbon Lattice 12
Terminal Groups That Produce High-Energy Surfaces 13
Terminal Groups That Produce Surfaces with Intermediate Energies 16
Terminal Groups That Produce Low Energy Surfaces 17
Use of Polar Surfaces as Sites for Covalent Attachment 17
1.4.2 Asymmetric Alkyl Disulfides 21
1.4.3 Alkanethiols Containing Ringed Structures 21
Alkanethiols Terminating in Bulky Tail Groups 23
Alkanethiols Terminating in Electroactive Bulky Tail Groups 26
Alkanethiols Containing Internal Aromatic Functional Groups 30
1.4.4 Aromatic Thiols 33
1.4.5 Polymeric Organosulfur Compounds 33
1.4.6 Organosulfur Analogs of Biological Oligomers 35
1.5 Summary and Outlook 36

The Kinetics and Thermodynamics of Monolayer Formation: In Situ Measurement of Alkanethiol Adsorption onto Gold

D. S. Karpovich, H. M. Schessler, and G. J. Blanchard

2.1 Introduction 44
2.2 Thermodynamics of Alkanethiol/Gold Monolayer Formation 45
2.2.1 Background 45
2.2.2 Measurement of Monolayer Formation 48
Operation of QCMs in Solution 48
Details of the Experimental Apparatus 50
Conductivity Measurements 52
2.2.3 Chemistry of the Gold/Thiol Interface 52
The Adsorption and Desorption Reactions 52
2.2.4 Experimental Data, Reaction Kinetics, and Their Relationship to the Thermodynamic Quantities 55
2.2.5 Implications of the Thermodynamic Results 65
2.3 Aliphatic Chain Annealing 65
2.4 Rigidity of the Aliphatic Environment on a Timescale Relevant to Chemical Interactions 70
2.4.1 The Need For Dynamical Information on SAMs 70
2.4.2 Measurement of Chromophore Dynamics 70
2.4.3 Dynamical Measurement Results 71
2.4.4 Implications of the Dynamical Data 76
2.5 Summary 77

Atomic Force Microscopy Studies of Self-Assembled Monolayers of Thiols

Gang-yu Liu, Song Xu, and Sylvain Cruchon-Dupeyrat

3.1 Introduction 82
3.2 Structure Characterization 83
3.2.1 *n*-Alkanethiol Monolayers 83
Order in SAMs 83
Domains and Depressions in SAMs 86
3.2.2 SAMs with Other Chains or Terminal Groups 86
SAMs of Thiodiethyl Alkyl Decanoate 87
SAMs of Fluorinated Thiols 88
SAMs with Other Terminal Groups 88
3.2.3 SAMs of Thiol Mixtures/Asymmetric Disulfide 89
3.3 Self-Assembly Kinetics and Mechanism 89
3.4 Nanomechanical Properties 93
3.4.1 Elastic Compliance 93
3.4.2 Adhesive and Frictional Forces 96
3.5 Nanofabrication 98
3.5.1 Displacement of Thiols under Ambient Laboratory Conditions 99
3.5.2 Displacement of Thiols in Liquid 100
Fabrication in Pure Solvents 100
Fabrication in Thiol Solutions 101
3.6 Concluding Remarks 102

X-ray and He Atom Diffraction Studies of Self-Assembled Monolayers

Paul Fenter

4.1 Introduction 112
4.1.1 Toward a Molecular Perspective 112
4.1.2 Diffraction Probes 114
4.1.3 Experimental Details 117
4.1.4 The Question of X-ray-Induced Damage 119
4.2 Self-Assembled Monolayer Structures 121
4.2.1 C_{18}/Au(001): Complementarity of He and X-ray Diffraction 121
4.2.2 Three Closely Related SAMs and Their Structures 125
4.3 Order and Phase Behavior of C_n/Au(111) 130
4.3.1 Order and Disorder in SAMs 131
4.3.2 Hydrocarbon Tilt Structure vs. Chain Length 132
4.3.3 Interface Structure of C_{10} and C_{18} Monolayers 135
4.3.4 (n,T) Phase Behavior of C_n/Au(111) 138
4.3.5 Structure of Submonolayer Phases ($\Theta < 1$) 142
4.4 Conclusions 144

NMR Spectroscopy of Self-Assembled Monolayers

Linda Reven and Lucy Dickinson

5.1 Introduction 150
5.2 Solid-State NMR Techniques 151
5.2.1 Cross-Polarization and Magic-Angle Spinning (CPMAS) 151
5.2.2 Relaxation Parameters 152
5.2.3 Line Shape Measurements 153
5.3 Alkanethiols on Gold Nanoparticles 154
5.3.1 ^{13}C Chemical Shifts: Conformation and Surface Bonding 154
5.3.2 Chain Dynamics of Alkanethiols 156
5.3.3 Functionalized Alkanethiols on Gold Nanoparticles 159
5.4 Alkylphosphonates on Metal Oxides 163
5.4.1 ^{31}P Solid-State NMR Studies of Alkylphosphonate Monolayers and Multilayers 164
Surface Bonding 164
Surface Reactivity 165
5.4.2 Dynamics of Alkylphosphonate Monolayers 166
5.4.3 Order–Disorder Transitions in Alkylphosphonate Monolayers 168
5.5 Alkylsiloxane SAMs 169
5.5.1 Solid-State NMR of Octadecylsiloxane Monolayers on Silica 170
5.6 Summary 175

The Structure of Alkanethiol Films on Liquid Mercury: An X-Ray Study

M. Deutsch, O. M. Magnussen, B. M. Ocko, M. J. Regan, and P. S. Pershan

6.1 Introduction 180
6.2 Experimental 185
6.2.1 Samples and Sample Cell 185
6.2.2 Surface X-ray Measuring Techniques 186
6.2.3 Modeling and Fitting 186
6.3 Results 188
6.3.1 Monolayers 188
Reflectivity 188
GIXD 192
6.3.2 Multilayers 193
Reflectivity 193
GIXD 197
6.4 Discussion 198
6.5 Conclusion 200

Simulations of Self-Assembled Monolayers of Thiols on Gold

J. Ilja Siepmann and Ian R. McDonald

7.1 Introduction 206
7.2 Monte Carlo or Molecular Dynamics? 207
7.3 Force Fields 209
7.3.1 Hautman-Klein United-Atom (HKUA) Model 209
7.3.2 Mar-Klein All-Atom (MKAA) Model 210
7.3.3 Gerdy-Goddard Dimer (GGD) Model 211
7.3.4 Force Fields for More Complex SAMs 211
7.4 Properties of Pure Monolayers 211
7.4.1 Translational, Orientational, and Conformational Order 212
7.4.2 Dynamics 214
7.4.3 Temperature Effects 215
7.4.4 Chain Length Effects 216
7.4.5 Finite-Size Effects 217
7.5 Structure of Mixed Monolayers 218
7.6 Prediction of Contact Angles 220
7.7 Mechanical Relaxation 221
7.8 SAMs on Colloidal Gold Particles 222
7.9 Conclusions 222

Microcontact Printing of SAMs

Joe Tien, Younan Xia, and George M. Whitesides

8.1 Introduction 228
8.2 Microcontact Printing 230
8.3 Patterned Hydrophobic SAMs as Ultrathin Resists 240
8.4 Two-Component Patterned SAMs as Templates........ 244
8.5 Conclusions 250
Author Index 255
Subject Index 273

ISBN 0-12-533025-1